Carpentry
&
Construction

Carpentry & Construction

MARK R. MILLER
Associate Professor
Texas A&M University–Kingsville
Kingsville, TX

REX MILLER
Professor Emeritus
State University College at Buffalo
Buffalo, New York

GLENN E. BAKER
Professor Emeritus
Texas A&M University
College Station, TX

4th Edition

McGraw-Hill

New York   Chicago   San Francisco   Lisbon   London
Madrid   Mexico City   Milan   New Delhi   San Juan
Seoul   Singapore   Sydney   Toronto

The **McGraw·Hill** Companies

Library of Congress Cataloging-in-Publication Data

Miller, Mark R.
    Carpentry & construction / Mark R. Miller, Rex Miller, Glenn E. Baker—4th ed.
      p. cm.
    Includes index.
    ISBN 0-07-144008-9
    1. Carpentry. 2. House construction. I. Miller, Rex II. Baker, Glenn E. III. Title.

TH5606.M52   2004
694—dc22   2004040301

3 4 5 6 7 8 9 0   QPD/QPD   0 9 8 7 6 5

ISBN 0-07-144008-9

*The sponsoring editor for this book was Larry Hager, the editing supervisor was
Caroline Levine, and the production supervisor was Sherri Souffrance. It was set in ITC
Century Light by Wayne Palmer of McGraw-Hill Professional's Hightstown, N.J.,
composition unit.*

McGraw-Hill books are available at special quantity discounts to use as premiums and sales
promotions, or for use in corporate training programs. For more information, please write to
the Director of Special Sales, McGraw-Hill Professional, Two Penn Plaza, New York, NY
10121. Or contact your local bookstore.

Contents

9 Installing Windows & Doors

10 Finishing Exterior Walls

15 Finishing the Interior

16 Special Construction Methods

Preface

Carpentry and Construction, Fourth Edition, is written for everyone who wants or needs to know about carpentry and construction. Whether remodeling an existing home or building a new one, the rewards from a job well done are manyfold.

This text can be used by students in vocational courses, technical colleges, apprenticeship programs, and construction classes in industrial technology programs. The home do-it-yourselfer will find answers to many questions that pop up in the course of getting a job done whether over a weekend or over a year's time.

In order to prepare this text, the authors examined courses of study in schools located all over the country. An effort was made to take into consideration the geographic differences and the special environmental factors relevant to a particular area.

Notice how the text is organized. The first chapter, "Starting the Job," presents the information needed to get construction started. The next chapter covers preparing the site. Then the footings and foundation are described. Once the roof is in place, the next step is the installation of windows and doors. When the windows and doors are in place, the exterior siding is applied. Next, heating and cooling are covered—all-important considerations for living quarters. Once the insulation is in place, the interior walls and ceilings are covered in detail before presenting interior finishing methods.

Special construction methods, maintenance, and remodeling, as well as careers in carpentry are then described. The building of solar houses and the design of solar heating are covered to keep the student and do-it-yourselfer up-to-date with the latest developments in energy conservation. Take a closer look at the steel framing used in more abundance today for private homes. Also note the use of foam and concrete to build homes of lasting quality that are almost completely impervious to tornado and hurricane damage.

No book can be complete without the aid of many people. The Acknowledgments that follow mention some of those who contributed to making this text the most current in design and technology techniques available to the carpenter. We trust you will enjoy using the book as much as we enjoyed writing it.

MARK R. MILLER
REX MILLER
GLENN BAKER

Acknowledgments

The authors would like to thank the following manufacturers for their generous efforts. They furnished photographs, drawings, and technical assistance. Without the donations of time and effort on the part of many people, this book would not have been possible. We hope this acknowledgment of some of the contributions will let you know that the field you are working in or about to enter is one of the best. (The name given below in parentheses indicates the abbreviated form used in the credit lines for the photographs appearing in this text.)

Abitibi Corporation; ALCOA; AFM Corporation; American Olean Tile Co.; American Plywood Association; American Polysteal Forms; American Standard, Inc.; Town of Amherst, New York (Town of Amherst, N.Y.); Andersen Corporation (Andersen); Armstrong Cork Company (Armstrong Cork); C. Arnold and Sons; Avalon Concepts; Beaver-Advance Company (Beaver-Advance); The Bilco Company (Bilco); Bird and Son, Inc. (Bird and Son); Black & Decker Manufacturing Company (Black & Decker); Boise-Cascade; Butler Manufacturing Company (Butler); Certain-Teed Corporation (Certain-Teed); California Redwood Association (California Redwood Assoc.); Conwed Corporation (Conwed); Corl Corporation; Dewalt, Div. of American Machine and Foundry Company (DeWalt); Dow Chemical Company (Dow Chemical); Dow-Corning Company (Dow-Corning); Duo-Fast Corporation (Duo-Fast); EMCO; Formica Corporation (Formica); Fox & Jacobs Corporation (Fox and Jacobs); General Products Corporation (General Products); Georgia-Pacific Corporation (Georgia-Pacific); Goldblatt Tools, Inc. (Goldblatt Tools); Gold Bond Building Products (Gold Bond); Grossman Lumber Company (Grossman Lumber); Gypsum Association (Gypsum.); Hilti Fastening Systems (Hilti); IRL Daffin Company (IRL Daffin); Kelly-Stewart Company, Inc. (Kelly-Stewart); Kenny Manufacturing Company (Kenny); Kirch Company (Kirch); Kohler Company (Kohler); Lennox Furnace Co.; Majestic; Manco Tape, Inc. (Manco Tape); Martin Industries (Martin); Masonite Corporation (Masonite); MFS; Milwaukee Electric Tool Company (Milwaukee Electric Tool); NAHB (National Association of Home Builders); National Aeronautical & Space Administration (NASA); National Homes; National Lock Hardware Company; National Oak Flooring Manufacturer's Association (National Oak Flooring Manufacturers); National Wood Manufacturers Association (National Wood Manufacturers); New York State Electric and Gas Corporation (NYSE & G); Novi; NuTone; Owens-Corning Fiberglas Corporation (Owens-Corning); Patent Scaffolding Company (Patent Scaffolding); Pella Windows and Doors, Inc. (Pella); Permograin Products; Plaskolite, Inc. (Plaskolite); Portland Cement Association (Portland Cement); Potlatch Corporation (Potlatch); Proctor Products, Inc. (Proctor Products); Red Cedar Shingle & Handsplit Shake Bureau (RCS & HSB); Reynolds Metals Products; Richmond Screw Anchor Company (Richmond Screw Anchor); Riviera Kitchens, Div. of Evans Products; Rockwell International, Power Tool Division (Rockwell); Sears, Roebuck and Company (Sears, Roebuck); Shakertown Corporation; Simplex Industries (Simplex); Stanley Tools Company (Stanley Tools); State University College at Buffalo; Southern Forest Products Association; TECO Products and Testing Corporation, Washington, D.C. 20015; Texas A & M University (Texas A & M); U.S. Gypsum Company (U.S. Gypsum); U.S. Department of Energy (U.S. Dept. of Energy); U.S. Bureau of Labor Statistics; U.S. Forest Service, Forest Products Laboratory (Forest Products Lab); United Steel Products; Universal

Fastenings Corporation; Universal Form Clamp Company (Universal Form Clamp); Valu, Inc. (Valu); Velux-American; Weiser Lock, Division of Norris Industries (Weiser Lock), Weslock; Western Wood Products Association (Western Wood Products); Weyerhauser Company (Weyerhauser); David White Instruments, Div. of Realist, Inc. (David White).

In addition, the authors would also like to thank Paul Consiglio of Buffalo, NY for some of the line drawings used.

AFM Corporation
American Polysteel Forms
American Standard, Inc.
Avalon Concepts
Clopay Building Products Co.
Congoleum Floors
Delta Building Products
Eljer Manufacturing Co.
Hy-Lite Block Windows
Kohler Company
L.J. Smith Stair Systems
Majestic Company
Nuconsteel Commercial Corp.
ODL, Inc.
Owens-Corning
Pittsburg-Corning
PlumbShop
Southeastern Metals Manufacturing Co.
Steel Framing, Inc.
Tri-Steel Structures, Inc.
Western Wood Products Association
Wilsonart International

1
CHAPTER

Starting the Job

BECAUSE CARPENTRY INVOLVES ALL KINDS of challenging jobs, it is an exciting industry. You will have to work with hand tools, power tools, and all types of building materials. You can become very skilled at your job. You get a chance to be proud of what you do. You can stand back and look at the building you just helped erect and feel great about a job well done.

One of the exciting things about being a carpenter is watching a building come up. You actually see it grow from the ground up. Many people work with you to make it possible to complete the structure. Being part of a team can be rewarding, too.

This book will help you do a good job in carpentry, whether you are remodeling an existing building or starting from the ground up. Because it covers all the basic construction techniques, it will aid you in making the right decisions.

You have to do something over and over again to gain skill. When reading this book, you might not always get the idea the first time. Go over it again until you understand. Then go out and practice what you just read. This way you can see for yourself how the instructions actually work. Of course, no one can learn carpentry by merely reading a book. You have to read, reread, and then do. This "do" part is the most important. You have to take the hammer or saw in hand and actually do the work. There is nothing like good, honest sweat from a hard day's work. At the end of the day you can say "I did that" and be proud that you did.

This chapter should help you build these skills:

• Select personal protective gear
• Work safely as a carpenter
• Measure building materials
• Lay out building parts
• Cut building materials
• Fasten materials
• Shape and smooth materials
• Identify basic hand tools
• Recognize common power tools

SAFETY

Figure 1-1 shows a carpenter using one of the latest means of driving nails: the compressed-air-driven nail driver, which drives nails into the wood with a single stroke. The black cartridge that appears to run up near the carpenter's leg is a part of the nailer. It holds the nails and feeds them as needed.

Fig. 1-1 *This carpenter is using an air-driven nail driver to nail these framing members.* (Duo-Fast)

As for safety, notice the carpenter's shoes. They have rubber soles for gripping the wood. This will prevent a slip through the joists and a serious fall. The steel toes in the shoes prevent damage to a foot from falling materials. The soles of the shoes are very thick to prevent nails from going through. The hard hat protects the carpenter's head from falling lumber, shingles, or other building materials. The carpenter's safety glasses cannot be seen in Fig. 1-1, but they are required equipment for the safe worker.

Other Safety Measures

To protect the eyes, it is best to wear safety glasses. Make sure your safety glasses are of tempered glass. They will not shatter and cause eye damage. In some instances you should wear goggles. This prevents splinters and other flying objects from entering the eye from under or around the safety glasses. Ordinary glasses aren't always the best, even if they are tempered glass. Just become aware of the possibilities of eye damage whenever you start a new job or procedure. See Fig. 1-2 for a couple of types of safety glasses.

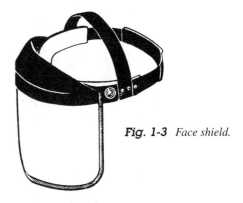

Fig. 1-3 Face shield.

Fig. 1-2 Safety glasses.

Sneakers are used only by roofers. Sneakers, sandals, and dress shoes do not provide enough protection for the carpenter on the job. Only safety shoes should be worn on the job.

Gloves Some types of carpentry work require the sensitivity of the bare fingers. Other types do not require the hands or fingers to be exposed. In cold or even cool weather gloves may be in order. Gloves are often needed to protect your hands from splinters and rough materials. It's only common sense to use gloves when handling rough materials.

Probably the best gloves for carpenter work are a lightweight type. A suede finish to the leather improves the gripping ability of the gloves. Cloth gloves tend to catch on rough building materials. They may be preferred, however, if you work with short nails or other small objects.

Body protection Before you go to work on any job, make sure your entire body is properly protected. The hard hat comes in a couple of styles. Under some conditions the face shield is better protection. See Fig. 1-3.

Is your body covered with heavy work clothing? This is the first question to ask before going onto the job site. Has as much of your body as practical been covered with clothing? Has your head been properly protected? Are your eyes covered with approved safety glasses or face shield? Are your shoes sturdy, with safety toes and steel soles to protect against nails? Are gloves available when you need them?

General Safety Rules

Some safety procedures should be followed at all times. This applies to carpentry work especially:

* Pay close attention to what is being done.
* Move carefully when walking or climbing.
* (Take a look at Fig. 1-4. This type of made-on-the-job ladder can cause trouble.) Use the leg muscles when lifting.

Fig. 1-4 A made-on-the-job ladder.

- Move long objects carefully. The end of a carelessly handled 2 × 4 can damage hundreds of dollars worth of glass doors and windows. Keep the workplace neat and tidy. Figure 1-5A shows a cluttered working area. It would be hard to walk along here without tripping. If a dumpster is used for trash and debris, as in Fig. 1-5B, many accidents can be prevented. Sharpen or replace dull tools.

(A)

(B)

Fig. 1-5 *(A) Cluttered work site; (B) A work area can be kept clean if a large dumpster is kept nearby for trash and debris.*

- Disconnect power tools before adjusting them.
- Keep power tool guards in place.
- Avoid interrupting another person who is using a power tool.
- Remove hazards as soon as they are noticed.

Safety on the Job

A safe working site makes it easier to get the job done. Lost time due to accidents puts a building schedule behind. This can cost many thousands of dollars and lead to late delivery of the building. If the job is properly organized and safety is taken into consideration, the smooth flow of work is quickly noticed. No one wants to get hurt. Pain is no fun. Safety is just common sense. If you know how to do something safely, it will not take any longer than if you did it in an unsafe manner. Besides, why would you deliberately do something that is dangerous? All safety requires is a few precautions on the job. Safety becomes a habit once you get the proper attitude established in your thinking. Some of these important habits to acquire are:

- Know exactly what is to be done before you start a job.
- Use a tool only when it can be used safely. Wear all safety clothing recommended for the job. Provide a safe place to stand to do the work. Set ladders securely. Provide strong scaffolding.
- Avoid wet, slippery areas.
- Keep the working area as neat as practical.
- Remove or correct safety hazards as soon as they are noticed. Bend protruding nails over. Remove loose boards.
- Remember where other workers are and what they are doing.
- Keep fingers and hands away from cutting edges at all times.
- Stay alert!

Safety Hazards

Carpenters work in unfinished surroundings. While a house is being built, there are many unsafe places around the building site. You have to stand on or climb ladders, which can be unsafe. You may not have a good footing while standing on a ladder. You may not be climbing a ladder in the proper way. Holding onto the rungs of the ladder is very unsafe. You should always hold onto the outside rails of the ladder when climbing.

There are holes that can cause you to trip. They may be located in the front yard where the water or sewage lines come into the building. There may be holes for any number of reasons. These holes can cause you all kinds of problems, especially if you fall into them or turn your ankle.

The house in Fig. 1-6 is almost completed. However, if you look closely you can see that some wood has been left on the garage roof. This wood can slide down and hit a person working below. The front porch has not been poured. This means stepping out of the front door can be a rather long step. Other debris around the yard can be a source of trouble. Long sliv-

Fig. 1-6 *Even when a house is almost finished, there can still be hazards. Wood left on a roof could slide off and hurt someone, and without the front porch, it is a long step down.*

ers of flashing can cause trouble if you step on them and they rake your leg. You have to watch your every step around a construction site.

Outdoor work Much of the time carpentry is performed outdoors. This means you will be exposed to the weather, so dress accordingly. Wet weather increases the accident rate. Mud can make a secure place to stand hard to find. Mud can also cause you to slip if you don't clean it off your shoes. Be very careful when it is muddy and you are climbing on a roof or a ladder.

Tools Any tool that can cut wood can cut flesh. You have to keep in mind that although tools are an aid to the carpenter, they can also be a source of injury. A chisel can cut your hand as easily as the wood. In fact, it can do a quicker job on your hand than on the wood it was intended for. Saws can cut wood and bones. Be careful with all types of saws, both hand and electric. Hammers can do a beautiful job on your fingers if you miss the nailhead. The pain involved is intensified in cold weather. Broken bones can be easily avoided if you keep your eye on the nail while you're hammering. Besides that, you will get the job done more quickly. And, after all, that's why you are there—to get the job done and do it right the first time. Tools can help you do the job right. They can also cause you injury. The choice is up to you.

In order to work safely with tools you should know what they can do and how they do it. The next few pages are designed to help you use tools properly.

USING CARPENTER TOOLS

A carpenter is lost without tools. This means you have to have some way of containing them. A toolbox is

very important. If you have a place to put everything, then you can find the right tool when it is needed. A toolbox should have all the tools mentioned here. In fact, you will probably add more as you become more experienced. Tools have been designed for every task. All it takes is a few minutes with a hardware manufacturer's catalog to find just about everything you'll ever need. If you can't find what you need, the manufacturers are interested in making it.

Measuring Tools

Folding rule When using the folding rule, place it flat on the work. The 0 end of the rule should be exactly even with the end of the space or board to be measured. The correct distance is indicated by the reading on the rule.

A very accurate reading may be obtained by turning the edge of the rule toward the work. In this position, the marked graduations of the face of the rule touch the surface of the board. With a sharp pencil, mark the exact distance desired. Start the mark with the point of the pencil in contact with the mark on the rule. Move the pencil directly away from the rule while making the mark.

One problem with the folding rule is that it breaks easily if it is twisted. This happens most commonly when it is being folded or unfolded. The user may not be aware of the twisting action at the time. You should keep the joints oiled lightly. This makes the rule operate more easily.

Pocket tape Beginners may find the pocket tape (Fig. 1-7) the most useful measuring tool for all types of work. It extends smoothly to full length. It returns quickly to its compact case when the return button is pressed. Steel tapes are available in a variety of lengths. For most carpentry a rule 6, 8, 10, or 12 feet long is used.

Fig. 1-7 *Tape measure. (Stanley Tools)*

Longer tapes are available. They come in 20-, 50-, and 100-foot lengths. See Fig. 1-8. This tape can be extended to 50 feet to measure lot size and the location of a house on a lot. It has many uses around a building site. A crank handle can be used to wind it up once you are finished with it. The hook on the end of the tape makes it easy for one person to use it. Just hook the tape over the end of a board or nail and extend it to your desired length.

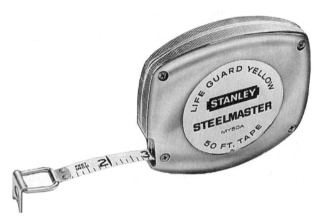

Fig. 1-8 *A longer tape measure.*

Saws

Carpenters use a number of different saws. These saws are designed for specific types of work. Many are misused. They will still do the job, but they would do a better job if used properly. Handsaws take quite a bit of abuse on a construction site. It is best to buy a good-quality saw and keep it lightly oiled.

Standard skew-handsaw This saw has a wooden handle. It has a 22-inch length. A 10-point saw (with 10 teeth per inch) is suggested for crosscutting. Crosscutting means cutting wood *across* the grain. The 26-inch-length, 5½-point saw is suggested for ripping, or cutting *with* the wood grain.

Figure 1-9 shows a carpenter using a handsaw. This saw is used in places where the electric saw cannot be used. Keeping it sharp makes a difference in the quality of the cut and the ease with which it can be used.

Backsaw The backsaw gets its name from the piece of heavy metal that makes up the top edge of the cutting part of the saw. See Fig. 1-10. It has a fine tooth configuration. This means it can be used to cut cross-grain and leave a smoother finished piece of work. This type of saw is used by finish carpenters who want to cut trim or molding.

Miter box As you can see from Fig. 1-11A, the miter box has a backsaw mounted in it. This box can be adjusted using the lever under the saw handle (see arrow). You can adjust it for the cut you wish. It can cut from 90° to 45°. It is used for finish cuts on moldings and trim materials. The angle of the cut is determined by the location of the saw in reference to the bed of the box. Release the clamp on the bottom of the saw support to adjust the saw to any degree desired. The wood is held with one hand against the fence of the box and the bed. Then the saw is used by the other hand. As you can see from the setup, the cutting should take place when the saw is pushed forward. The backward movement of the saw should be made with the pressure on the saw released slightly. If you try to cut on the backward movement, you will just pull the wood away from the fence and damage the quality of the cut.

Coping saw Another type of saw the carpenter can make use of is the coping saw (Fig. 1-12). This one can cut small thicknesses of wood at any curve or angle desired. It can be used to make sure a piece of paneling fits properly or a piece of molding fits another piece in the corner. The blade is placed in the frame with the teeth pointing toward the handle. This means it cuts only on the downward stroke. Make sure you properly support the piece of wood being cut. A number of blades can be obtained for this type of saw. The number of teeth in the blade determines the smoothness of the cut.

Hammers and Other Small Tools

There are a number of different types of hammers. The one the carpenter uses is the *claw* hammer. It has claws that can extract nails from wood if they have been put in the wrong place or have bent while being driven. Hammers can be bought in 20-ounce, 24-ounce, 28-ounce, and 32-ounce weights for carpentry work. Most carpenters prefer a 20-ounce. You have to work with a number of different weights to find out which will work best for you. Keep in mind that the hammer should be of tempered steel. If the end of the hammer has a tendency to splinter or chip off when it hits a nail, the pieces can hit you in the eye or elsewhere, causing serious damage. It is best to wear safety glasses whenever you use a hammer.

Nails are driven by hammers. Figure 1-13 shows the gage, inch, and penny relationships for the common box nail. The *d* after the number means *penny*. This is a measuring unit inherited from the English in the colonial days. There is little or no relationship between

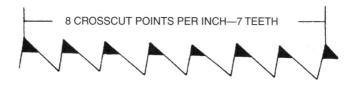

8 CROSSCUT POINTS PER INCH—7 TEETH

6 RIP POINTS PER INCH—5 TEETH

THE NUMBER OF POINTS PER INCH ON A HANDSAW
DETERMINES THE FINENESS OR COARSENESS
OF CUT. MORE POINTS PRODUCE A FINER CUT.

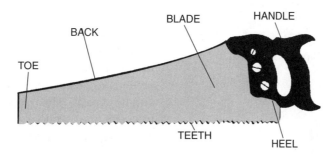

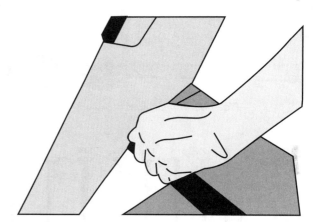

Fig. 1-9 *Using a handsaw.*

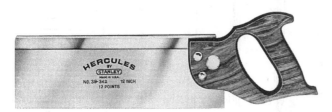

Fig. 1-10 *Backsaw.* (Stanley Tools)

penny and inches. If you want to be able to talk about it intelligently, you'll have to learn both inches and penny. The gage is nothing more than the American Wire Gage number for the wire that the nails were made from originally. Finish nails have the same measuring unit (penny) but do not have the large, flat heads.

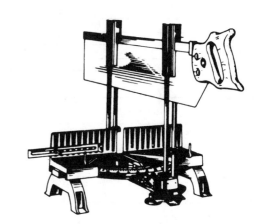

Fig. 1-11A Miter box. (Stanley Tools)

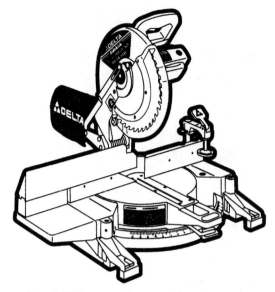

Fig. 1-11B *Powered compound miter saw.* (Delta)

Fig. 1-12 *Coping saw.* (Stanley Tools)

Nail set Finish nails are driven below the surface of the wood by a nail set. The nail set is placed on the head of the nail. The large end of the nail set is struck by the hammer. This causes the nail to go below the surface of the wood. Then the hole left by the countersunk nail is filled with wood filler and finished off with a smooth coat of varnish or paint. Figure 1-14 shows the nail set and its use.

The carpenter would be lost without a hammer. See Fig. 1-15. Here the carpenter is placing sheathing on rafters to form a roof base. The hammer is used to drive the boards into place, since they have to overlap slightly. Then the nails are driven by the hammer also.

In some cases a hammer will not do the job. The job may require a hatchet. See Fig. 1-16. This device can be used to pry and to drive. It can pry boards loose when they are improperly installed. It can sharpen posts to be driven at the site. The hatchet can sharpen the ends of stakes for staking out the site. It can also withdraw nails. This type of tool can also be used to drive stubborn sections of a wall into place when they are erected for the first time. The tool has many uses.

Scratch awl An awl is a handy tool for a carpenter. It can be used to mark wood with a scratch mark and to produce pilot holes for screws. Once it is in your tool box, you can think of a hundred uses for it. Since it does have a very sharp point, it is best to treat it with respect. See Fig. 1-17.

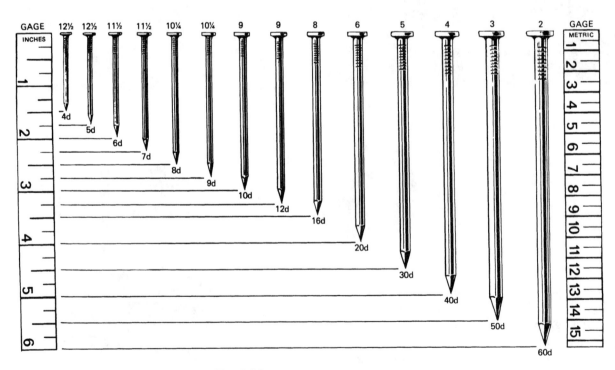

Fig. 1-13 *Nails.* (Forest Products Laboratory)

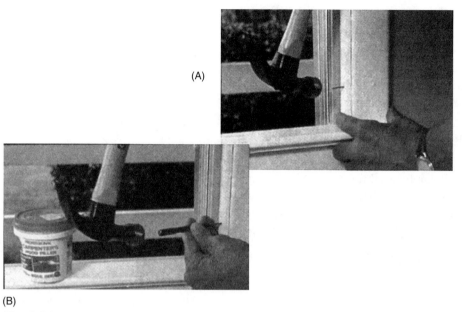

(A)

(B)

Fig. 1-14 *(A) Driving a nail with a hammer. (B) Finishing the job with a nail set to make sure the hammer doesn't leave an impression in the soft wood of the window frame.*

Fig. 1-15 *Putting on roof sheathing. The carpenter is using a hammer to drive the board into place.*

Fig. 1-16 *Hatchet.* (Stanley Tools)

Fig. 1-17 *Scratch awl.* (Stanley Tools)

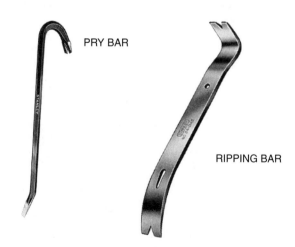

PRY BAR

RIPPING BAR

Fig. 1-18 *Wrecking bars.* (Stanley Tools)

Wrecking bar This device (Fig. 1-18) has a couple of names, depending on which part of the country you are in at the time. It is called a wrecking bar in some parts and a crowbar in others. One end has a chisel-sharp flat surface to get under boards and pry them loose. The other end is hooked so that the slot in the end can pull nails with the leverage of the long handle. This specially treated steel bar can be very helpful in prying away old and unwanted boards. It can be used to help give leverage when you are putting a wall in place and making it plumb. This tool has many uses for the carpenter with ingenuity.

Screwdrivers The screwdriver is an important tool for the carpenter. It can be used for many things other than turning screws. There are two types of screwdrivers. The standard type has a straight slot-fitting blade at its end. This type is the most common of screwdrivers. The Phillips-head screwdriver has a cross or X on the end to fit a screw head of the same design. Figure 1-19 shows the two types of screwdrivers.

Fig. 1-19 *Two types of screwdrivers.*

Squares

In order to make corners meet and standard sizes of materials fit properly, you must have things square. That calls for a number of squares to check that the two walls or two pieces come together at a perpendicular.

Try square The *try square* can be used to mark small pieces for cutting. If one edge is straight and the handle part of the square (Fig. 1-20) is placed against this straight edge, then the blade can be used to mark

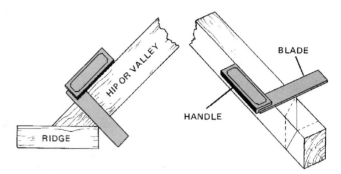

Fig. 1-20 *Use of a try square.* (Stanley Tools)

the wood perpendicular to the edge. This comes in handy when you are cutting 2 × 4s and want them to be square.

Framing square The framing square is a very important tool for the carpenter. It allows you to make square cuts in dimensional lumber. This tool can be used to lay out rafters and roof framing. See Fig. 1-21. It is also used to lay out stair steps.

Later in this book you will see a step-by-step procedure for using the framing square. The tools are described as they are called for in actual use.

Bevel A bevel can be adjusted to any angle to make cuts at the same number of degrees. See Fig. 1-22. Note how the blade can be adjusted. Now take a look at Fig. 1-23. Here you can see the overhang of rafters. If you want the ends to be parallel with the side of the house, you can use the bevel to mark them before they are cut off. Simply adjust the bevel so the handle is on top of the rafter and the blade fits against the soleplate below. Tighten the screw and move the bevel down the rafter to where you want the cut. Mark the angle along the blade of the bevel. Cut along the mark, and you have what you see in Fig. 1-23. It is a good device for transferring angles from one place to another.

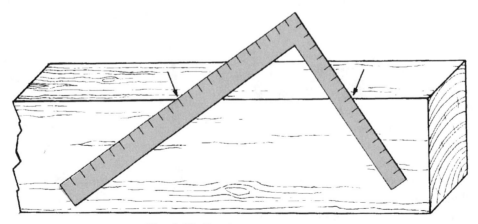

Fig. 1-21 *Framing square.* (Stanley Tools)

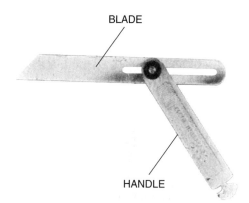

BLADE

HANDLE

Fig. 1-22 *Bevel. (Stanley Tools)*

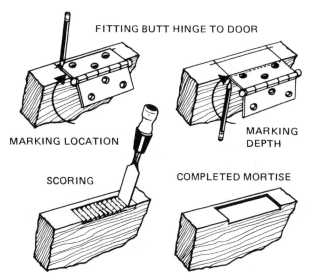

FITTING BUTT HINGE TO DOOR

MARKING LOCATION

MARKING DEPTH

SCORING

COMPLETED MORTISE

Fig. 1-24 *Using a wood chisel to complete a mortise.*

on the handle so the force of the hammer blows will not chip the handle. Other applications are up to you, the carpenter. You'll find many uses for the chisel in making things fit.

Plane Planes (Fig. 1-25) are designed to remove small shavings of wood along a surface. One hand holds the knob in front and the other the handle in back. The blade is adjusted so that only a small sliver of wood is removed each time the plane is passed over the wood. It can be used to make sure that doors and windows fit properly. It can be used for any number of wood smoothing operations.

Fig. 1-23 *Rafter overhang cut to a given angle.*

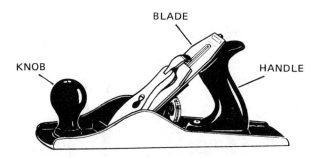

BLADE

KNOB

HANDLE

Fig. 1-25 *Smooth plane. (Stanley Tools)*

Chisel Occasionally you may need a wood chisel. It is sharpened on one end. When the other end is struck with a hammer, the cutting end will do its job. That is, of course, if you have kept it sharpened. See Fig. 1-24.

The chisel is commonly used in fitting or hanging doors. It is used to remove the area where the door hinge fits. Note how it is used to score the area (Fig. 1-24); it is then used at an angle to remove the ridges. A great deal of the work with the chisel is done by using the palm of the hand as the force behind the cutting edge. A hammer can be used. In fact, chisels have a metal tip

Dividers and compass Occasionally a carpenter must draw a circle. This is done with a compass. The compass shown in Fig. 1-26A can be converted to a divider by removing the pencil and inserting a straight steel pin. The compass has a sharp point that fits into the wood surface. The pencil part is used to mark the circle circumference. It is adjustable to various radii.

The dividers in Fig. 1-26A have two points made of hardened metal. They are adjustable. It is possible to use them to transfer a given measurement from the

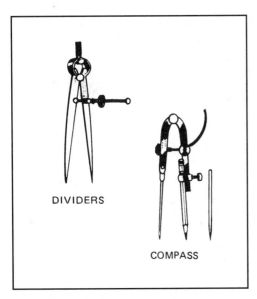

Fig. 1-26A *Dividers and compass*

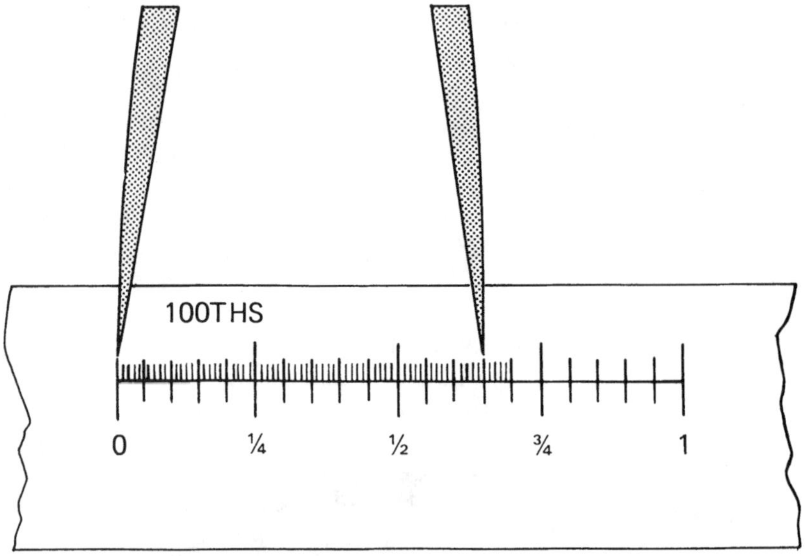

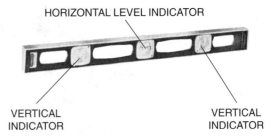

Fig. 1-26B *Dividers being used to transfer hundredths of an inch.*

framing square or measuring device to another location. See Fig. 1-26B.

Level In order to have things look as they should, a level is necessary. There are a number of sizes and shapes available. This one shown in Figure 1-27B is the most common type used by carpenters. The bubbles in the glass tubes tell you if the level is obtained. In Fig. 1-27A the carpenter is using the level to make sure the window is in properly before nailing it into place permanently.

If the vertical and horizontal bubbles are lined up between the lines, then the window is plumb, or verti-

HORIZONTAL LEVEL INDICATOR

VERTICAL INDICATOR

VERTICAL INDICATOR

Fig. 1-27B *A commonly used type of level.* (Stanley Tools)

cal. A plumb bob is a small, pointed weight. It is attached to a string and dropped from a height. If the bob is just above the ground, it will indicate the vertical direction by its string. Keeping windows, doors, and frames square and level makes a difference in fitting. It is much easier to fit prehung doors into a frame that is square. When it comes to placing panels of 4-×-8-foot plywood sheathing on a roof or on walls, squareness can make a difference as to fit. Besides, a square fit and a plumb door and window look better than those that are a little off. Figure 1-27C shows three plumb bobs.

Fig. 1-27A *Using a level to make sure a window is placed properly before nailing.* (Andersen)

Fig. 1-27C *Plumb bobs.* (Stanley Tools)

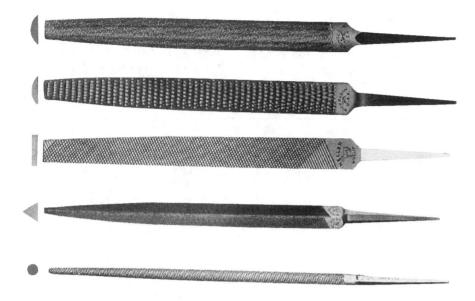

Fig. 1-28 Wood and cabinet files:
(A) Half-round; (B) Rasp; (C) Flat;
(D) Triangular; and (E) Round. (Millers
Falls Division, a division of Ingersol-Rand Co.)

Files A carpenter finds use for a number of types of files. The files have different surfaces for doing different jobs. Tapping out a hole to get something to fit may be just the job for a file. Some files are used for sharpening saws and touching up tool cutting edges. Figure 1-28 shows different types of files. Other files may also be useful. You can acquire them later as you develop a need for them.

Clamps C clamps are used for many holding jobs. They come in handy when placing kitchen cabinets by holding them in place until screws can be inserted and properly seated. This type of clamp can be used for an extra hand every now and then when two hands aren't enough to hold a combination of pieces till you can nail them. See Fig. 1-29.

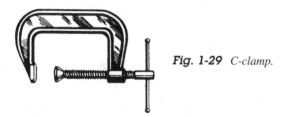

Fig. 1-29 C-clamp.

Cold chisel It is always good to have a cold chisel around. It is very much needed when you can't remove a nail. Its head may have broken off and the nail must be removed. The chisel can cut the nail and permit the separation of the wood pieces. See Fig. 1-30.

Fig. 1-30 Cold chisel. (Stanley Tools)

If a chisel of this type starts to "mushroom" at the head, you should remove the splintered ends with a grinder. Hammering on the end can produce a mushrooming effect. These pieces should be taken off since they can easily fly off when hit with a hammer. That is another reason for using eye protection when using tools.

Caulking gun In times of energy crisis, the caulking gun gets plenty of use. It is used to fill in around windows and doors and everywhere there may be an air leak. There are many types of caulk being made today. Another chapter will cover the details of the caulking compounds and their uses.

This gun is easily operated. Insert the cartridge and cut its tip to the shape you want. Puncture the thin plastic film inside. A bit of pressure will cause the caulk to come out the end. The long rod protruding from the end of the gun is turned over. This is so the serrated edge will engage the hand trigger. Remove the pressure from the cartridge when you are finished. Do this by rotating the rod so that the serrations are not engaged by the trigger of the gun.

Power Tools

The carpenter uses many power tools to aid in getting the job done. The quicker the job is done, the more valuable the work of the carpenter becomes. This is called productivity. The more you are able to produce, the more valuable you are. This means the contractor can make money on the job. This means you can have a job the next time there is a need for a good carpenter. Power tools make your work go faster. They also help you to do a job without getting fatigued. Many tools have been designed with you in mind. They are portable and operate from an extension cord.

Table 1-1 *Size of Extension Cords for Portable Tools*

Cord Length, Feet	Full-Load Rating of the Tool in Amperes at 115 Volts					
	0 to 2.0	2.10 to 3.4	3.5 to 5.0	5.1 to 7.0	7.1 to 12.0	12.1 to 16.0
	Wire Size (AWG)					
25	18	18	18	16	14	14
50	18	18	18	16	14	12
75	18	18	16	14	12	10
100	18	16	14	12	10	8
200	16	14	12	10	8	6
300	14	12	10	8	6	4
400	12	10	8	6	4	4
500	12	10	8	6	4	2
600	10	8	6	4	2	2
800	10	8	6	4	2	1
1000	8	6	4	2	1	0

If the voltage is lower than 115 volts at the outlet, have the voltage increased or use a much larger cable than listed.

The extension cord should be the proper size to take the current needed for the tool being used. See Table 1-1. Note how the distance between the outlet and the tool using the power is critical. If the distance is great, then the wire must be larger in size to handle the current without too much loss. The higher the number of the wire, the smaller the diameter of the wire. The larger the size of the wire (diameter), the more current it can handle without dropping the voltage.

Some carpenters run an extension cord from the house next door for power before the building site is furnished power. If the cord is too long or has the wrong size wire, it drops the voltage below 115. This means the saws or other tools using electricity will draw more current and therefore drop the voltage more. Every time the voltage is dropped, the device tries to obtain more current. This becomes a self-defeating phenomenon. You wind up with a saw that has little cutting power. You may have a drill that won't drill into a piece of wood without stalling. Of course the damage done to the electric motor is in some cases irreparable. You may have to buy a new saw or drill. Double-check Table 1-1 for the proper wire size in your extension cord.

Portable saw   This is the most often used and abused of carpenter's equipment. The electric portable saw, such as the one shown in Fig. 1-31, is used to cut all 2 × 4s and other dimensional lumber. It is used to cut off rafters. This saw is used to cut sheathing for roofs. It is used for almost every sawing job required in carpentry.

This saw has a guard over the blade. The guard should always be left intact. Do not remove the saw guard. If not held properly against the wood being cut, the saw can kick back and into your leg.

You should always wear safety glasses when using this saw. The sawdust is thrown in a number of direc-

Fig. 1-31 *Portable power saw. The favorite power tool of every carpenter. Note the blade should not extend more than ⅛" below the wood being cut. Also note the direction of the blade rotation.*

tions, and one of these is straight up toward your eyes. If you are watching a line where you are cutting, you definitely should have on glasses.

Table saw   If the house has been enclosed, it is possible to bring in a table saw to handle the larger cutting jobs. See Fig. 1-32. You can do ripping a little more safely with this type of saw because it has a rip fence. If a push stick is used to push the wood through and past the blade, it is safe to operate. Do not remove the safety guard. This saw can be used for both crosscut and rip. The blade is lowered or raised to the thickness of the wood. It should protrude about ¼ to ½ inch above the wood being cut. This saw usually requires a 1-horsepower motor. This means it will draw about 6.5 amperes to run and over 35 to start. It is best not to run the saw on an extension cord. It should be wired directly to the power source with circuit breakers installed in the line.

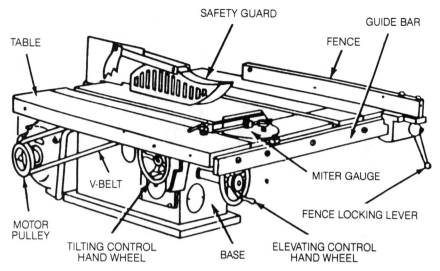

Fig. 1-32 *Table saw.* (Power Tool Division, Rockwell International)

Labels: SAFETY GUARD, GUIDE BAR, TABLE, FENCE, MITER GAUGE, V-BELT, FENCE LOCKING LEVER, MOTOR PULLEY, TILTING CONTROL HAND WHEEL, BASE, ELEVATING CONTROL HAND WHEEL

Radial arm saw This type of saw is brought in only if the house can be locked up at night. The saw is expensive and too heavy to be moved every day. It should have its own circuit. The saw will draw a lot of current when it hits a knot while cutting wood. See Fig. 1-33.

Fig. 1-33 *Radial arm saw.* (DeWalt)

In this model the moving saw blade is pulled toward the operator. In the process of being pulled toward you, the blade rotates so that it forces the wood being cut against the bench stop. Just make sure your left hand is in the proper place when you pull the blade back with your right hand. It takes a lot of care to operate a saw of this type. The saw works well for cutting large-dimensional lumber. It will crosscut or rip. This saw will also do miter cuts at almost any angle. Once you become familiar with it, the saw can be used to bevel crosscut, bevel miter, bevel rip, and even cut cir-

cles. However, it does take practice to develop some degree of skill with this saw.

Router The router has a high-speed type of motor. It will slow down when overloaded. It takes the beginner some time to adjust to *feeding* the router properly. If you feed it too fast, it will stall or burn the edge you're routing. If you feed it too slowly, it may not cut the way you wish. You will have to practice with this tool for some time before you're ready to use it to make furniture. It can be used for routing holes where needed. It can be used to take the edges off laminated plastic on countertops. Use the correct bit, though. This type of tool can be used to the extent of the carpenter's imagination. See Fig. 1-34.

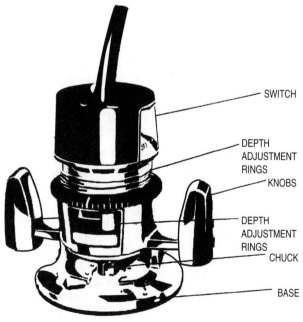

Fig. 1-34 *The hand-held router has many uses in carpentry.*

Labels: SWITCH, DEPTH ADJUSTMENT RINGS, KNOBS, DEPTH ADJUSTMENT RINGS, CHUCK, BASE

Using Carpenter Tools 15

Saw blades There are a number of saw blades available for the portable, table, or radial saw. They may be standard steel types or they may be carbide tipped. Carbide-tipped blades tend to last longer. See Fig. 1-35.

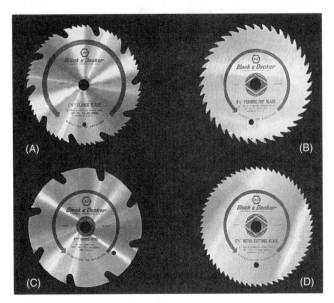

Fig. 1-35 *Saw blades. (A) Planer blade; (B) Framing rip blade; (C) Carbide tipped; (D) Metal cutting blade.* (Black & Decker)

Combination blades (those that can be used for both crosscut and rip) with a carbide tip give a smooth finish. They come in 7–7¼-inch diameter with 24 teeth. The arbor hole for mounting the blade on the saw is ¾ to ⅝ inch. A safety combination blade is also made in 10-inch-diameter size with 10 teeth and the same arbor hole sizes as the combination carbide-tipped blade.

The planer blade is used to crosscut, rip, or miter hard or soft woods. It is 6½ or 10 inches in diameter with 50 teeth. It too can fit anything from ¾- to ⅝-inch arbors.

If you want a smooth cut on plywood without the splinters that plywood can generate, you had better use a carbide-tipped plywood blade. It is equipped with 60 teeth and can be used to cut plywood, Formica, or laminated countertop plastic. It can also be used for straight cutoff work in hard or soft woods. Note the shape of the saw teeth to get some idea as to how each is designed for a specific job. You can identify these after using them for some time. Until you can, mark them with a grease pencil or marking pen when you take them off. A teflon coated blade works better when cutting treated lumber.

Saber saw The saber saw has a blade that can be used to cut circles in wood. See Fig. 1-36. It can be used to cut around any circle or curve. If you are making an inside cut, it is best to drill a starter hole first. Then, insert the blade into the hole and follow your mark. The saber saw is especially useful in cutting out

Fig. 1-36 *Saber saw.*

holes for heat ducts in flooring. Another use for this type of saw is cutting holes in roof sheathing for pipes and other protrusions. The saw blade is mounted so that it cuts on the upward stroke. With a fence attached, the saw can also do ripping.

Drill The portable power drill is used by carpenters for many tasks. Holes must be drilled in soleplates for anchor bolts. Using an electric power drill (Fig. 1-37A) is faster and easier than drilling by hand. This drill is capable of drilling almost any size hole through dimensional lumber. A drill bit with a carbide tip enables the carpenter to drill in concrete as well as bricks. Carpenters use this type of masonry hole to insert anchor bolts in concrete that has already hardened. Electrical boxes have to be mounted in drilled holes in the brick and concrete. The job can be made easier and can be more

Fig. 1-37A *Hand-held portable drill.*

efficiently accomplished with the portable power hand drill.

The drill has a tough, durable plastic case. Plastic cases are safer when used where there is electrical work in progress.

Carpenters are now using cordless electric drills (Fig. 1-37B). Cordless drills can be moved about the job without the need for extension cords. Improved battery technology has made the cordless drills almost as powerful as regular electric drills. The cordless drill has numbers on the chuck to show the power applied to the shaft. Keep in mind that the higher the number, the greater the torque. At low power settings, the chuck will slip when the set level of power is reached. This allows the user to set the drill to drive screws.

Fig. 1-37B *A cordless hand drill with variable torque*

Figure 1-37C shows a cordless drill and a cordless saw. This cordless technology is now used by carpenters and do-it-yourselfers. Cordless tools can be obtained in sets that use the same charger system (Fig. 1-37D). An

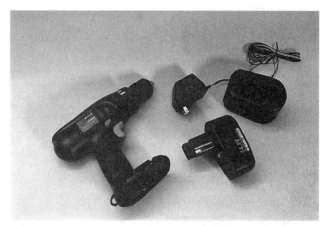

Fig. 1-37D *One charger can be used to charge saw and drill batteries of same voltage.*

extra set of batteries should be kept charging at all times and then "swapped out" for the discharged ones. This way no time is lost waiting for the battery to reach full charge. Batteries for cordless tools are rated by battery voltage. High voltage gives more power than low voltage.

As a rule, battery-powered tools do not give the full power of regular tools. However, most jobs don't require full power. Uses for electric drills are limited only by the imagination of the user. The cordless feature is very handy when mounting countertops on cabinets. Sanding discs can be placed in the tool and used for finishing wood. Wall and roof parts are often screwed in place rather than nailed. Using the drill with special screwdriver bits can make the job faster than nailing.

Sanders The belt sander shown in Fig. 1-38 and the orbital sanders shown in Figs. 1-39A and B can do almost any required sanding job. The carpenter needs the sander occasionally. It helps align parts properly, especially those that don't fit by just a small amount. The sander can be used to finish off windows, doors, counters, cabinets, and floors. A larger model of the

Fig. 1-37C *A cordless drill and a cordless saw using matching batteries.*

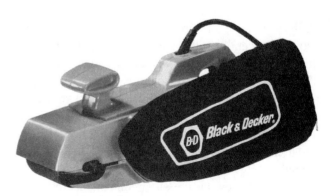

Fig. 1-38 *Belt sander.* (Black & Decker)

(A)

(B)

Fig. 1-39 *Orbital sanders: (A) dual action and (B) single action.*
(Black & Decker)

Fig. 1-40 *Air-powered nailer.* (Duo-Fast)

belt sander is used to sand floors before they are sealed and varnished. The orbital or vibrating sanders are used primarily to put a very fine finish on a piece of wood. Sandpaper is attached to the bottom of the sander. The sander is held by hand over the area to be sanded. The operator has to remove the sanding dust occasionally to see how well the job is progressing.

Nailers One of the greatest tools the carpenter has acquired recently is the nailer. See Fig. 1-40. It can drive nails or staples into wood better than a hammer. The nailer is operated by compressed air. The staples and nails are especially designed to be driven by the machine. See Tables 1-2 and 1-3 for the variety of fasteners used with this type of machine. The stapler or nailer can also be used to install siding or trim around a window.

The tool's low air pressure requirements (60 to 90 pounds per square inch) allow it to be moved from place to place. Nails for this machine (Fig. 1-40) are from 6d to 16d. It is magazine-fed for rapid use. Just pull the trigger.

FOLLOWING CORRECT SEQUENCES

One of the important things a carpenter must do is follow a sequence. Once you start a job, the sequence has to be followed properly to arrive at a completed house in the least amount of time.

Preparing the Site

Preparing the site may be expensive. There must be a road or street. In most cases the local ordinances require a sewer. In most locations the storm sewer and the sanitary sewer must be in place before building starts. If a sanitary sewer is not available you should plan for a septic tank for sewage disposal.

Figure 1-41 shows a sewer project in progress. This shows a street being extended. The storm sewer lines are visible, as is the digger. Trees had to be removed first by a bulldozer. Once the sewer lines are in, the roadbed or street must be properly prepared. Figure 1-42 shows the building of a street. Proper drainage is very important. Once the street is in and the curbs poured, it is time to locate the house.

Figure 1-43 shows how the curb has been broken and the telephone terminal box installed in the weeds. Note the stake with a small piece of cloth on it. This marks the location of the site.

Table 1-2 *Fine Wire Staples for a Pneumatic Staple Driver*

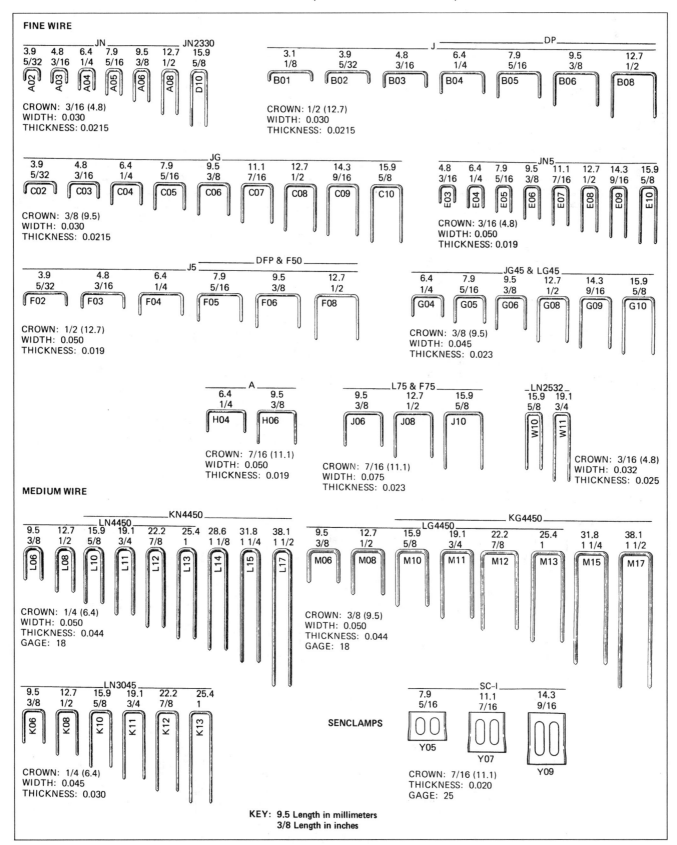

FINE WIRE

JN						JN2330
3.9 5/32	4.8 3/16	6.4 1/4	7.9 5/16	9.5 3/8	12.7 1/2	15.9 5/8
A02	A03	A04	A05	A06	A08	D10

CROWN: 3/16 (4.8)
WIDTH: 0.030
THICKNESS: 0.0215

	J		DP			
3.1 1/8	3.9 5/32	4.8 3/16	6.4 1/4	7.9 5/16	9.5 3/8	12.7 1/2
B01	B02	B03	B04	B05	B06	B08

CROWN: 1/2 (12.7)
WIDTH: 0.030
THICKNESS: 0.0215

			JG					
3.9 5/32	4.8 3/16	6.4 1/4	7.9 5/16	9.5 3/8	11.1 7/16	12.7 1/2	14.3 9/16	15.9 5/8
C02	C03	C04	C05	C06	C07	C08	C09	C10

CROWN: 3/8 (9.5)
WIDTH: 0.030
THICKNESS: 0.0215

			JN5				
4.8 3/16	6.4 1/4	7.9 5/16	9.5 3/8	11.1 7/16	12.7 1/2	14.3 9/16	15.9 5/8
E03	E04	E05	E06	E07	E08	E09	E10

CROWN: 3/16 (4.8)
WIDTH: 0.050
THICKNESS: 0.019

		J5	DFP & F50		
3.9 5/32	4.8 3/16	6.4 1/4	7.9 5/16	9.5 3/8	12.7 1/2
F02	F03	F04	F05	F06	F08

CROWN: 1/2 (12.7)
WIDTH: 0.050
THICKNESS: 0.019

		JG45 & LG45			
6.4 1/4	7.9 5/16	9.5 3/8	12.7 1/2	14.3 9/16	15.9 5/8
G04	G05	G06	G08	G09	G10

CROWN: 3/8 (9.5)
WIDTH: 0.045
THICKNESS: 0.023

A	
6.4 1/4	9.5 3/8
H04	H06

CROWN: 7/16 (11.1)
WIDTH: 0.050
THICKNESS: 0.019

L75 & F75		
9.5 3/8	12.7 1/2	15.9 5/8
J06	J08	J10

CROWN: 7/16 (11.1)
WIDTH: 0.075
THICKNESS: 0.023

LN2532	
15.9 5/8	19.1 3/4
W10	W11

CROWN: 3/16 (4.8)
WIDTH: 0.032
THICKNESS: 0.025

MEDIUM WIRE

	LN4450	KN4450						
9.5 3/8	12.7 1/2	15.9 5/8	19.1 3/4	22.2 7/8	25.4 1	28.6 1 1/8	31.8 1 1/4	38.1 1 1/2
L06	L08	L10	L11	L12	L13	L14	L15	L17

CROWN: 1/4 (6.4)
WIDTH: 0.050
THICKNESS: 0.044
GAGE: 18

		LG4450	KG4450				
9.5 3/8	12.7 1/2	15.9 5/8	19.1 3/4	22.2 7/8	25.4 1	31.8 1 1/4	38.1 1 1/2
M06	M08	M10	M11	M12	M13	M15	M17

CROWN: 3/8 (9.5)
WIDTH: 0.050
THICKNESS: 0.044
GAGE: 18

	LN3045				
9.5 3/8	12.7 1/2	15.9 5/8	19.1 3/4	22.2 7/8	25.4 1
K06	K08	K10	K11	K12	K13

CROWN: 1/4 (6.4)
WIDTH: 0.045
THICKNESS: 0.030

SENCLAMPS

	SC-I	
7.9 5/16	11.1 7/16	14.3 9/16
Y05	Y07	Y09

CROWN: 7/16 (11.1)
THICKNESS: 0.020
GAGE: 25

KEY: 9.5 Length in millimeters
3/8 Length in inches

Table 1-3 7-Digit Nail Ordering System

1st Digit: Diameter, Inches		2d Digit: Head		3d and 4th Digits: Length, Inches		5th Digit: Point		6th Digit: Wire Chem. and Finish		7th Digit: Finish	
A	0.0475	A	Brad	08	1/2	A	Diam.-reg.	A	Std. carbon-galv.	A	Plain
D	0.072	C	Flat	11	3/4	E	Chisel	E	Std. carb.	B	Sencote
E	0.0915	E	Flat/ring	13	1				"Weatherex"galv.	C	Painted
G	0.113		shank	15	1 1/4			G	Stainless steel	D	Painted and
H	0.120	F	Flat/screw	17	1 1/2				std. tensile		sencote
J	0.105		shank	19	1 3/4			H	Hardened high-		
K	0.131	Y	Slight-	20	1 7/8				carbon bright basic		
			headed	21	2						
U	0.080		pin	22	2 1/8			P	Std. carbon bright basic		
		Z	Headless pin	23	2 1/4						
				24	2 3/8						
				25	2 1/2						
				26	2 3/4						
				27	3	EXAMPLE: 10 1/4 ga. (K), flat head (C),					
				28	3 1/4	KC25AAA—2 1/2" (25), regular point (A), std. carb.					
				29	3 1/2	galvanized (A), plain, or uncoated (A) Senco-Nail					

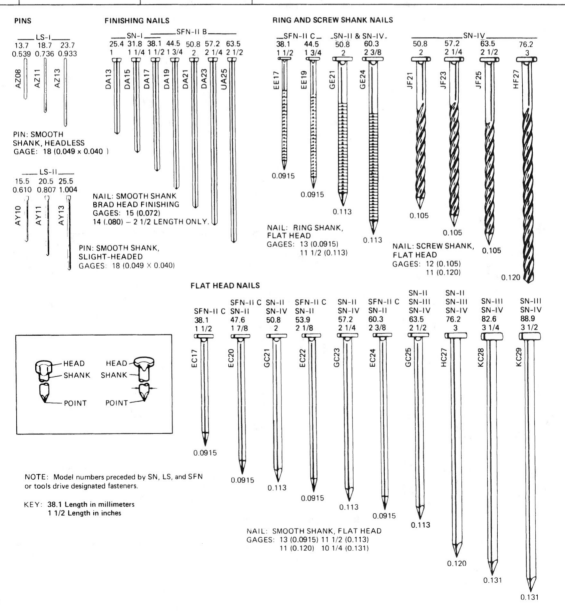

PINS

LS-I
13.7 18.7 23.7
0.539 0.736 0.933

AZ08 AZ11 AZ13

PIN: SMOOTH
SHANK, HEADLESS
GAGE: 18 (0.049 x 0.040)

LS-II
15.5 20.5 25.5
0.610 0.807 1.004

AY10 AY11 AY13

PIN: SMOOTH SHANK,
SLIGHT-HEADED
GAGES: 18 (0.049 × 0.040)

FINISHING NAILS

SN-I _____ SFN-II B _____
25.4 31.8 38.1 44.5 50.8 57.2 63.5
1 1 1/4 1 1/2 1 3/4 2 2 1/4 2 1/2

DA13 DA15 DA17 DA19 DA21 DA23 UA25

NAIL: SMOOTH SHANK
BRAD HEAD FINISHING
GAGES: 15 (0.072)
14 (.080) – 2 1/2 LENGTH ONLY.

RING AND SCREW SHANK NAILS

SFN-II C
38.1 44.5 50.8 60.3
1 1/2 1 3/4 2 2 3/8

SN-II & SN-IV

EE17 EE19 GE21 GE24

0.0915
0.0915
0.113
0.113

NAIL: RING SHANK,
FLAT HEAD
GAGES: 13 (0.0915)
11 1/2 (0.113)

SN-IV
50.8 57.2 63.5 76.2
2 2 1/4 2 1/2 3

JF21 JF23 JF25 HF27

0.105
0.105
0.105
0.120

NAIL: SCREW SHANK,
FLAT HEAD
GAGES: 12 (0.105)
11 (0.120)

FLAT HEAD NAILS

SFN-II C   SFN-II C   SN-II   SFN-II C   SN-II   SFN-II C   SN-II   SN-II   SN-III   SN-III
           SN-II   SN-IV   SN-II   SN-IV   SN-II   SN-III SN-III SN-IV   SN-IV
38.1   47.6   50.8   53.9   57.2   60.3   63.5   76.2   82.6   88.9
1 1/2   1 7/8   2   2 1/8   2 1/4   2 3/8   2 1/2   3   3 1/4   3 1/2

EC17 EC20 GC21 EC22 GC23 EC24 GC25 HC27 KC28 KC29

0.0915 0.0915 0.113 0.0915 0.113 0.0915 0.113 0.120 0.131 0.131

NAIL: SMOOTH SHANK, FLAT HEAD
GAGES: 13 (0.0915) 11 1/2 (0.113)
11 (0.120) 10 1/4 (0.131)

HEAD SHANK POINT
HEAD SHANK POINT

NOTE: Model numbers preceded by SN, LS, and SFN
or tools drive designated fasteners.

KEY: 38.1 Length in millimeters
1 1/2 Length in inches

Fig. 1-41 *Street being extended for a new subdivision.*

Fig. 1-42 *The beginning of a street.*

Fig. 1-43 *Locating a building site and removing the curb for the driveway.*

As you can see in Fig. 1-44, the curb has been removed. A gravel bed has been put down for the driveway.

The sewer manhole sticks up in the driveway. The basement has been dug. Dirt piles around it show how deep the basement really is. However, a closer look shows that the hole isn't too deep. That means the dirt will be pushed back against the basement wall to form a higher level for the house. This will provide drainage

Fig. 1-44 *Dirt from the basement excavation is piled high around a building site.*

away from the house when finished. See Fig. 1-45 for a look at the basement hole.

The Basement

In Fig. 1-46 the columns and the foundation wall have been put up. The basement is prepared in this case with courses of block with brick on the outside. This basement appears to be more of a crawl space under the first floor than a full stand-up basement.

Once the basement is finished and the floor joists have been placed, the flooring is next.

Fig. 1-45 *Hole for a basement.*

Fig. 1-46 *The columns and foundation walls will help support the floor parts.*

The Floor

Once the basement or foundation has been laid for the building, the next step is to place the floor over the joists. Note in Fig. 1-47 that the grooved flooring is laid in large sheets. This makes the job go faster and reinforces the floor.

Wall Frames

Once the floor is in place and the basement entrance hole has been cut, the floor can be used to support the wall frame. The 2 × 4s or 2 × 6s for the framing can be placed on the flooring and nailed. Once together, they are pushed into the upright position as in Fig. 1-48. For a two-story house, the second floor is placed on the first-story wall supports. Then the second-floor walls are nailed together and raised into position.

Fig. 1-47 *Carpenters are laying plywood subflooring with tongue-and-groove joints. This is stronger.* (American Plywood Association)

Fig. 1-48 *Wall frames are erected after the floor frame is built.*

Sheathing

Once the sheathing is on and the walls are upright, it is time to concentrate on the roof. See Fig. 1-49. The rafters are cut and placed into position and nailed firmly. See Fig. 1-50. They are reinforced by the proper horizontal bracing. This makes sure they are properly designed for any snow load or other loads that they may experience.

Roofing

The roofing is applied after the siding is on and the rafters are erected. The roofing is completed by applying the proper underlayment and then the shingles. If asphalt shingles are used, the procedure is slightly different from that for wooden shingles. Shingles and roofing are covered in another chapter in this book. Figure 1-51 shows the sheathing in place and ready for the roofing.

Fig. 1-49 *Beginning construction of the roof structure.* (Georgia-Pacific)

Fig. 1-50 *Framing and supports for rafters.*

Fig. 1-51 *Fiberboard sheathing over the wall frame.*

Fig. 1-53 *Siding applied to building. Note the pattern of the staples.*

Siding

After the roofing, the finishing job will have to be undertaken. The windows and doors are in place. Finish touches are next. The plumbing and drywall may already be in. Then the siding has to be installed. In some cases, of course, it may be brick. This calls for bricklayers to finish up the exterior. Otherwise the carpenter places siding over the walls. Figure 1-52 shows the beginning of the siding at the top left of the picture.

Figure 1-53 shows how the siding has been held in place with a stapler. The indentations in the wood show a definite pattern. The siding is nailed to the nail base underneath after a coating of tar paper (felt paper in some parts of the country) is applied to the nail base or sheathing.

Finishing

Exterior finishing requires a bit of caulking with a caulking gun. Caulk is applied to the siding that butts the windows and doors.

Finishing the interior can be done at a more leisurely pace once the exterior is enclosed. The plumbing and electrical work has to be done before the drywall or plaster is applied. Once the wallboard has been finished, the trim can be placed around the edges of the walls, floors, windows, and doors. The flooring can be applied after the finishing of the walls and ceiling. The kitchen cabinets must be installed before the kitchen flooring. There is a definite sequence to all these operations.

As you can imagine, it would be impossible to place roofing on a roof that wasn't there. It takes planning and following a sequence to make sure the roof is there when the roofing crew comes around to nail the shingles in place. The water must be there before you can flush the toilets. The electricity must be hooked up before you can turn on a light. These are reasonable things. All you have to do is sit down and plan the whole operation before starting. Planning is the key to sequencing. Sequencing makes it possible for everyone to be able to do a job at the time assigned to do it.

Fig. 1-52 *Siding applied on the top left side of the building.*

The Laser Level

The need for plumb walls and level mouldings as well as various other points straight and level is paramount in house building. It is difficult in some locations to establish a reference point to check for level windows, doors, and roofs, as well as ceilings and steps.

The laser level (Fig. 1-54) has eliminated much of this trouble in house building. This simple, easy-to-use tool is accurate to within ⅛ of an inch in 150 feet, and it has become less expensive recently so that even do-it-yourselfers can rent or buy one.

The laser level can generate a vertical reference plane for positioning a wall partition or for setting up forms. See Fig. 1-55. It can produce accurate height gauging and alignment of ceilings, mouldings, horizontal planes and can accurately locate doorways, windows, and thresholds for precision framing and finishing. See Fig. 1-56. The laser level can aid in leveling floors, both indoors and out. It can be used to check stairs, slopes, and drains. The laser beam is easy to use and accurate in locating markings for roof pitches, and it works well in hard-to-reach situations. See Fig. 1-57. The laser beam is generated by two AAA alkaline batteries that will operate for up to 16 hours.

Fig. 1-54 *Laserspirit level moves 360° horizontally and 360° vertically with the optional lens attachment. It sets up quickly and simply with only two knobs to adjust.* (Stabila®)

Fig. 1-55 *The laser level can be used to align ceilings, mouldings, and horizontal planes; it produces accurate locations for doorways, windows, and thresholds for precision framing and finishing.* (Stabila®)

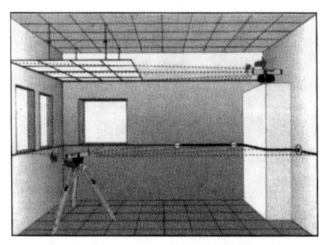

Fig. 1-56 *The laser level can be used for indoor or outdoor leveling of floors, stairs, slopes, drains and ceilings, and mouldings around the room.* (Stabila®)

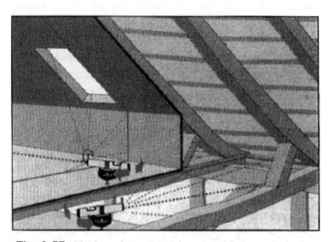

Fig. 1-57 *The laser beam is used to provide easy and accurate location markings on pitches and in hard-to-reach situations.* (Stabilia®)

The combination laser and spirit level quickly and accurately lays out squares and measures plumb. No protective eyewear is needed. The laser operates on a wavelength of 635 nm and can have an extended range up to 250 feet.

CHAPTER 1
STUDY QUESTIONS

1. What is carpentry involved with in terms of processes?

2. Can you learn carpentry from a book?

3. Why should a carpenter wear steel-toed shoes?

4. Why should a carpenter use safety glasses on the job?

5. Why is a hard hat suggested when a carpenter is working on site?

6. Why are gloves used on a carpentry job?

7. List at least six safety procedures that should be followed at all times.

8. List at least five important habits to acquire for safe working on the job.

9. Why would a carpenter need a folding rule?

10. What is the difference between a rip and a cross-cut saw?

11. What is a miter box?

12. What is a backsaw?

13. Why would a carpenter need a nail set?

14. What does a carpenter use a wrecking bar for?

15. What's the difference between a try square and a bevel?

16. Where does the carpenter use a caulking gun?

17. Why is the length of an extension cord important?

18. What's the difference between a table saw and a radial arm saw?

19. Why is following a sequence important?

20. How has the laser level aided in home construction?

2
CHAPTER

Preparing the Site

INITIAL PLANNING IS VITALLY IMPORTANT TO the success of the completed job. Before you begin any construction, you must plan the building on paper. Always keep in mind that it is much cheaper to make a mistake on paper than on site.

In this chapter, you will learn how to develop these skills:

- Locate the boundaries
- Lay out buildings
- Use the carpenter's level
- Prepare for the start of construction

Each has its importance. Locating the boundaries will allow you to properly lay out the building on the site.

The builder's level will show you how to make sure the building is level. No new leaning tower of Pisa is needed today. One is enough. It would be very hard to sell one today. That means you should choose the site so that the soil will support the weight of the building.

There is a basic sequence to follow in building. It should be followed for the benefit of those who are supposed to operate as part of the team.

BASIC SEQUENCE

The basic sequence involves the following operations:

1. Cruise the site and plan the job.
2. Locate the boundaries.
3. Locate the building area or areas.
4. Define the site work that is needed.
5. Clear any unwanted trees.
6. Lay out the building.
7. Establish the exact elevations.
8. Excavate the basement or foundation.
9. Provide for access during construction.
10. Start the delivery of materials to the site.
11. Have a crew arrive to start with the footings.

LOCATING THE BUILDING ON THE SITE

The proper location of a building is very important. It would be embarrassing and costly to move a building once it is built. That means a lot of things have to be checked first.

Property Boundaries

First, a clear deed to the land should be established. This can be done in the county courthouse. Check the records or have someone who is paid for this type of work do it. An abstract of the history of the ownership of the land is usually provided. In Iowa, for instance, the abstract traces ownership back to the Louisiana Purchase of 1803. In New York State the history of the land is traced by owners from the days of the Holland Land Company. Alabama can provide records back to the time the Creek or other Indians owned the land. Each state has its own history and its own procedure for establishing absolute ownership of land. It is best to have proof of this ownership before starting any construction project.

Surveyors should be called in to establish the limits of the property. A plot plan is drawn by the surveyors. This can be used to locate the property. Figure 2-1 is a plot plan showing the location of a house on a lot.

Sidewalks, utilities easements, and other things have to be taken into consideration. The location of the house may be specified by local ordinance. This type of ordinance will usually specify what clearance the house must have on each side. It will probably set the limits of setback from the street. You may also want to plan around trees. Since trees increase the

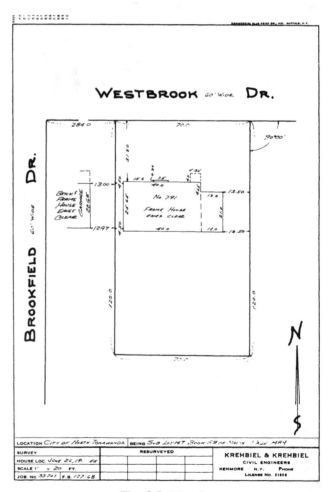

Fig. 2-1 *Plot plan.*

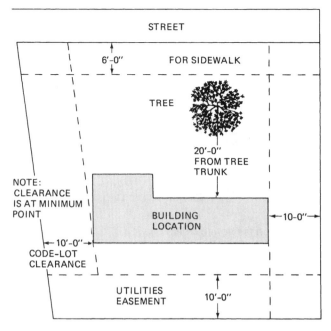

Fig. 2-2 Site location must be chosen carefully.

STREET

6'-0" FOR SIDEWALK

TREE

20'-0"
FROM TREE
TRUNK

NOTE:
CLEARANCE
IS AT MINIMUM
POINT

BUILDING
LOCATION

10-0"

10'-0"
CODE-LOT
CLEARANCE

UTILITIES
EASEMENT

10'-0"

Fig. 2-3 Rough-cleared lot. Only weeds need to be taken out as the basement is dug.

value of the property, it is important to save as many as possible. Figure 2-2 shows a sketch of some of these considerations.

An *easement* is the right of the utilities to use the space to furnish electric power, phone service, and gas to your location and to others nearby. This means you have given them permission to string wires or bury lines to provide their services. Keep in mind also the rights of the city or township to supply water and sewers. These may also cut across the property.

Laying Out the Foundation

Layout of the foundation is the critical beginning in house construction. It is a simple but extremely important process. It requires careful work. Make sure the foundation is square and level. You will find all later jobs, from rough carpentry through finish construction and installation of cabinetry, are made much easier.

1. Make sure your proposed house location on the lot complies with local regulations.
2. Set the house location, based on required setbacks and other factors, such as the natural drainage pattern of the lot. Level or at least rough-clear the site. See Fig. 2-3.
3. Lay out the foundation lines. Figure 2-4 shows the simplest method for locating these. Locate each outside corner of the house and drive small stakes into the ground. Drive tacks into the tops of the stakes. This is to indicate the outside line of the foundation wall. This is not the footings limit but

the outside wall limit. Next check the squareness of the house by measuring the diagonals, corner to corner, to see that they are equal. If the structure is rectangular, all diagonal measurements will be equal. You can check squareness of any corner by measuring 6 feet down one side, then 8 feet down the other side. The diagonal line between these two end points should measure exactly 10 feet. If it doesn't, the corner isn't truly square. See Fig. 2-5.

4. After the corners are located and squared, drive three 2×4 stakes at each corner as shown in Fig. 2-4. Locate these stakes 3 feet and 4 feet outside the actual foundation line. Then nail 1×6 batter boards horizontally so that their top edges are all level and at the same grade. Levelness will be checked later. Hold a string line across the tops of opposite batter boards at two corners. Using a plumb bob, adjust the line so that it is exactly over the tacks in the two corner stakes. Cut saw kerfs ¼ inch deep where the line touches the batter boards so that the string lines may be easily replaced if they are broken or disturbed. Figure 2-6 shows how carpenters in some parts of the country use a nail instead of the saw kerf to hold the thread or string. Figure 2-7 shows how the details of the location of the stake are worked out. This one is a 3 – 4 – 5 triangle, or 9 feet and 12 feet on the sides and 15 feet on the diagonal. If you use 6, 8, and 10 feet you get a 3 – 4 – 5 triangle also. This means $6 \div 2 = 3$, $8 \div 2 = 4$, and $10 \div 2 = 5$. In the other example, 9 feet $\div 3 = 3$, 12 feet $\div 3 = 4$, and 15 feet $\div 3 = 5$. So you have a 3 – 4 – 5 triangle in either measurement. Other combinations can be used but these are the most common. Cut all saw kerfs the same depth. This is because the string

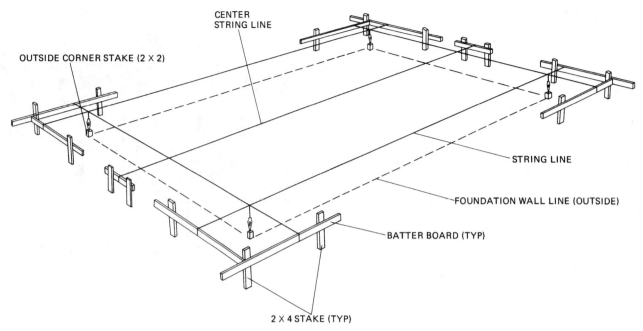

Fig. 2-4 *Staking out a basement.* (American Plywood Association)

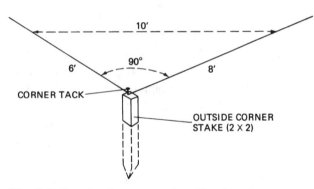

Fig. 2-5 *Squaring the corner and marking the point.* (American Plywood Association)

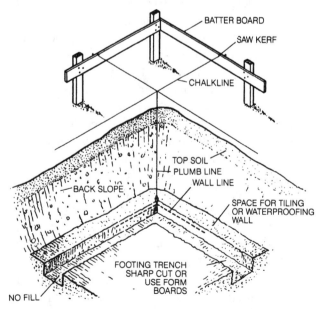

Fig. 2-6 *Note the location of the nails on the batter board.* (U.S. Department of Agriculture)

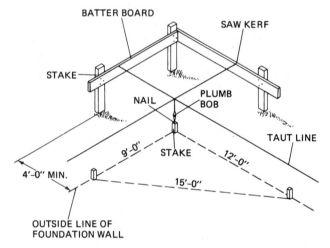

Fig. 2-7 *Staking and laying out the house.* (Forest Products Laboratory)

line not only defines the outside edges of the foundation but also will provide a reference line. This ensures uniform depth of footing excavation. When you have made similar cuts in all eight batter boards and strung the four lines in position, the outside foundation lines are accurately established.

5. Next, establish the lengthwise girder location. This is usually on the centerline of the house. Double-check your house plans for the exact position. This is because occasionally the girder will be slightly off the centerline to support an interior bearing wall. To find the line, measure the correct distance from the corners. Then install batter boards and locate the string line as before.

6. Check the foundation for levelness. Remember that the top of the foundation must be level around the entire perimeter of the house. The most accurate and simplest way to check this is to use a surveyor's level. This tool will be explained later in this chapter. The next best approach is to ensure that batter boards, and thus the string lines, are all absolutely level. You can accomplish this with a 10- to 14-foot-long piece of straight lumber. See Fig. 2-8A. Judge the straightness of the piece of lumber by sighting along the surface. Use this straightedge in conjunction with a carpenter's level. Laser levels (Fig. 2-8B) can also be used instead of a long board. Make sure the laser is level and at the right height. Then use the red dot to indicate the height of each leveling stake. Then drive temporary stakes around the house perimeter. The distance between them should not exceed the length of the straightedge. Then place one end of the straightedge on a batter board. Check for exact levelness. See Fig. 2-8. Drive another stake to the same height. Each time a stake is driven, the straightedge and level should be reversed end for end. This should ensure close accuracy in establishing the height of each stake with reference to the batter board. The final check on overall levelness comes when you level the last stake with the batter board where you began. If the straightedge is level here, then you have a level foundation baseline. During foundation excavation, the corner stakes and temporary leveling stakes will be removed. This stresses the importance of the level batter boards and string line. The corners and foundation levelness must be located using the string line.

THE BUILDER'S LEVEL

Practically all optical sighting and measuring instruments can be termed *surveying instruments*. Surveying, in its simplest form, simply means accurate measuring. Accurate measurements have been a construction requirement ever since humans started building things.

How Does It Work?

Even during the days of pyramid building, humans recognized the fact that the most accurate distant measurements were obtained with a perfectly straight line of sight. The basic principle of operation for today's modern instruments is still the same. A line of sight is a perfectly straight line. The line does not dip, sag, or curve. It is a line without weight and is continuous.

Any point along a level line of sight is exactly level with any other point along that line. The instrument itself is merely the device used to obtain this perfectly level line of sight for measurements.

Three Main Parts of a Builder's Level

1. *The Telescope* (Fig. 2-9). The telescope is a precision-made optical sighting device. It has a set of carefully ground and polished lenses. They produce a clear, sharp, magnified image. The magnification of a telescope is described as its power. An 18-power

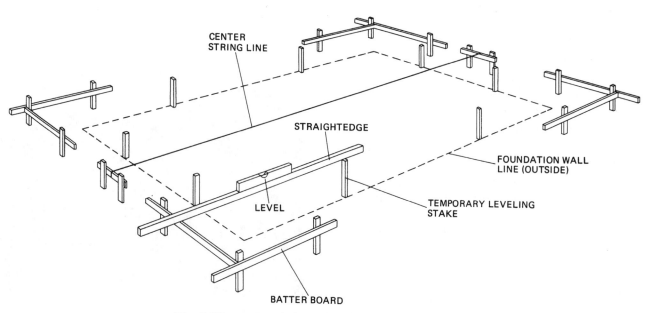

Fig. 2-8A *Leveling the batter boards.* (American Plywood Association)

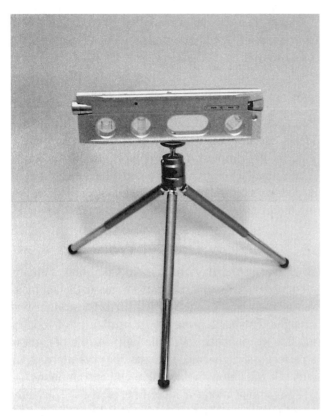

Fig. 2-8B *A laser level on a short tripod can be used on batter boards and leveling stakes.*

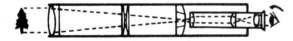

Fig. 2-9 *The telescope on an optical level.* (David White Instruments)

telescope will make a distant object appear 18 times closer than when viewed with the naked eye. Cross hairs in the telescope permit the object sighted on to be centered exactly in the field of view.

2. *The Leveling Vial* (Fig. 2-10). Also called the "bubble," the leveling vial works just like the familiar carpenter's level. However, it is much more sensitive and accurate in this instrument. Four leveling screws on the instrument base permit the user to center (level) the vial bubble perfectly and thus establish a level line of sight through the telescope. A vital first step in instrument use is leveling. Instrument vials are available in various degrees of sensitivity. In general, the more sensitive the vial, the more precise the results that may be obtained.

Fig. 2-10 *The leveling vial on an optical level.* (David White Instruments)

3. *The Circle* (Fig. 2-11). The perfectly flat plate upon which the telescope rests is called the circle. It is marked in degrees and can be rotated in any horizontal direction. With the use of an index pointer, any horizontal angle can be measured quickly. Most instruments have a *vernier scale*. An additional scale is subdivided. It divides degrees into minutes. There are 60 minutes in each degree. There are 360° in a circle.

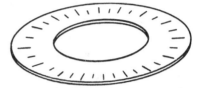

Fig. 2-11 *The circle on an optical level.* (David White Instruments)

Preparing the Instrument

Figure 2-12 shows a builder's level on site. Leveling the instrument is the most important operation in preparing the instrument for use.

Leveling the instrument First, secure the instrument to its tripod and proceed to level it as follows. Figure 2-13A shows the type of tripod used to support the instrument. The target pole is shown in Fig. 2-13B.

Place the telescope directly over one pair of opposite leveling screws. (See Fig. 2-14.) Turn the screws

Fig. 2-12 *Using the optical (or builder's) level on the job.* (David White Instruments)

(A)

(B)

Fig. 2-13 *(A) The tripod for an optical level; (B) the rod holder for use with an optical level.* (David White)

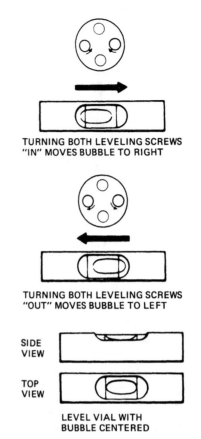

TURNING BOTH LEVELING SCREWS "IN" MOVES BUBBLE TO RIGHT

TURNING BOTH LEVELING SCREWS "OUT" MOVES BUBBLE TO LEFT

SIDE VIEW

TOP VIEW

LEVEL VIAL WITH BUBBLE CENTERED

Fig. 2-15 *Adjusting the leveling screws and watching the bubbles for level.* (David White Instruments)

directly under the scope in opposite directions at the same time (see step 5 in Fig. 2-14) until the level-vial bubble is centered. The telescope is then given a quarter (90°) turn. Place it directly over the other pair of leveling screws (step 2 of Fig. 2-14). The leveling operation is then repeated. Then recheck the other positions (steps 3 and 4 of Fig. 2-14) and make adjustments. This may not be necessary. Adjust if necessary. When leveling is completed, it should be possible to turn the telescope in a complete circle without any changes in the position of the bubble. See Fig. 2-15.

With the instrument leveled, you know that, since the line of sight is perfectly straight, any point on the line of sight will be exactly level with any other point. The drawing in Fig. 2-16 shows how exactly you can check the difference in height (elevation) between two points. If the rod reading at B is 3 feet and the reading at C is 4 feet, you know that point B is 1 foot higher than point C. Use the same principle to check if a row of windows is straight or if a foundation is level. Or you can check how much a driveway slopes.

Staking out a house Start at a previously chosen corner to stake out the house. Sight along line AB of Fig. 2-17 to establish the front of the house. Measure the desired distance to B and mark it with a stake.

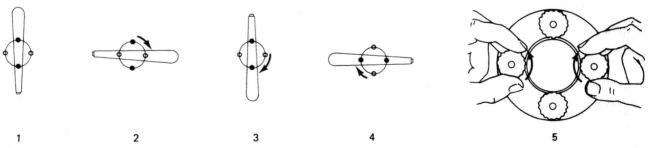

1   2   3   4   5

Fig. 2-14 *Adjusting the screws on the level-transit will level it. Note how it is leveled with two screws, then moved 90° and leveled again.* (David White Instruments)

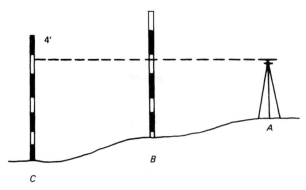

Fig. 2-16 *Finding the elevation with a level.* (David White Instruments)

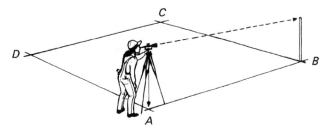

Fig. 2-17 *How to stake out a house on a building lot using a builder's level.* (David White Instruments)

Swing the telescope 90° by the circle scale. Mark the desired distance to *D*. This gives you the first corner. All the others are squared off in the same manner. You're sure all foundation corners are square, and all it took was a few minutes setup time. See Fig. 2-18.

This method eliminates the use of the old-fashioned string line-tape-plumb bob methods.

The Level-Transit

There are two types of levels used for building sites. The level and the level-transit are the two instruments used. The level has the telescope in a fixed horizontal position but can move sideways 360° to measure horizontal angles. It is usually all that is needed at a building site for a house. See Fig. 2-19.

A combination instrument is called a level-transit. The telescope can move in two directions. It can move up and down 45° as well as from side to side 360°. See Fig. 2-20. It can measure vertical as well as horizontal angles.

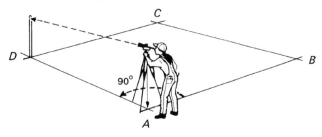

Fig. 2-18 *Squaring the other corner in laying out a building on a lot.* (David White Instruments)

Fig. 2-19 *Builder's level.* (David White Instruments)

Fig. 2-20 *Level-transit.* (David White Instruments)

A lock lever or levers permit the telescope to be securely locked in a true level position for use as a level. A full transit instrument, in addition to the features just mentioned, has a telescope that can rotate 360° vertically.

The level-transit is shown in operation in Fig. 2-21.

Using the Level and Level-Transit

Reading the circle and vernier The 360° circle is divided in quadrants (0 to 90°). The circle is marked by degrees and numbered every 10°. See Fig. 2-22.

To obtain degree readings it is only necessary to read the exact degree at the intersection of the zero index mark on the vernier and the degree mark on the circle (or on the vertical arc of the level-transit).

Fig. 2-21 *Using the level-transit on the site.* (David White Instruments)

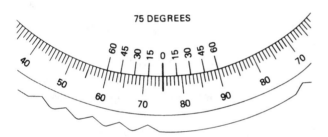

Fig. 2-22 *Reading the circle.* (David White Instruments)

For more precise readings, the vernier scale is used. See Fig. 2-23. The vernier lets you subdivide each whole degree on the circle into fractions, or minutes. There are 60 minutes in a degree. If the vernier zero does not line up exactly with a degree mark on the circle, note the last degree mark passed and, reading up the vernier scale, locate a vernier mark that coincides with a circle mark. This will indicate your reading in degrees and minutes.

Hanging the plumb bob To hang the plumb bob, attach a cord to the plumb bob hook on the tripod. Knot the cord as shown in Fig. 2-24.

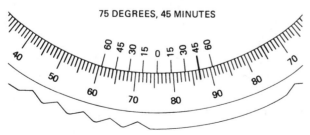

Fig. 2-23 *Reading the circle and vernier.* (David White Instruments)

Fig. 2-24 *To hang a plumb bob, attach a cord to the plumb bob hook on the tripod and knot the cord as shown here.* (David White Instruments)

If you are setting up over a point, attach the plumb bob. Move the tripod and instrument over the approximate point. Be sure the tripod is set up firmly again. Shift the instrument on the tripod head until the plumb bob is directly over the point. Then set the instrument leveling screws again to level the instrument.

Power The power of a telescope is rated in terms of magnification. It may be 24X or 37X. The 24X means the telescope is presenting a view 24 times as close as you could see it with the naked eye. Some instruments are equipped with a feature that lets you zoom in from 24X to 37X. It increases the effective reading range of the instrument more than 42 percent. It also permits greater flexibility in matching range, image, and light conditions. Use low power for brighter images in dim light. Since it gives a wider field of view, it is also handy in locating targets. Low power also provides better visibility for sighting through heat waves. See Fig. 2-25.

High power is used for sighting under bright light conditions. It is used for long-range sighting and for more precise rod readings.

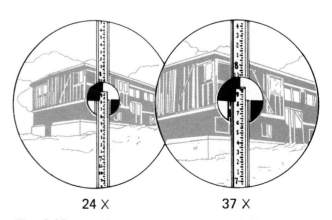

24 X   37 X

Fig. 2-25 *Variable instrument power is available.* (David White Instruments)

Rod Leveling rods are a necessary part of the transit leveling equipment. Rods are direct-reading with large graduations. All rods are equipped with a tough, permanent polyester film scale that will not shrink or expand. This is important when you consider the graduations can be $\frac{1}{100}$ of a foot. Figure 2-26 shows a leveling rod with the target attached at 4' 5¼". The target (Fig. 2-27) can be moved by releasing a small clamp in the back. Figure 2-28 shows a tape and the graduations. They are ⅛ inch wide and ⅛ inch apart. The tape is marked in feet, inches, and eighths of an inch. Feet are numbered in red. A three-section rod extends to 12 feet. A two-section rod extends to 8 feet 2 inches.

The rod holder is directed by hand signals from the surveyor behind the transit. The hand signals are easy to understand, since they motion in the direction of desired movement of the rod.

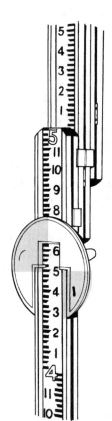

Fig. 2-26 *Leveling rod made of wood.* (David White Instruments)

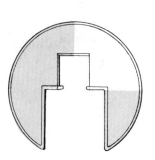

Fig. 2-27 *Target that fits wood rods.* (David White Instruments)

Fig. 2-28 *Tape face on the rod. This one is marked in feet, inches, and eighths.* (David White Instruments)

Establishing Elevations

Not all lots are flat. That means there is some kind of slope to be considered when digging the basement or locating the house. The level can help establish what these elevation changes are. From the grade line you establish how much soil will have to be removed for a basement. The grade line will also determine the location of the floor.

The bench mark is the place to start. A bench mark is established by surveyors when they open a section to development. This point is a reference to which the lot you are using is tied. The lot is so many feet in a certain direction from a given bench mark.

The bench mark may appear as a mark or point on the foundation of a nearby building. Sometimes it is the nearby sidewalk, street, or curb that is used as the level reference point.

The grade line is established by the person who designed the building. This line must be accurately established. Many measurements are made from this line. It determines the amount of earth removed from the basement or for the foundation footings.

Using the Leveling Rod

Use a leveling rod and set it at any point you want to check the elevation. Sight through the level or transit-level to the leveling rod. Take a reading by using the cross hair in the telescope. Move the rod to another point that is to be established. Now raise or lower the rod until the reading is the same as for the first point. This means the bottom of the rod is at the same elevation as the original point.

One person will hold the rod level. Another will move the target up or down till the cross hair in the telescope comes in alignment with that on the target. The difference between the two readings tells you what the elevation is.

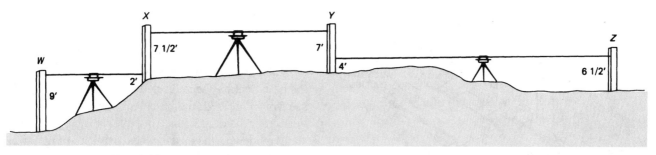

Fig. 2-29 *Getting the elevation when the two points cannot be viewed from a single point.*

Figure 2-29 shows how the difference in elevation between two points that are not visible from a single point is determined.

If point *Z* cannot be seen from point *W*, then you have to set the transit up again at two other points, such as *X* and *Y*. Take the readings at each location; then you will be able to determine how much of the soil has to be removed for a basement.

PREPARING THE SITE
Clearing

One of the first things to do in preparing the site for construction is to clear the area where the building will be located. Look over the site. Determine if there are trees in the immediate area of the house. If so, mark the trees to be removed. This can be done with a spray can of paint. Put an X on those to be removed or a line around them. In some cases, people have marked those that must go with a piece of cloth tied to a limb.

Also make sure those that are staying are not damaged when the heavy equipment is brought onto the site. Scarring trees can cause them to die later. Covering them more than 12 inches will probably kill them also. You have to cut off a part of the treetop. This helps it survive the covering of the roots.

Don't dig the sewer trench or the water lines through the root system of the trees to be saved. This can cause the tops of the trees to die later, and in some cases will kill them altogether.

Make a rough drawing of the location of the house and the trees to be saved. Make sure the persons operating the bulldozer and digger are made aware of the effort to save trees.

Cutting trees Keep in mind that removing trees can also be profitable. You can cut the trees into small logs for use in fireplaces. This has become an interest of many energy conservationists. The brush and undergrowth can be removed with a bulldozer or other type of equipment. Do not burn the brush or the limbs without checking with local authorities. There is always someone who is interested in hauling off the accumulation of wood.

Stump removal In some cases a tree stump is left and must be removed. There are a number of ways of doing this. One is to use a winch and pull it up by hooking the winch to some type of power takeoff on a truck, tractor, or heavy equipment. You could dig it out, but this can take time and too much effort in most cases.

The use of explosives to remove the stump is not permitted in some locations. Better check with the local police before setting off the blast.

The best way is to use the bulldozer to uproot the entire stump or tree. It all depends upon the size of the tree and the size of the equipment available for the job. Anyway, be sure the lot is cleared so the digging of the basement or footings can take place.

Excavation

A house built on a slab does not require any extensive excavation. One-piece or monolithic slabs are used on level ground and in warm climates. In cold climates, where the frostline penetrates deeper, or in areas where drainage is a problem, a two-piece slab has to be used. Figures 2-30 and 2-31 both show two-piece slab foundations.

Slab footings must rest beneath the frostline. This gives stability in the soil. The amount of reinforcement needed for a slab varies. The condition of the soil and the weights to be carried determine the reinforcement. Larger slabs and those on less stable soil need more reinforcement.

The top of the slab must be 8 inches above ground. This allows moisture under the slab to drain away from the building. It also gives you a good chance to spot termites building their tunnels from the earth to the floor of the house. The slab should always rest slightly above the existing grade. This is to provide for runoff water during a rainstorm.

Basements A basement is the area usually located underground. It provides most homes with a lot of

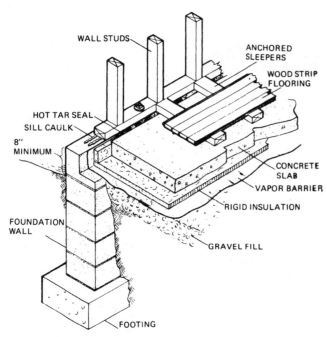

Fig. 2-30 *Two-piece slab with block foundation wall.* (Forest Products Laboratory)

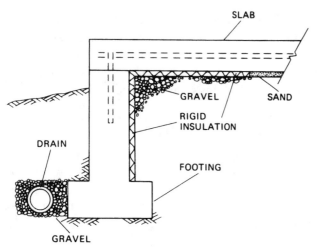

Fig. 2-31 *Two-piece slab with poured wall and footer.*

storage. In some, it is a place to do the laundry. It also serves as a place to locate the heating and cooling units. If a basement is desired, it must be dug before the house is started. The footings must be properly poured and seasoned. Seasoning should be done before poured concrete or concrete block is used for the wall. Some areas now use treated wood walls for a basement.

Figure 2-32 shows a basement dug for use in colder climates. Trenches from the street to the basement must be provided for the plumbing and water. Utilities may be buried also. If they are, the electric, phone, and gas lines must also be located in trenches or buried after the house is finished. It is a good idea to notify the utility companies so that they can schedule the installation of their services when you are ready for them.

Some shovel work may have to be done to dig the basement trenches for the sewer pipes. This is done after the basement has been leveled by machines. As you can see from Fig. 2-32, the basement may also need shovel work after the digger has left. Note the cave-ins and dirt slides evident in the basement excavation in Fig. 2-32.

The basement has to be filled later. Gravel is used to form a base for the poured concrete floor.

The high spots in the basement must be removed by shovel. Proper-size gravel should be spread after the sewer trench is filled. You may have to tamp the gravel to make sure it is properly level and settled. Do this before the concrete mixer is called for the floor job.

The footings have to be poured first. They are boxed in and poured before anything is done in the way of the basement floor. In some instances drain tile must be installed inside or outside the footing. The tile is allowed to drain into a sump. In other locations no drainage is necessary because of soil conditions.

PROVIDING ACCESS DURING CONSTRUCTION

The first thing to be established is who is to be on the premises. Check with your insurance company about liability insurance. This is in case someone is hurt on the location. Also decide who should be kept out. You also have to decide how access control is to be set up. It may be done with a fence or by an alert guard or dog. These things do have to be considered before the construction gets underway. If equipment is left at the site, who is responsible? Who will pay for vandalism? Who will repair damage caused by wind, hail, rain, lightning, or tornado?

Materials Storage

Where will materials for the job be stored? In Fig. 2-33 you can see how plywood is stored. What happens if someone decides to haul off some of the plywood? Who is responsible? What control do you have over the stored materials after dark?

Figure 2-34 shows plywood bundles broken open. This makes it easy for single sheets to disappear. With the current price of plywood, it becomes important to plan some type of storage facility on the site.

Some of the shingles in Fig. 2-35 may be hard to find if the wind gets to the broken bundle. What's to stop children playing on the site during off hours? They can also take the shingles and spread them over the landscape.

Fig. 2-32 *Excavation for a basement.*

Fig. 2-33 *Storing plywood on the site.*

Fig. 2-34 *Broken bundle of plywood sheathing.*

Storage of bricks can be a problem. See Fig. 2-36. They are expensive and can easily be removed by someone with a small truck. It is very important to have some type of on-site storage. It is also very important to make sure that materials are not delivered before they are needed. Some type of materials inventory has to be maintained. This may be worth a person's time. The location of the site is a major factor in the disappearance of materials. Location has a lot to do with the liability coverage needed from insurance companies.

Temporary buildings Some building sites have temporary structures to use as storage. In some cases the plans for the building are also stored in the tool shed. Covered storage is used in some locations where rain and snow can cause a delay by wetting the lumber, sand, or cement. If you are using drywall, you will need to keep it dry. In most instances it is not delivered to the site until the house is enclosed.

Some construction sheds are made on the scene. In other cases the construction shed may be delivered to the site on a truck. It is picked up and moved away once the building can be locked.

Mobile homes have been used as offices for supervisors. This usually is the case when a number of houses are made by one contractor and all are located in one row or subdivision.

A garage can be used as the headquarters for the construction. The garage is enclosed. It is closed off by

Fig. 2-35 *Broken bundle of shingles.*

Fig. 2-36 *Storing bricks on site.*

doors, so it can be locked at night. Since the garage is easy to close off, it becomes the logical place to take care of paper work. It also becomes a place to store materials that should not get wet.

When building a smaller house, the carpenter takes everything home at the end of the working day. The carpenter's car or truck becomes the working office away from home. Materials are scheduled for delivery only when actually needed. In larger projects some local office is needed, so the garage, tool shed, mobile home, or construction shack is used.

Storing construction materials   Storing construction materials can be a problem. It requires a great deal of effort to make sure the materials are on the job when needed. If delivery schedules are delayed, work has to stop. This puts people out of work.

If materials are stored on the job, make sure they are neatly arranged. This prevents accidents such as tripping over scattered materials. Sand should be delivered and placed out of the way. Keep it out of the normal traffic flow from the street to the building.

Everything should be kept in some order. This means you know where things are when you need them. Then you don't have to plow through a mound of

supplies just to find a box of nails. Everything should be laid out according to its intended order of use.

Lumber should be kept flat. This prevents warpage, cupping, and twisting. Plywood should be protected from rain and snow if it is interior grade. In any case it should not be allowed to become soaked. Keep it flat and covered.

Humidity control is important inside a house. This is especially true when you're working with drywall. It should be allowed to dry by keeping the windows open. Too much humidity can cause the wood to twist or warp.

Temporary Utilities

You will need electricity to operate the power tools. Power can be obtained by using a long extension cord from the house nearby. Or, you may have to arrange for the power company to extend a line to the side and put in a meter on a pole nearby.

Water is needed to mix mortar. The local line will have to be tapped, or you may have to dig the well before you start construction. It all depends upon where the building site is located. If the house is being built near another, you may want to arrange with the neighbors to supply water with a hose to their outside faucet. Make sure you arrange to pay for the service.

Waste Disposal

Every building site has waste. It may be human waste or paper and building-material wrappings. Human waste can be controlled by renting a Porta John or a Johnny-on-the-Spot. This can prevent the house from smelling like a urinal when you enter. The sump basin should not be used as a urinal. It does leave an odor to the place. Besides, it is unsanitary.

Waste paper can be burned in some localities. In others burning is strictly forbidden. You should check before you arrange to have a large bonfire for getting rid of the trash, cut lumber ends, paper, and loose shingles. There are companies that provide a trash-collecting service for construction areas. They leave the place *broom clean*. It leaves a better impression of the contractor when a building is delivered in order, without trash and wood pieces lying around. If you go to the trouble of building a fine home for someone, the least you can do is deliver it in a clean condition. After all, this is going to be a home.

Arranging Delivery Routes

Damage to the construction site by delivery trucks can cause problems later. You should arrange a driveway by putting in gravel at the planned location of the drive. Get permission and remove the curb at the entrance to the driveway. Make sure deliveries are made by this route. Pile the materials so that they are arranged in an orderly manner and can be reached when needed.

Concrete has to be delivered to the site for the basement, foundation, and garage floor. Be sure to allow room for the ready-mix truck to get to these locations. Lumber is usually strapped together. Make allowance for bundles of lumber to be dumped near the location where they will be needed.

Make sure the nearby plants or trees are protected. This may require a fence or stakes. Some method should be devised to keep the trucks, diggers, and earth movers from destroying natural vegetation.

Access to the building site is important. If this is the first house in the subdivision, or if it is located off the road, you have to provide for delivery of materials. You may have to put in a temporary road. This should be a road that can be traveled in wet weather without the delivery trucks becoming bogged down or stuck.

As you can see, it takes much planning to accomplish a building program that will come off smoothly. The more planning you do ahead of time, the less time will be spent trying to obtain the correct permissions and deliveries.

The key to a successful building program is planning. Make a checklist of the items that need attention beforehand. Use this checklist to keep yourself current with the delivery of materials and permissions.

It is assumed here the proper financial arrangements have been made before construction begins.

CHAPTER 2
STUDY QUESTIONS

1. What is a builder's level?
2. What is the difference between a builder's level and a level-transit?
3. Why is it important to have a clear title to a piece of land before building?
4. Who establishes the correct property limits?
5. What is an easement?
6. Why should a house be laid out square and level?
7. How do you set up a builder's level?
8. What is an elevation?
9. What is a vernier?
10. What is a plumb bob?
11. How is a plumb bob used by a carpenter?
12. How is the power of a telescope on a builder's level rated?
13. What is a leveling rod?
14. What is a target on a leveling rod?
15. What methods are used to remove stumps from a building site?
16. What does excavate mean?
17. Who has access to a house under normal conditions?
18. Why is some shovel work needed after the digger has excavated the basement?
19. What's the purpose of temporary buildings at a construction site?
20. Why would you want to arrange delivery routes to a building site?

3
CHAPTER

Laying Footings & Foundations

PEOPLE OFTEN THINK THAT FOOTINGS AND foundations are the same. Actually, the footing is the lowest part of the building and carries the weight. The foundation is the wall between the footing and the rest of the building. In this chapter you will learn how to:

- Design footings and foundations
- Locate corners and lines for forms
- Check the level of footing and foundation excavation
- Make the forms for footings
- Make the forms for foundations
- Reinforce the forms as required
- Mix or select concrete for usage
- Pour the concrete into the forms
- Finish concrete in the forms
- Embed anchor systems in forms
- Waterproof foundation walls if needed
- Make necessary drainage systems

INTRODUCTION

Footings bear the weight of the building. They spread the weight evenly over a wide surface. Figure 3-1 shows the three parts of a footing system. These parts are the bearing surface, the footing, and the foundation. The bearing surface must be located beneath the frostline on firm and solid ground. The frostline is the deepest level at which the ground will freeze in the wintertime. Moisture in the ground above the frostline will freeze and thaw. When it does, the ground moves and shifts. The movement will break or damage the footing or foundation. The location and construction of the

footing is very important. Think of the weight of all the lumber, concrete, stone, and furniture that must be supported by this layer. All of these must be supported without sinking or moving.

Footings may be made in several ways. There are flat footings, stepped footings, pillared footings, and pile footings. The flat footing, as in Fig. 3-2, is the easiest and simplest footing to make because it is all on one level. The stepped footing is used on sides of hills as in Fig. 3-3. The stepped footing is like a series of short flat footings at different levels, much like a flight of steps. By making this type of footing, no special digging (excavation) is needed. The third footing type is the pillared footing. See Fig. 3-4. The pillared foot-

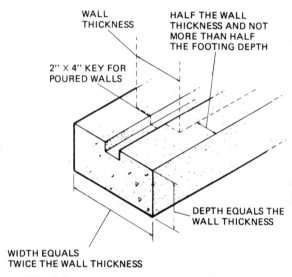

Fig. 3-2 *Regular flat footing.* (Forest Products Laboratory)

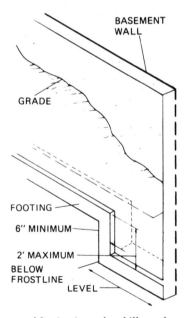

Fig. 3-3 *A stepped footing is used on hills or slopes.* (Forest Products Laboratory)

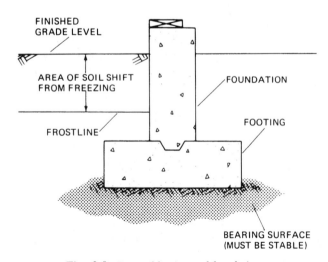

Fig. 3-1 *Parts of footing and foundation.*

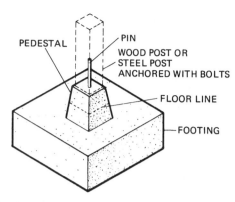

Fig. 3-4 *The pillar or post footing may be square or round.* (Forest Products Laboratory)

ing is used in many locations where the soil is evenly packed and little settling occurs. It consists of a series of pads, or feet. Columns are then built on the pads and the building rests upon the columns. Buildings with either flat or stepped footings usually have pillared footings in the center. This is because buildings are too wide to support the full weight without support in the middle areas.

Pile footings are the fourth type and are used where soil is loose, unstable, or very wet. As in Fig. 3-5, long columns are put into the ground. These are long enough to reach solid soil. The columns may be made of treated wood or concrete. The wooden piles are driven into the ground by pounding. The concrete piles are made by drilling a hole and filling it with concrete. Pads or caps are then put over the tops of the piles.

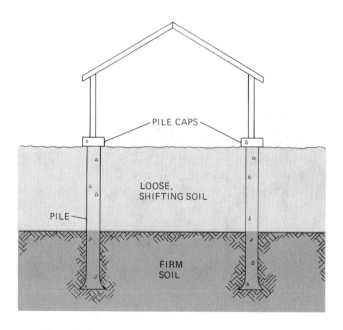

Fig. 3-5 *Pile footings reach through water or shifting soils.*

SEQUENCE

No matter what type of footing and foundation is used, a certain sequence should be followed. The sequence can change slightly according to the method involved. For example, the footing and foundation can be poured in one solid concrete piece. However, many footings are made separately and the concrete foundations are built on top of the footings. In both cases, the sequence is similar. The basic sequence is:

1. Find the amount of site preparation needed.
2. Lay out footing and foundation shape.
3. Excavate to proper depth.
4. Level the footing corners.
5. Build the footing forms.
6. Reinforce the forms as needed.
7. Estimate concrete needs.
8. Pour the concrete footing.
9. Build the foundation forms.
10. Reinforce the forms as needed.
11. Pour the concrete into forms.
12. Finish the concrete and embed anchors.
13. Remove the forms.
14. Waterproof and drain as required.

LAYING OUT THE FOOTINGS

Footings are the bottom of the building and must hold up the weight of the building. Two factors are involved in finding the correct shape and size. The first is the strength or solidness of the soil. The second factor is the width and depth of the footing for the weight of the building in that type of soil.

Soil Strength

Soil strength refers to how dense and solid the soil is packed. It also refers to how stable or unmoving the soil is. Some soils are very hard only when dry. Others keep the same strength whether they are wet or dry. In any condition, the soil must be dense and strong enough to support the weight of the building. When soil is soft, the footing is made wider to spread the weight over more surface. In this way, each surface unit holds up less weight. Figure 3-6 shows how much weight various soil types will support. Standard footings should not be poured on loose soil.

Type of Soil	Bearing Capacity (pounds per ft²)
Soft Clay loose dirt, etc.	2 000
Loose Sand hard clay, etc.	4 000
Hard Sand or Gravel	6 000
Partially Cemented Sand or Gravel soft stone, etc.	20 000

Fig. 3-6 *Bearing capacity of typical soils.*

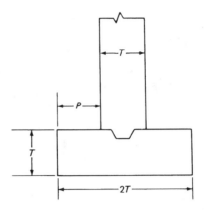

FLOORS	BASE-MENT	ALL WOOD FRAME		WOOD FRAME WITH MASONRY VENEER	
		T	P	T	P
1	None	6"	3"	6"	4"
1	Yes	6"	2"	6"	3"
2	None	6"	3"	6"	4"
2	Yes	6"	4"	8"	5"

NOTE: For soil with 2000 pounds per square foot (PSF) load capacity.

Fig. 3-7 *Typical footing size.*

Footing Width

The second factor is the width of the footing. As mentioned, the footing should be wider for soft soil. Figure 3-7 shows typical sizes for footings. As a rule, footings are about two times as wide as they are thick. The average footing is about 8 inches thick, and the footing is about the same thickness as the foundation wall.

Locating Footing Depth

Footings are laid out several inches below the frostline. For buildings with basements, place the top of the footing 12 inches below the frostline. For buildings that do not have basements, 4 to 6 inches below the frostline could be deep enough. Local building codes may give exact details.

Footings Under Columns

The footings and foundations that most people see support only the outside walls. But today most houses are wide, and support is needed in the center of a wide building. This support is from footings, pillars, or columns built in the center. Pillars or columns must have a footing just as the outside walls do. For houses with basements, the footings and pillars become part of the basement floor and walls. See Fig. 3-8. Many houses do not have basements. Instead, they have a crawl space between the ground and the floor. This crawl space provides access to the pipes and utilities. Pillars or columns built on footings are used for supports in the crawl spaces. The footings may be any

Fig. 3-8 *Footings in a basement later became a part of the basement floor.*

Fig. 3-9 *Footings and piers must be located in crawl spaces.*

shape—square, rectangular, or round. Figure 3-9 shows a site prepared in this manner.

Footings for either basements or crawl spaces are all similar. They should be below the frostline as in a regular footing. However, they carry a greater weight than do the outside footings. For this reason, they should be 2 to 3 feet square.

Special Strength Needs

Footings for heavier areas of a building such as chimneys, fireplaces, bases for special machinery, and other similar things should be wider and thicker. For chimneys in a one-story building, the footing should project at least 4 inches on each side. The chimneys on two-story buildings are taller and heavier. Therefore, the footing should project 6 to 8 inches on each side of the chimney. Figure 3-10 shows a foundation for a fireplace.

Fig. 3-10 *A special footing is used for fireplaces. It supports the extra weight.*

Reinforcement and Strength

Two things are done to the footing to make it stronger. First, it is reinforced with steel rods. Then, the footing is also matched or keyed so that the foundation wall will not shift or slide.

Reinforcement In most cases the footing should be reinforced with rods. These reinforcement rods are called *rebar*. Two or more pieces of rebar are used. The rebar should be located so that at least 3 inches of space for concrete is left around all edges. See Fig. 3-11.

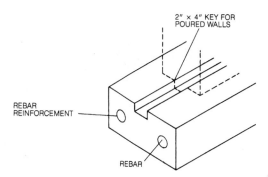

Fig. 3-11 *Footings may be reinforced. Note the key to keep the foundation from shifting.*

Keyed footings The best type of separate footing is keyed, as shown in Fig. 3-11. This means that the footing has a key or slot formed in the top. The slot is filled when the foundation is formed. The key keeps the foundation from sliding or moving off the footing. Without a key, freezing and thawing of water in the ground could force foundation walls off the footing.

EXCAVATING THE FOOTINGS

The procedure for locating the building on the lot was explained in Chapter 2. Batter boards were put up and lines were strung from them to show the location of the walls and corners.

Now the size, shape, and depth of footings must be decided.

Finding Trench Depth

Trenches or ditches must be dug, or excavated, for the footings and foundation. Ground that is extremely rough and uneven should be rough-graded before the excavation is begun. The topsoil that is removed can be piled at one edge of the building site. It should be used later when the ground is smoothed and graded around the building. Before the digging is started, determine how deep it is to be.

The trench at the lowest part of the site must be deep enough for the footing to be below the frostline. If the footing is to be 12 inches below the frostline, the trench at the lowest part must be deep enough for this. Figure 3-12 shows these depths. This lowest point becomes the level line for the entire footing. Elevations are taken at each corner to find out how deep the trenches are at each corner.

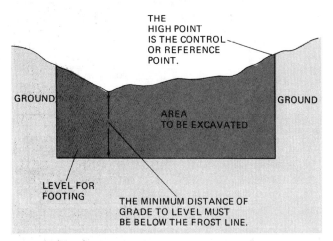

Fig. 3-12 *Footings must be below the frostline.*

Excavating for Deep Footings

Footings must be deep in areas where the frostline is deep. Deep footings are also needed when a basement will be dug.

Rough lines are drawn on the ground. They do not need to be very accurate, but the lines from the batter boards are used as guides. However, the rough line should be about 2 feet outside the line. See Fig. 3-13. The trench for the footing is dug much wider than the

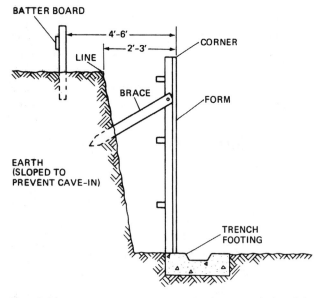

Fig. 3-13 *The trench is wider than the footing and sloped for safety.*

footing so that there is room to work. Since the footings are made of concrete, the molds (called *forms*) for the concrete must be built. Room is needed to build or put up the forms. Work that must be done after the footing or foundation is formed includes removing the forms, waterproofing the walls, and making proper drainage.

As the trench is dug, the depth is measured. When the trench has been dug to the correct depth, the machinery is removed. The forms are then laid out. For basements, the interior ground is also excavated.

Excavating for Shallow Footings

Rough lines are marked on the ground with chalk or shovel lines. These lines should be marked to show the width of the footing desired. Corner stakes are removed, and lines are taken from the batter boards. A trench is excavated to the correct depth.

The special footings for the interior are also excavated at this point. The excavation for the interior pad footings should be made to the same depth as that for the outside walls.

The concrete is poured directly into these trenches. Any reinforcement is made without forms in the excavation itself. The rebar is suspended with rebar stakes or metal supports called *chairs*. See Fig. 3-14.

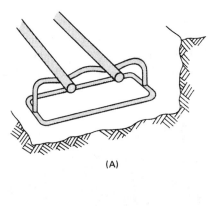

(A)

(B)

Fig. 3-14 *Chairs are used to hold rebar in place. (Richmond Screw Anchor)*

To form a key in this type of footing, stakes are driven along the edges as in Fig. 3-15. The board that forms the key in the footing is suspended in the center of the trench area.

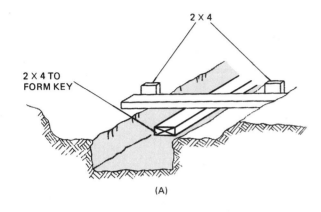

(A)

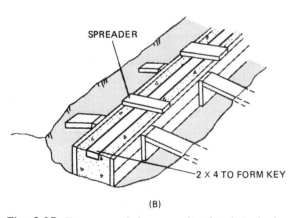

(B)

Fig. 3-15 *Keys are made by suspending 2 × 4s in the form. (A) Keys for trench forms; and (B) Keys for board forms.*

In many areas, concrete block wall is used on this type of footing. The blocks may be secured by inserting rebars into the footing area. The bricks or blocks are laid so that the rebar is centered in an opening in the block. The opening is then filled with mortar or concrete to secure the foundation against slipping.

The important thing to remember is that special forms are not used with shallow footings. Also, they may not be finished smooth. As a result they may appear very rough or unfinished. This is not important if they are the proper shape.

Slab Footings and Basements

Slab footings are used in areas where concrete floors are made. Slabs can combine the concrete floor and the footing as one unit. Slabs, basement floors, and other large concrete surfaces are detailed in another chapter. Basement floors are made separately from the footings and are done after the footings and basement walls are up.

BUILDING THE FORMS FOR THE FOOTINGS

After the excavation has been completed, the corners must be relocated. After the corners are relocated, the forms are built and leveled and the concrete is poured and allowed to harden. Then the forms are removed, and the foundation is erected. In many cases, the footing and the foundation are made as one piece.

Laying Out the Forms

After the excavation is complete, the first step is to relocate the corners and edges for the walls. To do this, the lines from the batter boards are restrung and a plumb bob is used to locate the corner points. The corner points and other reference points are marked with a stake. The stake is driven level for the top of the footing. This level is established by using a transit or a level. Refer to Chapter 2 for this procedure.

Nails

It is best to use double-headed, or duplex, nails for making the forms. Forms should be nailed with the nails on the outside. This means that the nails are not in the space where the concrete will be. This way the nailhead does not get embedded in the concrete and is left exposed. The double head allows the nail to be driven up tight; it will still be easy to pull out when the forms are taken apart.

Putting Up the Forms

With the corner stakes used for location and level, the walls of the forms are constructed. The amount that the footing is to project past the wall is determined. Usually this is one-half the thickness of the foundation wall. This dimension is needed because the corner indicates foundation corner and not footing corner. Stakes are driven outside the lines so that the form will be the proper width. The carpenter must allow for the width of the stake and the width of the boards used for the forms. See Fig. 3-16. Drive stakes as needed for support. As a rule, the distance between stakes is about twice the width of the footing. Nail the top board to the first stake and level the top board in two directions. For first direction, the top board is leveled with the corner stake. For the second direction, the top board is leveled on its length. See Fig. 3-17. After the top board is leveled, nail it to all stakes. Then nail the lower boards to the stakes. Both inside and outside forms are made this way.

If 1-inch-thick boards are used to build the form, stakes should be driven closer together. If boards 2

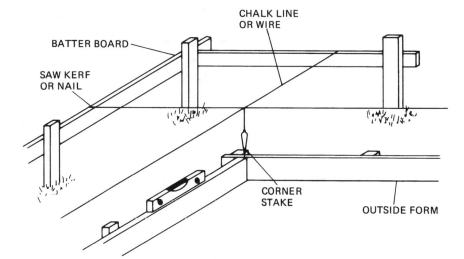

Fig. 3-16 *The footing corner is located and leveled.*

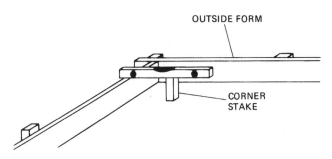

Fig. 3-17 *The form is leveled all around.*

inches thick are used, the stakes may be 4 to 6 feet apart. In both cases, the stakes are braced as shown in Fig. 3-18.

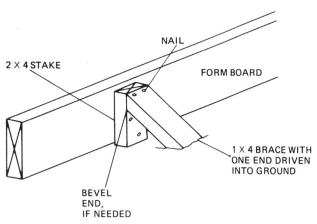

Fig. 3-18 *Bracing form boards.*

Loose dirt should be removed from under the footing form. It is best for the footing to be deeper than is needed. Never make a footing thinner than the specifications. Never fill any irregular hole or area with loose dirt. Always fill with gravel or coarse sand and tamp it firmly in place.

The keyed notch The key or slot in the footing is made with a board. The board is nailed to a brace that reaches across the top of the forms. The brace should be nailed in place at intervals of about 4 feet apart. Refer to Fig. 3-15 to see how the key is made.

Excavation for drains and utility lines Drain pipes and utility lines are sometimes located beneath the footings in a building. When this is done, trenches are dug underneath the footing forms. These trenches are usually dug by hand underneath the forms. After the drain pipes or utility lines are laid in place, the area is filled with coarse gravel or sand. This gravel or sand is tightly packed in place beneath the form.

Spacing the walls of the form The weight of the concrete can make the walls spread apart. To keep the walls straight, braces are used. The braces on the walls provide much support. Special braces called *spreaders* are also nailed across the top. Forms should be braced properly so that the amount of concrete ordered will fill the forms properly. Also, this practice ensures that excess concrete does not add extra weight to the building.

The forms should also be checked to make sure that there are no holes, gaps, or weak areas. These could let the concrete leak out of the form and thus weaken the structure. These leaks are called *blowouts*.

WORKING WITH CONCRETE

Before the concrete is ordered and poured, several things are done. The forms should be checked for the proper depth and level. Openings and trenches beneath the footing area for pipes and utility lines are made. These should be properly leveled and filled. The forms should be checked to make sure that they are properly braced and spaced. Finally, chalk lines and corner stakes should be removed from the forms.

Reinforcement

In most cases, reinforcement rods (rebar) are placed in the footing after the forms are finished. The amount of reinforcement is usually given in the plans. As it is laid, the rebar is tied in place. Soft metal wires, called *ties*, are twisted around the rebars. The carpenter must be sure that the footing conforms to the local building codes.

Specifying Concrete

Most concrete used today is made from cement, sand, and gravel mixed with water. The cement is the "glue" that hardens and holds or binds the materials. Most cement used today is portland cement. It is made from limestone that is heated, powdered, and mixed with certain minerals. When mixed with aggregates, or sand and gravel, it becomes concrete.

Concrete mixes can be denoted by three numbers, such as 1—2—3. This is the volume proportion of cement, sand, and gravel. 1—2—3 is the basic mix, but it is varied for strength, hardening speed, or other factors. However, it is recommended that concrete be specified by the water-to-cement ratio, aggregate size, and bags of concrete per cubic yard. See Table 3-1.

Most concrete today is delivered to the building site. Usually, the concrete is not mixed by the carpenters. It is delivered by concrete trucks from a concrete company. Figure 3-19 shows a transit-mix truck. The concrete is sold in units of cubic yards. The carpenter may need to make the order for concrete to the concrete company. To do so, the carpenter must be able to figure how much concrete to order.

Estimating Concrete Needs

A formula is used to estimate the volume of concrete needed. The basic unit for concrete is the cubic yard. A

Fig. 3-19 *A transit-mix truck delivers concrete to a site.*

cubic yard is made up of 27 cubic feet ($3 \times 3 \times 3$). To convert footing sizes, use the formula:

$$\frac{L'}{3} \times \frac{W''}{36} \times \frac{T''}{36} = \text{cubic yards}$$

where L' = length in feet
W'' = width in inches
T'' = thickness in inches

Example. A footing is 18 inches wide and 8 inches thick. It must support a building 48 feet long and 24 feet wide. The distance around the edges is called the perimeter. The perimeter is $(2 \times 48) + (2 \times 24)$, or 144 feet. This would be:

$$\frac{L'}{3} \times \frac{W''}{36} \times \frac{T''}{36} = \text{cubic yards}$$

$$= \frac{144}{3} \times \frac{18}{36} \times \frac{8}{36}$$

Table 3-1 *Concrete Use Chart*

Uses	Concrete, Bags per Cubic Yard	Sand, Pounds per Bag of Concrete	Gravel, Pounds per Bag of Concrete	Gravel Size, Average Diameter in Inches	Water, Gallons per Bag of Concrete	Consistency Slump
Footings, basement walls (8-inch), or foundation walls (8-inch thickness)	5.0	265	395	1½"	7	4 – 6 inches
Slabs, basement floors, sidewalks, etc. (4-inch thickness)	6.2	215	295	1"	6	4 – 6 inches
Basic 1-2-3 mixture (approximation only)	6.0	190	275	2"	5.5	2 – 4 inches

NOTES: 1. All figures are for slight to moderate ground water and medium-fineness sand.
2. All figures vary slightly.

and by cancellation,

$$\frac{48}{1} \times \frac{1}{2} \times \frac{2}{9} + \frac{96}{18} = 5.33 \text{ cubic yards}$$

The minimum amount that can be ordered is one cubic yard. After the first cubic yard, fractions can be ordered. The estimate is 5⅓ cubic yards. Often a little more is ordered to make sure enough is delivered.

Pouring the Concrete

To be ready, the carpenter sees to two things. First the forms must be done. Then the concrete truck must have a close access. The driver can move the spout to cover some distance. However, it may be necessary to carry the concrete an added distance. This can be done by pumping the concrete or by carrying it. Wheelbarrows, as in Fig. 3-20, are sometimes used.

Another method is to use a dump bucket carried by a crane. See Fig. 3-21.

The builder must spread, carry, and level the concrete. The truck will only deliver it to the site. The truck driver can remain only a few minutes. The driver is not allowed to help work the concrete. As the concrete is poured, it should be tamped. This is done with a board or shovel that is plunged into the concrete. See Fig. 3-22. Tamping helps get rid of air pockets. This makes the concrete solidly fill all the form.

For shallow footings, no smoothing or "finishing" need be done. For deep footings, the surface should be

Fig. 3-21 *A dump bucket is used to dump concrete into forms that trucks can't reach* (Universal Form Clamp)

Fig. 3-22 *Concrete is tamped into forms to get rid of air pockets.* (Portland Cement)

roughly leveled. This is done by resting the ends of a board across the top of the form. The board is then used to scrape the top of the concrete smooth and even with the form.

BUILDING THE FOUNDATION FORMS

The foundation is a wall between the footing and the floor of the building. It is often made of concrete.

Fig. 3-20 *Sometimes the concrete must be carried from the truck to the worksite.* (IRL Daffin)

However, it may also be made of concrete blocks, bricks, or stone. In some regions, foundation walls are made of treated plywood as well.

When concrete is used, special forms may be used for the foundations. These forms are easily put up and down. Often the footing and the foundation are made in one solid piece.

Builders also make the foundation in much the same way as they make the forms for the footings. After the form is removed, the lumber is used in framing the house.

In either method, the form should be spaced for the correct width. It must also be spaced to prevent the weight of the concrete from spreading the forms. The width of the form is important. A form that is not wide enough will not carry the weight of the building safely. A form that is too wide uses too much concrete. Too much concrete costs more and adds weight to the building. This weight can cause settling problems. However, spreading forms can also cause errors in pouring the concrete. Frequently, just enough concrete is ordered to fill the forms. If the forms are allowed to spread, more concrete is used. Thus enough concrete might not be delivered to the site.

Making the Forms

In making forms, several things must be considered. First, sections of a form must fit tightly together. This prevents leaks at the edges. Leaks can cause bubbles and air pockets in the concrete. This is called *honeycomb*. Honeycomb weakens the foundation wall.

When special forms are used and assembled to make the total form, they must be braced properly. Forms up to 4 feet wide are braced on the back side with studs. These forms are made from metal sheets or from plywood sheathing ¾ inch thick or thicker. For building walls higher than 4 feet, special braces called *wales* are used. See Fig. 3-23.

The sheathing is nailed to the studs and wales from the inside. The studs are laid out flat on the ground. The sheathing is then laid on the studs and nailed down. The assembled form is then erected and placed into position. It is spaced properly, and wales and braces are added.

Braces are erected every 4 to 6 feet. However, for extra weight or wall height, braces may be closer.

Joining the Forms Together

Edges and corners should be joined tightly so that no concrete leaks occur. When using plywood forms, join the edges together by nailing the plywood sheathing to the studs. Use 16d nails as in Fig. 3-24. When nailing the corners together, use the procedure also shown in Fig. 3-24. Again, 16d nails are used.

When special metal forms are used, the manufacturer's directions should be carefully followed.

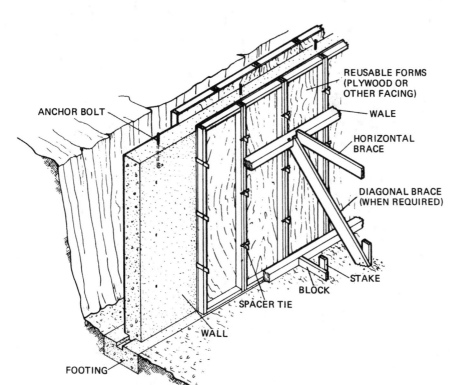

Fig. 3-23 Special braces called wales are sometimes needed. (Forest Products Laboratory)

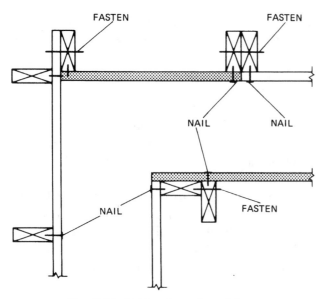

Fig. 3-24 *Nailing plywood panel forms.*

Spreaders

Spreaders are used on all forms to hold the walls apart evenly. Several types may be used. The spreaders may be made of metal at the site. Metal straps may be nailed in between the sections of the forms. However, most builders use spacers that have been made at a factory for that type of form.

After the concrete has hardened, the forms are removed. The spreader is broken off when the form is removed. Special notches are made in the rods to weaken them so that they will break at the required place. Figure 3-25 shows a typical spreader.

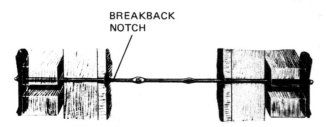

Fig. 3-25 *Special braces called spreaders keep the forms spaced apart. These rods are later broken off at the notches, called breakbacks. The rest of the spreader remains in the concrete.* (Richmond Screw Anchor)

Using Panel Forms

A panel form is a special form made up in sections. The forms may be used many times. Most are specially made by manufacturers. Each style has special connectors that enables the forms to be quickly and easily erected. By use of standard sizes such as 2-×-4- or 4-×-8-foot sections, walls of almost any size and shape can be erected quickly. The forms are made of metal or wood. The advantages are that they are quick and easy to use, they may be used many times, and they may be used on almost any size or shape of wall.

Panel forms must be braced and spaced just as forms constructed on the site are. To the builder making one building, there is little advantage to using such forms. They must be purchased, and they are not cheap. However, when they are used many times, the savings in time make them economical. Figure 3-26 shows an example.

Fig. 3-26 *Reusable forms are made from panels of plywood or metal. These special panel forms are assembled with special fasteners.* (Proctor Products)

One-Piece Forms

When the same style of footing and foundation is often used in the same type of soils, a one-piece form is used. This combines the footings and the foundation as one piece (Fig. 3-27). Several versions may be used. Some types allow a footing of any size to be cast with a foundation wall of any thickness. Some incorporate the footing and the foundation wall as a stepped figure. Others, as in Fig. 3-28, use a tapered design.

The one-piece form saves operational steps. Casting is quicker, it is easier, and it is done in one operation. Two-piece forms and the conventional processes require that the footing be cast and allowed to harden. The footing forms are then removed and the founda-

Fig. 3-27 *Panel forms can combine the footing and the foundation.* (Proctor Products)

Fig. 3-28 *Tapered form.* (Proctor Products)

tion forms erected, cast, allowed to harden, and removed. This takes several days and many hours of work. The one-piece form offers many advantages in the savings of time and cost of labor.

Special Forms

Certain types of form are used for shapes that are commonly used. A round form such as that in Fig. 3-29 is commonly used. This type of form may be used to cap pilings. It may also be used for the footings under the central foundation pillars.

Fig. 3-29 *Pier form.* (Proctor Products)

Other special forms include forms made of steel and cardboard. Steel forms may be used to cast square or round columns. They are normally used in construction of large projects such as bridges and dams. They are also used on large business buildings. The cardboard forms are made of treated paper and fibers. They are used one time and destroyed when they are removed. Figure 3-30 shows a pillar made with a cardboard form.

Fig. 3-30 *A round form made of cardboard.* (Proctor Products)

All such special forms allow time and labor to be saved. Little labor and time is needed to set up special forms. Reinforcement may be added as required. Also, such forms are available in many shapes and sizes.

Openings and Special Shapes

Openings for windows and doors are frequently required in concrete foundation walls. Also, special keys or notches are often needed. These hold the ends of support frames, joists, and girders. At times, utility and sewer lines run through a foundation. Special openings must also be made for these. It is very expensive to try to cut such openings into concrete once it is hardened and cured.

However, if portions of the forms are blocked off, concrete cannot enter these areas. This way almost any shape can be built into the wall before it is formed. This shape is called a *block-out* or a *buck*. The concrete is then poured and moves around these blocked-out areas. When the concrete hardens, the shape is part of the wall. This is quicker, cheaper, and easier. It also provides better strength to the wall and makes the forms used more versatile.

Of course, a carpenter can build a block-out of almost any shape in a form. First, one wall of the form is sheathed. The shape can then be framed out on that side. The inside of the shaped opening may be used for nails and braces. The outside, next to the concrete, should be kept smooth and well finished. However, it is expensive to pay a carpenter to frame special openings if they have to be repeated many times. It is better to use a form that can be used over again. Figure 3-31 shows an example.

Fig. 3-31 *Openings may be made with special forms that can be reused. (Proctor Products)*

When building a buck or block-out, first check the plans. Sometimes bucks are removed; other times they are left to form a wooden frame around the opening.

In either case, the size of the opening is the important thing. To determine rough opening sizes for windows in walls, see the chapter on building walls.

Buck keys Strips of wood are used along the sides of openings in concrete. These are used as a nail base to hold frames or units in the opening. These strips are called *keys*. See Fig. 3-32. If the buck is removed, the key is left in place. If the buck is left in place, the key holds the buck frame securely. Note how the key is undercut. The undercut prevents the key from being pulled out.

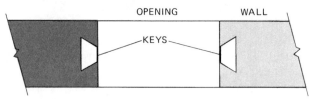

Fig. 3-32 *Keys are placed along the sides of openings as a nail base. Note undercut so that key cannot be removed.*

Bucks should be made from 2-inch lumber. The key can be made of either 1- or 2-inch lumber. The key needs to be only 1 or 2 inches wide. Usually, only the sides of the bucks are keyed.

Buck left as a frame First find the size of the opening. Next, cut the top and bottom pieces longer than the width. These pieces are usually 3 inches longer than the width. See Fig. 3-33. Next, cut the two sides to the same height as the desired opening. Nail with two or three 16d nails as shown. Note that the top piece goes over the sides. The desired size for the opening is the same as the size of the opening in the buck.

Buck removed First find the opening size. Next, cut the top and bottom pieces to the exact width of the opening. Then cut the two sides shorter. The amount is usually 3 inches (twice the thickness of the lumber used). See Fig. 3-34. Nail the frame together with 16d nails as shown.

Note that the opening size is the same as the outside dimensions of the buck. Also, the outside faces are oiled. This keeps the concrete from sticking to the sides of the buck. It also makes it easier to remove the buck.

Buck braces When the opening is large, the weight of the concrete can bend the boards. If the boards bend, the opening will not be the right size or shape.

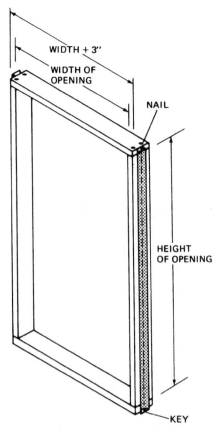

Fig. 3-33 *The buck may be left in the wall as the frame. In this case the opening in the buck is the desired size.*

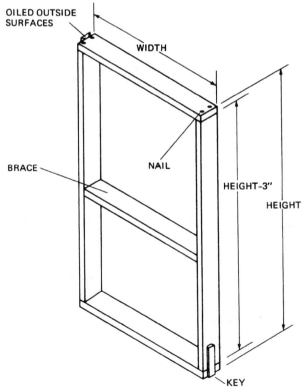

Fig. 3-34 *The buck may be removed. In this case the buck frame is the size desired.*

To prevent this, braces are placed in the opening. See Fig. 3-34. Note that the braces can run from side to side or from top to bottom.

Another type of form is used for porches, sidewalks, or overhangs. See Fig. 3-35. This allows porch supports to be part of the foundation. The earth is filled in later.

Fig. 3-35A *Special forms used for an overhang on a basement wall.*

Reinforcing Concrete Foundations

Concrete is very strong when compressed. However, it does not support weight without cracking. Even though it is very hard, it is also very brittle. In order to resist shifting soil, concrete should be reinforced. It is not a matter of whether the soil will shift. It is more a matter of how much the soil will shift.

It should be noted that sometimes the reinforcement is added before the forms are done. When the forms are tall, very narrow, or hard to get at, reinforcement is done first.

Concrete reinforcement is done in two basic ways. The first way is to use concrete reinforcement bars. The second way is to use mesh. Mesh is similar to a large screen made with heavy steel wire. The foundations

Fig. 3-35B *The overhang after the forms are removed.*

Fig. 3-35C *Basement walls are coated and waterproofed.*

Fig. 3-35D *Finally the concrete porch will be poured.*

may be reinforced by running rebars lengthwise across the top and at intervals up and down. Figure 3-36 shows some typical reinforcement.

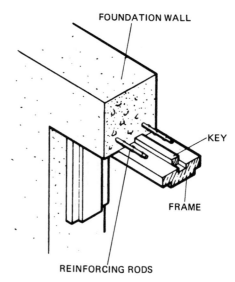

Fig. 3-36 *Foundations should be reinforced at the top.* (Forest Products Laboratory)

The amount and type of reinforcement used is determined by the soil and geographic location. The reinforcement is spaced and tied. There should be at least 3 inches of concrete around the reinforcement.

The carpenter should be sure that the foundation and reinforcement conform to the local building codes.

Estimating Concrete Volume

When the forms are complete, the amount of concrete needed is computed. The same formula as for footings is used:

$$\frac{L'}{3} \times \frac{W''}{36} \times \frac{T''}{36} = \text{cubic yards}$$

However, *W* in inches is replaced by the wall height *H* in feet, so that

$$\frac{L'}{3} \times \frac{H'}{3} \times \frac{T''}{36} = \text{cubic yards}$$

If the foundation is to be 8 inches thick and 8 feet high and the perimeter is 144 feet, then

$$\frac{144}{3} \times \frac{8}{3} \times \frac{8}{36} = \text{cubic yards}$$

and by cancellation,

$$\frac{48}{1} \times \frac{8}{3} \times \frac{2}{9} = \frac{16}{1} \times \frac{8}{1} \times \frac{2}{9}$$

$$= \frac{256}{9}$$

$$= 28.4 \text{ cubic yards}$$

Delivery and Pouring

Once the needs are estimated and the concrete has been ordered or mixed, the wall should be poured. If a transit-mix truck is used, the concrete is mixed and delivered to the site. The concrete truck should be backed as close as possible to the forms. As in Fig. 3-37, the concrete should be poured into the forms, tamped, and spread evenly. By doing this, air pockets and honeycombs are avoided.

Fig. 3-37 *Concrete is poured or pumped into the finished form.*

Finishing the Concrete

Two steps are involved in finishing the pouring of the concrete foundation wall. First, the tops of the forms must be leveled. Sometimes the concrete is poured to within 2 or 3 inches of the top. The concrete is then allowed to partially cure and harden. A concrete or grout with a finer mixture of sand may be used to finish out the top of the foundation.

Anchors are embedded in the concrete before it hardens completely. One end of each anchor bolt is threaded and the other is bent. See Fig. 3-38. The threaded end sticks up so that a sill plate may be bolted in place. As the concrete begins to harden, the bolts are slowly worked into place by being twisted back and forth and pushed down. Once they are embedded firmly, the concrete around them is troweled smooth. Figure 3-39 shows an anchor embedded in a foundation wall.

Fig. 3-38 *Anchor bolt.*

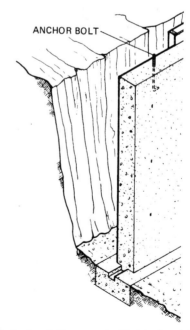

Fig. 3-39 *Anchor bolts are embedded in foundation walls and protrude from them.* (Forest Products Laboratory)

The forms are removed after the concrete cures and hardens. Low spots must be filled. A small spot can be shimmed with a wooden shingle. However, a larger area should be filled in with grout or mortar.

CONCRETE BLOCK WALLS

Concrete blocks are often used for basement and foundation walls. When the foundation is exposed, it may be faced with brick. See Fig. 3-40. Blocks need no form work and go up more quickly than brick or stone. The most common size is 7⅝ inches high, 8 inches wide, and 15⅝ inches long. The mortar joints are ⅜ inch wide. This gives a finished block size of 8 × 16 inches, and a wall 8 inches thick.

Fig. 3-40 *Concrete block foundations may be faced with brick or stone.*

The footing is rough and unfinished. This is because the mortar for the block is also used to smooth out the rough spots. No key is needed, but reinforcement rod should be used. Figure 3-41 shows a footing for a block wall.

Block walls should be capped with either concrete or solid block. Anchor bolts are mortared in the last row of hollow block. They then pass through the mortar joint of the solid cap. See Fig. 3-42.

A special pattern is sometimes used to lay the block. This is done when the wall will also be the visible finish wall. This pattern (see Fig. 3-42) is called a *stack bond*. It should be reinforced with small rebar.

PLYWOOD FOUNDATIONS

Plywood may be used for foundations or for basement walls in some regions. There are several advantages to

Fig. 3-41 *A footing for a concrete block foundation.*

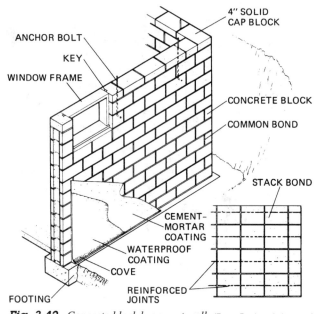

Fig. 3-42 *Concrete block basement wall.* (Forest Products Laboratory)

using plywood. First, it can be erected in even the coldest weather. It is fast to put up because no forms or reinforcement are required. For the owner, plywood makes a wall that is warmer and easier to finish inside. It also conserves on the energy required to heat the building.

The frame is formed with 2-inch studs located on 12- or 16-inch centers. The frame is then sheathed with plywood. Insulation is placed between the studs. The exterior of the wall is covered with plastic film, and building paper is lapped over the top part. The building paper is laid over the top of the rock fill and helps drainage. See Figs. 3-43 and 3-44.

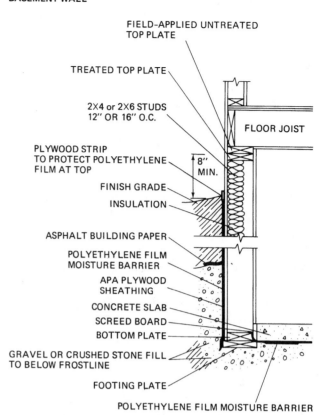

Fig. 3-44 *Cross section of plywood basement wall.* (American Plywood Association)

TYPICAL PANEL

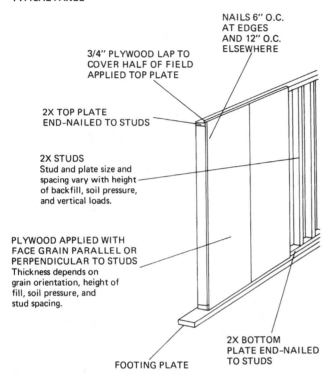

NOTE: Wood and plywood are treated.

Fig. 3-43 *A typical plywood foundation panel.* (American Plywood Association)

It is important to note that all lumber and plywood must be pressure-treated with preservative.

DRAINAGE AND WATERPROOFING

A foundation wall should be drained and waterproofed properly. If a wall is not drained properly, the water may build up and overflow the top of the foundation. Unless the wall is waterproofed, water may seep through it and cause damage to the foundation and footings. It may also cause damage to the interior of a basement. Proper drainage is ensured by placing drain pipes or drain tile around the outside edges of the footing. A gravel fill is used to place the tile slightly below the level of the top of the footing. Figure 3-45 shows the proper location of the drain pipe. The pipe is then covered with loose gravel and compacted slightly.

If the house has a basement, the foundation walls should be waterproofed. If the house has only a crawl space, no waterproofing is needed.

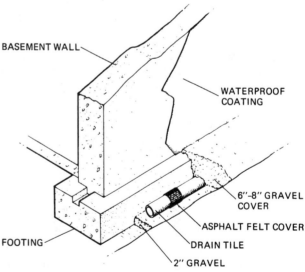

Fig. 3-45 *Basement walls should be coated and drained.* (Forest Products Laboratory)

Waterproofing Basement Walls

Three types of walls are commonly used today for basements. The most common is the solid cast concrete wall. However, concrete blocks are also used and so is plywood.

Concrete block foundations Concrete blocks should be plastered and waterproofed. Figure 3-42 shows the processes involved. First, the concrete wall is coated with a thin coat of plaster. This is called a *scratch coat*. After this has hardened, the surface is scratched so that the next coat will adhere more firmly. The second coat of plaster is applied thickly and smoothly over both the wall and the top of the footing. As shown in Fig. 3-42, this outside layer is then covered with a waterproof coating. Such a coating could include layers of bitumen, builder's felt, or plastic.

Waterproofing concrete walls No plaster is needed over a cast-concrete wall. The wall may be quickly coated with bitumen. However, plastic sheeting may also be applied to cover both the footing and the foundation in one piece. The most common process, as shown in Fig. 3-45, involves a bitumen layer. This bitumen layer is sometimes reinforced by a plastic panel which is then coated with another layer of bitumen.

TERMITES

Termite is the common name applied to white ants. They are neither all-white nor ants. In fact, they are closely related to cockroaches. Termites do, however, live in colonies somewhat in the same manner as ants do.

Most termite colonies are made up of three castes. The highest caste is the royal or reproductive group (Fig. 3-46A). The middle caste is the soldier. The worker is at the bottom of the social groupings (Fig. 3-46C). In every mature colony of termites, a group of young winged reproductives leave the parent nest, mate, and set out to found new colonies. Their wings are used only once; then they are broken off just before they seek a mate.

The worker caste is made up of small, blind, and wingless termites (Fig. 3-46B). They have pale or whitish soft bodies. Only their feet and heads are covered by a hard coating. The worker caste makes up the largest group within a colony.

The soldiers have very long heads in proportion to their bodies and are responsible for protecting the colony against its enemies, usually ants. The soldiers are also blind and wingless. Because of this, the workers do all the work. They enlarge the nest, search for food and water, and make tunnels. They also take care of the soldiers since they have to be fed individually.

Termite castes contain both female and male. The kings live as long as the queens. The queens become as long as the males—three inches—when they are full of eggs. They can lay many thousands of eggs a day for many days. The eggs for all castes appear the same.

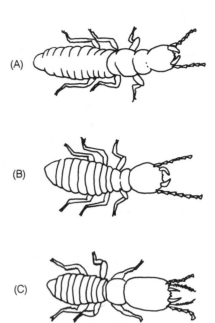

Fig. 3-46 (A) The supplementary queen leaves the colony to mate, then sets up a new colony. She breaks off her wings before mating; (B) The worker termite is small and has a pale soft body. Workers gather food for the colony; (C) The soldier termite is extremely large when compared to the worker. The soldier has a hard head and defends the colony.

Termites live in warm areas such as Africa, Australia, and the Amazon. They build nests as high as 20 feet, with the inside divided into chambers and galleries. They keep the king and queen separated in a closed cell. The workers carry the eggs away as fast as the queen lays them. The workers then care for the eggs until they hatch and then take care of the young until they grow up. Termites digest wood, paper, and other materials. They use cellulose for food. Most of the damage they do to homes and furniture is through tunneling. They have also been known to destroy books in search for cellulose to digest. They also do great damage to sugar cane and orange trees. They are considered a serious pest in many parts of the United States where they damage houses.

The best way to provide protection for a home is to follow some suggestions that have been developed through the years. There are about 2,000 different species of termites known. About 40 species live in the United States and two species in Europe. They do not build large mounds in the United States or Europe, but do most of their damage out of sight.

Types of Termites

Three groups of termites exist in the United States. They are grouped according to their habits. The *subterranean* (underground) termites are the smallest and the most destructive, for they nest underground. They

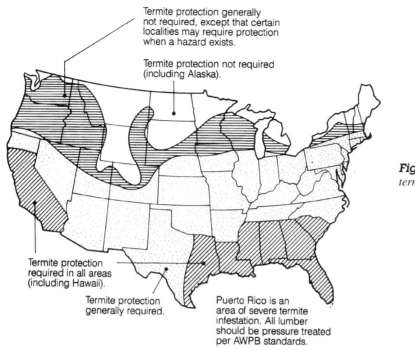

Termite protection generally
not required, except that certain
localities may require protection
when a hazard exists.

Termite protection not required
(including Alaska).

Termite protection
required in all areas
(including Hawaii).

Termite protection
generally required.

Puerto Rico is an
area of severe termite
infestation. All lumber
should be pressure treated
per AWPB standards.

Fig. 3-47 *Note where the termites are located in the U.S.*

extend their habitat for long distances into wood structures. The *damp wood* termites live only in very moist wood. This type causes trouble only on the Pacific Coast. The *dry wood* termites need very little moisture. They are found to be destructive in the Southwest. The damp wood and dry wood types do not have a distinctive working caste (Fig. 3-47).

Once termites get into the house, they eat books, cloth, and furniture. They also attack bridges, trestles, and other wooden structures. They do more damage each year in the United States than fire.

Termite Protection

Most termites cannot live without water. The best approach is to eliminate their source of moisture. This can be done by applying chemicals to the soil around the foundation of a building or impregnating the wood with chemicals that repel or kill termites. Creosote and other types of chemicals are used around the footings of buildings and termite shields are used to keep them from reaching the wood sill (Fig. 3-48). Metal shields are also placed on copper water pipe or soil pipe to prevent their climbing up the metal pipes to reach the wood of the building (see Fig. 3-49). The builder must take every precaution to prevent infestation.

Infestation can be prevented by using minimum space between joists and the soil in the crawl space (see Fig. 3-50). Keep the space to a minimum of 18 inches. The minimum recommended space between girders and the soil is 18 inches. The lowest wood

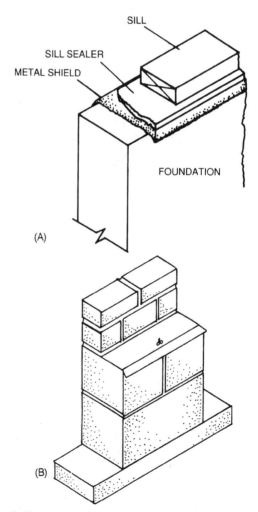

SILL

SILL SEALER

METAL SHIELD

FOUNDATION

(A)

(B)

Fig. 3-48 *(A) A metal shield is placed over the concrete foundation before the wood sill is applied; (B) Even masonry walls should have a termite shield in areas infested with termites.*

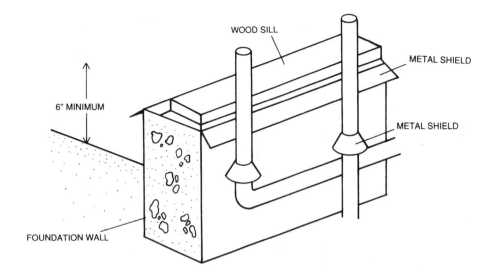

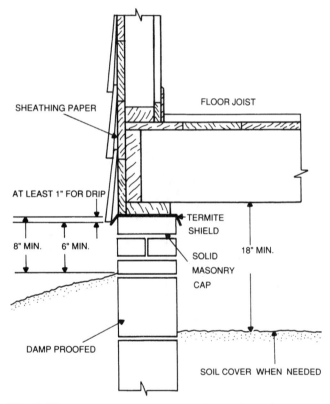

Fig. 3-49 *Note the placement of shield over pipes to protect the wooden floors from infestation.*

Fig. 3-50 *Note the minimum spacings for wood in reference to the grade.*

Fig. 3-51). Make sure there are no fine cracks in the concrete foundation or loose mortar in the cement blocks. Termites have been known to build clay tunnels around metal shields.

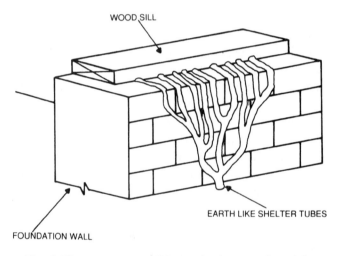

Fig. 3-51 *Termites are well known for their tunneling ability.*

Termites and Treated Wood

Decay and termite attack can occur when all four of the following conditions prevail: a favorable temperature (approximately 50–90 degrees F), a source of oxygen, a moisture content above 20 percent, and a source of food (wood fiber). If any one of these conditions is removed, infestation will not occur. Chemical preservatives eliminate wood as a food source.

PRESSURE-TREATED WOOD

Treatment of wood adds to its versatility. Treatment with chemical preservatives protects wood that is exposed to the elements, is in contact with the ground, or is used in areas of high humidity. The treatment of

member of the exterior of a house should be placed 6 inches or more above the grade. Metal termite shields should be used on each side of a masonry wall (see Fig. 3-48A). Metal shields should be at least 24-gage galvanized iron. Concrete should be compacted as it is placed in the forms. This makes sure rock pockets and honeycombs are eliminated.

After a house is completed make sure all wood scraps are removed. Any scraps buried in backfilling or during grading become possible pockets for termite colonies. From there they can tunnel into a house (see

wood allows you to build a project for outside exposure that will last a lifetime.

Properly treated, pressure-treated wood resists rot, decay and termites, and provides excellent service, even when exposed to severe conditions. It is wood that has been treated under pressure and under controlled conditions with chemical preservatives that penetrate deeply into the cellular structure.

Just because a wood is green in color doesn't mean it is pressure-treated wood. And just because it is pressure treated doesn't mean there is enough of the chemical deep enough in the wood to prevent decay and keep insects at bay. Only in recent years has treated wood been readily available from local lumber yards.

In order to obtain the properly treated wood you should use the American Wood Preservers Bureau's recommendations. First, you should consider the wood. Not all wood is created equal. Most wood species don't readily accept chemical preservatives. To assist preservative penetration, the American Wood Preservers Association standards require incising for all species except southern pine, ponderosa pine, and red pine. Incising is a series of little slits along the grain of the wood which assists chemical penetration and uniform retention. Depth of penetration is important in providing a chemical barrier thick enough so that any checking or splitting won't expose untreated wood to decay or insect attack.

To make sure the wood is strong enough for your intended use, you should always insist that the lumber you buy bears a lumber grade mark. See Fig. 3-52. This is typical of the symbols (grade marks) found on southern pine lumber.

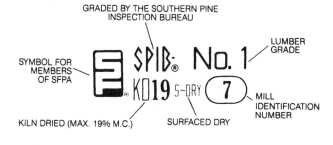

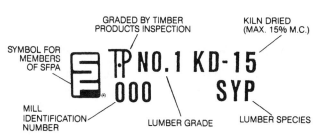

Fig. 3-52 *Two typical lumber grade marks.* (Southern Forest Products Association)

Preservatives

There are three types of wood preservatives used. *Waterborne preservatives* are used in residential, commercial, recreational, marine, agricultural, and industrial applications. *Creosote* and mixtures of creosote and coal tar in heavy oil are used for railroad ties, utility poles, piles and similar applications. *Pentachlorophenol* (Penta) in various solvents is used in industrial applications, utility poles, and some farm uses.

Waterborne preservatives are most commonly used in home construction due to their clean, odorless appearance. Wood treatment with waterborne preservatives can be stained or painted when dry. These preservatives also meet stringent EPA health guidelines. The preservative most commonly used in residential construction lumber is known by a variety of brand names, but in the trade it is known simply as CCA. The CCA stands for Chromates Copper Arsenate. It also accounts for the green (copper) color in the treated wood.

CCA is a waterborne preservative. The chromium salts combine with the wood sugar to form an insoluble compound that renders the CCA preservative nonleachable. Some estimate the life of properly treated wood to exceed 100 years because the chemicals do not leach out in time or with exposure to the elements. However, in order to make sure the wood is properly treated after it has been cut you must treat the exposed area with chemicals. It is a good idea to avoid field cuts and drilling in portions of treated wood that will be submerged in water.

Above-Ground and In-Ground Treatment

The standard retentions most commonly found in lumber yards are 0.25 (pounds of chemical per cubic foot of wood) for above-ground use and 0.40 for below- (or in ground contact with) ground use. The only difference is a slightly higher concentration of chemical in 0.40 below-ground treatments. So if the above-ground [.25] material is not available, below-ground [.40] can be used instead. However you should not use 0.25 material in contact with the ground.

Retentions (lbs./cu. ft.)	Uses
0.25	above-ground
0.40	ground contact
0.60	wood foundation
2.50	in salt water

Only FDN (foundation) treatment should be used for wood foundation lumber and plywood applications.

All pieces of lumber and plywood used for wood foundation applications must be identified by the FDN stamp (0.60 retention). That means the foundation should have a life longer than that of the rest of the structure.

Nails and Fasteners

Hot-dipped galvanized or stainless-steel nails and fasteners should be used to ensure maximum performance in treated wood. Such fasteners ensure permanence and prevent corrosion, which stains both the wood and its finishes. In structural applications, where a long service life is required, stainless steel, silicon, bronze, or copper fasteners are recommended. Smaller nails may be used with southern pine because of its greater nail-holding ability. Use 10d nails to fasten 2-inch dimensional lumber, 8d nails for fastening 1-inch boards to 2-inch dimensional lumber, and 6d nails for fastening 1-inch boards to 1-inch boards. The use of treated lumber adhesives can be considered for attaching deck boards to joists, in building fences, or in applications where the appearance of nailheads is not desired. For deck applications, fastening boards bark-side-up will help reduce surface checking and cupping. See Fig. 3-53.

Fig. 3-53 *The bark side of the lumber is exposed when using treated wood for a deck or other exposed surface.* (Southern Forest Products Association)

Handling and Storing Treated Wood

When properly handled, pressure-treated wood is not believed to be a health hazard. Treated wood should be disposed of by ordinary trash collection or burial. It should not be burned. Prolonged inhalation of sawdust from untreated and treated wood should be avoided. Sawing should be performed outdoors while wearing a dust mask. Eye goggles should be worn when power-sawing or machining. Before eating, drinking, or using tobacco products, areas of skin that have come in contact with treated wood should be washed thoroughly. Clothes accumulating preservatives and sawdust should be laundered before reuse and washed separately from other household clothing.

Care should be taken to prevent splitting or excessively damaging the surface of the lumber, since this could permit decay organisms to get past the chemical barrier and start deterioration from within. Treated lumber should be stacked and stored in the same manner as untreated wood. Treated wood will also weather. If it is stored outside and exposed to the sun and elements, the green color will eventually turn to the characteristic gray, just as natural brown or red-colored wood does.

CHAPTER 3
STUDY QUESTIONS

1. What is the lowest part of a building?
2. How is footing size determined?
3. What types of footings are used?
4. What is a form?
5. How deep should footings be for basements? For crawl spaces?
6. How are forms leveled?
7. What is a spreader?
8. How much concrete would be needed for a one-piece footing and foundation with these dimensions:

Footing thickness	= 8 inches
Width	= 18 inches
Perimeter	= 150 feet
Foundation thickness	= 8 inches
Height	= 9 feet

9. What should be used to fill holes for utility lines?
10. How is a footing key made?
11. How are anchor bolts embedded?
12. What two materials are commonly used to make foundation walls?
13. How should rebar be spaced?
14. How is rebar tied together?
15. How are foundations drained?
16. How are foundation walls waterproofed?
17. Why are double-headed nails preferred for building concrete forms?
18. What is a block-out?
19. What are the advantages of using panel forms?
20. What is a "basic" concrete volume mix?

4
CHAPTER

Pouring Concrete Slabs & Floors

ONCRETE SURFACES ARE USED FOR MANY things. Slabs combine footings, foundations, and subfloors in one piece. Concrete floors are common in basements and in baths. Concrete is used outdoors to form stairs, driveways, patios, and sidewalks. Carpenters build the forms and, in some cases, also help pour and finish the concrete. After this chapter you should be able to:

• Excavate

• Construct the forms

• Prepare the subsurface

• Lay drains and utilities

• Lay reinforcement

• Determine concrete needs

• Ensure correct pouring and surfacing

CONCRETE SLABS

Concrete slabs can combine footings, foundations, and subfloors as one piece. Slabs are cheaper to build than basements. In the past, basements were used as storage areas for furnaces, fuels, and ashes. Basements also held cooling and ventilation units and laundry areas. Today, these things are as easily built on the ground. However, in cold climates many people still prefer basements. They keep pipes from freezing and add warmth to the upper floors. They also provide storage space, play areas, and sometimes living areas.

Slabs are best used on level ground and in warm climates. They can be used where ground hardness is uneven. Slabs are good with split-level houses or houses on hills where the slab is used for the lower floor. See Fig. 4-1.

In the past, slabs did not make comfortable floors. They were very cold in the winter, and water easily condensed on them. Water could also seep up through the slab from the ground. Today's building methods can solve most of these problems.

Most heat energy is lost at the edges of a slab. This loss is reduced by using rigid insulation. See Fig. 4-2. In extreme cases a warmer floor is needed. Heating ducts may be built in the slab flooring itself. This is an efficient method and provides an even temperature in the building. A warmer type of floor can also be built over the concrete floor. This is discussed later in this chapter.

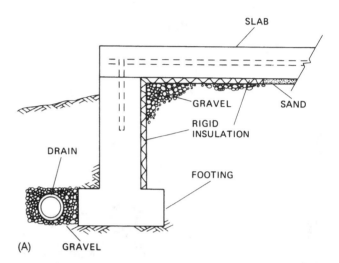

(B)

Fig. 4-2 (A) Insulation for a slab; (B) Forms for a slab. (Fox and Jacobs)

Sequence for Preparing a Slab

Slabs are used for many types of buildings and for several types of outdoor surfaces. In most cases, the general procedure is about the same. After the site is prepared and the building is located, this sequence is common:

1. Excavate.

2. Construct forms.

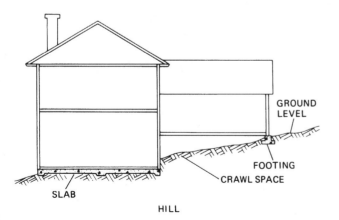

Fig. 4-1 A typical split-level home. The lower floor is a slab; the middle floor is a frame floor.

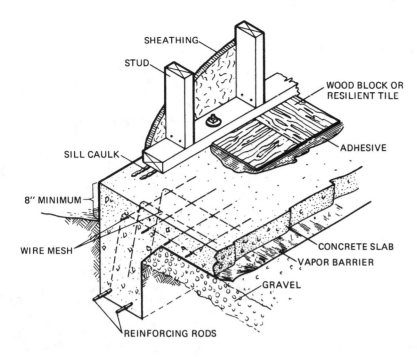

SHEATHING
STUD
WOOD BLOCK OR
RESILIENT TILE
SILL CAULK
ADHESIVE
8" MINIMUM
WIRE MESH
CONCRETE SLAB
VAPOR BARRIER
GRAVEL
REINFORCING RODS

Fig. 4-3 *Combined slab and foundation (thickened-edge slab).* (Forest Products Laboratory)

3. Prepare subsurface.
   a. Spread sand or gravel and level.
   b. Install drains, pipes, and utilities.
   c. Install moisture barrier.
4. Install reinforcement bar.
5. Construct special forms for lower levels, stairs, walks, etc.
6. Level tops of forms.
7. Determine concrete needs.
8. Pour concrete.
9. Tamp, level, and finish.
10. Embed anchors.
11. Cure and remove forms.

Types of Slabs

Slabs are built in two basic ways. Those called *monolithic* slabs are poured in one piece. They are used on level ground in warm climates.

In cold climates, however, the frostline penetrates deeper into the ground. This means that the footing must extend deeper into the ground to be below the frostline. Another method for slabs is often used where the footing is built separately. The two-piece slab is best for cold or wet climates. Figures 4-2 and 4-3 show both types of slabs.

Slab footings must rest beneath the freeze line. This gives the slab stability in the soil. Slabs should be reinforced, but the amount of reinforcement needed varies. It depends on soil conditions and weights to be carried. On dry, stable soil, slabs need little reinforce-

ment. Larger slabs or less stable soils need more reinforcement.

The top of the slab should be 8 inches above ground. If the slab is above the rest of the ground, moisture under the slab can drain away from the building. The ground around the slab should also be sloped for the best drainage. See Fig. 4-4.

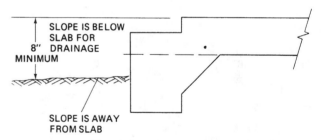

SLOPE IS BELOW
SLAB FOR
8" DRAINAGE
MINIMUM

SLOPE IS AWAY
FROM SLAB

Fig. 4-4 *Drainage of a slab.*

Excavate

At this point, the building has already been located and batter boards are in place. Lines on the batter boards show the location of corners and walls.

Now, trenches or excavations are made for the footings, drains, and other floor features. These must all be deeper than the rest of the slab. Footings are, of course, around the outside edges. However, a slab big enough for a house should also have central footings.

The locating and digging for slab footings is the same as for foundation footings. There is no difference at all when a two-piece slab is to be made. The main difference for a monolithic slab is how the forms are made.

The trenches can now be dug. Rough lines are used for guides. As a rule, inner trenches are dug first. The trenches for the outside footings are done last. This is easier when machines are moved around the site.

The excavations are then checked for depth level. Remember from Chapter 3 that the lowest point determines the depth. Trenches are also dug for drains and sewer lines inside the slab. Then, trenches are dug from the slab to the main utility lines. Sewer lines must connect the slab to the main sewer line, and so forth.

Construct the Forms

After the excavation is done, the corners are relocated. Lines are restrung on batter boards and the corners are plumbed. See Fig. 3-16. The footing forms for two-piece slabs are made like standard footing or foundation forms. Refer to Chapter 3.

Lumber is brought to the place where it is needed, and then forms are constructed. See Fig. 4-5. Monolithic forms are made like footing forms. The top board is placed and leveled first. See Fig. 3-16. It is leveled with the corner first. Then its length is leveled and the ends are nailed to stakes. As before, double-headed nails should be used from the outside. The remaining form boards are then nailed in place.

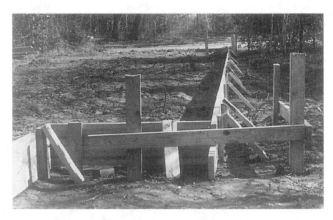

Fig. 4-6 *The forms are erected.*

Fig. 4-7 *Sand is dumped and spread in the form area.*

Fig. 4-5 *Forms are erected.* (Fox and Jacobs)

Fig. 4-8 *Excavations are made for outside footings. Plumbing is "roughed in."*

Another method may be used for monolithic slabs with shallow footings. It is a very fast and inexpensive way. The form boards are put up before excavating. See Fig. 4-6. Be sure to carefully check the plans. Next, the sand or gravel is dumped inside the form area. See Fig. 4-7. The sand is spread evenly over the form area. As in Fig. 4-8, the outside footings are then dug by hand. Chalk lines are strung from the forms.

They are then used as guides to dig the central footings. See Fig. 4-9. Trenches for drains and sewers are then dug.

Fig. 4-9 *Excavations are also made for footings in the slab.*

Prepare the Subsurface

The ground under any type of slab must be prepared for moisture control. Water must be kept from seeping up through a slab. The water must also be drained from under the slab. This preparation is needed because water from rain and snow will seep under the slab.

Outside moisture can be reduced by using good siding methods. The edges of the slab can be stepped for brick. The sheathing can overlap the slab edge on other types of siding. As mentioned before, heat energy can be lost easily from slabs. The main area of loss is around the edges. Rigid foam insulation can be put under the slab's edges. This will reduce the heat energy losses of the slab.

Subsurface preparation After excavation, drains, water lines, and utilities are "roughed in." These should be placed for areas such as the kitchen, baths, and laundry. Water lines may be run in ceilings or beneath the slab. If they are to be beneath the slab, soft copper should be used. Extra length should be coiled loosely to allow for slab movement. Metal or plastic pipe may be used for the drains. Conduit (metal pipe for electrical wires) should be laid for any electrical wires to go under the slab. Wires should never be laid without the conduit. All openings in the pipes are then capped. See Fig. 4-10. This keeps dirt and concrete from clogging them.

The various pipes are then covered with sand or gravel. Sand or gravel is dumped in the slab area and carefully smoothed and leveled. Chalk lines are strung across the forms to check the level. See Fig. 4-11. It is also wise to install a clean-out plug between the slab and the sewer line. See Fig. 4-12.

Lay a vapor seal Once the sand is leveled, the moisture barrier is laid. The terms vapor barrier, moisture barrier, vapor seal, and membrane mean the same

Fig. 4-10 *All openings in pipes are covered. This prevents them from clogging with dirt or concrete.*

Fig. 4-11 *The sand is leveled. Chalk lines are used as guides.*

Fig. 4-12 *Trenches for drains and sewers are also excavated. Cleanout plugs are usually outside the slab.*

Fig. 4-13 *The vapor barrier is then laid. The reinforcement is also laid.*

thing. As a rule, plastic sheets are used for moisture barriers. The moisture barrier is laid so that it covers the whole subsurface area. To do this, several strips of material will be used. The strips should overlap at least 2 inches at the edges.

Insulate the edges Figure 4-2 shows insulation for a slab. Insulation is laid after placement of the vapor barrier. The insulation is placed around the outside edges. This is called *perimeter insulation*. The insulation should extend to the bottom of the footing. It should extend into the floor area at least 12 inches. A distance of 24 inches is recommended. Rigid foam at least 1 inch thick is used. Perimeter insulation is not always used in warm climates. However, the moisture barrier should always be used.

Reinforcement Reinforcement should always be used. The amount of reinforcement rods (rebar) used in the footing should conform to local codes. The slab should also be reinforced with mesh. This mesh is made of 10-gage wire. The wires are spaced 4 to 6 inches apart. Figure 4-13 shows the reinforced form ready for pouring. Where the soil is unstable, more reinforcement is needed. The amount is usually given on the plans.

As the rebar is laid, it is tied in place. The soft metal ties are twisted around the rebars. See Fig. 4-14. The mesh may also be held off the bottom by metal stakes. See Fig. 4-15. Mesh may also be lifted into place as the concrete is poured. Chairs hold rebar in place for pouring. See Fig. 4-16.

Fig. 4-14 *Rebar is tied together with soft wire "ties."*

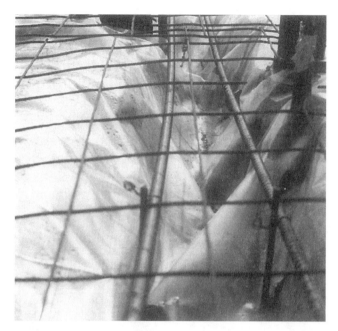

Fig. 4-15 *Short stakes hold mesh and rebar in place during pouring.*

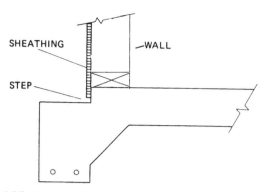

Fig. 4-17 *A stepped form aids drainage. The step also forms a base for brick siding.*

Fig. 4-16 *Chairs hold rebar in place for pouring.* (Fox and Jacobs)

Special Shapes

Special shapes are used for several reasons. A stepped edge, as in Fig. 4-17, helps drainage. It prevents rainwater from flowing onto the floor surface. Lower sur-

face areas are also common. They are used for garages, entry ways, and so forth.

Step edges are easily formed. A 2 × 4 or 2 × 6 is nailed to the top of the form. See Fig. 4-18A and B. To lower a larger surface, an extension form is used. The first form is used as a nailing base. The extension form is built inside the outer form. The lower area can then be leveled separately. See Fig. 4-19.

(A)

(B)

Fig. 4-18 *A stepped edge is formed by nailing a board to the form.* (Fox and Jacobs)

Fig. 4-19 *A lower surface is used for a garage. (See top left through bottom right.)*

(A)

(B)

(C)

(D)

(E)

Pouring the Slab

First, the corners of the forms are leveled. See Fig. 4-20. A transit is used for large slabs, but small areas are leveled with a carpenter's level. See Chapter 2 for the leveling process with a transit.

Diagonals are checked for squareness, and all dimensions are checked.

Most concrete today is delivered to the building site. Usually, the concrete is not mixed by the carpenters. It is delivered by transit-mix trucks from a concrete company. The concrete is sold in units of cubic yards. Before the concrete can be ordered, the amount of concrete needed must be determined.

Estimating volume A formula is used to estimate the volume of concrete needed. The following formula is used:

$$\frac{L'}{3} \times \frac{W'}{3} \times \frac{T''}{36} = \text{cubic yards}$$

For example, a slab has a footing 1 foot wide. The slab is to be 30 feet wide and 48 feet long. The slab is to be 6 inches thick. From the formula, the amount required is

$$\frac{30}{3} \times \frac{48}{3} \times \frac{6}{36} = \frac{8640}{324}$$

$$\frac{10}{1} \times \frac{16}{1} \times \frac{1}{6} = \frac{160}{6}$$

$$= 26\frac{2}{3}$$

But the perimeter footings are 18 inches deep. They are not just 6 inches deep. Thus, a portion 12 inches

Fig. 4-20 *Corners of forms are leveled before pouring the slab.*
(Portland Cement)

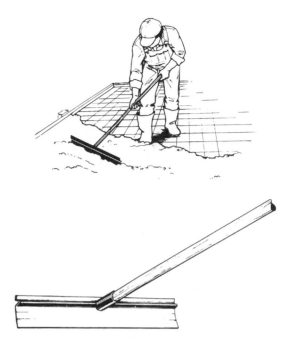

Fig. 4-21 *Concrete is spread evenly in the form.*

thick (18 minus 6) must be added. This additional amount of concrete is calculated as follows. The linear distance around the slab is

$$48 + 48 + 30 + 30 = 156$$

Then

$$\frac{L}{3} \times \frac{W}{3} \times \frac{T}{36} = \frac{1}{3} \times \frac{156}{3} \times \frac{12}{36}$$

$$= \frac{1}{3} \times \frac{52}{1} \times \frac{1}{3} = \frac{52}{9} = 5.7 \text{ cubic yards}$$

Thus, to fill the slab, the two elements are added:

$$\begin{array}{r} 26.7 \\ + 5.7 \\ \hline 32.4 \text{ cubic yards} \end{array}$$

Pouring To be ready, the carpenter sees to two things. First the forms must be done. Then the concrete truck must have a close access.

The builder must spread, carry, and level the concrete. The truck will only deliver it to the site. The truck driver can remain only a few minutes. The driver is not allowed to help work the concrete. As the concrete is poured, it should be spread and tamped. This is done with a board or shovel. See Fig. 4-21. The board or shovel is plunged into the concrete. Be careful not to cut the moisture barrier. Tamping helps get rid of air pockets. This makes the concrete solidly fill all the form.

After tamping, the concrete is leveled. This is done with a long board called a *strike-off*. See Fig. 4-22. The

(A)

(B)

Fig. 4-22 *(A) As the form is filled, it is leveled. This is done with a long board called a strike-off. (B) Using a "jitterbug" to remove trapped air from the concrete.* (Skrobarczk Properties)

ends rest across the top of the forms. The board is moved back and forth across the top. A short back-and-forth motion is used. If the board is not long enough, special supports are used. These are called *screeds*. A screed may be a board or pipe supported by metal pins. The screed is leveled with the tops of the forms. It is removed after the section of concrete is leveled. Any holes left by the screeds are then patched.

After leveling, the surface is treated using a "jitter-bug" to remove trapped air in the concrete. Then, it is floated. This is done after the concrete is stiff. However, the concrete must not have hardened. A finisher uses a float to tamp the surface gently. During tamping the float is moved across the surface. Large floats called *bull floats* are used. See Fig. 4-23. Floating lets the smaller concrete particles float to the top. The large particles settle. This gives a smooth surface to the concrete. Floats may be made of wood or metal.

Fig. 4-24 *A broom can be used to make a lined surface.* (Portland Cement)

Fig. 4-25 *Hand trowels may be used to smooth small surfaces.* (Portland Cement)

Fig. 4-23 *After leveling, the surface is smoothed by "floating" with a bull float.* (Portland Cement)

After floating, the finish is done. A rough, lined surface can be produced. A broom is pulled across the top to make lines. This surface is easier to walk on in bad weather. See Fig. 4-24. For most flooring surfaces, a smooth surface is desired. A smooth surface is made by troweling. For small surfaces a hand trowel may be used, as in Fig. 4-25. However, for larger areas, a power trowel will be used. See Fig. 4-26.

Now the surface has been finished. Next, the anchors for the walls are embedded. Remember, the concrete is not yet hard. Do not let the anchors interfere with joist or stud spacing. The first anchor is embedded at about one-half the stud spacing. This would be 8 inches for 16-inch spacings. Anchors are placed at four to eight foot intervals. Only two or three anchors are needed per wall.

Fig. 4-26 *A power trowel is used to finish large surfaces.* (Portland Cement)

The anchors are twisted deep into the concrete. The anchors are moved back and forth just a little. This settles the concrete around them. After the anchors are embedded, the surface is smoothed. A hand trowel is used. See Fig. 4-27.

The concrete is then allowed to harden and cure. Afterward, the form is removed. It takes about three

Fig. 4-27 *Anchor bolts are embedded into foundation wall.*

days to cure the concrete slab. During this time, work should not be done on the slab. When boards are used for the forms, they are saved. These boards are used later in the house frame. Lumber is not thrown away if it is still good. It is used where the concrete stains do not matter.

Expansion and Contraction

Concrete expands and contracts with heat and cold. To compensate for this movement, expansion joints are needed. Expansion joints are used between sections. The expansion joint is made with wood, plastic, or fiber. Joint pieces are placed before pouring. Such pieces are used between foundations and basement walls or between driveways and slabs. See Fig. 4-28. Often, the screed is made of wood. This can be left for the expansion joint.

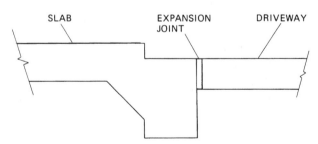

Fig. 4-28 *Expansion joints are used between large, separate pieces.*

Joints

Other joints are also used to control cracking. These, however, are shallow grooves troweled or cut into the concrete. They may be troweled in the concrete as it hardens. The joints may also be cut with a special saw blade after the concrete is hard. The joints help offset

and control cracking. Concrete is very strong against compression. However, it has little strength against bending or twisting. Note, for example, how concrete sidewalks break. The builder only tries to control breaks. They cannot be prevented. The joints form a weak place in concrete. The concrete will break at the weak point. But the crack will not show in the joint. The joint gives a better appearance when the concrete cracks. Figure 4-29 shows a troweled joint. Figure 4-30 shows a cut joint.

Fig. 4-29 *Joints may be troweled into a surface.* (Portland Cement)

Fig. 4-30 *Some joints may be cut with saws.*

CONCRETE FLOORS

Concrete is used for floors in basements and commercial buildings. The footings and foundations are built before the floor is made. An entire house may be built before the basement floor is made. See Fig. 4-31. The concrete can be poured through windows or floor openings.

Fig. 4-31 *Footings in a basement later become part of the basement floor.*

A concrete floor is made like a slab. It is also the way to make the two-piece slab. First, the ground is prepared. Drains, pipes, and utilities are positioned and covered with gravel. Next, coarse sand or gravel is leveled and packed. A plastic-film moisture barrier is placed to reach above the floor level. Perimeter insulation is laid as indicated. See Fig. 4-32. Reinforcement is placed and tied. Rigid foam insulation can also be used as an expansion joint. Finally, the floor is poured, tamped, leveled, and finished. Separate footings should be used to support beams and girders. Figure 4-33 shows this procedure.

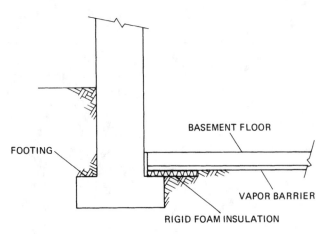

Fig. 4-32 *Rigid foam should be laid at the edges of the basement floor.*

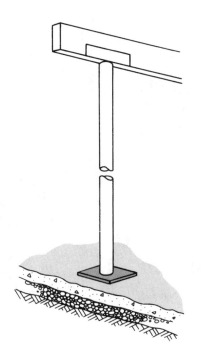

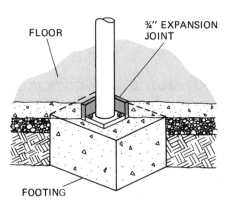

Fig. 4-33 *Floors should not be major supports. Separate footings should be used.*

Stairs

Entrances may be made as a part of the main slab. However, stairs used with slabs are poured separately. Separate stairs are used with slabs as an access in steep areas. They are also used with steep lawns. Forms for steps may be made by two methods. The first method uses short parallel boards. See Fig. 4-34. The second method uses a long stringer. Support boards are added as in Fig. 4-35. The top of the stair tread is left open in both types of forms. This lets the concrete step be finished. The bottoms of the riser forms are beveled. This lets a trowel finish the full surface. The stock used for building supports should be 2-inch lumber. This keeps the weight of the concrete from bulging the forms.

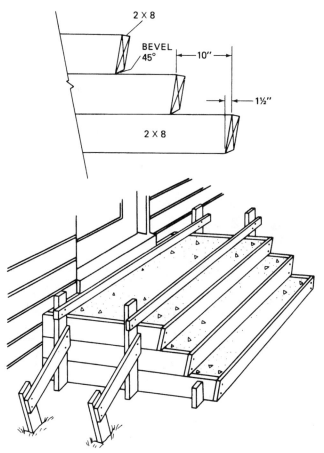

Fig. 4-34 *A form for steps.*

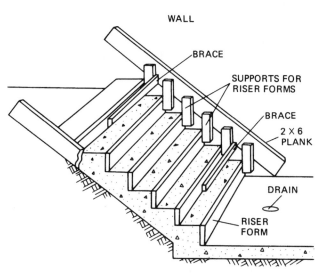

Fig. 4-35 *A form for steps against a wall.*

SIDEWALKS AND DRIVEWAYS

Sidewalks and driveways should not be a part of a slab. They must float free of the building. Where they touch, an expansion joint is needed. Sidewalks and driveways should slope away from the building. This lets water drain away from the building. Then, special methods are used for drainage. Often, a ledge about 1½ inches high

will be used. The floor of the building is then higher than the driveway. One side of the driveway is also raised. Water will then stop at the ledge and will flow off to the side. This keeps water from draining into the building.

Two-inch-thick lumber should be used for the forms. It should be four or six inches wide. The width determines slab thickness. Commercial forms may also be used. See Fig. 4-36.

Fig. 4-36 *Special forms may be used for sidewalks and driveways.* (Proctor Products)

Sidewalks

Sidewalks are usually 3 feet wide and 4 inches thick. Main walks or entry ways may be 4 feet or wider. No reinforcement is needed for sidewalks on firm ground, though it may be used for sidewalks on soft ground. A sand or gravel fill is used for support of sidewalks on wet ground. The earth and fill should be tamped solid. Figure 4-37 shows a sidewalk form. The slab is poured, leveled, and finished like other surfaces.

Fig. 4-37 *Sidewalk forms are made with 2-×-4 lumber.*

Driveways

Driveways should be constructed to handle great weight. Reinforcement rods or mesh should be used in driveways. Slabs 4 inches thick may be used for passenger cars. Six-inch-thick slabs are used where trucks are expected. The standard driveway is 12 feet wide. Double driveways are 20 feet wide. Other dimensions are shown in Fig. 4-38. The slab is poured, leveled, and finished like other surfaces.

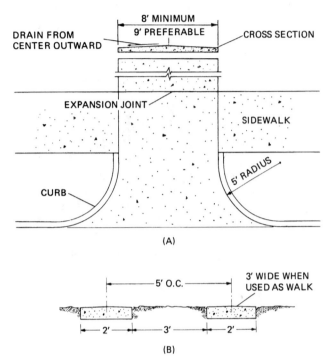

Fig. 4-38 *Driveway details: (A) Single-slab driveway; (B) Ribbon-type driveway.* (Forest Products Laboratory)

SPECIAL FINISHES & SURFACES

Concrete may be finished in several ways. Different surface textures are used for better footing or appearance. Also, concrete may be colored. It may also be combined with other materials.

Different Surface Textures

Different surface textures may be used for appearance. However, the most common purpose is for better footing and tire traction, especially in bad weather. Several methods may be used to texture the surface. The surface may be simply floated. This gives a smooth, slightly roughened surface. Floated surfaces are often used on sidewalks.

Brushing The surface may also be brushed. Brushing is done with a broom or special texture brush. The pattern may be straight or curved. For a straight pattern, the brush is pulled across the entire surface. Refer again to Fig. 4-24. For swirls, the brush can be moved in circles.

Pebble finish Pebbles can be put in concrete for a special appearance. See Fig. 4-39. The pebble finish is

not difficult to do. As the concrete stiffens, pebbles are poured on the surface. The pebbles are then tamped into the top of the concrete. The pebbles in the surface are leveled with a board or a float. Some hours later, the fine concrete particles may be hosed away.

Fig. 4-39 *A pebbled concrete surface.*

Color additives Color may be added to concrete for better appearance. The color is added as the concrete is mixed and is uniform throughout the concrete. The colors used most are red, green, and black. The surface of colored concrete is usually troweled smooth. Frequently, the surface is waxed and polished for indoor use.

Terrazzo A terrazzo floor is made in two layers. The first layer is plain concrete. The second layer is a special type of white or colored concrete. Chips of stone are included in the second, or top, layer of concrete. The top layer is usually about ½ to 1 inch thick. It is leveled but not floated. The topping is then allowed to set. After it is hardened, the surface is finished. Terrazzo is finished by grinding it with a power machine. This grinds the surface of both the stones and the cement until smooth. Metal strips are placed in the terrazzo to help control cracking. Both brass and aluminum strips are used. Figure 4-40 shows this effect. The result is a durable finish with natural beauty. The floor may be waxed and polished.

Terrazzo will withstand heavy foot traffic with little wear. It is often used in buildings such as post offices or schools. It is also easy to maintain.

Ceramic tile, brick, and stone Concrete may also be combined with ceramic tile or brick. The result is a better-appearing floor. The floor contrasts the concrete

Fig. 4-40 *A terrazzo floor is smooth and hard. It can be used in schools and public buildings.* (National Terrazzo and Mosaic Association)

Fig. 4-41 *A concrete floor is being placed over a plywood floor. The concrete will make the floor more durable.*

and the brick or tile. Also, the concrete may be finished in several ways. This gives more variety to the contrast. For example, pebble finish on the concrete may be used with special brick. Bricks are available in a variety of shapes, colors, and textures.

Ceramic tile comes in many sizes. The largest is 12 × 12 inches. Several shapes are also used. Glazed tile is used in bathrooms because the glaze seals water from the tile. Unglazed tile also has many uses, but it is not waterproof.

To set tile, stone, or bricks in a concrete floor, you must have a lowered area. The pieces are laid in the lowered area. The area between the pieces is filled with concrete, grout, or mortar.

Stone, tile, and brick are used. They add contrast and beauty and resist wear. They are easy to clean and resist oil, water, and chemicals.

Concrete over wood floors Concrete is also used for a surface over wood flooring. Figure 4-41 shows this kind of floor being made. A concrete topping on a floor has several advantages. These floors are harder and more durable than wood. They resist water and chemicals and may be used in hallways and rest rooms.

Wood over concrete Wood may be used for a finish floor over a concrete floor. The wooden floor is warmer in cold climates. Because it does not absorb heat as does concrete, energy can be saved. The use of wood can also improve the appearance of the floor.

Two methods are used for putting wood over a concrete floor. The first method is the older. A special glue, called *mastic*, is spread over the concrete. Then strips of wood are laid on the mastic. A wooden floor may be nailed to these strips. Figure 4-42A and B shows a cross section.

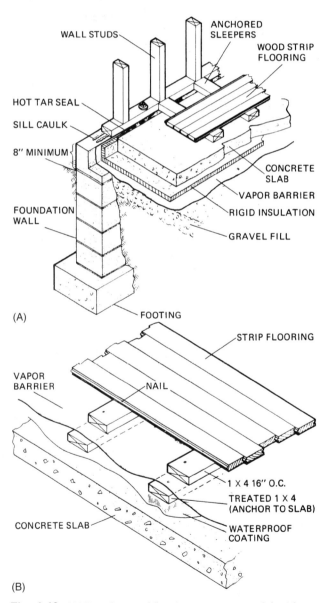

Fig. 4-42 *(A) Base for wood flooring on a concrete slab with vapor barrier under slab; (B) Base for wood flooring on a concrete slab with no vapor barrier under slab.* (Forest Products Laboratory)

For better energy savings, a newer method is used. In this method, rigid insulation is laid on the concrete. Mastic may be used to hold the insulation to the floor. See Fig. 4-43. The wooden floor is laid over the rigid insulation. Plywood or chipboard underlayment can also be used. The floor may then be carpeted. If desired, special wood surfaces can be used.

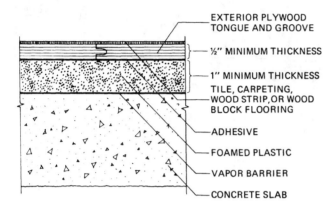

EXTERIOR PLYWOOD TONGUE AND GROOVE
½" MINIMUM THICKNESS
1" MINIMUM THICKNESS
TILE, CARPETING, WOOD STRIP, OR WOOD BLOCK FLOORING
ADHESIVE
FOAMED PLASTIC
VAPOR BARRIER
CONCRETE SLAB

NOTE: Vapor barrier may be omitted if an effective vapor barrier is in place beneath the slab.

Fig. 4-43 *A wood floor may be laid over insulation. The insulation is glued to the concrete floor.* (American Plywood Association)

ENERGY FACTORS

There are two ways of saving energy that are used with concrete floors. The first is to insulate around the edges of the slab. The second is to cover the floor with insulated material. Both methods have been mentioned previously.

CHAPTER 4
STUDY QUESTIONS

1. What does *monolithic* mean?
2. What are two ways of saving energy with concrete floors?
3. What is the sequence for pouring a slab?
4. Why are slab forms leveled just before pouring?
5. How is moisture beneath a slab drained?
6. How is moisture drained away from a building?
7. What thickness of lumber is best for making concrete forms?
8. What should be the thickness of a sidewalk?
9. What should be the thickness of a driveway for trucks?
10. How is energy saved when concrete floors are built?
11. How are step forms made?
12. What is the width of a single driveway?
13. Where should footings for a slab be located?
14. How is a large surface floated?
15. How is a large surface troweled?
16. How can concrete be finished for better footing?
17. How is a pebble surface made?
18. How much concrete would be needed for a plain slab floor 6 inches thick, 21 inches wide, and 36 feet long?
19. What happens to lumber used for forms?
20. How is cracking controlled? Is it really stopped?

5
CHAPTER

Building
Floor
Frames

IN THIS UNIT YOU WILL LEARN HOW TO BUILD frame floors. You will learn how to make floors over basements and crawl spaces. You will also learn how to make openings for stairs and other things. Overall, you will learn how to:

- Connect the floor to the foundation
- Place needed girders and supports
- Lay out the joist spacings
- Measure and cut the parts
- Put the floor frame together
- Lay the subflooring
- Build special framing
- Alter a standard floor frame to save energy

INTRODUCTION

Floors form the base for the rest of the building. Floor frames are built over basements and crawl spaces. Houses built on concrete slabs do not have floor frames. However, multilevel buildings may have both slabs and floor frames.

First the foundation is laid. Then the floor frame is made of posts, beams, sill plates, joists, and a subfloor. When these are put together they form a level platform. The rest of the building is held up by this platform. The first wooden parts are called the *sill plates*. The sill plates are laid on the edges of the foundations. Often, additional supports are needed in the middle of the foundation area. See Fig. 5-1. These are called *midfloor supports* and may take several forms. These supports may be made of concrete or masonry. Wooden posts and metal columns are also used. Wooden timbers called *girders* are laid across the central supports. Floor joists then reach (span) from the sill on the foundation to the central girder. The floor

joists support the floor surface. The joists are supported by the sill and girder. These in turn rest on the foundation.

Two types of floor framing are used on multistory buildings. The most common is the platform type. In platform construction each floor is built separately. The other type is called the balloon frame. In balloon frames the wall studs reach from the sill to the top of the second floor. Floor frames are attached to the long wall studs. The two differ on how the wall and floor frames are connected. These will be covered in detail later.

SEQUENCE

The carpenter should build the floor frame in this sequence:

1. Check the level of foundation and supports.
2. Lay sill seals, termite shields, etc.
3. Lay the sill.
4. Lay girders.
5. Select joist style and spacing.
6. Lay out joists for openings and partitions.
7. Cut joists to length and shape.
8. Set joists.
   a. Lay in place.
   b. Nail opening frame.
   c. Nail regular joists.
9. Cut scabs, trim joist edges.
10. Nail bridging at tops.
11. Lay subfloor.
12. Nail bridging at bottom.
13. Trim floor at ends and edges.
14. Cut special openings in floor.

SILL PLACEMENT

The sill is the first wooden part attached to the foundation. However, other things must be done before the sill is laid. When the anchors and foundation surface are adequate, a seal must be placed on the foundation. The seal may be a roll of insulation material or caulking. If a metal termite shield is used, it is placed over the seal. Next, the sill is prepared and fitted over the anchor bolts onto the foundation. See Fig. 5-2.

The seal forms a barrier to moisture and insects. Roll-insulation-type material, as in Fig. 5-3, may be used. The roll should be laid in one continuous strip with no joints. At corners, the rolls should overlap about 2 inches.

Fig. 5-1 *The piers and foundation walls will help support the floor frame.*

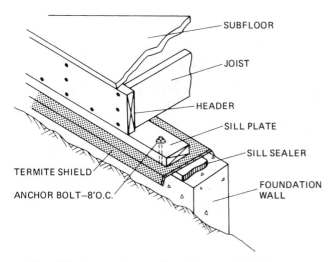

Fig. 5-2 *Section showing floor, joists, and sill placement.*

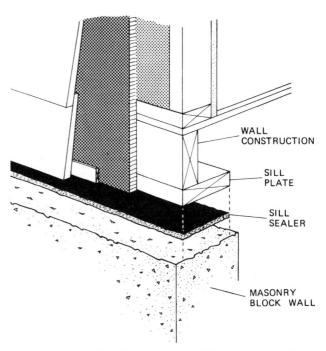

Fig. 5-3 *A seal fills in between the top of the foundation wall and the sill. It helps conserve energy by making the sill more weathertight.* (Conwed)

To protect against termites, two things are often done. A solid masonry top is used. Metal shields are also used. Some foundation walls are built of brick or concrete block that have hollow spaces. They are sealed with mortar or concrete on the top. A solid concrete foundation provides the best protection from termite penetration.

However, termites can penetrate cracks in masonry. Termites can enter a crack as small as ¹⁄₆₄ inch in width. Metal termite shields are used in many parts of the country. Figure 5-4 shows a termite shield installed.

Anchor the Sill

The sill must be anchored to the foundation. The anchors keep the frame from sliding from the foundation. They also keep the building from lifting in high winds. Three methods are used to anchor the sill to the foun-

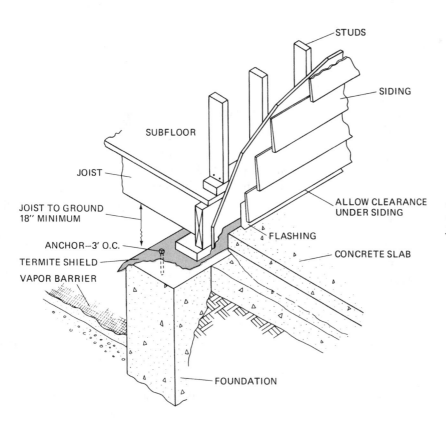

Fig. 5-4 *Metal termite shield used to protect wood over foundation.* (Forest Products Laboratory)

dation. The first uses bolts embedded in the foundation, as in Fig. 5-5. Sill straps are also used and so are special drilled bolt anchors. See Figs. 5-6 and 5-7. Special masonry nails are also sometimes used; however, they are not recommended for anchoring exterior walls. It is not necessary to use many anchors per wall. Anchors should be used about every 4 feet, depending upon local codes. Anchors may not be required on walls shorter than 4 feet.

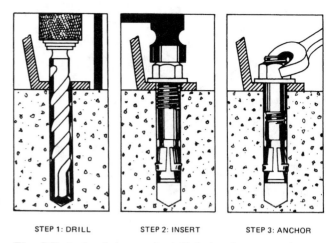

STEP 1: DRILL   STEP 2: INSERT   STEP 3: ANCHOR

Fig. 5-7 *Anchor holes may be drilled after the concrete has set.*
(Hilti-Fastening Systems)

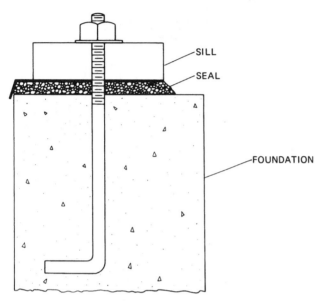

Fig. 5-5 *Anchor bolt in foundation.*

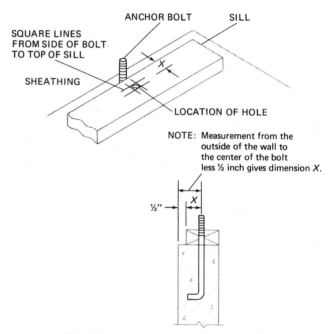

Fig. 5-8 *Locate the holes for anchor bolts.*

Fig. 5-6 *Anchor straps or clips can be used to anchor the sill.*

The anchor bolts must fit through the sill. The holes are located first. Washers and nuts are taken from the anchor bolts. The sill board is laid next to the bolts. See Fig. 5-8. Lines are marked using a framing square as a guide. The sheathing thickness is subtracted from one-half the width of the board. This distance is used to find the center of the hole for each anchor. The centers for the holes are then marked. As a rule the hole is bored ¼ inch larger than the bolt. This leaves some room for adjustments and makes it easier to place the sill.

Next, the sill is put over the anchors and the spacing and locations are checked. All sills are fitted and then removed. Sill sealer and termite shields are laid, and the sills are replaced. The washers and nuts are put on the bolts and tightened. The sill is checked for levelness and straightness. Low spots in the foundation can be shimmed with wooden wedges. However, it is best to use grout or mortar to level the foundation.

Special masonry nails may be used to anchor interior walls on slabs. These are driven by sledge hammers or by nail guns. The nail mainly prevents side slippage of the wall. Figure 5-9 shows a nail gun application.

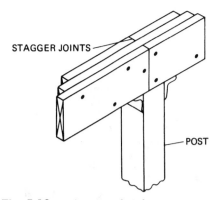

Fig. 5-10 *Built-up wood girder.* (Forest Products Laboratory)

Fig. 5-9 *A nail gun can be used to drive nails in the slab and for toenailing.* (Duo-Fast)

Setting Girders

Girders support the joists on one end. Usually the girder is placed halfway between the outside walls. The distance between the supports is called the *span*. The span on most houses is too great for joists to reach from wall to wall. Central support is given by girders.

Determine girder location Plans give the general spacing for supports and girders. Spans up to 14 feet are common for 2-×-10-inch or 2-×-12-inch lumber. The girder is laid across the leveled girder supports. A chalk line may be used to check the level. The support may be shimmed with mortar, grout, or wooden wedges. The supports are placed to equalize the span. They also help lower expense. The piers shown in Fig. 5-1 must be leveled for the floor frame.

The girder is often built by nailing boards together. Figure 5-10 shows a built-up girder. Girders are often made of either 2-×-10-inch or 2-×-12-inch lumber. Joints in the girder are staggered. The size of the girder and joists is also given on the plans.

There are several advantages to using built-up girders. First, thin boards are less expensive than thick ones. The lumber is more stable because it is drier. There is less shrinkage and movement of this type of girder. Wooden girders are also more fire-resistant than steel girders. Solid or laminated wooden girders take a long time to burn through. They do not sag or break until they have burned nearly through. Steel, on the other hand, will sag when it gets hot. It only takes a few minutes for steel to get hot enough to sag.

The ends of girders can be supported in several ways. Figure 5-11A and B shows two methods. The ends of girders set in walls should be cut at an angle. See Fig. 5-12. In a fire, the beam may fall free. If the ends are cut at an angle, they will not break the wall.

Metal girders should have a wooden sill plate on top. This board forms a nail base for the joists. Basement girders are often supported by post jacks. See Fig. 5-13. Post jacks are used until the basement floor is finished. A support post, called a *lolly column*, may be built beneath the girder. It is usually made from 2 × 4 lumber. Walls may be built beneath the girder. In many areas this is done so that the basement may later be finished out as rooms.

JOISTS

Joists are the supports under the floor. They span from the sill to the girder. The subfloor is laid on the joists.

Lay Out the Joists

Joists are built in two basic ways. The first is the *platform method*. The platform method is the more common method today. The other method is called *balloon framing*. It is used for two-story buildings in some areas. However, the platform method is more common for multistory buildings.

Joist spacing The most common spacing is 16 inches. This makes a strong floor support. It also allows the carpenter to use standard sizes. However, 12-inch and 24-inch spacing are also used. The spacing depends upon the weight the floor must carry. Weight comes from people, furniture, and snow, wind, and rain. Local building codes will often tell what joist spacings should be used.

Joist spacings are given by the distance from the center of one board to the center of the next. This is called the distance *on centers*. For a 16-inch spacing, it would be written, *16 inches O. C.*

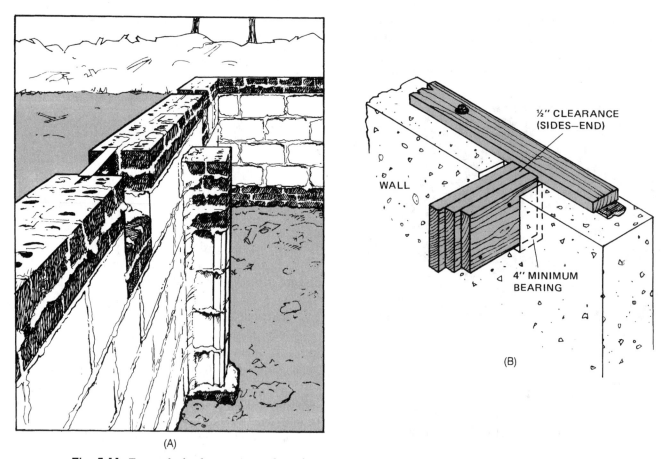

Fig. 5-11 *Two methods of supporting girder ends: (A) Projecting post; (B) Recessed pocket.* (Forest Products Laboratory)

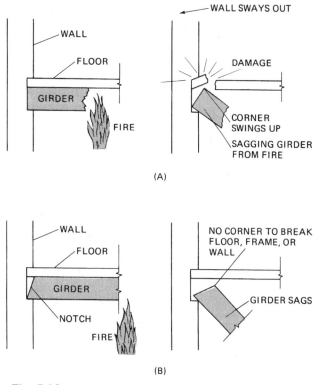

Fig. 5-12 *For solid walls, girder ends must be cut at an angle.*

Fig. 5-13 *A post jack supports the girder until a column or a wall is built.*

Modular spacings are 12, 16, or 24 inches O.C. These modules allow the carpenter to use standard-size sheet materials easily. The standard-size sheet is 48 × 96 inches (4 × 8 feet). Any of the modular sizes divides evenly into the standard sheet size. By using modules, the amount of cutting and fitting is greatly reduced. This is important since sheet materials are used on subfloors, floors, outside walls, inside walls, roof decks, and ceilings.

Joist layout for platform frames The position of the floor joists may be marked on a board called a *header*. The header fits across the end of the joists. See Fig. 5-14.

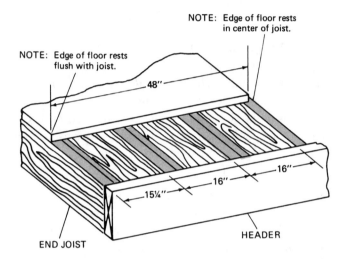

Fig. 5-14 *The positions of the floor joists may be marked on the header.*

Joist spacing is given by the distance between centers. However, the center of the board is a hard mark to use. It is much easier to mark the edge of a board. After all, if the centers are spaced right, the edges will be too!

The header is laid flat on the foundation. The end of the header is even with the end joist. The distance from the end of the header to the edge of the first joist is marked. However, this distance is not the same as the O.C. spacing. See Fig. 5-15. The first distance is always ¾ inch less than the spacing. This lets the edge of the flooring rest flush, or even, with the outside edge of the joist on the outside wall. This will make laying the flooring quicker and easier.

The rest of the marks are made at the regular O.C. spacing. See Fig. 5-15. As shown, an X indicates on which side of the line to put the joist.

Mark a pole It is faster to transfer marks than to measure each one. The spacing can be laid out on a board first. The board can then be used to transfer spacings. This board is called a *pole*. A pole saves time because measurements are done only once. To transfer the marks, the pole is laid next to a header. Use a square to

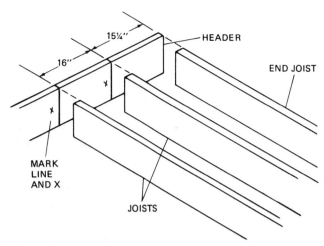

Fig. 5-15 *The first joist must be spaced ¾ inch less than the O.C. spacing used. Mark from end of header.*

project the spacing from the pole to the header. A square may be used to check the "square" of the line and mark.

Joists under walls Joists under walls are doubled. There are two ways of building a double joist. When the joist supports a wall, the two joists are nailed together. See Fig. 5-16. Pipes or vents sometimes go through the floors and walls. Then, a different method is used. See Fig. 5-17. The joists are spaced approximately 4 inches

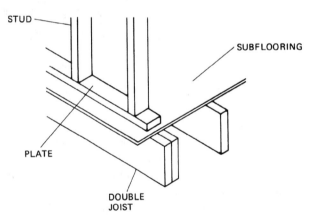

Fig. 5-16 *Joists are doubled under partitions.*

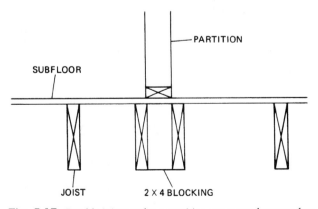

Fig. 5-17 *Double joists under a partition are spaced apart when pipes must go between them.*

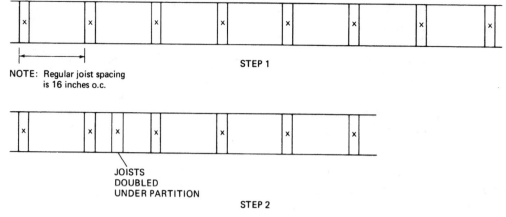

NOTE: Regular joist spacing is 16 inches o.c.

STEP 1

JOISTS DOUBLED UNDER PARTITION

STEP 2

Fig. 5-18 *Header joist layout. Next, add the joists for partitions.*

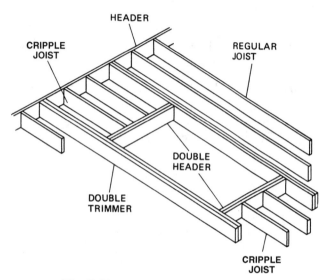

Fig. 5-19 *Frame parts for a floor opening.*

apart. This space allows the passage of pipes or vents. Figure 5-18 shows the header layout pole with a partition added. Special blocking should be used in the double joist. Two or three blocks are used. The blocking serves as a fire stop and as bracing.

Joists for openings Openings are made in floors for stairs and chimneys. Double joists are used on the sides of openings. They are called *double trimmers*. Double trimmers are placed without regard for regular joist spacings. Regular spacing is continued on each end of the opening. Short "cripple" joists are used. See Fig. 5-19. A pole can show the spacing for the openings. See Fig. 5-20.

Girder spacing The joists are located on the girders also. Remember, marks do not show centerlines of the joists. Centerlines are hard to use, so marks show the edge of a board. These marks are easily seen.

Balloon layout Balloon framing is different. See Fig. 5-21. The wall studs rest on the sill. The joists and the studs are nailed together as shown. However, the end joists are nailed to the end wall studs.

The first joist is located back from the edge. The distance is the same as the wall thickness. The second joist is located by the first wall stud.

The wall stud is located first. The first edge of the stud is ¾ inch less than the O.C. spacing. For 16 inches O.C., the stud is 15¼ inches from the end. A 2-inch

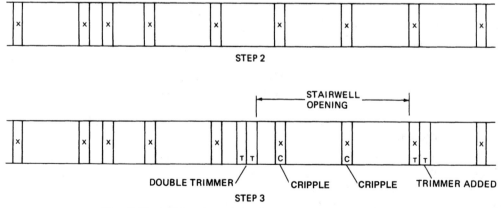

STEP 2

STAIRWELL OPENING

DOUBLE TRIMMER CRIPPLE CRIPPLE TRIMMER ADDED

STEP 3

Fig. 5-20 *Add the trimmers for the opening to the layout pole.*

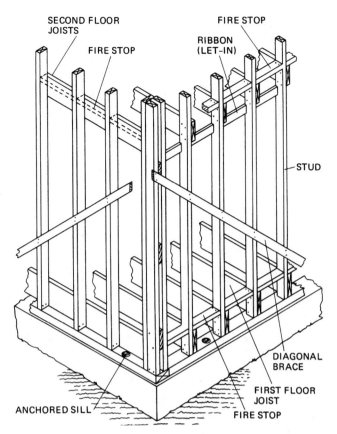

Fig. 5-21 *Joist and stud framing used in balloon construction.*
(Forest Products Laboratory)

stud will be 1½ inches thick. Thus, the edge of the first joist will be 16¾ inches from the edge.

Cut Joists

The joists span, or reach, from the sill to the girder. Note that joists do not cover the full width of the sill. Space is left on the sill for the joist header. See Fig. 5-4. For lumber 2 inches thick, the spacing would be 1½ inches. Joists are cut so that they rest on the girder. Four inches of the joist should rest on the girder.

The quickest way to cut joists is to cut each end square. Figure 5-22 shows square-cut ends. The joists

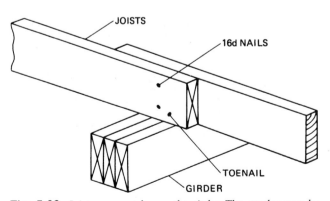

Fig. 5-22 *Joists may overlap on the girder. The overlap may be long or short.*

overlap across the girder. The ends rest on the sill with room for the joist header. This way the header fits even with the edge.

It is sometimes easier to put the joist header on after the joists are toenailed and spaced. Or, the joist header may be put down first. The joists are then just butted next to the header. But it is very important to carefully check the spacing of the joists before they are nailed to either the sill or the header.

Other Ways to Cut Joists

Ends of joists may be cut in other ways. The ends may be aligned and joined. Ends are cut square for some systems. For others, the ends are notched. Metal girders are sometimes used. Then, joists are cut to rest on metal girders.

End-joined joists The ends of the joists are cut square to fit together. The ends are then butted together as in Fig. 5-23. A gusset is nailed (10d) on each side to hold the joists together. Gussets may be made of either plywood or metal. This method saves lumber. Builders use it when they make several houses at one time.

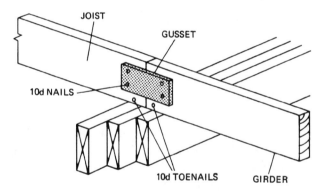

Fig. 5-23 *Joists may be butt-joined on the girder. Gussets may be used to hold them. Plywood subfloor may also hold the joists.*

Notched and lapped joists Girders may be notched and lapped. See Fig. 5-24. This connection has more interlocking but takes longer and costs more. First, the notch is cut on the end of the joists. Next a 2-×-4-inch joist support is nailed (16d) on the girder. Nails should be staggered 6 or 9 inches apart. The joists are then laid in place. The ends overlap across the girder.

Joist-girder butts This is a quick method. With it the top of the joist can be even with the top of the girder.

A 2-×-4-inch ledger is nailed to the girder with 16d common nails. See Fig. 5-25. The joist rests on the ledger and not on the girder. This method is not as strong.

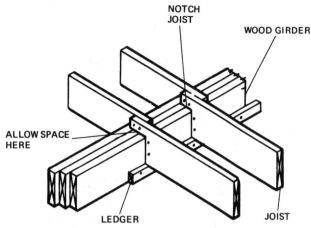

Fig. 5-24 *Joists can be notched and lapped on the girder.* (Forest Products Laboratory)

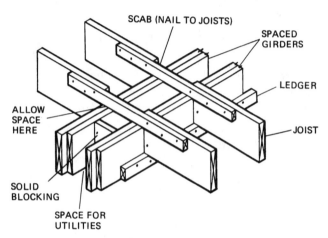

Fig. 5-25 *Joists may be butted against the girder. Note that the girder may be spaced for pipes, etc.* (Forest Products Laboratory)

Also, a board is used to join the girder ends. The board is called a *scab*. The scab also makes a surface for the subfloor. It is a 2-×-4-inch board. It is nailed with three 16d nails on each end.

Joist hangers Joist hangers are metal brackets. See Fig. 5-26. The brackets hold up the joist. They are nailed (10d) to the girder. The joist ends are cut square. Then, the joist is placed into the hanger. It is also nailed with 10d nails as in Fig. 5-26. Using joist hangers saves time. The carpenter need not cut notches or nail up ledgers.

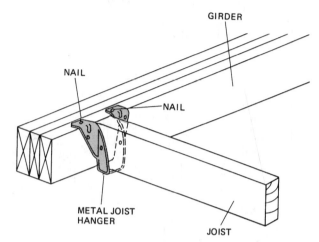

Fig. 5-26 *Using metal joist hangers saves time.*

Joists for metal girders Joists must be cut to fit into metal girders. See Fig. 5-27. A 2-×-4-inch board is first bolted to the metal girder. The ends of the joist are then beveled. This lets the joist fit into the metal girder. The joist rests on the board. The board also is a nail base for the joist. The tops of the joists must be scabbed. The scabs are made of 2-×-4-inch boards. Three 16d nails are driven into each end.

Setting the Joists

Two jobs are involved in setting joists. The first is laying the joists in place. The second is nailing the joists. The carpenter should follow a given sequence.

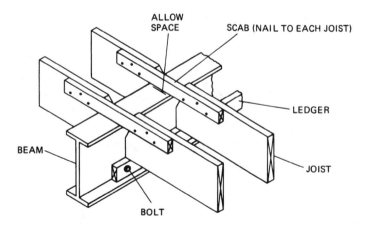

(A)

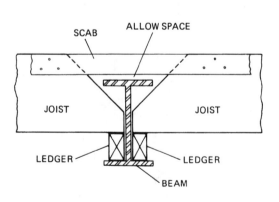

(B)

Fig. 5-27 *Systems for joining joists to metal girders.* (Forest Products Laboratory)

Lay the joists in place First, the header is toenailed in place. Then the full-length joists are cut. Then they are laid by the marks on the sill or header. Each side of the joist is then toenailed (10d) to the sill. See Fig. 5-28. Joists next to openings are not nailed. Next, the ends of the joists are toenailed (10d) on the girder. Then the overlapped ends of the joists are nailed (16d) together. See Fig. 5-22. These nails are driven at an angle, as in Fig. 5-29.

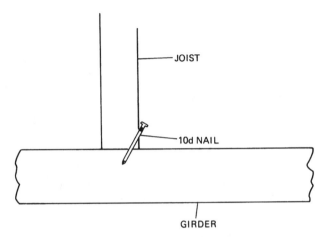

Fig. 5-28 *The joists are toenailed to the girders.*

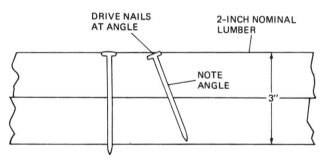

Fig. 5-29 *When nailing joists together, drive nails at an angle. This holds better and the ends do not stick through.*

Nail opening frame A special sequence must be used around the openings. The regular joists next to the opening should not be nailed down. The opening joists are nailed (16d) in place first. These are called *trimmers.*

Then, the first *headers* for openings are nailed (16d) in place. Note that two headers are used. For 2-×-10-inch joists, three nails are used. For 2-×-12 inch lumber, four nails are used. Figure 5-30 shows the spacing of the nails.

Next, short cripple or tail joists are nailed in place. They span from the first header to the joist header. Three 16d nails are driven at each end. Then, the second header is nailed (16d) in place. See Fig. 5-31.

The double trimmer joist is now nailed (16d) in place. These pieces are nailed next to the opening. The nails are alternated top and bottom. See Fig. 5-32. This

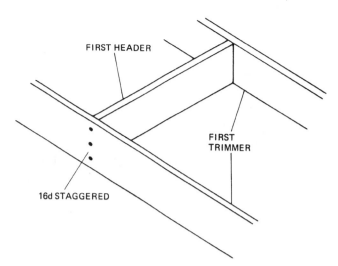

Fig. 5-30 *Nailing the first parts of an opening.*

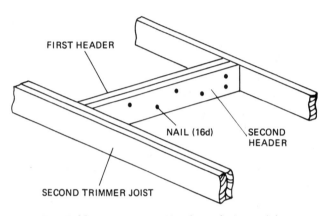

Fig. 5-31 *Add the second header and trimmer joists.*

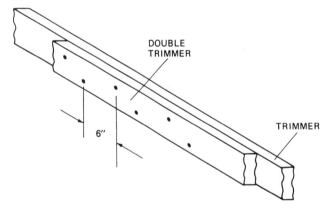

Fig. 5-32 *Stagger nails on double trimmer. Alternate nails on top and bottom.*

finishes the opening. The regular joists next to the opening are nailed in place. Finally, the header is nailed to the joist ends. Three 16d nails are driven into each joist.

Fire Stops

Fire stops are short pieces nailed between joists and studs. See Fig. 5-33. They are made of the same boards

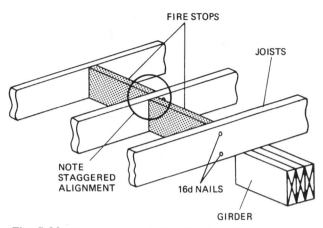

Fig. 5-33 *Fire stops are nailed in. They keep fire from spreading between walls and floors.*

as the joists. Fire stops keep fire from spreading between walls and floors. They also help keep joists from twisting and spreading. Fire stops are usually put at or near the girder. Two 16d nails are driven at each end of the stop. Stagger the boards slightly as shown. This makes it easy to nail them in place.

Bridging

Bridging is used to keep joists from twisting or bending. Bridging is centered between the girder and the header. For most spans, center bridging is adequate. For joist spans longer than 16 feet, more bridging is used. Bridging should be put in every 8 feet. This must be done to comply with most building codes.

Most bridging is cut from boards. It may be cut from either 1-inch or 2-inch lumber. Use the framing square to mark the angles as in Fig. 5-34. With this method, the angle may be found.

A radial arm saw may be used to cut multiple pieces. See Fig. 5-35. Also, a jig can be built to use a portable power saw. See Fig. 5-36.

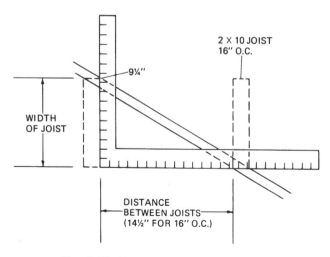

Fig. 5-34 *Use a square to lay out bridging.*

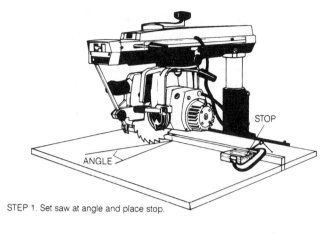

STEP 1. Set saw at angle and place stop.

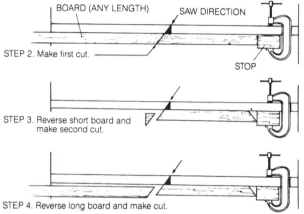

STEP 2. Make first cut.

STEP 3. Reverse short board and make second cut.

STEP 4. Reverse long board and make cut.

Fig. 5-35 *Radial arm saw set up for bridging.*

Special steel bridging is also used. Figure 5-37 shows an example. Often, only one nail is needed in each end. Steel bridging meets most codes and standards.

All the bridging pieces should be cut first. Nails are driven into the bridging before it is put up. Two 8d or 10d nails are driven into each end. Next, a chalk line is strung across the tops of the joists. This gives a line for the bridging.

The bridging is nailed at the top first. This lets the carpenter space the joists for the flooring when it is laid. The bottoms of the bridging are nailed after the flooring is laid.

The bridging is staggered on either side of the chalk line. This prevents two pieces of bridging from being nailed at the same spot on a joist. To nail them both at the same place would cause the joist to split. See Fig. 5-38.

SUBFLOORS

The last step in making a floor frame is laying the subflooring. Subflooring is also called *underlayment*. The subfloor is the platform that supports the rest of the structure. It is covered with a finish floor material in the living spaces. This may be of wood,

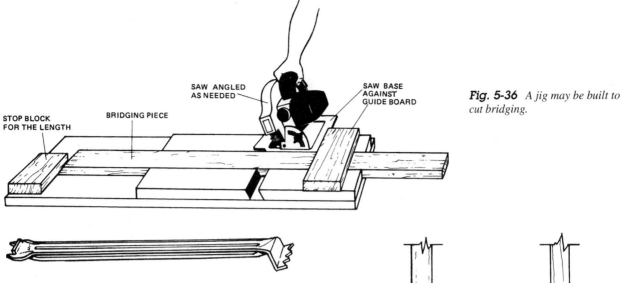

SAW ANGLED AS NEEDED

SAW BASE AGAINST GUIDE BOARD

BRIDGING PIECE

STOP BLOCK FOR THE LENGTH

Fig. 5-36 *A jig may be built to cut bridging.*

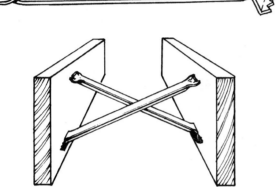

Fig. 5-37 *Most building codes allow steel bridging.*

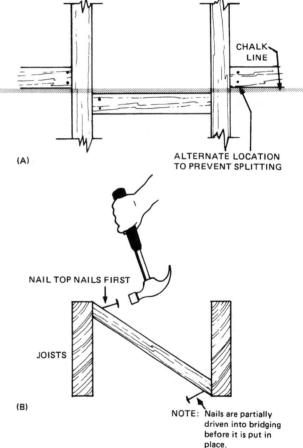

CHALK LINE

(A)

ALTERNATE LOCATION TO PREVENT SPLITTING

NAIL TOP NAILS FIRST

JOISTS

(B)

NOTE: Nails are partially driven into bridging before it is put in place.

Fig. 5-38 *Bridging pieces are staggered.*

carpeting, tile, or stone. However, the finish floor is added much later.

Several materials are commonly used for subflooring. The most common material is plywood. Plywood should be C—D grade with waterproof or exterior glues. Other materials used are chipboard, fiberboard, and boards.

Plywood Subfloor

Plywood is an ideal subflooring material. It is quickly laid and takes little cutting and trimming. It may be either nailed or glued to the joists. Plywood is very flat and smooth. This makes the finished floor smooth and easy to lay. Builders use thicknesses from ½- to ¾-inch plywood. The most common thicknesses are ½ and ⅝ inch. The FHA minimum is ½-inch-thick plywood.

Plywood as subflooring has fewer squeaks than boards. This is because fewer nails are required. The squeak in floors is caused when nails work loose. Table 5-1 shows minimum standards for plywood use.

Chipboard and Fiberboard

As a rule, plywood is stronger than other types of underlayment. However, both fiberboard and chipboard

are also used. Chipboard underlayment is used more often. The minimum thickness for chipboard or fiberboard is ⅝ inch. This thickness must also be laid over 16-inch joist spacing. Both chipboard and fiberboard are laid in the same manner as plywood. In any case, the ends of the large sheets are staggered. See Fig. 5-39.

Laying Sheets

The same methods are used for any sheet subflooring. Nails are used most often, but glue is also used. An

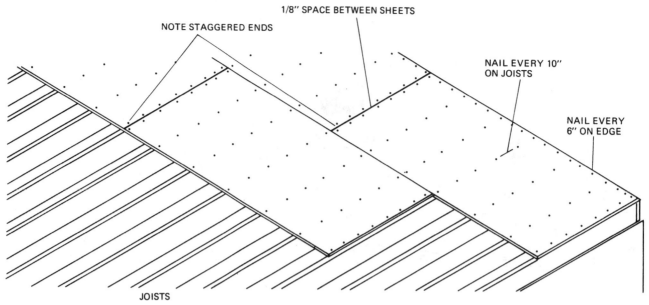

Fig. 5-39 *The ends of subfloor sheets are staggered.*

Table 5-1 *Minimum Flooring Standards*

Single-Layer (Resilient) Floor			
Joists, Inches O.C.	Minimum Thickness, Inches	Common Thickness, Inches	Minimum Index
12	19/32	5/8	24/12
16	5/8	5/8 or 3/4	32/16
24	3/4	3/4	48/24
Subflooring with Finish Floor Layer Applied			
Joists, Inches O.C.	Minimum Thickness, Inches	Common Thickness, Inches	Minimum Index
12	1/2	1/2 or 5/8	32/16
16	1/2	5/8	32/16
24	3/4	3/4	48/24

NOTES: 1. C-C grade underlayment plywood.
2. Each piece must be continuous over two spans.
3. Sizes can vary with span and depth of joists in some locations.

outside corner is used as the starting point. The long grain or sheet length is laid across the joists. See Fig. 5-40. The ends of the different courses are staggered. This prevents the ends from all lining up on one joist. If they did, it could weaken the floor. By staggering the end joints, each layer adds strength to the total floor. The carpenter must allow for expansion and contraction. To do this, the sheets are spaced slightly apart. A paper match cover may be used for spacing. Its thickness is about the correct space.

Nailing The outside edges are nailed first with 8d nails. Special "sinker" nails may be used. The outside

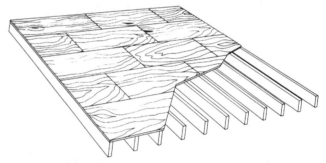

Fig. 5-40 *The long grain runs across the joists.*

nails should be driven about 6 inches apart. Nails are driven into the inner joists about 10 inches apart. See Fig. 5-41. Power nailers can be used to save time, cost, and effort. See Fig. 5-42.

Gluing Gluing is now widely used for subflooring. Modern glues are strong and durable. Glues, also called adhesives, are quickly applied. Glue will not squeak as will nails. Figure 5-43 shows glue being applied to floor joists. Floors are also laid with tongue-

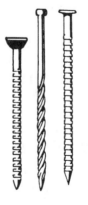

Fig. 5-41 *Flooring nails. Note the "sinker" head on the first nail.*

Fig. 5-42 *Using power nailers saves time and effort.* (Duo-Fast)

Fig. 5-43 *Subflooring is often glued to the joists. This makes the floor free of squeaks.* (American Plywood Association)

Fig. 5-44 *Plywood subflooring may also have tongue-and-groove joints. This is stronger.* (American Plywood Assocation)

Fig. 5-45 *A buffer board is used to protect the edges of tongue-and-groove panels.* (American Plywood Association)

and-groove joints (Fig. 5-44). Buffer boards are used to protect the edges of the boards as the panels are put in place. (See Fig. 5-45.)

Board Subflooring

Boards are also used for subflooring. There are two ways of using boards. The older method lays the boards diagonally across the joists. Figure 5-46 shows this. This way takes more time and trimming. It takes a longer time to lay the floor, and more material is wasted by trimming. However, diagonal flooring is still used. It is preferred where wood board finish flooring will be used. This way the finish flooring may be laid at right angles to the joists. Having two layers that run in different directions gives greater strength.

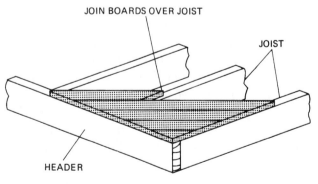

Fig. 5-46 *Diagonal board subfloors are still used today.*

Today, board subflooring is often laid at right angles to the joists. This is appropriate when the finish floor will be sheets of material.

Either way, two kinds of boards are used. Plain boards are laid with a small space between the boards. It allows for expansion. End joints must be made over a joist for support. See Fig. 5-47. Grooved boards are also used. See Fig. 5-48. End joints may be made at any point with grooved boards.

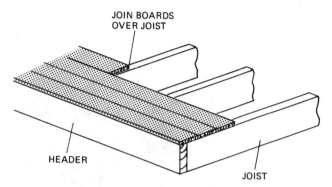

Fig. 5-47 *Plain board subfloor may be laid across joists. Joints must be made over a joist.*

Fig. 5-48 *Grooved flooring is laid across joists. Joints can be made anywhere.*

Nailing Boards are laid from an outside edge toward the center. The first course is laid and nailed with 8d nails. Two nails are used for boards 6 inches wide or less. Three nails are used for boards wider than 6 inches.

The boards are nailed down untrimmed. The ends stick out over the edge of the floor. This is done for both grooved and plain boards. After the floor is done, the ends are sawed off. They are sawed off even with the floor edges.

SPECIAL JOISTS

The carpenter should know how to make special joists. Several types of joists are used in some buildings. Spe-

cial joists are used for overhangs and sunken floors. A sunken floor is any floor lower than the rest. Sunken floors are used for special flooring such as stone. Floors may also be lowered for appearance. Special joists are also used to recess floors into foundations. This is done to make a building look lower. This is called the *low-profile* building.

Overhangs

Overhangs are called *cantilevers*. They are used for special effects. Porches, decks, balconies, and projecting windows are all examples. Figure 5-49A shows an example of projecting windows. Figure 5-49B is a different type of bay. However, both rest on overhanging floor joist systems. Overhangs are also used for "garrison" style houses. When a second floor extends over the wall of the first, it is called a garrison style. See Fig. 5-50A and B.

The longest projection without special anchors is 24 inches. Windows and overhangs seldom extend 24 inches. A balcony, however, would extend more than 24 inches. Thus, a balcony would need special anchors.

Overhangs with joist direction Some overhangs project in the same direction as the floor joists. Little extra framing is needed for this. This is the easiest way to build overhangs. In this method, the joists are simply made longer. Blocking is nailed over the sill with 16d nails. Figure 5-51 shows blocking and headers for this type of overhang. Here the joists rest on the sill. Some overhangs extend over a wall instead of a foundation. Then, the double top plate of the lower wall supports the joists.

Overhangs at angles to joist direction Special construction is needed to frame this type of overhang. It is similar to framing openings in the floor frame. Stringer joists form the base for the subflooring. Stringer joists must be nailed to the main floor joists. See Fig. 5-52. They must be inset twice the distance of the overhang. Two methods of attaching the stringers are used. The first method is to use a wooden ledger. However, this ledger is placed on the top. See Fig. 5-53. The other method uses a metal joist hanger. Special anchors are needed for large overhangs such as rooms or decks.

Sunken Floors

Subfloors are lowered for two main reasons. A finish floor may be made lower than an adjoining finish floor for appearance. Or the subfloor may be made lower to accommodate a finish floor of a different material. The different flooring could be stone, tile, brick, or concrete.

(A)

(B)

Fig. 5-49 *(A) A bay window rests on overhanging floor joists.* (American Plywood Association); *(B) A different type of bay. Both types rest on overhanging floor joist systems.*

(A)

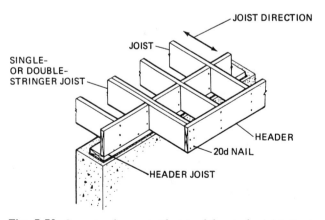

(B)

Fig. 5-50 *(A) Joist protections from the overhang for a garrison type second story; (B) The finished house.*

These materials are used for appearance or to drain water. However, they are thicker than most finish floors. To make the floor level, special framing is done to lower the subfloor.

The sunken portion is framed like a special opening. First, header joists are nailed (16d) in place. See

Fig. 5-51 *Some overhangs simply extend the regular joist.* (Forest Products Laboratory)

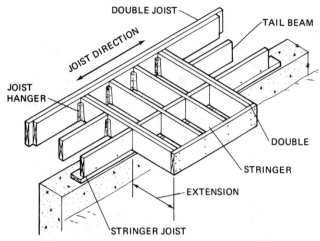

Fig. 5-52 *Frame for an overhang at an angle to the joists.* (Forest Products Laboratory)

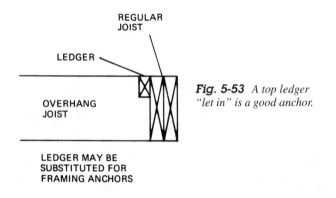

Fig. 5-53 *A top ledger "let in" is a good anchor.*

Fig. 5-54. The headers are not as deep, or wide, as the main joists. This lowers the floor level. To carry the load with thinner boards, more headers are used. The headers are added by spacing them closer together. Double joists are nailed (16d) after the headers.

Low Profiles

The lower profile home has a regular size frame. However, the subfloor and walls are joined differently. Figure 5-55 shows the arrangement. The sill is below the top of the foundation. The bottom plate for the wall is attached to the foundation. The wall is not nailed to the subfloor. This makes the joists below the common foundation level. The building will appear to be lower than normal.

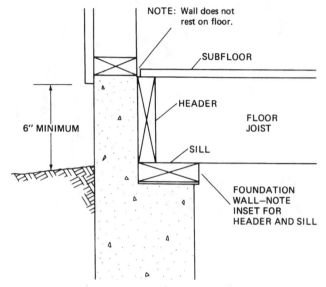

Fig. 5-55 *Floor detail for a low-profile house.*

ENERGY FACTORS

Most energy is not lost through the floor. The most heat is lost through the ceiling. This is because heat rises. However, energy can be saved by insulating the floor. In the past, most floors were not insulated. Floors over basements need not be insulated. Floors over enclosed basements are the best energy savers.

Floors over crawl spaces should be insulated. The crawl space should also be totally enclosed. The foundation should have ventilation ports. But, they should

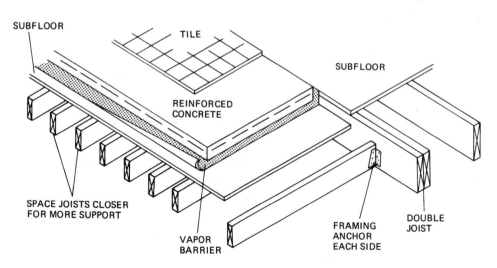

Fig. 5-54 *Details of frame for a sunken floor.*

be closed in winter. The most energy is saved by insulating certain areas. Floors under overhangs and bay windows should be insulated. Floors next to the foundation should also be insulated. The insulation should start at the sill or header. It should extend 12 inches into the floor area. See Fig. 5-56. The outer corners are the most critical areas. But, for the best results, the whole floor can be insulated. Roll or bat insulation is placed between joists and supported. Supports are made of wood strips or wire. Nail (6d) them to the bottom of the joists. See Fig. 5-57.

Fig. 5-57 *Insulation between floor joists should be supported.*

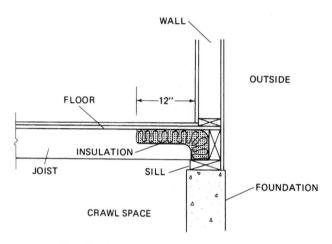

Fig. 5-56 *Insulate the outside floor edges.*

Moisture Barriers

Basements and slabs must have moisture barriers beneath them. Moisture barriers are not needed under a floor over a basement. However, floors over crawl spaces should have moisture barriers. The moisture barrier is laid over the subfloor. See Fig. 5-58. A moisture barrier may be added to older floors below the joists. This may be held in place by either wooden strips or wires. Six-mil plastic or builder's felt is used. See Fig. 5-59.

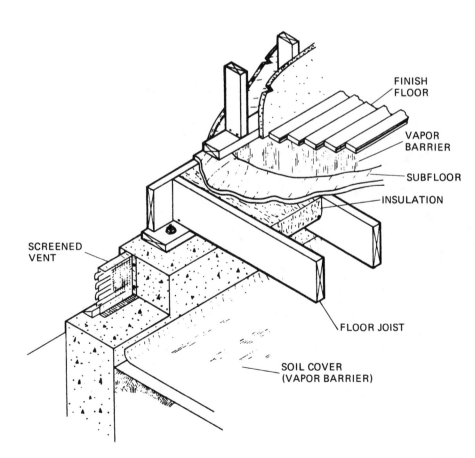

Fig. 5-58 *A moisture barrier is laid over the subfloor above a crawl space.* (Forest Products Laboratory)

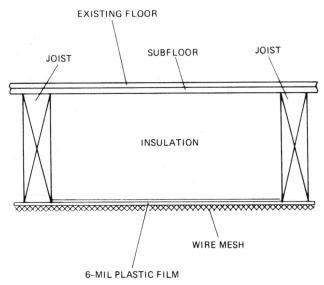

Fig. 5-59 A moisture barrier may also be added beneath floors.

Energy Plenums

A *plenum* is a space for controlled air. The air is pressurized a little more than normal. Plenum systems over crawl spaces allow air to circulate beneath floors. This maximizes the heating and cooling effects. Figure 5-60 shows how the air is circulated. Doing this keeps the temperature more even. Even temperatures are more efficient and comfortable.

The plenum must be carefully built. Insulation is used in special areas. See Fig. 5-61. A hatch is needed for plenum floors. The hatch gives access to the plenum area. Access is needed for inspection and servicing. There are no outside doors or vents to the plenum.

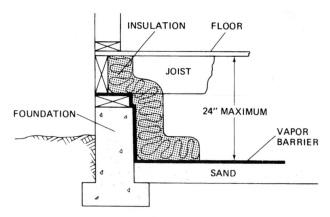

Fig. 5-61 Section view of the energy plenum.

The plenum arrangement offers an advantage to the builder: A plenum house can be built more cheaply. There are several reasons for this. The circulation system is simpler. No ducts are built beneath floors or in attics. Common vents are cut in the floors of all the rooms. The system forces air into the sealed plenum

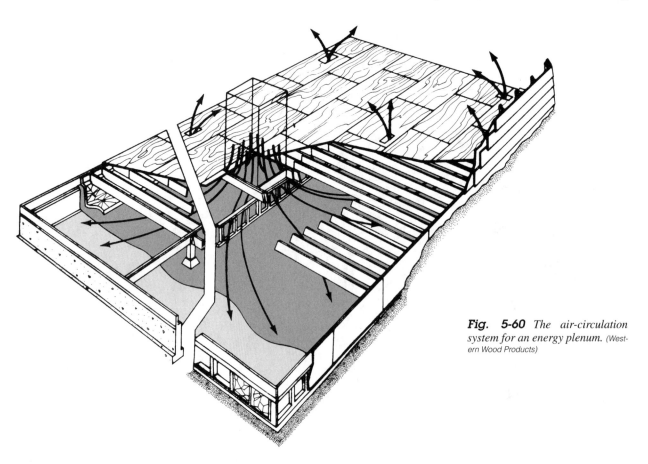

Fig. *5-60* The air-circulation system for an energy plenum. *(Western Wood Products)*

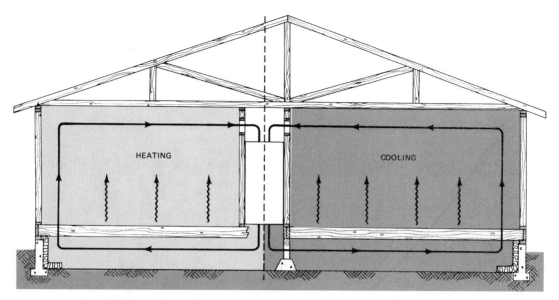

Fig. 5-62 *Conditioned air is circulated through the plenum.* (Western Wood Products)

(see Fig. 5-62). The air does not lose energy in the insulated plenum. The forced air then enters the various rooms from the plenum. The blower unit is in a central portion of the house. The blower can send air evenly from a central area. The enclosed louvered space lets the air return freely to the blower.

Rough plumbing is brought into the crawl space first. Then the foundation is laid. Fuel lines and cleanouts are located outside the crawl space. The minimum clearance in the crawl space should be 18 inches. The maximum should be 24 inches. This size gives the greatest efficiency for air movement.

Foundation walls may be masonry or poured concrete. Special treated plywood foundations are also used. Proper drainage is essential. After the foundation is built, the sill is anchored. Standard sills, seals, and termite shields may be used. The plenum area must be covered with sand. Next, a vapor barrier is laid over the ground and extends up over the sill. See Fig. 5-61. This completely seals the plenum area (Fig. 5-62). Then insulation is laid. Either rigid or batt insulation may be used. It should extend from the sill to about 24 inches inside the plenum. The most energy loss occurs at foundation corners. The insulation covers these corners. Then, the floor joists are nailed to the sill. The joist header is nailed (16d) on and insulated. Subflooring is then nailed to the joists. This completes the plenum. After this, the building is built as a normal platform.

CHAPTER 5
STUDY QUESTIONS

1. What is the procedure for building a floor?
2. List the parts of a frame floor.
3. Why are regular joists by openings nailed last?
4. What does a girder do?
5. What is a crawl space?
6. What holds the wooden frame to the foundation?
7. Where is joist spacing marked?
8. Why is the first joist spaced differently?
9. What is a post jack?
10. What are the two joist (framing) methods?
11. What are common joist spacings?
12. What size lumber is used for joists?
13. What is the easiest way to cut joists?
14. What are the advantages of using joist hangers?
15. What sequence is used to nail the opening frames?
16. Why are fire stops and bridging used?
17. What is done to frame overhangs?
18. What is done to frame sunken floors?
19. How can floors be changed to save energy?
20. What is an energy plenum?

6

Framing Walls

OW TO BUILD A WALL FRAME IS THE TOPIC of this unit. How to cut the parts for the wall is covered, then how to connect the wall to the rest of the building. Why the parts are made as they are is also explained. You will learn how to:

- Lay out wall sections
- Measure and cut the parts
- Assemble and erect wall sections
- Join wall sections together
- Change a standard wall frame to save energy and materials

INTRODUCTION

There are two ways of framing a building. The most common is called the *western platform method*. The other is the *balloon method*. In most buildings today, the western platform method is used.

In the western platform method, walls are put up after the subfloor has been laid. Walls are started by making a frame. The frame is made by nailing boards to the tops and bottoms of other boards. The top and bottom boards are called *plates*. The vertical boards are called *studs*. The frame must be made very strong because it holds up the roof. After the frame is put together, it is raised and nailed in place. The wall frames are put up one at a time. A roof can be built next. Then the walls are covered and windows and doors are installed. Figure 6-1 shows a wall frame in place. Note that the roof is not on the building yet.

Fig. 6-1 *Wall frames are put up after the floor is built.*

Sometimes the first covering for the wall is added before the wall is raised. This first covering is called *sheathing*. It is very easy to nail the sheathing on the frame while the frame is flat on the floor. Doing this makes the job quicker and reduces problems with

keeping the frame square. Another advantage is that no scaffolds will be needed to reach all the areas. This reduces the possibility of accidents and saves time in moving scaffolds. However, most builders still nail sheathing on after the wall is raised.

The wall is attached to the floor in most cases. When the floor is built over joists, the wall is nailed to the floor. See Fig. 6-2. When the wall is built on a slab, it is anchored to the slab. See Fig. 6-3. As a rule, walls for both types of floors are made the same way.

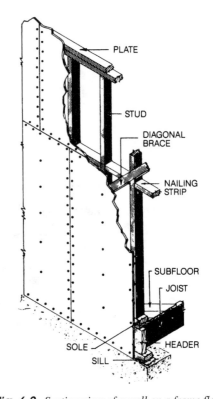

Fig. 6-2 *Section view of a wall on a frame floor.*

Walls that help hold up the roof, or the next floor, are made first. As a rule, all outside, or exterior, walls do this. These walls are also referred to as *load-bearing* walls.

Inside walls are called *partitions*. They can be load-bearing walls, too. However, not all partitions carry loads. Interior walls that do not carry loads may be built after the roof is up. Interior walls that do not carry loads are also called *curtain walls*.

After the walls are put up, the roof is built. The walls are then covered. The first cover is the sheathing. Putting on the sheathing after the roof lets a builder get the building waterproofed or weatherproofed a little sooner. The siding is put on much later.

Wall sections are made one at a time. The longest outside walls are made first. The end walls are made next. However, the sequence can be changed to fit the

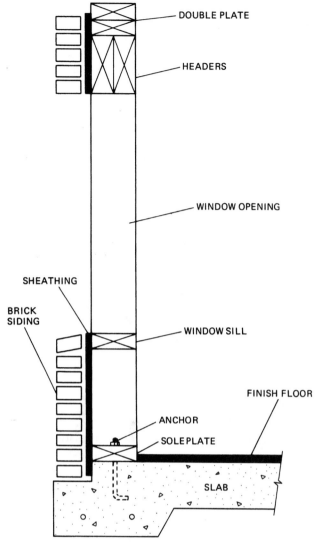

Fig. 6-3 *Section view of a wall on a slab.*

Labels on figure: DOUBLE PLATE, HEADERS, WINDOW OPENING, SHEATHING, BRICK SIDING, WINDOW SILL, FINISH FLOOR, ANCHOR, SOLEPLATE, SLAB

job. There are many ways to make walls. This chapter will show the most common method.

SEQUENCE

The general sequence for making wall frames is:

1. Lay out the longest outside wall section.
2. Cut the parts.
3. Nail the parts together.
4. Raise the wall.
5. Brace it in place.
6. Lay out the next wall.
7. Repeat the process.
8. Join the walls.
9. Do all outside walls.
10. Do all inside walls.

11. Build the roof.
12. Sheath the outside walls.
13. Install outside doors and windows.
14. Cut soleplates from inside door openings.

WALL LAYOUT

All the parts of the wall must be planned. The carpenter must know beforehand what parts to cut and where to nail. The carpenter plans the construction by making a *wall layout* on boards. One layout is done on the top and bottom boards (plates). Another layout is done for the wall studs.

Plate Layout

Soleplates First, select pieces of 2-inch lumber for the bottom of the wall. The bottom part of the wall is called a *soleplate*. The soleplate is laid along the edges of the floor where the walls will be. No soleplate is put across large door openings. Sometimes, the soleplate can be made across small doors. After the wall is put up, the soleplate is sawed out.

Top plate After a soleplate has been laid, another piece is laid beside it. It will be the top part of the wall. This piece is called a *top plate*. The soleplate and the top plate are laid next to each other with a flat side up. Figure 6-4 shows how the soleplate and the top plate are laid so that you can measure and mark both the top and bottom plates at the same time. Because it keeps both top and bottom locations aligned, this method ensures accuracy.

Sole- and top plates are often spliced. The splice must occur over the center of a full stud. See Fig. 6-5. Otherwise, the wall section will be weakened.

Stud Layout

Several types of studs are used in walls. The studs that run from the soleplate to the top plate are called *full studs*. Studs that run from the soleplate to the top of a rough opening are called *trimmer studs*. Short studs that run from either plate to a header or a sill are called *cripple studs*. Figure 6-6 shows a part of a wall section.

Spacing full studs Most full studs are spaced a standard distance apart. This standard distance is an even part of the sizes of plywood, sheathing, and other building materials. The studs act both as roof support and as a nailing base for the sheathing. The most common standard distance is 16 inches. Studs spaced 16 inches apart are said to be 16 inches on *centers* (O.C.). Another common spacing is 24 inches O.C. However,

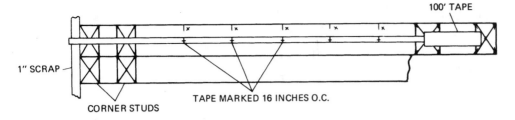

CORNER STUDS

1" SCRAP

TAPE MARKED 16 INCHES O.C.

100' TAPE

NAIL A PIECE OF 1 INCH SCRAP AT THE END. HOOK
THE TAPE OVER THE END. MARK THE INTERVAL.
THEN MAKE A SMALL X TO THE RIGHT OF THE MARK.

Fig. 6-4 Lay sole- and top
plates for marking.

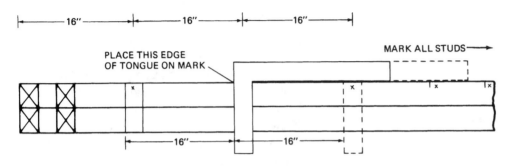

PLACE THIS EDGE
OF TONGUE ON MARK

MARK ALL STUDS ⟶

USE A 2 INCH SCRAP TO MARK WIDTHS. THEN
MARK STUD LOCATIONS.

Fig. 6-5 Top plates are spliced over a stud.

lines on centers are not used to show where boards are placed. An easier way is shown in Fig. 6-4. The stud is located by using the left end of the wall as a starting point. Nail a 1-inch scrap piece there. Then the distance between centers is measured from the block on the outside corner of the wall. A mark is made, and the X is made to the right of the mark.

After all of the marks are made, a square is used to line in the locations. This method is easier because the measurement is taken from the side of the stud. The distance is the same whether it is center-to-center or side-to-side.

Space the rough openings The next step in spacing is to find where the windows and doors are to be made. The openings for windows and doors are framed no matter what the stud spacing is. The openings are called rough openings (R.O.). The size of rough openings may be shown in different ways. The actual size may be shown in the plan. However, many plans show *window schedules*, which often list the window as a number, like "2442." This means that the window sash opening is 2 feet 4 inches wide and 4 feet 2 inches high. The rough openings are larger. Wooden windows require larger R.O.s than metal ones. It is best to refer to specifications. However, when they are not available, carpenters can use a rule of thumb: Wooden windows are usually written, for example, 24 × 42. The R.O. should be 3 inches wider (2'7") and 4 inches higher (4'6"). Metal frames are usually written without the X. The R.O. is 2 inches wider and 3 inches higher. For a 2442 metal window, the R.O. would be 2 feet 6 inches wide and 4 feet 5 inches high.

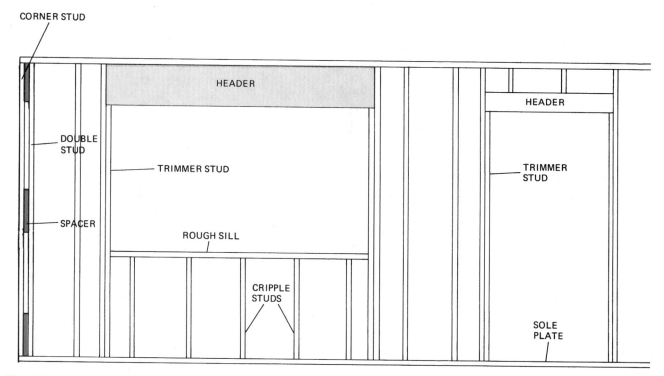

Fig. 6-6 *A wall section and parts. Note the large header over the opening. It is sometimes larger than needed, to eliminate the need for top cripples. This saves labor costs because it takes longer to cut and nail cripples.*

To find the locations of rough openings, measure the distance from the corner or end of the wall to the center of the rough opening. Make a mark called a *centerline* on the soleplate. From the centerline measure half of the rough opening width on each side. Make another mark at each side of the centerline for the rough opening. Mark a line, as in Fig. 6-7, to show the thickness of a stud. This thickness goes on the outside of the opening. Note that the distance between the lines must

be the width of the opening. Mark a T in the regular stud spaces. This tells you that a *trimmer* stud is placed there. On the outside of the space for the trimmer stud, lay out another thickness. Mark an X from one corner of the opening to the other as shown in the figure. The X is used to show where a full stud is placed. The T is used to show where a trimmer stud is placed. The trimmer stud does not extend from the sole- to the top plate. Thus, it is shown only on the soleplate.

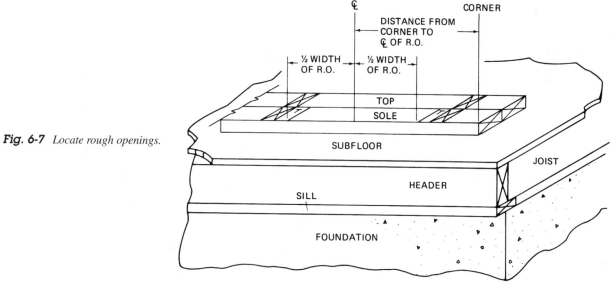

Fig. 6-7 *Locate rough openings.*

Corner Studs

A strong way of nailing the walls together is needed. To accomplish this, a double stud is used for one corner.

Corner studs on the first wall The spacing of the studs starts at one corner. Another stud space is marked at the second corner. The regular spacing is not used at corners. A stud is laid out at each corner, or end, of the wall.

To make the corner stronger, another stud is placed in the corner section. See Fig. 6-8A. The second corner stud is spaced from the first with spacer blocks. This is called a *built-up corner*. It is done to give the corner greater strength and to make a nail base on the inside. A nail base is needed on the inside to nail the inside wall covering in place. After the end stud is marked on the corner, mark the thickness of one more stud on the plate. Mark it with an S for *spacer*. Next, lay out the stud as shown in Fig. 6-8B. Mark it with an X to indicate a full stud.

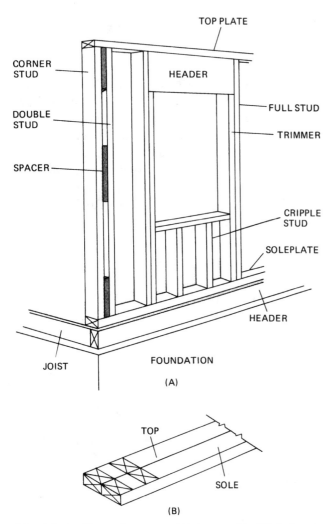

Fig. 6-8 *Stud layout for corners: (A) the assembly; (B) the layout.*

Corner studs for the second wall The second wall does not need double studs. It is laid out in the regular way. The walls are nailed together as in Fig. 6-9.

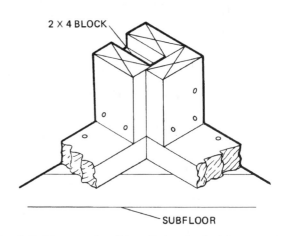

Fig. 6-9 *The second wall is nailed to the double corner.* (Forest Products Laboratory)

Partition Studs

As mentioned earlier, inside walls are called partitions. The partitions must be solidly nailed to the outside walls. To make a solid nail base, special studs are built into the exterior walls. See Fig. 6-10. The most common method is to place two studs as shown in Fig. 6-10. The studs are placed 1½ inches apart. The space is made just like corners on the first wall. The corner of the partition can now rest on the two studs. The two special studs act as a nail base for the wall. Also, ¾ inch of each stud is exposed. This makes a nail base for the interior wall covering. Other methods of joining partitions to the outside walls are shown in Figs. 6-11 and 6-12.

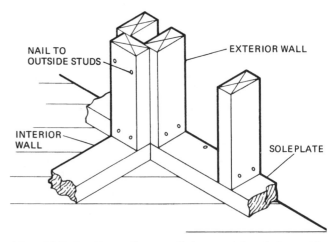

Fig. 6-10 *Partition walls are nailed to special studs in outside walls.* (Forest Products Laboratory)

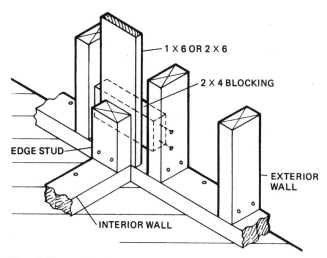

Fig. 6-11 *Another way to join partitions to the outside walls.* (Forest Products Laboratory)

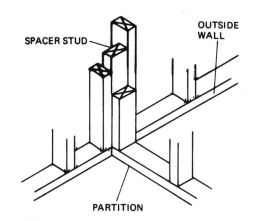

Fig. 6-12 *A third way to join partitions to the outside walls.*

Find Stud Length

Before cutting any studs, the carpenter must find the proper lengths. There are two ways of doing this. One way lets the carpenter measure the length. In the other way, the carpenter must compute it.

Make a master stud pattern A carpenter can make a master stud pattern. This is the way that stud lengths are measured. A 2-inch board just like the studs is used. A side view of the wall section is drawn. The side view is full-size and will show the stud lengths.

The carpenter starts by laying out the distance between the floor and the ceiling height. The distances in between are then shown. Rough openings are added. This then shows the lengths of trimmers and cripples. The pattern also lets the carpenter check the measurements before cutting the pieces. The stud pattern is also called a story pole or rod. As a rule, it is done for only one floor of the house. Figure 6-13 shows a master stud pattern.

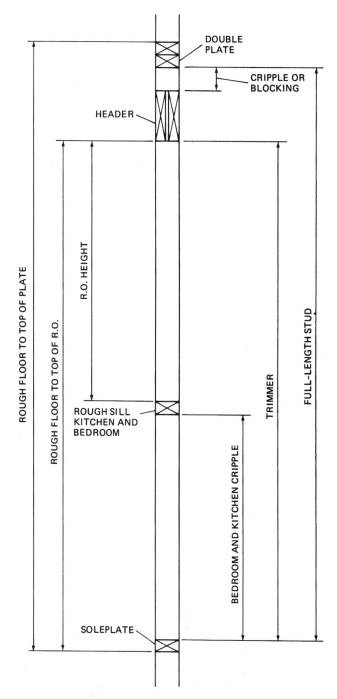

Fig. 6-13 *Make a master stud pattern by drawing it full size on a board.*

Compute stud length Another way to find stud lengths is to compute them. The carpenter must do some arithmetic and check it very carefully. This method is shown in Fig. 6-14.

The usual finished floor-to-ceiling distance is 8 feet ½ inch. The thicknesses of the finish floor and ceiling material are added to this dimension. The floor and ceiling thicknesses are commonly ½ inch each. The distance is written all in inches and adds up to 97½

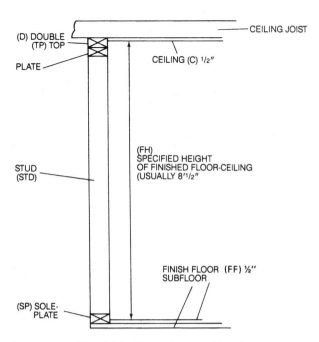

Fig. 6-14 *Computing a stud length.*

inches. The thickness of the sole- and top plates is subtracted. For one soleplate and a double top plate, this thickness is 4½ inches (3 times 1½ inches). The remainder is the stud length. For this example, the stud length is 93 inches. This is a commonly used length.

Using precut studs Sometimes the studs are cut to length at the mill and delivered to the site. When this is done, the carpenter does not make any measurements or cuts for full studs. The standard length for such precut studs is 93 inches. The carpenter who orders such materials should be very careful to specify "precision end trim" (P.E.T.) for lumber. See Fig. 6-15.

Fig. 6-15 *Precut studs ready at a site.*

Frame Rough Openings

The locations of full and trimmer studs for the rough opening are shown on the soleplates. The lengths of these can be found from the size of the rough opening.

The story pole is used for reference. The width of the R.O. sets the distance between the trimmers. See Fig. 6-6. A full stud is used on the outside of the trimmer.

Trimmer Studs

Trimmer studs extend from the soleplate to the top of the rough opening. They provide support for the header. The header must support the wall over the opening. It is important that the header be solidly held. The trimmers give solid support to the ends of the headers. The length of trimmers is the distance from the soleplate to the header.

Header Size

The size of the header is determined by the width of the rough opening. Table 6-1 shows the size for a typical opening width. As in Fig. 6-3, two header pieces are used over the rough opening. In some cases, the headers may be large enough to completely fill the space between the rough opening and the top plate. See Fig. 6-16. However, doing this will make the wall

Table 6-1 Size of Lumber for Headers

Width of Rough Opening	Minimum Size Lumber	
	One Story	Two or More
3'0"	2 × 4	2 × 4
3'6"	2 × 4	2 × 6
4'0"	2 × 6	2 × 6
4'6"	2 × 6	2 × 6
5'0"	2 × 6	2 × 6
6'0"	2 × 6	2 × 8
8'0"	2 × 8	2 × 10
10'0"	2 × 10	2 × 12
12'0"	2 × 12	2 × 12

Fig. 6-16 *Solid headers may be used instead of cripples.*

section around the header shrink and expand at a different rate than the rest of the wall. A solid header section is also harder to insulate.

The sill The bottom of the rough opening is framed in by a sill. The sill does not carry a load. Thus, it can simply be nailed between the trimmer studs at the bottom of the rough opening. The sill does not require solid support beneath the ends; however, it is common to use another short trimmer for this. See Fig. 6-17.

Fig. 6-18 *A radial arm saw can be used on the site.*

Fig. 6-17 *A trimmer may be used under both the header and the sill.*

Cripples Cripple studs are short studs. They join the header to the top plate. They also join the sill to the soleplate. They continue the regular stud spacing. This makes a nail base for sheathing and wallboard.

CUTTING STUDS TO LENGTH

After the wall parts are laid out, carpenters must find how many studs are needed. They must do this for each length. A great deal of time can be saved if all studs are cut at one time. Saws are set and full studs are cut first. Then the settings are changed and all trimmer studs are cut. Next, cripple studs are cut. Headers and sills should be cut last.

Cutting Tips

Most cuts to length are done with power saws. Two types are used. Radial arm saws can be moved to a location. See Fig. 6-18. Portable circular saws can be used almost anyplace. See Fig. 6-19. With either

Fig. 6-19 *Portable circular saws can be used almost anywhere.*

type of saw, special setups can be used. Pieces can be cut to the right length without measuring each one.

Portable circular saw Sawing several pieces to the same length is done with a special jig. See Fig. 6-20. Two pieces of stud lumber are nailed to a base. Enough space is left between them for another stud. A stop block is nailed at one end. A guide board is then nailed across the two outside pieces. Care should be taken because the guide is for the saw frame. The blade cuts a few inches away.

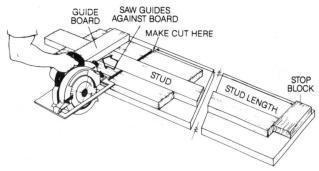

Fig. 6-20 A jig for cutting studs with a portable circular saw.

Radial arm saw A different method is used on a radial arm saw. No marking is done. Figure 6-21 shows how to set the stop block. This is quicker and easier. The piece to be cut is simply moved to touch the stop block. This sets the length. The piece is held against the back guide of the saw. The saw is then pulled through the work.

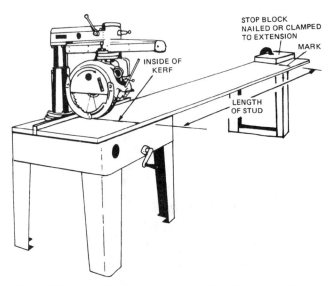

Fig. 6-21 A stop block is used to set the length with a radial arm saw. This way all the studs can be cut without measuring. This saves much time and money.

WALL ASSEMBLY

Once all pieces are cut, the wall may be assembled. Headers should be assembled before starting. See Figs. 6-22 and 6-23. The soleplate is moved about 4 inches away from the edge of the floor. It is laid flat with the stud markings on top. Then it is turned on edge. The markings should face toward the middle of the house. Then the top plate is moved away from the soleplate. The distance should be more than the length of a full stud. The soleplate and top plate markings should be aligned. They must point toward each other.

Fig. 6-22 Use ½-inch plywood for spaces between headers.

Fig. 6-23 Nailers can be used to assemble headers.

The straightest studs are selected for the corners. They are put at the corners just as they will be nailed. Next, a full stud is laid at every X location between the sole- and top plates. Figure 6-24 shows carpenters doing this.

Fig. 6-24 Laying the studs in place on the subfloor.

Nailing Studs to Plates

All studs are laid in position. The soleplate and top plate are tapped into position. Make sure that each stud is within the marks.

Corner studs Before nailing, corner spacers are cut. Three spacers are used at each corner. Each spacer should be about 16 inches long. Spacers are put between the two studs at the corners. One spacer should be at the top, one should be at the bottom, and one should be in the center. Two 16d nails should be used on each side. See Fig. 6-25. After the corner studs are nailed together, they are nailed to the plates. Two 16d nails are used for each end of each stud.

Full studs All the full studs are nailed in place at the X marks. Two 16d nails are used at each end. Figures 6-26 and 6-27 show how.

Trimmer studs Next, the trimmer studs are laid against the full studs. The spacing of the rough openings is checked. The trimmers are then nailed to the studs from the trimmer side. See Fig. 6-28. Use 10d nails. The nails should be staggered and 16 inches apart. Staggered means that one nail is near the top edge and the next nail is near the bottom edge. See Fig. 6-29.

Headers and sills The headers are nailed in place next. The headers are placed flush with the edges of the studs. The headers are nailed in place with 16d nails. The nails are driven through the studs into the ends of the headers.

The sills are nailed in place after the headers. Locate the position of the sill. Toenail the ends of the sill to the trimmers. Also, another trimmer may be used. See Fig. 6-30.

Cripple studs Next, the cripples are laid out. They are nailed in place from the soleplate with two 16d nails. Then, two 16d nails are used to nail the sill to the ends of the cripples. Finally, the top cripples are nailed into place. See Fig. 6-31.

CORNER BRACES

Corner braces should be put on before the wall is raised. The bracing prevents the wall frame from swaying sideways. Two methods are commonly used.

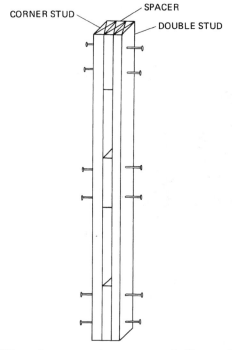

Fig. 6-25 *Nailing spacers into corner studs. Use two 16d nails on each side of spacer.*

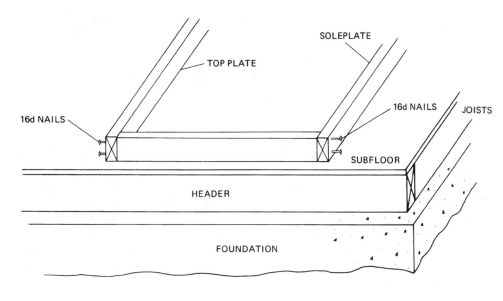

Fig. 6-26 *Nailing full studs to plates.*

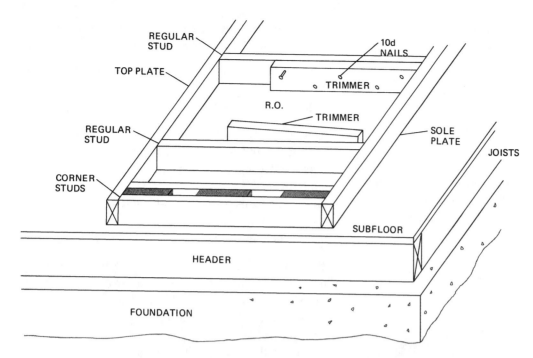

Fig. 6-27 *Trimmers are nailed into the rough opening.*

Fig. 6-28 *Nailing studs for wall frames.*

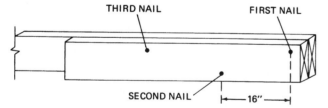

Fig. 6-29 *In stagger nailing, one nail is near the top and the other is near the bottom.*

Plywood Corner Braces

The first method uses plywood sheets as shown in Fig. 6-32. It is best to nail the plywood on before the wall is erected. Plywood bracing costs more but takes little time and effort. If used, it should be the same thickness as the wall sheathing. Plywood should not be used where "energy" sheathing is used. This will be discussed later.

Diagonal Corner Braces

The second method is to make a board brace. The brace is made from 1-×-4-inch lumber. It is "let in" or recessed into the studs. The angle may be any angle, but 45° is common. Braces are "let in" to the studs so that the outside surface of the wall will stay flat. See Fig. 6-33.

To make a diagonal brace, select the piece of wood to be used and nail it temporarily in place across the outside. Mark the layout on each stud. Also mark the angle on the ends of the brace. Remove the brace and gage the depth of the cut across each stud. Make the cuts with a saw. Knock out the wood between the cuts with a chisel. Then, trim off the ends of the brace. Place the brace in the "let in" slots. Check for proper fit. If the fit is good, nail the brace in place. Use two 8d nails at each stud.

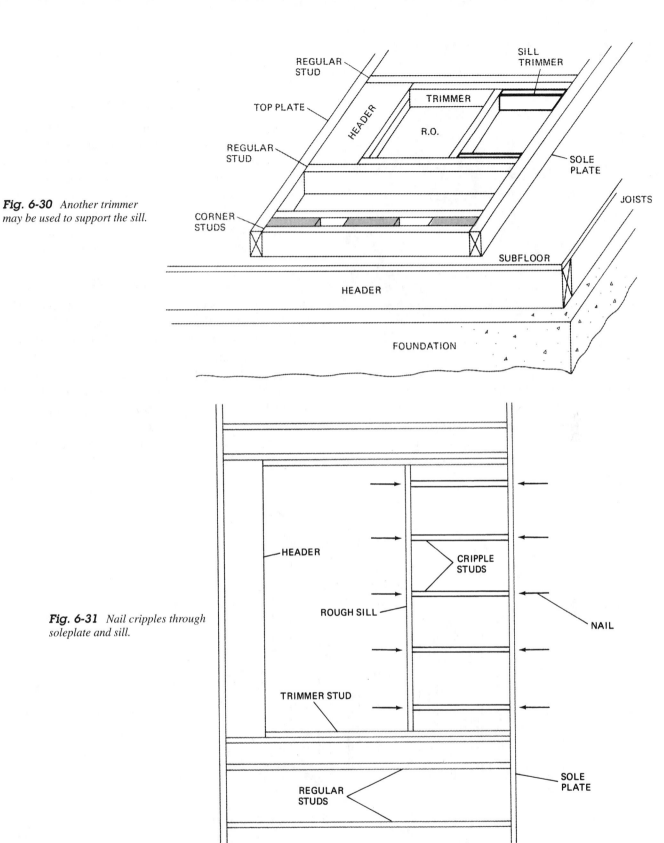

Fig. 6-30 *Another trimmer may be used to support the sill.*

Fig. 6-31 *Nail cripples through soleplate and sill.*

ERECT THE WALLS

Once the wall section is assembled, it is raised upright. The wall may be raised by hand. See Fig. 6-34. Special wall jacks may also be used, as in Fig. 6-35. Figure 6-36 shows another method, raising the wall with a forklift unit. This is common when a large wall must be placed over anchor bolts on a slab. For slabs, the anchor holes in the soleplate should be drilled before the

Fig. 6-32 *Plywood is used for corner bracing.*

is usually kept until the roof is on. See Figs. 6-38 and 6-39.

Wall Sheathing

Wall sheathing may be put on a wall before it is raised. The advantage is that this reduces the amount of lifting and holding. However, it slows down getting the roof up. It is also harder to make extra openings for vents and other small objects. Wall sheathing may be nailed or stapled. Manufacturer's recommendations should be followed. Sheathing is made from wood, plywood, fiber, fiberboard, plastic foam, or gypsum board.

To Raise the Wall

Before the wall is raised, a line should be marked. The line should show the inside edge of the wall. It is made with the chalk line on the subfloor. This shows the position of the wall.

Before the wall is raised, all needed equipment should be ready. Enough people or equipment to raise the wall should be ready. Also the tools to plumb and brace the wall should be ready.

When all are ready, the wall may be raised. The top edge of the wall is picked up first. Lower parts are

wall is raised (Fig. 6-37). See the section on anchoring sills in Chapter 5. Walls that have been raised must be braced. The brace is temporary. The brace is used to hold the wall upright at the proper angle. Wind will easily blow walls down if they are not braced. Bracing

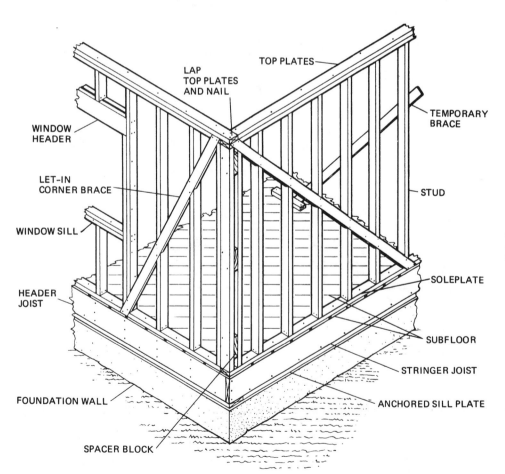

Fig. 6-33 *Boards are used for corner bracing.* (Forest Products Laboratory)

Fig. 6-34 *Raising a wall section by hand.* (American Plywood Association)

Fig. 6-35 *Raising a wall with wall jacks.* (Proctor Products)

Fig. 6-36 *Raising a wall with a fork-lift unit.*

Fig. 6-37 *Wall sections are nailed to slabs after being erected.* (Fox and Jacobs)

Fig. 6-38 *Once raised, walls are held in place with temporary braces.*

grasped to raise it into a vertical position. See Figs. 6-34 through 6-36. The wall is held firmly upright. As it is held, the wall is pushed even with the chalk line on the floor.

Some walls are built erect, not raised into position. As a rule, these walls are built on slabs. The layout and cutting are the same.

However, the soleplate is bolted to the slab first. Next studs are toenailed to the soleplate. The top plate

Fig. 6-39 *Nailing a temporary brace.* (Fox and Jacobs)

can be put on next. Then the wall is braced. Openings and partition bases may then be built.

Put Up a Temporary Brace

After the wall is positioned on the floor, it is nailed in place. See Fig. 6-40. Then one end is plumbed for vertical alignment. A special jig can be used for the level. See Fig. 6-41. After the wall is plumb, a brace is nailed. See Fig. 6-39.

Each wall is treated in the same manner. After the first wall, each added wall will form a corner. The corners are joined by nailing as in Fig. 6-42. It is important that the corners be plumb.

After several walls have been erected, the double plate is added at the top. Figure 6-42 shows a double plate in place. Note that the double plate overlaps on corners to add extra strength.

INTERIOR WALLS

Interior walls, or partitions, are made in much the same manner as outside walls. However, the carpenter must remember that studs for inside walls may be longer. This is because inside walls are often curtain walls. A curtain wall does not help support the load of the roof. Because of this a double plate might not be used. But, the top of the wall must be just as high.

Most builders wish to make the building weathertight quickly. Therefore the first partitions that are made are load-bearing partitions.

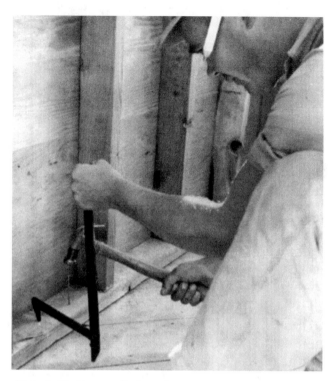

Fig. 6-40 *Once walls have been pulled into place, they are nailed to the subfloor.* (Proctor Products)

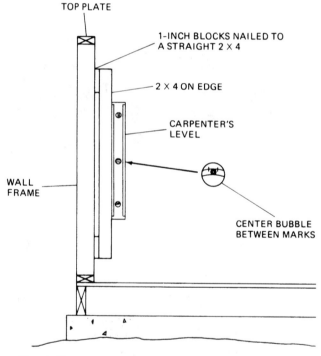

Fig. 6-41 *Use a spanner jig to plumb corners with the level.*

When roof trusses are used, load-bearing partitions may not be necessary. Trusses distribute loads so that inside support is not needed. However, many carpenters make all walls alike. This lets them cut all studs the same. It also lets them use any wall for support.

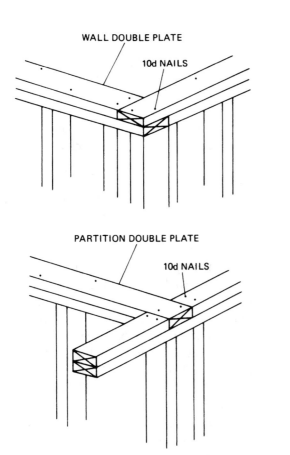

Fig. 6-42 *Double plates are added after several walls have been erected.*

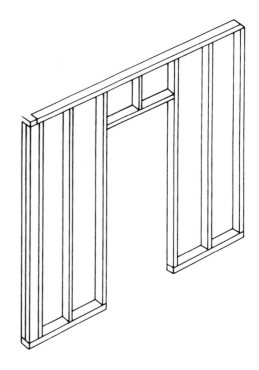

Fig. 6-43 *Single boards are also used as headers on inside partitions.*

Locate Soleplates for Partitions

To locate the partitions, the centerlines are determined from the plans. The centerline is then marked on the floor with the chalk line. Plates are then laid out.

Studs

Studs and headers are cut. The partition walls are done in the same way as outside walls. As a rule, the longer partitions are done first. Then the short partition walls are done. Last are the shortest walls for the closets.

Corners

Corners and wall intersections are made just as for outside walls. The size and amount of blocking can be reduced. The purpose of blocking is to provide nail surfaces. These are needed at inside and outside corners. They are a base for nailing wall covering.

Headers and Trimmers

Headers are not required for rough openings in curtain walls. Often, openings are framed with single boards. See Fig. 6-43. The header for inside walls is much like a sill. Trimmers are optional. They provide more sup-port. They are recommended when single-board headers are used.

Soleplate

The soleplate is not cut out for door openings. It is made as one piece. It is cut away after the wall is raised. See Fig. 6-44.

Special Walls

Several conditions may call for special wall framing. Walls may need to be thicker to enclose plumbing. Drain pipes may be wider than the standard 3½-inch-thick wall. Thickness may be added in two ways: Wider studs may be used, or extra strips may be nailed to the edges of studs. Other special needs are sound-proofing and small openings.

Soundproofing

Most household noise is transmitted by sound waves vibrating through the air. Blocking the air path with a standard interior wall or ceiling reduces the sound somewhat, but not completely.

The reason? Vibrations still travel through solids, particularly when the materials provide a continuous path. Standard interior walls and metal air ducts allow sound waves to continue their transmission.

So an effective sound control system must not only block the sound path, it needs to break the vibration

POINT TO CUT OUT SOLE PLATE

Fig. 6-44 At door openings, the partition soleplates are cut out after the wall is up.

path. Further noise reduction is accomplished with a system that absorbs the sound.

The Sound Transmission Class (STC) rating is a measurement—expressed in decibels—that describes a structure's ability to resist sound transmission. The higher the STC rating, the greater the structure's ability to limit the transmission of sound.

By incorporating combinations of all three elements for proper sound control, it is possible to raise the STC rating of a standard interior wall from 4 to 12 decibels, depending on the wall's construction.

Special insulation is available for sound control. When installed inside interior walls, the acoustic batts provide extra material to help block the sound path. The fiberglass insulation has tiny air pockets that absorb sound energy.

To break the vibration path, the insulation manufacturer recommends two options: One approach is to install staggered wall studs. The sole- and top plates are made wider. Regular studs are used. The studs are staggered from one side to the other. See Fig. 6-45. Insulation can now be woven between the studs. Sound is transmitted better through solid objects. But, with staggered studs, there is no solid part from wall to wall. In this way, there is no solid bridge for easy sound passage.

Another alternative is to install resilient channels between drywall and studs. For floors and ceilings, resilient channels should be installed in addition to the insulation batts. Figure 6-46A illustrates this using wood studs, 16 inches on center with ½-inch drywall on each side and one thickness of fiberglass acoustic batt, which is 3.5 inches thick. Figure 6-46B shows a better option. Single wood studs are used 16 inches on

WALL DETAIL	DESCRIPTION	STC RATING
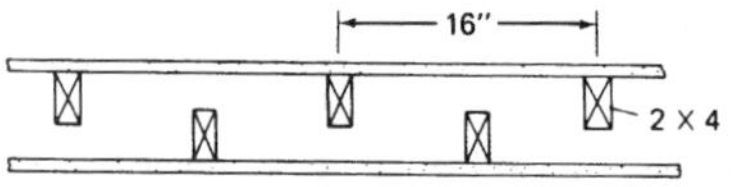 16" 2 × 4	1/2-INCH GYPSUM WALLBOARD	45
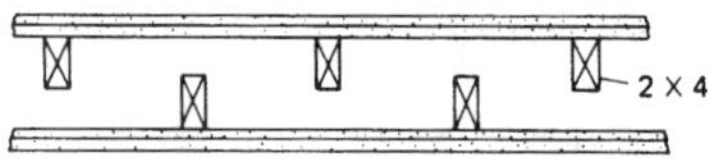 2 × 4	5/8-INCH GYPSUM WALLBOARD (DOUBLE LAYER EACH SIDE)	53
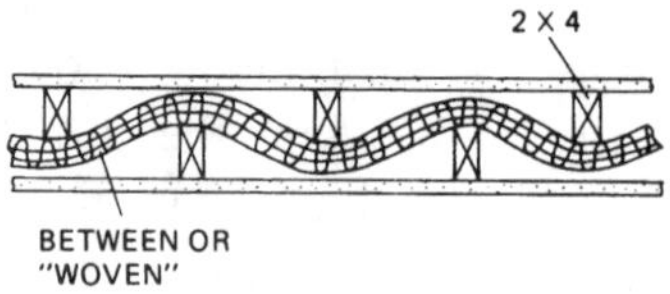 2 × 4 BETWEEN OR "WOVEN"	1/2-INCH GYPSUM WALLBOARD 1 1/2-INCH FIBROUS INSULATION	60
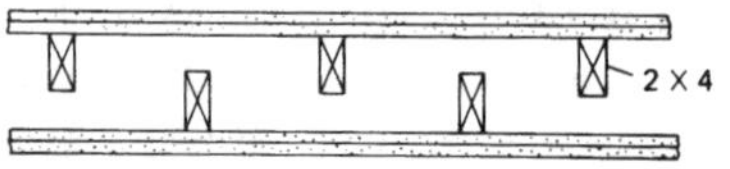 2 × 4	1/2-INCH SOUND DEADENING BOARD (NAILED) 1/2-INCH GYPSUM WALLBOARD (LAMINATED)	51

Fig. 6-45 Sound insulation of double walls. (Western Wood Products)

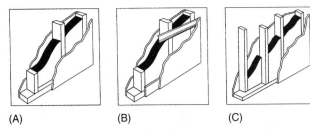

(A)   (B)   (C)

Fig. 6-46 *(A) Single wood studs, 16 inches on center. (Owens-Corning); (B) Single wood studs, 16 inches on center with resilient channel. (Owens-Corning); (C) Staggered wood studs, 16 inches on center. (Owens-Corning)*

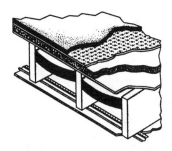

Fig. 6-48 *Standard carpet and pad with ⅜-inch particle board surface, ⅝-inch plywood subfloor, and a single sheet of ½-inch gypsum board mounted on resilient channels spaced 24 inches. (Owens-Corning)*

center with a resilient channel and ½-inch drywall on each side. A 3.5-inch thick insulation batt is used. Resilient channels help reduce noise by dissipating sound energy and decreasing sound transmission through the framing. For the wall, install resilient channels spaced 24 inches horizontally over 16 inches on-center framing. Another, better option is shown in Fig. 6-46C. Here staggered wood studs placed 16 inches on-center with ½-inch gypsum board on each side with one thickness of fiberglass acoustic batt, 3.5 inches thick. Keep in mind that the insulation batts inserted between the walls are not intended as a thermal barrier, but for sound control.

There are other sources of sound or noise transmission that also must be addressed if the problem is to be minimized. Figure 6-47 shows the electrical wiring can cause sound transmission if holes for the wire are not caulked. Do not place light switches and outlets back to back. Place wall fixtures a minimum of 24 inches apart. Light switches should be spaced at least 36 inches apart. Note also how elastic caulk is used to fill in around the electrical receptacle box.

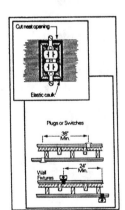

Fig. 6-47 *Light switches and receptacles. (Owens-Corning)*

Wood joist floors, Fig. 6-48, usually have a standard carpet and pad with ⅜-inch plywood or particle board subfloor. Single layer ½-inch gypsum ceiling mounted on resilient channels spaced every 24 inches also improves the sound deadening quality of the ceil-

ing and floor. This can be improved more by adding a 3.5-inch-thick fiberglass acoustic batts.

Doors with a solid wood core will improve the noise control. See Fig. 6-49. Install a threshold closure at the bottom of the door to reduce sound transmission. Ducts can be insulated with fiberglass batts to aid in maintaining constant air temperatures while minimizing sound transmission. Duct work featuring a smooth acrylic coating on the inner surface for easy cleaning and maintenance also improves the sound qualities. See Fig. 6-50.

Caulking also can decrease the sound level. Caulking around the perimeter of drywall panels, plumbing fixtures, pipes, and wall plates can reduce

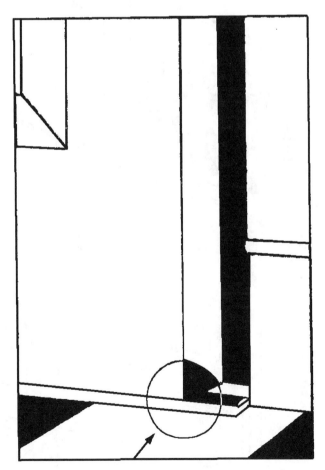

Fig. 6-49 *Doors need a threshold closure to keep out noise. (Owens-Corning)*

Fig. 6-50 *Installing fiberglass on ductwork.* (Owens-Corning)

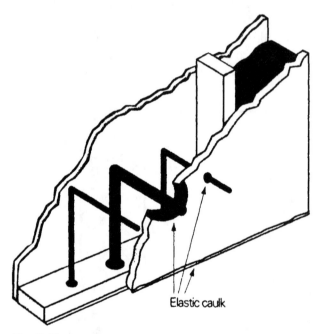

Elastic caulk

Fig. 6-51 *Caulking around plumbing helps seal out noise.* (Owens-Corning)

the incidence of unwanted noises considerably. See Fig. 6-51.

Small openings Small openings are often needed in walls. They are needed for air ducts, plumbing, and drains. Recessed cabinets, such as bathroom medicine cabinets, also need openings. These openings are framed for strength and support. See Fig. 6-52.

SHEATHING

Several types of sheathing are used. Sheathing is the first layer put on the outside of a wall. Sheathing makes the frame still and rigid and provides insulation. It will also keep out the weather until the building is finished.

There are five main types of sheathing used today: fiberboard, gypsum board, boards, plywood, and rigid foam. Corner bracing is needed for all types except plywood and boards.

Fiberboard Sheathing

The most common type of sheathing is treated fiberboard. It should be applied vertically for best bracing and strength characteristics. It is usually ½ inch thick. Plywood ½ inch thick can be used for corner bracing. The ½-inch fiber sheathing and plywood fit flush for a smooth surface. The exterior siding can easily be attached. Fiber sheathing should be fastened with roofing nails. Nails 1½ inches long are spaced 3 to 6 inches apart. Nails should always be driven at least ⅜ inch from the edge. If plywood is not used, diagonal corner braces are required. See Fig. 6-53A to C.

Gypsum Sheathing

Sheathing made from gypsum material is also used. See Fig. 6-54. This gypsum sheathing is not the same as the sheets used on interior walls. This gypsum sheathing is treated to be weather-resistant. The most common thickness is ½ inch. A ½-inch-thick plywood

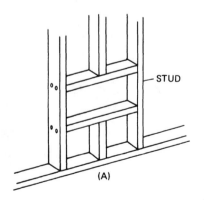

STUD

(A)

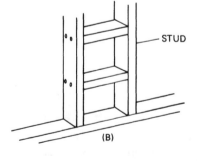

STUD

(B)

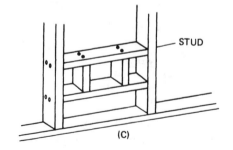

STUD

(C)

Fig. 6-52 *Framing for small openings in walls.*

Fig. 6-53 *Corner bracing; (A) Fiberboard sheathing with plywood corner braces; (B) Fiberboard sheathing with inlet board corner braces—seen from the outside; (C) Fiberboard sheathing with inlet board corner braces—seen from the inside.*

Fig. 6-54 *Gypsum sheathing is inexpensive and fire resistive.*

corner brace may be used. The sheathing should be fastened with roofing nails. Nails 1¾ inches long should be used. The nails should be 4 inches apart.

Plywood Sheathing

Plywood is also used for sheathing. When it is used, no corner bracing is required. Plywood is both strong and fire-resistant. It can be nailed up quickly with little cutting. A moisture barrier must be added when plywood is used. However, plywood and board sheathing are both expensive.

When plywood is used, it should be at least ⁵⁄₁₆ inch thick. Half-inch thickness is recommended. Exterior siding can be nailed directly to ½-inch plywood. Plywood sheets can be applied vertically or horizontally. The sheets should be fastened with 6d or 8d nails. The nails should be 6 to 12 inches apart.

Energy Sheathing

Two types of energy sheathing are commonly used. The first is a special plastic foam called *rigid foam*. See Fig. 6-55. Its use can greatly reduce energy consumption. It

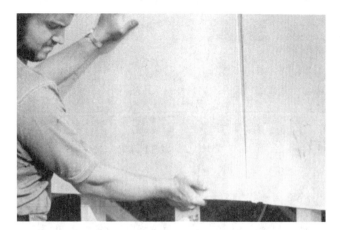

Fig. 6-55 *Rigid polystyrene foam sheathing provides more insulation.* (Dow Chemical)

is roughly equivalent to 3 more inches of regular wall insulation. It is grooved on the sides and ends. It is fitted together horizontally and is nailed in place with 1¼-inch nails spaced 9 to 12 inches apart. Joints may be made at any point. Rigid foam may also be covered with a shiny foil on one or both sides. See Fig. 6-56. The foil surface further reduces energy losses. It does so because the shiny surface reflects heat. Also, the foil prevents air from passing through the foam. However, foam burns easily. A gypsum board interior wall should be used with the foam. The gypsum wall reduces the fire hazard.

Fig. 6-56 *Rigid polystyrene foam may be coated with reflective foil to increase its effectiveness as an insulation sheathing.*

The second type is a special fiber. It is also backed on both sides with foil. Its fibers do not insulate as well as foam. It does prevent air movement better than plain foam. The foil surface is an effective reflector. Also, it is more fire-resistant and costs less. See Fig. 6-57.

Boards

Boards are still used for sheathing. However, diagonal boards are seldom used. Plain boards may be used. Boards with both side and end tongue-and-groove

Fig. 6-57 *Another style of energy efficient sheathing is similar to hardboard but is coated with reflective foil.* (Simplex)

joints are used. For grooved boards, joints need not occur over a stud. This saves installation time. Carpenters need not cut boards to make joints over studs. However, for plain boards joints should be made over studs. A special moisture barrier should be added on the outside. Builder's felt is used most often. It is nailed in place with 1-inch nails through metal disks. The bottom layers are applied first. See Fig. 6-58.

FACTORS IN WALL CONSTRUCTION

A carpenter may learn the procedure for making a wall. However, the carpenter may not know why walls are made as they are. Each part of a wall frame has a specific role. The corner pieces help tie the walls together. The double top plates also add strength to corners. The top plate is doubled to help support the weight of the rafters and ceiling. A rafter or a ceiling joist between studs is held up by the top plate. A single top plate could eventually sag and bend.

Standard Spacing

The spacing of wall members is also important. Standard construction materials are 4 feet wide and 8 feet long. Standard finish floor-to-ceiling heights are 8 feet ½ inch. The extra ½ inch lets pieces 8 feet long be used without binding.

Further, using even multiples of 4 or 8 means that less cutting is needed. Buildings are often designed in multiples of 4 feet. Two-foot roof overhangs can be used. The overhang at each end adds up to 4 feet. This

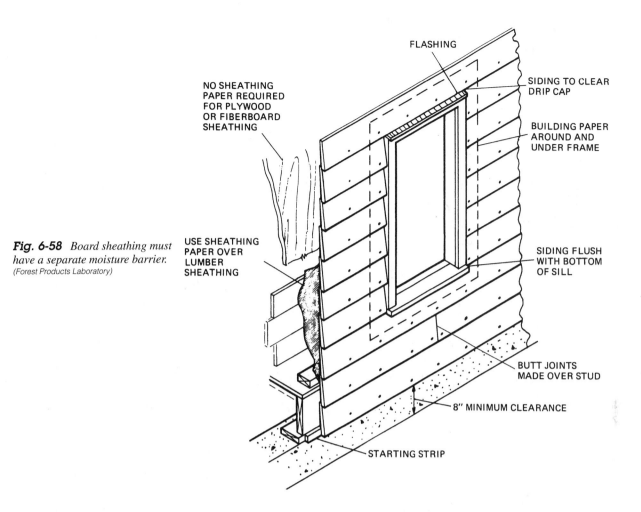

Fig. 6-58 *Board sheathing must have a separate moisture barrier.* (Forest Products Laboratory)

FLASHING

NO SHEATHING
PAPER REQUIRED
FOR PLYWOOD
OR FIBERBOARD
SHEATHING

SIDING TO CLEAR
DRIP CAP

BUILDING PAPER
AROUND AND
UNDER FRAME

USE SHEATHING
PAPER OVER
LUMBER
SHEATHING

SIDING FLUSH
WITH BOTTOM
OF SILL

BUTT JOINTS
MADE OVER STUD

8" MINIMUM CLEARANCE

STARTING STRIP

reduces the cutting done on siding, floors, walls, and ceilings, thereby reducing time and costs.

Notching and Boring

Whenever a hole or notch is cut into a structural member, the structural capacity of the piece is weakened and a portion of the load supported by the cut member must be transferred properly to other joists. It is best to design and frame a project to accommodate mechanical systems from the outset, as notching and boring should be avoided whenever possible. However, unforeseen circumstances sometimes arise during construction.

If it is necessary to cut into a framing member, the following figures (6-59, 6-60, 6-61) provide a guide for doing so in the least destructive manner. These drawings comply with the requirements of the three major model building codes: Uniform (UBC), Standard (SBC), and National (BOCA), and the CABO one- and

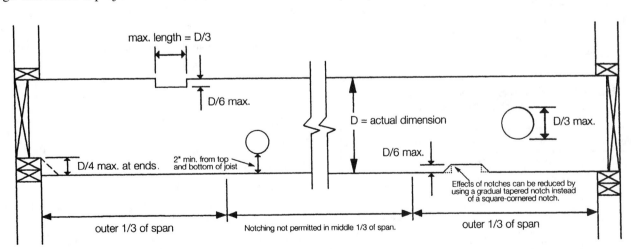

max. length = D/3

D/6 max.

D = actual dimension

D/3 max.

D/4 max. at ends.

2" min. from top and bottom of joist

D/6 max.

Effects of notches can be reduced by using a gradual tapered notch instead of a square-cornered notch.

outer 1/3 of span

Notching not permitted in middle 1/3 of span.

outer 1/3 of span

Fig. 6-59 *Placement of cuts in floor joists.* (Western Wood Products Association)

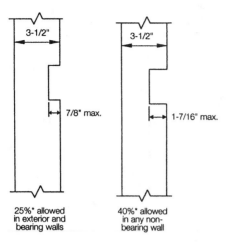

Fig. 6-60 *Notches in 2 × 4 studs.* (Western Wood Products Assocation)

3-1/2"

7/8" max.

25%* allowed
in exterior and
bearing walls

3-1/2"

1-7/16" max.

40%* allowed
in any non-
bearing wall

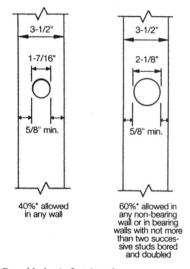

3-1/2"

1-7/16"

5/8" min.

40%* allowed
in any wall

3-1/2"

2-1/8"

5/8" min.

60%* allowed in
any non-bearing
wall or in bearing
walls with not more
than two succes-
sive studs bored
and doubled

Fig. 6-61 *Bored holes in 2 × 4 studs.* (Western Wood Products Association)

two-family dwelling code. Figure 6-59 shows the placement of cuts in floor joists. Notches and holes in 2 × 4 studs are shown in Figs. 6-60 and 6-61. Table 6-2 presents maximum sizes for cuts in floor joists.

The Western Wood Products Association provides information on the woods harvested in their area and used for studs and floor joists. Notching and boring holes in wood are a common practice for use as means

Table 6-2 *Maximum Sizes for Cuts in Floor Joists (Western Wood Products Association)*

Joist Size	Max. Hole	Max Notch Depth	Max. End Notch
2x4	none	none	none
2x6	1-1/2"	7/8"	1-3/8"
2x8	2-3/8"	1-1/4"	1-7/8"
2x10	3"	1-1/2"	2-3/8"
2x12	3-3/4"	1-7/8"	2-7/8"

of installing plumbing, electrical wiring, security systems, and sound systems. However, there are some very good reasons for not indiscriminately drilling a hole where you want. There can be serious results when the lumber being used for its strength is weakened by the placement of the hole or notch.

For instance, when structural wood members are used vertically to carry loads in compression, the same engineering procedure is used for both studs and columns. However, differences between studs and columns are recognized in the model building codes for conventional light-frame residential construction.

The term *column* describes an individual major structural member subjected to axial compression loads, such as columns in timber-frame or post-and-beam structures. The term *stud* describes one of the members in a wall assembly or wall system carrying axial compression loads, such as 2 × 4 studs in a stud wall that includes sheathing or wall board. The difference between columns and studs can be further described in terms of the potential sequences of failure.

Columns function as individual major structural members. Consequently, failure of a column is likely to result in partial collapse of a structure (or complete collapse in extreme cases due to the domino effect). However, studs function as members in a system. Due to the system effects (load sharing, partial composite action, redundancy, load distribution, etc.), studs are much less likely to fail and result in a total collapse than are columns.

Notching or boring into columns is not recommended and rarely acceptable; however, model codes established guidelines for allowable notching and boring into studs used in a stud-wall system.

Figures 6-60 and 6-61 illustrate the maximum allowable notching and boring of 2×4 studs under all model codes except BOCA. BOCA allows a hole one third the width of the stud in all cases.

Bored holes shall not be located in the same cross section of a stud as a cut or notch.

It is important to recognize the point at which a notch becomes a rip, such as when floor joists at the entry of a home are ripped down to allow underlayment for a floor.

Ripping wide dimension lumber lowers the grade of the material, and is unacceptable under all building codes.

When a sloped surface is necessary, a non-structural member can be ripped to the desired slope and fastened to the structural member in a position above the top edge. Do not rip the structural member.

Modular Standards

Currently, a new modular system of construction is being used more. The modular system uses stud and rafter systems 24 inches O.C. The wider spacing does not provide as much support for wall sheathing. However, the ability to support the roof is not reduced much. A building with studs on 16-inch centers is very strong. Yet the difference in strength between 16- and 24-inch centers is very small. The advantage of using the 24-inch modular system is that it saves on costs. For a three-bedroom house, the cost of wall studs may be reduced about 25 percent. Moreover, the costs of labor are also reduced. This saves the time that would be used for cutting and nailing that many more studs.

The savings are even more in terms of money and wages paid. This saving is made because fewer resources are used with almost the same results. In the modular system, several ways of reducing material use are employed. Floor joists are butted and not lapped. Building and roof size are exact multiples of 4 or 8 feet. Also, single top plates are used on partitions.

Energy

Energy is also a matter of concern today. Insulated walls save energy used for heating and cooling. The amount of insulation helps determine the efficiency. A 6-inch-thick wall can hold more insulation than a 4-inch wall. It will reduce the energy used by about 20 percent. However, it takes more material to build such a wall. Canadian building codes often specify 6-inch walls.

The double-wall system of frame buildings is a better insulator than a solid wall. Wood is good insulation when compared with other materials. But the best insulation is a hollow space filled with material that does not conduct heat. A wall should have three layers. An outside, weatherproof wall is needed. This layer stops rain and wind. A thick layer of insulation is next. This helps keep heat energy from being conducted through the wall. The inside wall is the third layer. It helps reduce air movement. It also helps seal and hold the insulation in place.

More insulation effect can also be added by adding a shiny surface. See Figs. 6-56 and 6-57.

Energy sheathing adds insulation in two ways: it insulates in the same way as regular insulation, and it also covers the studs. This way, the stud does not conduct heat directly through the wall. See Fig. 6-63.

The use of headers also affects wall construction. Headers reduce time and construction costs; however,

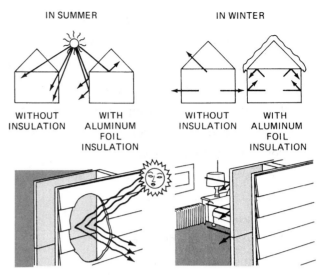

Fig. 6-62 *Reflected heat makes the house cooler in summer and warmer in winter.* (U.S. Gypsum)

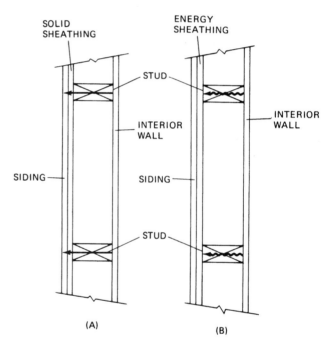

Fig. 6-63 *(A) Solid sheathing, stud and wall are a solid path. This conducts energy loss straight through a wall; (B) Energy sheathing forms a barrier. There is no solid path for energy.*

they are also difficult to insulate. Truss headers or a single header is better. This allows more insulation to be used.

The builder and the carpenter can alter a wall frame to save energy. Walls can be made thicker so that more insulation can be used. Thicker walls also let insulation cover studs in the same way as for sound insulation. Energy sheathing can be used to cover the studs. It adds insulation without requiring thick walls. Reflective foil also makes it more efficient. See Fig. 6-62.

CHAPTER 6
STUDY QUESTIONS

1. Which is usually built first, the wall or the floor?
2. On what piece are studs spaced?
3. What size nail is used to join studs to plates?
4. What size nail is used to join headers to studs?
5. How are headers supported?
6. When are soleplates cut for interior doors?
7. When are headers made?
8. What is a curtain wall?
9. Why is corner bracing done?
10. How are corners braced?
11. How are corners plumbed?
12. Why are walls braced temporarily?
13. What is used for sheathing?
14. How can you tell when a wall is in place?
15. How can walls be changed to save energy?
16. What is a story pole?
17. What is needed for sound control in a house?
18. What kind of doors improve sound control in a house?
19. What happens to a structural member when a hole is drilled in it?
20. What happens to dimensional lumber when it is ripped?

7
CHAPTER

Building
Roof Frames

THE FRAMING FOR A ROOF IS DETERMINED by the choice of roof style. There are a number of types of rooflines. Some of them are the gable, mansard, shed, hip, and gambrel. Each has identifying characteristics. Each style presents unique problems. Rafters and sheathing will be shaped according to the roofline desired.

In this unit you will learn how to build roof frames. You will learn how to make roofs of different types. You will also learn how to make openings for chimneys and soil pipes. Things you can learn to do are:

- Design a particular type of roof rafter
- Mark off and cut a roof rafter
- Put roof trusses in place
- Lay sheathing onto a roof frame
- Build special combinations of framing for roofs
- Identify the type of plywood needed for a particular roof

INTRODUCTION

The roof is an important part of any building. It is needed to keep out the weather and to control the heat and cooling provided for human comfort inside. There are many types of roofs. Each serves a purpose. Each one is designed to keep the inside of the house warm in the winter and cool in the summer. The roof is designed to keep the house free of moisture, whether rain, snow, or fog.

The roof is made of rafters and usually supported by ceiling joists. When all the braces and forms are put together, they form a roof. In some cases roofs have to be supported by more than the outside walls. This means some of the partitions must be weight-bearing. However, with truss roofs it is possible to have a huge open room without supports for the roof.

Figure 7-1 shows various roof shapes. The shape can affect the type of roofing materials used. The various shapes call for some special details. Roofs have to withstand high winds and ice and snow. The weight of snow can cause a roof to collapse unless it is properly designed. It is necessary to make sure the load on the roof can be supported. This calls for the proper size rafters and decking.

Shingles on a roof protect it from rain, wind, and ice. They have to withstand many years of exposure to all types of weather conditions.

The hip roof with its hip and hip jack rafters is of particular concern. It is one of the most popular types. The common rafter is the simplest in the gable roof. The

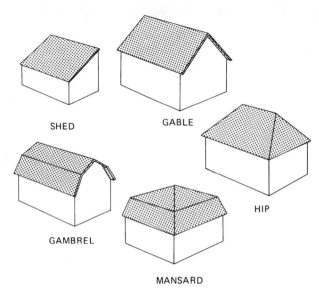

Fig. 7-1 *Various roof shapes or styles.*

mansard roof and the gambrel roof present some interesting problems with some very interesting solutions. All this will be explained in detail later in this chapter.

Various valleys, ridge boards, and cripple rafters will be described in detail here.

SEQUENCE

The carpenter should build the roof frame in this sequence:

1. Check the plans to see what type of roof is desired.
2. Select the ceiling joist style and spacing.
3. Lay out ceiling joists for openings.
4. Cut ceiling joists to length and shape.
5. Lay out regular rafter spacing.
6. Lay out the rafters and cut to size.
7. Set the rafters in place.
8. Nail the rafters to the ridge board.
9. Nail the rafters to the wall plate.
10. Figure out the sheathing needed for the roof.
11. Apply the sheathing according to specifications.
12. Attach the soffit.
13. Put in braces or lookouts where needed. See Fig. 7-2.
14. Cut special openings in the roof decking.

There may be a need for cutting dormer rafters after the rest of the roof is finished. Structural elements may be installed where needed. Truss roof rafters will need special attention as to spacing and nailing. Double-check to make sure they fit the manufacturer's specifications.

Fig. 7-2 *Braces (lookouts) are put in where they are needed.* (Duo-Fast)

ERECTING TRUSS ROOFS
Truss Construction

Trussed rafters are commonly used in light frame construction. They are used to support the roof and ceiling loads. Such trusses are designed, fabricated, and erected to meet the design and construction criteria of various building codes. They efficiently use the excellent structural properties of lumber. They are spaced either 16 inches on centers (O.C.) or in some cases 24 inches O.C. See Fig. 7-3.

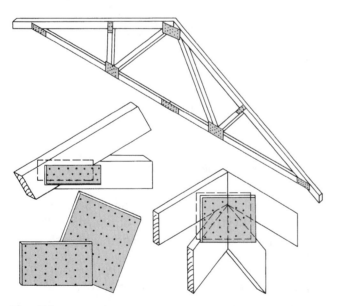

Fig. 7-3 *W-type truss roof and metal plates used to make the junction points secure.*

Truss Disadvantages

You should keep in mind that the truss type of construction does have some disadvantages. The attic space is limited by the supports that make up the truss.

Truss types of construction may need special equipment to construct them. In some instances it is necessary to use a crane to lift the trusses into position on the job site.

Roof framing The roof frame is made up of rafters, ridge board, collar beams, and cripple studs. See Fig. 7-4. In gable roof-construction, all rafters are precut to the same length and pattern. Figure 7-5 shows a gable roof. Each pair of rafters is fastened at the top to a ridge board. The ridge board is commonly a 2 × 8 for 2 × 6 rafters. This board provides a nailing area for rafter ends. See Fig. 7-4.

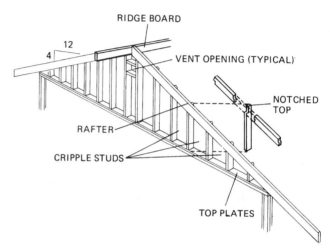

Fig. 7-4 *Gable-type rafter with cripple studs and ridge board. Note the notched top in the cripple stud.* (American Plywood Association)

Fig. 7-5 *Two-story house with gable roof. Note the bay windows with hip roof.*

Getting started Getting started with erection of the roof framing is the most complicated part of framing a house. Plan it carefully. Make sure you have all materials on hand. It is best to make a "dry run" at ground level. The erection procedure will be much easier if you have at least two helpers. A considerable amount of temporary bracing will be required if the job must be done with only one or two persons.

Steps in framing a roof Take a look at Fig. 7-6. It shows two persons tipping up trusses. They are tipped up and nailed in place one at a time. Two people, one working on each side, will get the job done quickly. This is one of the advantages of trussed-roof construction. The trusses are made at a lumber yard or some other location. They are usually hauled to the construction site on a truck. Here they are lifted to the roof of the building with a crane or by hand. In some cases the sheet metal bracket shown in Fig. 7-7 holds the truss to the wall plate. Figure 7-8 shows how the metal bracket is used to fasten the truss in place. In Fig. 7-9 toenailing is used to fasten the truss to the wall plate.

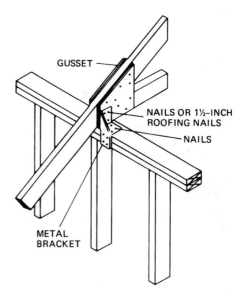

Fig. 7-8 *Using a nailing bracket to attach a truss to the wall plate and nailing into the gusset to attach the rafter.*

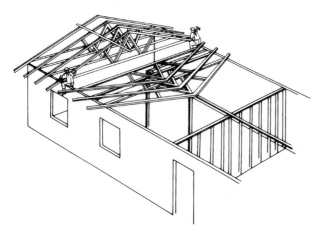

Fig. 7-6 *Roof trusses are placed on the walls, then tipped up and nailed in place.*

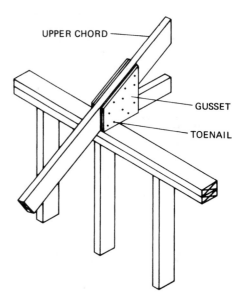

Fig. 7-9 *Toenailing the truss rafter to the wall plate.*

Fig. 7-7 *Rafter truss attached to the wall plate with a sheet-metal bracket.*

Advantages of trusses Manufacturers point out that the truss has many advantages. In Fig. 7-10 the conventional framing is used. Note how bearing walls are required inside the house. With the truss roof you can place the partitions anywhere. They are not weight-bearing walls. With conventional framing it is possible for ceilings to sag. This causes cracks. The truss has

supports to prevent sagging ceilings. See Fig. 7-11. Notice how the triangle shape is obvious in all the various parts of the truss. The triangle is a very strong structural form.

Details of trusses There are a number of truss designs. A W-type is shown in Fig. 7-12. Note how the 2 × 4s are brought together and fastened by plywood that is both nailed and glued. In some cases, as shown in Fig. 7-3, steel brackets are used. Whenever plywood is used for the gussets, make sure the glue fits the climate. In humid parts of the country, where the attic may be damp, the glue should be able to take the humidity. The manufacturer will inform you of the glue's use and how it will perform in humid climates. The

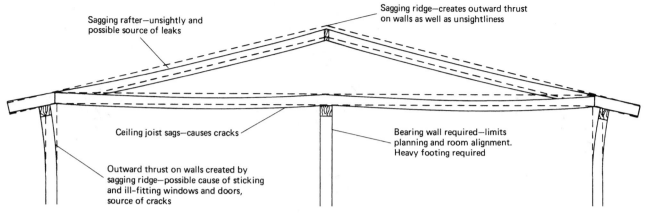

Fig. 7-10 *Disadvantages of conventional roof framing.*

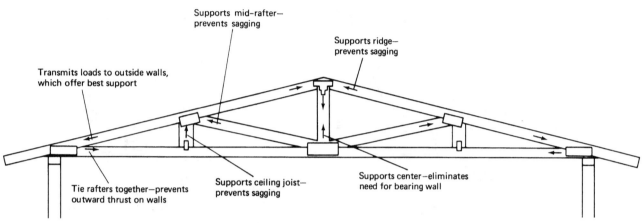

Fig. 7-11 *Advantages of roof trusses.*

glue should not lose its strength when the weather turns humid. Most glue containers have this information on them. If not, check with the manufacturer before making trusses. In most instances, the manufacturer of trusses is very much aware of the glue requirements for a particular location.

Figure 7-13 shows how the three suggested designs are different. The truss in Fig. 7-13A is known as the W type. Note the W in the middle of the truss. In Fig. 7-13B the king post is simple. It is used for a low-pitch roof. The scissors type is shown in Fig. 7-13C.

The W type is the most popular. It can be used on long spans. It can also be made of low-grade lumber. The scissor type is used on houses with sloping living room ceilings. It can be used for a cathedral ceiling. This truss is cheaper to make than the conventional type of construction for cathedral ceilings. King-post trusses are very simple. They are used for houses. This truss is limited to about a 26-foot span—that is, if 2 × 4s are used for the members of the truss. Compare this with a W type, which could be used for a span of 32 feet. King type is economical for use in medium spans. It is also useful in short spans. See Fig. 7-14.

The type of truss used depends upon the wind and snow. The weight applied to a roof is an important factor.

Make sure the local codes allow for truss roofs. In some cases, the FHA has to inspect them before insuring a mortgage with them in the house.

Lumber to use in trusses The lumber used in construction of trusses must be that which is described in Table 7-1. The moisture content should be from 7 to 16 percent. Glued surfaces have to be free of oil, dirt, or any foreign matter. Each piece must be machine-finished, but not sanded.

Lumber with roughness in the gusset area cannot be used. Twisted, cupped, or warped lumber should not be used either. This is especially true if the twist or cup is in the area of the gusset. Keep the intersecting surfaces of the lumber within $\frac{1}{32}$ inch.

Glue for trusses For dry or indoor locations use casein-type glue. It should meet Federal Specification MMM—A—125, Type 11. For wet conditions use resorcinol-type glue. Military specifications are MIL—A—46051 for wet locations. If the glue joint is

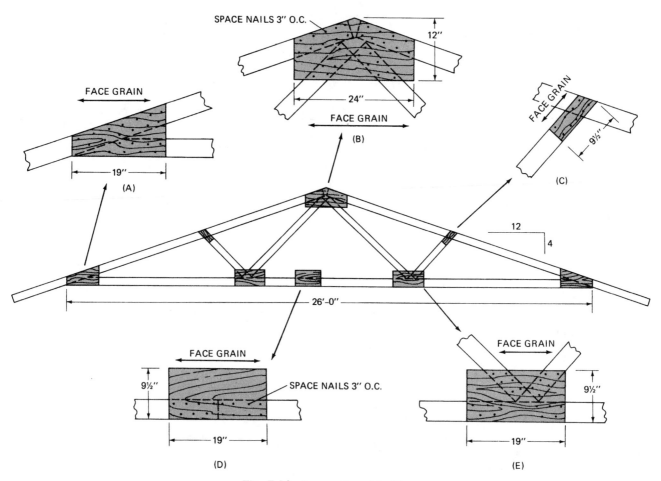

Fig. 7-12 *Construction of the W truss.*

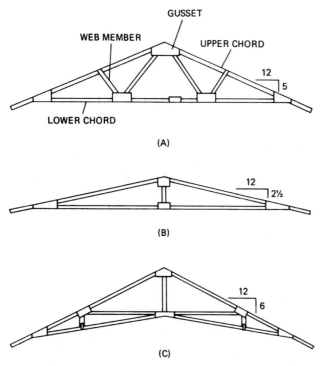

Fig. 7-13 *Wood trusses: (A) W type; (B) King post type; (C) Scissors type.*

exposed to the weather or used at the soffit, use the resorcinol-type glue.

Load conditions for trusses Table 7-2 shows the loading factors needed in designing trusses. Note the 30 pounds per square foot (psf) and 40 psf columns. Then look up the type of gusset—either beveled-heel or square-heel type. Standard sheathing or C–C Ext—APA grade plywood is the type used here for the gussets and sheathing (Table 7-3). C-D plywood is often used for roof sheathing.

Covering the trusses Once the trusses are in place, cover them with sheathing or plywood. This will make a better structure once the sheathing is applied. The underlayment is applied and the roofing attached properly. The sheathing or plywood makes an excellent nail base for the shingles. See Chapter 8 for applying shingles.

THE FRAMING SQUARE

In carpentry it is necessary to be able to use the framing square. This device has a great deal of informa-

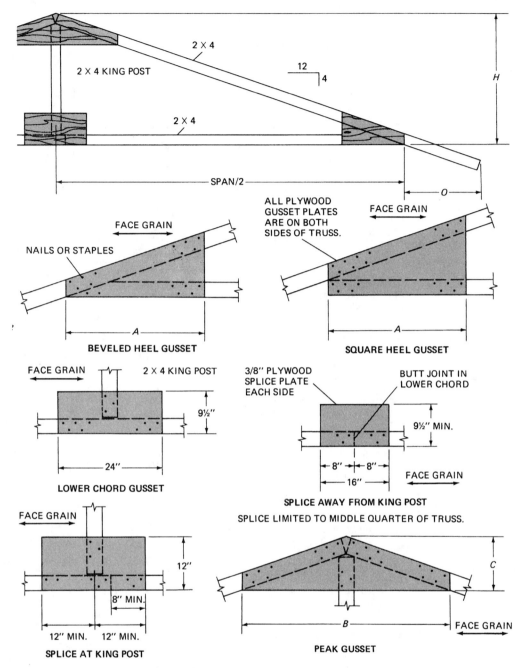

Fig. 7-14 *Construction of the king post truss.*

tion stamped on its body and tongue. See Fig. 7-15 for what the square can do in terms of a right angle. Figure 7-16 shows the right angle made by a framing square. The steel square or carpenter's square is made in the form of a right angle. That is, two arms (the body and the tongue of the square) make an angle of 90°.

Note in Fig. 7-15 how a right triangle is made when points A, B, and C are connected. Figure 7-16 shows the right triangle. A right triangle has one angle which is 90°.

Parts of the Square

The steel square consists of two parts—the body and the tongue. The body is sometimes called the *blade*. See Fig. 7-17.

Body The body is the longer and wider part. The body of the Stanley standard steel square is 24 inches long and 2 inches wide.

Tongue The tongue is the shorter and narrower part and usually is 16 inches long and 1½ inches wide.

Table 7-1 *Chord Code Table for Roof Trusses*

Chord Code	Size	Grade and Species Meeting Stress Requirements	Grading Rules	f	t//	c//
1	2 × 4	Select structural light framing WCDF	WCLIB	1950	1700	1400
		No. 1 dense kiln-dried Southern pine	SPIB	2000	2000	1700
		1.8E	WWPA	2100	1700	1700
2	2 × 4	1500f industrial light framing WCDF	WCLIB	1500	1300	1200
		1500f industrial light framing WCH	WCLIB	1450	1250	1100
		No. 1 2-inch dimension Southern pine	SPIB	1450	1450	1350
		1.4E	WWPA	1500	1200	1200
3	2 × 4	1200f industrial light framing WCDF	WCLIB	1200	1100	1000
		1200f industrial light framing WCH	WCLIB	1150	1000	900
		No. 2 2-inch dimension Southern pine	SPIB	1200	1200	900
4	2 × 6	Select structural J&P WCDF	WCLIB	1950	1700	1600
		Select structural J&P Western larch	WWPA	1900	1600	1500
		No. 1 dense kiln-dried Southern pine	SPIB	2000	2000	1700
		1.8E	WWPA	2100	1700	1700
5	2 × 6	Construction grade J&P WCDF	WCLIB	1450	1300	1200
		Construction grade J&P WCH	WCLIB	1450	1250	1150
		Structural J&P Western larch	WWPA	1450	1300	1200
		No. 1 2-inch dimension Southern pine	SPIB	1450	1450	1350
		1.4E	WWPA	1500	1200	1200
6	2 × 6	Standard grade J&P WCDF	WCLIB	1200	1100	1050
		Standard grade J&P WCH	WCLIB	1150	1000	950
		Standard structural Western larch	WWPA	1200	1100	1050
		No. 2 2-inch dimension Southern pine	SPIB	1200	1200	900
7	2 × 4	Select structural light framing WCDF	WCLIB	1900	1900	1400
		Select structural light framing Western larch	WWPA	1900	1900	1400
		No. 1 dense kiln-dried Southern pine	SPIB	2050	2050	1750
		1.8E	WWPA	2100	1700	1700
8	2 × 4	1500f industrial light framing WCDF	WCLIB	1500	1500	1200
		Select structural light framing WCH	WCLIB	1600	1600	1100
		Select structural light framing WH	WWPA	1600	1600	1100
		1500f industrial light framing Western larch	WWPA	1500	1500	1200
		No. 1 2-inch dimension Southern pine	SPIB	1500	1500	1350
		1.4E	WWPA	1500	1200	1200
9	2 × 4	1200f industrial light framing WCDF	WCLIB	1200	1200	1000
		1200f industrial light framing Western larch	WWPA	1200	1200	1000
		1500f industrial light framing WCH	WCLIB	1500	1500	1000
		1500f industrial light framing WH	WWPA	1500	1500	1000
		No. 2 2-inch dimension Southern pine	SPIB	1200	1200	900
10	2 × 6	Select structural J&P WCDF	WCLIB	1900	1900	1500
		Select structural J&P Western larch	WWPA	1900	1900	1500
		No. 1 dense kiln-dried Southern pine	SPIB	2050	2050	1750
		1.8E	WWPA	2100	1700	1700
11	2 × 6	Construction grade J&P WCDF	WCLIB	1500	1500	1200
		Construction grade J&P Western larch	WWPA	1500	1500	1200
		Select structural J&P WCH	WCLIB	1600	1600	1200
		Select structural J&P WH	WWPA	1600	1600	1200
		No. 1 2-dimension Southern pine	SPIB	1500	1500	1350
		1.4E	WWPA	1500	1200	1200
12	2 × 6	Standard grade J&P WCDF	WCLIB	1200	1200	1000
		Standard grade J&P Western larch	WWPA	1200	1200	1000
		Standard grade J&P WCH	WCLIB	1200	1200	1000
		Standard grade J&P WH	WWPA	1200	1200	1000
		No. 2 2-inch dimension dense Southern pine	SPIB	1400	1400	1050

Heel The point at which the body and the tongue meet on the outside edge of the square is called the *heel*. The intersection of the inner edges of the body and tongue is sometimes also called the heel.

Face The face of the square is the side on which the manufacturer's name, Stanley in this case, is stamped, or the visible side when the body is held in the left hand and the tongue is held in the right hand. See Fig. 7-17.

Back The back is the side opposite the face. See Fig. 7-18.

The modern scale usually has two kinds of marking: scales and tables.

Table 7-2 *Designs When Using Standard Sheathing as C-C EXT*

Loading Condition, Total Roof Load, psf	Span	Beveled-Heel Gusset							Square-Heel Gusset						
		Dimensions, Inches					Chord code		Dimensions, Inches						
		A	B	C	H	O	Upper	Lower	A	B	C	H	O	Upper	Lower
30 psf (20 psf live load, 10 psf dead load) on upper chord and 10 psf dead load on lower chord. Meets FHA requirements.	20'8"	32	48	12	45⅛	44	2	3	19	32	12	48¾	44	2	3
	22'8"	32	48	12	49⅛	48	1	2	19	32	12	52¾	48	1	3
	24'8"	48	60	16	53⅛	48	1	2	24	48	12	56¾	48	1	3
	26'8"	48	72	16	57⅛	48	1	2	32	60	16	60¾	48	1	2
40 psf (30 psf live load, 10 psf dead load) on upper chord and 10 psf dead load on lower chord.	20'8"	32	48	12	45⅛	43	8	9	19	32	12	48¾	48	7	9
	22'8"	32	60	16	49⅛	48	7`	8	19	48	12	52¾	48	7	9
	24'8"								32	60	16	56¾	48	7	9

Table 7-3 *Plywood Veneer Grades*

Grade	Description
N	Special order "natural finish" veneer. Select all heartwood or all sapwood. Free of open defects. Allows some repairs.
A	Smooth and paintable. Neatly made repairs permissible. Also used for natural finish in less demanding applications.
B	Solid surface veneer. Circular repair plugs and tight knots permitted.
C	Knotholes to 1". Occasional knotholes ½" larger permitted providing total width of all knots and knotholes within a specified section does not exceed certain limits. Limited splits permitted. Minimum veneer permitted in Exterior type plywood.
C Plugged	Improved C veneer with splits limited to ⅛" in width and knotholes and borer holes limited to ¼" by ½".
D	Permits knots and knotholes to 2½" in width, and ½" larger under certain specified limits. Limited splits permitted.

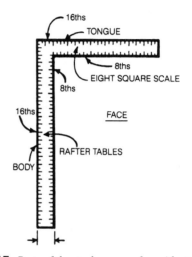

Fig. 7-17 *Parts of the steel square—face side.* (Stanley Tools)

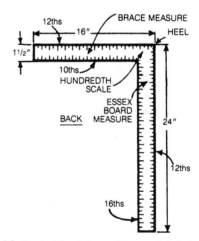

Fig. 7-18 *Back side of the steel square labeled.* (Stanley Tools)

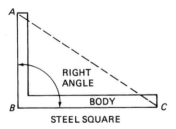

Fig. 7-15 *The steel square has a 90° angle between the tongue and the body.* (Stanley Tools)

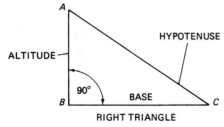

Fig. 7-16 *The parts of a right triangle.* (Stanley Tools)

Scales The scales are the inch divisions found on the outer and inner edges of the square. The inch graduations are in fractions of an inch. The Stanley square has the following graduations and scales:

Face of body—outside edge	Inches and sixteenths
Face of body—inside edge	Inches and eighths

Face of tongue— outside edge	Inches and sixteenths
Back of tongue— inside edge	Inches and eighths
Back of body— outside edge	Inches and twelfths
Back of body— inside edge	Inches and sixteenths
Back of tongue— outside edge	Inches and twelfths
Back of tongue— inside edge	Inches and tenths

Hundredths scale This scale is located on the back of the tongue, in the corner of the square, near the brace measure. The hundredths scale is 1 inch divided into 100 parts. The longer lines indicate 25 hundredths, while the next shorter lines indicate 5 hundredths. With the aid of a pair of dividers, fractions of an inch can be obtained. See Fig. 7-19 for the location of the hundredths scale.

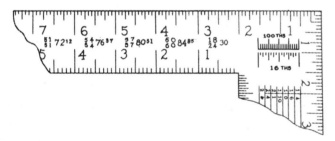

Fig. 7-19 *Location of the hundredths scale on a square.* (Stanley Tools)

One inch graduated in sixteenths is below the hundredths scale on the latest squares, so that the conversion from hundredths to sixteenths can be made at a glance without the need to use dividers. This comes in handy when you are determining rafter lengths using the figures of the rafter tables, where hundredths are given.

Rafter scales These tables will be found on the face of the body and will help you to determine rapidly the lengths of rafters as well as their cuts.

The rafter tables consist of six lines of figures. Their use is indicated on the left end of the body. The first line of figures gives the lengths of common rafters per foot of run. The second line gives the lengths of hip-and-valley rafters per foot of run. The third line gives the length of the first jack rafter and the differences in the length of the others centered at 16 inches. The fourth line gives the length of the first jack rafter and the differences in length of the others spaced at 24-inch

centers. The fifth line gives the side cuts of jacks. The sixth line gives the side cuts of hip-and-valley rafters.

Octagon scale The octagon or "eight-square" scale is found along the center of the face of the tongue. Using this scale a square timber may be shaped into one having eight sides, or an "octagon."

Brace scale This table is found along the center of the back of the tongue and gives the exact lengths of common braces.

Essex board measure This table is on the back of the body and gives the contents of any size lumber.

Steel Square Uses

The steel square has many applications. It can be used as a try square or to mark a 90° line along a piece of lumber. See Fig. 7-20.

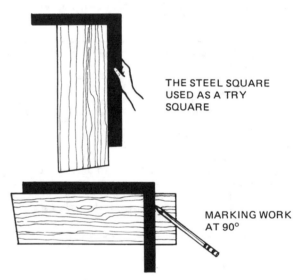

THE STEEL SQUARE USED AS A TRY SQUARE

MARKING WORK AT 90°

Fig. 7-20 *Uses of the steel square.* (Stanley Tools)

The steel square can also be used to mark 45° angles and 30–60° angles. See Fig. 7-21.

In some instances you may want to use the square for stepping off the length of rafters and braces. See Fig. 7-22. Another use for the square is the laying out of stairsteps. Figure 7-23 shows how this is done.

However, one of the most important roles the square plays in carpentry is the layout of roof framing. Here it is used to make sure the rafters fit the proper angles. The length of the rafters and the angles to be cut can be determined by the use of the framing square. Rafter cuts are shown in the following sections.

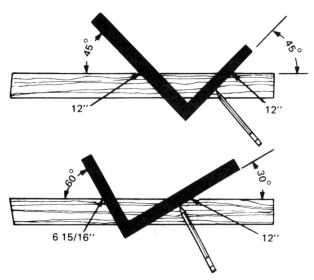

Fig. 7-21 *Using the steel square for marking angles.* (Stanley Tools)

Fig. 7-22 *Using the steel square to step off the length of rafters and braces.* (Stanley Tools)

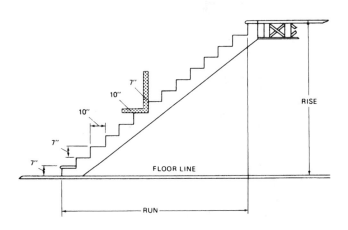

Fig. 7-23 *Using the square to lay out stairs.* (Stanley Tools)

ROOF FRAMING

There are a number of types of roofs. A great variety of shapes is prevalent, as you can see in any neighborhood. Some of the most common types will be identified and worked with here.

Shed roof The shed roof is the most common type. Easy to make, it is also sometimes called the lean-to roof. It has only a single slope. See Fig. 7-24.

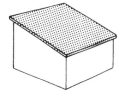

Fig. 7-24 *Lean-to or shed roof.* (Stanley Tools)

Gable or pitch roof This is another type of roof that is commonly used. It has two slopes meeting at the center or ridge, forming a gable. It is a very simple form of roof, and perhaps the easiest to construct. See Fig. 7-25.

Fig. 7-25 *Gable roof.* (Stanley Tools)

Gable-and-valley or hip-and-valley roof This is a combination of two intersecting gable or hip roofs. The valley is the place where two slopes of the roof meet. The roofs run in different directions. There are many modifications of this roof, and the intersections usually are at right angles. See Figs. 7-26 and 7-27.

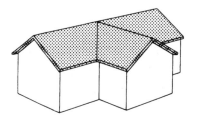

Fig. 7-26 *Gable-and-valley roof.* (Stanley Tools)

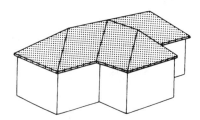

Fig. 7-27 *Hip-and-valley roof.* (Stanley Tools)

Hip roof This roof has four sides, all sloping toward the center of the building. The rafters run up diagonally to meet the ridge, into which the other rafters are framed. See Fig. 7-28.

Roof Terms

There are a number of terms you should be familiar with in order to work with roof framing. Each type of

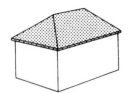

Fig. 7-28 Hip roof. (Stanley Tools)

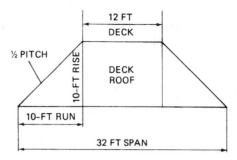

Fig. 7-30 Location of the deck roof. (Stanley Tools)

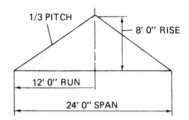

Fig. 7-31 One-third pitch. (Stanley Tools)

roof has its own particular terms; however, some of the common terms are:

Span The span of a roof is the distance over the wall plates. See Fig. 7-29.

Run The run of a roof is the shortest horizontal distance measured from a plumb line through the center of the ridge to the outer edge of the plate. See Fig. 7-29. In equally pitched roofs, the run is always equal to half the span or generally half the width of the building.

Rise The rise of a roof is the distance from the top of the ridge and of the rafter to the level of the foot. In figuring rafters, the rise is considered as the vertical distance from the top of the wall plate to the upper end of the measuring line. See Fig. 7-29.

Deck roof When rafters rise to a deck instead of a ridge, the width of the deck should be subtracted from the span. The remainder divided by 2 will equal the run. Thus, in Fig. 7-30 the span is 32 feet and the deck is 12 feet wide. The difference between 32 and 12 is 20 feet, which divided by 2 equals 10 feet. This is the run of the common rafters. Since the rise equals 10 feet, this is a ½-pitch roof.

Pitch The pitch of a roof is the slant or the slope from the ridge to the plate. It may be expressed in several ways:

1. The pitch may be described in terms of the ratio of the total width of the building to the total rise of the roof. Thus, the pitch of a roof having a 24-foot span with an 8-foot rise will be 8 divided by 24, which equals ⅓ pitch. See Fig. 7-31.

2. The pitch of a roof may also be expressed as so many inches of vertical rise to each foot of horizontal run. A roof with a 24-foot span and rising 8 inches to each foot of run will have a total rise of 8 × 12 = 96 inches or 8 feet. Eight divided by 24 equals ⅓. Therefore, the roof is ⅓ pitch. See Fig. 7-31.

Note that in Fig. 7-32 the building is 24 feet wide. It has a roof with a 6-foot rise. What is the pitch of the roof? The pitch equals 6 divided by 24, or ¼.

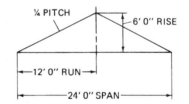

Fig. 7-32 One-fourth pitch. (Stanley Tools)

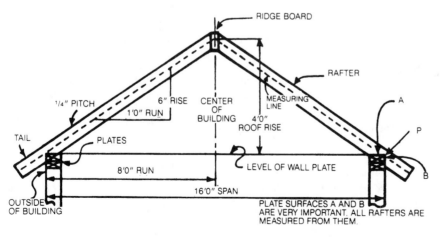

Fig. 7-29 Span, run, rise, and pitch. (Stanley Tools)

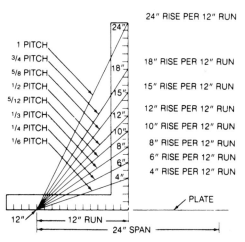

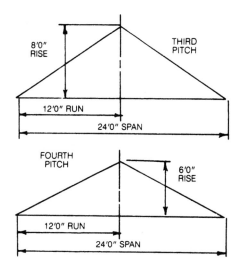

Fig. 7-33 *Principal roof pitches. (Stanley Tools)*

Principal roof pitches Figure 7-33 shows the principal roof pitches. They are called ½ pitch, ⅓ pitch, or ¼ pitch, as the case may be, because the height from the level of the wall plate to the ridge of the roof is one-half, one-third, or one-quarter of the total width of the building.

Keep in mind that roofs of the same width may have different pitches, depending upon the height of the roof.

Take a look at Fig. 7-34. This will help you interpret the various terms commonly used in roof construction.

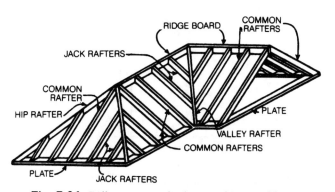

Fig. 7-34 *Different types of rafters used in a roof frame.*

Principal Roof Frame Members

The principal members of the roof frame are the plates at the bottom and the ridge board at the top. To them the various rafters are fastened. See Fig. 7-34.

Plate The plate is the roof member to which rafters are framed at their lower ends. The top, *A*, and the outside edge of the plate, *B*, are the important surfaces from which rafters are measured in Fig. 7-29.

Ridge board The ridge board is the horizontal member that connects the upper ends of the rafters on one side to the rafters on the opposite side. In cheap con-

struction the ridge board is usually omitted. The upper ends of the rafters are spiked together.

Common rafters A common rafter is a member that extends diagonally from the plate to the ridge.

Hip rafters A hip rafter is a member that extends diagonally from the corner of the plate to the ridge.

Valley rafters A valley rafter is one that extends diagonally from the plate to the ridge at the line of intersection of two roof surfaces.

Jack rafters Any rafter that does not extend from the plate to the ridge is called a jack rafter. There are different kinds of jacks. According to the position they occupy, they can be classified as hip jacks, valley jacks, or cripple jacks.

Hip jack A hip jack is a jack rafter with the upper end resting against a hip and the lower end against the plate.

Valley jack A valley jack is a jack rafter with the upper end resting against the ridge board and the lower end against the valley.

Cripple jack A cripple jack is a jack rafter with a cut that fits in between a hip-and-valley rafter. It touches neither the plate nor the ridge.

All rafters must be cut to proper angles so that they will fit at the points where they are framed. These different types of cuts are described below.

Top or plumb cut The cut of the rafter end which rests against the ridge board or against the opposite rafter is called the top or plumb cut.

Bottom or heel cut The cut of the rafter end which rests against the plate is called the bottom or heel cut. The bottom cut is also called the foot or seat cut.

Side cuts Hip-and-valley rafters and all jacks, besides having top and bottom cuts, must also have their sides at the end cut to a proper bevel so that they will fit into the other members to which they are to be framed. These are called side cuts or cheek cuts. All rafters and their cuts are shown in Fig. 7-35.

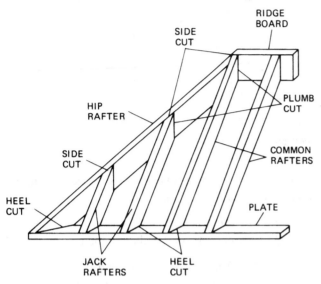

Fig. 7-35 *Rafter cuts.* (Stanley Tools)

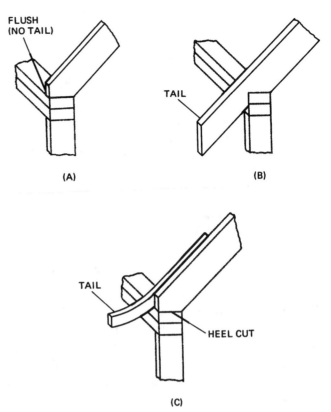

Fig. 7-36 *Tails or overhangs of rafters. (A) Flush tail (no tail); (B) Full tail; (C) Separate curved tail.*

RAFTERS
Layout of a Rafter

The *measuring line* of a rafter is a temporary line on which the length of the rafter is measured. This line runs parallel to the edge of the rafter and passes through the point *P* on the outer edge of the plate. Point *P* is where cut-lines *A* and *B* converged. This is the point from which all dimensions are determined. See Fig. 7-29.

Length The length of a rafter is the shortest distance between the outer edge of the plate and the center of the ridge line.

Tail That portion of the rafter extending beyond the outside edge of the plate is called the tail. In some cases it is referred to as the cave. The tail is figured separately and is not included in the length of the rafter as mentioned above. See Fig. 7-29.

Figure 7-36 shows the three variations of the rafter tail. Figure 7-36(A) shows the flush tail or no-tail. The rafter butts against the wall plate with no overhang. In Fig. 7-36(B), the full tail is shown. Note the overhang. Figure 7-36(C) shows the various shapes possible. This one indicates a separate tail that has been curved. It is nailed onto the no-tail or flush rafter.

All the cuts for the various types of common rafters are made at right angles to the sides of the rafter.

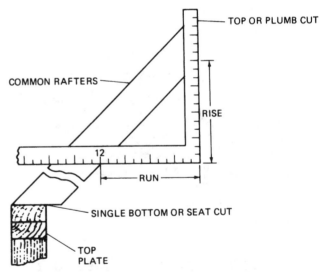

Fig. 7-37 *Using the steel square to mark off the top or plumb cut.*

Figure 7-37 shows how the framing square is used to lay out the angles. Find the 12-inch mark on the square. Note how the square is set at 12 for laying out the cut.

The distance 12 is the same as 1 foot of run. The other side of the square is set with the edge of the stock to the rise in inches per foot of run. In some cases the tail is not cut until after the rafter is in place. Then it is cut to match the others and aligns better for the fascia board.

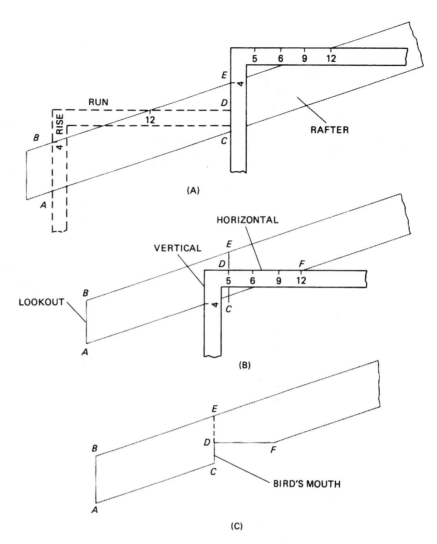

Fig. 7-38 *Using the steel square to lay out rafter lookout and bird's mouth.*

Figure 7-38 shows a method of using the square to lay out the bottom and lookout cuts. If there is a ridge board, you have to deduct one-half the thickness of the ridge from the rafter length. Figure 7-38 shows how to place the square for marking the rafter lookout. Scribe the cut line as shown in Fig. 7-38(A). The rise is 4 and the run is 12. Then move the square to the next position and mark from C to E. The distance from B to E is equal to the length of the lookout. Move the square up to E (with the same setting). Scribe line CE. On this line, lay off CD. This is the length of the vertical side of the bottom cut. Now apply the same setting to the bottom edge of the rafter. This is done so that the edge of the square cuts D. Scribe DF. See Fig. 7-38(B). This is the horizontal line of the bottom cut. In making this cut, the wood is cut out to the lines CD and DE. See Fig. 7-39(A) for an example of rafters cut this way. Note how the portable handsaw makes cuts beyond the marks.

In Figs. 7-39(B) and 7-40 you can see the rafters in place. Note the 90° angle. The rafter and the ridge should meet at 90°. The rafter in Fig. 7-40 has not had

Fig. 7-39A *Notice the saw cuts past the bird's mouth.*

the lookout cut. The overhang is slightly different from the type just shown. Can you find the difference?

Length per foot of run The rafter tables on the Stanley squares are based on the rise per foot run, which means that the figures in the tables indicate the length

Fig. 7-39B *Rafters in place. They are nailed to the ridge board, the wall plate, and the ceiling joists. The other side of the roof is already covered with plywood sheathing.*

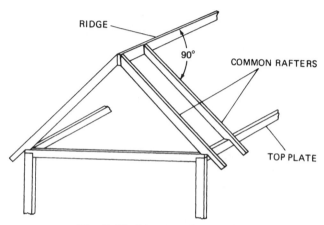

Fig. 7-40 *Common rafters in place.*

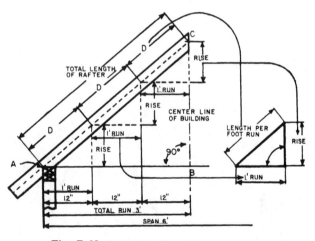

Fig. 7-41 *Length per foot of run.* (Stanley Tools)

of rafters per 1-foot run of common rafters for any rise of roof. This is shown in Fig. 7-41.

The roof has a 6-foot span and a certain rise per foot. The figure may be regarded as a right triangle *ABC*, having for its sides the run, the rise, and the rafter.

The run of the rafter has been divided into three equal parts, each representing 1-foot run.

It will be noted that by drawing vertical lines through each division point of the run, the rafter also will be divided into three equal parts *D*.

Since each part *D* represents the length of rafter per 1-foot run and the total run of the rafter equals 3 feet, it is evident that the total length of rafter will equal the length *D* multiplied by 3.

The reason for using this per-foot-run method is that the length of any rafter may be easily determined for any width of building. The length per foot run will be different for different pitches. Therefore, before you can find the length of a rafter, you must know the rise of roof in inches or the rise per foot of run.

> RULE: To find the rise per foot run, multiply the rise by 12 and divide by the length of the run.

The factor *12* is used to obtain a value in inches. The rise and run are expressed in feet. See Figs. 7-42 and 7-43.

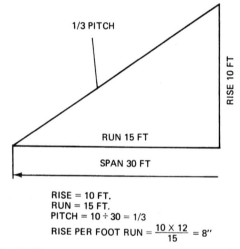

RISE = 10 FT.
RUN = 15 FT.
PITCH = 10 ÷ 30 = 1/3
$$\text{RISE PER FOOT RUN} = \frac{10 \times 12}{15} = 8''$$

Fig. 7-42 *Finding the rise per foot of run.* (Stanley Tools)

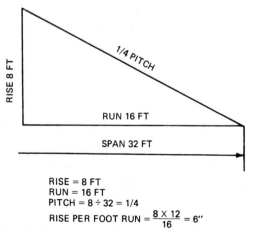

RISE = 8 FT
RUN = 16 FT
PITCH = 8 ÷ 32 = 1/4
$$\text{RISE PER FOOT RUN} = \frac{8 \times 12}{16} = 6''$$

Fig. 7-43 *Finding the rise per foot of run.* (Stanley Tools)

The rise per foot run is always the same for a given pitch and can be easily remembered for all ordinary pitches.

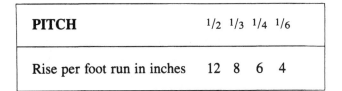

PITCH	1/2	1/3	1/4	1/6
Rise per foot run in inches	12	8	6	4

The members of a firmly constructed roof should fit snugly against each other. Rafters that are not properly cut make the roof shaky and the structure less stable. Therefore, it is very important that all rafters are the right lengths and that their ends are properly cut. This will provide a full bearing against the members to which they are connected.

Correct length, proper top and bottom cuts, and the right side or cheek cuts are the very important features to be observed when framing a roof.

Length of Rafters

The length of rafters may also be obtained in other ways. There are three in particular which can be used:

1. Mathematical calculations
2. Measuring across the square
3. Stepping off with the square

The first method, while absolutely correct, is very impractical on the job. The other two are rather unreliable and quite frequently result in costly mistakes.

The tables on the square have eliminated the need for using the three methods just mentioned. These tables let the carpenter find the exact length and cuts for any rafter quickly, and thus save time and avoid the possibility of errors.

Common Rafters

The common rafter extends from the plate to the ridge. Therefore, it is evident that the rise, the run, and the rafter itself form a right triangle. The length of a common rafter is the shortest distance between the outer edge of the plate and a point on the centerline of the ridge. This length is taken along the *measuring line*. The measuring line runs parallel to the edge of the rafter and is the hypotenuse or the longest side of a right triangle. The other sides of the triangle are the run and the rise. See Fig. 7-44.

The rafter tables on the face of the body of the square include the outside edge graduations on both

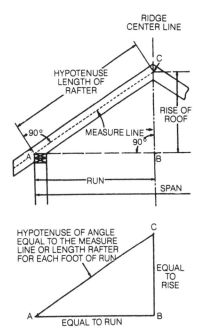

Fig. 7-44 *Measuring the line for a common rafter.* (Stanley Tools)

the body and the tongue, which are in inches and sixteenths of an inch.

The length of rafters The lengths of common rafters are found on the first line, indicated as *Length of main rafters per foot run*. There are seventeen of these tables, beginning at 2 inches and continuing to 18 inches. Figure 7-45 shows the square being used.

RULE: To find the length of a common rafter, multiply the length given in the table by the number of feet of run.

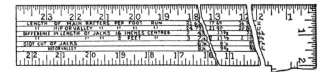

Fig. 7-45 *Note the labels for tables on the steel square.* (Stanley Tools)

For example, if you want to find the length of a common rafter where the rise of roof is 8 inches per foot run, or one-third pitch, and the building is 20 feet wide, you first find where the table is. Then on the inch line on the top edge of the body, find the figure that is equal to the rise of the roof, which in this case will be 8. On the first line under the figure 8 will be found 14.42. This is the length of the rafter in inches per foot run for this particular pitch. Examine Fig. 7-46.

The building is 20 feet wide. Therefore, the run of the rafter will be 20 divided by 2, which equals 10 feet.

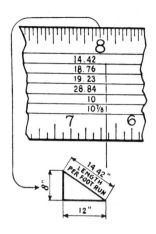

Fig. 7-46 *Finding the proper number on the square.* (Stanley Tools)

Since the length of the rafter per 1-foot run equals 14.42 inches, the total length of the rafter will be 14.42 multiplied by 10, which equals 144.20 inches, or 144.20 divided by 12, which equals 12.01 feet, or for all practical purposes, 12 feet. Check Fig. 7-46.

Top and bottom cuts of the common rafter The top or plumb cut is the cut at the upper end of the rafter where it rests against the opposite rafter or against the ridge board. The bottom cut or heel cut is the cut at the lower end which rests on the plate. See Fig. 7-47.

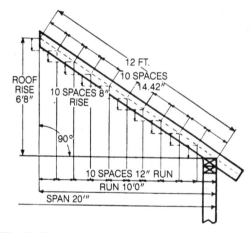

Fig. 7-47 *Finding the length of the rafter.* (Stanley Tools)

The top cut is parallel to the centerline of the roof, and the bottom cut is parallel to the horizontal plane of the plates. Therefore, the top and bottom cuts are at right angles to each other.

RULE: To obtain the top and bottom cuts of a common rafter, use 12 inches on the body and the rise per foot run on the tongue. Twelve inches on the body will give the horizontal cut, and the figure on the tongue will give the vertical cut.

To illustrate the rule, we will examine a large square placed alongside the rafter as shown in Fig. 7-48. Note that the edge of the tongue coincides with the top cut of the rafter. The edge of the blade coincides with the heel cut. If this square were marked in feet, it would show the run of the rafter on the body and the total rise on the tongue. Line *AB* would give the bottom cut and line *AC* would give the top cut.

However, the regular square is marked in inches. Since the relation of the rise to 1-foot run is the same as that the total rise bears to the total run, we use 12 inches on the blade and the rise per foot on the tongue to obtain the respective cuts. The distance 12 is used as a unit and is the 1-foot run, while the figure on the other arm of the square represents the rise per foot run. See Figs. 7-49 and 7-50.

Actual length of the rafter The rafter lengths obtained from the tables on the square are to the centerline of the ridge. Therefore, the thickness of half of the ridge board should always be deducted from the obtained total length before the top cut is made. See Fig. 7-51. This deduction of half the thickness of the ridge is measured at right angles to the plumb line and is marked parallel to this line.

Figure 7-52 shows the wrong and right ways of measuring the length of rafters. Diagram D shows the measuring line as the edge of the rafter, which is the case when there is no tail or eave.

Cutting the rafter After the total length of the rafter has been established, both ends should be marked and allowance made for a tail or eave. Don't forget to allow for half the thickness of the ridge.

Both cuts are obtained by applying the square so that the 12-inch mark on the body and the mark on the tongue that represents the rise are at the edge of the stock.

All cuts for common rafters are made at right angles to the side of the rafter.

For example, a common rafter is 12 feet 6 inches, and the rise per foot run is 9 inches. Obtain the top and bottom cuts. See Fig. 7-53.

Points *A* and *B* are the ends of the rafter. To obtain the bottom or seat cut, take 12 inches on the body of the square and 9 inches on the tongue. Lay the square on the rafter so that the body will coincide with point *A* or the lower end of the rafter. Mark along the body of the square and cut.

To obtain the top cut, move the square so that the tongue coincides with point *B*. This is the upper end of the rafter. Mark along the tongue of the square.

Deduction for the ridge The deduction for half the thickness of the ridge should now be measured. Half

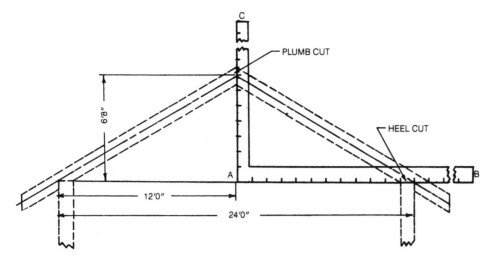

Fig. 7-48 *Using the steel square to check plumb and heel cuts.* (Stanley Tools)

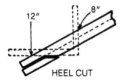

HEEL CUT

Fig. 7-49 *Using the steel square to lay out the heel cut.* (Stanley Tools)

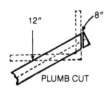

PLUMB CUT

Fig. 7-50 *Using the steel square to lay out the plumb cut.* (Stanley Tools)

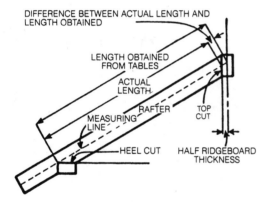

Fig. 7-51 *How to find the difference between the actual length and the length obtained.* (Stanley Tools)

Fig. 7-52 *Right and wrong ways of measuring rafters.* (Stanley Tools)

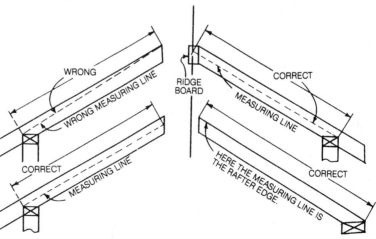

the thickness of the ridge is 1 inch. One inch is deducted at right angles to the top cut mark or plumb line, which is point *C*. A line is then drawn parallel to the top cut mark, and the cut is made. You will notice that the allowance for half the ridge measured along the measuring line is 1¼ inches. This will vary according to the rise per foot run. It is therefore important to measure for this deduction at right angles to the top cut mark or plumb line.

Measuring rafters The length of rafters having a tail or eave can also be measured along the back or top edge instead of along the measuring line, as shown in Fig. 7-54. To do this it is necessary to carry a plumb line to the top edge from *P*, and the measurement is started from this point.

Occasionally in framing a roof, the run may have an odd number of inches; for example, a building might have a span of 24 feet 10 inches. This would mean a run of 12'5". The additional 5 inches can be added easily without mathematical division after the length obtained from the square for 12 feet of run is measured. The additional 5 inches is measured at right

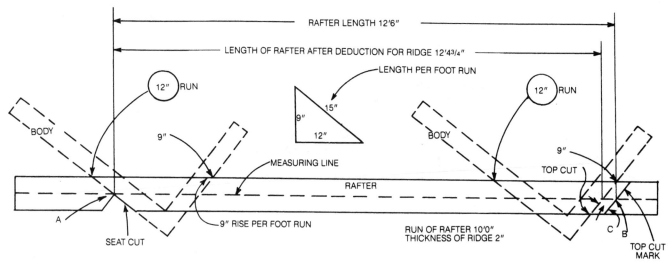

Fig. 7-53 *Applying the square to lay out cuts.* (Stanley Tools)

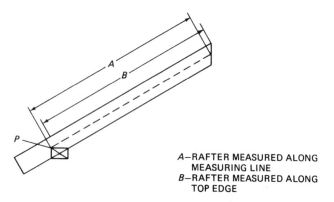

A—RAFTER MEASURED ALONG
MEASURING LINE
B—RAFTER MEASURED ALONG
TOP EDGE

Fig. 7-54 *Two places to measure a rafter.* (Stanley Tools)

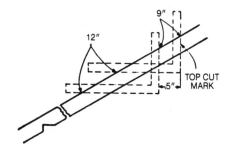

Fig. 7-55 *Adding extra inches to the length of a rafter.* (Stanley Tools)

angles to the last plumb line. See Fig. 7-55 for an illustration of the procedure.

Hip-and-Valley Rafters

The hip rafter is a roof member that forms a hip in the roof. This usually extends from the corner of the building diagonally to the ridge.

The valley rafter is similar to the hip, but it forms a valley or depression in the roof instead of a hip. It also extends diagonally from the plate to the ridge. Therefore, the total rise of the hip-and-valley rafters is the same as that of the common rafter. See Fig. 7-56.

The relation of hip-and-valley rafters to the common rafter is the same as the relation of the sides of a right triangle. Therefore, it will be well to explain here one of the main features of right triangles.

In a right triangle, if the sides forming the right angle are 12 inches each, the hypotenuse, or the side opposite the right angle, is equal to 16.97 inches. This is usually taken as 17 inches. See Fig. 7-57.

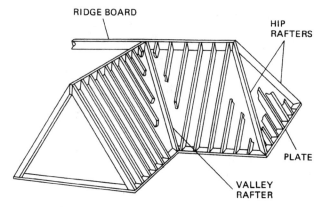

Fig. 7-56 *Hip-and-valley rafters.* (Stanley Tools)

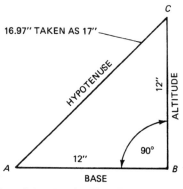

Fig. 7-57 *In a right triangle, 12-inch base and 12-inch altitude produces an isosceles triangle. This means the hypotenuse is 16.97 inches.* (Stanley Tools)

The position of the hip rafter and its relation to the common rafter is shown in Fig. 7-58, where the hip rafter is compared to the diagonal of a square prism. The prism (as shown in Fig. 7-58) has a base of 5 feet square, and its height is 3 feet 4 inches.

D is the corner of the building
BC is the total rise of the roof
AC is the common rafter
DB is the run of the hip rafter
DC is the hip rafter

It should be noted that the figure *DAB* is a right triangle whose sides are the portion of the plate *DA*, the run of the common rafter *AB*, and the run of the hip rafter *DB*. The run of the hip rafter is opposite right angle *A*. The hypotenuse is the longest side of the right triangle.

If we should take only 1 foot of run of common rafter and a 1-foot length of plate, we would have a right triangle *H*. The triangle's sides are each 12 inches long. The hypotenuse is 17 inches, or more accurately 16.97 inches. See Figs. 7-57 and 7-59.

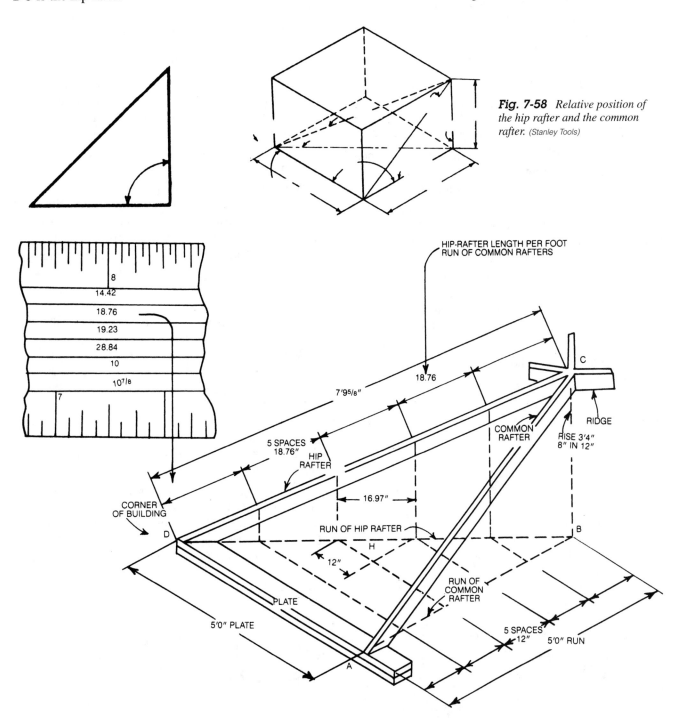

Fig. 7-58 *Relative position of the hip rafter and the common rafter.* (Stanley Tools)

Fig. 7-59 *Finding the length of the hip rafter. Note the location of the 18.76 on the square.* (Stanley Tools)

The hypotenuse in the small triangle *H* in Fig. 7-59 is a portion of the run of the hip rafter *DB*. It corresponds to a 1-foot run of common rafter. Therefore, the run of hip rafter is always 16.97 inches for every 12 inches of run of common rafter. The total run of the hip rafter will be 16.97 inches multiplied by the number of feet run of common rafter.

Length of hip-and-valley rafters Lengths of the hip-and-valley rafters are found on the second line of the rafter table. It is entitled *Length of hip or valley rafters per foot run*. This means that the figures in the table indicate the length of hip-and-valley rafters per foot run of common rafters. See Fig. 7-45.

> RULE: To find the length of a hip or valley rafter, multiply the length given in the table by the number of feet of the run of common rafter.

For example, find the length of a hip rafter where the rise of the roof is 8 inches per foot of run, or one-third pitch. The building is 10 feet wide. See Fig. 7-58.

Proceed as in the case of the common rafters. That is, find on the inch line of the body of the square the figure corresponding to the rise of roof—which is 8. On the second line under this figure is found 18.76. This is the length of hip rafter in inches for each foot of run of common rafter for one-third pitch. See Fig. 7-59.

The common rafter has a 5-foot run. Therefore, there are also five equal lengths for the hip rafter, as may be seen in Fig. 7-59. We have found the length of the hip rafter to be 18.76 inches per 1 foot run. Therefore, the total length of the hip rafter will be 18.76×5 = 93.80 inches. This is 7.81 feet, or for practical purposes 7'9¹³⁄₁₆" or 7'9⅝".

Top and bottom cuts The following rule should be followed for top and bottom cuts.

> RULE: To obtain the top and bottom cut for hip or valley rafters, use 17 inches on the body and the rise per foot run on the tongue. Seventeen on the body will give the seat cut, and the figure on the tongue will give the vertical or top cut. See Fig. 7-60.

Measuring hip-and-valley rafters The length of all hip-and-valley rafters must always be measured along the center of the top edge or back. Rafters with a tail or eave are treated like common rafters, except that the measurement or measuring line is at the center of the top edge.

Deduction from hip or valley rafter for ridge The deduction for the ridge is measured in the same way as for the common rafter (see Fig. 7-53), except that half the diagonal (45') thickness of the ridge must be used.

Side cuts In addition to the top and bottom cuts, hip-and-valley rafters must also have side or check cuts at the point where they meet the ridge.

These side cuts are found on the sixth or bottom line of the rafter tables. It is marked *Side cut hip or valley—use*. The figures given in this line refer to the graduation marks on the outside edge of the body. See Fig. 7-45.

The figures on the square have been derived by determining the figure to be used with 12 on the tongue for the side cuts of the various pitches by the following method. From a plumb line, the thickness of the rafter is measured and marked as at *A* in Fig. 7-61. A line is then squared across the top of the rafter and the diagonal points connected as at *B*. The line *B* or side cut is obtained by marking along the tongue of the square.

> RULE: To obtain the side cut for hip or valley rafters, take the figure given in the table on the body of the square and 12 inches on the tongue. Mark the side cut along the tongue where the tongue coincides with the point on the measuring line.

For example, find the side cut for a hip rafter where the roof has 8 inches rise per foot run or pitch. See Fig. 7-62.

Figure 7-62 represents the position of the hip rafter on the roof. The rise of the roof is 8 inches to the foot. First, locate the number 8 on the outside edge of the body. Under this number in the bottom line you will find 10⅞. This figure is taken on the body and 12

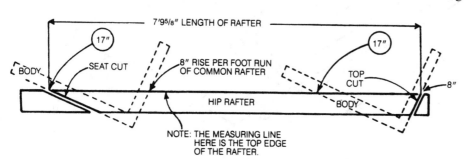

Fig. 7-60 *Using the square to lay out top and seat cuts on a hip rafter.* (Stanley Tools)

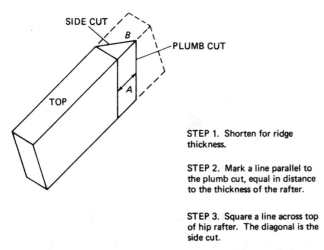

STEP 1. Shorten for ridge thickness.

STEP 2. Mark a line parallel to the plumb cut, equal in distance to the thickness of the rafter.

STEP 3. Square a line across top of hip rafter. The diagonal is the side cut.

Fig. 7-61 *Making side cuts so that the hip will fit into the intersection of rafters.* (Stanley Tools)

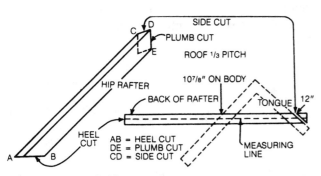

Fig. 7-62 *Hip rafter cuts.* (Stanley Tools)

inches is taken on the tongue. The square is applied to the edge of the back of the hip rafter. The side cut *CD* comes along the tongue.

In making the seat cut for the hip rafter, an allowance must be made for the top edges of the rafter. They would project above the line of the common and jack rafters if the corners of the hip rafter were not removed or backed. The hip rafter must be slightly lowered. Do this by cutting parallel to the seat cut. The distance varies with the thickness and pitch of the roof.

It should be noted that on the Stanley square the 12-inch mark on the tongue is always used in all angle cuts—top, bottom, and side. This leaves the worker with only one number to remember when laying out side or angle cuts. That is the figure taken from the fifth or sixth line in the table.

The side cuts always come on the right-hand or tongue side on rafters. When you are marking boards, these can be reversed for convenience at any time by taking the 12-inch mark on the body and using the body references on the tongue.

Obtain additional inches in run of hip or valley rafters by using the explanation given earlier for common rafters. However, use the diagonal (45°) of the ad-

ditional inches. This is approximately 7⅟₁₆ inches for 5 inches of run. This distance should be measured in a similar manner.

Jack Rafters

Jack rafters are *discontinued* common rafters. They are common rafters cut off by the intersection of a hip or valley before reaching the full length from plate to ridge.

Jack rafters lie in the same plane as common rafters. They usually are spaced the same and have the same pitch. Therefore, they also have the same length per foot run as common rafters have.

Jack rafters are usually spaced 16 inches or 24 inches apart. Because they rest against the hip or valley equally spaced, the second jack must be twice as long as the first one, the third three times as long as the first, and so on. See Fig. 7-63.

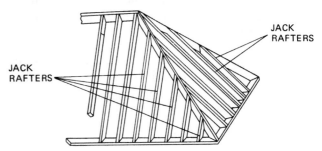

Fig. 7-63 *Location of jack rafters.* (Stanley Tools)

Length of jack rafters The lengths of jacks are given in the third and fourth lines of the rafter tables on the square. They are indicated:

Third line: Difference in length of jacks—16 inch centers

Fourth line: Difference in length of jacks—2 foot centers

The figures in the table indicate the length of the first or shortest jack, which is also the difference in length between the first and second jacks, between the second and third jacks, and so on.

> RULE: To find the length of a jack rafter, multiply the value given in the tables by the number indicating the position of the jack. From the obtained length, subtract half the diagonal (45°) thickness of the hip or valley rafter.

For example, find the length of the second jack rafter. The roof has a rise of 8 inches to 1 foot of run of common rafter. The spacing of the jacks is 16 inches.

On the outer edge of the body, find the number 8, which corresponds to the rise of the roof. On the third line under this figure find 19¼. This means that the first jack rafter will be 19¼ inches long. Since the length of the second jack is required, multiply 19¼ by 2, which equals 38½ inches. From this length half the diagonal (45°) thickness of the hip or valley rafter should be deducted. This is done in the same manner that the deduction for the ridge was made on the hip rafter.

Proceed in the same manner when the lengths of jacks spaced on 24-inch centers are required. It should be borne in mind that the second jack is twice as long as the first one. The third jack is three times as long as the first jack, and so on.

Top and bottom cuts for jacks Since jack rafters have the same rise per foot run as common rafters, the method of obtaining the top and bottom cuts is the same as for common rafters. That is, take 12 inches on the body and the rise per foot run on the tongue. Twelve inches will give the seat cut. The figure on the tongue will give the plumb cut.

Side cut for jacks At the end where the jack rafter frames to the hip or valley rafter, a side cut is required. The side cuts for jacks are found on the fifth line of the rafter tables on the square. It is marked: *Side cut of jacks—use*. See Fig. 7-45.

> RULE: To obtain the side cut for a jack rafter, take the figure shown in the table on the body of the square and 12 inches on the tongue. Mark along the tongue for side cut.

For example, find the side cut for jack rafters of a roof having 8 inches rise per foot run, or ⅓ pitch. See Figs. 7-64 and 7-65. Under the number 8 in the fifth line of the table find 10. This number taken on the outside edge of the body and 12 inches taken on the tongue will give the required side cut.

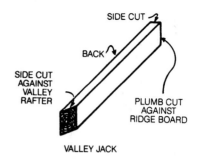

Fig. 7-65 *Valley jack rafter cuts.* (Stanley Tools)

BRACE MEASURE

In all construction there is the need for some braces to make sure certain elements are held securely. See Fig. 7-120(A). The brace measure table is found along the center of the back of the tongue of the carpenter's square. It gives the lengths of common braces. See Fig. 7-66.

Fig. 7-66 *Brace measure table on back side of steel square tongue.* (Stanley Tools)

For example, find the length of a brace whose run on post and beam equals 39 inches. See Fig. 7-67. In the brace table find the following expression:

$$\begin{matrix} 39 \\ & 55.15 \\ 39 \end{matrix}$$

This means that with a 39-inch run on the beam and a 39-inch run on the post, the length of the brace will be 55.15 inches. For practical purposes, use 55⅛ inches.

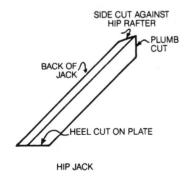

Fig. 7-64 *Hip jack rafter cuts.* (Stanley Tools)

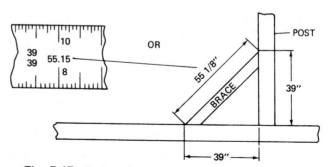

Fig. 7-67 *Cutting a brace using a square table.* (Stanley Tools)

Braces may be regarded as common rafters. Therefore, when the brace run on the post differs from the run on the beam, their lengths as well as top and bottom cuts may be determined from the figures given in the tables of the common rafters.

ERECTING THE ROOF WITH RAFTERS

Rafters are cut to fit the shape of the roof. Roofs are chosen by the builder or planner. The design of the rafter is determined by the pitch, span, and rise of the type of roof chosen. The gable roof is simple. It can be made easily with a minimum of difficult cuts. In this example we start with the gable type and then look at other variations of rooflines. See Fig. 7-68 for an example of the gable roof.

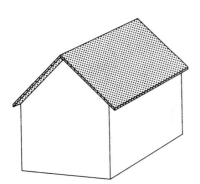

Fig. 7-68 *Gable roof.*

Rafter Layout

One of the most important tools used to lay out rafters is the carpenter's square. All rafters must be cut to the proper angle or bevel. They fasten to the wall plate or to the ridge board. In some cases there is an overhang. This overhang must be taken into consideration when the rafter is cut. Gable siding, soffits, and overhangs are built together (Fig. 7-69).

Fig. 7-69 *Gable siding, soffits, and gable overhangs can be built together. (Fox and Jacobs)*

Raising Rafters

Mark rafter locations on the top plate of the side walls. The first rafter pair will be flush with the outside edge of the end wall. See Fig. 7-70. Note the placement of the gable end studs. The notch in the gable end stud is made to fit the 2 × 4, or whatever thickness of rafter you are using. Space the first interior rafter 24 inches, measured from the end of the building to the center of the rafter. In some cases 16 inches O.C. is used for spacing. All succeeding rafter locations are measured 24 inches center to center. They will be at the sides of ceiling joist ends. See Fig. 7-71.

Next, mark rafter locations on the ridge board. Allow for the specified gable overhang. To achieve the required total length of ridge board, you may have to

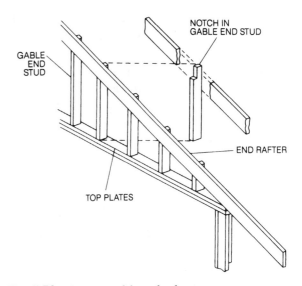

Fig. 7-70 *Placement of the end rafter. (American Plywood Association)*

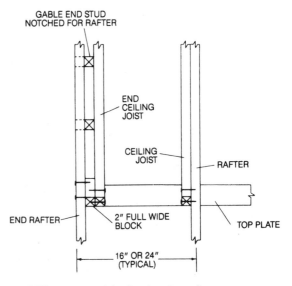

Fig. 7-71 *Spacing of the first interior rafter. (American Plywood Association)*

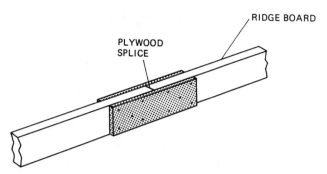

Fig. 7-72 *Method of splicing a ridge board.* (American Plywood Association)

splice it. See Fig. 7-72. Do not splice at this time. It is easier to erect it in shorter sections, then splice it after it is in place.

Check your house plan for roof slope. For example, a 4-inch rise in 12 inches of run is common. It is usually considered the minimum for asphalt or wood shingles.

Lay out one pair of rafters as previously shown. Mark the top and bottom angles and seat-cut location. Make the cuts and check the fit by setting them up at floor level. Mark this set of rafters for identification and use it as a pattern for the remainder.

Cut the remaining rafters. For a 48-foot house with rafters spaced 24 inches O.C. you will need 24 pairs cut to the pattern (25 pairs counting the pattern). In addition, you will need two pairs of fascia rafters for the ends of the gable overhang. See Fig. 7-73. Since they cover the end of the ridge board, they must be longer than the pattern rafters by half the width of the ridge board. Fascia rafters have the same cuts at the top and bottom as the regular rafters. However, they do not have a seat cut.

Build temporary props of 2 × 4s to hold the rafters and ridge board in place during roof framing installation. The props should be long enough to reach from the top plate to the bottom of the ridge board. They should be fitted with a plywood gusset at the bottom. When the props are installed, the plywood gusset is nailed temporarily to the top plate or to a ceiling joist. The props are also diagonally braced from about midpoint in both directions to maintain true vertical (check with a plumb bob). See Fig. 7-74.

Move the ridge board sections and rafters onto the ceiling framing. Lay plywood panels over the ceiling joists for a safe footing. First erect the ridge board and the rafters nearest its ends. See Fig. 7-74. If the ridge

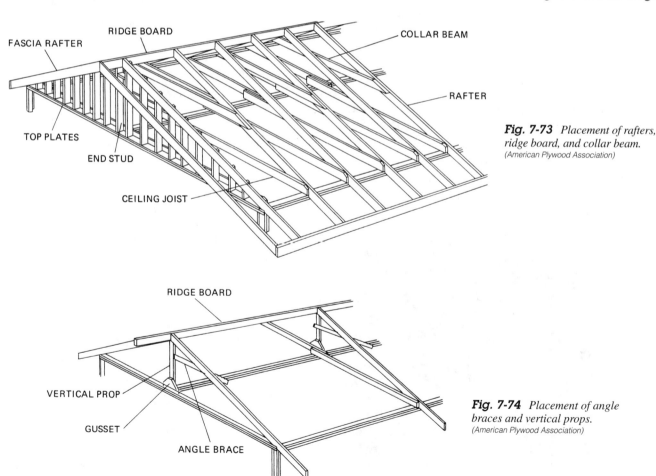

Fig. 7-73 *Placement of rafters, ridge board, and collar beam.* (American Plywood Association)

Fig. 7-74 *Placement of angle braces and vertical props.* (American Plywood Association)

of the house is longer than the individual pieces of ridge board, you'll find it easier to erect each piece separately, rather than splice the ridge board full-length first. Support the ridge board at both ends with the temporary props. Toenail the first rafter pair securely to the ridge board using at least two 8d nails on each side. Then nail it at the wall. Install the other end rafter pair in the same manner.

Make the ridge board joints, using plywood gussets on each side of the joint. Nail them securely to the ridge board.

Check the ridge board for level. Also check the straightness over the centerline of the house.

After the full length of the ridge board is erected, put up the remaining rafters in pairs. Nail them securely in place. Check them occasionally to make sure the ridge board remains straight. If all rafters are cut and assembled accurately, the roof should be self-aligning.

Toenail the rafters to the wall plate with l0d nails. Use two per side. Also nail the ceiling joists to the rafters. For a 24-foot-wide house, you will need (our 16d nails) at each lap. In high-wind areas, it is a good idea to add metal-strap fasteners for extra uplift resistance. See Fig. 7-75.

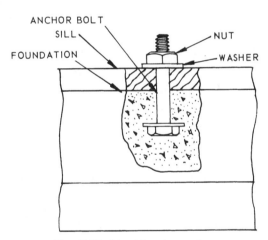

Fig. 7-75 *Metal framing anchors.*

Cut and install 1 × 6 collar beams at every other pair of rafters (4 feet O.C.). See Fig. 7-73. Nail each end with four 8d nails. Collar beams should be in the upper third of the attic crawl space. Remove the temporary props.

Square a line across the end wall plate directly below the ridge board. If a vent is to be installed, measure half its width on each side of the center mark to locate the first stud on each side. Mark the positions for the remaining studs at 16 inches O.C. Then measure and cut the studs. Notch the top end to fit under

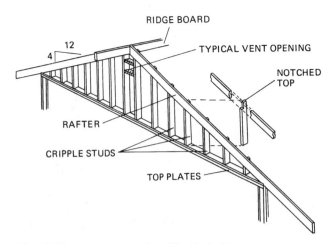

Fig. 7-76 *Vent openings should be blocked in.* (American Plywood Association)

the rafter so that the stud bottom will be flush with the top plate. Cut the cripple studs and headers to frame in the vent opening. See Fig. 7-76.

Cut and install fascia board to the correct length of the ridge board. Bevel the top edge to the roof slope. Nail the board to the rafter ends. Then, install fascia rafters. Fascia rafters cover the end of the ridge board. See Fig. 7-73. Where the nails will be exposed to weather, use hot-dipped galvanized or other nonstaining nails.

SPECIAL RAFTERS

There are some rafters needed to make special roof shapes. The mansard roof and the hip roof both call for special rafters. Jack rafters are needed for the hip roof. This type of roof may also have valleys and have to be treated especially well. Dormers call for some special rafters, too. For bay windows and other protrusions, some attention may have to be given to rafter construction.

Dormers

Dormers are protrusions from the roof. They stick out from the roof. They may be added to allow light into an upstairs room. Or, they may be added for architectural effect. See Fig. 7-77. Dormers may be made in three types. They are:

1. Dormers with flat, sloping roofs that have less slope than the roof in which they are located. This can be called a shed-type dormer (Fig. 7-78).

2. Dormers with roofs of the gable type at right angles to the roof (Fig. 7-79). No slope in this one.

3. The two types can be combined. This is called the hip-roof dormer.

Fig. 7-77 *A dormer.*

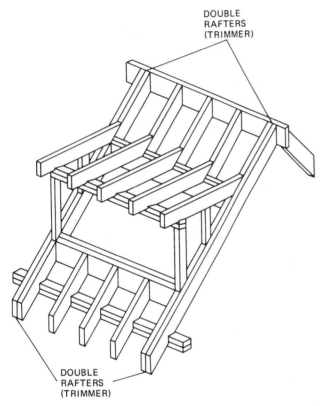

Fig. 7-78 *Shed dormer.*

Bay Windows

Bay windows are mostly for decoration. They add to the architectural qualities of a house. They stick out from the main part of the house. This means they have to have special handling. The floor joists are extended out past the foundation the required amount. A band is then used to cap off the joist ends. See Fig. 7-80. Take a closer look at the ceiling joists and rafters for the bay. The rafters are cut according to the rise called for on the plans. Cuts and lengths have already been discussed. No special problems should be presented by

this method of framing. In order to make it easier, it is best to lay out the rafter plan at first so that you know which are the common and which are the hip rafters. In some cases you may need a jack rafter or two depending upon the size of the bay. See Fig. 7-81.

CEILING JOISTS

Ceiling joists serve a number of purposes. They keep the wall from falling inward or outward. They anchor the rafters to the top plate. Ceiling joists also hold the finished ceiling in place. The run of the joist is important. The distance between supports for a joist determines its size. In some cases the ceiling joist has to be spliced. Figure 7-82 shows one method of splicing. Note how the splice is made on a supporting partition. Figure 7-83 shows how the joists fit on top of the plate. In some cases it is best to tie the joist down to the top plate by using a framing bracket. Figure 7-84 shows how one type of bracket is used to hold the joist. This helps in high-wind areas.

The ceiling joists in Fig. 7-85 (A) and (B) have been trimmed to take the angle of the rafter into consideration once the rafter is attached to the top plate and joist.

The size of the joist is determined by the local code. However, there are charts that will give you some idea of what size piece of dimensional lumber to use. Table 7-4 indicates some allowable spans for ceiling joists. These are given using non-stress graded lumber.

There are some special arrangements for ceiling joists. In some cases it is necessary to interrupt the free flow of lines represented by joists. For example, a chimney may have to be allowed for. An attic opening may be called for on the drawings. These openings have to be reinforced to make sure the joists maintain their ability to support the ceiling and some weight in the ceiling at a later time. See Fig. 7-86.

Figure 7-87 shows how framing anchors are used to secure the joists to the double header. This method can be used for both attic openings and fireplace openings.

OPENINGS

As mentioned before, fireplaces do come out through the roof. This must be allowed for in the construction process. The floor joists have to be reinforced. The area around the fireplace has to be strong enough to hold the hearth. However, what we're interested in here is how the fireplace opening comes through the rafters. See Fig. 7-88. The roofing here has been boxed

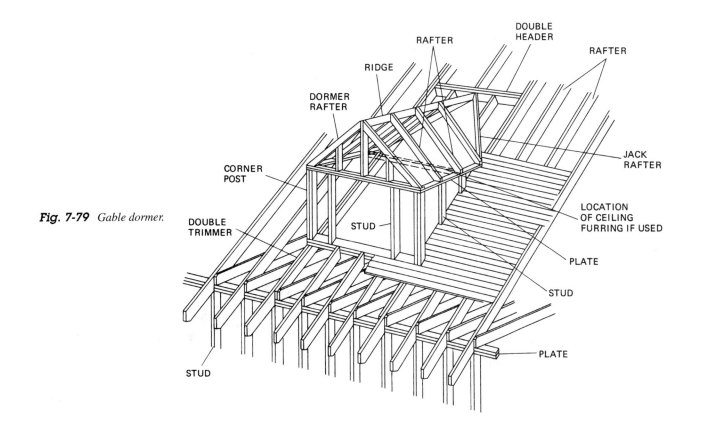

Fig. 7-79 *Gable dormer.*

RIDGE

RAFTER

DOUBLE HEADER

RAFTER

DORMER RAFTER

CORNER POST

JACK RAFTER

DOUBLE TRIMMER

STUD

LOCATION OF CEILING FURRING IF USED

PLATE

STUD

STUD

PLATE

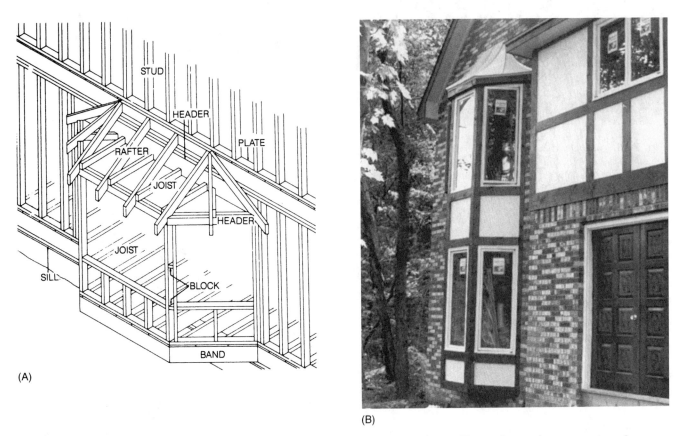

STUD

HEADER

PLATE

RAFTER

JOIST

HEADER

JOIST

SILL

BLOCK

BAND

(A)

(B)

Fig. 7-80 *(A) Bay window framing; (B) Two bay windows stacked for a two-story house. The metal cover does not require rafters.*

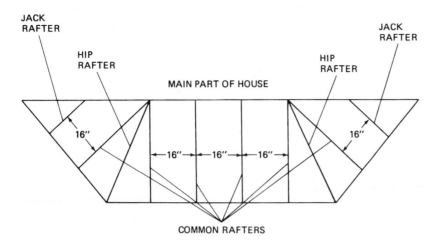

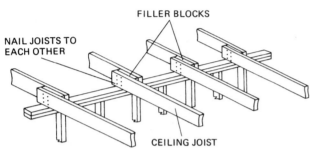

Fig. 7-81 *Rafter layout for a bay window.*

Fig. 7-82 *Ceiling joist splices are made on a supporting partition.*

(A)

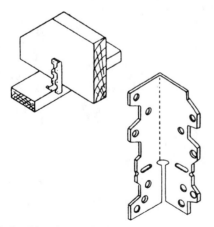

Fig. 7-83 *Looking up toward the ceiling joists. Notice how these are supported on the wall plate and are not cut or tapered. Two nails are used to toenail the joists to the plate.*

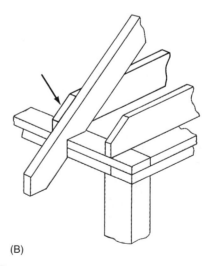

(B)

Fig. 7-84 *Steel bracket used to hold ceiling joist to the top plate. Note how it is bent to fit.*

Fig. 7-85 *(A) Ceiling joists in place on the top plate. Note the cuts on the ends of the joists. (B) Set the first ceiling joist on the inside of the end of the wall. This will allow the ceiling drywall to be nailed to it later.*

Table 7-4 Ceiling Joists

Size of Ceiling Joists, Inches	Spacing of Ceiling Joists, Inches	Maximum Allowable Span			
		Group I	Group II	Group III	Group IV
2 × 4	12 16	11'6" 10'6"	11'0" 10'0"	9'6" 8'6"	5'6" 5'0"
2 × 6	12 16	18'0" 16'0"	16'6" 15'0"	15'6" 14'6"	12'6" 11'0"
2 × 8	12 16	24'0" 21'6"	22'6" 20'6"	21'0" 19'0"	19'0" 16'6"

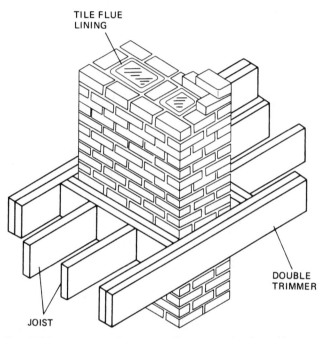

Fig. 7-86 Blocking in the joists to allow an opening for a chimney.

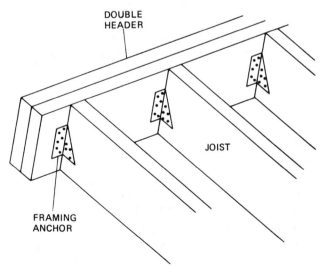

Fig. 7-87 Using framing anchors to hold the tail joists to the header.

Fig. 7-88 Arrows show the openings in the roof for the chimney.

off to allow the fireplace to come through. The chimney at the top has the flashing ready for installation as soon as the bricks are laid around the flue.

Other openings are for soil pipes. These are used for venting the plumbing system. A hole in the plywood deck is usually sufficient to allow their exit from the inside of the house. See Fig. 7-89.

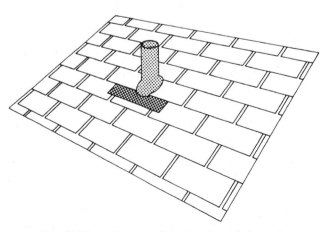

Fig. 7-89 Soil pipe stack coming through the roof.

DECKING

A number of types of roof deckings are used. One is plywood. It is applied in 4-×-8-foot sheets. Plywood adds structural strength to the building. Plywood also saves time, since it can be placed on the rafters rather quickly.

Boards of the 1-×-6- or 1-×-8-inch size can be used for sheathing. This decking takes a longer time to apply. Each board has to be nailed and sawed to fit. This type of decking adds strength to the roof also.

Another type of decking is nothing more than furring strips applied to the rafters. This is used as a nail base for cedar shingles.

Plywood Decking

Roof sheathing, the covering over rafters or trusses, provides structural strength and rigidity. It makes a solid base for fastening the roofing material.

A roof sheathing layout should be sketched out first to determine the amount of plywood needed to cover the rafters. See Fig. 7-90.

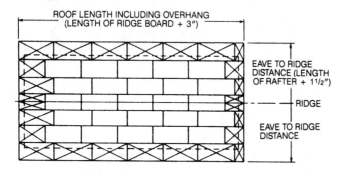

NOTE: For "open soffits" all panels marked with an "x" must be EXT-DFPA "soffit" panels.

Fig. 7-90 *Roof sheathing layout for a plywood deck.* (American Plywood Association)

Draw your layout. It may be a freehand sketch, but it should be relatively close to scale. The easiest method is to draw a simple rectangle representing half of the roof. The long side will represent the length of the ridge board. Make the short side equal to the length of your rafters, including the overhangs. If you have open soffits, draw a line inside the ends and bottoms. Use a dotted line as shown in Fig. 7-90. This area is to be covered by *exterior* plywood. Remember that this is only half of the roof. Any cutting of panels on this side can be planned so that the cut-off portions can be used on the other side. If your eave overhang is less than 2 feet and you have an open soffit, you may wish to start with a half panel width of soffit plywood. Figure 7-91 shows

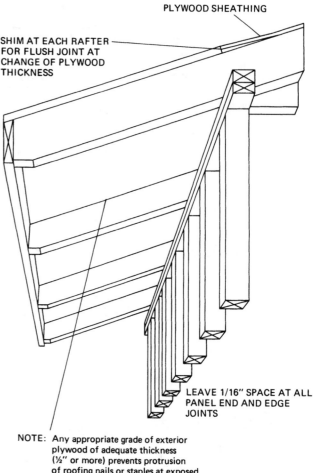

NOTE: Any appropriate grade of exterior plywood of adequate thickness (½" or more) prevents protrusion of roofing nails or staples at exposed underside, and carries design roof load.

Fig. 7-91 *Open soffit.* (American Plywood Association)

the open soffit. Figure 7-92 shows the boxed soffit. Otherwise you will probably start with a full 4-×-8-foot sheet of plywood at the bottom of the roof and work upward toward the ridge. This is where you may have to cut the last row of panels. Stagger panels in succeeding rows.

Complete your layout for the whole roof. The layout shows panel size and placement as well as the number of sheathing panels needed. This is shown in Fig. 7-90.

If your diagram shows that you will have a lot of waste in cutting, you may be able to reduce scrap by slightly shortening the rafter overhang at the eave, or the gable overhang.

An example is shown in Fig. 7-90, where nearly half of the panels are "soffit" panels. In such a case, rather than using shims to level up soffit and interior sheathing panels, you may want to use interior sheathing panels of the same thickness as your soffit panels, even though they might then be a little thicker than the minimum required.

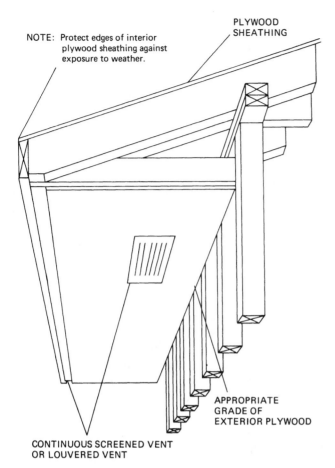

NOTE: Protect edges of interior plywood sheathing against exposure to weather.

PLYWOOD SHEATHING

APPROPRIATE GRADE OF EXTERIOR PLYWOOD

CONTINUOUS SCREENED VENT OR LOUVERED VENT

Fig. 7-92 *Boxed soffit. (American Plywood Association)*

Fig. 7-93 *Using a stapler to fasten plywood decking to the rafters. (American Plywood Association)*

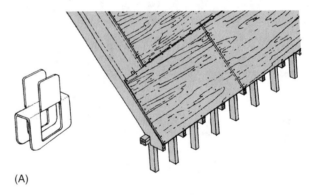

(A)

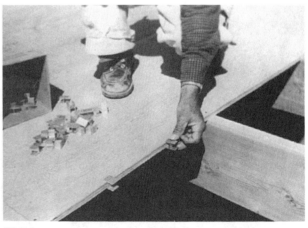

(B)

Fig. 7-94 *Plywood clips reinforce the surface area where the sheathing butts.*

Cut panels as required, marking the cutting lines first to ensure square corners.

Begin panel placement at any corner of the roof. If you are using special soffit panels, remember to place them best or textured side down.

Fasten each panel in the first course (row), in turn, to the roof framing using 6d common smooth, ring-shank, or spiral-threaded nails. Space nails 6 inches O.C. along the panel ends and 12 inches O.C. at intermediate supports.

Leave a $\frac{1}{16}$-inch space at panel ends and $\frac{1}{8}$ inch at edge joints. In areas of high humidity, double the spacing.

Apply the second course, using a soffit half panel in the first (overhang) position. If the main sheathing panels are thinner than the soffit sheathing, install small shims to ease the joint transition. See Fig. 7-91 for location of the shims.

Apply the remaining courses as above.

Note that if your plans show closed soffits, the roof sheathing will all be the same grade thickness. To apply plywood to the underside of closed soffits, use non-staining-type nails.

Figure 7-93 shows plywood decking being applied with a stapler. Figure 7-94A and B shows an H clip for plywood support along the long edges. This gives extra support for the entire length of the panels.

Figure 7-95 shows the erection of the sidewalls to a house. In Fig. 7-96 the plywood sheathing has been placed on one portion of the rafters. Note the rig (arrow) that allows a sheet of plywood to be passed up from the ground to roof level. The sheet is first placed on the rack. Then it is taken by the person on the roof and moved over to the needed area.

Fig. 7-95 *Erection of sidewalls to a house.*

Fig. 7-96 *Plywood sheathing placed on one portion of rafters. Note the ladder made for plywood lifting.*

Figure 7-97 shows some boxed soffit. Note the louvers already in the board. The temporary supports hold the soffit in place until final nailing is done and it can be supported by the fascia board. The fascia has a groove along its entire length to allow the soffit to slide into it.

Boards for Decking

Lumber may be used for roof decking. In fact, it is necessary in some special-effects ceilings. It is needed where the ceiling is exposed and the underside of the decking is visible from below or inside the room.

Roof decking comes in a variety of sizes and shapes. See Fig. 7-98. A 2° angle is cut in the lumber decking ends to ensure a tight joint (Fig. 7-99). This type of decking is usually nailed, and so the nails must be concealed. This requires nailing as shown in Fig. 7-100. Eight-inch spikes are usually used for this type of nailing. Note the chimney opening in Fig. 7-101.

Figure 7-102 illustrates the application of 1-×-6 or 1-×-8 sheathing to the rafters. Note the two nails used to hold the boards down. The common nail is used here. In the concealed nailing it takes a finishing nail to be completely concealed.

Once the decking is in place, it is covered by an underlayment of felt paper. This paper is then covered by shingles as selected by the builder.

Shingle Stringers

For cedar shingles the roof deck may be either spaced or solid. The climatic conditions determine if it is a solid deck or one that is spaced. In areas where there are blizzards and high humidity, the spaced deck is not used. In snow-free areas, spaced sheathing is practical.

Fig. 7-97 *Louvered soffit held in place with temporary braces.*

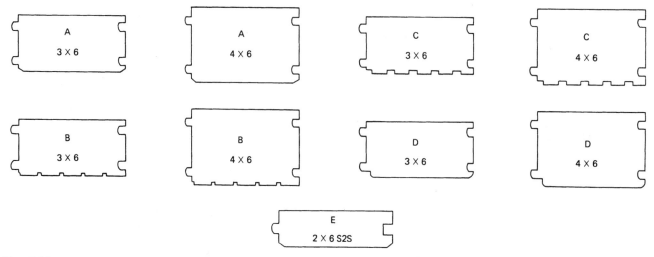

Fig. 7-98 *Different sizes and shapes of roof decking made of lumber. (A) Shows the regular V-jointed decking; (B) Indicates the straited decking; (C) Shows the grooved type; (D) Illustrates the eased joint or bullnosed type; (E) Indicates single tongue-and-groove with a V joint.*

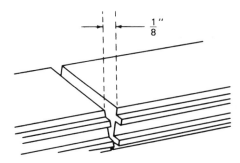

Fig. 7-99 *A 2° angle is cut in the lumber decking ends. This ensures a tight face joint on the exposed ceiling below.*

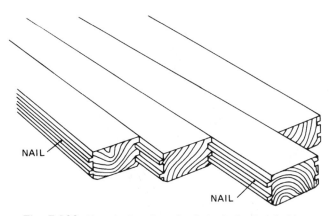

Fig. 7-100 *Note the location of nails in the lumber decking.*

Use 1 × 6s spaced on centers equal to the weather exposure at which the shakes are to be laid. However, the spacing should not be over 10 inches. In areas where wind-driven snow is encountered, solid sheathing is recommended. See Fig. 7-102 for an example of spaced sheathing.

Roof pitch and exposure Handsplit shakes should be used on roofs where the slope or pitch is sufficient to ensure good drainage. Minimum recommended

pitch is ⅙ or 4-in-12 (4-inch vertical rise for each 12-inch horizontal run), although there have been satisfactory installations on lesser slopes. Climatic conditions and skill and techniques of application are modifying factors.

Maximum recommended weather exposure is 10 inches for 24-inch shakes and 7½ inches for 18-inch shakes. A superior three-ply roof can be achieved at slight additional cost if these exposures are reduced to 7½ inches for 24-inch shakes and 5½ inches for 18-inch shakes.

Figure 7-103 shows the shakes in place on spaced sheathing. Note how the amount of exposure to the weather makes a difference in the spacing of the sheathing. Figure 7-104(A) and (B) gives a better view of the spaced sheathing and how the roofing is applied to it.

CONSTRUCTING SPECIAL SHAPES

The gambrel shape is familiar to most people, since it is the favorite shape for barns. It consists of a double-slope roof. This allows for more space in the attic or upper story. More can be stored there. In modern home designs, this type of roof has been used to advantage. It gives good headroom for an economical structure with two stories. This design was brought to the United States by Germans in the early days of the country.

Gambrel-Shaped-Roof Storage Shed

A storage shed will give you an idea of the simplest way to utilize the gambrel-shaped roof. Examine the details and then obtain the equipment and supplies needed. The bill of materials lists the supplies that are

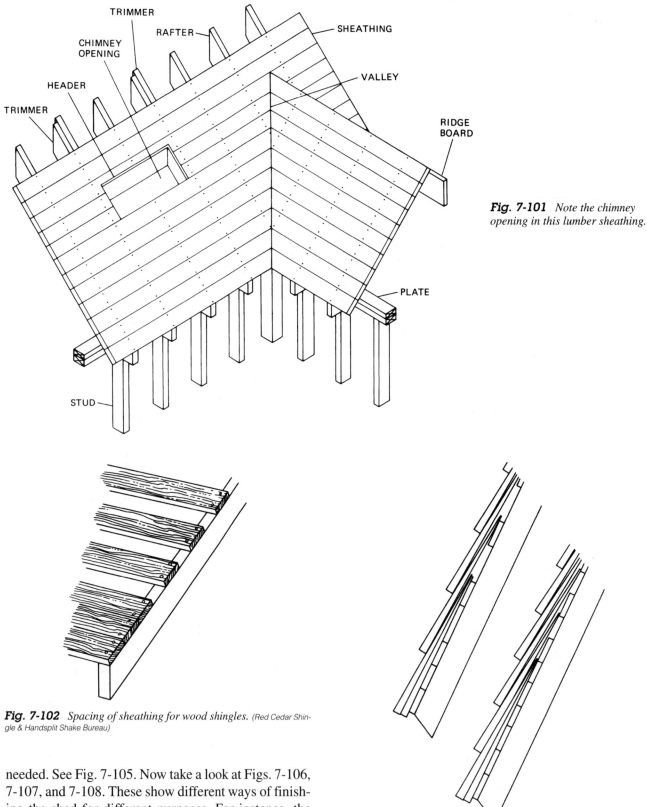

Fig. 7-101 Note the chimney opening in this lumber sheathing.

TRIMMER
CHIMNEY OPENING
RAFTER
HEADER
TRIMMER
SHEATHING
VALLEY
RIDGE BOARD
PLATE
STUD

Fig. 7-102 Spacing of sheathing for wood shingles. (Red Cedar Shingle & Handsplit Shake Bureau)

needed. See Fig. 7-105. Now take a look at Figs. 7-106, 7-107, and 7-108. These show different ways of finishing the shed for different purposes. For instance, the structure can be covered with glass, Plexiglas, or polyvinyl as in Fig. 7-107 and made into a greenhouse. Then there is the rustic look shown in Fig. 7-106 and the contemporary look shown in Fig. 7-108.

Fig. 7-103 Handsplit shakes should be used on roofs where the slope is sufficient to ensure good drainage. Two different exposures to the weather are shown. Note the spacing of the sheathing under the shakes. (Red Cedar Shingle & Handsplit Shake Bureau)

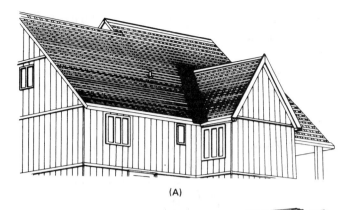

(A)

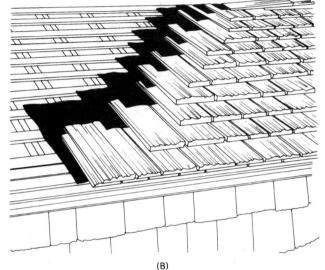

(B)

Fig. 7-104 *(A) Roof ready for application of shingles. (B) Cedar shingle being applied to prepared sheathing.* (Red Cedar Shingle & Handsplit Shake Bureau)

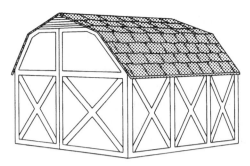

Fig. 7-106 *Rustic shed design.* (TECO)

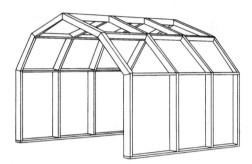

Fig. 7-107 *A greenhouse can be made by covering the frame with plastic.* (TECO)

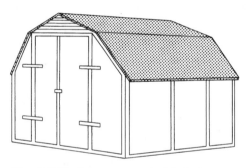

Fig. 7-108 *Contemporary shed design.* (TECO)

Fig. 7-105 *Bill of materials for a storage shed.* (TECO Products and Testing Corporation, Washington, DC 20015)

BILL OF MATERIALS			
QUAN.	**DESCRIPTION**	**QUAN.**	**DESCRIPTION**
28	2″ x 4″ x 8 FT. LONG	40	TECO C-7 PLTS.
9	4′ x 8′ x ½″ PLYWD.	30	TECO JOIST HGR.
2	1″ x 4″ x 6 FT. LONG	12	TECO ANGLES
1 ROLL	ROOFING FELT	30	TECO A-5 PLTS.
1 GAL	ROOF. CEMENT	3	3 BUTT HINGES
1 GAL	BARN-RED PAINT	1	HASP & LOCK
5#	6d COM. NAILS	10 BG.	90# CONC. MIX
2#	12d COM. NAILS		OR
2#	½″ ROOF. NAILS	4	6″ x 8″ x 8′ RAIL TIE

Frame layout Note the dimensions of the shed. It is 7 feet high and 8 feet wide. A detail of the framing angle is shown in the frame-to-sill detail (Fig. 7-109). Note the spacing of the slopes of the roof. Figure 7-110 indicates how the vertical stud member and the roof member are attached with metal plates. Figure 7-111 indicates how an 18° angle is used for cutting the studs and roof members.

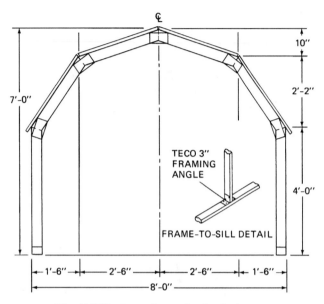

Fig. 7-109 *Frame layout for the shed.* (TECO)

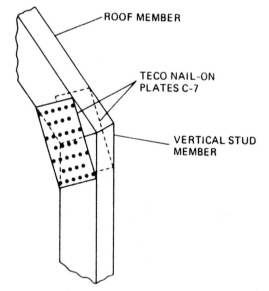

Fig. 7-110 *Detail of connection of vertical stud member and roof member.* (TECO)

Frame-cutting instructions Mark off 18° angles on the 2 × 4s. See Fig. 7-111. Cut to length. Note the exact length required for the roof member and the vertical stud. When you have cut the required number of studs and roof members, place the sections on a hard,

flat surface. A driveway or sidewalk can be used. Nail the metal plates equally into each member and flip the frame over. Then nail metal plates on this side. Make four frames for a shed with an 8-foot depth. You can change the number of frames to match your length or depth requirements. Just add a frame for every 2 feet 8 inches of additional depth.

Roof framing plan Figure 7-112 shows the roof framing layout. Such a layout will eliminate any problems you might have later if you did not properly plan your project. The purlin is extended in length every time you extend the depth of the shed by 2 feet 6 inches. Note how the purlins are attached to the vertical stud member and the roof member. See Fig. 7-113. Once you have attached all the purlins, you have the standing frame. It is time to consider other details. Figure 7-114 shows how a vent is built into the rear elevation. It can be a standard window or constructed from 1 × 2s with a polyvinyl backing.

Door details You have to decide upon the design of the doors to be used. Figure 7-115 shows how the door is constructed for both rustic and contemporary styles. Cut the angles at 18° when making the door for the rustic style. Use the same template used for the studs and roof members.

The contemporary door is nothing more than squared-off corners. Figure 7-116 shows the placement of the hasps used on the door. The plywood sheathing for the roof and the shingles should follow the instructions in Chapter 8.

Mansard Roofs

The mansard roof has its origins in France. The mansard usually made in the United States is slightly different from the French style. Figure 7-117 shows the American style of mansard. It uses a roof rafter with a steep slope for the side portion and one with a very low slope for the top. Standard 2-×-4 or 2-×-6 framing lumber is used to make these rafters. Plywood sheathing is applied over the rafters, and roofing is usually applied over the entire surface. This is supposed to make the house look lower. It effectively lowers the "belt-line" and makes the roof look closer to the ground. Shingles have to be chosen for a steep slope so that they are not blown off by high winds.

Figure 7-118 shows a mansard roof with cedar shakes. The steep slope portion adds to the effect when covered by cedar shingles. It is best to use the wooden shakes for the low slope as well; however, in some cases the slope may be too low and other shingles may be needed to do the job properly.

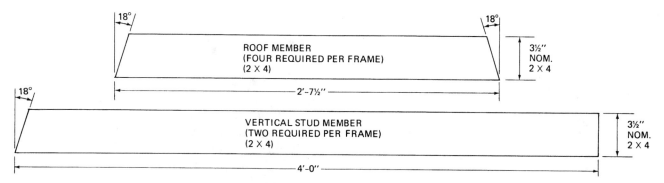

Fig. 7-111 *Frame cutting instructions.* (TECO)

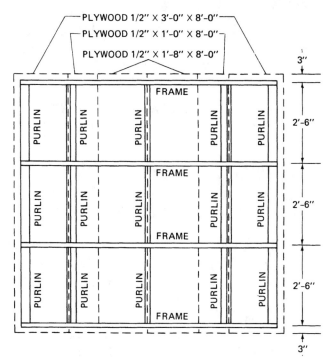

Fig. 7-112 *Roof-framing plan for the shed.* (TECO)

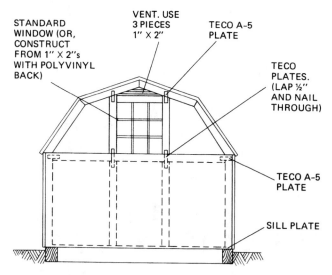

Fig. 7-114 *Rear elevation of the shed. Note how the vent is built in.* (TECO)

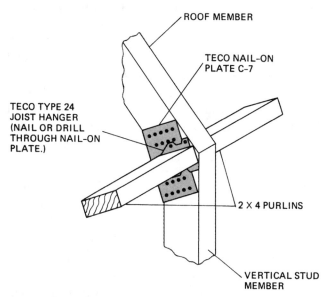

Fig. 7-113 *Attachment of purlins to vertical stud member and roof member.* (TECO)

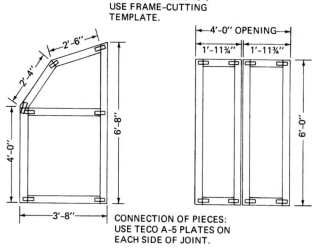

Fig. 7-115 *Door details for the shed.* (TECO)

Figure 7-119 shows how the French made the mansard roof truss. It was used on hotels and some homes. This style was popular during the nineteenth century. Note the elaborate framing used in those days to hold the various angles together. At that time they

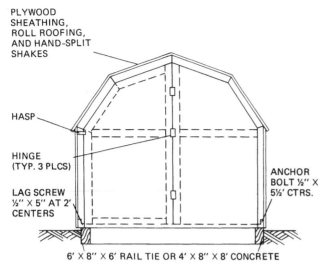

Fig. 7-116 *Front elevation of the shed.* (TECO)

PLYWOOD
SHEATHING,
ROLL ROOFING,
AND HAND-SPLIT
SHAKES

HASP

HINGE
(TYP. 3 PLCS)

LAG SCREW
½″ × 5″ AT 2′
CENTERS

ANCHOR
BOLT ½″ ×
5½′ CTRS.

6′ × 8″ × 6′ RAIL TIE OR 4′ × 8″ × 8′ CONCRETE

Fig. 7-118 *Handsplit shakes used on a mansard roof.* (Red Cedar Shingle & Handsplit Shake Bureau)

Fig. 7-117 *Mansard roof.*

used wrought-iron straps instead of today's metal (steel) brackets and plates.

Post-and-Beam Roofs

The post-and-beam type of roof is used for flat or low-slope roofs. This type of construction can use the roof decking as the ceiling below. See Fig. 7-120(A). The exposed ceiling or roof decking has to be finished. That means the wood used for the roof deck has to be surfaced and finished on the inside surface. Post-and-beam methods don't use regular rafters. See Fig. 7-120(B). The rafters are spaced at a greater distance than in conventional framing. This calls for larger dimensional lumber in the rafters. These are usually exposed also and need to be finished according to plan.

ROOF LOAD FACTORS

Plywood roof decking offers builders some of their most attractive opportunities for saving time and money. Big panels really go down fast over large areas. They form a smooth, solid base with a minimum of joints. Waste is minimal, contributing to the low in-place cost. It is frequently possible to cut costs still further by using fewer rafters with a somewhat thicker panel for the decking. For example, use ¾-inch plywood over framing 4 feet O.C. Plywood and trusses are often combined in this manner. For recommended spans and plywood grades, see Table 7-5.

Plywood roof sheathing with conventional shingle roofing Plywood roof sheathing under shingles provides a tight deck with no wind, dust, or snow infiltration and high resistance to racking. Plywood has stood up for decades under asphalt shingles, and it has performed equally well under cedar shingles and shakes.

Plywood sheathing over roof trusses spaced 24 inches O.C. is widely recognized as the most economical construction for residential roofs and has become the industry standard.

Design Plywood recommendations for plywood roof decking are given in Table 7-4. They apply for the following grades: C—D INT APA, C—C EXT APA, Structural II and II C—D INT APA, and Structural I and II C—C EXT APA. Values assume 5 pounds per square foot dead load. Uniform load deflection limit is ¹⁄₁₈₀ of the span under live load plus dead load, or ¹⁄₂₄₀ under live load only. Special conditions, such as heavy

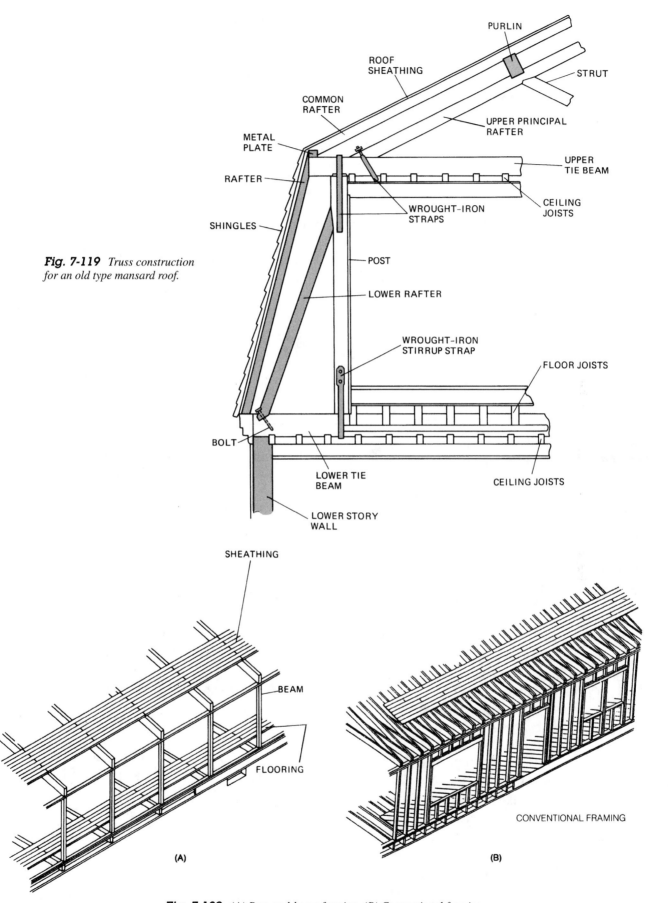

Fig. 7-119 *Truss construction for an old type mansard roof.*

PURLIN

STRUT

ROOF
SHEATHING

COMMON
RAFTER

UPPER PRINCIPAL
RAFTER

METAL
PLATE

UPPER
TIE BEAM

RAFTER

CEILING
JOISTS

SHINGLES

WROUGHT-IRON
STRAPS

POST

LOWER RAFTER

WROUGHT-IRON
STIRRUP STRAP

FLOOR JOISTS

BOLT

LOWER TIE
BEAM

CEILING JOISTS

LOWER STORY
WALL

SHEATHING

BEAM

FLOORING

(A)

CONVENTIONAL FRAMING

(B)

Fig. 7-120 *(A) Post-and-beam framing. (B) Conventional framing.*

Table 7-5 *Plywood Roof Decking*

Identi- fication Index	Plywood Thickness, Inches	Maximum Span, Inches	Unsupported Edge—Max. Length, Inches	Allowable Live Loads, psf Spacing of Supports Center to Center, Inches									
				12	16	20	24	30	32	36	42	48	60
12/0	5/16	12	12	150									
16/0	5/16, 3/8	16	16	160	75								
20/0	5/16, 3/8	20	20	190	105	65							
24/0	3/8, 1/2	24	24	250	140	95	50						
32/16	1/2, 5/8	32	28	385	215	150	95	50	40				
42/20	5/8, 3/4, 7/8	42	32		330	230	145	90	75	50	35		
48/24	3/4, 7/8	48	36		300	190	120	105	65	45	35		
2•4•1	1 1/8	72	48				390	245	215	135	100	75	45
1 1/8" Grp. 1 and 2	1 1/8	72	48				305	195	170	105	75	55	35
1 1/4" Grp. 3 and 4	1 1/4	72	48				355	225	195	125	90	85	40

concentrated loads, may require constructions in excess of these minimums. Plywood is assumed continuous across two or more spans, and applied face grain across supports.

Application Provide adequate blocking, tongue-and-groove edges, or other edge support such as plyclips when spans exceed maximum length for unsupported edges. See Fig. 7-121 for installation of plyclips. Use two plyclips for 48-inch or greater spans and one for lesser spans.

Space panel ends 1/16 inch apart and panel edges 1/8 inch apart. Where wet or humid conditions prevail, double the spacings. Use 6d common smooth, ring-shank, or spiral-thread nails for plywood 1/2 inch thick or less. Use 8d nails for plywood to 1 inch thick. Use 8d ringshank or spiral nails or 10d common smooth for 2 • 4 • 1, 1 1/8 inch and 1 1/4 inch panels. Space nails 6 inches at panel edges and 12 inches at intermediate supports, except where spans are 48 inches or more. Then space nails 6 inches at all supports.

Plywood nail holding Extensive laboratory and field tests, reinforced by more than 25 years experience, offer convincing proof that even 5/16-inch plywood will hold shingle nails securely and permanently in place, even when the shingle cover is subjected to hurricane-force winds.

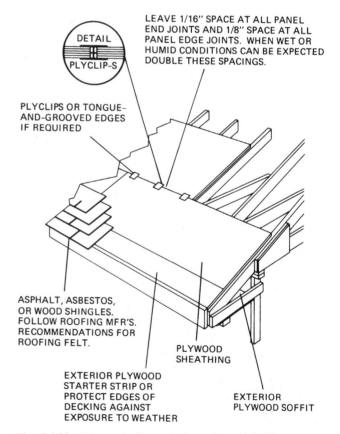

DETAIL
PLYCLIP-S

LEAVE 1/16" SPACE AT ALL PANEL END JOINTS AND 1/8" SPACE AT ALL PANEL EDGE JOINTS. WHEN WET OR HUMID CONDITIONS CAN BE EXPECTED DOUBLE THESE SPACINGS.

PLYCLIPS OR TONGUE-AND-GROOVED EDGES IF REQUIRED

ASPHALT, ASBESTOS, OR WOOD SHINGLES. FOLLOW ROOFING MFR'S. RECOMMENDATIONS FOR ROOFING FELT.

EXTERIOR PLYWOOD STARTER STRIP OR PROTECT EDGES OF DECKING AGAINST EXPOSURE TO WEATHER

PLYWOOD SHEATHING

EXTERIOR PLYWOOD SOFFIT

Fig. 7-121 *Using a plyclip to reinforce plywood decking.* (American Plywood Assocation)

The maximum high wind pressure or suction is estimated at 25 psf except at the southern tip of Florida, where wind pressures may attain values of 40 to 50 psf. Because of shape and height factors, however, actual suction or lifting action even in Florida should not exceed 25 psf up to 30-foot heights. Thus any roof sheathing under shingles should develop at least that much withdrawal resistance in the nails used.

Plywood sheathing provides more than adequate withdrawal resistance. A normal wood-shingled room will average more than 6 nails per square foot. Each nail need carry no more than 11 pounds. Plywood sheathing only 5/16 inch thick shows a withdrawal resistance averaging 50 pounds for a single 3d shingle nail in laboratory tests and in field tests of wood shingles after 5 to 8 years' exposure. In addition, field experience shows asphalt shingles consistently tear through at the nail before the nail pulls out of the plywood.

Figure 7-122 shows the markings found on plywood used for sheathing. Note the interior and exterior glue markings. APA stands for the American Plywood Association.

LAYING OUT A STAIR

So far you have used the framing square to lay out rafters. There is also another use for this type of instrument. It can be used to lay out the stairs going to the basement or going upstairs in a two-story house.

Much has been written about stairs. Here we only lay out the simplest and most useful of the types available. The fundamentals of stair layout are offered here.

1. Determine the height or rise. This is from the top of the floor from which the stairs start, to the top of the floor on which they are to end. See Fig. 7-123.

2. Determine the run or distance measured horizontally.

3. Mark the total rise on a rod or a piece of 1-×-2-inch furring to make a so-called *story pole*. Divide the height or rise into the number of risers desired. A simple method is to lay out the number of risers wanted by spacing off the total rise with a pair of compasses. It is common to have this result in fractions of an inch. For example, a total rise of 8 feet 3 ¾ inches or 99¾ inches divided by 14 = 7.125 or 7⅛-inch riser. This procedure is not necessary in the next step because the horizontal distance, or run, is seldom limited to an exact space as is the case with the rise.

4. Lay out or space off the number of treads wanted in the horizontal distance or run. There is always one

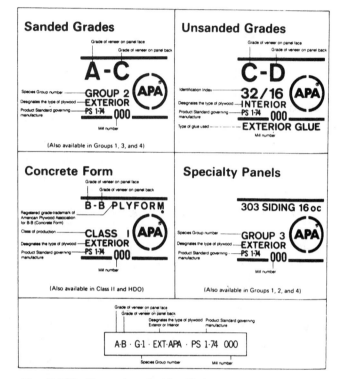

Fig. 7-122 *Plywood grades identified.* (American Plywood Association)

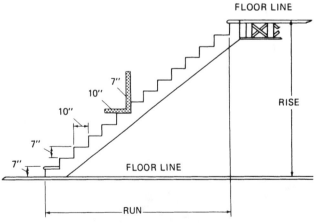

NOTE: The stairs shown with approximately the above dimensions of the riser and tread are considered easy or comfortable to climb.

Fig. 7-123 *Laying out stairs with the steel square.* (Stanley Tools)

less tread than there are risers. If there are 14 risers in the stair, there are only 13 treads. For example, if the tread is 10 inches wide and the riser is 7 inches, the stair stringer would be laid out or "stepped off" with the square, ready for cutting as shown in Fig. 7-123. The thickness of the tread should be deducted from the first riser as shown. This is in order to have this first step the same height as all the others.

ALUMINUM SOFFIT

So far the soffit has been mentioned as the covering for the underside of the overhang. This has been shown to be covered with a plywood sheathing of ¼-inch thickness or with a cardboard substance about the thickness of the plywood suggested. The cardboard substitute is called Upson board. This is because the Charles A. Upson Company makes it. If properly installed and painted, it will last for years. However, it should not be used in some climates.

One of the better materials for soffit is aluminum. More and more homes are being fitted with this type of maintenance-free material. This particular manufacturer no longer makes this roll-type soffit.

Material Availability

Aluminum soffit can be obtained in 50-foot rolls with various widths. They can be obtained (Fig. 7-124) in widths of 12, 18, 24, 30, 36, and 48 inches. These are pushed or pulled into place as shown in Fig. 7-125. The hip roof with an overhang all around the house would require soffit material pulled in as shown in Fig. 7-126. The runners supporting the material are shown in Fig. 7-127. Covering the ends is important to ensure a neat job. Corner trim and fascia closure are available to help give the finished job a look not unlike that of an all-wood soffit.

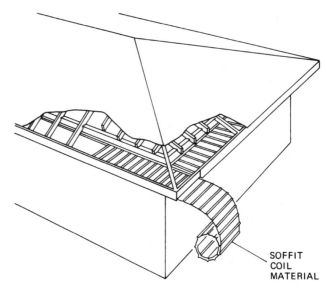

Fig. 7-125 *Method of inserting the aluminum soffit material.* (Reynolds Metals Product)

Figure 7-128 shows how the fascia runner, the frieze runner, and corner trim are located for ease of installation of the soffit coil.

After the material has been put in place and the end cuts have been made, the last step is to insert a plastic liner to hold the aluminum in place. This prevents rattling when the wind blows. The material can be obtained with a series of holes prepunched. This will serve as ventilation for the attic. See Fig. 7-129.

Figure 7-130 gives more details on the installation of the runners that support the soffit material. The fascia runner is notched at points *b* for about 1½ inches maximum. Then the tab is bent upward and against the inside of the fascia board. Here it is nailed to the board for support. Take a look at *c* in Fig. 7-130 to see how the tab is bent up. Note how the width of the channels is the soffit coil width plus at least ⅜ inch and not more than ⅞ inch. This allows for expansion by the aluminum. Aluminum will expand in hot weather.

Cutting the runner to desired lengths can be done by cutting the channel at *a* and *b* of Fig. 7-131. Then bend the metal back and forth along a line such as at *c* until it breaks. Of course you can use a pair of tin snips to make a clean cut.

Figure 7-132 shows how the soffit is installed with a brick veneer house. Part (A) shows how the frieze runner is located along the board. Insert (B) shows how the runner is nailed to the board.

When the fascia board is more than 1 inch wide, it is necessary to place 1-inch aluminum strips as shown in Fig. 7-133. The tabs are hooked in between the fold in the runner. Then the runner is brought under the fas-

WIDTHS

12″   18″   24″   30″   36″   48″

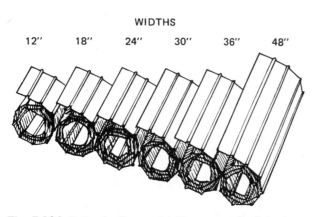

Fig. 7-124 *Rolls of soffit material. These are made of aluminum.*

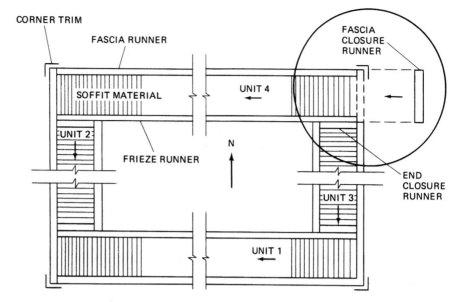

Fig. 7-126 *Steps in installing aluminum soffit in a hip roof with an overhang all around.* (Reynolds Metals Products)

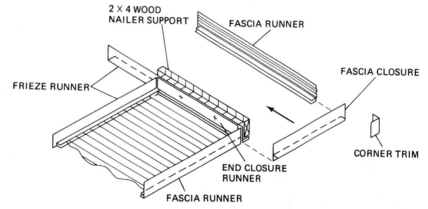

Fig. 7-127 *Closing off the ends of the soffit with aluminum.* (Reynolds Metal Products)

cia board and bent back and nailed. This will allow the runner to expand when it is hot. Do not nail the overlapping runners to one another.

In some instances it is necessary to use a double-channel runner. This is done so that there will be no sagging of the soffit material. See Fig. 7-134. Note how the frieze runner and double-channel runner are located. Note the gravel stop on this flat roof. In some parts of the country more overhang is needed because it gives more protection from the sun.

Figure 7-135 shows how the H-molding joint works to support the two soffit materials as they are unrolled into the channel molding. Note the location of the vent strip, when needed.

As was mentioned before, the aluminum soffit makes for a practically maintenance-free installation. More calls will be made for this type of finish. The carpenter will probably have to install it, since the carpenter is responsible for the exterior finish of the building and the sealing of all the openings. With the advent of aluminum siding, it is only natural that the soffit be aluminum.

METAL CONNECTORS

Every year many houses are destroyed when the force of high winds cause roofs to fly off and walls to collapse. One of the methods used by builders located in high-wind areas such as along the shores of lakes, bays, oceans, and gulfs is the metal connector. Various fasteners are designed to increase the structural strength of homes built to withstand hurricanes, and in some instances, tornadoes and earthquakes.

Figure 7-136 shows how the use of metal connectors can increase the chances for a house to escape a hurricane's full force. The connectors are illustrated in the next few pages. Take a close look at the encircled number and then refer to the following pages for an illustration on where and how the connector is utilized to its best advantage.

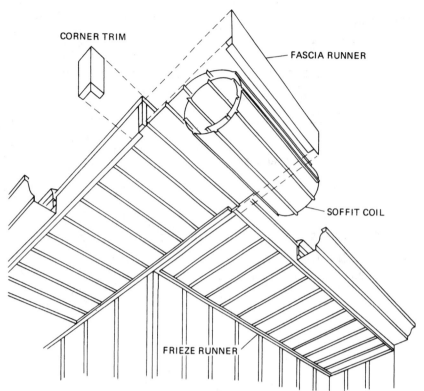

CORNER TRIM

FASCIA RUNNER

SOFFIT COIL

FRIEZE RUNNER

Fig. 7-128 *Installing soffit coil in the overhang space. Note the corner trim to give a finished appearance.* (Reynolds Metals Products)

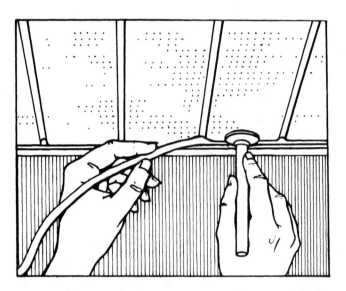

Fig. 7-129 *Finishing up the job with a vinyl insert to hold the aluminum in place.* (Reynolds Metals Products)

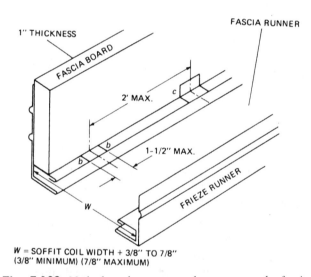

1" THICKNESS

FASCIA BOARD

FASCIA RUNNER

2' MAX.

c

1-1/2" MAX.

b

b

FRIEZE RUNNER

W

W = SOFFIT COIL WIDTH + 3/8" TO 7/8"
(3/8" MINIMUM) (7/8" MAXIMUM)

Fig. 7-130 *Method used to support the runner on the fascia.* (Reynolds Metals Products)

High winds can be combatted by building a house that will withstand the force of Mother Nature. The only successful way to combat these forces is to employ proper construction techniques ahead of time to ensure the integrity of the structure. The best method is to use an uninterrupted load path from the roof members to the foundation. Metal connectors are engineered to satisfy the necessary wind uplift load requirements. See Fig. 7-137.

Figure 7-138 shows how foundations are prepared with metal connectors to cause the sole plate and studs to be permanently and solidly anchored to prevent damage by high winds. Figure 7-139 illustrates how connectors are used for a Stem Wall System. A wood-to-wood type of construction enhanced by metal connectors is shown in Fig. 7-140. Note the allowable loads tables that give the nail size and uplift in pounds per square inch (psi).

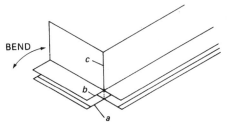

Fig. 7-131 *Bending and breaking the runner material.* (Reynolds Metals Products)

A poured masonry header has the rafters attached by hangers anchored in the concrete. See Fig. 7-141.

Second floor problems can be solved by tying the first and second floors together with metal connectors as shown in Fig. 7-142. A variety of fasteners are illustrated. Regular truss-to-top plate construction is shown reinforced in Fig. 7-143 while the top plate-to-stud reinforcement is illustrated by

Fig. 7-132 *Soffit on a brick veneer house. (A) Locating the frieze runner along the board. (B) Using the quarter-round type of frieze runner.* (Reynolds Metals Products)

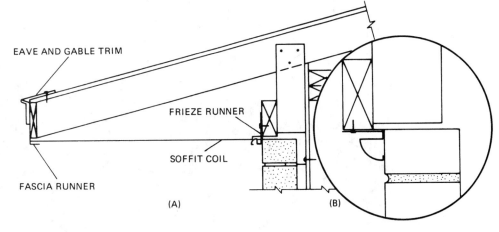

ALTERNATE USE OF QUARTER ROUND FRIEZE RUNNER

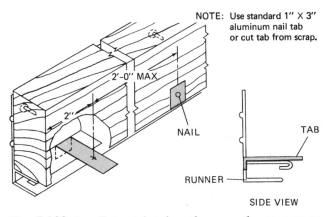

Fig. 7-133 *Installing a tab to keep the runner free to move as aluminum expands on hot days.* (Reynolds Metals Products)

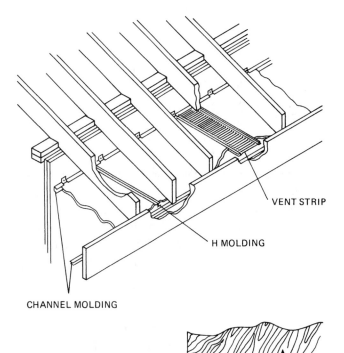

Fig. 7-135 *Installing the H-molding joint.* (Reynolds Metals Products)

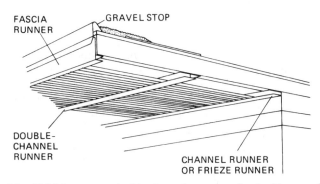

Fig. 7-134 *Using a double-channel runner and a double row of aluminum soffit material.* (Reynolds Metals Products)

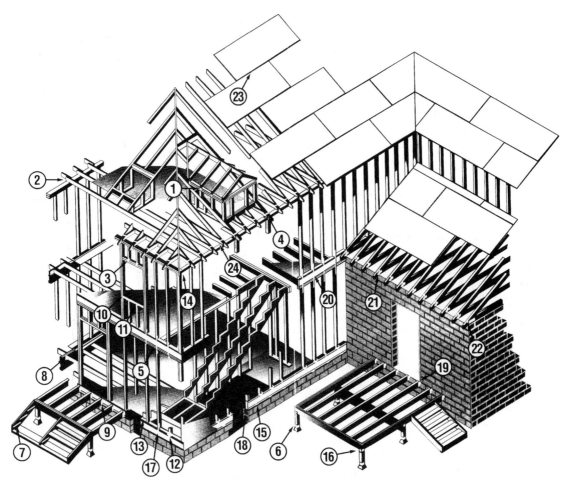

Fig. 7-136 *Location of metal connectors.* (SEMCO)

Fig. 7-144. These metal fasteners or connectors are standard requirements in areas buffeted by high winds, hurricanes, and earthquakes. The additional costs can often be offset by lower insurance rates on the finished house.

There are charts that show the expected winds in a given area. Figure 7-145 is a map of the United States with the wind speeds indicated. The wind load calculations placed on a structure are determined by a number of factors. Chief among these is the wind speed ratings for the location of the structure.

CHAPTER 7
STUDY QUESTIONS

1. List at least five types of roof lines.
2. What is the difference between a hip roof and a mansard roof?
3. What are the parts of a roof frame?

4. Where are trussed rafters commonly used?
5. What is a disadvantage of a truss roof?
6. What is an advantage of the truss type of construction?
7. What is used to cover trusses?
8. What is a framing square?
9. Identify the following parts of a steel square:
   a. Body
   b. Tongue
   c. Heel
   d. Face
10. How can you make use of the hundredths scale on a square?
11. What is the difference between the octagon and brace scales on a square?
12. Identify the following terms:
    a. Shed roof
    b. Gable or pitch roof
    c. Valley roof

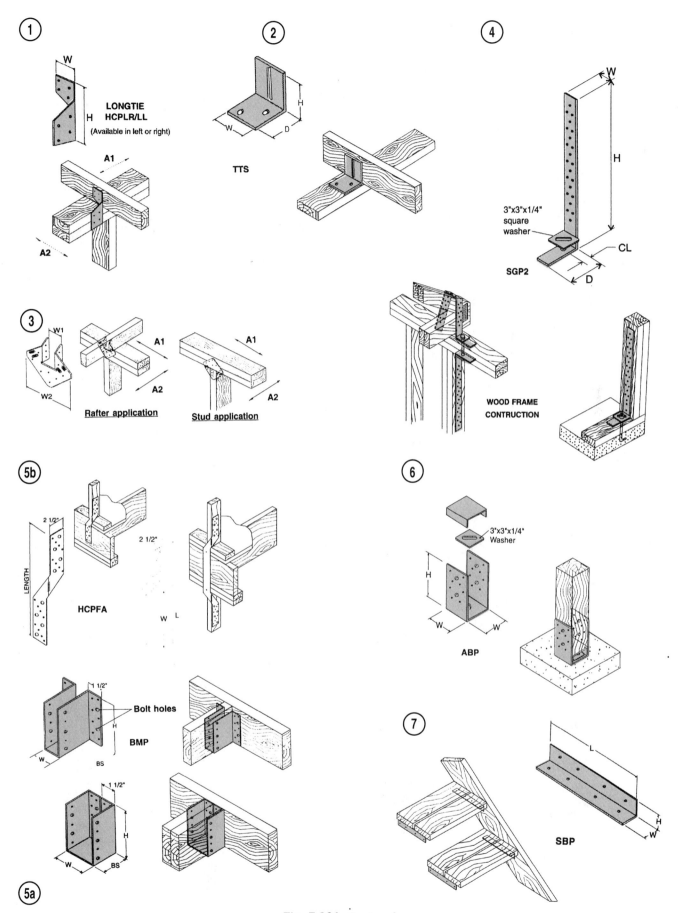

Fig. 7-136 Continued.

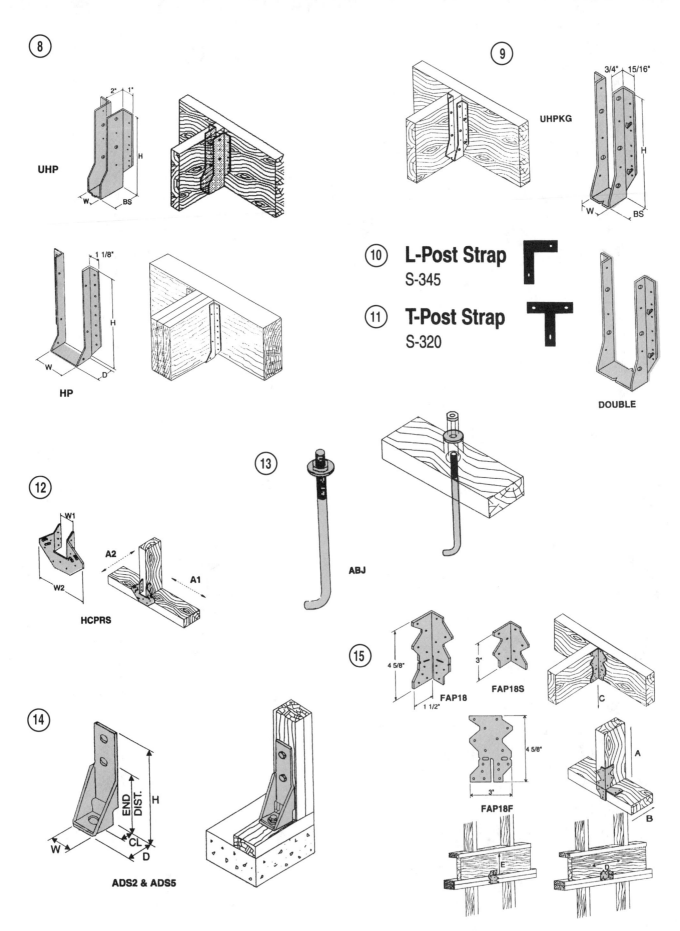

Fig. 7-136 *Continued.*

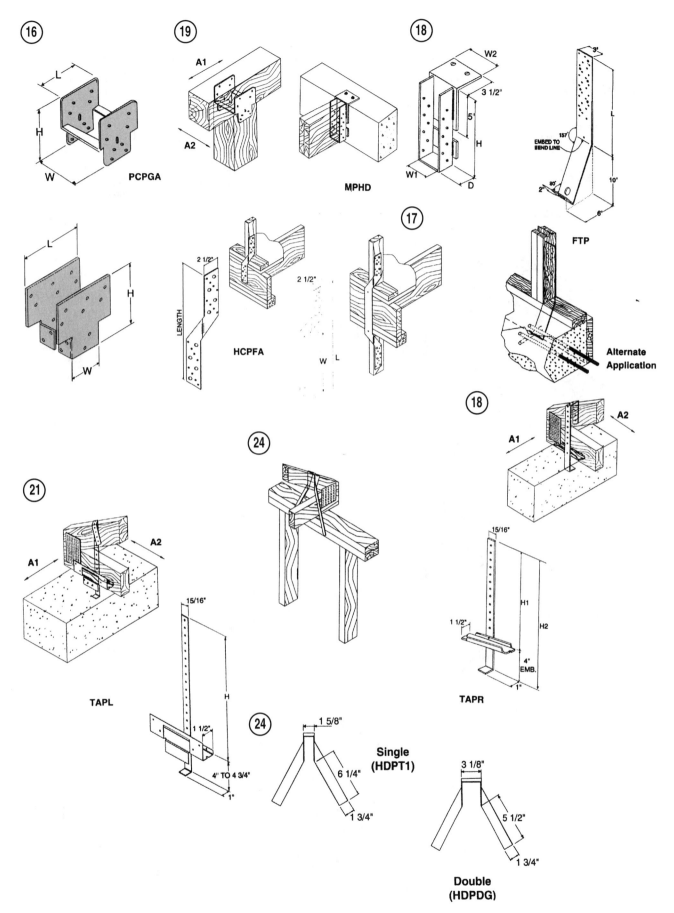

16

L
H
W

PCPGA

19

A1
A2

2 1/2"

LENGTH

HCPFA

2 1/2"

W L

17

18

W2
3 1/2"
5"
H
W1
D

MPHD

3"
L
157°
EMBED TO
BEND LINE
80°
2"
10"
6"

FTP

Alternate
Application

18

A2
A1

15/16"
H1
H2
1 1/2"
4"
EMB.
1"

TAPR

21

A1
A2

TAPL

15/16"
H
1 1/2"
4" TO 4 3/4"
1"

24

24

1 5/8"
6 1/4"
1 3/4"

Single
(HDPT1)

3 1/8"
5 1/2"
1 3/4"

Double
(HDPDG)

Fig. 7-136 Continued.

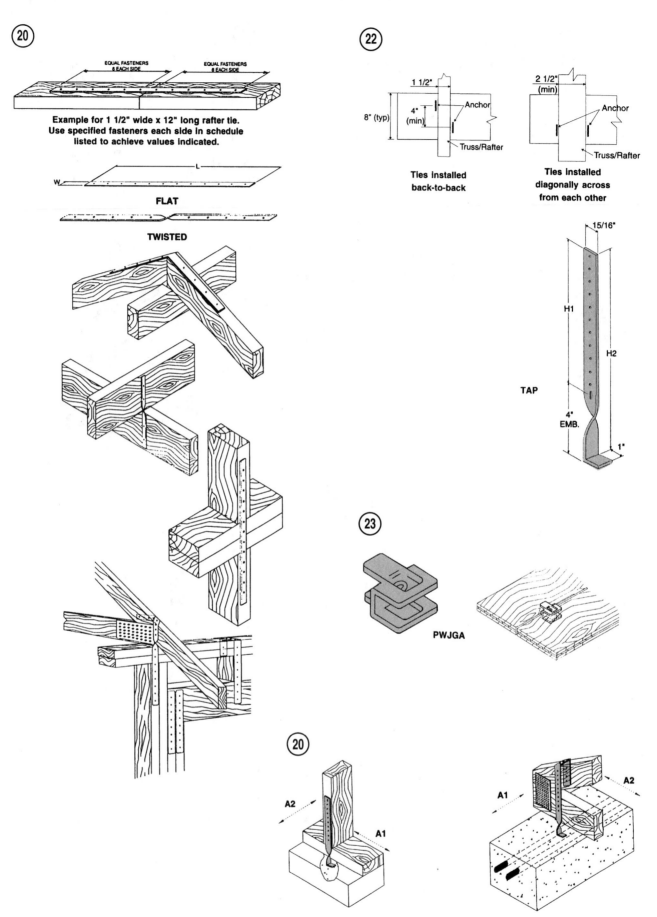

Example for 1 1/2" wide x 12" long rafter tie.
Use specified fasteners each side in schedule
listed to achieve values indicated.

FLAT

TWISTED

Ties installed
back-to-back

Ties installed
diagonally across
from each other

TAP

PWJGA

Fig. 7-136 *Continued.*

① **Hurricane Anchor** This tie adds increased resistance to wind uplift. The tie eliminates toe nailing utilizing correctly located nail holes for fast, easy, and strong attachment of rafters and trusses to plates and studs. They are made of 20-18 gage galvanized steel. They can be installed on each side of the rafter for twice the loads when the rafter thickness is a minimum of 2.5 inches or diagonally when rafter is 1.5 inches.

② **90° Bracket** Used for tying trusses to non-load bearing walls.

③ **Rafter Clip/Stud Plate** A fast, economical tie to secure rafters or trusses to wall studs and top plates, and from studs to sill plate.

④ **Girder Truss Strap** Ideal for girder truss connections when there is a high uplift load requirement. Can be used for wood-to-wood application or concrete-to-wood application. A washer plate adds increased resistance to wind uplift.

⑤ **A. Beam Support** These face mount supports are designed to provide full support of the top chord preventing joist rotation. Bolt holes are also provided for additional load capacity. **B. Floor Tie Anchor** These ties are designed especially for use with floors constructed above grade, as a connection between first or second floor level to studs. They are designed for engineered floor systems with larger clear spans of 21 inches and 24 inches. They can be fastened with bolts or nails.

⑥ **Heavy Duty Post Anchor** Post base is to elevate post above concrete to allow for ventilation. Heavy-duty design permits higher uplift loads and simplified installation with 0.5-inch anchor bolts that fit through pre-punched holes, slotted to permit adjustment to align for off-position anchor bolts.

⑦ **Staircase Bracket** These brackets are designed to simplify and reinforce stair construction.

⑧ **Joist Support** The regular joist support is made of 18 gage and the Heavy-Duty Joist Hanger is made to support headers, joists, and trusses. It is made from 14-gage galvanized steel.

⑨ **Kwik Grip Joist Support** These supports are designed for quick installation in place for easy nailing. The offset nail hole placement is for secure positive nailing. Precision formed high-strength 18-gage galvanized is used for long life.

⑩ **L-Post Strap** Can be used to tie the window framing together.

⑪ **T-Post Strap** Can be used at perpendicular junction points for cripple studs and the rough sill.

⑫ **Stud Plate/Rafter Clip** Used for tying studs to sill plate and top plates, or rafters to top plates.

⑬ **Anchor Bolt** 6-inch minimum embedment with 3000 psi concrete will resist 1,635 lbs. Wind unlift loads are based on the shear capacity of No. 2 Southern Pine. Compression perpendicular to grain 565 psi.

⑭ **Hold Down Anchor** Ideal for shear walls and vertical posts.

⑮ **Universal Framing Anchor** This is a multi-purpose anchor for almost any wood connection task. It anchors rafters and roof trusses to plates, and it anchors floor and ceiling joists to headers and solid blocking to plates. The 90° framing angles can be used to join posts to beams and make other right-angle connections.

⑯ **Post Cap** Simplifies installing 4×4 wood posts to wood beams and trusses. Makes a full-strength, positive connection between the post top and lateral beams or trusses. It is considerably stronger than random toe-nailing and spiking, and is less time-consuming than drilling and lag bolting. Post caps are of the split design to offer maximum flexibility in application. Direct-load path-side plates maximizes load capacities. Toe-nail slots speed plum and level adjustments.

⑰ **Floor Tie Anchor** As previously noted, these ties are designed especially for use with floors constructed above grade, as a connection between first or second floor level to studs. Nails or bolts can be used.

⑱ **Foundation Tie** There is a prepunched hole in the foot of the tie to increase concrete grip and allow alternate rebar rod support. Prepunched holes in the bend are there for nails to hold it in the form board. Ties should have a minimum of 3 inches in the concrete.

⑲ **Heavy-Duty Masonry Beam Hanger** Designed to work with standard block wall or concrete tie beam construction. Eliminates the need of constructing special seats to support floor joists.

⑳ **Ratter Tie** Tie straps meet a variety of application and design load conditions and specifications. Use rafter ties when tying rafters to plate, anchoring studs to sill, or framing over girders and bearing partitions.

㉑ **Lateral Truss Anchor** This anchor is designed to meet the lateral and uplift load demands for hurricane-resistant construction. It provides a custom connection to wood for trusses or rafters. An attached beam seat-plate eliminates treated sill or moisture-barrier installation. The riveted plate on the seat-plate design prevents truss movement parallel and perpendicular to the wall.

㉒ **Truss Anchor** Accommodates diverse design requirements for concrete-to-wood installation, allows a 4-inch embedment in concrete. You can use two anchors installed, one on each side of the rafter, for twice the load per single rafter thickness. Minimum edge distance is 2 inches.

㉓ **Galvanized Plywood Clip** This clip is designed for easy and fast application to sheathing edges. It gives a snug self-lock and that keeps the clip firmly in position throughout panel placement. It eliminates unreliable wood edge blocking. This type of clip is also available in aluminum. The clip gives an automatic spacing to the plywood or sheathing.

㉔ **Truss Tie Down Strap (Gun Tie)** This strap provides additional increased resistance to wind uplift to secure rafters or trusses to top plates. It eliminates fastening through the truss nail plates and requires no truss nailing. It is designed to allow gun nailing for quick installation. The clip is made of 20-gage steel and the installer should wear eye protection. Fasteners are placed on each side of the truss into the top and bottom layer of the double top-plate members in equal quantity. Fasteners should be no less than 0.25 inches from the edge of the strap and placed no less than 0.375 inches from the edge of the framing member. **True Tie** is the same share as the *Gun Tie*, but it has 18-gage galvanized metal.

Fig. 7-136 *Continued.*

PRODUCT CODE	GAUGE	FASTENER SCHEDULE		ALLOWABLE LOADS	
				WIND / EARTHQUAKE	
		HEADER / PLATE	STUD	UPLIFT 133%	UPLIFT 160%
ABJBL10W	---	---	---	1635	1635
FOP41	12	12-16d	---	2190	2465
HCPFA	16	8-16d	8-16d	1200	1415
HCPSA	18	---	16-16d	1200	1415
HCPRS	18	5-8d	6-8d	540	540
CLP5W	18	11-8d	6-8d	530	540
SRP121630F	12	---	18-16d	2815	3380
RS150	16	---	11-10d	1645	1645
FAP18F	18	6-8d	6-8d	765	915
ADS2	12	(1) 5/8"	(2) 5/8"	2775	3330

PRODUCT CODE	GAUGE	FASTENER SCHEDULE		ALLOWABLE LOADS	
				WIND / EARTHQUAKE	
		STUD	PLATE	UPLIFT 133%	UPLIFT 160%
HCPLR	18	4-8d	4-8d	510	520
FAP18	18	6-8d	6-8d	745	745
HCPRS	18	6-10d	5-10d	540	540
CLP5W	18	6-10d	11-10d	540	540
TPP4	20	8-10d	8-10d	1335	1335

PRODUCT CODE	GAUGE	FASTENER SCHEDULE		ALLOWABLE LOADS			
				LATERAL		WIND / EARTHQUAKE	
		TRUSS / RAFTER	SEAT PLATE OR BEAM	PERP. TO WALL	PARAL. TO WALL	UPLIFT 133%	UPLIFT 160%
SGP2	14	14-16d	---	---	---	1455	1455
TAPL12	14&20	11-16d	4-10dx1 1/2"	1405	1405	1950	1950
TAP16	14	11-16d	---	595	210	1950	1950
TAPR216	14&20	11-16d	---	595	210	1950	1950
HDA6	1/4"	(2) 3/4"	(2) 3/4"	---	---	4256	4256

PRODUCT CODE	GAUGE	FASTENER SCHEDULE			ALLOWABLE LOADS			
					LATERAL		WIND / EARTHQUAKE	
		TRUSS / RAFTER	PLATE	STUD	PERP. TO WALL	PARAL. TO WALL	UPLIFT 133%	UPLIFT 160%
HDPT2	18	---	12-16d	---	450	450	1915	2300
RT10	20	5-8dx1 1/2"	8-8d	5-8dx1 1/2"	95	115	555	555
HCPLR	18	4-8d	4-8d	4-8d	95	145	510	520
HCPRF	18	6-10d	6-10d	6-10d	395	235	540	540
RTPGA818T	14	9-16d	---	9-16d	---	---	1360	1635
HCPFA	16	---	8-16d	8-16d	---	---	1200	1415
TPP4	20	---	---	8-10d	---	---	1290	1335

PRODUCT CODE	GAUGE	FASTENER SCHEDULE		ALLOWABLE LOADS	
				WIND / EARTHQUAKE	
		STUD / TOP	SILL	UPLIFT 133%	UPLIFT 160%
TAP18	14	(12) 16d		1950	1950
ADS2	12	(2) 5/8"	(1) 5/8"	2775	3330
FA3	16	(4) 8dX1 1/2"	(2) 8dX1 1/2"	1155	1155
FTP42*	12	(22) 16d		4050	4050
FAS118	18	(4) 8dX1 1/2"	(4) 8dX1 1/2"	755	755
SGP2	14	(14) 16d		1455	1455
ABJBL10W	---			1635	1635

Fig. 7-137 *Characteristics of metal connectors.* (SEMCO)

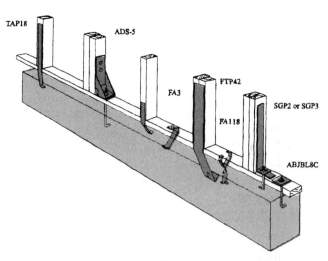

Fig. 7-138 *Wood-to-concrete foundation connections.* (SEMCO)

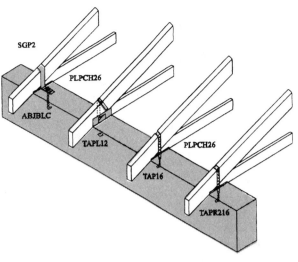

Fig. 7-141 *Poured masonry header with connectors holding the trusses in place.* (SEMCO)

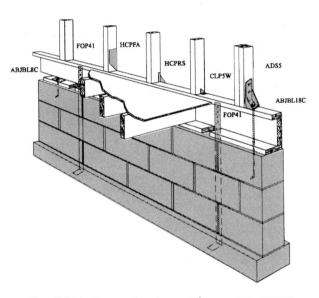

Fig. 7-139 *Stem wall system metal connectors.* (SEMCO)

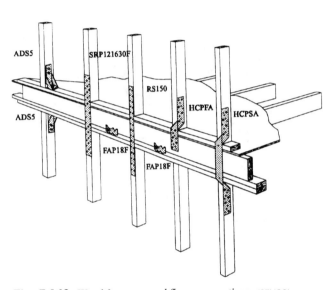

Fig. 7-142 *Wood frame second floor connections.* (SEMCO)

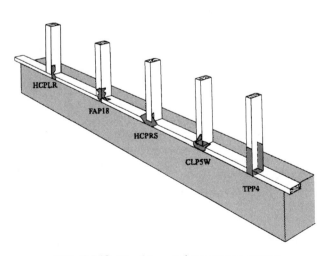

Fig. 7-140 *Wood-to-wood connectors.* (SEMCO)

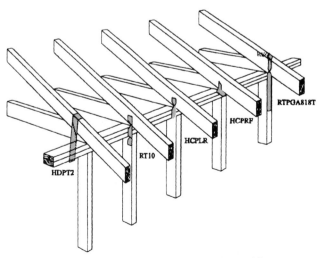

Fig. 7-143 *Top plate, truss or rafter connections with connectors making the truss to top plate a little more secure.* (SEMCO)

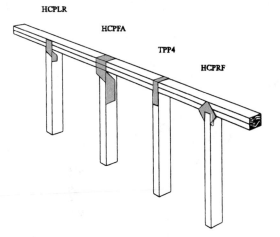

Fig. 7-144 *Various connectors used in the top plate-to-stud construction.* (SEMCO)

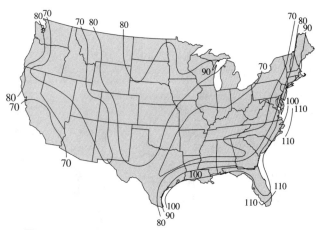

Fig. 7-145 *Basic Wind Speed Map for Continental USA.*

13. What do the following terms mean:
    a. Span
    b. Run
    c. Rise
    d. Pitch

14. Identify the following roof frame members:
    a. Plate
    b. Ridge board
    c. Common rafters
    d. Hip rafters
    e. Jack rafters
    f. Valley rafters

15. What part of a rafter is the tail?

16. What is the difference between a valley rafter and a hip rafter?

17. How do you describe a jack rafter?

18. What is a bay window?

19. What method does a builder use to minimize high-wind damage to a house?

20. In what areas of the country are metal fasteners or connectors standard requirements?

8
CHAPTER

Covering Roofs

IN THIS UNIT YOU WILL LEARN HOW THE carpenter covers roofs, how the roof is prepared for shingles, and how to place the shingles on the prepared surface. In addition, details for applying an asphalt shingle roof are given. Skills covered include how to:

• Prepare a roof deck for shingles
• Apply shingles to a roof deck
• Estimate shingles needed for a job
• Figure slope of a roof

INTRODUCTION

Roofing shingles are made in different sizes and shapes. Roofs have different angles and shapes. Pipes stick up through the roof. Roofing has to be fitted. There are crickets also to be fitted. (A cricket fits between a chimney and the roof.) A number of fine details are presented by roof shapes. All these have to be considered by the roofer.

Some industrial and commercial buildings use roll shingles. This material requires a slightly different approach. Asphalt shingles are safer than wooden shingles. The asphalt shingles will resist fire longer. Because of fire regulations, wooden shingles are not allowed in some sections of the country.

BASIC SEQUENCE

The carpenter should apply a roof in this order:

1. Check the deck for proper installation.
2. Decide which shingles to use for the job.
3. Estimate the amount needed for the job.
4. Apply drip strips.
5. Place the underlayment.
6. Nail the underlayment.
7. Start the first course of shingles.
8. Continue other courses of shingles.
9. Cut and install flashings:
   a. valleys
      (1) open
      (2) closed-cut
      (3) woven
   b. soil pipe flashing
   c. chimney flashing
   d. other flashings
10. Cover ridges.
11. Cover all nailheads with cement.
12. Glue down tabs, if needed.

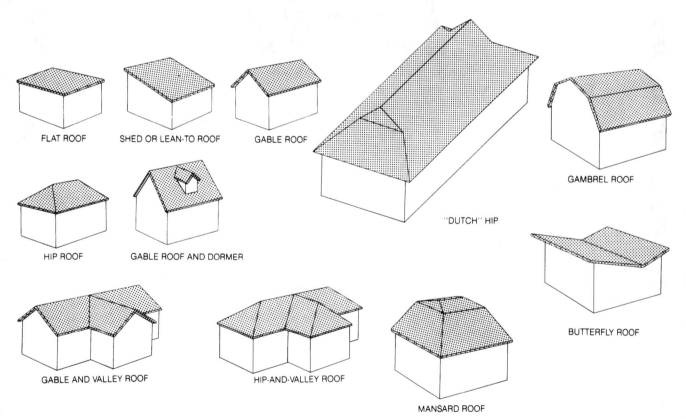

Fig. 8-1 *Various types of roof shapes.*

Types of Roofs

There are a number of roof types; each is classified according to its shape. Figure 8-1 shows the different types. Each type presents roofing problems, and different methods are used to cover the decking, ridges, and drip areas.

The *mansard roof* presents some unique problems. Figure 8-2 shows a mansard roof. Note that the dark area is covered with shingles. The angles presented by the various vertical and sloping sections need special bracing. Attaching the shingles also requires attention to details on the vertical sections. (Refer back to 7-119 for an illustration showing truss construction on a mansard style roof.) The attached garage in Fig. 8-2 has a hip roof. Figure 8-3 also shows a hip roof. Note how the entrance is also a hip, but shorter.

Fig. 8-2 *Mansard roof with hip on garage.*

Fig. 8-3 *Hip-and-valley roof.*

The *gable roof* is a common type. See Fig. 8-4. It is a simple roof that is easy to build. Figure 8-5 shows a variety of gable roofs. Each is a complete unit. The garage shows the angles of this type of roof very well.

Drainage Factors

The main purpose of a roof is to protect the inside of a building. This is done by draining the water from the

Fig. 8-4 *Gable roof.*

Fig. 8-5 *Gable roof with add-ons.*

roof. The water goes onto the ground or into the storm sewer. Some parts of the country allow the water to be dropped onto the earth below. Other sections require the collected water to be moved to a storm sewer. The main idea is to prevent water seepage. Roof water should not seep back into a basement.

Ice is frequently a problem in colder climates. Ice forms and makes a dam for melting snow. See Fig. 8-6A. Water backs up under the shingles and leaks into the ceiling below. This problem can be caused by insufficient insulation. The lack of soffit ventilation will also cause leaks. Heating cables can be installed to prevent ice dams. Leaking can be prevented by adding ventilation. If insulation can be added, this too should be done. See Fig. 8-6B.

The point where rooflines come together is called a *valley*. See Fig. 8-7A, B, and C. Valleys direct water to the drain. This keeps it out of the house. They need special attention during roofing.

Eave troughs and downspouts drain water from the roof. It is drained into gutters. Downspouts carry the water to the ground. Downspouts connect to other pipes. That piping sometimes goes to the street storm

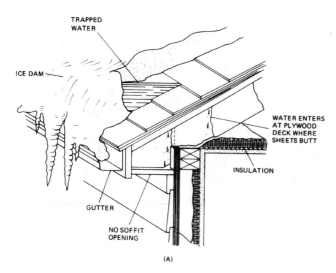

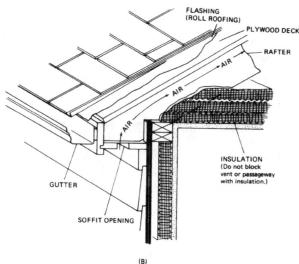

Fig. 8-6 *(A) Water leakage caused by ice dam. (B) Using ventilation and insulation to prevent leaks.*

sewer. This eliminates seepage into the basement or under the slab. See Fig. 8-8A. In most cases a splash block is located under the downspout to disperse the water. See Fig. 8-8B.

Roofing Terms

There are a number of roofing terms used by roofers and carpenters. You should become familiar with the terms; you will then be able to talk with roofing salespersons.

Square   Shingles needed to cover 100 square feet of roof surface. That means 10 feet *square*, or 10 feet by 10 feet.

Exposure   Distance between exposed edges of overlapping shingles. Exposure is given in inches. See Fig. 8-9. Note the 5-inch and 3-inch exposures.

Head lap   Distance between the top of the bottom shingle and the bottom edge of the one covering it. See Fig. 8-10.

Top lap   Distance between the lower edge of an overlapping shingle and the upper edge of the lapping shingle. See Fig. 8-10. Top lap is measured in inches.

Side lap   Distance between adjacent shingles that overlap. Measured in inches.

Valley   Angle formed by two roofs meeting. The internal part of the angle is the valley.

Rake   On a gable roof, the inclined edge of the surface to be covered.

Flashing   Metal used to cover chimneys and around things projecting through the roofing. Used to keep the weather out.

Underlayment   Usually No. 15 or No. 30 felt paper applied to a roof deck. It goes between the wood and the shingle.

Ridge   The horizontal line formed by the two rafters of a sloping roof being nailed together.

Hip   The external angle formed by two sides of a roof meeting.

Roofing is part of the exterior building. The carpenter is called upon to place the covering over a frame. This frame is usually covered by plywood. Plywood comes in 4-×-8-foot sheets and can be quickly installed onto the rafters. Sheathing may be 1-×-6-inch or 1-×-10-inch boards. Sheathing takes longer to install than plywood. The frame is covered with a tar or felt paper. This paper goes on over the sheathing. The paper allows moisture to move from the wood upward. Moisture then escapes under the shingles. This prevents a buildup of moisture. If the weather is bad, moisture can freeze. This forms a frost under the shingles.

Shingles of wood, asphalt, asbestos, and Fiberglas are used for roofing. Tile and slate were once commonly used; however, they are rather expensive to install. Copper, galvanized iron, and tin are also used as roof coverings.

Commercial buildings may use a built-up roof, which has a number of layers. This type uses a gravel topping or cap sheet. Asphalt-saturated felt is mopped down with hot asphalt or tar. See Fig. 8-8C. Choice of roofing is determined by three things: cost, slope, and life expected. In some local climates (wind, rain, snow), flat roofs may have to be rejected.

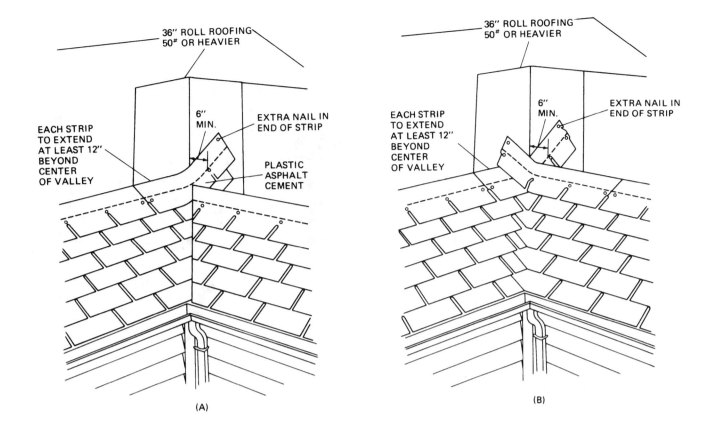

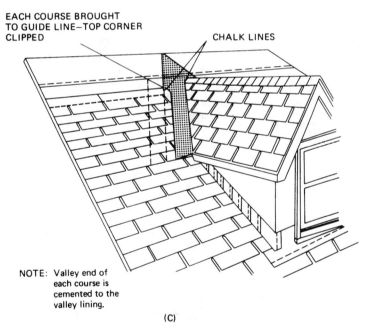

Fig. 8-7 *(A) Closed-cut valley; (B) Woven valley; (C) Open valley.*

In certain applications, such as homes, appearance is another important consideration. Shingles are used most frequently for homes. See Fig. 8-8D. In some areas, cedar shingles are not permitted because wood burns too easily. Once aged, however, it becomes more fire-resistant.

Pitch

Drainage of water from a roof surface is essential. This means that pitch should be considered. The pitch or slope of a roof deck determines the choice of shingle. It also determines drainage.

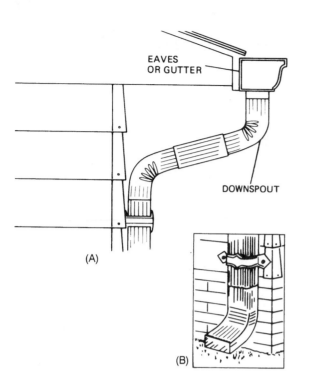

Fig. 8-8 (A) Eave trough and downspout. (B) Downspout elbow turns water away from the basement. (C) Preparing a flat roof. Asphalt-saturated felt is mopped down with hot asphalt or tar. (D) Applying a shingle roof. The shingles are packaged so that they can be placed in a location convenient for the roofer.

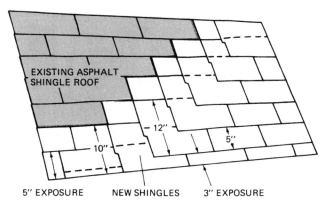

Fig. 8-9 *Exposure is the distance between the exposed edges of overlapping shingles.* (Bird and Son)

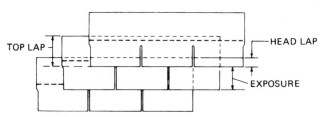

Fig. 8-10 *Head lap and top lap.* (Bird and Son)

Pitch limitations are shown in Fig. 8-11. Any shingle may safely be used on roofs with normal slopes. Normal is 4 inches rise or more per horizontal foot. An exception exists for square-butt strip shingles. They may be used on slopes included in the shaded area in Fig. 8-11.

When the pitch is less than 4 inches per foot, it is best to use roll roofings. In the range of 4 inches down to 1 inch per foot, the following rules apply:

- Roll roofing may be applied by the exposed nail method if the pitch is not lower than 2 inches per foot.

- Roll roofings applied by the concealed nail method may be used on pitches down to, but not below, 1 inch per foot. This is true if (1) they have at least 3 inches of top lap, and (2) they have double coverage roofing with a top lap of 19 inches.

Any of the above may be applied on a deck with a pitch steeper than the stated minimum. Pitch is given as a fraction. For example, a roof has a rise of 8 feet and a run of 12 feet. Then its pitch is

$$\frac{8}{2 \times 12} \text{ or } \frac{8}{24} \text{ or } \frac{1}{3}$$

Pitch is equal to the rise divided by 2 times the run. Or

$$\text{Pitch} = \frac{\text{rise}}{2 \times \text{run}}$$

Slope

Slope is how fast the roof rises from the horizontal. See Fig. 8-12. Slope is equal to the rise divided by the run. Or

$$\text{Slope} = \frac{\text{rise}}{\text{run}}$$

Slope and *pitch* are often used interchangeably. However, you can see there is a difference. Some roofers' manuals use them as if they were the same.

Before a roof can be applied, you have to know how many shingles are needed. This calls for estimating the area to be covered. First determine the number of square feet. Then divide the number of square feet by 100 to produce the number of squares needed.

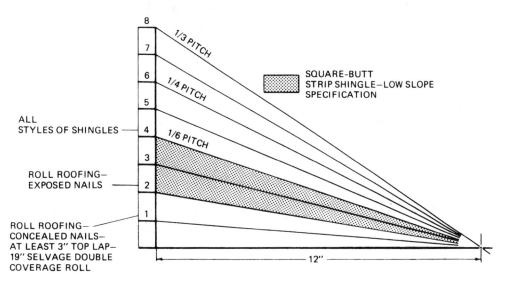

Fig. 8-11 *Minimum pitch requirements for different asphalt roofing products.* (Bird and Son)

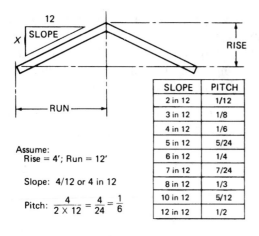

SLOPE	PITCH
2 in 12	1/12
3 in 12	1/8
4 in 12	1/6
5 in 12	5/24
6 in 12	1/4
7 in 12	7/24
8 in 12	1/3
10 in 12	5/12
12 in 12	1/2

Assume:
Rise = 4'; Run = 12'

Slope: 4/12 or 4 in 12

Pitch: $\dfrac{4}{2 \times 12} = \dfrac{4}{24} = \dfrac{1}{6}$

$$\text{Slope} = \frac{\text{rise}}{\text{run}} \qquad\qquad \text{Pitch} = \frac{\text{rise}}{2 \times \text{run}}$$

Fig. 8-12 *Slope, pitch, and run of a roof.*

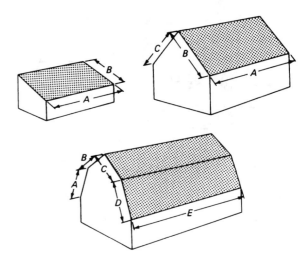

Fig. 8-13 *Simple roof types and their dimensions: Shed, gable, and gambrel.* (Bird and Son)

ESTIMATING ROOFING QUANTITIES

Roofing is estimated and sold in squares. *A square of roofing is the amount required to cover 100 square feet.* To estimate the required amount, you have to compute the total area to be covered. This should be done in square feet. Then divide the amount by 100. This determines the number of squares needed. Some allowance should be made for cutting and waste. This allowance is usually 10 percent. If you use 10 percent for waste and cutting, you will have the correct number of shingles. A simple roof with no dormers will require less than 10 percent.

Complicated roofs will require more than 10 percent for cutting and fitting.

Estimating Area

The areas of simple surfaces can be computed easily. The area of the shed roof in Fig. 8-13 is the product of the eave line and the rake line ($A \times B$). The area of the simple gable roof in Fig. 8-13 equals the sum of the two rakes B and C multiplied by the eave line A. A gambrel roof is estimated by multiplying rake lines A, B, C, and D by eave line E. See Fig. 8-13.

Complications arise in roofs such as the one in Fig. 8-14. Ells, gables, or dormers can cause special problems. Obtain the lengths of eaves, rakes, valleys, and ridges from drawings or sketches. Measuring calls for dangerous climbing. You may want to estimate without climbing. To do this:

1. The pitch of the roof must be known or determined.
2. The horizontal area in square feet covered by the roof must be computed.

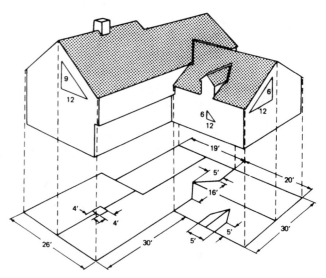

Fig. 8-14 *Complicated dwelling roof shown in perspective and plan views.* (Bird and Son)

Pitch is shown in Fig. 8-15. The pitch of a roof is stated as a relationship between rise and span. If the span is 24'0" and the rise is 8'0", the pitch will be 8/24 or ⅓. If the rise were 6'0", then the pitch would be 6/24 or ¼. The ⅓-pitch roof rises 8 inches per foot of horizontal run. The ¼-pitch roof rises 6 inches per foot of run.

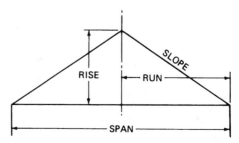

Fig. 8-15 *Pitch relations.*

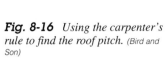

Fig. 8-16 *Using the carpenter's rule to find the roof pitch.* (Bird and Son)

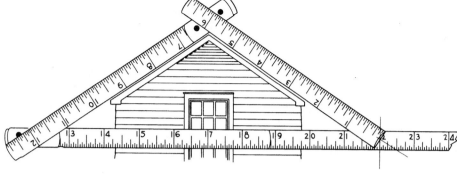

You can determine the pitch of any roof without leaving the ground. Use a carpenter's folding rule in the following manner.

Form a triangle with the rule. Stand across the street or road from the building. Hold the rule at arm's length. Align the roof slope with the sides of the rule. Be sure that the base of the triangle is held horizontal. It will appear within the triangle as shown in Fig. 8-16. Take a reading on the base section of the rule. Note the reading point shown in Fig. 8-17. Locate in the top line, headed *Rule Reading* in Fig. 8-18, the point nearest your reading. Below this point is the pitch and the rise per foot of run. Here the reading on the rule is 22. Under 22 in Fig. 8-18, the pitch is designated as this is a rise of 8 inches per foot of horizontal run.

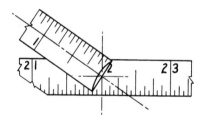

READING POINT

Fig. 8-17 *Reading the carpenter's rule to find the point needed for the pitch figures.* (Bird and Son)

Horizontal Area

Figure 8-14 is a typical dwelling. The roof has valleys, dormers, and variable-height ridges. Below the perspective the total ground area is covered by the roof. All measurements needed can be made from the ground. Or they can be made within the attic space of the house. No climbing on the roof is needed.

Computation of Roof Areas

Make all measurements. Draw a roof plan. Determine the pitches of the various elements of the roof. Use the carpenter's rule. The horizontal areas can now be quickly worked out.

Include in the estimate only those areas having the same pitch. The rise of the main roof is 9 inches per foot. That of the ell and dormers is 6 inches per foot.

The horizontal area under the 8-inch-slope roof will be

$$
\begin{array}{rl}
26 \times 30 = & 780 \text{ square feet} \\
19 \times 30 = & \underline{570} \text{ square feet} \\
\text{Total} & 1350 \text{ square feet}
\end{array}
$$

Less

$$
\begin{array}{rl}
8 \times 5 = & 40 \text{ (triangular area under ell roof)} \\
4 \times 4 = & \underline{16} \text{ (chimney)} \\
& 56 \text{ square feet}
\end{array}
$$

1350 − 56 = 1294 square feet total

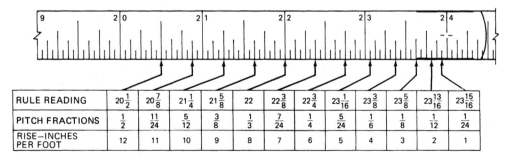

RULE READING	$20\frac{1}{2}$	$20\frac{7}{8}$	$21\frac{1}{4}$	$21\frac{5}{8}$	22	$22\frac{3}{8}$	$22\frac{3}{4}$	$23\frac{1}{16}$	$23\frac{3}{8}$	$23\frac{5}{8}$	$23\frac{13}{16}$	$23\frac{15}{16}$
PITCH FRACTIONS	$\frac{1}{2}$	$\frac{11}{24}$	$\frac{5}{12}$	$\frac{3}{8}$	$\frac{1}{3}$	$\frac{7}{24}$	$\frac{1}{4}$	$\frac{5}{24}$	$\frac{1}{6}$	$\frac{1}{8}$	$\frac{1}{12}$	$\frac{1}{24}$
RISE—INCHES PER FOOT	12	11	10	9	8	7	6	5	4	3	2	1

Fig. 8-18 *Reading point converted to pitch.* (Bird and Son)

The area under the 6-inch-rise roof will be

$20 \times 30 = 600$ square feet
$8 \times 5 = \underline{40}$ (triangular area projecting over the
640 main house)

Duplications

Sometimes one element of a roof projects over another. Add duplicated areas to the total horizontal area. If the eaves in Fig. 8-14 project only 4 inches, there will be

1. A duplication of 2 (7 × ⅓) or 4⅔ square feet under the eaves of the main house. This is where they overhang the rake of the ell section.
2. A duplication under the dormer eaves of 2 (5 × ⅓) or 3⅓ square feet.

3. A duplication of 9½ × ⅓ or 3¹⁄₁₆ square feet under the eaves of the main house. This is where they overhang the rake of the ell section.

The total is 11⅙ or 12 square feet. Item 2 should be added to the area of the 6-inch-pitch roof. Items 1 and 3 should be added to the 9-inch-pitch roof. The new totals will be 640 + 4 or 644 for the 6-inch pitch and 1294 + 8 or 1302 for the 9-inch pitch.

Converting Horizontal to Slope Areas

Now convert horizontal areas to slope areas. Do this with the aid of the *conversion table*, Table 8-1. Horizontal areas are given in the first column. Corresponding slope areas are given in columns 2 to 12.

Table 8-1 *Conversion Table*

Rise, Inches per Foot of Horizontal Run	1	2	3	4	5	6	7	8	9	10	11	12
Pitch, Fractions	1/24	1/12	1/8	1/6	5/24	1/4	7/24	1/3	3/8	5/12	11/24	1/2
Conversion Factor	1.004	1.014	1.031	1.054	1.083	1.118	1.157	1.202	1.250	1.302	1.356	1.414
Horizontal, area in Square Feet or length in feet												
1	1.0	1.0	1.0	1.1	1.1	1.1	1.2	1.2	1.3	1.3	1.4	1.4
2	2.0	2.0	2.1	2.1	2.2	2.2	3.2	2.4	2.5	2.6	2.7	2.8
3	3.0	3.0	3.1	3.2	3.2	3.2	3.5	3.6	3.8	3.9	4.1	4.2
4	4.0	4.1	4.1	4.2	4.3	4.5	4.6	4.8	5.0	5.2	5.4	5.7
5	5.0	5.1	5.2	5.3	5.4	5.6	5.8	6.0	6.3	6.5	6.8	7.1
6	6.0	6.1	6.2	6.3	6.5	6.7	6.9	7.2	7.5	7.8	8.1	8.5
7	7.0	7.1	7.2	7.4	7.6	7.8	8.1	8.4	8.8	9.1	9.5	9.9
8	8.0	8.1	8.3	8.4	8.7	8.9	9.3	9.6	10.0	10.4	10.8	11.3
9	9.0	9.1	9.3	9.5	9.7	10.1	10.4	10.8	11.3	11.7	12.2	12.7
10	10.0	10.1	10.3	10.5	10.8	11.2	11.6	12.0	12.5	13.0	13.6	14.1
20	20.1	20.3	20.6	21.1	21.7	22.4	23.1	24.0	25.0	26.0	27.1	28.3
30	30.1	30.4	31.0	31.6	32.5	33.5	34.7	36.1	37.5	39.1	40.7	42.4
40	40.2	40.6	41.2	42.2	43.3	44.7	46.3	48.1	50.0	52.1	54.2	56.6
50	50.2	50.7	51.6	52.7	54.2	55.9	57.8	60.1	62.5	65.1	67.8	70.7
60	60.2	60.8	61.9	63.2	65.0	67.1	69.4	72.1	75.0	78.1	81.4	84.8
70	70.3	71.0	72.2	73.8	75.8	78.3	81.0	84.1	87.5	91.1	94.9	99.0
80	80.3	81.1	82.5	84.3	86.6	89.4	92.6	96.2	100.0	104.2	108.5	113.1
90	90.4	91.3	92.8	94.9	97.5	100.6	104.1	108.2	112.5	117.2	122.0	127.3
100	100.4	101.4	103.1	105.4	108.3	111.8	115.7	120.2	125.0	130.2	135.6	141.4
200	200.8	202.8	206.2	210.8	216.6	223.6	231.4	240.4	250.0	260.4	271.2	282.8
300	301.2	304.2	309.3	316.2	324.9	335.4	347.1	360.6	375.0	390.6	406.8	424.2
400	401.6	405.6	412.4	421.6	433.2	447.2	462.8	480.8	500.0	520.8	542.4	565.6
500	502.0	507.0	515.5	527.0	541.5	559.0	578.5	601.0	625.0	651.0	678.0	707.0
600	602.4	608.4	618.6	632.4	649.8	670.8	694.2	721.2	750.0	781.2	813.6	848.4
700	702.8	709.8	721.7	737.8	758.1	782.6	809.9	841.4	875.0	911.4	949.2	989.8
800	803.2	811.2	824.8	843.2	864.4	894.4	925.6	961.6	1000.0	1041.6	1084.8	1131.2
900	903.6	912.6	927.9	948.6	974.7	1006.2	1041.3	1081.8	1125.0	1171.8	1220.4	1272.6
1000	1004.0	1014.0	1031.0	1054.0	1083.0	1118.0	1157.0	1202.0	1250.0	1302.0	1356.0	1414.0

The total area under the 9-inch rise is 1302 square feet. Under the column headed 9 (for 9-inch rise) on the conversion table is found:

	Horizontal Area		Slope Area
Opposite	1000	is	1250.0
Opposite	300	is	375.0
Opposite	00	is	00.0
Opposite	2	is	2.5
Totals	1302		1627.5

The total area under the 6-inch rise is 644 square feet.

	Horizontal Area		Slope Area
Opposite	600	is	670.8
Opposite	40	is	44.7
Opposite	4	is	4.5
Totals	644		720.0

The total area for both pitches is 1627.5 + 720 = 2347.5 square feet.

Now, add a percentage for waste. This should be 10 percent. That brings the total area of roofing required to 2582 square feet. Divide 2582 by 100 and get 25.82 or, rounded, *26 squares*.

One point about this method should be emphasized. The method is possible because of one fact. Over any given horizontal area, at a given pitch, a roof will always contain the same number of square feet regardless of its design. A shed, a gable, and a hip roof, with or without dormers, will each require exactly the same square footage of roofing—that is, if each is placed over the same horizontal area with the same pitch.

Accessories

Quantities of starter strips, edging strips, ridge shingles, and valley strips all depend upon linear measurements. These measurements are taken along the eaves, rake ridge, and valley. Eaves and ridge are horizontal. The rakes and valleys run on a slope. Quantities for the horizontal elements can be taken off the roof plan. True length of rakes and valleys must be taken from conversion tables.

LENGTH OF RAKE

Determine the length of the rake of the roof. Measure the horizontal distance over which it extends. In this case the rakes on the ends of the main house span distances are 26 and 19 feet. More rake footage is 26 + 19 + 13 + 3½ = 61½ feet.

Refer to Table 8-1 under the 9-inch-rise column. Opposite the figures in column 1 find the length of the rake.

	Horizontal Run	Length of Rake	
	60	75.00	
	1	1.3	
	0.5	0.6	
Totals	61.5	76.9	(actual length of the rake)

Use the same method and apply it to the rake of the ell. This will indicate its length, including the dormer, to be 39.1 inches. Add these amounts to the total length of eaves. The figure obtained can be used for an estimate of the amount of edging needed.

Hips and Valleys

Hip-and-valley lengths can be determined. Use the run off the common rafter. Then refer to the hip-and-valley table, Table 8-2.

Common rafter run is one-half the horizontal distance that the roof spans. This determines the length of a valley. The run of the common rafter should be taken at the lower end of the valley.

Figure 8-14 shows the portion of the ell roof that projects over the main roof. It has a span of 16 feet at the lower end of the valley. Therefore, the common rafter at this point has a run of 8'0".

There are two valleys at this roof intersection. Total run of the common rafter is 16'0". Refer to Table 8-2. Opposite the figures in the column headed *Horizontal*, find the linear feet of valleys. Then check the column under the pitch involved.

One of the intersecting roofs has a rise of 6 inches. The other has a rise of 9 inches. Length for each rise must be found. The average of the two is then taken. This gives a close approximation of the true length of the valley.

Thus,

Horizontal	6-inch rise	9-inch rise
10	15	16
6	9	9.6
16	24	25.6

24 + 25.6 = 49.6

49.6 ÷ 2 = 24.8 length of valleys

Table 8-2 Hip-and-Valley Conversions

Rise, Inches per Foot of Horizontal Run	4	5	6	7	8	9	10	11	12	14	16	18
Pitch { Degrees	18°26′	22°37′	26°34′	30°16′	33°41′	36°52′	39°48′	42°31′	45°	49°24′	53°8′	56°19′
Fractions	1/6	5/24	1/4	7/24	1/3	3/8	5/12	11/24	1/2	7/12	2/3	3/4
Conversion Factor	1.452	1.474	1.500	1.524	1.564	1.600	1.642	1.684	1.732	1.814	1.944	2.062
Horizontal Length in Feet												
1	1.5	1.5	1.5	1.5	1.6	1.6	1.6	1.7	1.7	1.8	1.9	2.1
2	2.9	2.9	3.0	3.0	3.1	3.2	3.3	3.4	3.5	3.6	3.9	4.1
3	4.4	4.4	4.5	4.6	4.7	4.8	4.9	5.1	5.2	5.4	5.8	6.2
4	5.8	5.9	6.0	6.1	6.3	6.4	6.6	6.7	6.9	7.3	7.8	8.2
5	7.3	7.4	7.5	7.6	7.8	8.0	8.2	8.4	8.7	9.1	9.7	10.3
6	8.7	8.8	9.0	9.1	9.4	9.6	9.9	10.1	10.4	10.9	11.7	12.4
7	10.2	10.3	10.5	10.7	10.9	11.2	11.5	11.8	12.1	12.7	13.6	14.4
8	11.6	11.8	12.0	12.2	12.5	12.8	13.1	13.5	13.9	14.5	15.6	16.5
9	13.1	13.3	13.5	13.7	14.1	14.4	14.8	15.2	15.6	16.3	17.5	18.6
10	14.5	14.7	15.0	15.2	15.6	16.0	16.4	16.8	17.3	18.1	19.4	20.6
20	29.0	29.5	30.0	30.5	31.3	32.0	32.8	33.7	34.6	36.3	38.9	41.2
30	43.6	44.2	45.0	45.7	46.9	48.0	49.3	50.5	52.0	54.4	58.3	61.9
40	58.1	59.0	60.0	61.0	62.6	64.0	65.7	67.4	69.3	72.6	77.8	82.5
50	72.6	73.7	75.0	76.2	78.2	80.0	82.1	84.2	86.6	90.7	97.2	103.1
60	87.1	88.4	90.0	91.4	93.8	96.0	98.5	101.0	103.9	108.8	116.6	123.7
70	101.6	103.2	105.0	106.7	109.5	112.0	114.9	117.9	121.2	127.0	136.1	144.3
80	116.2	117.9	120.9	121.9	125.1	128.0	131.4	134.7	138.6	145.1	155.5	165.0
90	130.7	132.7	135.0	137.2	140.8	144.0	147.8	151.6	155.9	163.3	175.0	185.6
100	145.2	147.4	150.0	152.4	156.4	160.0	164.2	168.4	173.2	181.4	194.4	206.2

Dormer Valleys

The run of the common rafter at the dormer is 2.5 feet. Check Table 8-2. It is found that:

Horizontal	6-inch rise
2.0	3.0
0.5	0.75
2.5	3.75 (length of valley)

Two such valleys will total 7.5 feet.

The total length of valley will be 24.8 + 7.5 = 32.3 feet. Use these figures to estimate the flashing material required.

ROOFING TOOLS

Most roofing tools are already in the carpenter's toolbox. Tools needed for roofing are shown in Fig. 8-19.

Roof brackets Can be used to clamp onto ladder.

Ladders A pair of sturdy ladders with ladder jacks are needed. Shingles are placed on the roof using a hoist on the delivery truck. These ladders come in handy for side roofing.

Staging These are planks for the ladder jacks. They hold the roofer or shingles. They are very useful on mansard roof jobs.

Apron The carpenter's apron is very necessary. It holds the nails and hammer. Other small tools can fit into it. It saves time in many ways. It keeps needed tools handy.

Hammer The hammer is a necessary device for roofing. It should be a balanced hammer for less wrist fatigue.

Chalk and line This combination is needed to draw guidelines. Shingles need alignment. The chalk marks are needed to make sure the shingles line up.

Tin snips Heavy-duty tin snips are needed for trimming flashing. They can also be used for trimming shingles.

Kerosene A cleaner is needed to remove tar from tools. Asphalt from shingles can be removed from tools with kerosene.

Tape measure A roofer has to make many measurements. This is a necessary tool.

Utility knife A general-purpose knife is needed for close trimming of shingles.

Putty knife This is used to spread roofing cement.

Carpenter's rule This makes measurements and also serves to determine the pitch of a roof. See Fig. 8-16.

Stapler Some new construction roofing can use a stapler. This device replaces the hammer and nails.

Fig. 8-19 Carpenter's tools needed for roofing: (Bird and Son) (A) Planking support; (B) Ladder with planking; (C) Claw hammer; (D) Carpenter's apron; (E) Chalk and cord; (F) Snips; (G) Tape measure; (H) Kerosene; (I) Stapler; (J) Carpenter's folding rule; (K) Utility knife; (L) Putty knife.

SAFETY

Working on a roof can be dangerous. Here are a few helpful hints. They may save you broken bones or pulled muscles.

1. Wear sneakers or rubber-soled shoes.
2. Secure ladders and staging firmly.
3. Stay off wet roofs.
4. Keep away from power lines.
5. Don't let debris accumulate underfoot.
6. Use roofing brackets, planks if the roof slopes 4 inches or more for every 12 inches of horizontal run.

APPEARANCE

How the finished job looks is important. Here are a few precautions to improve roof appearance.

1. Avoid shingling in extremely hot weather. Soft asphalt shingles are easily marred by shoes and tools.

2. Avoid shingling when the temperature is below 40°F. Cold shingles are stiff and may crack.

3. Measure carefully and snap the chalk line frequently. Roof surfaces aren't always square. You'll want to know about problems to come so that you can correct them.

4. Start at the rear of the structure. If you've never shingled before, this will give you a chance to gain experience before you reach the front.

APPLYING AN ASPHALT ROOF

Asphalt roofing products will serve well when they are correctly applied. Certain fundamentals must be considered. These have to do with the deck, flashing, and application of materials.

Roof Problems

A number of roof problems are caused by defects in the deck. A nonrigid deck may affect the lay of the roofing. Poorly seasoned deck lumber may warp. This can cause cocking of the shingle tabs. It can also cause wrinkling and buckling of roll roofing.

Improper ventilation can have an effect similar to that of green lumber. The attic area should be ventilated. This area is located directly under the roof deck. It should be free of moisture. In cold weather, be sure the interiors are well ventilated. This applies when plaster is used in the building. A positive ventilation of air is required through the building during roofing. This can usually be provided by opening one or two windows. Windows in the basement or on the first floor can be opened. This can create a positive draft through the house. Open windows at opposite ends of the building. This will also create a flow of air. The moving air has a tendency to dry out the roof deck. It helps to eliminate excess moisture. Condensation under the roofing can cause problems.

Deck construction Wood decks should be made from well-seasoned tongue-and-groove lumber that is more than 6 inches wide. Wider sheathing boards are more likely to swell or shrink, producing a buckling of the roof material. Sheathing should be tightly matched. It should be secured to the rafter with at least two 8d nails. One should be driven through the edge of the board. The other should be driven through the board face. Boards containing too many resinous areas should be rejected. Boards with loose knots should not be used. Do not use badly warped boards.

Figure 8-20 shows how a wood roof deck is constructed. In most cases today, 4-×-8-foot sheets of plywood are used as sheathing. The plywood goes over the rafters. C—D grade plywood is used.

Underlayment Apply one layer of no. 15 asphalt-saturated felt over the deck as an underlayment if the deck has a pitch of 4 inches per foot or greater. The felt should be laid horizontally. See Fig. 8-20. Do not use no. 30 asphalt felt. Do not use any tar-saturated felt. Laminated waterproof papers should not be used either. Do not use any vapor barrier-type material. Lay each course of felt over the lower course. Lap the courses 4 inches. Overlap should be at least 2 inches where ends join. Lap the felt 6 inches from both sides over all hips and ridges.

Apply underlayment as specified for low-slope roofs, where the roof slope is less than 4 inches per

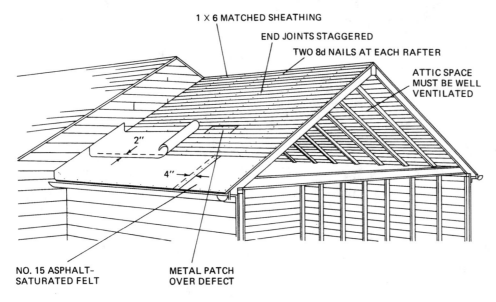

1 X 6 MATCHED SHEATHING

END JOINTS STAGGERED

TWO 8d NAILS AT EACH RAFTER

ATTIC SPACE MUST BE WELL VENTILATED

2"

4"

NO. 15 ASPHALT-SATURATED FELT

METAL PATCH OVER DEFECT

Fig. 8-20 *Features of a good wood roof deck.* (Bird and Son)

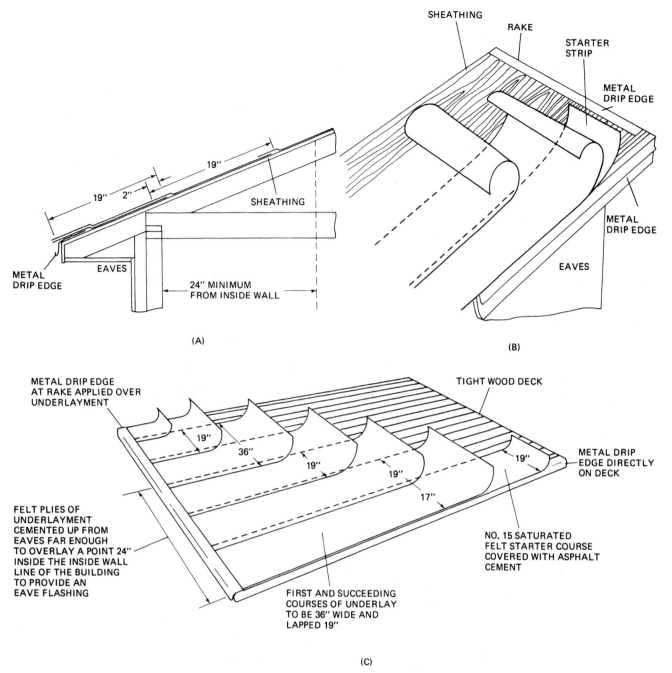

Fig. 8-21 *(A) Eaves flashing for a low-slope roof. (B) Placing of sheathing and drip edge. (C) Placement of the underlayment for a shingle roof.* (Bird and Son)

foot and not below 2 inches. Check the maker's suggestions. See Fig. 8-21. Felt underlayment performs three functions:

1. It ensures a dry roof for shingles. This avoids buckling and distortion of shingles. Buckling may be caused by shingles being placed over wet roof boards.

2. Felt underlay prevents the entrance of wind driven rain onto the wood deck. This may happen when shingles are lifted up.

3. Underlay prevents any direct contact between shingles and resinous areas. Resins may cause chemical reactions. These could damage the shingles.

Plywood decks Plywood decks should meet the Underwriters' Laboratories standards. Standards are set according to grade and thickness. Design the eaves, rake, and ridge to prevent problems. Openings through the deck should be made in such a way that the plywood will not be exposed to the weather. See Fig. 8-22.

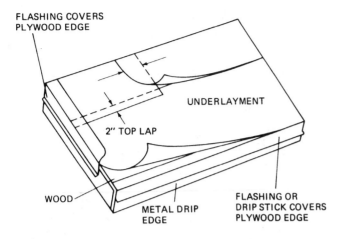

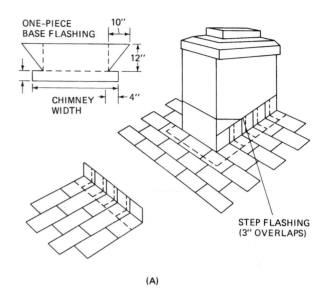

(A)

Fig. 8-22 *Preparing the roof deck for shingling.* (Bird and Son)

Nonwood deck materials Nonwood materials are sometimes used in decks. Such things as fiberboard, gypsum, concrete plank, and tile are nonwoods used for decks. These materials have their own standards. Check with the manufacturer for suggestions.

Flashings Roofs are often complicated by intersections with other roofs. Some adjoining walls have projections through the deck. Chimneys and soil stacks create leakage problems. Special attention must be given to protecting against the weather here. Such precautions are commonly called *flashing*. Careful attention to flashing is critical. It helps provide good roof performance. See Fig. 8-23.

Valleys Valleys exist where two sloping roofs meet at an angle. This causes water runoff toward and along the joint. Drainage concentrates at the joint. This makes the joint an easy place for water to enter. Smooth, unobstructed drainage must be provided. It should have enough capacity to carry away the collected water.

There are three types of valleys: open, woven, and closed-cut. See Fig. 8-24.

Each type of valley calls for its own treatment. Figure 8-25 shows felt being applied to a valley. A 36-inch-wide strip of 15# asphalt-saturated felt is centered in the valley. It is secured with nails. They hold it in place until shingles are applied. Courses of felt are cut to overlay the valley strip. The lap should not be less than 6 inches. Eave flashing is then applied.

PUTTING DOWN SHINGLES

Before you put the shingles down, you need an underlayment. See Fig. 8-26. Covering the underlayment is

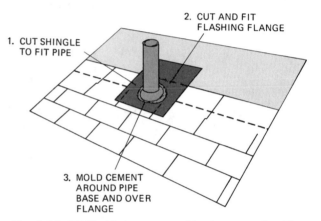

Fig. 8-23 *(A) Flashing patterns and in place around a chimney* (Bird and Son). *(B) Flashing around a soil pipe.* (Bird and Son)

a sheet of saturated felt or tar paper. Table 8-3 shows characteristics of typical asphalt rolls. It is best not to put down the first shingle until you know what is available. Study Table 8-4 to check the characteristics and sizes of typical asphalt shingles. Remember the # symbol means pound or lb., as in 285# or 285 pounds per square of shingles. A *square* covers an area of 100 square feet.

Nails

Nails used in applying asphalt roofings are large-headed and sharp-pointed. Some are hot-galvanized steel. Others are made of aluminum. They may be barbed or otherwise deformed on the shanks. Figure 8-27 shows three types of asphalt nails.

Roofing nails should be long enough to penetrate through the roofing material. They should go at least ¾ inch into the wood deck. This requires that they be of the lengths indicated in Table 8-5.

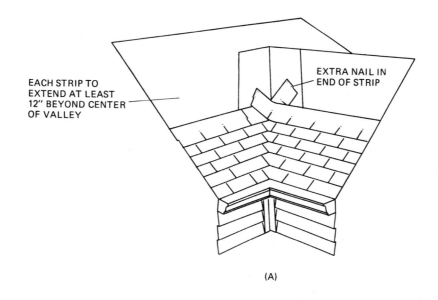

(A)

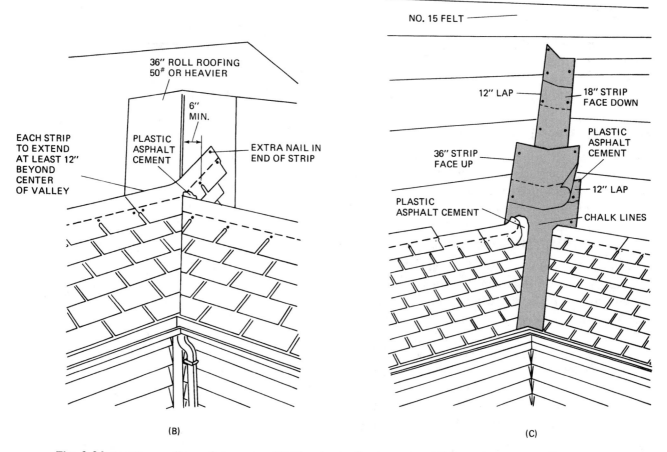

(B)   (C)

Fig. 8-24 *(A) Woven valley roof (Bird and Son); (B) Closed-cut valley (Bird and Son); (C) Preparing an open valley. (Bird and Son)*

Number of Nails The number of nails required for each shingle type is given by its maker. Manufacturer's recommendations come with each bundle.

Use 2-inch centers in applying roll roofing. This means 252 nails are needed per square. If 3-inch centers are used, then 168 nails are needed per square.

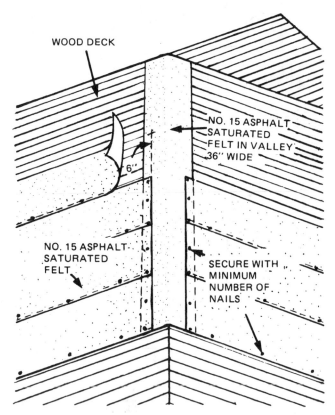

Fig. 8-25 *Felt underlay centered in the valley before valley linings are applied.* (Bird and Son)

Fasteners for Nonwood Materials

Gypsum products, concrete plank and tile, fiberboard, or unusual materials require special fasteners. This type of deck varies with its manufacturer. In such cases follow the manufacturer's suggestions.

Shingle Selection

There are a number of types of shingles available. They may be used for almost any type of roof. Various colors are used to harmonize with buildings. Some are made for various weather conditions. White and light colors are used to reflect the sun's rays. Pastel shingles are used to achieve a high degree of reflectivity. They still permit color blending with siding and trim. Fire and wind resistance should be considered. Simplicity of application makes asphalt roofings the most popular in new housing. They are also rated high for reroofing.

Farm buildings There is no one kind of asphalt roofing for every job. Building types are numerous. The style of roof on the farmhouse may affect the choice of roof on other buildings. They probably should have the same color roofing. This would make the group harmonize. A poultry laying house or machine storage shed near the house calls for a roof like the farm house.

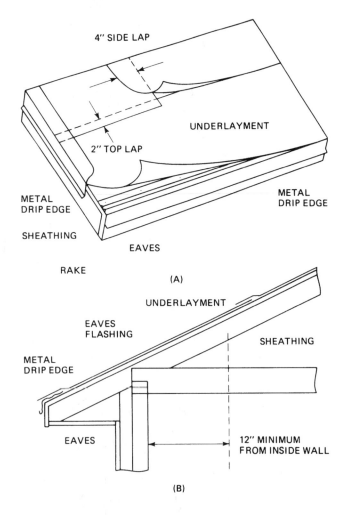

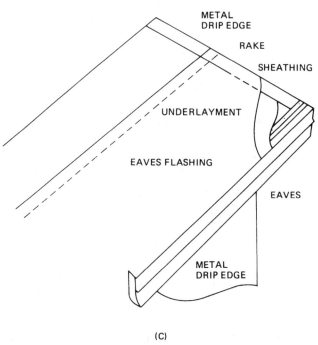

Fig. 8-26 *(A) Underlayment and drip edge; (B) Eaves flashing for a roof; (C) Underlayment, eaves flashing, and metal drip edge.* (Bird and Son)

Table 8-3   Typical Asphalt Rolls

1	2		3	4		5	6		7
Product	Approximate Shipping Weight		Squares Per Package	Length, Feet	Width, Inches	Side or End Lap, Inches	Top Lap, Inches	Exposure, Inches	Underwriters' Listing
	Per Roll	Per Square							
Mineral surface roll double coverage	75# to 90#	75# to 90#	1	36 38	36 38	6	2 4	34 32	C
			Available in some areas in 9/10 or 3/4 square rolls.						
Mineral surface roll	55# to 70#	55# to 70#	1/2	36	36	6	19	17	C
Coated roll	50# to 65#	50# to 65#	1	36	36	6	2	34	None
Saturated felt	60# 60# 60#	15# 20# 30#	4 3 2	144 108 72	36 36 36	4 to 6	2	34	None

An inexpensive roll roofing might be used for an isolated building.

The main idea is to select the right product for the building. The main reason for selecting a roofing material is protection of the contents of a building. The second reason for selecting a roofing is low maintenance cost.

Staples Staples may be used as an alternative to nails. This is only for new buildings. Staples must be zinc-coated. They should be no less than 16 gage. A semiflattened elliptical cross section is preferred. They should be long enough to penetrate ¾ inch into the wood deck. They must be driven with *pneumatic* (air-driven) staplers. The staple crown must bear tightly against the shingle. However, it must not cut the shingle surface. Use four staples per shingle. See Fig. 8-28. The crown of the staple must be parallel to the tab edge. Position it as shown in Fig. 8-28. Figure 8-29 shows how shingles are nailed. Figure 8-30 shows how the shingles are overlapped.

Cements

Six types of asphalt coatings and cements are:

1. Plastic asphalt cements
2. Lap cements
3. Quick-setting asphalt adhesives
4. Asphalt water emulsions
5. Roof coatings
6. Asphalt primers

Methods of softening The materials are flammable. Cement should be applied to a dry, clean surface. It

Table 8-4 *Typical Asphalt Shingles*

Product*	Configuration	Per Square			Size		Exposure, Inches	Underwriters' Listing
		Approximate Shipping Weight	Shingles	Bundles	Width, Inches	Length Inches		
Wood appearance strip shingle more than one thickness per strip 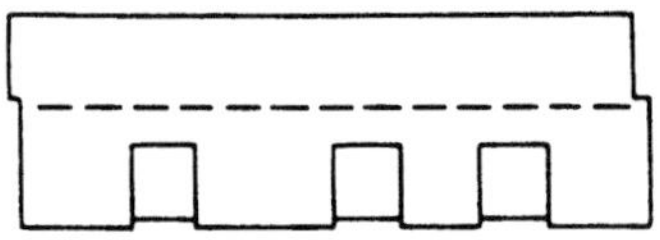 Laminated or job-applied	Various edge, surface texture, and application treatments	285# to 390#	67 to 90	4 or 5	11½ to 15	36 or 40	4 to 6	A or C, many wind-resistant
Wood appearance strip shingle single thickness per strip 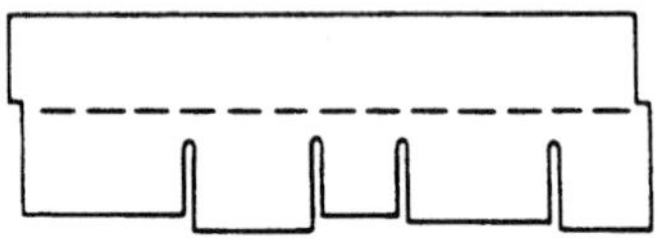	Various edge, surface texture, and application treatments	Various, 250# to 350#	78 to 90	3 or 4	12 or 12¼	36 or 40	4 to 5⅛	A or C, many wind-resistant
Self-sealing strip shingle 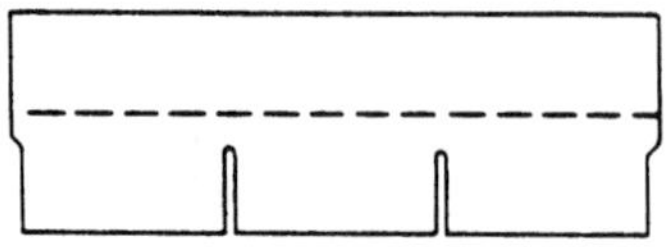	Conventional three-tab	205# to 240#	78 or 80	3	12 or 12¼	36	5 or 5⅛	A or C, all wind-resistant
	Two- or four-tab	Various, 215# to 325#	78 or 80	3 or 4	12 or 12¼	36	5 or 5⅛	
Self-sealing strip shingle 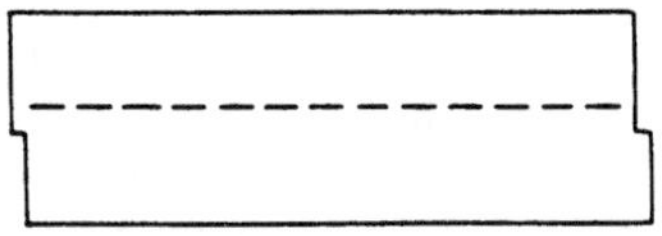 No cutout	Various edge and texture treatments	Various, 215# to 290#	78 to 81	3 or 4	12 or 12¼	36 or 36¼	5	A or C, all wind-resistant
Individual lock-down 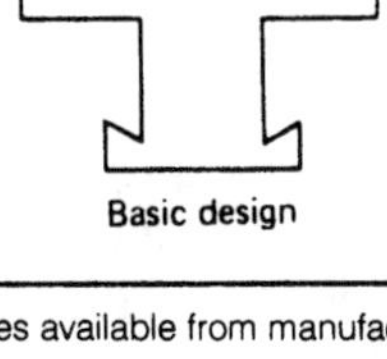 Basic design	Several design variations	180# to 250#	72 to 120	3 or 4	18 to 22¼	20 to 22½		C, many wind-resistant

*Other types available from manufacturers in certain areas of the country. Consult your regional Asphalt Roofing Manufacturers Association manufacturer.

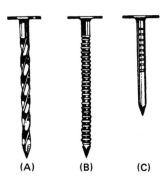

Fig. 8-27 *(A) Screw-threaded nail; (B) Annular threaded nail; (C) Asphalt shingle nail—smooth.*

Table 8-5 *Recommended Nail Length*

Purpose	Nail Length, Inches
Roll roofing on new deck	1
Strip or individual shingles—new deck	1 1/4
Reroofing over old asphalt roofing	1 1/4 to 1 1/2
Reroofing over old wood shingles	1 3/4

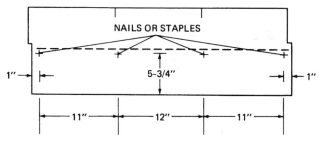

Fig. 8-28 *Nailing or stapling a strip asphalt shingle.* (Certain-Teed)

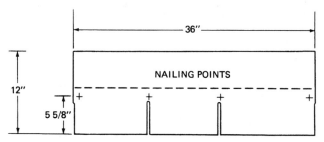

Fig. 8-29 *Nailing points on a strip shingle.* (Bird and Son)

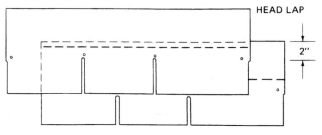

Fig. 8-30 *Overlap of shingles.* (Bird and Son)

should be troweled or brushed vigorously to remove air bubbles. The material should flow freely. It should be forced into all cracks and openings. An emulsion may be applied to damp or wet surfaces. It should not be applied in an exposed location. It should not be rained on for at least 24 hours. Emulsions are water soluble.

Uses Plastic asphalt cements are used for flashing cements. They are so processed that they will not flow at summer temperatures. They are elastic after setting. This compensates for normal expansion and contraction of a roof deck. They will not become brittle at low temperatures.

Lap cements come in various thicknesses. Follow the manufacturer's suggestions. Lap cement is not as thick as plastic cement. It is used to make a watertight bond. The bond is between lapping elements of roll roofing. It should be spread over the entire lapped area. Nails used to secure the roofing should pass through the cement. The shank of the nail should be sealed where it penetrates the deck material.

Seal down the free tabs of strip shingles with quick-setting asphalt adhesive. It can also be used for sealing laps of roll roofing.

Quick-setting asphalt adhesive is about the same thickness as plastic-asphalt cement. However, it is very adhesive. It is mixed with a solvent that evaporates quickly. This permits the cement to set up rapidly.

Roof coatings are used in spray or brush thickness. They are used to coat the entire roof. They can be used to resurface old built-up roofs. Old roll roofing or metal roofs can also be coated.

Asphalt water emulsions are a special type of roof coating made with asphalt and sometimes mixed with other materials. Because they are emulsified with water, they can freeze. Be sure to store them in a warm location. They should not be rained on for at least 24 hours.

Masonry primer is very fluid. Apply it with a brush or by spray. It must be thin enough to penetrate rapidly into the surface pores of masonry. It should not leave a continuous surface film. Thin, if necessary, by following instructions on the can.

Asphalt primer is used to prepare the masonry surface. It should bond well with other asphalt products. These are found on built-up roofs. Other products are plastic-asphalt cement or asphalt coatings.

Starter Course

Putting down the shingles isn't too hard—that is, if you have the roof deck in place. It should be covered

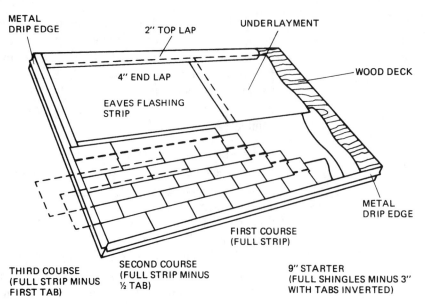

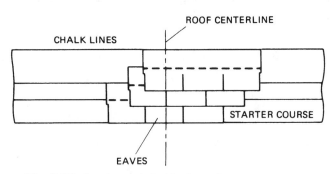

METAL
DRIP EDGE
2" TOP LAP
UNDERLAYMENT
4" END LAP
EAVES FLASHING
STRIP
WOOD DECK
METAL
DRIP EDGE
FIRST COURSE
(FULL STRIP)
THIRD COURSE
(FULL STRIP MINUS
FIRST TAB)
SECOND COURSE
(FULL STRIP MINUS
½ TAB)
9" STARTER
(FULL SHINGLES MINUS 3"
WITH TABS INVERTED)

Fig. 8-31 *Starting asphalt shingles at the rake.* (Bird and Son)

by the proper underlayment. The eaves should be properly prepared. Refer back to Fig. 8-26.

Starting at the rake Use only the upper portion of the asphalt shingle. Cut off the tabs. Position it with the adhesive dots toward the eaves. The starter course should overhang the eaves and rake edges by ¼ inch. Nail it in a line 3 to 4 inches above the eaves. See Fig. 8-31.

Start the first course with a full strip. Overhang the drip edges at the eaves and rake by ¼ inch. Nail the strip in place. Drive the nails straight. The heads should be flush with the surface of the shingle.

Snap a chalk line along the top edge of the shingle. The line should be parallel with the eaves. Snap several others parallel with the first. Make them 10 inches apart. Use the lines to check alignment at every other course. Snap lines parallel with the rake at the shingle cutouts. Use the lines to check cutout alignment.

Start the second course with a full strip less 6 inches. This means half a tab is missing. Overhang the cut edge at the rake. Nail the shingle in place.

Start the third course with a full strip less a full tab.

Start the fourth course with half a strip.

Continue to reduce the length of the first shingle in each course by an additional 6 inches. The sixth course starts with a 6-inch strip.

Return to the eaves. Apply full shingles across the roof, finishing each course. Dormer, chimney, or vent pipe instructions are another matter. They will be found later in this chapter.

For best color distribution, lay at least four strips in each row. Do this before repeating the pattern up the roof.

Start the *seventh course* with a full shingle. Repeat the process of shortening. Each successive course of

shingles is shortened by an additional 6 inches. This continues to the twelfth course.

Return to the seventh course. Apply full shingles across the roof.

Starting at the Center (Hip Roof)

Snap a vertical chalk line at the center of the roof. See Fig. 8-32.

ROOF CENTERLINE
CHALK LINES
STARTER COURSE
EAVES

Fig. 8-32 *Starting asphalt shingles at the center.* (Bird and Son)

Put down starter strips along the eaves. Do this in each direction from the chalk line. Go slightly over the centerlines of the hips. Overhang the eaves by ¼ inch.

Align the butt edge of a full shingle with the bottom edge of the starter strip. Also, align it with its center tab centered on the chalk line.

Snap a chalk line along the top of the shingle parallel with the eaves. Snap several others parallel with the first. They should be 10 inches apart. Use the lines to check the alignment of alternate courses. Finish the first course with full shingles. Extend the shingles part way over the hips.

Finish the remaining courses with full shingles.

Valleys

There are three types of valleys. One is the open type. Here, the saturated felt can be seen after the shingles are applied. Another type is the woven valley. This one has the shingles woven. There is no obvious valley line. The other is the closed-cut valley. This one has a straight line where the roofs intersect.

Open valleys Use mineral-surfaced-material roll roofing for this valley. Match or contrast the color with that of the roof covering. The open valley method is shown in Fig. 8-33. The felt underlay is centered in the valley before shingles are applied. See Fig. 8-34.

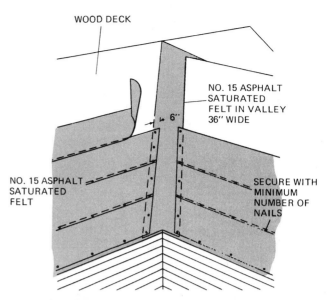

Fig. 8-34 *Felt underlay centered in the valley before valley linings are applied.* (Bird and Son)

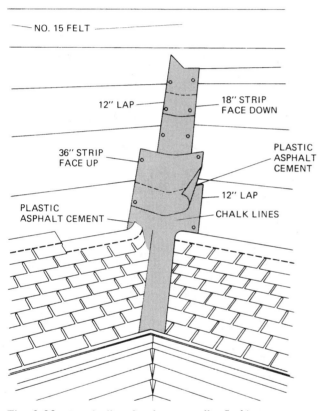

Fig. 8-33 *Use of roll roofing for open valley flashing.* (Bird and Son)

Center an 18-inch-wide layer of mineral-surfaced roll roofing in the valley. The surfaced side goes down. Cut the lower edge to conform to and be flush with the eave flashing strip. The ends of the upper segments overlap the lower segments in a splice. The ends are secured with plastic asphalt cement. See Fig. 8-33. Use only enough nails, 1 inch in from each edge, to hold the strip smoothly. Press the roofing firmly in place into the valley as you nail. Place another strip 36 inches wide on top of the first strip. This is placed surfaced side up. Center it in the valley. Nail it in the same manner as the underlying 18-inch strip.

Do this before the roofing is applied. Snap two chalk lines the full length along the valley, one line on each side of the valley. They should be 6 inches apart at the ridge, or, 3 inches when measured from the center of the valley. The marks diverge at the rate of $\frac{1}{8}$ inch per foot as they approach the eaves. A valley of 8 feet in length will be 7 inches wide at the eaves. One 16 feet long will be 8 inches wide at the eaves. The chalk line serves as a guide in trimming the last unit to fit the valley. This ensures a clean, sharp edge. The upper corner of each end shingle is clipped. See Fig. 8-34. This keeps water from getting in between the courses. The roofing material is cemented to the valley lining. Use plastic asphalt cement.

Woven and closed-cut valleys Some roofers prefer woven or closed-cut valleys. These are limited to strip-type shingles. Individual shingles cannot be used. Nails may be required at or near the center of the valley lining. Avoid placing a nail in an overlapped shingle too close to the center of the valley. It may sometimes be necessary to cut a strip short. That is done if it would otherwise end near the center. Continue from this cut end over the valley with a full strip. These methods increase the coverage of the shingles throughout the length of the valleys. This adds to the weather resistance of the roofs at these points.

Woven valleys There are two methods of weaving the shingles. See Fig. 8-35. They can be applied on both roof areas at the same time. This means you weave each course, in turn, over the valley. Or, you may cover each roof area first. Do this to a point about

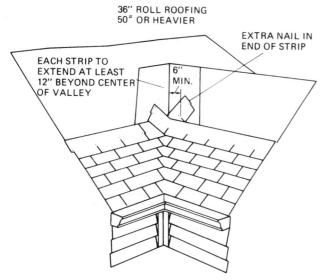

Fig. 8-35 *Weaving each course in turn to make a woven valley.* (Bird and Son)

Fig. 8-36 *Worker installing shingles. Note that the first strip is at the bottom. Also note the felt strip in the valley.* (Fox and Jacobs)

3 feet from the center of the valley. Then weave the valley shingles in later.

In the first method, lay the first course. Place it along the eaves of one roof area up to and over the valley. Extend it along the adjoining roof area. Do this for a distance of at least 12 inches. Then lay the first course along the eaves of the intersecting roof area. Extend it over the valley. It goes on top of the previously applied shingle. The next courses go on alternately. Lay along one roof area and then along the other. Weave the valley shingles over each other. See Fig. 8-35. Make sure that the shingles are pressed tightly into the valley. Nail them in the normal manner. No nails are located closer than 6 inches to the valley centerline. Two nails are located at the end of each terminal strip. See Fig. 8-35.

Closed-cut valleys For a closed-cut valley, lay the first course of shingles along the eaves of one roof area up to and over the valley. Extend it along the adjoining roof section. The distance is at least 12 inches. Follow the same procedure when applying the next courses of shingles. See Fig. 8-36. Make sure that the shingles are pressed tightly into the valley. Nail in the normal manner, except that no nail is to be located closer than 6 inches to the valley centerline. Two nails are located at the end of each terminal strip. See Fig. 8-37.

Apply the first course of shingles. Do this along the eaves of the intersecting roofs. Extend it over previously applied shingles. Trim a minimum of 2 inches up from the centerline of the valley. Clip the upper corner of each end shingle. This prevents water from getting between courses. Embed the end in a 3-inch-wide strip of plastic asphalt cement. Other courses are applied and completed. See Fig. 8-37.

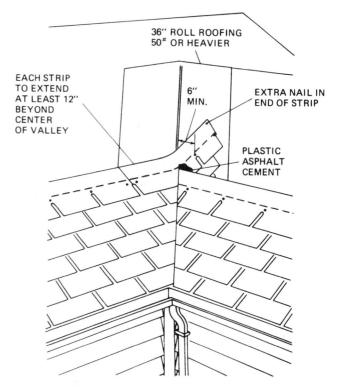

Fig. 8-37 *Closed-cut valley.* (Bird and Son)

An open valley for a dormer roof A special treatment is needed where an open valley occurs at a joint between the dormer roof and the main roof through which it projects. See Fig. 8-38.

First apply the underlay. Main roof shingles are applied to a point just above the lower end of the valley. The course last applied is fitted. It is fitted close to and flashed against the wall of the dormer. The wall is under the projecting edge of the dormer eave. The first strip of valley lining is then applied. Do this the same way as for the open valley. The bottom end is cut so that it extends ¼ inch below the edge of the dormer deck. The lower edge of the section lies on the main deck. It projects at least 2 inches below the joining

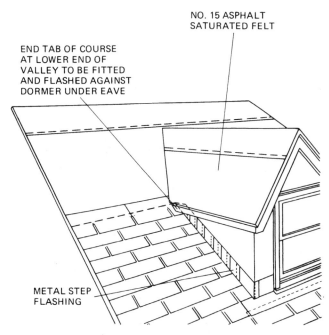

Fig. 8-38 An open valley for a dormer roof. Shingles have been laid on main roof up to lower end of the valley. (Bird and Son)

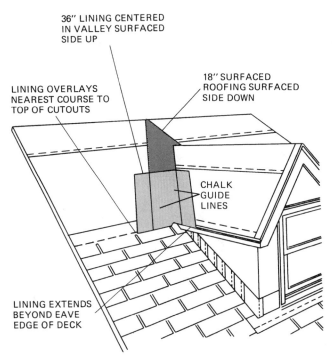

Fig. 8-39 Application of valley lining for a dormer roof. (Bird and Son)

roofs. Cut the second or upper strip on the dormer side. It should match the lower end of the underlying strip. Cut the side that lies on the main deck. It should overlap the nearest course of shingles. This overlap is the same as the normal lap of one shingle over another. It depends on the type of shingle being applied. In this case it extends to the top of the cutouts. This is a 12-inch-wide three-tab square-butt strip shingle.

The lower end of the lining is then shaped. See Fig. 8-39. It forms a small canopy over the joint between the two decks.

Apply shingles over the valley lining. The end shingle in each course is cut. It should conform to the guidelines. Bed the ends in a 3-inch-wide strip of plastic asphalt cement. Valley construction is completed in the usual manner. See Fig. 8-40.

Flashing Against a Vertical Wall

Step flashing is used when the rake of a roof abuts a vertical wall. It is best to protect the joint by using metal flashing shingles. They are applied over the end of each course of shingles.

The flashing shingles are rectangular in shape. They are from 5 to 6 inches long. They are 2 inches wider than the exposed face of the roofing shingles. When used with strip shingles laid 5 inches to the weather, they are 6 to 7 inches long. They are bent so as to extend 2 inches out over the roof deck. The remainder goes up the wall surface. Each flashing shingle is placed just uproof from the exposed edge of the

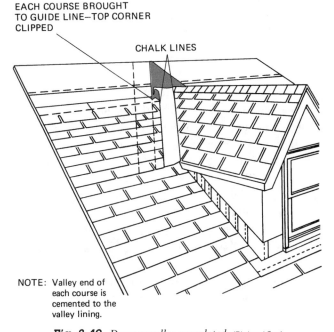

Fig. 8-40 Dormer valley completed. (Bird and Son)

single which overlaps it. It is secured to the wall sheathing with one nail in the top corner. See Fig. 8-41. The metal is 7 inches wide. The roof shingles are laid 5 inches to the weather. Each element of flashing will lap the next by 2 inches. See Fig. 8-41.

The finished siding is brought down over the flashing to serve as a cap flashing. However, it is held far enough away from the shingles. This allows the

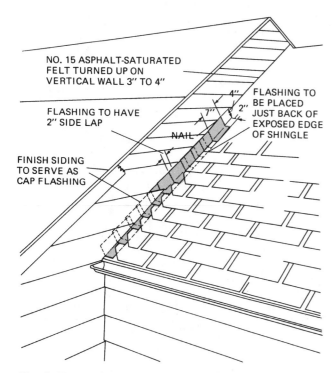

NO. 15 ASPHALT-SATURATED FELT TURNED UP ON VERTICAL WALL 3" TO 4"

FLASHING TO HAVE 2" SIDE LAP

FINISH SIDING TO SERVE AS CAP FLASHING

4"

7"

2"

NAIL

FLASHING TO BE PLACED JUST BACK OF EXPOSED EDGE OF SHINGLE

Fig. 8-41 *Use of metal flashing shingles to protect the joint between a sloping roof and a vertical wall.* (Bird and Son)

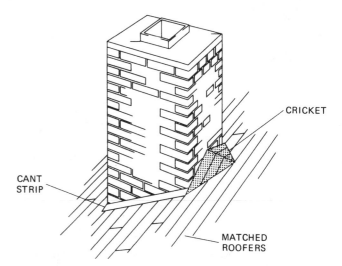

CRICKET

CANT STRIP

MATCHED ROOFERS

Fig. 8-42 *Cricket or saddle built behind the chimney.* (Bird and Son)

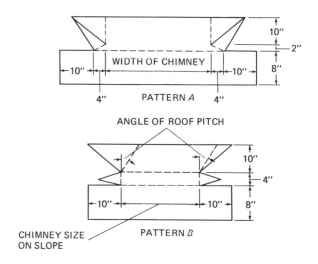

WIDTH OF CHIMNEY

10"

10"

10"

2"

8"

4"

PATTERN *A*

4"

ANGLE OF ROOF PITCH

10"

4"

10"

10"

8"

CHIMNEY SIZE ON SLOPE

PATTERN *B*

ends of the boards to be painted. Paint excludes dampness and prevents rot.

Chimneys

Chimneys are usually built on a separate foundation. This avoids stresses and distortions due to uneven settling. It is subject to differential settling. Flashing at the point where the chimney comes through the roof calls for something that will allow movement without damage to the water seal. It is necessary to use base flashings. They should be secured to the roof deck.

The counter or cap flashings are secured to the masonry. Figures 8-42 through 8-46 show how roll roofing is used for base flashing. Metal is used for cap flashing.

Apply shingles over the roofing felt up to the front face of the chimney. Do this before any flashings are placed. Make a saddle or cricket. See Fig. 8-42. This sits between the back face of the chimney and the roof deck. The cricket keeps snow and ice from piling up. It also deflects downflowing water around the chimney.

Apply a coat of asphalt primer to the brick work. This seals the surface. This is where plastic cement will later be applied. Cut the base flashing for the front. Cut according to the pattern shown in Fig. 8-43 (pattern *A*). This one is applied first. The lower section is laid over the shingles in a bed of plastic asphalt cement. Secure the upper vertical section against the ma-

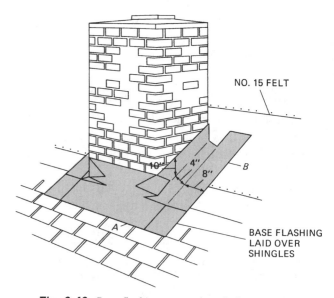

NO. 15 FELT

10"

4"

8"

B

A

BASE FLASHING LAID OVER SHINGLES

Fig. 8-43 *Base flashings cut and applied.* (Bird and Son)

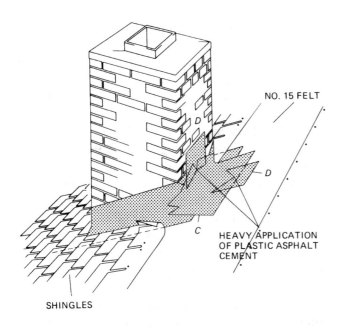

Fig. 8-44 *Flashing over the cricket in the rear of the chimney.*
(Bird and Son)

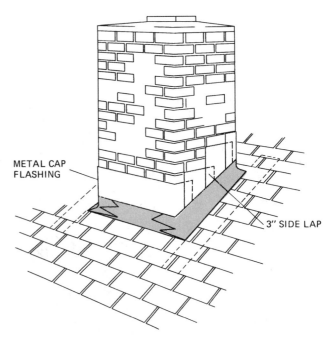

Fig. 8-46 *Metal cap flashing applied to cover the base flashing.*
(Bird and Son)

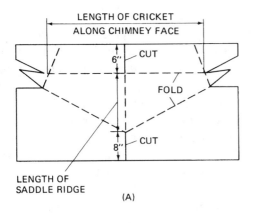

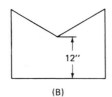

Fig. 8-45 *Flashing patterns.* (Bird and Son)

sonry with the same cement. Secure the upper vertical section against the masonry with the cement. Nails can also be used here, driven into the mortar joints. Bend the triangular ends of the upper section around the corners of the chimney. Cement in place.

Cut the side base flashings next. Use pattern *B* of Fig. 8-43. Bend them to shape and apply them as shown. Embed them in plastic asphalt cement. Turn the triangular ends of the upper section around the chimney corners. Cement them in place over the front base flashing.

Figure 8-44 shows the cutting and fitting of base flashings over the cricket. The cricket consists of two triangular pieces of board. The board is cut to form a ridge. The ridge extends from the centerline of the chimney back to the roof deck. The boards are nailed to the wood deck. They are also nailed to one another along the ridge. This is done before felt underlayment is applied. Cut the base flashing. See Fig. 8-45A. Bend it to cover the entire cricket. Extend it laterally to cover part of the side base flashing. Cut a second rectangular piece of roofing. See Fig. 8-45B. Make a cutout on one side to conform to the rear angle of the cricket. Set it tightly in plastic asphalt cement. Center it over that part of the cricket flashing extending up to the deck. This piece provides added protection where the ridge of the cricket meets the deck. Cut a second similar rectangular piece of flashing. Cut a V from one side. It should conform to the pitch of the cricket. Place it over the cricket ridge and against the chimney. Embed it in plastic asphalt cement.

Use plastic asphalt cement generously. Use it to cement all standing portions of the base flashing to the brickwork.

Cap flashings are shown in Fig. 8-46. They are made of sheet copper, 16-ounce or heavier. You can also make the caps of 24-gage galvanized steel. If steel is used, it should be painted on both sides.

Brickwork is secured to the cap flashing in Figs. 8-46 and 8-47. These drawings show a good method. Rake the mortar joint to a depth of 1½ inches. Insert the

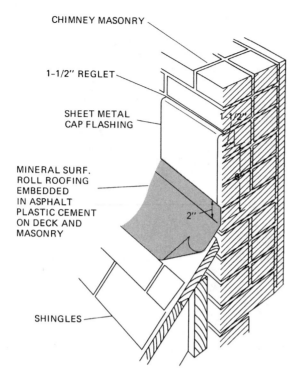

Fig. 8-47 *Method of securing cap flashing to the chimney.* (Bird and Son)

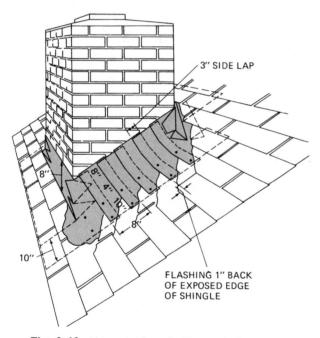

Fig. 8-48 *Alternative base flashing method.* (Bird and Son)

bent back edge of the flashing into the cleared space between the bricks. It is under slight spring tension. It cannot easily be dislodged. Refill the joint with portland cement mortar. Or you can use plastic asphalt cement. This flashing is bent down to cover the base flashing. The cap lies snugly against the masonry.

The front unit of the cap flashing is one continuous piece. On the sides and rear, the sections are of similar

size. They are cut to conform to the locations of brick joints. The pitch of the roof is also needed. The side units lap each other. See Fig. 8-46. This lap is at least 3 inches. Figure 8-48 shows another way to flash a sloping roof abutting a vertical masonry wall. This is known as the *step flashing* method. Place a rectangular piece of material measuring 8 × 22 inches over the end tab of each course of shingles. Hold the lower edge slightly back of the exposed edge of the covering shingle. Bend it up against the masonry. Secure it with suitable plastic asphalt cement. Drive nails through the lower edge of the flashing into the roof deck. Cover the nails with plastic asphalt cement. Repeat the operation for each course. Flashing units should be wide enough to lap each other at least 3 inches. The upper one overlays the lower one each time.

Asphalt roofing can be used for step flashing. It simply replaces the base flashing already shown. Metal cap flashings must be applied in the usual manner. The metal cap completes a satisfactory job.

Soil Stacks

Most building roofs have pipes or ventilators through them. Most are circular in section. They call for special flashing methods. Asphalt products may be successfully used for this purpose. Figures 8-49 through 8-54 show a step-by-step method of flashing for soil pipes. A soil pipe is used as a vent for plumbing. It is made of cast iron or copper. The pipe gets its name from being buried in the soil as a sanitary sewer pipe.

An alternative procedure for soil stacks can be used. Obtain noncorrodible metal pipes. They should have adjustable flanges. These flanges can be applied as a flashing to fit any roof pitch.

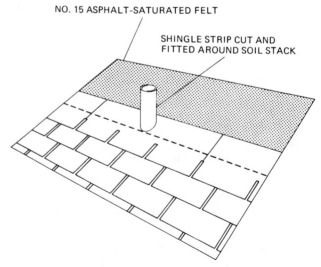

Fig. 8-49 *Roofing is first applied up to the soil pipe and fitted around it.* (Bird and Son)

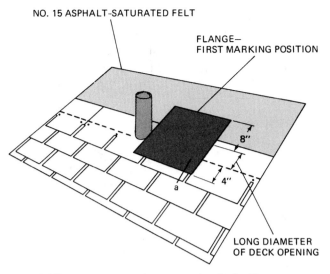

Fig. 8-50 First step in marking an opening for flashing. (Bird and Son)

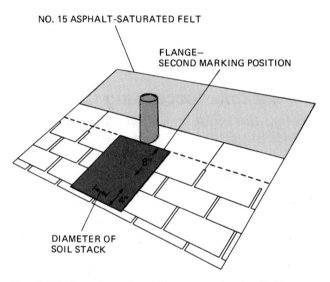

Fig. 8-51 Second step in marking an opening for flashing. (Bird and Son)

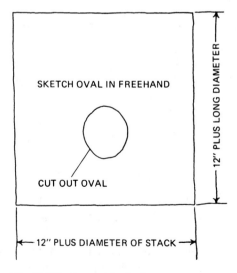

Fig. 8-52 Cut oval in the flange. (Bird and Son)

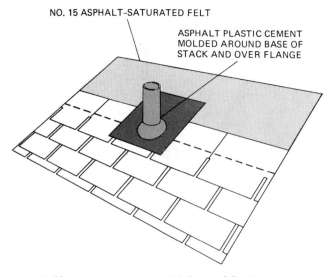

Fig. 8-53 Cement the collar molded around the pipe. (Bird and Son)

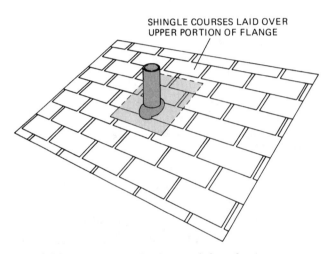

Fig. 8-54 Shingling completed past and above the pipe. (Bird and Son)

STRIP SHINGLES

Prepare the deck properly before starting to apply strip shingles. First place an underlayment down.

Deck Preparation

Metal drip edge Use a metal drip edge, made of noncorrodible, nonstaining metal. Place it along the eaves and rakes. See Figs. 8-55 and 8-56. The drip edge is designed to allow water runoff to drip free into a gutter. It should extend back from the edge of the deck not more than 3 inches. Secure it with nails spaced 8 to 10 inches apart. Place the nails along the edge. Drip edges of other materials may be used. They should be of approved types.

Underlayment Place a layer of no. 15 asphalt-saturated felt down. Cover the entire deck as shown. See Fig. 8-20.

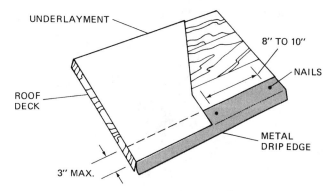

Fig. 8-55 *Application of the metal drip edge at eaves directly onto the deck.* (Bird and Son)

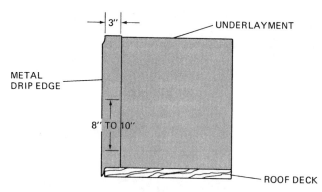

Fig. 8-56 *Application of the metal drip edge at the rakes over the underlayment.* (Bird and Son)

Chalk lines On small roofs, strip shingles may be laid from either rake. On roofs 30 feet or longer, it is better to start them at the center, then work both ways from a vertical line. This ensures better vertical alignment. It also provides for meeting and matching above a dormer or chimney. Chalk lines are used to control shingle alignment.

Eaves flashing Eaves flashing is required in cold climates. (January daily average temperatures of 25 degrees F or less call for eaves flashing.) In cold climates there is a possibility of ice forming along the eaves. If this happens, it causes trouble. Flashing should be used if there is doubt. Ice jams and water backup should be avoided. They can cause leakage into the ceiling below.

There are two flashing methods to prevent leakage. The methods depend on the slope of the roof. Possible severe icing conditions is another factor in choice of method.

Normal slope is 4 inches per foot or over. Install a course of 90-pound mineral-surfaced roll roofing. Or, apply a course of smooth roll roofing. It should not be less than 50 pounds. Install it to overhang the underlay and metal drip edge from ¼ to ⅜ inch. It should extend up to the roof. Cover a point at least 12 inches inside

the building's interior wall line. For a 36-inch eave overhang, the horizontal lap joint must be cemented. It should be located on the roof deck extending beyond the building's exterior wall line. See Fig. 8-57.

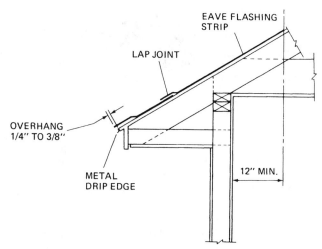

Fig. 8-57 *Eave flashing strip for normal-slope roof. This means 4 inches per foot or over.* (Bird and Son)

First and Succeeding Courses

Start the first course with a full shingle. Succeeding courses are started with full or cut strips. It depends upon the style of shingles being applied. There are three major variations for square-butt strip shingles:

1. Cutouts break at joints on the thirds. See Fig. 8-58.
2. Cutouts break at joints on the halves. See Fig. 8-59.
3. Random spacing. See Fig. 8-60.

Random spacing can be done by removing different amounts from the rake tab of succeeding courses. The amounts are removed according to the following scheme:

1. The width of any rake tab should be at least 3 inches.
2. Cutout centerlines should be located at least 3 inches laterally from other cutout centerlines. This means both the course above and the course below.
3. The rake tab widths should not repeat closely enough to cause the eye to follow a cutout alignment.

Ribbon Courses

Use a ribbon course to strengthen the horizontal roof lines. It adds a massive appearance that some people prefer. See Fig. 8-61.

One method involves special starting procedures. This is repeated every fifth course. Some people prefer this method.

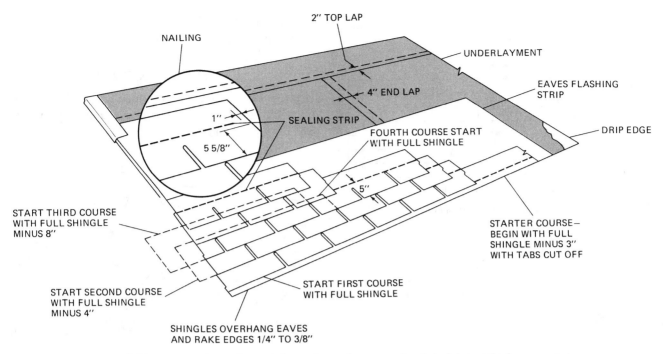

Fig. 8-58 *Applying three-tab square butt strips so that cutouts break the joints at thirds.* (Bird and Son)

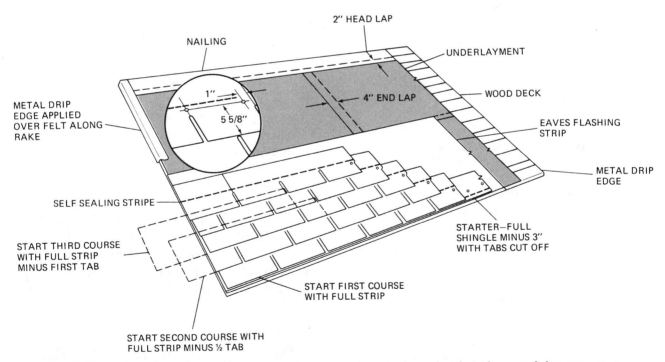

Fig. 8-59 *Applying three-tab square butt strips so that cutouts are centered over the tabs in the course below.* (Bird and Son)

1. Cut 4 inches off the top of a 12-inch-wide strip shingle. This will give you an unbroken strip 4 inches by 36 inches. You also get a strip 8 inches by 36 inches. Both strips contain the cutouts.

2. Lay the 4-×-36-inch strip along the eave.

3. Cover this with the 8-×-36-inch strip. The bottom of the cutouts is laid down to the eave.

4. Lay the first course of full (12-×-36-inch) shingles. It goes over layers *B* and *C*. The bottom of the cutouts is laid down to the eave. See Fig. 8-62.

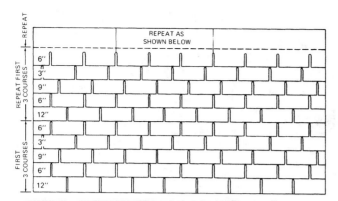

Fig. 8-60 *Random spacing of three-tab square butt strips.* (Bird and Son)

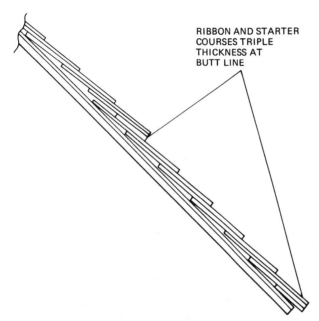

Fig. 8-61 *Ribbon courses. Side view.* (Bird and Son)

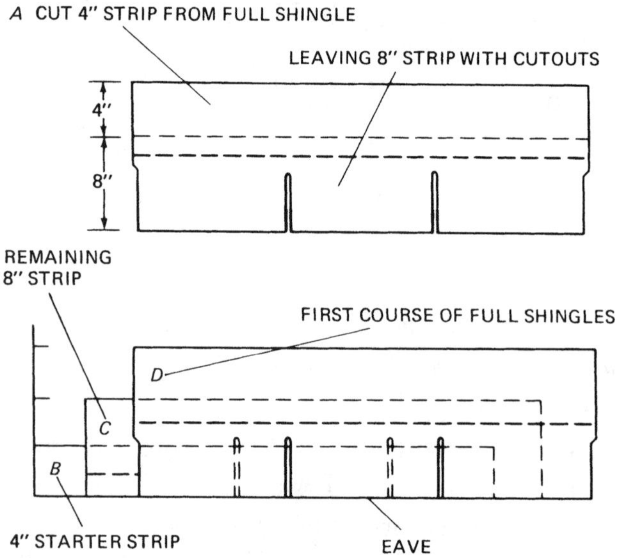

A CUT 4" STRIP FROM FULL SHINGLE

LEAVING 8" STRIP WITH CUTOUTS

4"

8"

REMAINING 8" STRIP

FIRST COURSE OF FULL SHINGLES

D

C

B

4" STARTER STRIP

EAVE

Fig. 8-62 *Laying the ribbon courses.* (Bird and Son)

218   Covering Roofs

Cutouts should be offset. This is done according to thirds, halves, or random spacing.

Wind Protection

High winds call for specially designed shingles. Cement the free tabs for protection against high winds. See Figs. 8-63 and 8-64.

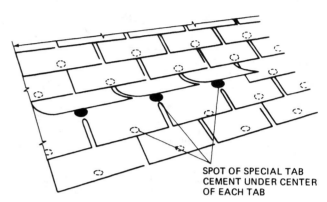

SPOT OF SPECIAL TAB CEMENT UNDER CENTER OF EACH TAB

Fig. 8-63 *Location of tab cement under square butt tabs.* (Bird and Son)

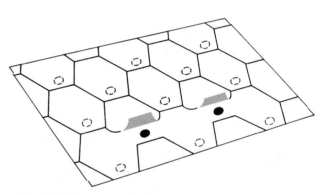

Fig. 8-64 *Location of tab cement under hex tabs.* (Bird and Son)

With a putty knife or caulking gun, apply a spot of quick-setting cement on the underlaying shingle. The cement should be about the size of a half-dollar. Press the free tab against the spot of cement. Do not squeeze the cement beyond the edge of the tab. Don't skip or miss any shingle tabs. Don't bend tabs back farther than needed.

Two- and Three-Tab Hex Strips

Nail two- and three-tab hex strips with four nails per strip. Locate the nails in a horizontal line 5¼ inches above the exposed butt edge.

Figure 8-65 shows how the two-tab strip is applied. Use one nail 1 inch back from each end of the strip. One nail is applied ¾ inch back from each angle of the cutouts.

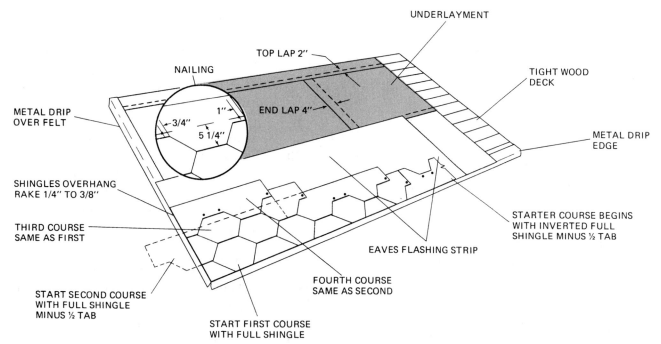

Fig. 8-65 *Application of two-tab hex strips.* (Bird and Son)

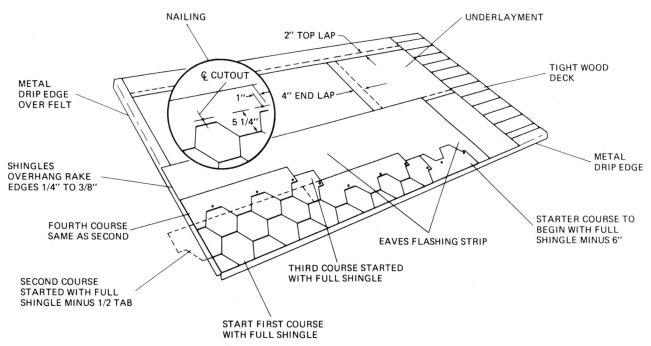

Fig. 8-66 *Application of three-tab hex strips.* (Bird and Son)

Three-tab shingles require one nail 1 inch back from each end. One nail is centered above each cutout. See Fig. 8-66.

Hips and Ridges

Use hip and ridge shingles to finish hips and ridges. They are furnished by shingle manufacturers. You can cut them from 12-×-36-inch square-butt strips. They should be at least 9 × 12 inches. One method of applying them is shown in Fig. 8-67.

Bend each shingle lengthwise down the center. This gives equal exposure on each side of the hip or ridge. Begin at the bottom of a hip. Or you can begin at one end of a ridge. Apply shingles over the hip or ridge. Expose them by 5 inches. Note the direction of the prevailing winds. This is important when you are placing ridge shingles. Secure each shingle with one

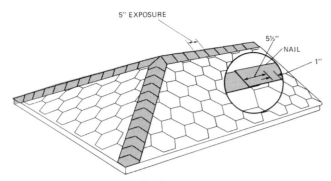

Fig. 8-67 *Hip-and-ridge shingles applied with hex strips.* (Bird and Son)

nail on each side. The nail should be 5½ inches back from the exposed end. It should be 1 inch up from the end. Never use metal ridge roll with asphalt roofing products. Corrosion may discolor the roof.

STEEP-SLOPE AND MANSARD ROOFS

New rooflines have caused changes recently. The mansard roofline requires some variations in shingle application. See Fig. 8-68.

Fig. 8-68 *Mansard roof.*

Excessive slopes give reduced results with factory self-sealing adhesives. This becomes obvious in colder or shaded areas.

Maximum slope for normal shingle application is 60° or 21 inches per foot.

Shingles on a steeper slope should be secured to the roof deck with roofing nails. See the maker's suggestions. Nails are placed 5⅝ inches above the butt. They should not be in or above the self-sealing strip.

Quick-setting asphalt adhesives should be applied as spots. The spots should be about the size of a quarter. They should be applied under each shingle tab. Do this immediately upon installation. Ventilation is needed to keep moisture-laden air from being trapped behind sheathing.

INTERLOCKING SHINGLES

Lockdown shingles are designed for windy areas. They have high resistance to high winds. They have an integral locking device. They can be classified into five groups. See Fig. 8-69.

These shingles generally do not require the use of adhesives. They may require a restricted use of cement. This is needed along the rakes and eaves. That is where the locking device may have to be removed.

Lock-type shingles can be used for both new and old buildings. Roof pitch is not too critical with these shingles. The designs can be classified into types 1, 2, 3, 4, and 5. Type 5 is a strip shingle with two tabs per strip.

Figure 8-70 shows how locking shingles work. Nail placement is suggested by the manufacturer. Nail placement is important for good results.

Figure 8-71 shows how the drip edge should be placed. The drip edge is designed to allow water runoff to drip free into gutters. The drip edge should not extend more than 3 inches back from the edge of the deck. Secure it with appropriate nails. Space the nails 8 to 10 inches apart along the inner edge.

Use the manufacturer's suggestions for starter course placement. Chalk lines will be very useful. Because it is very short, this type of shingle needs chalk lines for alignment.

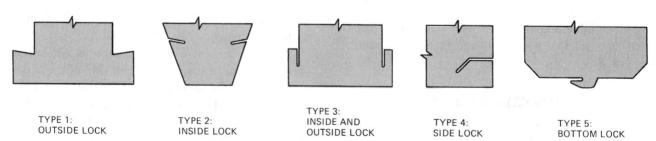

TYPE 1:
OUTSIDE LOCK

TYPE 2:
INSIDE LOCK

TYPE 3:
INSIDE AND
OUTSIDE LOCK

TYPE 4:
SIDE LOCK

TYPE 5:
BOTTOM LOCK

Fig. 8-69 *Locking devices used in interlocking shingles.* (Bird and Son)

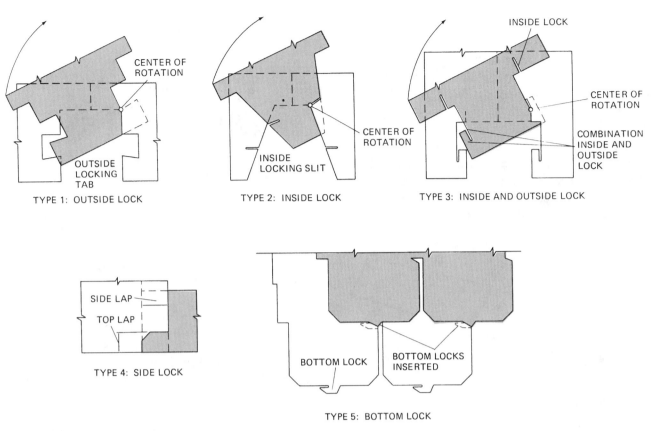

Fig. 8-70 *Methods of locking shingles types 1 through 5. In each case only the locking device is shown.* (Bird and Son)

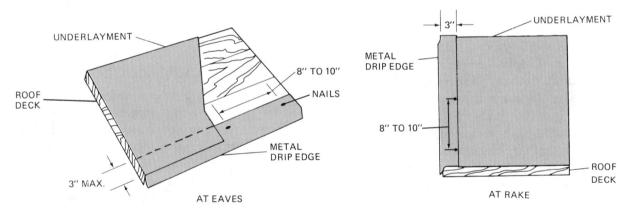

Fig. 8-71 *Placement of the metal drip edge for interlocking shingles.* (Bird and Son)

Hips and Ridges

Hips and ridges require shingles made for that purpose. However, you can cut them from standard shingles. They should be 9 × 12 inches. Either 90-pound mineral-surfaced roofing or shingles can be used. See Fig. 8-72.

In cold weather, warm the shingles before you use them. This keeps them from cracking when they are bent to lock. Do not use metal hip or ridge materials. Metal may become corroded and discolor the roof.

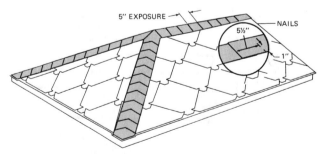

Fig. 8-72 *Hip-and-ridge application of interlocking shingles.* (Bird and Son)

Time is an important element in any job. You can speed up the roofing job when you don't have to cut and fit the hips and ridges with shingles you are using to cover the roof. You can speed things up and become more efficient by using pre-cut fiberglass high-profile hip and ridge caps. High profile means the cap sticks up higher than usual and gives a more "finished" quality to the roof. The high profile adds texture and shadow line to improve the appearance and enhance the beauty of any roof. They come in boxes with 48 pieces that cover 30 lineal feet of finished hip or ridge when applied with 8 inches to the weather. See Fig. 8-73.

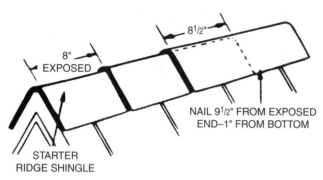

Fig. 8-73 *Back nail each shingle with two nails 9.5 inches from the exposed end.* (Ridglass)

The ridge and hip caps are back nailed with two nails 9.5 inches from the exposed end. See Fig. 8-74. Then continue to apply the caps with an 8-inch exposure. All nails should be covered by the succeeding shingle by at least 1 inch. See Fig. 8-75. In high-wind areas, seal down each installed ridge unit with elastomer adhesive or face nail each shingle with two nails, one on each side as shown in Fig. 8-75. In general, use 11- or 12-gage galvanized roofing nails with $\frac{7}{16}$-inch heads that are long enough to penetrate the roof deck by ¾ inch. When applying the Ridglass® caps in cold weather (under 50° F), unpack the carton on the roof and allow the shingles to warm up before application.

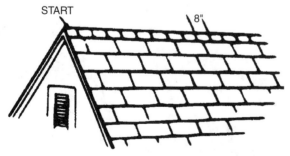

Fig. 8-74 *Continue to apply Ridglass with 9-inch exposure. All nails should be covered by the succeeding shingle by at least 1 inch.* (Ridglass)

Fig. 8-75 *Seal down each installed ridge unit with elastomeric adhesive and face nail each shingle with two nails.* (Ridglass)

These caps are made of fiberglass, not paper. They are SBS modified asphalt with no granule loss. There is no cutting, folding, or fabrication on the roof and matching colors are available for all manufacturers' shingles.

ROLL ROOFING

Roofing in rolls in some instances is very economical. Farm buildings and sheds are usually covered with inexpensive roll roofing. It is easier to apply and cheaper than strip shingles.

Do not apply roll roofing when the temperature is below 45 degrees F. If it is necessary to handle the material at lower temperatures, warm it before unrolling it. Warming avoids cracking the coating.

The sheet should be cut into 12-foot and 18-foot lengths. Spread them in a pile on a smooth surface until they flatten.

Windy Locations

Roll roofings are recommended for use in windy locations. Apply them according to the maker's suggestions. Use the pattern edge and blind nail. This means the nails can't be seen after the roofing is applied. Blind nailing can be used with 18-inch-wide mineral-surfaced or 65-pound smooth roofing. This can be used in windy areas. The 19-inch selvage double-coverage roll roofing is also suited for windy places.

Use concealed nailing rather than exposed nailing to apply the roll roofing. This ensures maximum life in service.

Use only lap cement or quick-setting cement. It should be cement the maker of the roofing suggests. Cements should be stored in a warm place until you are ready to use them. Place the unopened container in hot water to warm. Never heat asphalt cements directly over a fire. Use 11- or 12-gage hot-dipped galvanized nails. Nails should have large heads. This means at least ⅜-inch-diameter heads. The shanks should be ⅞ to 1 inch long. Use nails long enough to penetrate the wood below.

Exposed Nails—Parallel to the Rake

Exposed nailing, parallel to the eaves, is shown in Fig. 8-76. Figure 8-77 also shows the exposed nail method. It is parallel to the rake in this case. The overhang is ¼ to ⅜ inch over the rake. End laps are 6 inches wide and cemented down. Stagger the nails in rows, 1 inch apart. Space the nails on 4-inch centers in each row. Stagger all end laps. Do not have an end lap in one course over or adjacent to an end lap in the preceding course.

Hips and Ridges

For the method used to place a cap over the hips and ridge, see Fig. 8-78. Butt and nail sheets of roofing as they come up on either side of a hip or ridge. Cut strips of roll roofing 12 inches wide. Bend them lengthwise through their centers. Snap a chalk line guide parallel to the hip or ridge. It should be 5½ inches down on each side of the deck.

Cement a 2-inch-wide band on each side of the hip or ridge. The lower edge should be even with the chalk line. Lay the bent strip over the hip or ridge. Embed it in asphalt lap cement.

Secure the strip with two rows of nails. One row is placed on each side of the hip or ridge. The rows should be ¾ inch above the edges of the strip. Nails are spaced on 2-inch centers. Be sure the nails penetrate the cemented portion. This seals the nail hole with some of the asphalt.

WOOD SHINGLES

Wood shingles are the oldest method of shingling. In early U.S. history, pine and other trees were used for

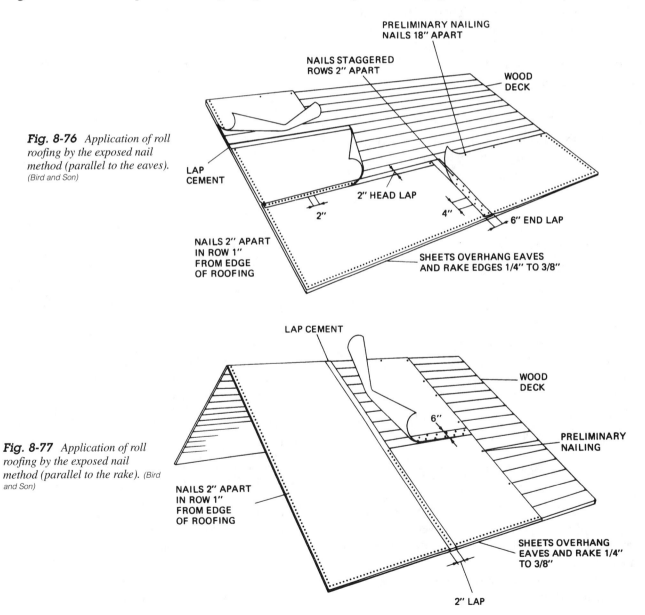

Fig. 8-76 *Application of roll roofing by the exposed nail method (parallel to the eaves).* (Bird and Son)

Fig. 8-77 *Application of roll roofing by the exposed nail method (parallel to the rake).* (Bird and Son)

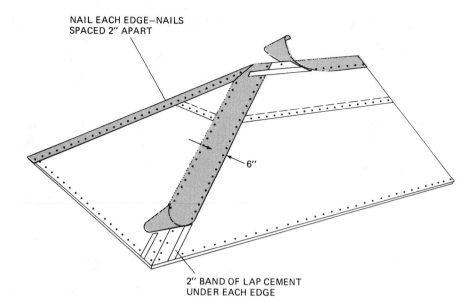

NAIL EACH EDGE—NAILS
SPACED 2" APART

6"

2" BAND OF LAP CEMENT
UNDER EACH EDGE

Fig. 8-78 *Hip-and-ridge application of roll roofing.* (Bird and Son)

shingles. Then the western United States discovered the cedar shingle. It is resistant to water and rot. If properly cared for, it will last at least 50 years. Application of this type of roofing material calls for some different methods.

Sizing Up the Job

You need to know a few things before ordering these shingles. First, find the pitch of your roof. See Fig. 8-79. Simply measure how many inches it rises for every foot it runs.

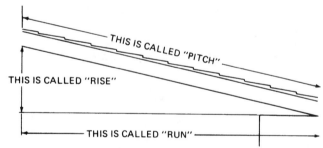

THIS IS CALLED "PITCH"

THIS IS CALLED "RISE"

THIS IS CALLED "RUN"

Fig. 8-79 *Figuring the pitch of a roof.* (Red Cedar Shingle & Handsplit Shake Bureau)

Remember that a square contains four bundles. It will cover 100 square feet of roof area. See Fig. 8-80.

Roof Exposure

Exposure refers to the area of the shingle that contacts the weather. See Fig. 8-81. Exposure depends upon roof pitch. A good shingle job is never less than three layers thick. See Table 8-6 for important information about shingles.

There are three lengths of shingles: 16 inches, 18 inches, and 24 inches.

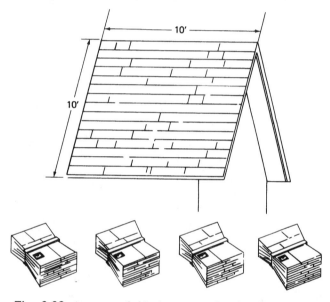

Fig. 8-80 *A square of shingle contains four bundles.* (Red Cedar Shingle & Handsplit Shake Bureau)

If the roof pitch is 4 inches in 12 inches or steeper (*three-ply roof*):

• For 16-inch shingles, allow a 5-inch exposure.
• For 18-inch shingles, allow a 5½-inch exposure.
• For 24-inch shingles, allow a 7½-inch exposure.

If the roof pitch is less than 4 inches in 12 inches but not below 3 inches in 12 inches (*four-ply roof*):

• For 16-inch shingles, allow a 3¾-inch exposure.
• For 18-inch shingles, allow a 4¼-inch exposure.
• For 24-inch shingles, allow a 5¾-inch exposure.

If the roof pitch is less than 3 inches in 12 inches, cedar shingles are not recommended. These exposures

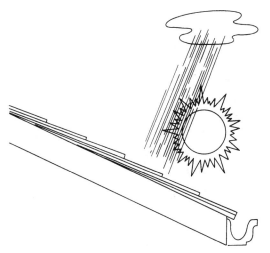

Fig. 8-81 *Shingle exposure to the weather.* (Red Cedar Shingle & Handsplit Shake Bureau)

are for no. 1 grade shingles. In applying no. 3 shingles, make sure you check with the manufacturer.

Estimating Shingles Needed

Determine the ground area of your house. Include the eaves and cornice overhang. Do this in square feet. If the roof pitch found previously:

- Rises 3 in 12, add 3 percent to the square foot total.
- Rises 4 in 12, add 5½ percent to the square foot total.
- Rises 5 in 12, add 8½ percent to the square foot total.
- Rises 6 in 12, add 12 percent to the square foot total.
- Rises 8 in 12, add 20 percent to the square foot total.
- Rises 12 in 12, add 42 percent to the square foot total.

Table 8-6 *Summary of Sizes, Packing, and Coverage of Wood Shingles*

Shake Type, Length, and Thickness, Inches	No. of Courses per Bundle	No. of Bundles per Square	Approximate Coverage (in Square Feet) of One Square, When Shakes are Applied with ½-Inch Spacing, at Following Weather Exposures (in Inches):								
			5½	6½	7	7½	8½	10	11½	14	16
18 × ½ medium resawn	9/9[a]	5[b]	55[c]	65	70	75[d]	85[e]	100[f]			
18 × ¾ heavy resawn	9/9[a]	5[b]	55[c]	65	70	75[d]	85[e]	100[f]			
24 × ⅜ handsplit	9/9[a]	5		65	70	75[g]	85	100[h]	115[i]		
24 × ½ medium resawn	9/9[a]	5		65	70	75[c]	85	100[j]	115[i]		
24 × ¾ heavy resawn	9/9[a]	5		65	70	75[c]	85	100[j]	115[i]		
24 × ½ to ⅝ tapersplit	9/9[a]	5		65	70	75[c]	85	100[j]	115[i]		
18 × ⅜ true-edge straight-split	14[k] straight	4								100	112[i]
18 × ⅜ straight-split	19[k] straight	5	65[c]	75	80	90[j]	100[i]				
24 × ⅜ straight-split	16[k] straight	5		65	70	75[c]	85	100[j]	115[i]		
15 starter-finish course	9/9[a]	5	Use supplementary with shakes applied with not over 10-inch weather exposure.								

[a]Packed in 18-inch wide frames.

[b]Five bundles will cover 100 square feet of roof area when used as starter-finish course at 10-inch weather exposure; six bundles will cover 100 square feet wall area when used at 8½-inch weather exposure; seven bundles will cover 100 square feet roof area when used at 7½-inch weather exposure. [m]

[c]Maximum recommended weather exposure for three-ply roof construction.

[d]Maximum recommended weather exposure for two-ply roof construction; seven bundles will cover 100 square feet of roof area when applied at 7½-inch weather exposure. [m]

[e]Maximum recommended weather exposure for sidewall construction; six bundles will cover 100 square feet when applied at 8½-inch weather exposure. [m]

[f]Maximum recommended weather exposure for starter-finish course application; five bundles will cover 100 square feet when applied at 10-inch weather exposure. [m]

[g]Maximum recommended weather exposure for application on roof pitches between 4 in 12 an 8 in 12.

[h]Maximum recommended weather exposure for application on roof pitches of 8 in 12 and steeper.

[i]Maximum recommended weather exposure for single-coursed wall construction.

[j]Maximum recommended weather exposure for two-ply roof construction.

[k]Packed in 20-inch-wide frames.

[l]Maximum recommended weather exposure for double-coursed wall construction.

[m]All coverage based on ½-inch spacing between shakes.

- Rises 15 in 12, add 60 percent to the square foot total.
- Rises 18 in 12, add 80 percent to the square foot total.

Divide the number you have found by 100. The answer is the number of shingle "squares" you should order to cover your roof if the pitch is 4 inches in 12 inches or steeper. If the roof is of lesser pitch, allow one-third more shingles to compensate for reduced exposure.

Also, add 1 square for every 100 linear feet of hips and valleys.

Tools of the Trade

A shingler's hatchet speeds the work. See Fig. 8-82. Sneakers or similar traction shoes make the job safer. A straight board keeps your rows straight and true.

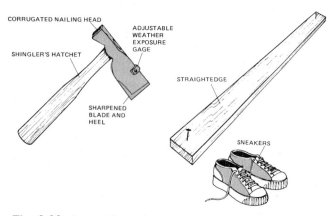

Fig. 8-82 *Tools of the trade.* (Red Cedar Shingle & Handsplit Shake Bureau)

APPLYING THE SHINGLE ROOF

Begin with a double thickness of shingles at the bottom edge of the roof. See Fig. 8-83. Let the shingles protrude over the edge to assure proper spillage into the eaves-trough or gutter. See *A* in Fig. 8-83.

Figure 8-84 shows how the nails are placed so that the next row above will cover the nails by not more than 1 inch. Use the board as shown in Fig. 8-85. Use

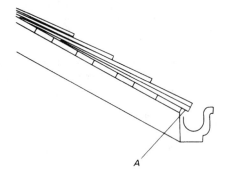

Fig. 8-83 *Applying a double thickness of shingles at the bottom edge of the roof.* (Red Cedar Shingle & Handsplit Shake Bureau)

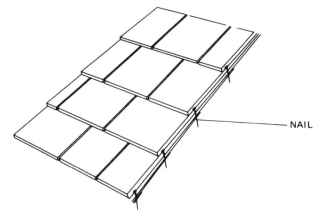

Fig. 8-84 *Covering nails in the previous course.* (Red Cedar Shingle & Handsplit Shake Bureau)

it as a straightedge to line up rows of shingles. Tack the board temporarily in place as a guide. It makes the work faster and the results look professional.

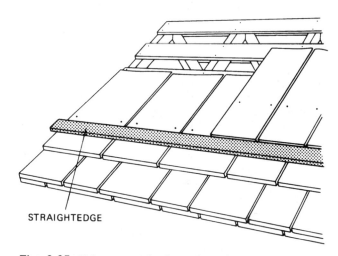

Fig. 8-85 *Using a straightedge to keep the ends lined up.* (Red Cedar Shingle & Handsplit Shake Bureau)

In Fig. 8-86 you can see the location of the nails. They should be placed no farther than ¾ inch from the edge of the shingle.

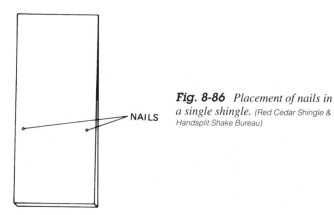

Fig. 8-86 *Placement of nails in a single shingle.* (Red Cedar Shingle & Handsplit Shake Bureau)

Figure 8-87 shows how the shingles are spaced ¼ inch apart to allow for expansion. Other simple rules are also shown in Fig. 8-88.

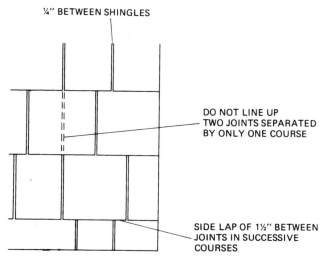

Fig. 8-87 *Spacing of shingle between courses.* (Red Cedar Shingle & Handsplit Shake Bureau)

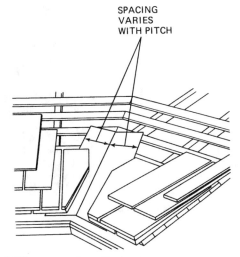

Fig. 8-88 *Spacing for valleys.* (Red Cedar Shingle & Handsplit Shake)

Valleys and Flashings

Extend the valley sheets beneath shingles. They should extend 10 inches on either side of the valley center. This is the case if the roof pitch is less than 12 inches in 12 inches. For steeper roofs, the valley sheets should extend at least 7 inches. See Fig. 8-88.

Most roof leaks occur at points where water is channeled for running off the roof. Or they occur where the roof abuts a vertical wall or chimney. At these points, use pointed metal valleys and flashings to assist the shingles in keeping the roof sound and dry. Suppliers will provide further information on which of

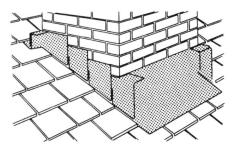

Fig. 8-89 *Flashing around a chimney.* (Red Cedar Shingle & Handsplit Shake Bureau)

the various metals to use. Figure 8-89 shows the flashing installed around a chimney.

Shingling at Roof Junctures

Apply the final course of shingles at the top of the wall. Install metal flashing (26-gage galvanized iron, 8 inches wide). Cover the top 4 inches of the roof slope. Bend the flashing carefully; avoid fracturing or breaking it. Make sure the flashing covers the nails that hold the final course. Apply a double starter course at the eave. Allow for a 1½-inch overhang of the wall surface. Complete the roof in the normal manner. See Fig. 8-90 for the convex juncture.

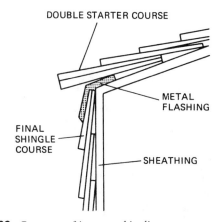

Fig. 8-90 *Convex roof juncture shingling.* (Red Cedar Shingle & Handsplit Shake Bureau)

For the concave juncture, apply the final course of shingles as shown in Fig. 8-91. Install the metal flashing to cover the last 4 inches of roof slope and bottom 4 inches of wall surface. Make sure the flashing covers the nails that hold the final course. Apply a double starter course at the bottom of the wall surface. Complete the shingling in the normal manner.

Before applying the final course of shingles, install 12-inch-wide flashing. This is to cover the top 8 inches of roof. Bend the remaining 4 inches to cover the top portion of the wall. See Fig. 8-92 for the treatment of apex junctures. Complete the roof shingling to

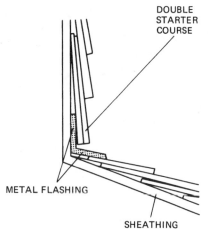

Fig. 8-91 *Concave roof junction shingling.* (Red Cedar Shingle & Handsplit Shake Bureau)

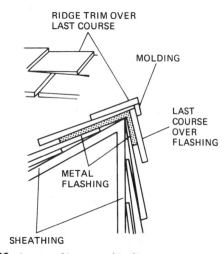

Fig. 8-92 *Apex roof juncture shingling.* (Red Cedar Shingle & Handsplit Shake Bureau)

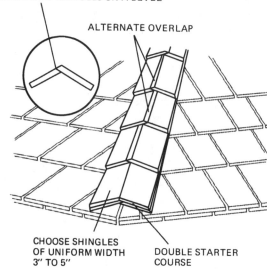

Fig. 8-93 *Applying overlap-type roofing ridges.* (Red Cedar Shingle & Handsplit Shake Bureau)

Fig. 8-94 *Applying a factory-assembled hip or ridge unit.* (Red Cedar Shingle & Handsplit Shake Bureau)

cover the flashing. Allow the shingle tips to extend beyond the juncture. Complete the wall shingling. Trim the last courses to fit snugly under the protecting roof shingles. Apply a molding strip to cover the top most portion of the wall. Trim the roof shingles even with the outer surface of the molding. Apply a conventional shingle "ridge" across the top edge of the roof. This is done in a single strip without matching pairs.

Applying Shingles to Hips and Ridges

The alternative overlap-type hip and ridge can be built by selecting uniform-width shingles. Lace them as shown in Fig. 8-93.

Time can be saved if factory-assembled hip-and-ridge units are used. These are shown in Fig. 8-94.

Nails for Wooden Shingles

Rust-resistant nails are very important. Zinc-coated or aluminum nails can be used. Don't skimp on nail quality. (See Fig. 8-95.)

CHAPTER 8 STUDY QUESTIONS

1. How many courses of shingles are used with wood shingles?
2. Sketch a gable roof shape on a simple building.
3. What is the main purpose of a roof?
4. How does soffit ventilation affect ice dams?

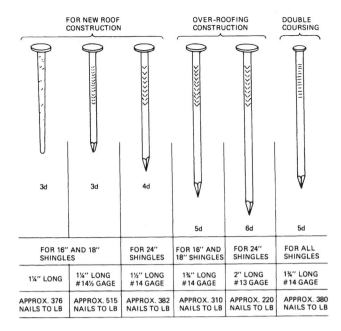

Fig. 8-95 *Nails used for wood shingles*

FOR NEW ROOF CONSTRUCTION			OVER-ROOFING CONSTRUCTION		DOUBLE COURSING
3d	3d	4d	5d	6d	5d
FOR 16" AND 18" SHINGLES	FOR 24" SHINGLES	FOR 16" AND 18" SHINGLES	FOR 24" SHINGLES	FOR ALL SHINGLES
1¼" LONG	1¼" LONG #14½ GAGE	1½" LONG #14 GAGE	1¾" LONG #14 GAGE	2" LONG #13 GAGE	1¾" LONG #14 GAGE
APPROX. 376 NAILS TO LB	APPROX. 515 NAILS TO LB	APPROX. 382 NAILS TO LB	APPROX. 310 NAILS TO LB	APPROX. 220 NAILS TO LB	APPROX. 380 NAILS TO LB

5. What is a valley?

6. What are eaves troughs used for?

7. How are plywood and sheathing different when used on a roof to provide a deck?

8. What are three types of shingles that can be used on a roof?

9. How many shingles are there in a square?

10. What is meant by shingle exposure?

11. What is meant by underlayment?

12. What is the difference between pitch and slope?

13. What is a woven valley?

14. How does a woven valley differ from a closed cut valley?

15. What is staging?

16. How long should roofing nails be?

17. What is galvanized iron used for when roofing a house?

18. What are asphalt-water emulsions?

19. Why is plastic asphalt cement used in roofing?

20. Why can't one type of asphalt roofing be used for every job?

9
CHAPTER

Installing Windows & Doors

WINDOWS AND DOORS PLAY AN important part in any type of house or building. They allow the air to circulate. They also allow passage in and out of the structure. Doors open to allow traffic in a planned manner. Windows are closed or open in design. There may be open and closed combinations, too. The design of a window or door is dictated by the building's use.

Buildings like those in Fig. 9-1 use the window as part of the design. The shape of a window may add to or detract from the design. The designer must be able to determine which is the right window for a building. The designer must also be able to choose a door that is architecturally compatible. Doors and windows come in many designs. However, they are limited in their function. This means some standards are set for the design of both. Most residential doors, for instance, are 6 feet 8 inches in height. If you are taller than that, you have to duck to pass through. Since most people are shorter, it is a safe height to use for doors.

Windows should be placed so that they will allow some view from either a standing or a sitting position. In most instances the window top and the door top are even. This way they look better from the outside.

In this unit you will learn how the carpenter installs windows and doors. You will learn how windows and doors are used to enhance a design. You will learn the various sizes and shapes of windows. You will learn the different sizes and hinging arrangements of doors. Locks will be presented so that you can learn how to install them.

Details for the placement of windows and doors are given. Things you will learn to do are:

- Prepare the window for installation
- Shim the window if necessary
- Secure the window in its opening
- Level and check for proper operation of the window
- Prepare a door for installation
- Shim the door if necessary
- Secure the door in its opening
- Level and check for proper operation

BASIC SEQUENCE

The carpenter should install a window in this order:

1. Check for proper window opening.
2. Uncrate the prehung window.
3. Remove the braces, if called for by the manufacturer.
4. Place builder's paper or felt between window and sheathing.
5. Place the window in the opening and check for level and plumb.
6. Attach the window at the corner or place one nail in the casing or flange (depending on the window design).
7. Check for proper operation of the window.
8. Secure the window in its opening.

The carpenter should install a door in this order:

1. Check the opening for correct measurements.
2. Uncrate the prehung door. If the door is not prehung, place the molding and trim in place first. Attach the door hinges by cutting the gains and screwing in the hinges.
3. Check for plumb and level.
4. Temporarily secure the door with shims and nails.
5. Check for proper operation.
6. Secure the door permanently.
7. Install the lock and its associated hardware.

Fig. 9-1 *Windows can add to or detract from a piece of architecture.* (Western Wood Product)

TYPES OF WINDOWS

There are many types of windows. Each hinges or swings in a different direction. They may be classified as:

1. Horizontal sliding windows
2. Awning picture windows
3. Double-hung windows
4. Casement windows

Most windows are made in factories today, ready to be placed into a rough opening when they arrive at the site. There are standards for windows. For instance, Commercial Standard 190 is shown in Fig. 9-2. The window requires a number of features, which are pointed out in the drawing. It must be weatherstripped to prevent air infiltration in excess of 0.75 cubic feet per minute per perimeter foot. This is under a 25 mph wind.

No more than two species of wood can be used in a unit. The wood used is ponderosa pine or a similar type of pine. Spruce, cedar, redwood, and cypress may also be used in the window frame, sill, and sash.

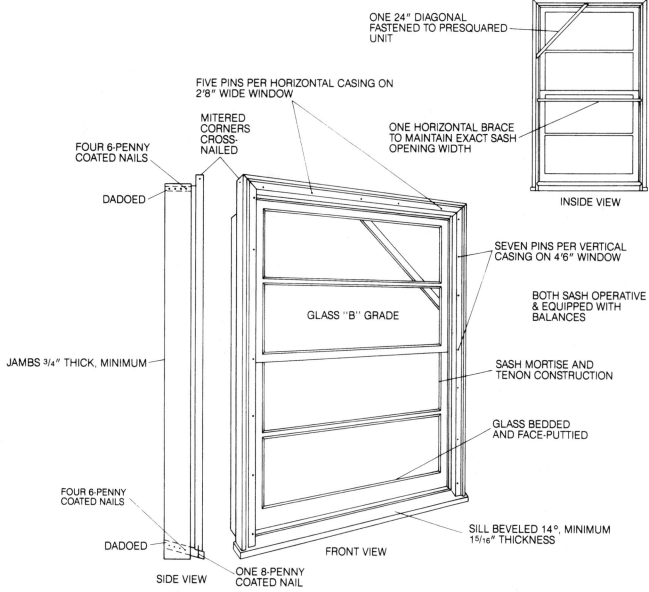

1. Weather-stripped to prevent air infiltration in excess of 0.75 cubic feet per minute per perimeter foot, under 25 MPH wind pressure.
2. No more than two species of wood per unit.
3. Chemically treated in accordance with NWMA minimum standards.
4. Finger-jointing permitted. See Commercial Standard 190, Par. 3.1.7.
5. Sash manufactured under Commercial Standard 163.
6. Ease of operation. See Commercial Standard 190, Par. 3.1.5.

Fig. 9-2 *Standard features of a window required to meet Commercial Standard 190.* (C. Arnold & Sons)

The wood has to be chemically treated in accordance with NWMA (National Window Manufacturers Association) minimum standards. Finger jointing is permitted. The sash has to be manufactured under Commercial Standard 163. Ease of operation is also spelled out in the written standard. Note how braces are specified to hold the window square and equally distant at all points. These braces are removed once the window has been set in place.

Horizontal sliding windows This type of window is fitted with a vinyl coating. The wood is not exposed at any point, which means less maintenance in the way of painting or glazing. The window is trimmed in vinyl so that it can be nailed into place on the framing around the rough opening. Figure 9-3 shows the window and details of its operation.

Figure 9-4 shows how the rough opening is made for a window. In this case the framing is on a 24-inch O.C. spacing. The large timber over the opening has to be large enough to support the roof without a stud where the window is placed. This prevents the window from buckling, which would stop the window from sliding. The cripple studs under the window opening are continuations of the studs that would be there normally. They are placed there to properly support the opening and to remove any weight from the window frame.

Sliding windows are available in a number of sizes. Figure 9-5 shows the possibilities. To find the overall unit dimension for a window which has a nonsupporting mullion, add the sum of the unit dimensions and subtract 2 inches. The mullion is the vertical bar between windows in a frame which holds two or more windows. The overall rough-opening dimension is equal to ¾ inch less than the overall unit dimension.

Double-hung window This window gets its name from the two windows that slide past one another. In this case they slide vertically. See Fig. 9-6. This is the most common type used today. The window shown is coated with plastic (vinyl) and can be easily attached to a stud through holes already drilled into the plastic around the frame. This plastic is called the flashing or the flange.

The double-hung window can be installed rather easily. See Fig. 9-7. The distance between the side jambs is checked to make sure they are even. Once the window is in the opening, place a shim where necessary. See Fig. 9-8. Note the placement of the nails here. Notice that this window does not have a vinyl flange around it. That is why the nails are placed as shown in Fig. 9-8. Figure 9-9 shows how shims are

placed under the raised jamb legs and at the center of the long sills of a double window. Figure 9-10 shows a sash out of alignment. See the arrow. The sashes will not be parallel if the unit is out of square.

(A)

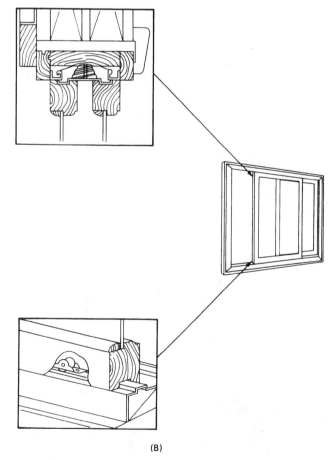

(B)

Fig. 9-3 *(A) Horizontal sliding window; (B) Details.* (Andersen)

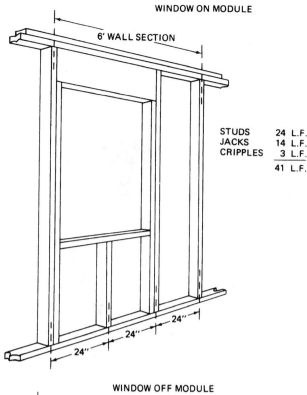

WINDOW ON MODULE

6' WALL SECTION

STUDS 24 L.F.
JACKS 14 L.F.
CRIPPLES 3 L.F.
_____
41 L.F.

WINDOW OFF MODULE

6' WALL SECTION

STUDS 32 L.F.
JACKS 14 L.F.
CRIPPLES 6 L.F.
_____
52 L.F.

(23% MORE FRAMING REQUIRED)

L.F. = LINEAR FEET

Fig. 9-4 *Window openings in a house frame. Window on module and off module with 24" centers.* (American Plywood Association)

Once the window is in place and properly seated, you can pack insulation between the jambs and the trimmer studs. See Fig. 9-11. Figure 9-12 shows how 1¾-inch galvanized nails are placed through the vinyl

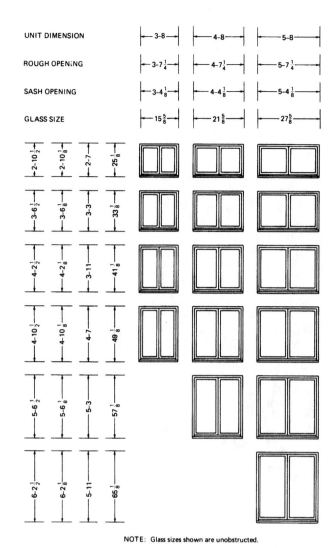

Fig. 9-5 *Sliding window sizes.* (Andersen)

NOTE: Glass sizes shown are unobstructed.

anchoring flange. This flange is then covered by the outside wall covering. The nails are not exposed to the weather.

Double-hung windows can be bought in a number of sizes. Figure 9-13 shows some of the sizes. In Fig. 9-14 the windows are all in place. The upperstory windows will be butted by siding. The downstairs windows are sitting back inside the brick. The upper windows will have part of the trim sticking out past the exterior siding.

Once the windows are in place and the house completed, the last step is to put the window dividers in place. They snap into the small holes in the sides of the window. The design may vary. See Fig. 9-15. The plastic grill patterns can be changed to meet the needs of the architectural style of the house. They can be removed by snapping them out. This way it is easier to clean the window pane. Figure 9-16 shows a house with the diamond light pattern installed in windows that swing out.

Fig. 9-6 *Double-hung window.* (Andersen)

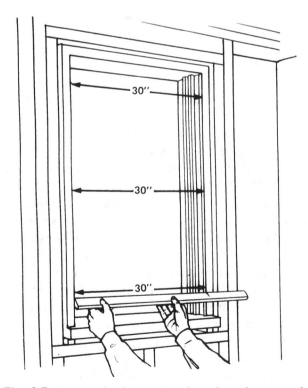

Fig. 9-7 *Measure the distance in at least three places to make sure the window is square.* (Andersen)

Fig. 9-8 *Shim the window where necessary. Side jambs are nailed through the shims.* (Andersen)

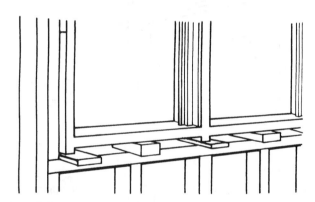

Fig. 9-9 *Shims raise the jamb legs and keep the window square.* (Andersen)

Casement window The casement window is hinged so that it swings outward. The whole window opens, allowing for more ventilation. See Fig. 9-17. This particular type has a vinyl flange for nailing it to the frame opening. It can be used as a single or in groups. This type of window can more easily be made weatherproof if it opens outward instead of inward.

Plastic muntins can be added to give a varied effect. They can be put into the windows as shown in Fig. 9-16. The diamond light muntins divide the glass space into small diamonds which resemble individual panes of glass.

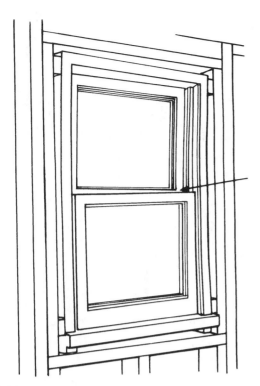

Fig. 9-10 *If the unit is not square, the sash rails will not be parallel. This can be spotted by eye.* (Andersen)

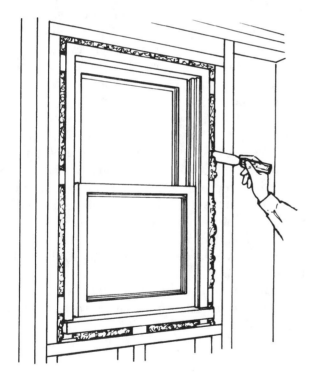

Fig. 9-11 *Loose insulation batting can be placed around the window to prevent drafts.* (Andersen)

The crank is installed so that the window opens outward with a twist of the handle. Figure 9-18 shows some of the ways this type of window may be operated.

Screens are mounted on the inside. Storm windows are mounted on the outside as in Fig. 9-19. In most

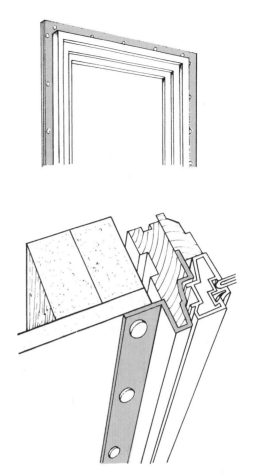

Fig. 9-12 *Installing the plastic flange around a prehung window with 1½-inch galvanized nails.* (Andersen)

cases, however, there is a thermopane used for insulation purposes. The thermopane is a double sheet of glass welded together with an air space between. This is then set into the sash and mounted as one piece of glass. See Fig. 9-20.

Multiple units are available. They may be movable or stationary. They may have one stationary part in the middle and two movable parts on the ends. Various combinations are available, as shown in Fig. 9-21.

Awning picture window This type of window has a large glass area. It also has a bottom panel which swings outward. A crank operates the bottom section. As it swings out, it has a tendency to form an awning effect—thus the name for this type. See Fig. 9-22. A number of sizes and combinations are available in this type of window. See Fig. 9-23. The fixed sash with an awning sash is also available in multiple units. Glass sizes are given in Fig. 9-23. If you need to find the overall basic unit dimension, add the basic unit to 2 ⅞ inch. The rough opening is the sum of the basic units plus ½ inch.

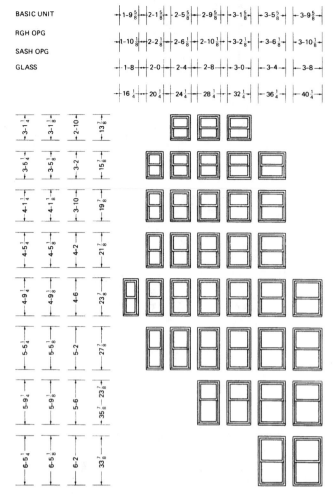

Fig. 9-13 Various sizes of double-hung windows. (Andersen)

Fig. 9-14 Double-hung windows in masonry (bottom floor) and set for conventional sliding (top floor).

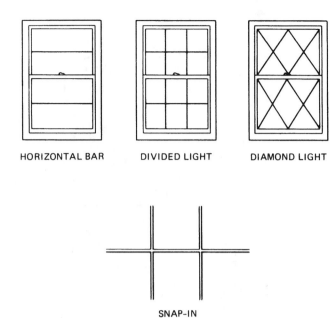

Fig. 9-15 Plastic dividers make possible different windowpane treatments. (Andersen)

Figure 9-24 shows the inswinging hopper type and the outswinging awning type of casement window. The awning sash type, shown in Fig. 9-22, has a bottom sash that swings outward. You have to specify which type of opening you want when you order. Specifying bottom-hinged or top-hinged is the quickest way to order.

Figure 9-25 shows the various sizes of hopper- and awning-type casements available. They can be stacked vertically. If this is done, the overall unit dimension for stacked units is the sum of the basic units plus ¾ inch for two units high, plus ¼ inch for three units high, and less 1¼ inches for four units high. To find the overall basic unit width of multiple units, add the basic unit dimensions plus 2⅞ inches to the total. To find the rough opening width, add the basic unit width plus ½ inch.

Figure 9-26 shows how a number of units may be stacked vertically. All units in this case open for maximum ventilation.

PREPARING THE ROUGH OPENING FOR A WINDOW

It is important that you consult the window manufacturer's specifications before you make the rough opening.

Fig. 9-16 *Diamond light pattern installed with plastic dividers.*

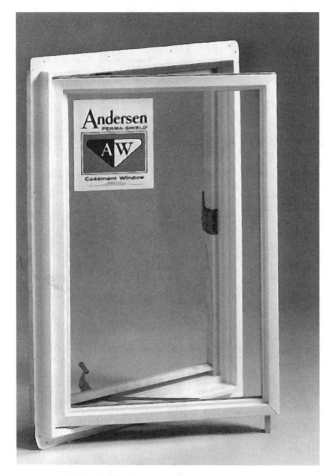

Fig. 9-17 *Casement window. (Andersen)*

Installation techniques, materials, and building codes vary according to area. Contact the window dealer for specific recommendations.

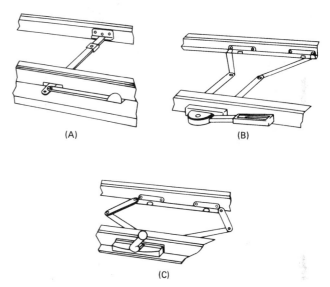

Fig. 9-18 *Methods of operating casement windows: (A) Standard push bar; (B) Lever lock; (C) Rotary gear.* (Andersen)

The same rough opening preparation procedures are used for wood and Perma-Shield windows. Figures 9-27 and 9-28 show the primed wood window and the Perma-Shield windows made by Andersen. These will be the windows discussed here. The instructions for installation will show how a manufactured window is installed.

Figure 9-29 shows how the wood casement window operates. Note the operator and its location. Figure 9-30 gives the details of the Perma-Shield Narroline window.

Figure 9-31 shows some of the possible window arrangements available from a window manufacturer. There is a window for almost any use. Select the window

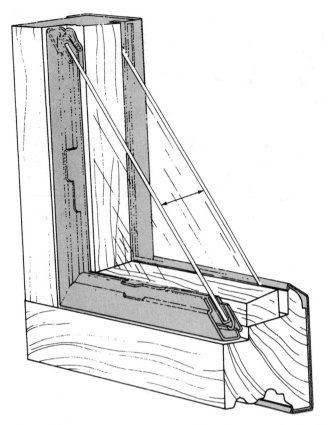

Fig. 9-19 *Double glass insulation. A ¹³/₁₆-inch air space is placed between panes.* (Pella)

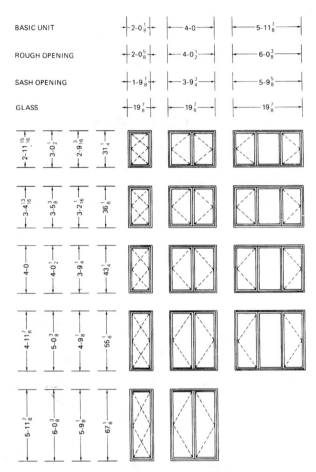

Fig. 9-21 *Various sizes of casement windows.* (Andersen)

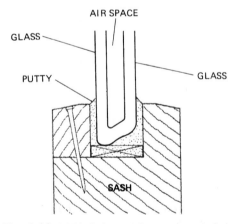

Fig. 9-20 *Welded glass or thermopane windows.*

needed and follow the instructions or similar steps for your window.

Brick veneer with a frame backup wall is similar in construction to the frame wall in the following illustrations.

When the opening must be enlarged, make certain the proper size header is used. Contact the dealer for the proper size header. To install a smaller size window, frame the opening as in new installation.

Fig. 9-22 *Picture window with awning bottom.* (Andersen)

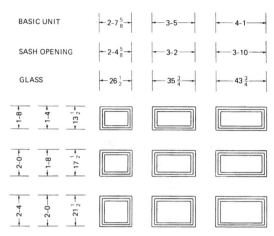

BASIC UNIT	2-10	3-10
SASH OPENING	2-7	3-7
GLASS	$28\frac{3}{4}$	$40\frac{3}{4}$

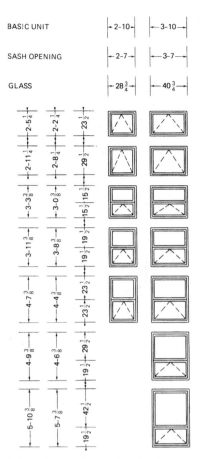

Fig. 9-23 *Various sizes of the fixed sash with awning sash windows.* (Andersen)

IN SWINGING HOPPER

OUT SWINGING AWNING

Fig. 9-24 *In swinging hopper and the out swinging awning types of window. These are casement windows.*

Steps in Preparing the Rough Opening

In some remodeling jobs, this must be done:

1. Lay out the window-opening width between regular studs to equal the window rough opening width plus

BASIC UNIT	$2\text{-}7\frac{5}{8}$	3-5	4-1
SASH OPENING	$2\text{-}4\frac{5}{8}$	3-2	3-10
GLASS	$26\frac{1}{2}$	$35\frac{3}{4}$	$43\frac{3}{4}$

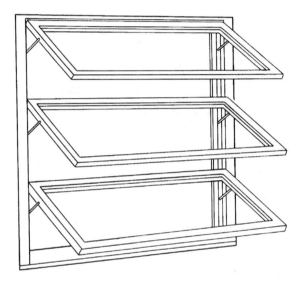

Fig. 9-25 *Various sizes of out swinging and in swinging windows.*
(Andersen)

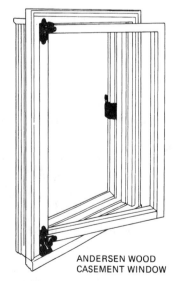

Fig. 9-26 *Vertical stacking of out swinging windows.*

ANDERSEN WOOD
CASEMENT WINDOW

Fig. 9-27 *Primed-wood window.* (Andersen)

Fig. 9-28 *Perma-Shield window.* (Andersen)

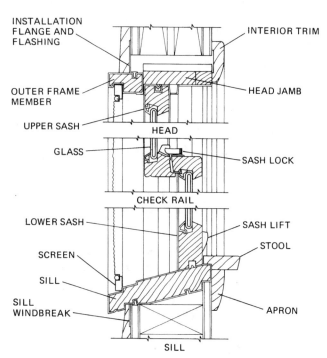

Fig. 9-30 *Details of the Perma-Shield window.* (Andersen)

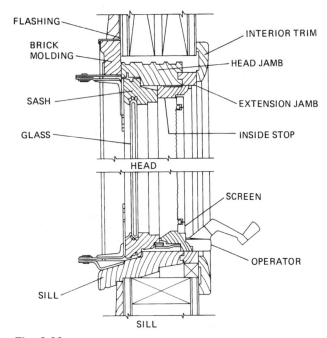

Fig. 9-29 *Details of the primed-wood casement window.* (Andersen)

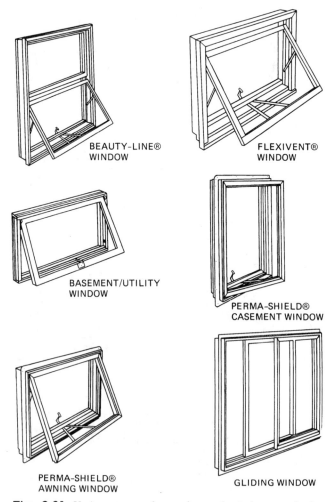

Fig. 9-31 *Various types of manufactured windows ready for quick installation.* (Andersen)

the thickness of two regular studs. See Fig. 9-32. Normally, in new construction the rough opening is already there, so all you have to do is install the window in it.

2. Cut two pieces of window header material to equal the rough opening of the window plus the thickness of two jack or trimmer studs. Nail the two header members together using an adequate spacer so that the header thickness equals the width of the jack or trimmer stud. See Fig. 9-33.

3. Cut the jack or trimmer studs to fit under the header for support. Nail the jack or trimmer studs to the regular studs. See Fig. 9-34.

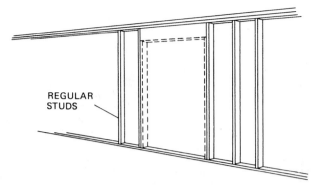

Fig. 9-32 *How to locate a window rough opening.* (Andersen)

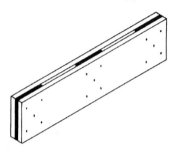

Fig. 9-33 *Making the header outside of the window opening.* (Andersen)

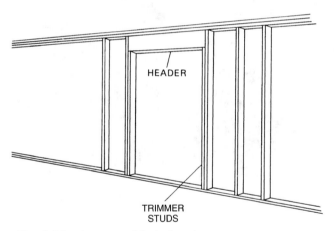

HEADER

TRIMMER STUDS

Fig. 9-34 *Placement of the jack studs.* (Andersen)

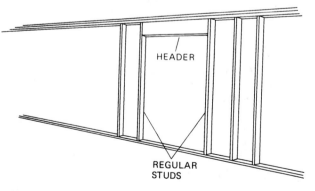

HEADER

REGULAR STUDS

Fig. 9-35 *Placing the header where it belongs.* (Andersen)

4. Position the header at the desired height between the regular studs. Nail through the regular studs into the header to hold the header in place until the next step is completed. See Fig. 9-35.

5. Measure the rough opening height from the bottom of the header to the top of the rough sill. Cut 2-×-4-inch cripples and the rough sill to the proper length. See Fig. 9-36. The rough sill length is equal to the rough opening width of the window. Assemble the cripples by nailing the rough sill to the ends of the cripples.

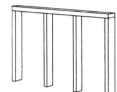

Fig. 9-36 *Assembling the cripples for easy placement.* (Andersen)

6. Fit the rough sill and cripples between the jack studs. See Fig. 9-37. Toenail the cripples to the bottom plate and the rough sill to the jack studs at the sides. See the round insert in Fig. 9-37.

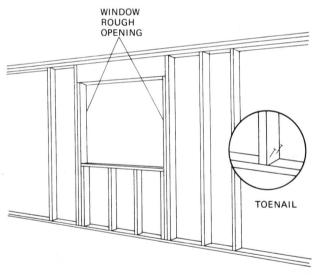

WINDOW ROUGH OPENING

TOENAIL

Fig. 9-37 *Placing the rough sill and cripples between the jack studs. Note the insert showing the toenailing.* (Andersen)

7. Apply the exterior sheathing (fiberboard, plywood, etc.) flush with the rough sill, header, and jack or trimmer stud framing members. See Fig. 9-38.

INSTALLING A WOOD WINDOW

The installation of a wood window is slightly different from that of a Perma-Shield window. However, there are many similarities. The following steps will show you how the windows are installed in the

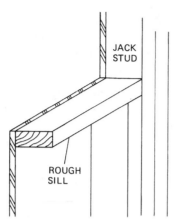

Fig. 9-38 *Applying the exterior sheathing.* (Andersen)

rough opening you just made from the preceding instructions.

1. Set the window in the opening from the outside with the exterior window casing overlapping the exterior sheathing. Locate the unit on the rough sill and center it between the side framing members (jack studs). Use 3½-inch casing nails and partially secure one corner through the head casing. See Fig. 9-39. Drive the nail at a slight upward angle, through the head casing into the header. See Fig. 9-40.

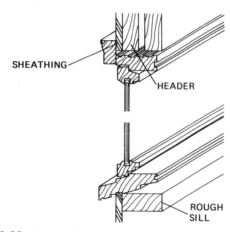

Fig. 9-39 *Placing the window in the rough opening.* (Andersen)

2. Level the window across the casing and nail it through the opposite corner with a 3½-inch casing nail. It may be necessary to shim the window under the side jambs at the sill to level it. This is done from the interior. See Fig. 9-41A and B.

3. Plumb (check the vertical of) the side jamb on the exterior window casing and drive a nail into the lower corner. See Fig. 9-42. Complete the installation by nailing through the exterior casing with 3½-inch nails. Space the nails about 10 inches apart.

Fig. 9-40 *Uses 3½-inch nails to partially secure one corner through the head casing.* (Andersen)

(A)

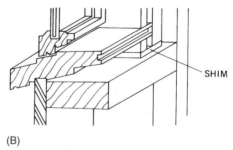

(B)

Fig. 9-41 *(A) Level the window across the casing and nail through the opposite corner* (Andersen). *(B) Location of the shim that holds the window level.*

4. Before you finally nail in the window, make sure you check the sash to see that it operates easily.

5. Apply a flashing with the rigid portion over the head casing. See Fig. 9-43. Secure this flashing

Fig. 9-42 *Check for plumb and nail the lower corner.*

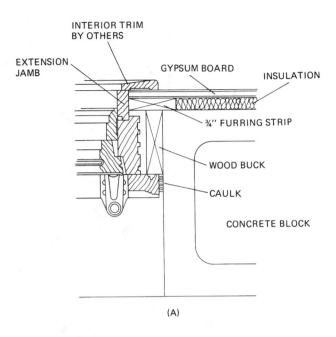

(A)

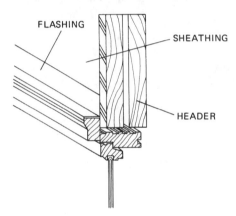

Fig. 9-43 *Apply the flashing with the rigid part over the head casing.* (Andersen)

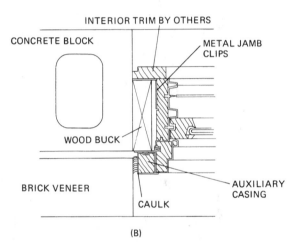

(B)

Fig. 9-44 *Details of a window installation in a masonry or brick veneer wall. (A) Wood window; (B) Perma-Shield window.* (Andersen)

with 1-inch nails through the flexible vinyl into the sheathing. Do not nail into the head casing.

6. Caulk around the perimeter of the exterior casing after the exterior siding or brick is applied.

Masonry or brick veneer wall This type of window can be installed in masonry wall construction. Fasten the wood buck to the masonry wall and nail the window to the wood buck using the procedures just shown for frame wall construction.

In Fig. 9-44A you see the wood window installed in a masonry wall. Figure 9-44B shows a Perma-Shield window installed with metal jam clips in a masonry wall with brick veneer. The metal jam clips and the auxiliary casing are available when specified.

Keep in mind that when brick veneer is used as an exterior finish, adequate clearance must be left for caulking between the window sill and the masonry.

This will prevent damage and bowing of the sill. The bowing is caused by the settling of the structural member. Shrinkage will also cause damage. Shrinkage takes place as the rough-in lumber dries after enclosure and the heat is turned on in the house.

Installing Windows by Nailing the Flange to the Sheathing

A simple procedure is used to place windows into the rough openings left in the framing of the house for such purposes. See Fig. 9-45. Most windows come with a flange that can be nailed to the sheathing or window framing. See Fig. 9-46. This eliminates cold air seepage in the winter and some noise generated by

Fig. 9-45 *A double-hung window mounted by nailing the flange to the sheathing and structural frame of the house.*

Nail

Nail

Nail

Fig. 9-46 *Using a light-weight plastic "glass block" window that fits as a unit and can be mounted by nailing the flange to the frame.*

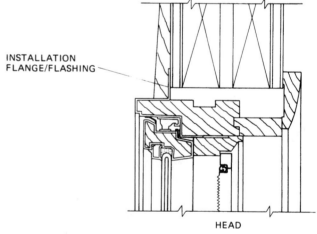

INSTALLATION FLANGE/FLASHING

HEAD

Fig. 9-47 *Details of how the flange or flashing is covered by the exterior wall covering.*

brisk winds. Examine Fig. 9-47 for details of how the flange or flashing is covered by the exterior wall covering.

In most instances, it is possible for the window unit to be handed out the opening to an outside carpenter who can nail it in place with little effort while the inside carpenter holds the unit in place or levels it with shims.

Figure 9-48 shows a house with windows installed using the methods just described.

Fig. 9-48 *Windows help make the house attractive.*

SKYLIGHTS

Having windows in the walls is not enough today. Houses also have windows in the ceiling. This new demand has been popular where more light is needed and an air of openness is desired. The skylight seems to be the answer to the demands of today's lifestyles. Bathrooms and kitchens are the rooms most often fitted with skylights.

Skylights can be installed when the house is built or they can be added later. In our examples here, we have selected the second approach since it does encompass both methods and can be easily adapted for original construction. Describing how to do it in original construction, however, would not necessarily serve those who want to install skylights after the house is built.

There are four basic types of skylights shown here, ranging from flush-mount to venting types. Figure 9-49 shows the flush-mount type. The low-profile flush-mount skylight provides the most economical solution to skylight installation. The flush-mount includes two heavy gage domes that are formed to provide a built-in deck mounting flange. This flange reduces installation error and allows fast and easy roof attachment when used with the mounting clips pre-packaged in the carton. The frameless/seamless feature eliminates potential leakage and provides airtight reliability. This one is designed specifically for residential use. It is designed to be installed on a pitched roof of 20° or more. It is available in four roof opening sizes.

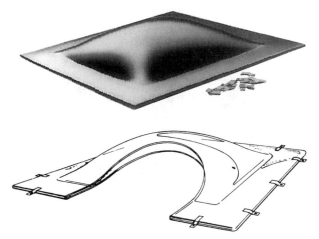

Fig. 9-49 *Flush-mount skylight.* (Novi)

Figure 9-50 is a curb-mount model and is ideal for locations where water, leaves, or snow collect on a roof, making an elevated skylight desirable. It is designed to utilize a wood curb put in place by the installer. This type can be installed on either a flat or pitched roof. It also is available in four roof opening sizes.

Fig. 9-50 *Curb-mount skylight.* (Novi)

Figure 9-51 shows a self-curbing type of skylight, which eliminates on-roof curb construction. This feature allows simple, time-saving installation while eliminating the potential for leakage around a wood curb. The premanufactured curbing is fully insulated and includes an extra-wide deck mounting flange with predrilled holes. This unit is easily installed by placing it directly over the roof opening and fastening it through the flange. It also has built-in condensation

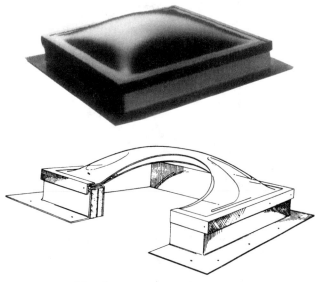

Fig. 9-51 Self-mount skylight. (Novi)

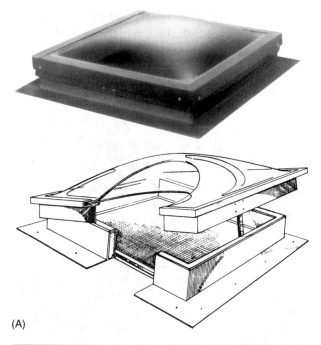

(A)

channels that get rid of unwanted moisture for maintenance-free operation. It, too, can be installed on a flat or pitched roof and comes in four roof opening sizes.

If you want something in the venting type, Fig. 9-52 has it. A built-in crank system allows operation with a hand crank or optional extension pole. The venting system is chain driven for better control and provides airtight closure when required. It can be installed on a flat or pitched roof and comes in two roof opening sizes. This one not only provides light but also gives you a way to allow for hot air to escape in the summer.

Installing the Skylight

Determine the roof opening location for the skylight on an inside ceiling or attic surface. See Fig. 9-53. Roof openings should be positioned between rafters whenever possible to keep rafter framing to a minimum. Take a look at Fig. 9-54 for framing diagrams for both 16-inch O.C. and 24-inch O.C. rafter spacing.

In a room with a cathedral ceiling, use a drill and wire probe to determine rafter spacing. When working in an attic, position the roof opening so that it is relative to the proposed opening. See Fig. 9-53. Make sure the installation area does not have plumbing or electrical wires inside the opening area.

Preparing the Roof Opening

From inside the house, square the finished roof opening dimensions for the skylight between the roof rafters. If a roof rafter does not cross the proposed opening, mark the four corner points of the roof opening. See Fig. 9-55.

(B)

Fig. 9-52 (A) Venting self-mount skylight. (Novi) (B) Using a crank pole to open the skylight. (Velux-America)

When a rafter must be cut away from the opening, measure 1½ inch beyond the finished roof opening dimensions on the upper and lower sides of the proposed opening and mark the four corner points. See Fig. 9-55B. This extra measurement allows for lumber that will frame the openings.

Fig. 9-53 *Measuring the opening for the skylight.*

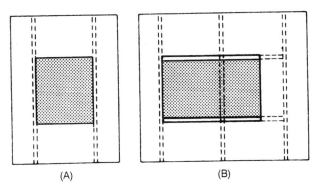

Fig. 9-55 *(A) Roof rafter does not cross the proposed opening. (B) Rafter must be cut to make room for the opening.* (Novi)

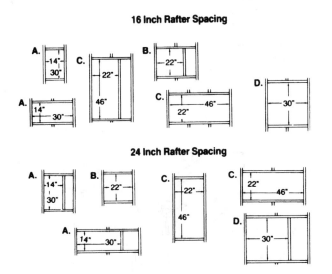

16 Inch Rafter Spacing

24 Inch Rafter Spacing

MODEL	KEY	OVERALL SIZE	FINISHED ROOF OPENING DIMENSION
16 x 32	A	24" x 40"	14" x 30"
24 x 24	B	32" x 32"	22" x 22"
24 x 48	C	32" x 56"	22" x 46"
32 x 32	D	40" x 40"	30" x 30"

Fig. 9-54 *Framing diagrams.* (Novi)

Cutting the Roof Opening

Drive a nail up through the roof at each designated corner point. Go to the roof and remove the shingles covering the proposed opening as indicated by the nails. Remove the shingles 12 to 14 inches beyond the proposed opening at the top and both sides, leaving the bottom row of shingles in place. If the roof felt underlays the shingles, it is not necessary to remove it from the installation area.

Fig. 9-56 *Cutting the hole in the roof.* (Novi)

Draw connecting lines between the nails and cut a hole in the roof using a saber saw or circular saw set to a depth of about one inch. See Fig. 9-56.

The rafter sections that remain in the opening must be cut away perpendicular to the roof deck surface with a handsaw. Temporary rafter supports should be installed to maintain structural alignment of the rafters.

Framing the Roof Opening

Frame the roof opening at the top and bottom by cutting two sections of lumber to fit between the existing rafters under the roof deck. Refer to the framing diagrams in Fig. 9-54.

Lumber used in framing should be the same size as the existing roof rafter lumber.

When working with a hole such as shown in Fig. 9-55, position each header under the roof deck to align with the top and bottom edges of the roof opening and secure it with nails.

If a rafter has been removed from the opening, shown in Fig. 9-55B, secure each header between the rafters and into the cut rafter ends with nails. See Fig. 9-57. Nail a 1½-inch-wide sheathing patch over the top of each header to make the opening level with the roof.

For installations made on a pitched roof you have to rotate the skylight until the runoff diverter strip is positioned at the top side of the roof openings as shown in Fig. 9-59.

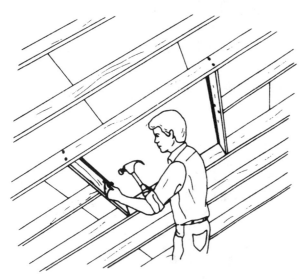

Fig. 9-57 *Framing the roof opening. (Novi)*

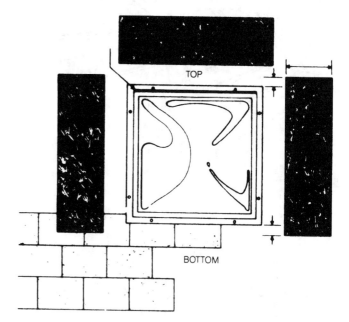

Install a side header where the roof opening does not align with an existing roof rafter. The finished rafter frame should align with the finished roof opening dimension.

Mounting the Skylight

Working on the roof, apply a layer of roof mastic (¼ inch deep) around the outer edge of the roof opening, 3 to 4 inches wide. Keep the mastic 1 inch away from the edge of the roof opening to prevent oozing. See Fig. 9-58.

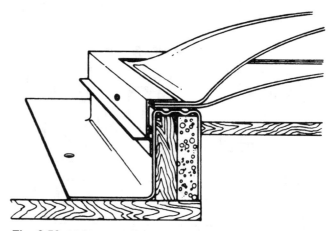

Fig. 9-59 *Making sure the diverter strip is properly located. (Novi)*

Set the skylight into the mastic to align with the roof opening. Be sure that the lower skylight flange overlaps the bottom row of shingles by at least 1 inch. Secure the skylight to the roof with screws using the predrilled flange holes.

Sealing the Installation

Select a flashing material that will seal three sides of the installation area. You may want to use metal, aluminum, or asphalt.

If you use asphalt, select a minimum 30-pound (#) rolled asphalt. Begin by cutting three sheets 5 inches

Fig. 9-58 *Spreading mastic around the opening. (Novi)*

longer than each exterior deck flange located at the top and both sides of the skylight. The width of each sheet should be cut to measure 8 to 10 inches. See Fig. 9-59.

Apply a layer of roof mastic to cover the exterior deck flange and roof deck at the top and both sides of the skylight. Spread the mastic to completely cover the deck flange and 8 to 10 inches of the roof deck around three sides of the installation. See Fig. 9-60.

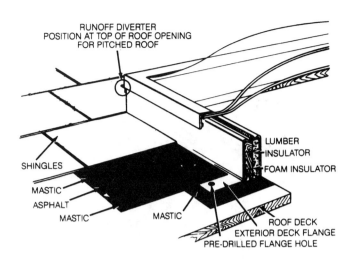

Fig. 9-60 *Applying mastic to cover the exterior deck flange.* *(Novi)*

Fig. 9-61 *Applying asphalt or tar paper or roofing paper.* *(Novi)*

Working with either side of the skylight, center the asphalt over the deck flange and press it into the mastic. Asphalt should extend 2½ inches over the flange at the top and bottom for adequate coverage. Continue to apply asphalt to the opposite side, and then to the top. Asphalt at the top must overlap asphalt on both sides. See Fig. 9-61.

Replacing the Shingles

After the asphalt has been set into place, apply a layer of mastic to completely cover the asphalt around the

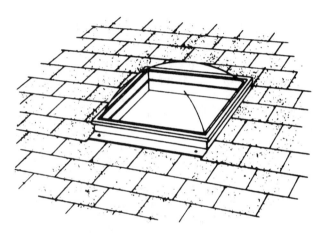

Fig. 9-62 *Replacing the shingles.* *(Novi)*

skylight. Starting at the bottom, replace each row of shingles. Trim the shingles around the skylight where necessary. See Fig. 9-62.

Preparing the Ceiling Opening

Drop a plumb line from each inside corner of the framed roof opening to the ceiling and mark four corner points. See Fig. 9-63. Remove the insulation between the joists 4 to 6 inches beyond the proposed ceiling opening.

If you want a ceiling opening that is larger than the roof opening or if an angled shaft is desired, some more measuring will have to be done at this time.

To angle the base of the skylight shaft beyond a parallel ceiling opening, pull the plumb line taut to the floor at the desired angle and mark each point. Using a tape measure and carpenter's square, determine the exact size and location of the proposed ceiling opening.

Fig. 9-63 *Dropping a plumb line to find the ceiling opening.* (Novi)

Tap a nail through the ceiling at each corner point. Find the locator nails from the room below and draw connecting lines between each point. Cut through the ceiling along the lines and remove the section. See Fig. 9-64.

Fig. 9-64 *Removing the ceiling section.* (Novi)

Framing the Ceiling Opening

The ceiling opening may be framed using procedures that apply to the framing of any roof opening. Using the same dimensional lumber as existing ceiling joists, cut the headers to fit between the joists and secure them in place with nails. The finished inside frame should align with the ceiling opening.

Constructing the Light Shaft

The light shaft is constructed with bevel cut 2-×-4-inch lumber hung vertically from the corners of the roof frame to the corners of the ceiling frame. Right angles are formed using two 2 × 4s at each corner of the proposed shaft to provide a nailing surface for the shaft liner. Additional 2 × 4 nailers are spaced to reinforce the shaft frame. See Fig. 9-65.

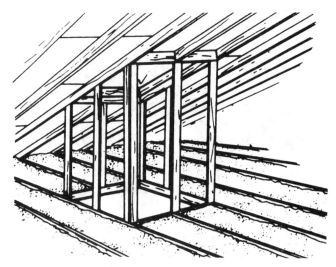

Fig. 9-65 *Framing the light shaft.* (Novi)

Cut the drywall or plywood to the size of the inside shaft walls and nail into place. The shaft may be insulated from the attic for maximum efficiency. See Fig. 9-66. The interior light shaft surfaces may be finished to match the room decor. Place trim around the opening in the ceiling and finish to match the room decor.

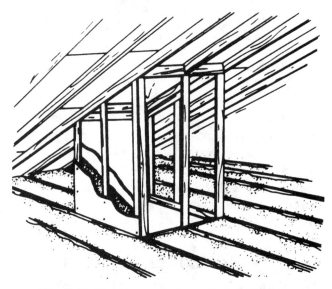

Fig. 9-66 *Finishing or closing up the light shaft.* (Novi)

OPERATION AND MAINTENANCE OF SKYLIGHTS

Condensation

Drops of condensation may appear on the inner dome surfaces with sudden temperature changes or during periods of high humidity. These droplets are condensed moisture and do not indicate a water leak from outside moisture. Condensation will evaporate as conditions of temperature and humidity normalize.

Figure 9-67 shows how light shaft installations can be used to present the light from the skylight to various parts of the room below. In Figs. 9-68 and 9-69 you will find how the original installations are made in houses under construction. The details and basic sizes are given, along with the roof pitch/slope chart. These will help you plan the installation from the start.

Care and Maintenance

If the dome is made of plastic, the outer dome surface may be polished with paste wax for added protection from outdoor conditions. If it is made of glass, you may want to wash it before installation and then touch up the finger marks after it is in place. Roofing mastic can be removed with rubbing alcohol or lighter fluid. Avoid petroleum-based or abrasive cleaners, especially on the clear plastic domes. Roof inspection should be conducted every two years to determine potential loosening of screws, cracked mastic, and other weather related problems that may result from the normal exposure to outdoor conditions.

Tube-type Skylights

The newer tube-type skylights can be installed during construction or after. They are designed to provide maximum light throughput from a relatively small unit. They are right for areas where a larger, standard skylight may not be practical. See Fig. 9-70. The tube shaft can be designed to reflect 95 percent of available sunlight. The low profile ceiling diffuser spreads natural light evenly through interior space. Early morning and late afternoon light can be captured and used by the dome to provide good illumination even during winter months in northern locations. See Fig. 9-71.

The tube-type skylight comes in kit form with everything needed, including illustrated instructions for the do-it-yourselfer. It installs in a few hours with basic hand tools. There is no framing, dry-walling, mudding, or painting required. It is available in both 10-inch and 14-inch diameters and therefore fits easily between standard 16-inch and 24-inch on-center rafters. See Fig. 9-72.

Most people are concerned with skylights because they have heard of them leaking, especially during the winter with snow pileup and melting. The illustrated skylight has a one-piece roof flashing that eliminates leaks. Flashing is specific to the roof type and ensures a perfect fit. The 14-inch spreads light up to 300 square feet. There is also an electric light kit available that makes the skylight into a standard light fixture at night and during dark periods of the day. It is designed to work from a wall switch and is a UL-approved installation. See Fig. 9-73.

Installation To start, locate the diffuser position on the ceiling. Check the attic for any obstructions or wiring. Locate the position on the roof for flashing and

SUGGESTED LIGHT SHAFT INSTALLATIONS

Where a roof window is installed above a flat ceiling, a light shaft will be needed. Typical installations are shown below. Flaring the shaft will give broader light distribution. Shaft construction by others.

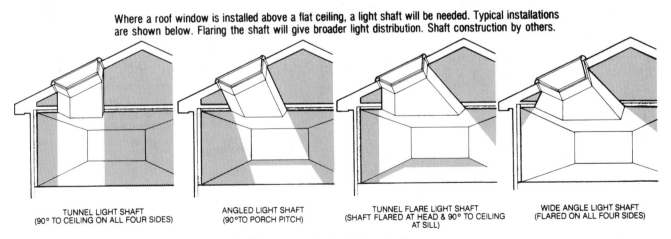

TUNNEL LIGHT SHAFT
(90° TO CEILING ON ALL FOUR SIDES)

ANGLED LIGHT SHAFT
(90°TO PORCH PITCH)

TUNNEL FLARE LIGHT SHAFT
(SHAFT FLARED AT HEAD & 90° TO CEILING AT SILL)

WIDE ANGLE LIGHT SHAFT
(FLARED ON ALL FOUR SIDES)

Fig. 9-67 *Suggested light shaft installations.* (Andersen)

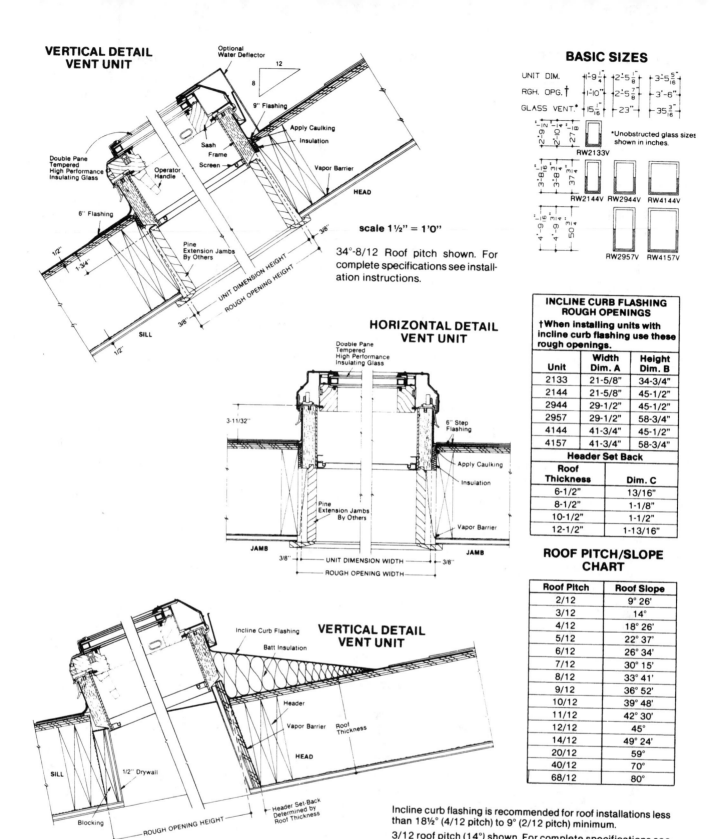

VERTICAL DETAIL VENT UNIT

scale 1½" = 1'0"

34°-8/12 Roof pitch shown. For complete specifications see installation instructions.

HORIZONTAL DETAIL VENT UNIT

VERTICAL DETAIL VENT UNIT

Incline curb flashing is recommended for roof installations less than 18½° (4/12 pitch) to 9° (2/12 pitch) minimum.

3/12 roof pitch (14°) shown. For complete specifications see installation instructions.

BASIC SIZES

*Unobstructed glass sizes shown in inches.

INCLINE CURB FLASHING ROUGH OPENINGS

†When installing units with incline curb flashing use these rough openings.

Unit	Width Dim. A	Height Dim. B
2133	21-5/8"	34-3/4"
2144	21-5/8"	45-1/2"
2944	29-1/2"	45-1/2"
2957	29-1/2"	58-3/4"
4144	41-3/4"	45-1/2"
4157	41-3/4"	58-3/4"

Header Set Back

Roof Thickness	Dim. C
6-1/2"	13/16"
8-1/2"	1-1/8"
10-1/2"	1-1/2"
12-1/2"	1-13/16"

ROOF PITCH/SLOPE CHART

Roof Pitch	Roof Slope
2/12	9° 26'
3/12	14°
4/12	18° 26'
5/12	22° 37'
6/12	26° 34'
7/12	30° 15'
8/12	33° 41'
9/12	36° 52'
10/12	39° 48'
11/12	42° 30'
12/12	45°
14/12	49° 24'
20/12	59°
40/12	70°
68/12	80°

Fig. 9-68 *Roof window vent unit, in place.* (Andersen)

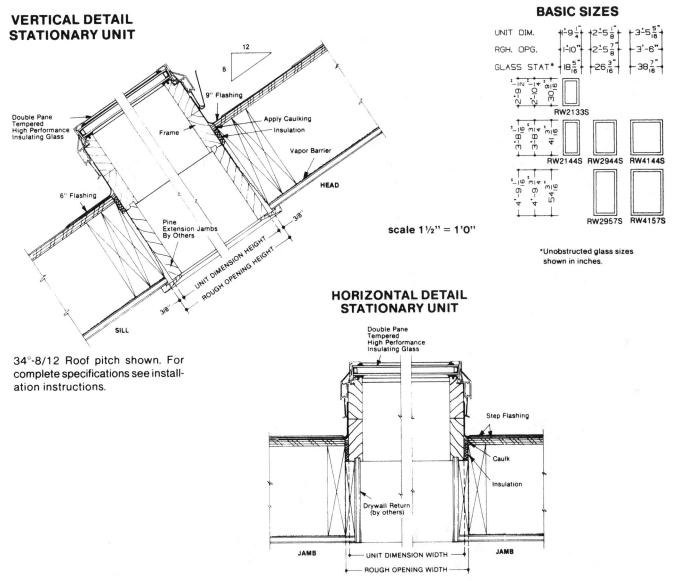

VERTICAL DETAIL STATIONARY UNIT

Double Pane Tempered High Performance Insulating Glass

9" Flashing

Frame

Apply Caulking

Insulation

Vapor Barrier

HEAD

6" Flashing

Pine Extension Jambs By Others

3/8"

3/8"

UNIT DIMENSION HEIGHT

ROUGH OPENING HEIGHT

SILL

12
8

scale 1½" = 1'0"

34°-8/12 Roof pitch shown. For complete specifications see installation instructions.

BASIC SIZES

UNIT DIM.	1'-9¼"	2'-5⅛"	3'-5 5/16"
RGH. OPG.	1'-10"	2'-5⅞"	3'-6"
GLASS STAT*	18 5/16	26 3/16	38 7/16

RW2133S

RW2144S   RW2944S   RW4144S

RW2957S   RW4157S

*Unobstructed glass sizes shown in inches.

HORIZONTAL DETAIL STATIONARY UNIT

Double Pane Tempered High Performance Insulating Glass

Step Flashing

Caulk

Insulation

Drywall Return (by others)

JAMB   UNIT DIMENSION WIDTH   JAMB

ROUGH OPENING WIDTH

Fig. 9-69 *Roof window, stationary unit, in place.* (Andersen)

dome. If the skylight is being installed in new construction, you can make sure plumbing and electrical take the skylight into consideration during the construction phase. Measure and cut an opening in the roof. Loosen shingles and install the flashing. (In new construction, it may be best to install the flashing before shingles are in place.) Insert the adjustable tube. Attach the dome. See Fig. 9-74. Measure and cut an opening in the ceiling. Install the ceiling trim ring. Attach the diffuser. In the attic, assemble, adjust, and install the tubular components. In colder climates it is necessary to insulate the tube shaft.

TERMS USED IN WINDOW INSTALLATION

Now is a good time to review the terms associated with the installation of a window. This will make it possible for you to understand the terminology when you work with a crew installing windows.

Plumb The act of checking the vertical line of a window when installing it in a rough opening.

Level The act of checking the horizontal line of a window when installing it in a rough opening.

Regular stud A vertical frame member that runs from the bottom plate on the floor to the top plate at the ceiling. In normal construction, this is a 2 × 4 approximately 8 feet long.

Jack or trimmer stud A vertical frame member that forms the window rough opening at the sides and supports the header. It runs from the bottom plate at the floor to the underside of the header.

Header A horizontal framing member located over the window rough opening supported by the jack studs.

Fig. 9-70 *Skylight installation. (ODL Inc.)*

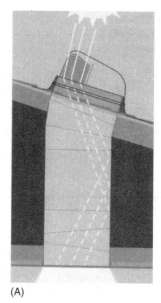

(A)

Fig. 9-71A *The dome above the roof line. (ODL Inc.)*

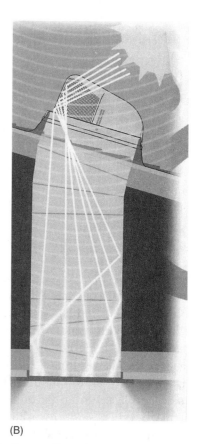

(B)

Fig. 9-71B *Dome reflects the sunlight coming from any angle throughout the day in any season. (ODL Inc.)*

Depending upon the span, headers usually are double 2 × 6s, 2 × 8s, or 2 × 10s in frame wall construction, or steel I beams in heavier masonry construction.

Rough sill A horizontal framing member, usually a single 2 x 4, located across the bottom of the window rough opening. The window unit rests on the rough sill.

Cripples Short vertical framing members spaced approximately 16 inches O.C., located below the rough sill across the width of the rough opening. Also used between the header and the top plate, depending upon the size of the headers required.

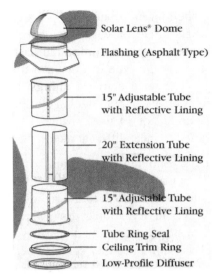

- Solar Lens® Dome
- Flashing (Asphalt Type)
- 15" Adjustable Tube with Reflective Lining
- 20" Extension Tube with Reflective Lining
- 15" Adjustable Tube with Reflective Lining
- Tube Ring Seal
- Ceiling Trim Ring
- Low-Profile Diffuser

Fig. 9-72 *Exploded view of the skylight.* (ODL Inc.)

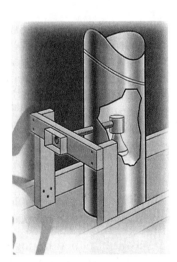

Fig. 9-73 *Conversion of the skylight to a light fixture.* (ODL Inc.)

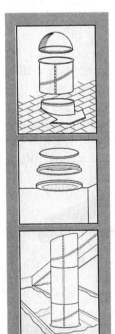

Fig. 9-74 *Installation of the skylight.* (ODL Inc.)

Shim An angled wood member—wedge shaped—used as a filler at the jamb and sill. (Wood shingle makes a good shim.)

Wood buck A structural wood member secured to a masonry opening to provide an installation frame for the window unit.

PREHUNG DOORS
Types of Doors

Exterior doors are made in many sizes and shapes. See Fig. 9-75. They may be solid with a glass window. They may have an X shape at the bottom, in which case they are referred to as a *cross-buck*. These doors are made in a factory and crated and shipped to the site. There they are unpacked and placed in the proper opening. There is little to do with them other than level them and nail them in place. The hardware is already mounted on the door.

Figure 9-76 shows a door that was prehung and shipped to the site. Note how it sticks out from the sheathing so that the siding can be applied and butted to the side jamb. Doors are chosen for their contribution to the architecture of the building. They must harmonize with the design of the house. Figure 9-77 shows a door that adds to the design of the house. The door facing or trim adds to the column effect of the porch.

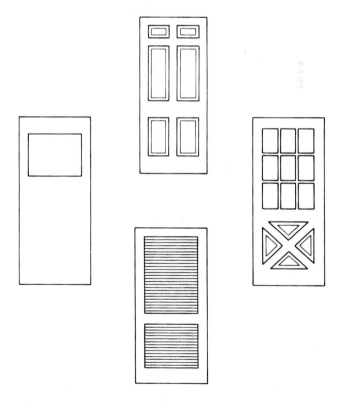

Fig. 9-75 *Various door designs.* (National Woodwork Manufacturers)

Fig. 9-76 *A prehung exterior door with three panels of glass. Note how the trim sticks out sufficiently for the siding to butt against it.*

Fig. 9-77 *The proper door can do much to improve the looks of the house.*

Flush doors Flush doors are made of plywood or some facing over a solid core. The core may be made of a variety of materials. In some instances where the door is used inside, the inside of the door is nothing more than a mesh or strips. See Fig. 9-78. Wood is usually preferred to metal for exterior doors of homes. Wood is nature's own insulator. While metal readily conducts heat and cold, wood does not. Wood is 400 times more effective an insulator than steel and 1800 times more effective than aluminum.

Stock wood flush doors come in a wide variety of sizes, designs, and shapes. Standard wood door frames will accommodate wood combination doors, storm doors, and screen doors without additional framing expense.

Panel doors This type of door has solid vertical members, rails, and panels. Many types are available. See Figs. 9-79 and 9-80. The amount of wood and glass varies. Many people want a glass section in the

front door. The four most popular types are clear, diamond obscure, circle obscure, and amber Flemish. Figure 9-80 shows some of the decorative variations in doors. Note how the type of door can improve the architecture. The main entrance may be highlighted with sidelights on one or both sides of the door. See Fig. 9-81. These are 12 or 14 inches wide. The panels of glass are varied to meet different requirements.

Sliding doors Sliding doors are just what the name suggests. They usually have tempered or safety glass. They can slide to the left or to the right. You have to specify a right- or left-sliding door when ordering. Most have insulating construction. They have two panes of glass with a dead-air space in between. See Fig. 9-82. In Fig. 9-83 you can see the sizes available. Also note the arrow which indicates the direction in which the door slides.

French doors This is usually two or more doors grouped to open outward onto a patio or veranda. They have glass panes from top to bottom. They may be made of metal or wood. Later in this chapter you will see two- and three-window groupings mounted step by step.

INSTALLING AN EXTERIOR DOOR

The door frame has to be installed in an opening in the house frame before the door can be hung. Figure 9-84 shows the parts of the door frame. Note the way it goes together. Figure 9-85 shows the frame in position with a wire on the right. It has been pulled through for the installation of the doorbell push button. Note the spacing of the hinges. The door in the background is a six-panel type that is already hung. The siding has not yet been butted against the door casing. It will be placed as close as possible and then caulked to prevent moisture from damaging the wood over a period of time.

Figure 9-86 shows the general information you need to be able to identify the parts of a door. It is important that the door be fitted so that there is a uniform ⅛-inch clearance all around to allow for free swing. Allow ½ to ¾ inch at the bottom for a better fit with carpet.

There are nine steps to installing an exterior door:

1. See Fig. 9-86B. Trim the height of the door. Many doors are made with extra long stiles or horns B. Before proceeding, cut the horns off the *top* of the door, even with the top of the top rail. When cutting, start with the saw at the outside edge to avoid splintering the edges of the door.

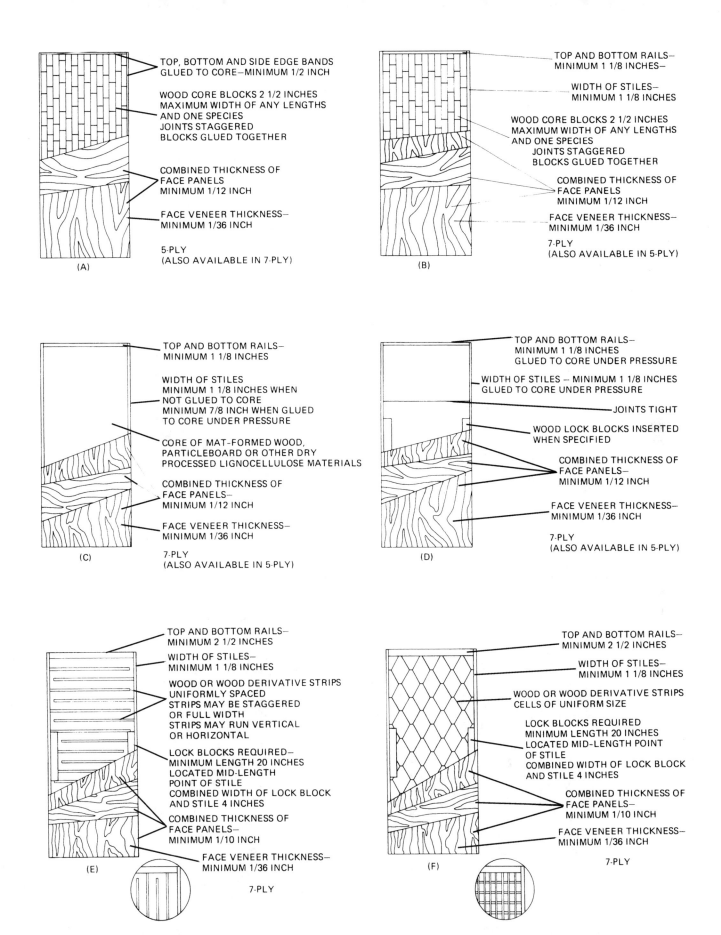

Fig. 9-78 *Various types of materials are used to fill the interior space in flush doors.* (National Woodwork Manufacturers)

Fig. 9-79 *Add-on panels and lights give designers options with flush doors.* (General Products)

2. See Fig. 9-86C. On the inside of the room, place the door into the opening upside down, and against the door jamb. Keep the hinge stile tight against the hinge jamb A1 of the door frame. This edge must be kept straight to ensure that hinges will be parallel when installed. Place two ¼-inch blocks under the door. These will raise the door to allow for ⅛-inch clearance at both top and bottom when cut. Mark the door at C, the top of the door frame opening, for cutting.

3. See Fig. 9-86D. After cutting the door to the proper height, place the door into the opening again—this

CHATEAU 8-PANEL

WILLIAMSBURG 6-PANEL

COLONIAL 9-LIGHT

CROSS BUCK 9-LIGHT

Fig. 9-80 *The four most popular door styles. They have clear, diamond obscure, circle obscure, and amber Flemish safety glass.*
(General Products)

Fig. 9-81 *Sidelights are designed for fast installation as integrated units in wood or plastic models. This insulated safety glass comes in 12-inch or 14-inch widths. (General Products)*

time right side up, with ⅛-inch blocks under the door for clearance. With the door held tightly against the hinge jamb of the door frame, have someone mark a pencil line along the lock stile of the door from outside the door opening (line 1 to 1A), holding a ⅛-inch block between the pencil and the door frame. This will automatically allow for the necessary ⅛-inch clearance needed.

4. See Fig. 9-86E. Trim the width of the door. If the amount of wood to be removed from the door (line 1 to 1A) is more than ¼ inch, it will be necessary to trim both edges of the door. Trim one-half the width of the wood to be removed from each edge of the door. Use a smooth or jack plane.

Fig. 9-82 *Sliding door.*

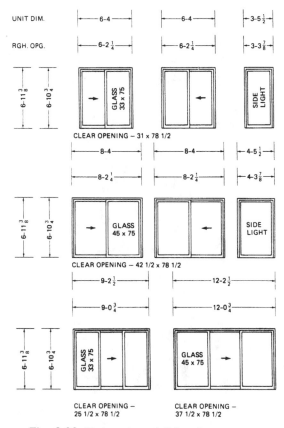

Fig. 9-83 *Various sizes of sliding doors. (Andersen)*

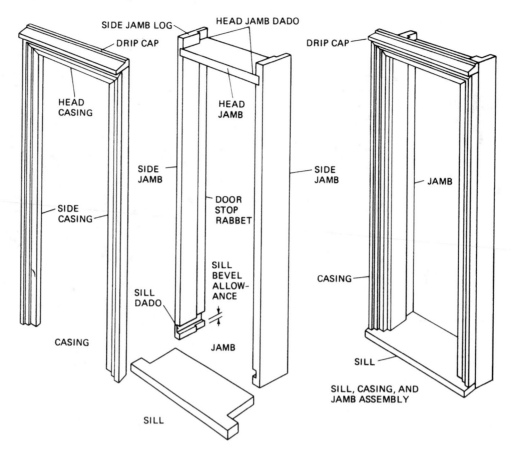

Fig. 9-84 *Assembling a door frame (left to right).*

Fig. 9-85 *Prehung door in place. Note the wires hanging out on the left side. They indicate where an outside light will go.*

5. See Fig. 9-86F. Bevel the lock stile of the door. The lock stile of the door should be planed to about a 3° angle so that it will clear the door frame when the door is swung shut.

6. See Fig. 9-86G. Install the hinges. Be sure that markings for hinges are uniform for all hinges used. The mortises (cutouts) for hinges should be of uniform depth (thickness of the hinge). Measure 7 inches down from the top of the door and 7⅛ inches down from the underside of the door frame top. Mark the locations of the top edge of the upper hinge. Placement of hinges will be unnecessary if the door is prehung.

7. See Fig. 9-86H. The bottom hinge is 9 inches up from the door bottom (9⅛ inches up from the threshold). The middle hinge is centered in the door height. Attach the hinge leafs to the door and door frame. Hang the door in the opening. If the mortises are cut properly but the hinges still bind, the frame jamb may be distorted or bowed. It may be necessary to place a thin shim under one edge of the frame hinge leaf to align it parallel and relieve the binding.

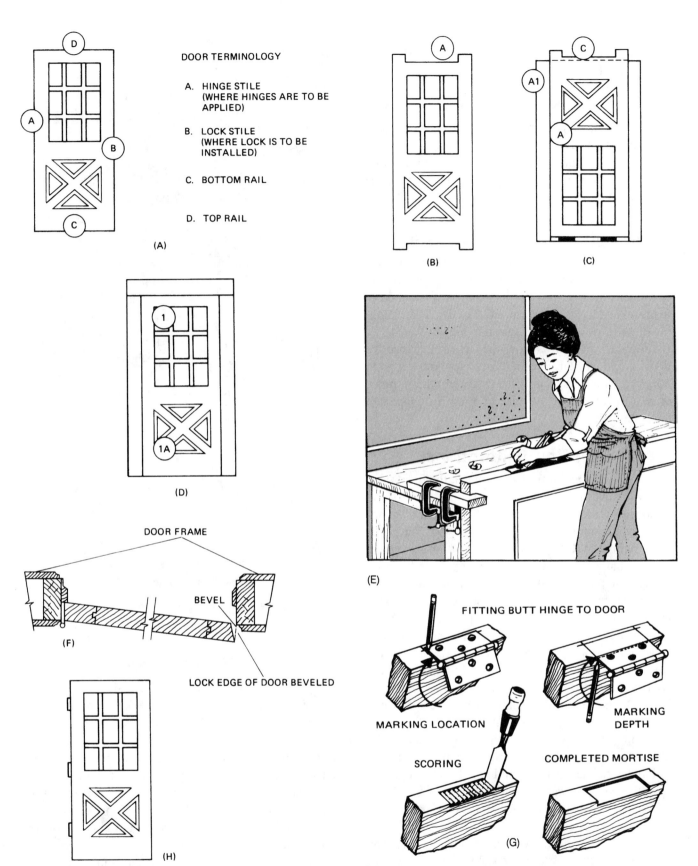

Fig. 9-86 *(A) Terms used with doors* (Grossman Lumber). *(B) Trim the door height. (C) Mark the door to fit the frame. (D) Mark the other end of the door. (E) Trim the width of the door. (F) Bevel the lock stile of the door. (G) Install the hinges. (H) Check the height of the hinges.*

8. Install the lockset. Because of the variety of styles of locksets, there is no one way to install them. This subject will be discussed in detail later in this chapter. Each lockset comes with a complete set of instructions. The best advice is to follow these instructions.

9. Finish the door. Care must be taken to paint or seal all door surfaces. Top, bottom, edges, and faces should be sealed and painted. Weather and moisture can hurt a door and decrease its performance. A properly treated door will give many years of satisfactory service.

Hanging a Two-Door System

In some cases the two-door system is the main entrance. See Fig. 9-87. In other cases the two door system may be French doors. See Fig. 9-88. This type of door comes in pieces and has to be assembled. The details for hanging of this type of door are shown in Fig. 9-89.

Figure 9-90 shows how the active and inactive doors are identified first. In most cases both doors do not open. This is especially the case when the entrance door is involved. In the case of French doors, both doors can open.

Handing Instructions

With the assembly kit furnished, one of the first things you will want to check is the "hand" of the door. Hand of doors is always determined by the *outside*. Inswinging doors are more common. See Fig. 9-91. The right-hand symbol is RH, and it means the door swings on hinges that are mounted on the right. If the door swings out, it is a right-hand reverse door and the symbol is RHR. It is still hinged on the right looking at it from the outside.

The left-hand symbol is LH. That means that on an inswinging door the hinges are on the left. This is most convenient for persons who are right-handed. If the door is outswinging, the left-hand reverse symbol is LHR. Doors usually swing into a wall where they can rest against the adjoining wall. They usually swing back only 90°. The traffic pattern also determines the way a door swings. Doors swing out in most commercial, industrial, and school buildings. This lets a person open the door outward so that it will be easy to leave the building if there is a fire. Safety is the prime consideration in this case.

Figure 9-92 shows how energy conservation has entered the picture. The figure shows how the top and bottom of the door are fitted to make sure air does not leak through the door. The door may be made of metal. Because metal conducts heat and cold, the door is insulated. See Fig. 9-93.

Fig. 9-87 *Double doors installed. The one on the left is the active door.*

Fig. 9-88 *Double door ready for assembly.* (General Products)

Metal Doors

Metal doors may also be used for residential houses. In some cases they are used to replace old or poorly fitting wooden doors. Figure 9-94 shows how doors with metal frames are designed for ease of installation. In Fig. 9-95 you can see how the metal frame is attached to concrete, wood, and concrete blocks.

One of the advantages of metal doors is their fire resistance. Frames for 2'8" and 3'0" doors carry a 1½-hour label. The frames for 3'6" and double doors are not labeled as to fire resistance.

In the case of metal frames, the frame has to be installed before the walls are constructed. The frame requires a rough opening 4½ inches wider and 2¼ inches higher than nominal. The stock frame is usually 5¾ inches wide.

Step 1
Attach Astragal to Inactive Door

Remove plastic filler plates from deadbolt and lock locations. Remove appropriate metal knock-out at deadbolt location.

Place door on edge, place astragal with *notch to door top* as shown. Compress outer flange against face of door, install bolt retainer spring at top and bottom of astragal, and secure astragal with five (5) self-tapping screws using power driver or drill.

Place two nylon screw bosses under the strike route and secure strike with 2 No. 8 screws provided.

Strike has tab that can be adjusted to assure proper closing while hanging door.

Place the bolt assembly in top and bottom of astragal. Adjust the bolt retainer spring to proper position and secure bolt retainer with Allen wrench provided.

Snap in 2 ea. channel closures in bolt recess above and below strike location. (proper lengths provided).

Tape one pile pad to interior face of inactive door for installation on astragal after door is installed in opening.

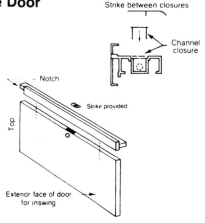

Step 2
Attach Sweep

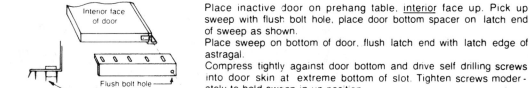

Place inactive door on prehang table, <u>interior</u> face up. Pick up sweep with flush bolt hole, place door bottom spacer on latch end of sweep as shown.

Place sweep on bottom of door, flush latch end with latch edge of astragal.

Compress tightly against door bottom and drive self drilling screws into door skin at extreme bottom of slot. Tighten screws moderately to hold sweep in up position.

Check operation of bolt assembly.
Turn door over (Interior face down)
Place active door on table (interior face up) and install sweep in same manner as first door.
Turn door over and proceed with frame prehanging. (Step 3)

Step 3
Attach Hinge Jambs

Place hinge jambs beside edge of doors as shown and attach hinges to doors with No. 10 machine screws.

Apply caulking tape to jambs at threshold locations. Make sure it follows contour of vinyl threshold part.

Fig. 9-89 *Details for hanging a two-door system.*

INSTALLING FOLDING DOORS

Folding doors are used to cover closets with any number of interesting patterns. They may be flush, and plain or mirrored. They may have two panels or four panels. The sizes and door widths vary to suit the particular application. See Table 9-1.

Figure 9-96 shows the openings and the details of fitting the metal bifold door.

Figure 9-97 shows the details of the four panels to be installed and different panel styles available. Note the names given to the parts so you can follow the installation instructions.

1. Carefully center the top track lengthwise in the finished opening. This should suit both flush and recessed mountings. Attach the track with No. 10 × 1¼-inch screws in the provided holes. The bifold

Step 4

Installing Header and Threshold

Stick ⅛" PAK-WIK spacers—2 ea. on header jambs, and 3 ea. on the astragal side at locations marked "X", to maintain ⅛" clearance between doors and jambs. Place header jamb against header stop block. Line up doors with header jamb and press doors firmly against it. Raise jambs up, make sure they are flush with header jamb at top corners. Drive 3 ea. 2¼" long staples in each corner of frame.

Assemble vinyl and aluminum threshold parts. Place threshold in frame and secure with #10 x 1½" screws through pre-drilled holes. Back edge to be flush with frame. (Make sure pile is firmly in contact with threshold and weatherstripping.) If not, remove and reposition.

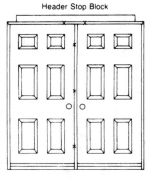

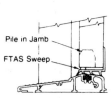

Step 5

Attaching Brickmold

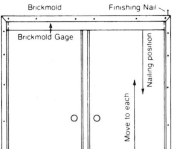

NOTE: If door is to be outswing, proceed with bracing shown in Step 6. Then turn unit upside down and install bolt strike as in Step 6. Proceed with brickmold in Step 5. NOTE: Outswing frame requires ⅝" longer header brickmold than inswing. Place brickmold gage at top and bottom of jambs. Position miter joints of jamb and header brickmold. Align fit of mitered corners and properly space reveal. Tack each corner, nail header brickmold, move gage down jambs and nail brickmold. Use 6 ea. No. 10 x 2½" finishing nails as shown, drive 2 ea. No. 10 x 2½" finishing nails in two corners as shown.

Step 6

Installing Flush Bolt Strike and Bracing Frame

Turn unit upside down and replace on table. Place pencil mark at flush bolt locations on header and threshold. Open inactive door, place thin dab of putty at bolt pencil marks on header and threshold. Close inactive door. Move bolts to mark the putty. Open inactive door. Center punch top and bottom bolt locations with nail. Remove putty. Drill ⅝" hole in header and threshold. Place bolt strike on header and align with drilled hole. Install 2 ea. No. 6 x 1" screws provided. Close inactive door and secure bolts. Close active door. Tack corner braces as shown. Tack strip of wood across frame, approximately 12" above threshold as shown. Use 8d coated box nails. On outswing unit, cut bracing to fit between brickmold.

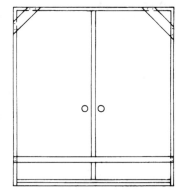

Fig. 9-89 *Continued.*

track is assembled for four-panel door installation. The track may be separated at the center without the use of tools when a two-panel door is installed. See Fig. 9-98. Knobs, screws, and rubber stops are packaged for two-panel installation.

2. Place the bottom track, either round edge or square edge, toward the room. Plumb the groove with the top track. See Fig. 9-99. Screw the track to the floor with ½-inch screws or fasten it to a clean floor with 3M double-coated tape no. 4432.

3. On all two-panel sections, lower the bottom pivot rod until it projects ½ inch below the edge of the door. Make it 1¼ inch if the carpet is under the door.

4. Attach the doorknobs.

5. Lift one door set. Insert the bottom pivot rod (threaded) into the bottom pivot bracket. Pull down the top spring-loaded pivot rod. Insert it into the pivot bracket in the top rack. Insert the top and bottom nylon glide rod tips into the track (Fig. 9-100).

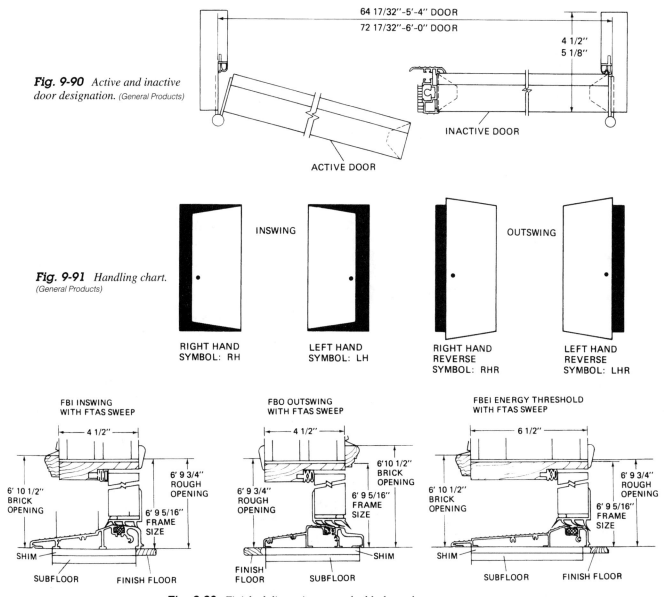

Fig. 9-90 *Active and inactive door designation.* (General Products)

64 17/32"–5'-4" DOOR
72 17/32"–6'-0" DOOR
4 1/2"
5 1/8"

INACTIVE DOOR

ACTIVE DOOR

Fig. 9-91 *Handling chart.* (General Products)

INSWING

OUTSWING

RIGHT HAND
SYMBOL: RH

LEFT HAND
SYMBOL: LH

RIGHT HAND
REVERSE
SYMBOL: RHR

LEFT HAND
REVERSE
SYMBOL: LHR

FBI INSWING
WITH FTAS SWEEP

4 1/2"

6' 10 1/2"
BRICK
OPENING

6' 9 3/4"
ROUGH
OPENING

6' 9 5/16"
FRAME
SIZE

SHIM

SUBFLOOR   FINISH FLOOR

FBO OUTSWING
WITH FTAS SWEEP

4 1/2"

6' 10 1/2"
BRICK
OPENING

6' 9 3/4"
ROUGH
OPENING

6' 9 5/16"
FRAME
SIZE

SHIM

FINISH
FLOOR   SUBFLOOR

FBEI ENERGY THRESHOLD
WITH FTAS SWEEP

6 1/2"

6' 10 1/2"
BRICK
OPENING

6' 9 3/4"
ROUGH
OPENING

6' 9 5/16"
FRAME
SIZE

SHIM

SUBFLOOR   FINISH FLOOR

Fig. 9-92 *Finished dimensions on a double-hung door.* (General Products)

6. Install the second door set the same way.

7. Insert the rubber stop in the center of the top and bottom tracks. Make sure that the stop seats firmly in the track. For a two-panel installation, cut the rubber stop to the proper length.

8. Because of their design, the bifold doors are rigid enough to operate smoothly without a full bottom track. This permits better carpeting in the closet. Saw off a 4-inch section of the bottom track. Place this on the floor. Use a plumb bob to pivot the bottom points with the top pivot. See Fig. 9-101. Fasten the section to the floor with two ½-inch screws. Remove the bottom glide rods from the doors. Single-track installation is not recommended for 8'0"-high doors, 7'0"-wide four-panel doors, or 3'6"-wide two-panel doors.

Final adjustments To raise or lower the doors to the desired height, turn the threaded bottom pivot rod with a screwdriver. Make sure the doors are even and level across the top. Tighten the locknut.

Doors should close snugly against the rubber stop. For horizontal (lateral) alignment, loosen the screw holding the top or bottom pivot brackets in the track. Adjust the door in or out. Retighten the screw.

Keep all glides and track free from paint and debris. The aluminum track is already lubricated to ensure smooth operation. Occasionally repeat the lubrication with silicon spray, paraffin, or soap. This keeps door operation free and easy.

These instructions are for a particular make of door. However, most manufacturers' instructions are basically the same. There are some minor adjustments

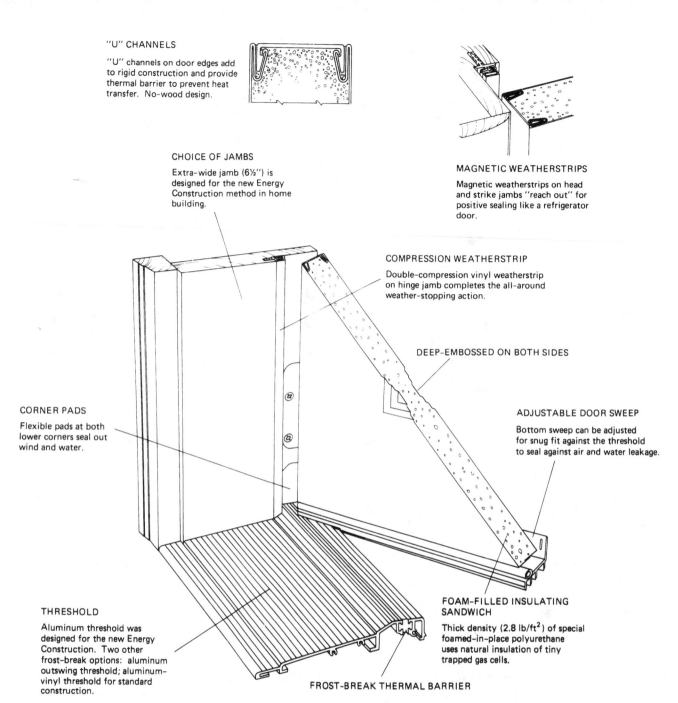

"U" CHANNELS

"U" channels on door edges add to rigid construction and provide thermal barrier to prevent heat transfer. No-wood design.

CHOICE OF JAMBS

Extra-wide jamb (6½") is designed for the new Energy Construction method in home building.

MAGNETIC WEATHERSTRIPS

Magnetic weatherstrips on head and strike jambs "reach out" for positive sealing like a refrigerator door.

COMPRESSION WEATHERSTRIP

Double-compression vinyl weatherstrip on hinge jamb completes the all-around weather-stopping action.

DEEP-EMBOSSED ON BOTH SIDES

CORNER PADS

Flexible pads at both lower corners seal out wind and water.

ADJUSTABLE DOOR SWEEP

Bottom sweep can be adjusted for snug fit against the threshold to seal against air and water leakage.

THRESHOLD

Aluminum threshold was designed for the new Energy Construction. Two other frost-break options: aluminum outswing threshold; aluminum-vinyl threshold for standard construction.

FOAM-FILLED INSULATING SANDWICH

Thick density (2.8 lb/ft^2) of special foamed-in-place polyurethane uses natural insulation of tiny trapped gas cells.

FROST-BREAK THERMAL BARRIER

Fig. 9-93 *Insulated metal door for commercial, industrial, or residential use. Note the energy-saving features.* (General Products)

you will have to make for each manufacturer. Make sure you follow the manufacturer's recommendations.

DOOR AND WINDOW TRIM
Interior Door Trim

Most inside or interior doors have two hinges. They usually come in a complete package. Once they are set in place, the casing has to be applied. See Fig. 9-102 for the location of the casing around an interior door. Note

how the jamb is installed. The stop is attached with nails and has a bevel cut at the bottom of the door. It prevents the door from swinging forward more than it should.

In Fig. 9-103 you will find two of the most commonly used moldings applied to the trim of a door. These two are colonial and ranch casing moldings. These are the names you use when ordering them. They are ordered from a lumberyard or mill.

Figure 9-104 shows how a molded casing is mitered at the corner. It is secured with a nail through the 45° cut. In the other part of this figure you see the

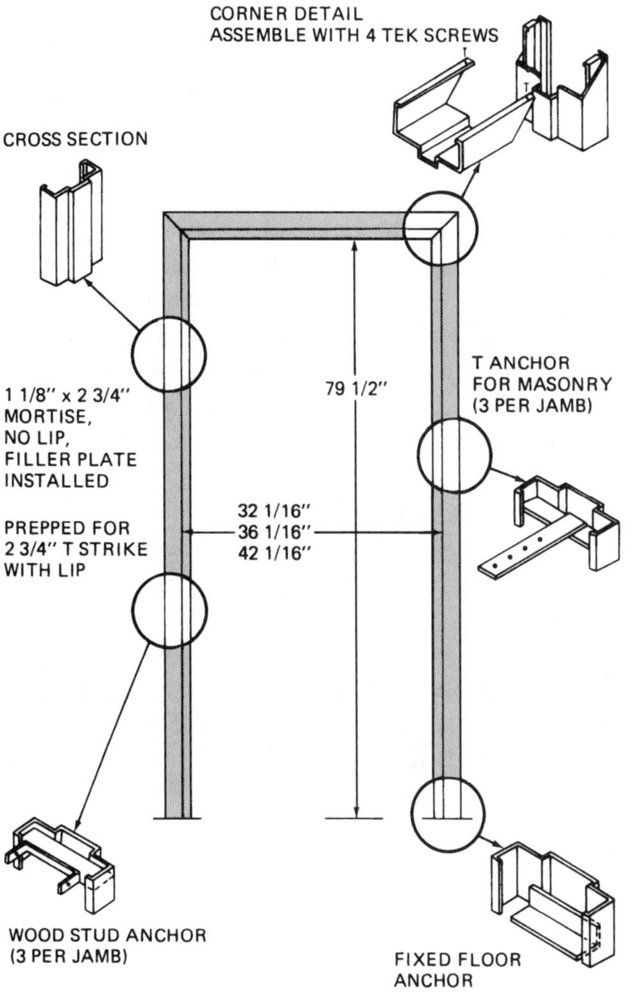
CORNER DETAIL
ASSEMBLE WITH 4 TEK SCREWS
CROSS SECTION
1 1/8" x 2 3/4"
MORTISE,
NO LIP,
FILLER PLATE
INSTALLED
PREPPED FOR
2 3/4" T STRIKE
WITH LIP
79 1/2"
32 1/16"
36 1/16"
42 1/16"
T ANCHOR
FOR MASONRY
(3 PER JAMB)
WOOD STUD ANCHOR
(3 PER JAMB)
FIXED FLOOR
ANCHOR

Fig. 9-94 *Putting together a metal frame for a door.* (General Products)

butt joint. This is where the casing meets at the side and top. Notice the way the nail is placed to hold the two pieces securely. Also notice the other nail locations. Why do you need to drill the nail hole for the toenailed side?

Table 9-1 ***Finished Opening Sizes for Bifold Doors.***

Door Width	Number of Panels	Door opening, Inches* 6′8″	Door opening, Inches* 8′0″	Actual Door Width, Inches
1′6″	2	18 1/2 × 80 3/4	18 1/2 × 95 1/4	17 7/16
2′0″	2	24 1/2 × 80 3/4	24 1/2 × 95 1/4	23 7/16
2′6″	2	30 1/2 × 80 3/4	30 1/2 × 95 1/4	29 7/16
3′0″	2	36 1/2 × 80 3/4	36 1/2 × 95 1/4	35 7/16
3′0″	4	36 1/2 × 80 3/4	36 × 95 1/4	35
3′6″	2	42 1/2 × 80 3/4	42 1/2 × 95 1/4	41 7/16
4′0″	4	48 × 80 3/4	48 × 95 1/4	47
5′0″	4	60 × 80 3/4	60 × 95 1/4	59
6′0″	4	72 × 80 3/4	72 × 95 1/4	71
7′0″	4	84 × 80 3/4	84 × 95 1/4	83

*Finished opening width shown provides 1/2 inch clearance each side of door. Finished opening width may be reduced by 1/2 inch provided finished opening is square and plumb. This will require cutting track. Finished opening heights shown provide 3/8 inch clearance between door and track—top and bottom. (This makes 7/8 inch between door and floor.) Doors can be raised to have 1 1/8 inch clearance door to floor without increasing opening height.

Installation of the strike plate in the side jamb is shown in Fig. 9-105. It has to be routed or drilled out. This allows the door locking mechanism to move into the hole. Figure 9-106 shows how the strike plate is mounted onto the door jamb.

Window Trim

Windows have to be trimmed. This completes the installation job. See Fig. 9-107. There are a couple of ways to trim a window. Take a look at Fig. 9-108 and note the difference. Shown is a trimmed window with casing at the bottom instead of a stool and apron. This is a quicker and simpler method of finishing a window. There is no need for a stool to overlap the apron or casing in some instances. This is the choice of the architect or the owner of the home. There are problems with the apron and stool method. The apron and stool will pull away from the inside casing. This leaves a gap of up to ¼ inch. It can become unsightly in time.

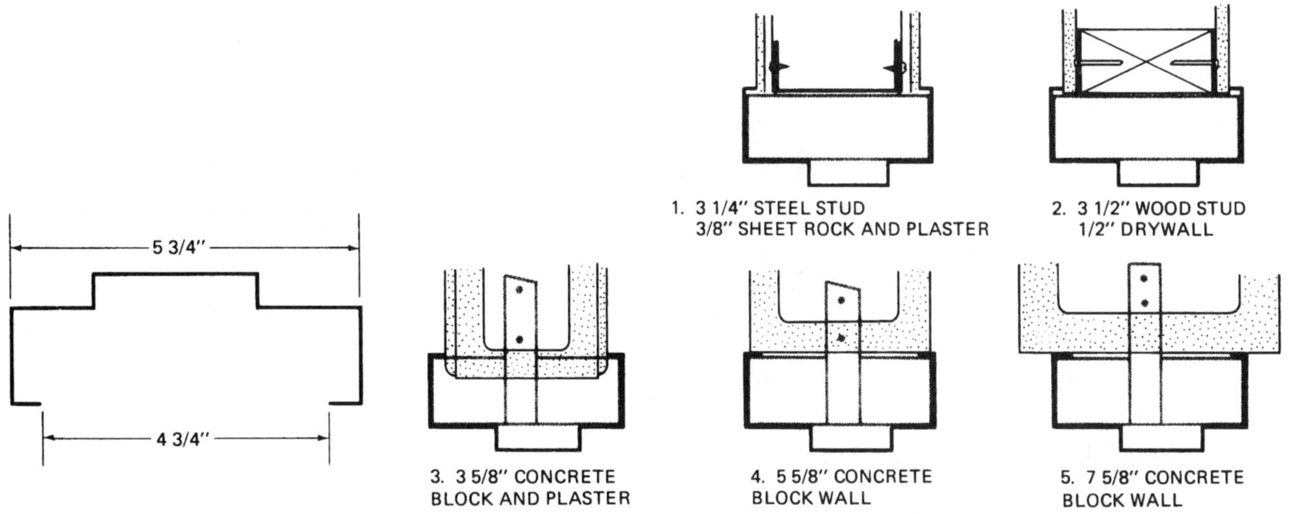
5 3/4"
4 3/4"
1. 3 1/4" STEEL STUD
3/8" SHEET ROCK AND PLASTER
2. 3 1/2" WOOD STUD
1/2" DRYWALL
3. 3 5/8" CONCRETE
BLOCK AND PLASTER
4. 5 5/8" CONCRETE
BLOCK WALL
5. 7 5/8" CONCRETE
BLOCK WALL

Fig. 9-95 *Typical installations with metal door frames.* (General Products)

Figure 9-109 shows some of the moldings that can be used in trimming windows, doors, or panels. These moldings are available in prepared lengths of 8 feet and 12 feet. Generally speaking, the simpler the molding design, the easier it is to clean. Many depressions or designs in a piece of wood can allow it to pick up dust. Some are very difficult to clean.

INSTALLING LOCKS

There are seven simple steps to installing a lock in a door. Figure 9-110 shows them in order.

In some cases you might want to reverse the lock. This may be the case when you change the lock from one door to another. The hand of the door might be different. In Fig. 9-111 you can see how simple it is to change the hand of the lock. In some cases you may have bought the lock without noticing how it should fit. This way you are able to make it fit in either direction.

There are a number of locks available. Figure 9-112 shows how 18 different locksets can be replaced by National Lock's locksets or lever sets. Figure 9-113 shows some of the designs available for strikes. The strike is always supplied with the lockset. Figure 9-114 shows the latch bolts. They may have round or square corners. They may have or may not have deadlock capability.

Entrance handle locks Most homes have an elaborate front door handle. In Fig. 9-115 you can see two of the types of escutcheons used to decorate the doorknob.

Door handles also become something of a decorative item. They come in a number of styles. Each lock manufacturer offers a complete line. See Fig. 9-116 for an illustration of two such handles. These handles are usually cast brass.

A number of lockset designs are available for entrance and interior doors. They may lock, then require a nail or pin to be opened. Or they may require a key. See Fig. 9-117.

Auxiliary locks Auxiliary locks are those placed on exterior doors to prevent burglaries. They are called deadlocks. They usually have a 1-inch bolt that projects

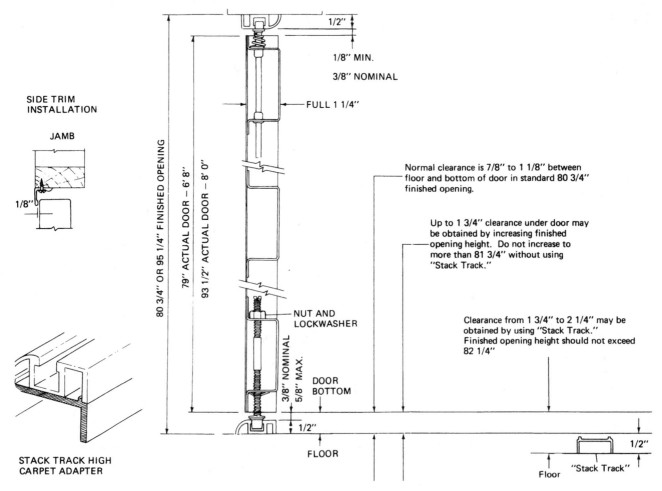

Fig. 9-96 *Installation of the bifold door.* (General Products)

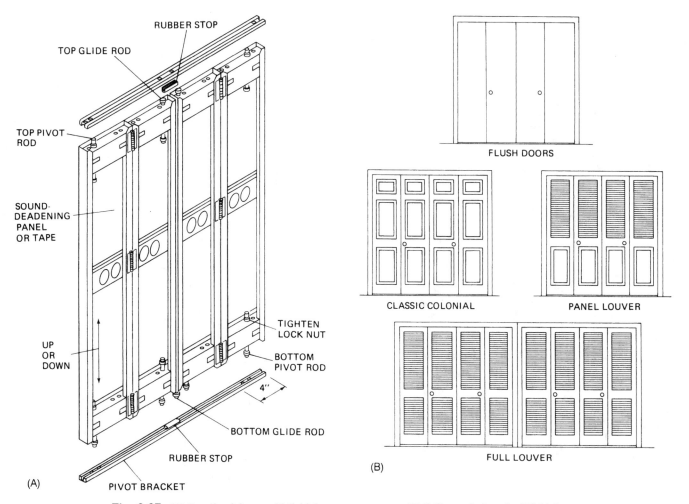

RUBBER STOP

TOP GLIDE ROD

TOP PIVOT
ROD

SOUND-
DEADENING
PANEL
OR TAPE

UP
OR
DOWN

TIGHTEN
LOCK NUT

BOTTOM
PIVOT ROD

4"

BOTTOM GLIDE ROD

RUBBER STOP

PIVOT BRACKET

(A)

FLUSH DOORS

CLASSIC COLONIAL

PANEL LOUVER

FULL LOUVER

(B)

Fig. 9-97 (A) Details of the metal bifold door. (General Products) (B) Different designs for bifold doors.

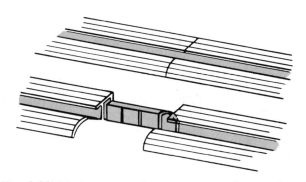

Fig. 9-98 Vinyl connectors let you snap apart four-panel track instantly to install two-panel bifolds. (General Products)

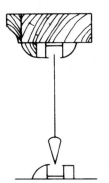

Fig. 9-99 Using a plumb bob to make sure the tracks line up. (General Products)

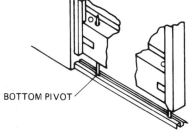

BOTTOM PIVOT

Fig. 9-100 Pivoting the bottom pin in the track. (General Products)

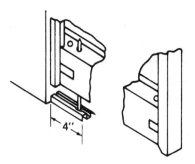

4"

Fig. 9-101 You can remove most of the bottom track if you don't want it on the floor. This lets carpeting run straight through to the closet wall . (General Products)

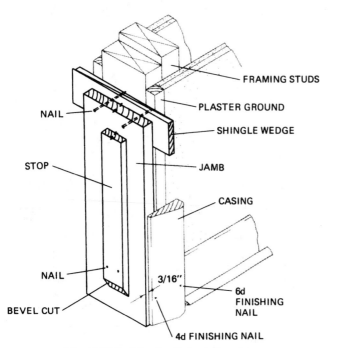

Fig. 9-102 Trim details for a door frame.

Labels: FRAMING STUDS, PLASTER GROUND, SHINGLE WEDGE, JAMB, CASING, 6d FINISHING NAIL, 4d FINISHING NAIL, 3/16", BEVEL CUT, NAIL, STOP, NAIL

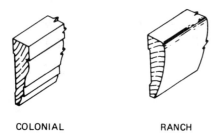

COLONIAL   RANCH

Fig. 9-103 Two popular types of molding used for trim around a door.

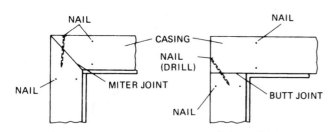

Labels: NAIL, NAIL, CASING, NAIL (DRILL), MITER JOINT, BUTT JOINT, NAIL, NAIL

Fig. 9-104 Two methods of joining trim over a door.

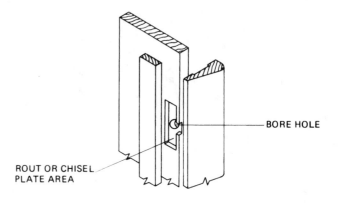

Labels: BORE HOLE, ROUT OR CHISEL PLATE AREA

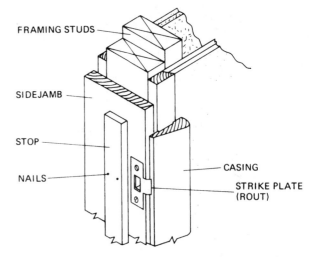

Labels: FRAMING STUDS, SIDEJAMB, STOP, NAILS, CASING, STRIKE PLATE (ROUT)

Fig. 9-105 Installing a strike plate for the lockset.

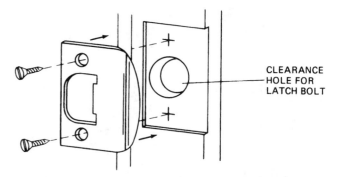

Labels: CLEARANCE HOLE FOR LATCH BOLT

Fig. 9-106 Installing the strike plate on the door jamb.

past the door. It fits into the door jamb (Fig. 9-118). In Fig. 9-119 you can see three of the various deadlock designs used. The key side, of course, goes on the outside of the door.

A standard type of lock is exploded for you in Fig. 9-120. Note the names of the parts. These locks have keys that fit into a cylinder. The keys lock or unlock the latch bolt.

Exposed brass, bronze, or aluminum parts are buffed or brushed. They are protected with a coat of lacquer. Aluminum is brushed and anodized.

Construction keying There are a couple of methods used for keying locksets. One of them makes it possible for a construction supervisor to get into a number of buildings with one key. Once the building is occupied, the lock is converted. The lock's key and no other will then operate it. See Fig. 9-121.

Builders use a short four-pin tumbler key to operate the lock during the construction period. A nylon wafer is inserted in the keyway at the factory. This blocks operation of the fifth and sixth tumblers. Accidental

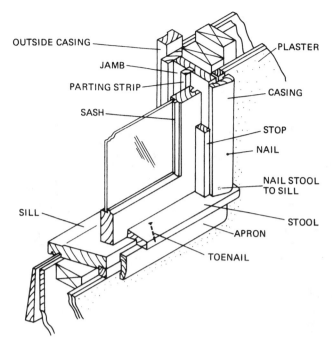

Fig. 9-107 *Window trim. Note the apron and stool.*

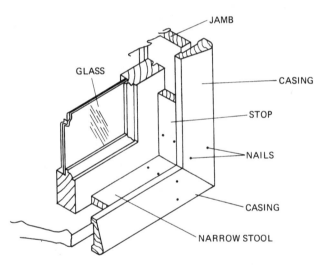

Fig. 9-108 *Window trim. Note the absence of the apron.*

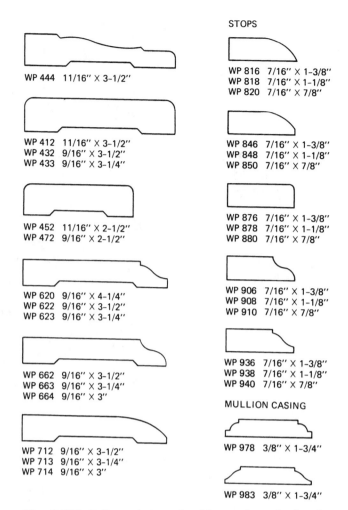

Fig. 9-109 *Different designs of molding used as trim for windows, doors, and other parts of the house.*

deactivation of the builder's key is unlikely. This is because a conscious effort to apply a 10- to 15-pound force is required to dislodge the nylon wafer the first time the five-tumbler key is used. The owner's keys are packed in specially marked, sealed envelopes.

When construction is completed, the unit is ready for occupancy. The homeowner inserts the regular five-pin tumbler key to move the nylon wafer. This makes the fifth tumbler operative. It automatically deactivates the four-pin tumbler arrangement, making the builder's key useless. Now the locksets can be operated only by the owner's keys.

Other lockmakers have different methods for this key operation. See Fig. 9-122. Figure 9-123 shows another method of key operation of locks. In this case the whole cylinder of the new lock is removed. The construction worker inserts a different cylinder. This cylinder works with the master key. Once the job is finished, the original cylinder is reinserted. The construction worker's key no longer operates this lock. Only the homeowner can operate the lock.

STORM DOORS AND WINDOWS

In most windows thermopane is used. It consists of two pieces of glass welded together with an air space. (See Fig. 9-20.) Some windows have the space evacuated so that a vacuum exists inside. This cuts down on the transfer of heat from the inside of a heated building to the cold outside, or the reverse during the summer. In some cases, as shown in Chapter 15, another piece of glass or plastic is added. It fits over the twin panes of glass. See Fig. 9-124.

Metal transfers heat faster than wood. Wood is 400 times more effective than steel as an insulator. It is 1800 times more effective than aluminum.

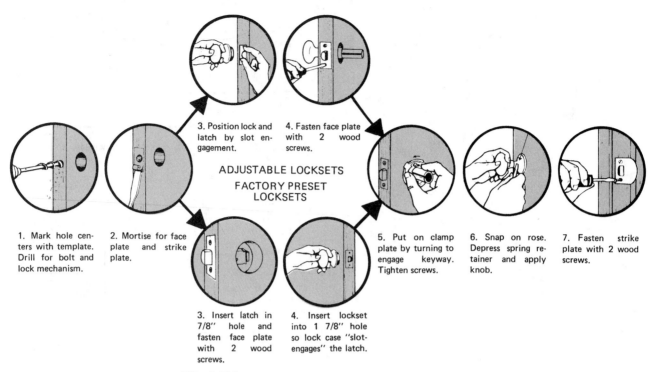

1. Mark hole centers with template. Drill for bolt and lock mechanism.

2. Mortise for face plate and strike plate.

ADJUSTABLE LOCKSETS

3. Position lock and latch by slot engagement.

4. Fasten face plate with 2 wood screws.

FACTORY PRESET LOCKSETS

3. Insert latch in 7/8" hole and fasten face plate with 2 wood screws.

4. Insert lockset into 1 7/8" hole so lock case "slot-engages" the latch.

5. Put on clamp plate by turning to engage keyway. Tighten screws.

6. Snap on rose. Depress spring retainer and apply knob.

7. Fasten strike plate with 2 wood screws.

Fig. 9-110 *Seven simple steps to install a lockset.* (National Lock)

Apply lock to door with key to outside as shown above. (See installation instruction sheet.) Turn turnbutton or pushbutton on inside of door to locked position.

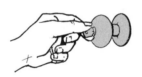

Insert key in lock. Turn key 30° either left or right. Do not retract latch bolt.

Keeping key in 30° position, remove lock and knob. First depress knob retainer pin. Then pull on key and knob together. Do not pull the knob out ahead of the key.

Be sure plug is in locked position. Then remove key approximately halfway out of plug. Turn entire plug, key, and cylinder in knob so bitting of key is up.

Apply knob on cam by engaging tab on knob with slot in tube. Keep key in vertical position. Push knob on tube while rotating key and plug gently. You will feel the proper engagement of locating tab on lock plug with slot in tube.

Depress knob retainer pin and push on knob as far as possible.

Push key all the way in keyway. Turn key to 30° position and push knob until retainer pin engages slot in knob shank.

Fig. 9-111 *Reversing the hand of a lock.* (National Lock)

Storm doors come in a wide variety of shapes and designs. They usually have a combination of screen and glass. The glass is removed in the summer and a screen wire panel is inserted in its place. This way the storm door serves year-round. See Fig. 9-125. They are delivered prehung and ready for installation. All that has to be done to install them is to level the door and add screws in the holes around the edges. A door closer is added to make sure the door closes after use. In some cases a spring adjustment device is added to the top so that the door closer and the door are protected from wind gusts. Most storm doors are made of metal. However, they are available in wood or plastic.

EASY LOCKSET INSTALLATION

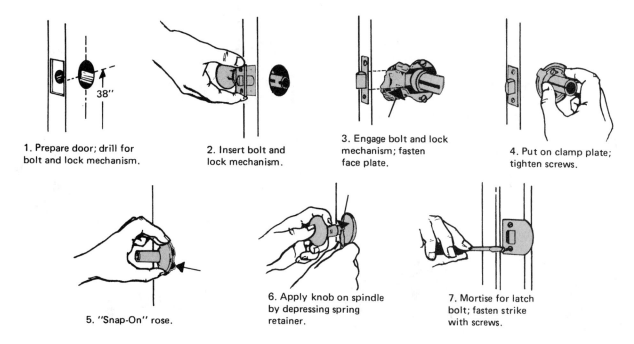

1. Prepare door; drill for bolt and lock mechanism.

2. Insert bolt and lock mechanism.

3. Engage bolt and lock mechanism; fasten face plate.

4. Put on clamp plate; tighten screws.

5. "Snap-On" rose.

6. Apply knob on spindle by depressing spring retainer.

7. Mortise for latch bolt; fasten strike with screws.

EASY LOCKSET REPLACEMENT

With just a screwdriver, the following 18 residential lockset brands can be replaced by National Lock locksets or lever sets.
ARROW COMET CORBIN DONNER ELGIN HARLOC KWIKSET LOCKWOOD MEDALIST NATIONAL RUSSWIN SARGENT SCHLAGE TROJAN TROY WEISER WESLOCK YALE

Fig. 9-112 *Replacement of a lockset.* (National Lock)

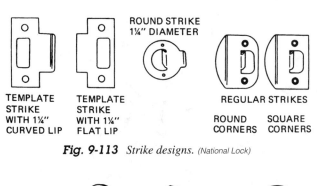

Fig. 9-113 *Strike designs.* (National Lock)

TEMPLATE STRIKE WITH 1¼" CURVED LIP

TEMPLATE STRIKE WITH 1¼" FLAT LIP

ROUND STRIKE 1¼" DIAMETER

REGULAR STRIKES

ROUND CORNERS SQUARE CORNERS

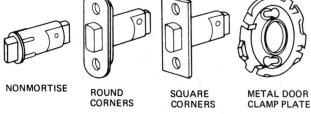

NONMORTISE

ROUND CORNERS

SQUARE CORNERS

METAL DOOR CLAMP PLATE

Fig. 9-114 *Latch bolt designs.* (National Lock)

Standard sizes are for openings from 35¾ to 36⅜ inches wide and from 79¾ to 81¼ inches high. There is an extender Z bar available for openings up to 37⅛ inches.

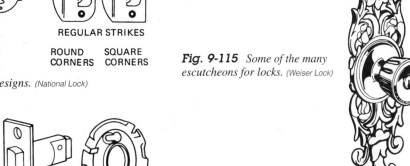

Fig. 9-115 *Some of the many escutcheons for locks.* (Weiser Lock)

According to the company's testing lab, this plastic (polypropylene) door has 45 percent more heat retention than an aluminum door. See Fig. 9-126.

INSTALLING A SLIDING DOOR

Sliding doors are a common addition to a house today. The doors slide open so that the patio can be reached

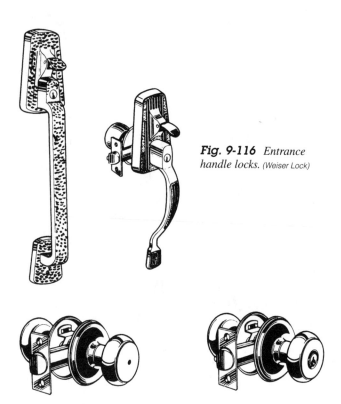

Fig. 9-116 *Entrance handle locks.* (Weiser Lock)

Fig. 9-118 *A deadlock projects out from the door and fits into the door jamb to make a secure door.* (Weiser Lock)

Fig. 9-117 *Locksets for interior and exterior doors.* (Weiser Lock)

Fig. 9-119 *Auxiliary locks. Chain and bolt, and two types of deadbolts.* (Weiser Lock)

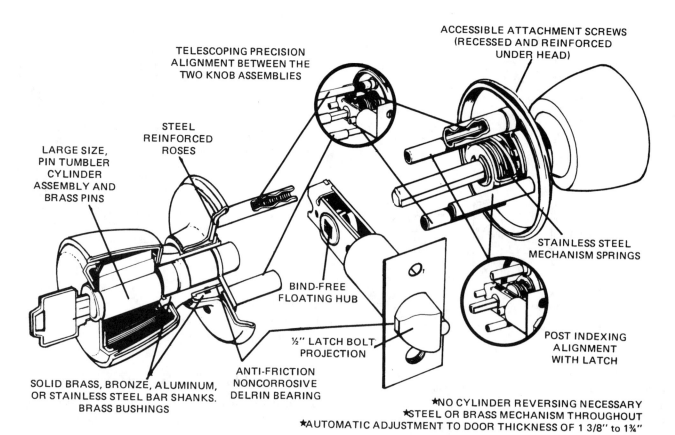

TELESCOPING PRECISION ALIGNMENT BETWEEN THE TWO KNOB ASSEMBLIES

ACCESSIBLE ATTACHMENT SCREWS (RECESSED AND REINFORCED UNDER HEAD)

STEEL REINFORCED ROSES

LARGE SIZE, PIN TUMBLER CYLINDER ASSEMBLY AND BRASS PINS

STAINLESS STEEL MECHANISM SPRINGS

BIND-FREE FLOATING HUB

½" LATCH BOLT PROJECTION

POST INDEXING ALIGNMENT WITH LATCH

SOLID BRASS, BRONZE, ALUMINUM, OR STAINLESS STEEL BAR SHANKS. BRASS BUSHINGS

ANTI-FRICTION NONCORROSIVE DELRIN BEARING

★NO CYLINDER REVERSING NECESSARY
★STEEL OR BRASS MECHANISM THROUGHOUT
★AUTOMATIC ADJUSTMENT TO DOOR THICKNESS OF 1 3/8" to 1¾"

Fig. 9-120 *Exploded view of a lockset.* (Weiser Lock)

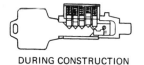 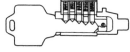

DURING CONSTRUCTION   AFTER CONSTRUCTION

Fig. 9-121 *Construction keying of locksets. (National Lock)*

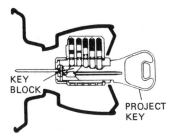

KEY BLOCK

PROJECT KEY

Lock cylinder is operated by the special "project key." The last two pins in the cylinder are held inoperative by the key block.

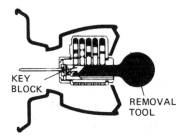

KEY BLOCK

REMOVAL TOOL

The special "project key" is canceled out by removal of the key block. A key block removal tool is furnished with the master keys for the locks. Simply push the removal tool into the keyway. Upon withdrawal, the key block will come out of the keyway. Thereafter, the "project key" no longer will operate the lock cylinder.

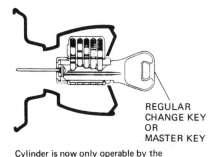

REGULAR CHANGE KEY OR MASTER KEY

Cylinder is now only operable by the regular change key or master key.

Fig. 9-122 *Another method of construction keying. (Weiser Lock)*

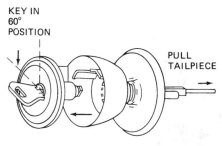

KEY IN 60° POSITION

PULL TAILPIECE

REMOVE REGULAR KEY CYLINDER

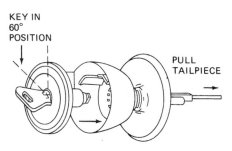

KEY IN 60° POSITION

PULL TAILPIECE

INSERT CONSTRUCTION CYLINDER

Fig. 9-123 *Some types of construction cylinders have to be removed. (Weiser Lock)*

Fig. 9-124 *Examples of thermopane windows. (Andersen)*

easily. It takes some special precautions to make sure the sliding doors will operate correctly. Most of these doors are made by a manufacturer like Andersen. They require a minimum of effort on the part of the carpenter. However, some very special steps are required. This portion of the chapter will deal primarily with the primed-wood type of gliding door and the Andersen Perma-Shield® type of gliding door.

The primed-wood gliding door (see Fig. 9-127) does not have a flange around it for quick installation. It requires some special attention. You will get an idea of how it fits into the rough opening from Fig. 9-128.

The Perma-Shield® gliding door is slightly different from the primed-wood type. See Fig. 9-129. In Fig. 9-130 you will find the details of the Perma-Shield door so that you can see the differences between the two types.

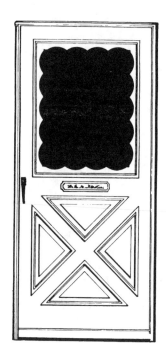

Fig. 9-125 *Storm door. (EMCO)*

Installation of both types of doors requires a rough opening in the frame structure of the house or building. The rough opening is constructed the same way for both types of doors.

Preparation of the Rough Opening

Installation techniques, materials, and building codes vary from area to area. Contact your local material supplier for specific recommendations for your area.

The same rough opening procedures can be used for both doors. There are, however, variations to note in gliding door installation procedures. These will be looked at more fully as we go along here.

If you need to enlarge the opening, make sure you use the proper-size header. Header size is usually given by the manufacturer of the door, or you can obtain it from your local supplier.

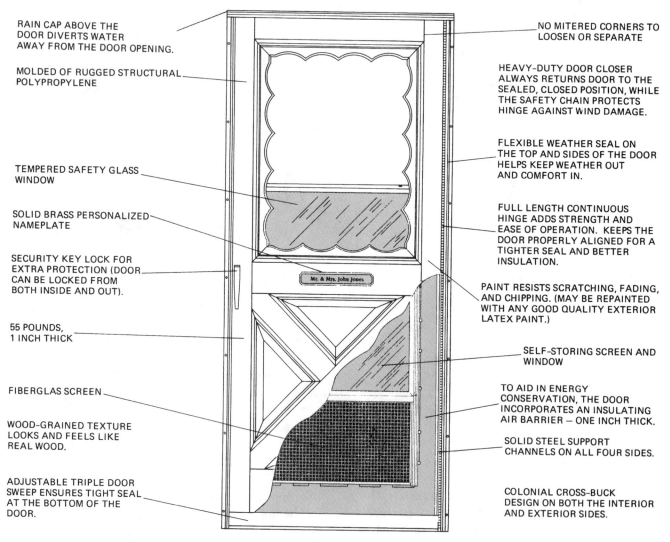

RAIN CAP ABOVE THE DOOR DIVERTS WATER AWAY FROM THE DOOR OPENING.

MOLDED OF RUGGED STRUCTURAL POLYPROPYLENE

TEMPERED SAFETY GLASS WINDOW

SOLID BRASS PERSONALIZED NAMEPLATE

SECURITY KEY LOCK FOR EXTRA PROTECTION (DOOR CAN BE LOCKED FROM BOTH INSIDE AND OUT).

55 POUNDS, 1 INCH THICK

FIBERGLAS SCREEN

WOOD-GRAINED TEXTURE LOOKS AND FEELS LIKE REAL WOOD.

ADJUSTABLE TRIPLE DOOR SWEEP ENSURES TIGHT SEAL AT THE BOTTOM OF THE DOOR.

NO MITERED CORNERS TO LOOSEN OR SEPARATE

HEAVY-DUTY DOOR CLOSER ALWAYS RETURNS DOOR TO THE SEALED, CLOSED POSITION, WHILE THE SAFETY CHAIN PROTECTS HINGE AGAINST WIND DAMAGE.

FLEXIBLE WEATHER SEAL ON THE TOP AND SIDES OF THE DOOR HELPS KEEP WEATHER OUT AND COMFORT IN.

FULL LENGTH CONTINUOUS HINGE ADDS STRENGTH AND EASE OF OPERATION. KEEPS THE DOOR PROPERLY ALIGNED FOR A TIGHTER SEAL AND BETTER INSULATION.

PAINT RESISTS SCRATCHING, FADING, AND CHIPPING. (MAY BE REPAINTED WITH ANY GOOD QUALITY EXTERIOR LATEX PAINT.)

SELF-STORING SCREEN AND WINDOW

TO AID IN ENERGY CONSERVATION, THE DOOR INCORPORATES AN INSULATING AIR BARRIER — ONE INCH THICK.

SOLID STEEL SUPPORT CHANNELS ON ALL FOUR SIDES.

COLONIAL CROSS-BUCK DESIGN ON BOTH THE INTERIOR AND EXTERIOR SIDES.

Fig. 9-126 *Energy saving plastic storm door. (EMCO)*

Preparation of the rough opening should follow these steps:

1. Lay out the gliding-door opening width between the regular studs to equal the gliding-door rough opening width plus the thickness of two regular studs. See Fig. 9-131.

2. Cut two pieces of header material to equal the rough opening width of the gliding door plus the

Fig. 9-127 *Primed-wood gliding door.* (Andersen)

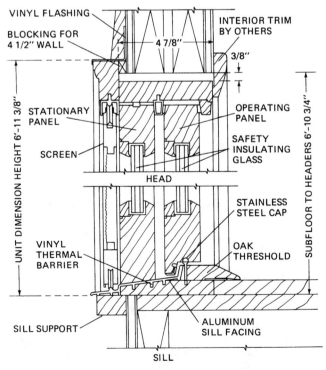

Fig. 9-128 *Details of the primed-wood gliding door.* (Andersen)

Fig. 9-129 *Perma-Shield® gliding door.* (Andersen)

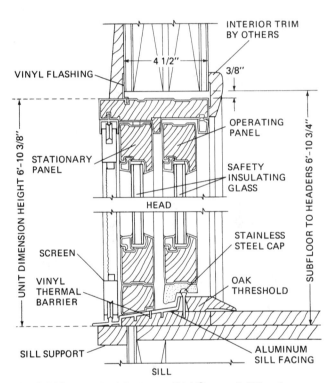

Fig. 9-130 *Details of the Perma-Shield® type of gliding door.* (Andersen)

thickness of two trimmer studs. Nail two header members together using an adequate spacer so that the header thickness equals the width of the trimmer stud. See Fig. 9-132.

3. Position the header at the proper height between the regular studs. Nail through the regular studs into the header to hold the header in place until the next step is completed. See Fig. 9-133.

4. Cut the jack or trimmer studs to fit under the header. This will support the header. Nail the jack or trimmer studs to the regular studs. See Fig. 9-134.

5. Apply the exterior sheathing (fiberboard, plywood, etc.) flush with the header and jack stud members. See Fig. 9-135.

Installation of a Wood Gliding Door

Keep in mind that all these illustrations are as viewed from the outside. Be sure the subfloor is level and the rough opening is plumb and square before installing the gliding-door frame. If you follow these steps closely, the door should be properly installed.

1. Run caulking compound across the opening to provide a tight seal between the door sill and the floor. Remove the shipping skids from the sill of the frame if the gliding door has been shipped set up. Follow the instructions included in the package if the frame is not set up. See Fig. 9-136.

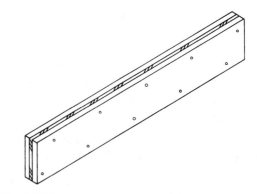

Fig. 9-132 *Header for a gliding-door opening.* (Andersen)

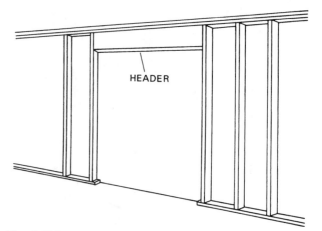

Fig. 9-133 *Placement of the header in the rough opening.* (Andersen)

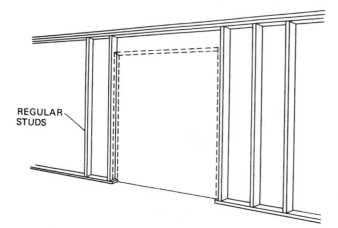

Fig. 9-131 *Layout of a rough opening for a gliding door.* (Andersen)

2. Position the frame in the opening from the outside. See Fig. 9-137. Apply pressure to the sill to properly distribute the caulking compound. The sill must be level. Check carefully and shim if necessary.

3. After leveling the sill, secure it to the floor by nailing along the inside edge of the sill with 8d coated nails spaced approximately 12 inches apart. See Fig. 9-138.

4. The jamb must be plumb and straight. Temporarily secure it in the opening with 10d casing nails through each side casing into the frame members. Using a straightedge, check the jambs for bow and shim. Shim solidly (five per jamb) between side jambs and jack studs.

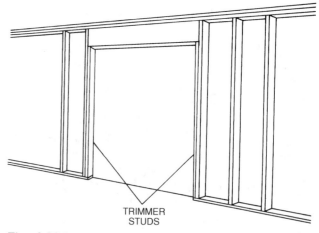

Fig. 9-134 *Placement of jack or trimmer studs in the rough opening.* (Andersen)

5. Complete the exterior nailing of the unit in the opening by nailing through the side and head casings into the frame members with 10d casing nails. See Fig. 9-139.

6. Position the flashing on the head casing and secure it by nailing through the vertical leg. The vertical center brace may now be removed from the frame.

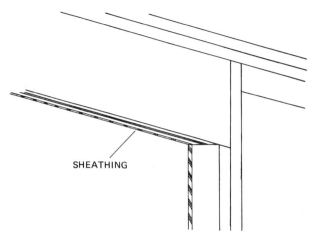

Fig. 9-135 *Application of exterior sheathing over the header.* (Andersen)

SHEATHING

Fig. 9-138 *Securing the sill to the floor by nailing.*

Fig. 9-136 *Running a bead of caulking to seal the sill and door for a sliding door.*

Fig. 9-139 *Exterior nailing of the unit.* (Andersen)

Fig. 9-137 *Leveling the sill.*

Be sure to remove and save the head and sill brackets. See Fig. 9-140.

7. Apply the treated wood sill support under the protruding metal sill facing. See Fig. 9-141. Install it tight to the underside of the metal sill with 10d casing nails.

Fig. 9-140 *Securing the flashing on the head casing by nailing.*

Fig. 9-141 *Applying the sill support under the metal sill facing.*

8. Position the stationary door panel in the outer run. Be sure the bottom rail is straight with the sill. Force the door into the run of the side jamb with a 2-×-4 wedge. See Fig. 9-142. Check the position by aligning the screw holes of the door bracket with the holes in the sill and head jamb. Repeat the above procedure for stationary panels of the triple door (if one is used here). Before the left-hand stationary panel is installed in a triple door, be sure to remove the screen bumper on the sill. Keep in mind, however, that only a double door is shown here.

9. Note the mortise in the bottom rail for a bracket. Secure the bracket with No. 8 one-inch flathead screws through the predrilled holes. See Fig. 9-143

Fig. 9-142 *Positioning the stationary door panel in the outer run by using a 2-×-4 wedge.*

for details. Align the bracket with the predrilled holes in the head jamb and secure it with No. 8 one-inch flathead screws. See Fig. 9-144. Repeat the procedure for the stationary panels of a triple door. The head stop is now removed if the unit has been shipped set up.

10. Apply security screws. Apply the two 1½-inch No. 8 flathead painted head screws through the parting stop into the stationary door top rail. See Fig. 9-145. Repeat for the stationary panels of a triple door.

11. Place the operating door on the rib of the metal sill facing, and tip the door in at the top. See Fig. 9-146. Position the head stop and apply with 1⁵⁄₁₆-inch No. 7 screws. See Fig. 9-147.

12. Check the door operation. If the door sticks or binds or is not square with the frame, locate the two adjustment sockets on the outside of the bottom rail. See Fig. 9-148. Simply remove the caps, insert the screwdriver, and turn to raise or lower the door. Replace the caps firmly.

Fig. 9-143 *Securing the bottom bracket with a screw.* (Andersen)

Fig. 9-144 *Securing the top bracket with a screw.* (Andersen)

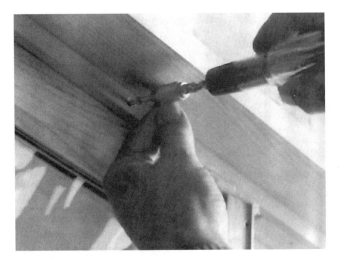

Fig. 9-145 *Applying the security screws.* (Andersen)

Fig. 9-146 *Placing the operating door on the rib of the metal sill facing.* (Andersen)

Fig. 9-147 *Positioning the head stop.* (Andersen)

13. If it is necessary to adjust the "throw" of the latch on two-panel doors, turn the adjusting screw to move the latch in or out. See Fig. 9-149. The lock may be adjusted on triple doors by loosening the screw to move the lock plate. See Fig. 9-150.

Masonry or Brick-Veneer Wall Installation of a Gliding Door

Gliding doors can be installed in masonry wall construction. Fasten a wood buck to the masonry wall and nail the sliding door to the wood buck using the procedures shown for frame wall construction.

Fig. 9-148 *Adjusting the door for square.* (Andersen)

Fig. 9-149 *The throw of the door is adjusted by this screw.* (Andersen)

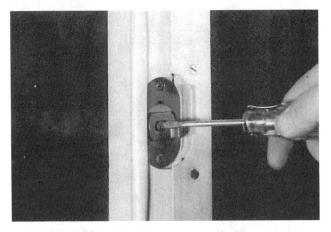

Fig. 9-150 *Lock adjustment on a triple door.* (Andersen)

In Fig. 9-151 a wood gliding door is installed in a masonry wall. Figure 9-152 shows a Perma-Shield® door installed with metal wall plugs or extender plugs in the masonry wall with brick veneer. Metal wall plugs or extender plugs and auxiliary casing can be specified when the door is ordered.

Keep in mind that when brick veneer is used as an exterior finish, adequate clearance must be left for caulking between the frame and the masonry. This will prevent damage and bowing caused by shrinkage and settling of the structural lumber.

Installation of a Perma-Shield Gliding Door

The Perma-Shield type of door is installed in the same way as has just been described, with some exceptions. The exceptions follow:

1. Note that wide vinyl flanges which provide flashing are used at the head and side jambs. See Fig. 9-153A. Locate the side member flush with the bottom of the sill with an offset leg pointing toward the inside of the frame. Tap with the hammer using a wood block to firmly seat the flashing in the groove. See Fig. 9-153B. Apply the head member similarly. Overlap the side flange on the outside.

2. After securing the frame to the floor with nails through the sill, apply clamps to draw the flanges tightly against the sheathing. See Fig. 9-154.

3. Temporarily secure the door in the opening with 10d casing nails through each side casing into the frame members. Using a straightedge, check the jambs for bow and shim. The jamb must be plumb

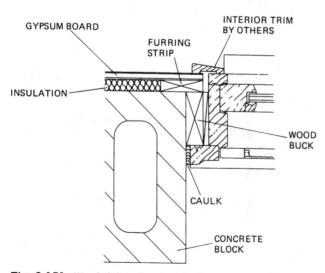

Fig. 9-151 *Wood gliding door installed in masonry wall.* (Andersen)

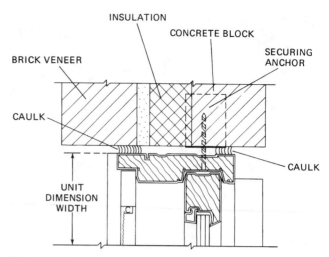

Fig. 9-152 *Perma-Shield® door installed with metal wall plugs or extender plugs in masonry wall with a brick veneer.* (Andersen)

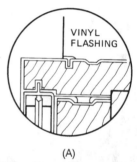

(A)

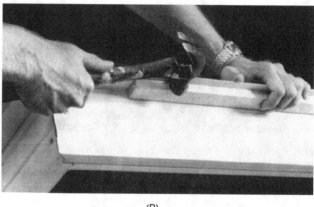

(B)

Fig. 9-153 *(A) Location of the vinyl flashing after it has been applied* (Andersen). *(B) Using a wooden block to apply the vinyl flashing.* (Andersen)

and straight. Shim solidly using five shims per jamb between the side jambs and the jack studs.

4. Side members on the Perma-Shield® gliding doors have predrilled holes to receive 2½-inch No. 10 screws. See Fig. 9-155. Shim at all screw holes between the door frame and the studs. Drill pilot holes into the studs. Secure the door frame to the studs with screws.

Fig. 9-154 *Clamps draw the flanges tightly against the sheathing.* (Andersen)

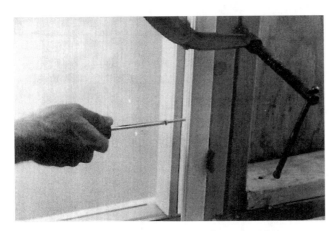

Fig. 9-155 *Securing the door frame to the studs with screws.* (Andersen)

5. The head jamb also has predrilled holes to receive the 2½-inch No. 10 screws. Shim at all screw holes between the door frame and header. Drill the pilot holes into the header. Insert the screws through the predrilled holes and draw up tightly. Do not bow the head jamb. See Fig. 9-156.

Figure 9-157 shows a Perma-Shield® door completely installed. Interior surfaces of the panel and frame should be primed before or immediately after installation for protection.

INSTALLING THE GARAGE DOOR

Garage doors are available in metal and wood. Wood doors are available in a high-grade hemlock/fir frame construction and recessed hardboard or raised redwood panels. Rough sawn, flush wood doors are also available. They come ready to be primed, painted, or stained to match the house finish. Steel doors come with a primer and need to be painted with a second coat to match the owner's preference in color or finish.

Fig. 9-156 *Placing screws in predrilled holes in the door header.* (Andersen)

Fig. 9-157 *Finished installation of a Perma-Shield® gliding door.*

Figures 9-158 and 9-159 show the types of springs used to aid in the raising of the garage door. The torsion springs are usually used for heavy doors for a two-car garage. A garage door with extension springs is usually involved with single-car garage doors.

A word of caution is usually sufficient: to avoid installation problems that could result in personal injury or property damage, use only the track specified and supplied with the door unit.

Some large doors weigh as much as 400 pounds when the spring tension is released. A single door weighs up to 200 pounds, and two people should work on it to prevent damage.

One of the primary concerns about installing a door is the headroom. Headroom is the space needed above the top of the door for the door, the overhead tracks, and the springs. Measure to check that there is

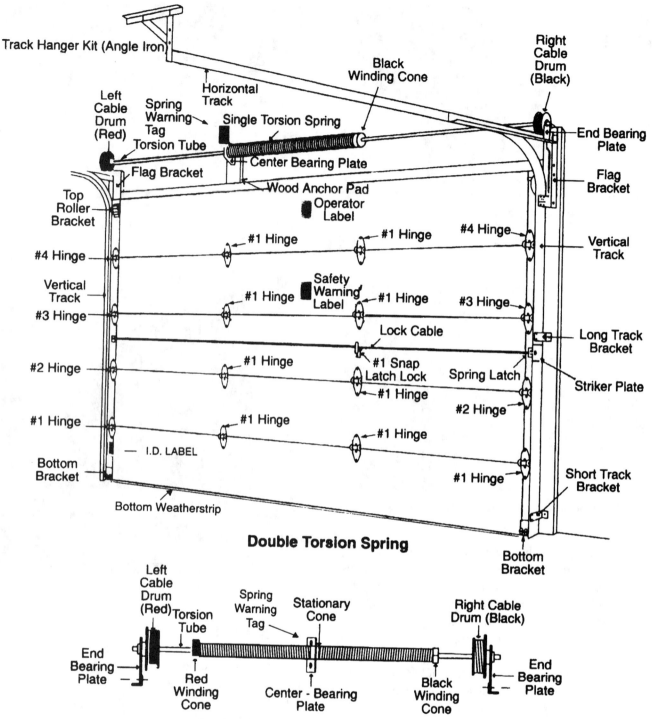

Double Torsion Spring

Fig. 9-158 A garage door with torsion springs. *(Clopay)*

no obstructions in your garage within that space. The normal space requirement is shown in Table 9-2. The backroom distance is measured from the back of the door into the garage, and should be at least 18 inches more than the height of the garage door. A minimum sideroom of 3.75 inches (5.5 inch EZ-Set Spring®) should be available on each side of the door on the interior wall surface to allow for attachment of the vertical track assembly. See Fig. 9-160.

Track radius is another important concern in the installation of the track. The radius of the track can be determined by measuring the dimension R in Fig. 9-161. If the dimension R measures 11 to 12 inches, then you have a 12-radius track. If R equals 14 to 15 inches, then you have a 15-inch radius track. See Fig. 9-161. About 3 inches of additional headroom height at the center plus additional backroom is needed to install an automatic garage door opener.

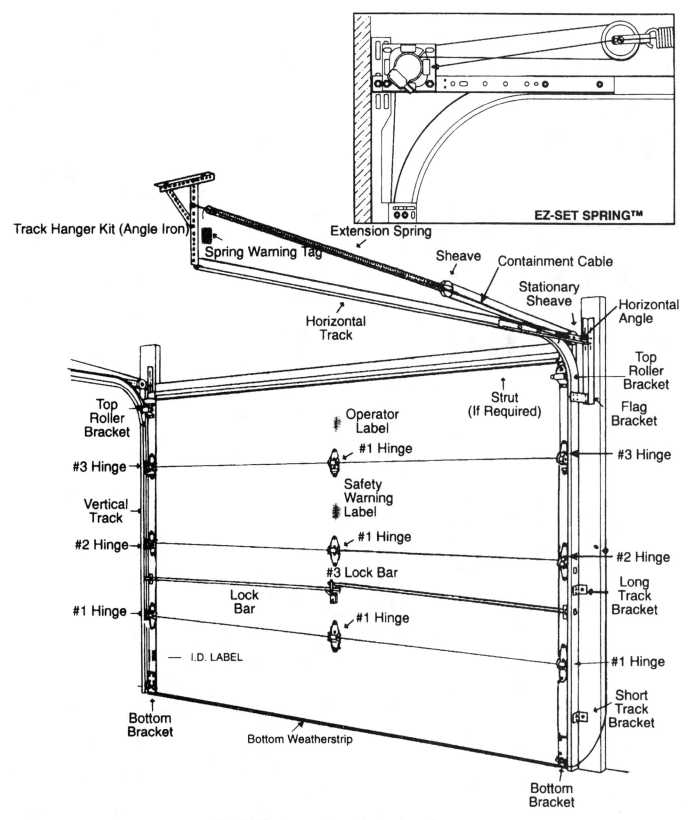

EZ-SET SPRING™

Track Hanger Kit (Angle Iron)

Spring Warning Tag

Extension Spring

Sheave

Containment Cable

Stationary Sheave

Horizontal Angle

Horizontal Track

Top Roller Bracket

Flag Bracket

Top Roller Bracket

Operator Label

#1 Hinge

Strut (If Required)

#3 Hinge

#3 Hinge

Vertical Track

Safety Warning Label

#2 Hinge

#1 Hinge

#2 Hinge

Long Track Bracket

#3 Lock Bar

Lock Bar

#1 Hinge

#1 Hinge

I.D. LABEL

#1 Hinge

Short Track Bracket

Bottom Bracket

Bottom Weatherstrip

Bottom Bracket

Fig. 9-159 *A garage door with extension springs.* (Clopay)

Check the door opener instructions to be sure. If the headroom is low there are a few options to compensate. For instance, a double track low headroom reduces the headroom requirement to 4.5 inches on

EZ-Set Spring® and extension springs, (9.5 inches on front mount torsion springs). Instructions are provided with the track. There is a low headroom conversion kit that reduces the required headroom to 4.5

Table 9-2 *Headroom Requirement Chart*

SPRING TYPE	TRACK RADIUS	HEADROOM REQUIRED
EZ-Set Spring™/Extension	12"	10"
EZ-Set Spring™/Extension	15"	12"
Torsion	12"	12"
Torsion	15"	14"

with any other low headroom option. This is used in place of the existing top roller. Instructions are included with the kit.

The next step is to prepare the opening. Figure 9-163 shows the rough opening for the door and the necessary additions. An option is stop molding featuring a built-in weather seal as shown in Fig. 9-164.

Next, prepare for installing the door sections. Spread the hardware on the floor in groups so that you can easily find the parts. Assemble as the directions require. One thing to keep in mind is if the door is going to be equipped with an automatic garage door opener, make sure that the door is always unlocked when the operator is being used. This will avoid damage to the door. Instructions come with the door in the form of a pamphlet with detailed line drawings.

Assemble and install the track. Follow the directions for the specific door being used. Pay special attention and use adequate length screws to fasten the

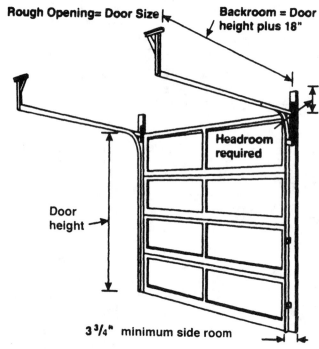

Fig. 9-160 *Required headroom.* (Clopay)

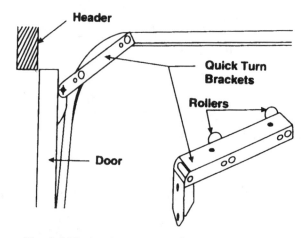

Fig. 9-162 *Quick Turn bracket for low headroom.* (Clopay)

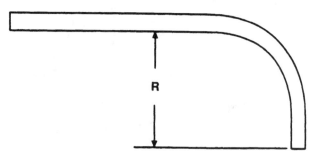

Fig. 9-161 *Track radius measurement.* (Clopay)

inches. This option is designed to modify the standard track. Instructions are provided in the kit. Another way to reduce the headroom requirement is to use the Quick Turn bracket. See Fig. 9-162. The Quick-Turn bracket cannot be used in conjunction

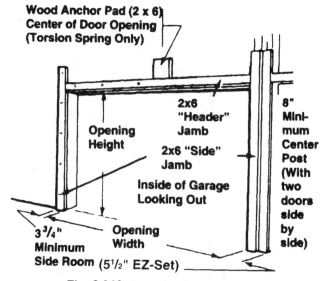

Fig. 9-163 *Preparing the opening.* (Clopay)

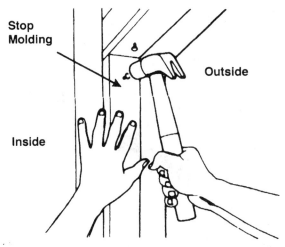

Fig. 9-164 *Door stop molding.* (Clopay)

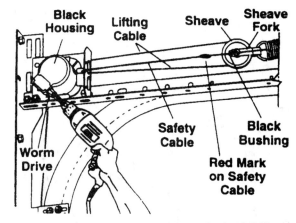

Fig. 9-165 *Adjusting the spring tension with a hand drill.* (Clopay)

rear track hangers into the trusses. A door may fall and cause serious injury if not properly secured.

Attaching springs Lifting cables and springs can be dangerous when installed incorrectly. It is very important to follow the instructions closely when installing the springs. Garage door springs can cause serious injury and property damage if they break under tension and are not secured with safety cables.

Keep your head well below the track when the spring is under tension or being tensioned because springs are dangerous when they are fully or partially wound. The first time the door is opened, make sure the door doesn't fall. This might happen if the tracks are not correctly aligned or the rear track hangers are not strong enough. Proceed slowly and carefully and follow the directions provided by the manufacturer. Both springs should be adjusted equally for proper operation.

The EZ-Set Spring® by Clopay can be adjusted by using a ⅜-inch drill using the ¼-inch hex driver provided in the door kit. Make sure the ¼-inch hex driver is inserted completely into the worm drive. The spring is tensioned by operating the drill in the clockwise direction. See Fig. 9-165.

Extension springs are used on single doors and have two springs, such as in Fig. 9-166. The door with less weight uses a single spring on each side of the door. The heavier door may use the double spring on each side. If your door was supplied with four extension springs, take notice of the color coding on the ends of the springs. If there are two color codes, be sure to use one of each on either side so that the spring tension is equal on both sides.

Torsion spring installation Torsion springs can be very dangerous if they are improperly installed or mis-

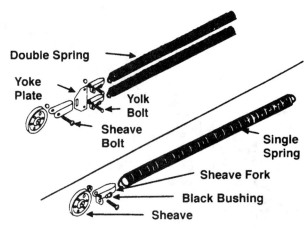

Fig. 9-166 *Single- and double-extension springs.* (Clopay)

handled. Do not attempt to install them alone unless you have the right tools and reasonable mechanical aptitude or experience, and you follow instructions very carefully. It is important to firmly and securely attach the torsion spring assembly to the frame of the garage. See Fig. 9-167.

Attaching an automatic opener When installing an automatic garage door opener, make sure to follow the manufacturer's installation and safety instructions carefully. Remove the pull-down rope and unlock or remove the lock. If attaching an operator bracket to the wood anchor pad, make sure the wooden anchor pad is free of cracks and splits and is firmly attached to the wall. Always drill pilot holes before attaching lag screws.

To avoid damage to the door, you must reinforce the top section of the door in order to provide a mounting point for the opener to be attached. Failure to reinforce the door, as illustrated in Fig. 9-168, will void the manufacturer's warranty. Note: All reinforcing angles are to be attached with #14 × ⅝-inch sheet metal screws at the reinforcement back-up plate locations.

Torsion Spring Installation

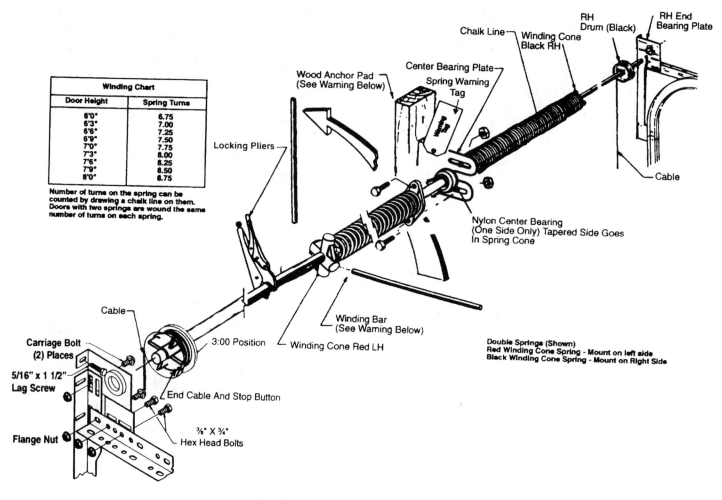

Winding Chart	
Door Height	Spring Turns
6'0"	6.75
6'3"	7.00
6'6"	7.25
6'9"	7.50
7'0"	7.75
7'3"	8.00
7'6"	8.25
7'9"	8.50
8'0"	8.75

Number of turns on the spring can be counted by drawing a chalk line on them. Doors with two springs are wound the same number of turns on each spring.

Locking Pliers

Wood Anchor Pad (See Warning Below)

Center Bearing Plate

Spring Warning Tag

Chalk Line

Winding Cone Black RH

RH Drum (Black)

RH End Bearing Plate

Cable

Nylon Center Bearing (One Side Only) Tapered Side Goes In Spring Cone

Cable

Carriage Bolt (2) Places

5/16" x 1 1/2" Lag Screw

Flange Nut

3:00 Position

End Cable And Stop Button

⅜" X ¾" Hex Head Bolts

Winding Bar (See Warning Below)

Winding Cone Red LH

Double Springs (Shown)
Red Winding Cone Spring - Mount on left side
Black Winding Cone Spring - Mount on Right Side

Fig. 9-167 *Torsion spring installation.* (Clopay)

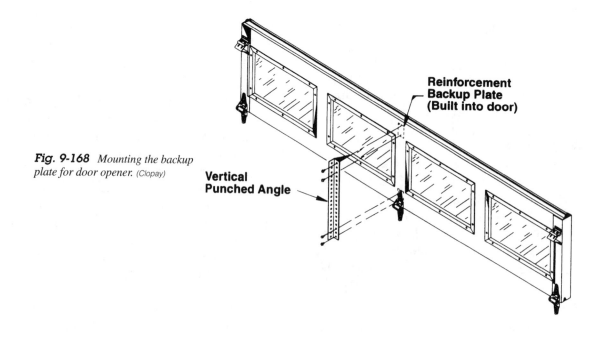

Reinforcement Backup Plate (Built into door)

Vertical Punched Angle

Fig. 9-168 *Mounting the backup plate for door opener.* (Clopay)

Cleaning and painting Before painting the door, it must be free of dirt, oils, chalk, waxes, and mildew. The prepainted surfaces can be cleaned of dirt, oils, chalk, and mildew with a diluted solution of trisodium phosphate. Trisodium phosphate is available over the counter at most stores under the name of Soilax®, in many laundry detergents without fabric softener additives, and in some general purpose cleaners. Check the label for trisodium phosphate content. The recommended concentration is ⅓ cup of powder to 1.5 to 2 gallons of water. After washing the door, always rinse well with clear water and allow it to dry.

The steel door can be painted with a high-quality flat latex exterior grade paint. Because all paints are not created equal, the following test needs to be performed. Paint should be applied on a small area of the door (following the instruction on the paint container), allowed to dry, and evaluated prior to painting the entire door. Paint defects to look for are blistering and peeling. An additional test is to apply a strip of masking tape over the painted area and peel back, checking to see that the paint adheres to the door and not to the tape.

After satisfactorily testing a paint, follow the directions on the container and apply it to the door. Be sure to allow adequate drying time should you decide to apply a second coat.

Window frames and inserts can be painted with a high-quality latex paint. The plastic should first be lightly sanded to remove any surface gloss.

ENERGY FACTORS

There are some coatings for windows and door glass that reflect heat. A thin film is placed over the glass area. This thin film reflects up to 46 percent of the solar energy. Only 23 percent is admitted. That means 77 percent of the solar energy is reflected or turned away. Look at Fig. 9-169 to see the extent of this ability to absorb energy and reflect it.

A thin vapor coating of aluminum prevents solar radiation from passing through glass. It does this by reflecting it back to the exterior. The temperature of the glass is not raised significantly. The coating minimizes undesirable secondary radiation from the glass. Visible light is reduced. However, the level of illumination remains acceptable. During the winter this coating reflects long-wave radiation and keeps the heat in the room.

The film is easily applied to existing windows. See Fig. 9-169. Step 1 calls for spraying the entire window with cleaner. Scrape every square inch of the window with a razor blade. This removes the paint, varnish,

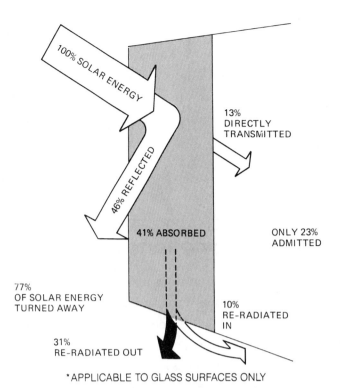

Fig. 9-169 *Energy-saving film for windows.* (Kelly-Stewart)

and excess putty. Wipe the window clean and dry. Step 2 calls for respraying the window with cleaner. Squeegee the entire window. Use vertical strokes from top to bottom of the window. Use paper towels to dry the edges and corners.

Step 3 calls for clear water to be sprayed onto the glass area. Remove the separator film. See Fig. 9-170. Test for the tacky side of film by folding over and touching to itself. The tacky side will stick slightly. Position the tacky side of the film against the top of the glass. Smooth the film by carefully pulling on the edges, or gently press the palms of your hands against the film. Slide the sheet of film easily to remove large wrinkles.

In step 4, spray the film, which is now on the glass, with water. Standing at the center, use a squeegee in gentle, short vertical and horizontal strokes to flatten the film. Slowly work out the wrinkles. Work out the bubbles and remove excess water from under the film. Be careful not to crease or wrinkle the film. Trim the edges with a razor blade. Wipe the excess water from the edges. The film will appear hazy for about two days until the excess water evaporates.

This is shown here to illustrate the possibilities of making your home an energy saver. Films and other devices will be forthcoming as we look forward to conserving energy.

1 2 3 4

Fig. 9-170 *Step-by-step application of energy-saving film.* (Kelly-Stewart)

CHAPTER 9
STUDY QUESTIONS

1. What purpose do windows serve?
2. What purpose does a door serve?
3. What is the height of a door?
4. Where are windows made today?
5. What is meant by a pre-hung door?
6. What is a double-hung window?
7. What is a casement window?
8. What is an awning-type window?
9. What is meant by a hopper-type window?
10. What is a flush door?
11. Why does a door need a 3° angle on the lock stile?
12. What determines the hand of a door?
13. What does LHR mean?
14. What is a construction key?
15. How are the locks in a house changed after the house is completed?
16. How are storm doors installed?
17. Where are storm windows installed?
18. What type of garage door uses extension springs?
19. Why is headroom important in garage door installation?
20. How much can a double garage door weigh?

10
CHAPTER

Finishing Exterior Walls

EXTERIOR WALLS ARE FINISHED BY TWO basic processes. The first is by covering the wall with a wood or wood product material called siding. The other is to cover the wall with a masonry material such as brick, stone, or stucco. The carpenter will install the exterior siding and will sometimes prepare the exterior wall for the masonry materials. However, people employed in the "trowel trades" install the masonry materials.

Exterior siding is applied over the wall sheathing. It adds protection against weather, strengthens the walls, and also gives the wall its final appearance or beauty. Siding may be made of many materials. Often, more than one type of material is used. Wood and brick could be combined for a different look. Other materials may also be combined. This chapter will help you learn these new skills:

- Prepare the wall for the exterior finish
- Estimate the amount of siding needed
- Select the proper nails for the procedure
- Erect scaffolds
- Install flashing and water tables to help waterproof the wall
- Finish the roof edges
- Install exterior siding
- Trim out windows and doors

INTRODUCTION

Before the exterior is finished, windows and doors must be installed. Also, the roof should be up and the sheathing should be on the walls. Then, to finish the exterior, three things are done. First, the cornices and rakes around the roof are enclosed. Next, the siding is applied. Finally, the finish trim for windows and doors is installed.

The *cornice* is the area beneath the roof overhang. This area is usually enclosed or boxed in. Figure 10-1 shows a typical cornice. In many areas, the cornice is boxed in as part of the roofing job. The cornice is often painted before siding is installed. This is particularly likely when brick or stone is used.

Carpenters install several types of exterior siding. Most types of siding are made of wood, plywood, or wood fibers. Some types of siding are made of plastic and metal. These are usually formed to look like wood siding.

Siding is the outer part of the house that people see. Beauty and appearance are important factors. However, siding is also selected for other reasons. The builder may want a siding that can be installed quickly and easily. This means that it costs less to install. The owner may want a siding that requires little maintenance. Both will want a siding that will not rot or warp. Whether insects will attack the siding is another factor.

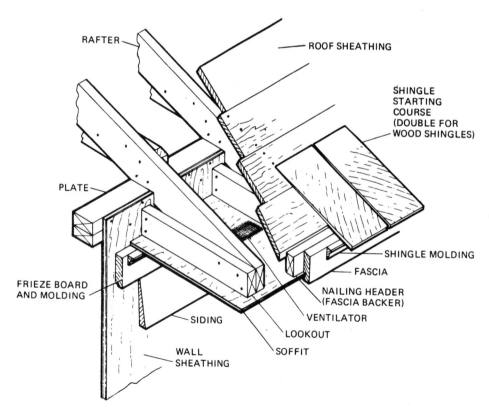

Fig. 10-1 *Typical box cornice framing.* (Forest Products Laboratory)

TYPES OF SIDING

Siding is made of plywood, wood boards, wood fibers, various compositions, metal, or plastics. See Fig. 10-2A to D. However, the shape is the main factor that determines how the siding is applied. Wood fiber is made in two main shapes, "boards" and sheets or panels, which often look like boards. However, the sheets are put up as whole sheets. Fiber panels can also be made to look like other types of siding.

Plywood panels can also be made to look like other types of siding. Figure 10-2B shows a house with plywood siding that looks like individual boards. Plywood can also be made in "board" strips. These are applied like boards.

Siding is also made from shingles and shakes. Shingles are made from wither wood or asbestos mineral compounds. Actually, both types of shingle siding are applied in much the same manner.

(C)

(A)

(B)

(D)

Fig. 10-2 *(A) Cedar boards are used both horizontally and diagonally for this siding.* (Potlatch) *(B) Plywood siding has many "looks" and styles. This looks like boards* (American Plywood Association) *(C) This siding combines rough brick, smooth stucco, and boards. (D) Here stone siding is applied over frame construction.*

SEQUENCE FOR SIDING

Sequence is determined by the type of siding, the type of roof, and the type of sheathing (if any) used on the building. How high the building is also affects the sequence. To work on high places, carpenters erect scaffolds.

In most cases, board siding is put on from bottom to top. However, if scaffolds have to be nailed to the wall, the siding on the bottom could be damaged. In that case the sequence can be changed. The scaffolds can be put up and the siding can be put on the bottom. This way the siding is not damaged by scaffold nails or bolts.

However, many scaffolds can stand alone. They need not be nailed to the wall. The common sequence for finishing an exterior wall is:

1. Prepare for the job. Make sure that the windows are installed, the vapor barrier is in place, the nails are selected, and the amount of siding is estimated.
2. Erect necessary scaffolds.
3. Install the flashing and water tables.
4. Finish the roof edges.
5. Install siding on the upper gable ends and on upper stories.
6. Install siding on the sides.
7. Finish the corners.
8. Trim windows and doors.

PREPARE FOR THE JOB

Several things should be done before siding is put on the wall. Windows and doors should be properly installed. Rough openings should have been moisture-shielded. Then any spaces between window units and the wall frame are blocked in. An air space should be present. Some types of sheathing are also good moisture barriers. However, other types are not. These must have a moisture barrier installed. Figure 10-3 shows a window that has been blocked properly so that siding may now be installed.

Vapor Barrier

One of the reasons for installing a vapor barrier is to protect the outside walls from vapor coming from inside the house. Paint problems on the siding of the house can arise from too much moisture escaping from inside the house. The main purpose of the barrier is to prevent water vapor from entering the enclosed wall space where condensation might occur and cause rot and odor problems. The warm side of the wall should

Fig. 10-3 *A window unit blocked for brick siding.*

be as vapor-tight as possible. The asphalt- or tar-saturated felt papers are not vapor-proof, but can be used on the sheathing for they are water-repellent.

Vapor barriers are available on insulation. This means the barrier side of the insulation should be installed toward the warm or inside of the house. In some instances the entire inside walls and ceiling are covered with a plastic (polyethylene) film to ensure a good vapor barrier. Heated air contains moisture. It travels from inside the building to the outside to blister and bubble paint on the siding. Damp air can also increase the susceptibility to rot and damage.

Sheathing is usually covered with some type of tar saturated paper to improve on the siding's ability to shed water and at the same time improve the energy efficiency of the house by preventing air leakage through the walls. Reflective energy barriers are also used. See Fig. 10-4. They help keep energy on one side of a wall. They can cause the inside of the house to be warmer in the winter and cooler in the summer.

To put up an air barrier, start at the bottom. Nail the top part in place. Most air-barrier materials will be applied in strips. The strips should shed water to the outside. See Fig. 10-5. To do this, the bottom strip is installed first. Each added strip is overlapped. Moisture barriers should also be lapped into window openings. See Chapter 9 on installing windows for this procedure.

Fig. 10-4 *Foil surfaces reflect energy.*

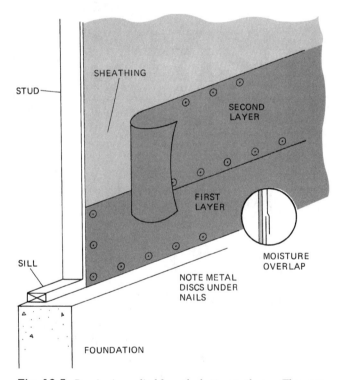

Fig. 10-5 *Barrier is applied from the bottom to the top. The overlap helps shed moisture.*

Vapor barriers are made from plastic films, metal foil, and from builder's felt (also called tar paper). Builder's felt and some plastic films are used most frequently for these purposes.

Nail Selection

Several methods of nailing can be used. Some siding may be put up by any of several methods. Other siding should be put up with special fasteners. The important thing is that the nail must penetrate into something solid in order to hold. For example, nails driven into fiber sheathing will not hold. Siding over this type of sheathing must be nailed at the studs. If the studs are placed on 16-inch centers, the nails are driven at 16-inch intervals. Likewise, splices in the siding should be made over studs. Also, the nail must be long enough to penetrate into the stud.

However, if plywood or hardboard sheathing is used, then nails can be driven at any location. The plywood will hold even a short nail.

Strips of wood are also used to make a nail base. These strips are placed over the sheathing. Then they are nailed to the studs through the sheathing. Nail strips are used for several types of shingled siding. See Fig. 10-6.

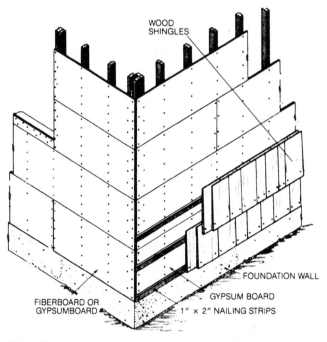

Fig. 10-6 *Nail strips can be used over sheathing.* (Forest Products Laboratory)

The nail should enter the nail base for at least ½ inch. A nonstructural sheathing such as fiber sheathing cannot be the nail base. With fiber or foam sheathing, the nail must be longer. The nail must go through siding and the sheathing and ½ inch into a stud.

Also, the type of nail should be considered. When natural wood is exposed, a finishing or casing-head nail would look better. The head of the nail can be set beneath the surface of the wood. The head will not be seen. This way the nail will not detract from the appearance. However, if the wood is to be painted, a common or a box-head nail may be used. Coated nails are preferred for composition and mineral siding. These nails are coated with zinc to keep them from rusting. This makes them more weather-resistant.

For some siding, the nail is driven at an angle. This means that it must be longer than the straight line distance. This type of nailing is frequently done in grooved and edged siding. The nails are driven in the grooves and edges. This way the siding is put on so that the nails are not exposed to the weather or the eye of the viewer.

Estimate the Amount of Siding Needed

Many people think that "one by six" (1-×-6-inch) boards are 1 inch thick and 6 inches wide. However, a 1-×-6 board is actually ¾ inch thick by 5½ inches wide. Also, most board siding overlaps. Rabbet, bevel, and drop sidings all overlap. When overlapped, each board would not expose 5½ inches. Each board would expose only about 5 inches.

Several things must be known to estimate the amount of siding needed. First, you must determine the height and width of the wall. Then find the type of siding and the sizes of windows or doors. Consider the following example:

Siding:	1-×-8-inch bevel siding
Overlap:	1½ inches
Wall height:	8 feet
Wall length:	40 feet
Windows and Doors:	Two windows, each 2 × 4 feet
	One door, 8 × 3 feet

First, find the total area to be covered. To do this, multiply the length of the wall times the height.

$$\text{Area} = 40 \times 8$$
$$= 320 \text{ square feet}$$

Next, subtract the area of the doors and windows from the area of the wall. To do this:

$$
\begin{aligned}
\text{Area of windows} &= \text{width} \times \text{length} \\
&= 2 \times 4 \\
&= 8 \text{ square feet} \times \text{number of windows} \\
&= 16 \text{ square feet} \\
\text{Door area} &= \text{width} \times \text{length} \\
&= 3 \times 8 \\
&= 24 \text{ square feet} \\
\text{Total opening area} &= 16 \text{ square feet} \\
&\quad + 24 \text{ square feet} \\
\text{Total} &\quad \overline{ 40} \text{ square feet}
\end{aligned}
$$

In order to get enough siding to cover this area, the percentage of overlap must be considered. Also, there

is always some waste in cutting boards so that they join at the proper place. When slopes and corners are involved, there is more waste. The amount of overlap for 8-inch siding lapped 1¼ inches is approximately 17 percent. However, it is best to add another 15 percent to this for waste. Thus, about 32 percent would be added to the total requirements. Thus, to side the wall would require 320 – 40 = 280 square feet. However, enough siding should be ordered to cover 280 square feet plus 32 percent (90 board feet). A total of 370 board feet should be ordered. Table 10-1 shows allowances for different types of siding.

Table 10-1 *Allowances that Must Be Added to Estimate Siding Needs*

Type	Size, Inches	Amount of Lap, Inches	Allowance, Percent (Add)
Bevel siding	1 × 4	3/4	45
	1 × 6	1	35
	1 × 8	1¼	32
	1 × 10	1½	30
	1 × 12	1½	25
Drop siding (shiplap)	1 × 4		30
	1 × 6		20
	1 × 8		17
Drop siding (matched)	1 × 4		25
	1 × 6		17
	1 × 8		15

To estimate the area for gable ends, the same procedure is used. Find the length and the height of the gable area. Then multiply the two dimensions for the total. However, since slope is involved, as shown in Fig. 10-7, only one-half of this figure is required. The allowance for waste and overlap is based upon this halved figure.

The amount of board siding required for an entire building may be estimated. First, the total area of all the walls is found. Then the areas of all the gable sections are found. These areas are added together. The allowances are made on the total figure.

Ordering paneled siding Paneled or plywood siding is sold in sheets. The standard sheet size is 4 feet wide and 8 feet long. The standard height of the wall is 8 feet. Thus a panel fully covers a 4-foot length of wall. To estimate the amount needed, find the length of the wall. Then divide by 4, the width of a panel. For example, a wall is 40 feet long. The number of panels is 40 divided by 4. Thus, it would take 10 panels to side this wall with plywood paneling. If, however, the length of the wall is 43 feet, then 11 panels would be

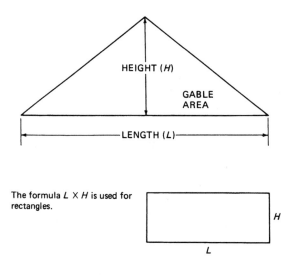

The formula L × H is used for rectangles.

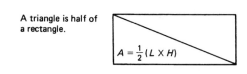

A triangle is half of a rectangle.

$$A = \frac{1}{2}(L \times H)$$

Its formula is $\frac{1}{2}(L \times H)$
A gable is two triangles.

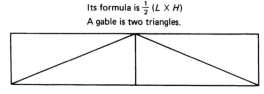

Fig. 10-7 *Estimating gable areas.*

from the total area. Shingles vary in length and width. The size of the shingle must be considered when estimating. Also, shingles come in packages called *bundles*. Generally, four bundles are approximately 1 "square" of coverage. This means that four bundles will cover about 100 square feet of surface area. However, the amount of shingle exposed determines the actual coverage. If only 4 inches of a 16-inch-long shingle is exposed, the coverage will be much less. Also, if more of the shingle is exposed, a greater area can be covered. For walls, more of the shingle can be exposed than for roofs. Table 10-2 shows the coverage of 1 square of shingles when different lengths are exposed.

ERECTING SCAFFOLDS

Many areas on a building are high off the ground. When carpenters must work up high, they put up scaffolds. Figure 10-8 shows a typical pump jack scaffold. Scaffolds are also sometimes called *stages*.

A scaffold must be strong enough to hold up the weight of everything on it. This includes the people, the tools, and the building materials. There are several types of scaffolds that can be used. The type used depends upon the number of workers involved and the weight of the materials. Also, how high the scaffold will be is a factor. How long the scaffold will remain up must be considered.

Job-Built Scaffolds

Scaffolds are often built from job lumber. The maximum distance above the ground should be less than 18 feet. Three main types of job scaffolds can be built. Supports for the platform should be no more than 10 feet apart, and less is desirable.

required. Only whole panels may be ordered. No allowances for windows and doors are made. Sections of panels are cut out for windows and doors.

Estimating shingle coverage To find the amount of shingles needed, the actual wall area should be found. The areas of the doors and windows should be deducted

Table 10-2 *Coverage of Wood Shingles for Varying Exposures*

Length and Thickness[b]	Approximate Coverage of Four Bundles or One Carton, [a]Square Feet									
	Weather Exposure, Inches									
	5½	6	7	7½	8	8½	9	10	11	11½
Random-width and dimension 16″ × 5/2″	110	120	140	150[c]	160	170	180	200	220	230
18″ × 5/2¼″	100	109	127	136	145½	154½[c]	163½	181½	200	209
24″ × 4/2″	—	80	93	100	106½	113	120	133	146½	153[c]

[a] Nearly all manufacturers pack four bundles to cover 100 square feet when used at maximum exposures for roof construction; rebutted-and-rejoined and machine-grooved shingles typically are packed one carton to cover 100 square feet when used at maximum exposure for double-coursed sidewalls.
[b] Sum of the thickness, e.g. 5/2″ means 5 butts equal 2″.
[c] Maximum exposure recommended for single-coursed sidewalls.

Fig. 10-8 *Pumpjack scaffold for light, low work such as bricklaying, painting, and the application of siding.*

Double-pole scaffolds This style of scaffold is shown in Fig. 10-9. The poles form each support section. They should be made from 2-×-4-inch pieces of lumber. A 1-×-6 board is nailed near the bottom of both poles for a bottom brace. Three 12d nails should be used in each end of the l-×-6-inch board. The main support for the working platform is called a *ledger*. It is nailed to the poles at the desired height. It should be made of 2-×-4-inch or 2-×-6-inch lumber. It is nailed with three 16d common nails in each end of the ledger. Sway braces should be nailed between each end pole

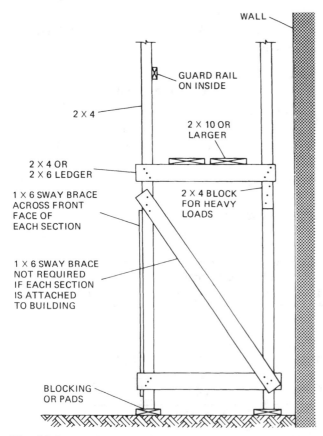

WALL

GUARD RAIL
ON INSIDE

2 × 4

2 × 10 OR
LARGER

2 × 4 OR
2 × 6 LEDGER

1 × 6 SWAY BRACE
ACROSS FRONT
FACE OF
EACH SECTION

2 × 4 BLOCK
FOR HEAVY
LOADS

1 × 6 SWAY BRACE
NOT REQUIRED
IF EACH SECTION
IS ATTACHED
TO BUILDING

BLOCKING
OR PADS

Fig. 10-9 *A double-pole scaffold can stand free of the building. Guard rails are needed above 10'0".*

as shown in Fig. 10-9. The sway braces can be made of 1-inch lumber. They are nailed as needed.

After two or more of the pole sections have been formed on the ground, they can be erected. Place a small piece of wood under the leg of the pole to provide a bearing surface. This keeps the ends of the pole from sinking into the soft earth. The sections should be held in place by the carpenters. Then another carpenter nails braces between each pole section. These are nailed diagonally, as are sway braces. They may be made of l-×-6-inch lumber. They are nailed with three 14d common nails at each end.

When the ledgers are more than 8 feet above the ground, a guard rail should be used. The guard rail is nailed to the pole about 36 inches above the platform. Note that the guard rail is nailed to the inside of the pole. This way a person can lean against the guard rail without pushing it loose.

The double-pole type of scaffold can be freestanding. That is, it need not be attached to the building. However, carpenters often nail a short board between the scaffold and the wall. This holds the scaffold safely. If this is done, the board should be nailed near the top of the scaffold.

Single-pole scaffolds Single-pole scaffolds are similar to double-pole scaffolds. However, the building forms one of the "poles." Blocks are nailed to the building. They form a nail base for the brace and ledgers. Figure 10-10 shows a single pole scaffold. Sway braces are not needed on the pole sections. However, sway braces should be used between each section of the scaffold.

Wall brackets Wall brackets are often made on the job. See Fig. 10-11. Wall brackets are most often used when the distance above the ground is not very great. They are quicker and easier to build than other scaffold types. Of course, they are also less expensive.

On the ends, the wall brackets are nailed to the outside corners of the wall. On intersections, nail blocks are used, similar to the single-pole scaffold blocks. Wall brackets must be nailed to solid wall members. It is best to use 20d common nails to fasten them in place. Figure 10-12 shows another type of wall bracket. This type is sturdier than the other.

Factory Scaffolds

Today, many builders use factory-made scaffolds. These have several advantages for a builder. They are quick and easy to erect. They are strong and durable and can be easily taken down and reused.

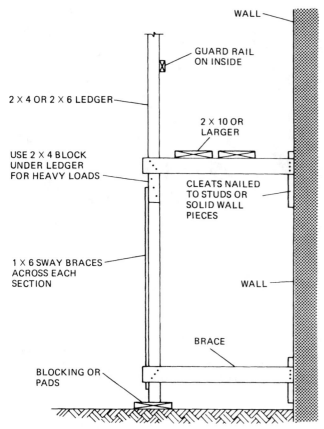

Fig. 10-10 *A single-pole scaffold must be attached to the building. Guard rails are needed above 10'0".*

WALL

GUARD RAIL
ON INSIDE

2 X 4 OR 2 X 6 LEDGER

2 X 10 OR
LARGER

USE 2 X 4 BLOCK
UNDER LEDGER
FOR HEAVY LOADS

CLEATS NAILED
TO STUDS OR
SOLID WALL
PIECES

1 X 6 SWAY BRACES
ACROSS EACH
SECTION

WALL

BRACE

BLOCKING OR
PADS

No lumber is used. Also, no cutting and nailing is needed to erect the scaffold. Thus, no lumber is ruined or made unusable for the building. The metal scaffold parts are easily stored and carried. They are not affected by weather and will not rot. There are several different types of scaffolds.

Double-pole scaffold sections The double-pole type of scaffold features a welded steel frame. It includes sway braces, base plates, and leveling jacks. See Figs. 10-13 and 10-14. Two or more sections can be used to gain greater distance above the ground. Special pieces provide sway bracing and leveling. Guard rails and other scaffold features may also be included.

Wall brackets Wall brackets, as in Figs. 10-15 through 10-17, are common. No nail blocks are needed on the walls for these metal brackets. These brackets are nailed or bolted directly to the wall. Be sure that they are nailed to a stud or another structural member. Use 16 or 20d common nails. After the nails are driven in place, be sure to check the heads. If the nail heads are damaged, remove them and renail the bracket with new nails. Remember that it is the nail head that holds the bracket to the wall. A damaged nail

Fig. 10-11 *A typical low-level wall bracket.*

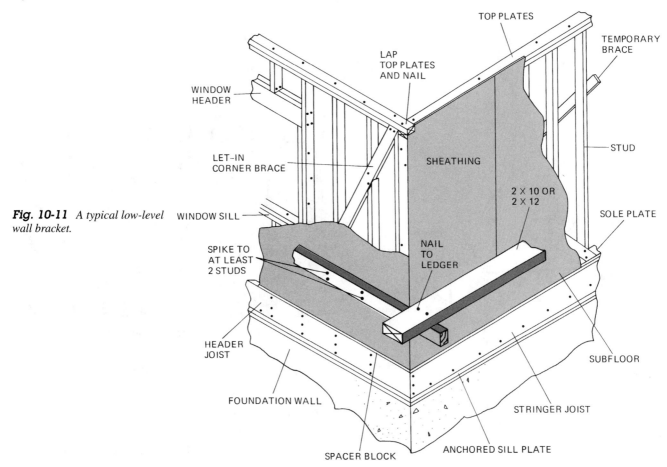

TOP PLATES

TEMPORARY
BRACE

LAP
TOP PLATES
AND NAIL

WINDOW
HEADER

STUD

LET-IN
CORNER BRACE

SHEATHING

WINDOW SILL

2 X 10 OR
2 X 12

SOLE PLATE

SPIKE TO
AT LEAST
2 STUDS

NAIL
TO
LEDGER

HEADER
JOIST

SUBFLOOR

FOUNDATION WALL

STRINGER JOIST

SPACER BLOCK

ANCHORED SILL PLATE

Fig. 10-12 *A job-made wall-bracket scaffold. It is dangerous to work on this kind of scaffold in high places without safety equipment.*

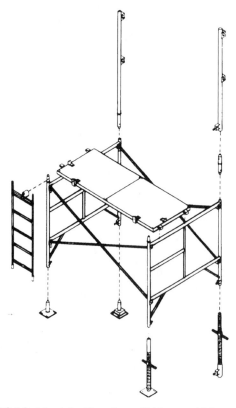

Fig. 10-13 *Metal double-pole scaffolds are widely used.* (Beaver-Advance)

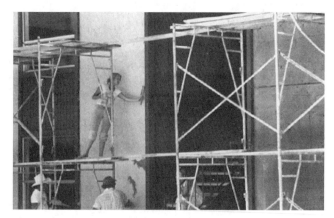

Fig. 10-14 *Metal sections can be combined in several ways to get the right height and length.*

EASY TO ERECT

SIMPLE INSTALLATION

SIMPLE REMOVAL

EASY STORAGE

SHIPPING/STORAGE CONFIGURATION

Fig. 10-15 *Metal wall brackets are quick to put up and take down.* (Richmond Screw Anchor)

Fig. 10-16 *Wall brackets can be used on concrete forms.*

head may break. A break could let the entire scaffold fall.

Trestle jack Trestle jacks are used for low platforms. They are used both inside and outside. They can be moved very easily. Trestle jacks are shown in Fig. 10-18.

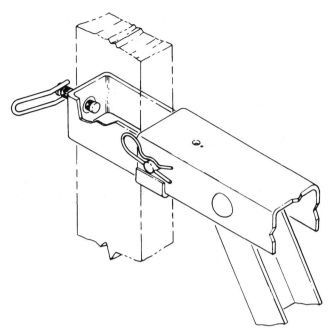

Fig. 10-17 *Guard rails may be added through an end bracket.*
(Richmond Screw Anchor)

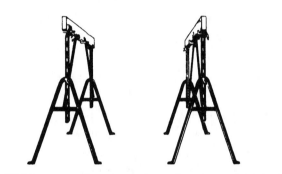

Fig. 10-18 *Two trestle jacks can form a base for scaffold planks.*
(Patent Scaffolding Co.)

Two trestle jacks should be used at each end of the section. A ledger, made of 2-×-4-inch lumber, is used to connect the two trestle jacks. Platform boards are then placed across the two ledgers. Platform boards should always be at least 2 inches thick.

As you can see, it takes four trestle jacks for a single section. Trestle jacks can be used on uneven areas, but they provide for a platform height of only about 24 inches. However, this is ideal for interior use.

Ladder jacks Ladder jacks, as in Fig. 10-19, hang a platform from a ladder. They are most suitable for repair jobs and for light work where only one carpenter is on the job. Two types of jacks are used. The type shown in Fig. 10-19 puts the platform on the outside of the ladder. The type shown in Fig. 10-20 places the platform below or on the inside of the ladder.

Ladder Use

Using ladders safely is an important skill for a carpenter. A ladder to be erected is first laid on the ground. The bottom end of the ladder should be near the building. The top end is raised and held overhead. The carpenter gets directly beneath the ladder. The hands are then moved from rung to rung. As the top end of the ladder is raised, the carpenter walks toward the building. Thus, the ladder is raised higher and higher with every step. Care is taken to watch where the top of the ladder is going. The top of the ladder is guided to its proper position. Then the ladder is leaned firmly against some part of the building. The base of the ladder is

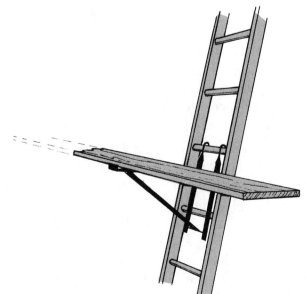

Fig. 10-19 *Outside ladder jacks.*

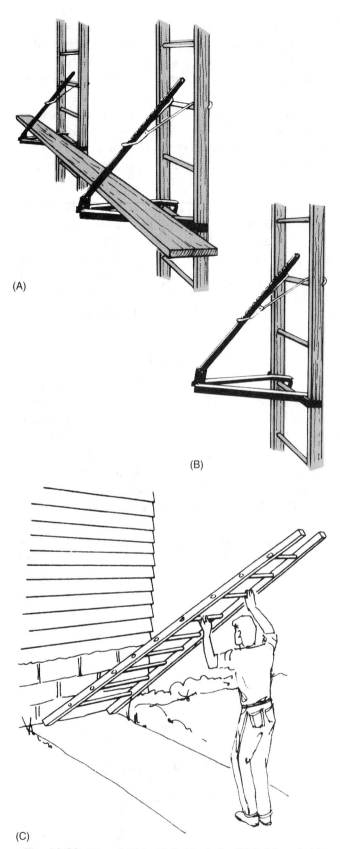

(A)

(B)

(C)

Fig. 10-20 *(A) and (B) Inside ladder jacks. (C) Raising a ladder can be a two-man job. In fact, a heavier ladder should have two men on it. However, if you have single ladder that's not too heavy, you can place the end of the ladder against the house or some obstruction and walk it up one rung at a time.*

made secure on solid ground or concrete. Both ends of the bottom should be on a firm base.

Both wooden and aluminum ladders are commonly used by carpenters. Aluminum ladders are light and easy to manage. However, they have a tendency to sway more than wooden ladders. They are very strong and safe when used properly. Wooden ladders do not sway or move as much, but they are much heavier and harder to handle.

Wooden ladders are sometimes made by carpenters. A ladder is made from clear, straight lumber. It should have been well seasoned or treated. The sides are called rails and the steps are rungs. Joints are cut into the rails for the rungs. Boards should never be just nailed between two rails.

Ladder Safety

1. Ladder condition should be checked before use.
2. The ladder should be clean. Grease, oil, or paint on rails or rungs should be removed.
3. Fittings and pulleys on extension ladders should be tight. Frayed or worn ropes and lines should be replaced.
4. The bottom ends of the ladder must rest firmly and securely on a solid footing.
5. Ladders should be kept straight and vertical. Never climb a ladder that is leaning sideways.
6. The bottom of the ladder should be one-fourth the height from the wall. For example, if the height is 12 feet, the bottom should be 3 feet from the wall.

Scaffold Safety

1. Scaffolds should be checked carefully before each use.
2. Design specifications from the manufacturer should be followed. State codes and local safety rules should also be followed.
3. Pads should be under poles.
4. Flimsy steps on scaffold platforms or ladders should never be used. Height should only be increased with scaffold of sound construction.
5. For platforms, planking that is heavy enough to carry the load and span should be used.
6. Platform boards should hang over the ledger at least 6 inches. This way, when boards overlap, the total overlap should be at least 12 inches.
7. Guard rails and toe boards should be used.

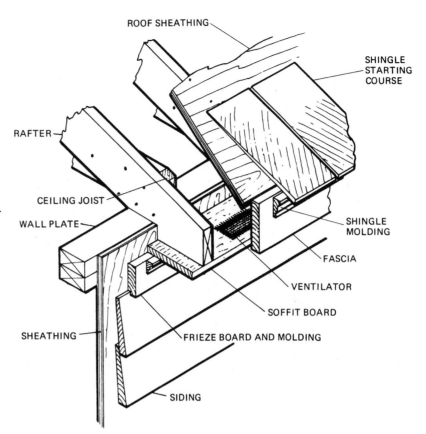

Fig. 10-21 *Narrow box cornice. A closed overhang is called a cornice.* (Forest Products Laboratory)

Labels in figure: ROOF SHEATHING, SHINGLE STARTING COURSE, RAFTER, CEILING JOIST, WALL PLATE, SHINGLE MOLDING, FASCIA, VENTILATOR, SOFFIT BOARD, SHEATHING, FRIEZE BOARD AND MOLDING, SIDING

8. Scaffolds should never be put up near power lines without proper safety precautions. The electric service company can be consulted for advice when a procedure is not known.

9. All materials and equipment should be taken off before a platform or a scaffold is moved.

FINISHING ROOF EDGES

Most roofs have an overhang. This portion is called the *eave* of the roof. If eaves are enclosed, they are also called a *cornice*. See Fig. 10-21. Usually, the edges of roofs are finished when the roof is sheathed. However, building sequences do vary from place to place. Two methods of finishing the eaves are commonly employed. These are the open method and the closed method. There are several versions of the closed method.

Open Eaves

A board is usually nailed across the ends of the rafters when the roof is sheathed. See Fig. 10-22. This board is called the *fascia*. The fascia helps to brace and strengthen the rafters. Fascia should be joined or spliced as shown in Fig. 10-23. However, the fascia is not needed structurally. Some types of open-eave construction do not use fascia.

Open eaves expose the area where the rafters and joists rest on the top plate of a wall. This area should be sealed either by a board or by the siding. See Fig. 10-24. Sealing this area helps prevent air currents from entering the wall. It also helps keep insects and small animals from entering the attic area. However, to com-

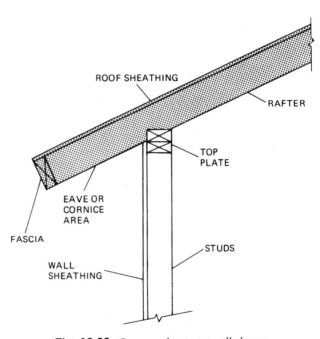

Labels in figure: ROOF SHEATHING, RAFTER, TOP PLATE, EAVE OR CORNICE AREA, FASCIA, STUDS, WALL SHEATHING

Fig. 10-22 *Open overhangs are called eaves.*

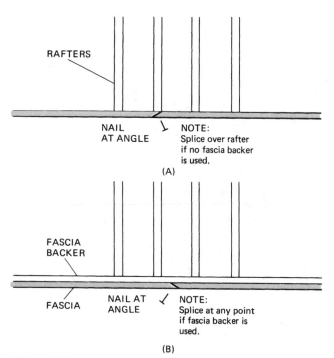

Fig. 10-23 *Splicing the fascia. (A) No fascia backer is used. (B) A fascia backer is used.*

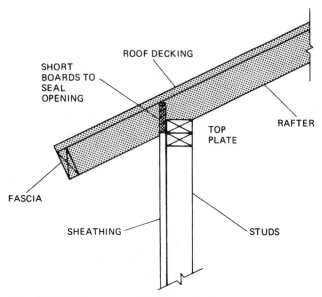

Fig. 10-24 *Open eaves should be sealed and vented properly.*

pletely seal this restricts airflow. This airflow is important to keep the roof members dry. It also prevents rotting from moisture. Airflow will also cool a building. To allow airflow, vents should be installed. These vents are backed with screen or hardware cloth to keep animals and insects from entering.

Enclosed Cornices

Eaves are often enclosed for a neater appearance. Enclosed eaves are called *cornices*. Some houses do not have eaves or cornices. These are called *close* cornices. See Fig. 10-25.

There are several ways of enclosing cornices. The two most common types are the standard slope cornice and the flat cornice. Both types seal the cornice area with panels called *soffits*. The panels are usually made of plywood or metal. Note that both types should have some type of ventilation. Special cornice vents are used. However, when plywood is used for the soffits, an opening is often left as in Fig. 10-26. This provides a continuous vent strip. This strip is covered with some type of grill or screen.

Standard slope cornice The standard slope cornice is the simplest and quickest to make. It is shown in Fig. 10-27A. The soffit panel is nailed directly to the underside of the rafters. If ¼-inch paneling is used, a 6d box nail is appropriate. Rust-resistant nails are recommended. Casing nails, when used, should be set and covered before painting. Panels should join on a rafter. In this way, each end of each panel has firm support. One edge of the panel is butted against the fascia. The edge next to the wall is closed with a piece of trim called a *frieze*.

Standard flat cornices The standard flat cornice has a soffit that is flat or horizontal with the ground. It is not sloped with the angle of the roof.

A nail base must be built for the soffit. Special short joists are constructed. These are called *lookouts*. See Fig. 10-27B. The flat cornice should be vented, as are other types. Either continuous strips or stock vents can be used.

The lookouts need a nail base on both the wall and the roof. They are nailed to the tail of the rafter for the roof support. On the wall, they can be nailed to the top plate or to a stud. More often than not, neither the top plate nor the stud can be used. Then a ledger is nailed to the studs through the sheathing. See Fig. 10-28.

Either 16d or 20d common nails should be used to nail the ledger to the wall. The bottom of the ledger should be level with the bottom of the rafters.

Next, find the correct length needed for the lookouts. Cut the lookouts to length with both ends cut square. Drive nails into one end of the lookout. The nails are driven until the tips just barely show through the lookout. Then they can be held against the rafter and driven down completely. Butt the other end of the lookout against the ledger. Then toenail the lookout on each side using an 8d or 12d common nail.

Soffit panels should be cut to size. Cornice vents should be cut or spaced next. Cornice vents can be attached to soffits before mounting. After the soffits are

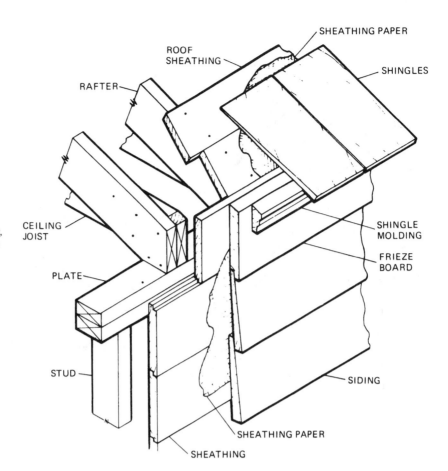

Fig. 10-25 *With close cornices, the roof does not project over the walls.* (Forest Products Laboratory)

Labels in figure: ROOF SHEATHING, SHEATHING PAPER, RAFTER, SHINGLES, CEILING JOIST, SHINGLE MOLDING, FRIEZE BOARD, PLATE, STUD, SIDING, SHEATHING PAPER, SHEATHING

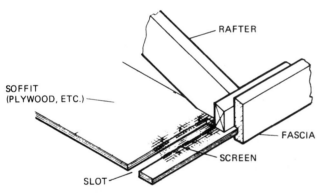

Fig. 10-26 *A strip opening may be used for ventilation instead of cornice vents.* (Forest Products Laboratory)

Labels in figure: RAFTER, SOFFIT (PLYWOOD, ETC.), FASCIA, SCREEN, SLOT

ready, they are nailed in place. For ¼-inch panels, a 6d common nail or box nail can be used.

Soffit panels should be joined on a solid nail base. The ends should join at the center of a lookout. One edge is butted against the fascia. The inner edge is sealed with a frieze board or a molding strip.

Many builders use prefabricated soffit panels. Sometimes fascia is grooved for one edge of the soffit panel. When prefabricated soffits are used, a vent is usually built into the panel.

Closed rakes The rake is the part of the roof that hangs over the end of a gable. See Fig. 10-29. When a cornice is closed, the rake should also be closed. However, most rakes are not vented. This is because they are not connected with the attic space as are the cornices.

It is common to add the gable siding before the rake soffit is added. Then a 2-×-4-inch nailing block is nailed to the end rafter. See Fig. 10-30. Use 16d common nails. A fly rafter is added and supported by the fascia and the roof sheathing.

The soffit is then nailed to the bottom of the nail block and the fly rafter. Frieze boards or bed molding are then added to finish the soffit.

Solid rake Today many roofs do not extend over the ends of the gables. In this type of roof, a fascia is nailed directly onto the last rafter. The roof is then finished over the fascia. This becomes what is called a *solid rake*. See Fig. 10-31.

Siding the Gable Ends

On many buildings, cornices are painted before any siding is installed. In other cases the gable siding is installed first, then both the cornice and gables are painted. After these are done, siding is installed. This is common when two different types of siding are used. For example, wood siding is put on the gables

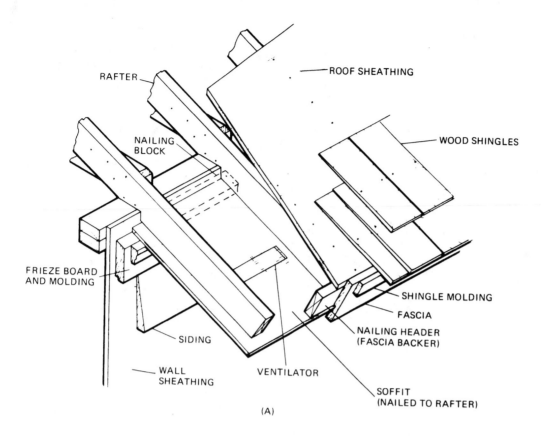

RAFTER

ROOF SHEATHING

NAILING BLOCK

WOOD SHINGLES

FRIEZE BOARD AND MOLDING

SHINGLE MOLDING

FASCIA

NAILING HEADER (FASCIA BACKER)

SIDING

WALL SHEATHING

VENTILATOR

SOFFIT (NAILED TO RAFTER)

(A)

Fig. 10-27 (A) Standard slope cornice; (B) Standard flat cornice. (Forest Products Laboratory)

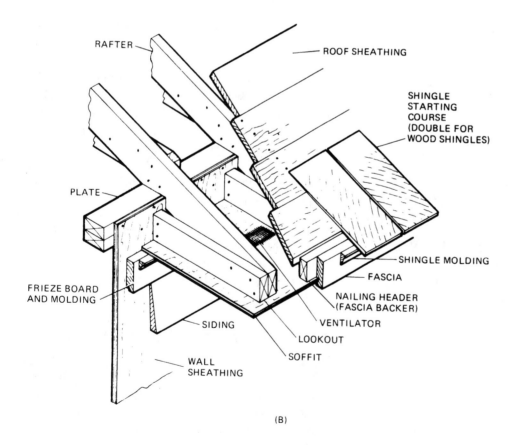

RAFTER

ROOF SHEATHING

SHINGLE STARTING COURSE (DOUBLE FOR WOOD SHINGLES)

PLATE

SHINGLE MOLDING

FASCIA

NAILING HEADER (FASCIA BACKER)

FRIEZE BOARD AND MOLDING

VENTILATOR

SIDING

LOOKOUT

SOFFIT

WALL SHEATHING

(B)

Fig. 10-28 *Lookouts are nailed first to the rafter tail. Then they are toenailed to a solid part of the wall.*

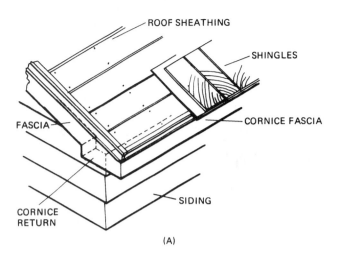

(A)

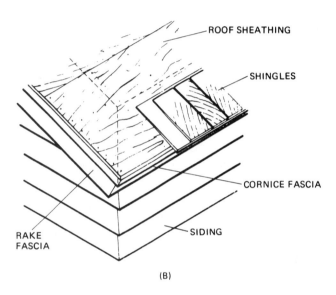

(B)

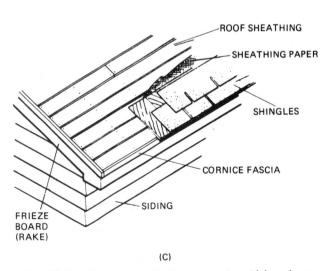

(C)

Fig. 10-29 *Closed rakes: (A) Narrow cornice with boxed return; (B) Narrow box cornice and closed rake; (C) Wide overhang at cornice and rake.* (Forest Products Laboratory)

and painted, then a brick siding is laid. See Figs. 10-32 and 10-33.

Gable walls may be sided in several ways. Gable siding may be the same as that on the rest of the walls. In this case, the gable and the walls are treated in the same way. However, gable walls are often covered with different siding. Brick exterior walls topped by wooden gable walls are common. Different types of wood and fiber sidings may also be combined. See Figs. 10-34 and 10-35. Different textures, colors, and directions are used to make pleasing contrasts.

It is important to apply the gable siding in such a way that water is shed properly from gable to wall. For brick or stone, the gable must be framed to overhang the rest of the wall. See Fig. 10-36. This overhang provides an allowance for the thickness of the stone or brick.

For wood, plywood, or fiber siding, special drainage joints are used. Also metal strips and separate wooden moldings are used.

Framed overhangs Gable ends can be framed to overhang a wall. See Fig. 10-36. Remember that this must be done to allow for the thickness of a brick wall.

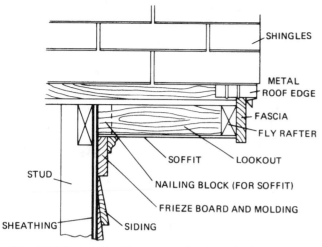

Fig. 10-30 *A nailing block is nailed to the end rafter for the soffit.* (Forest Products Laboratory)

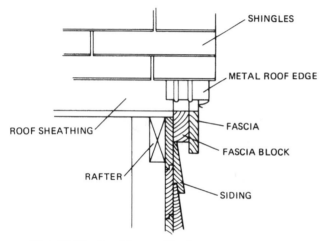

Fig. 10-31 *Detail for a solid rake.* (Forest Products Laboratory)

Fig. 10-32 *Here the gable siding is added and painted before brick siding is added.*

Fig. 10-33 *A wall and gable prepared for a combination of diagonal siding and brick veneer. Note that the wood has been painted before the brick has been applied.*

Several methods are used. One of the most common is to extend the top plate as in Fig. 10-37A. This puts the regular end rafter just over the wall for the overhang. A short (2-foot) piece is nailed to the top plate. The rafter bird's mouth is cut 1½ inch deeper than the others. A solid header is added on the bottom as shown in Fig. 10-37B. The gable is framed in the usual manner. Gable siding is done before wall siding (Figs. 10-38 and 10-39).

Drainage joints Drainage joints are made in several ways. As mentioned before, they are important because water must be shed from gable to wall properly. Otherwise, water would run inside the siding and damage the wall.

Drainage joints are used wherever two or more pieces are used, one above the other. Figure 10-40 shows the most common drainage joints. In one method, molded wooden strips are used. These strips are called *drip caps.*

Some siding, particularly plywood panels, has a rabbet joint cut on the ends. These are overlapped to stop water from running to the inside.

Metal flashing is also used. This flashing may be used alone or with drip caps.

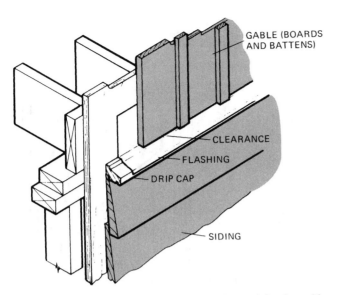

Fig. 10-34 *Different types of siding can be used for the gable and the wall.* (Forest Products Laboratory)

Fig. 10-35 *Smooth gable panels contrast with the brick siding.*

Fig. 10-36 *This gable is framed to overhang the wall.*

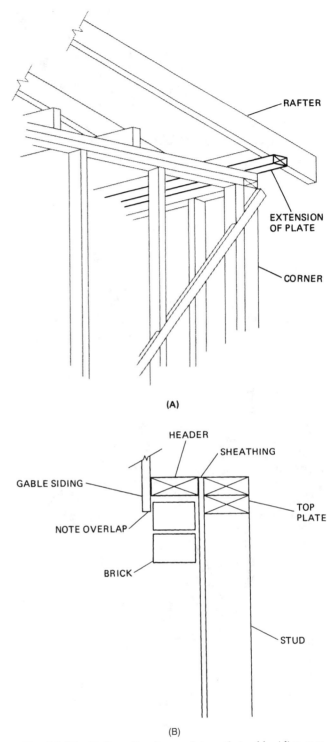

(A)

(B)

Fig. 10-37 *(A) Extending the top plate so that gable siding overhangs a brick or masonry wall; (B) With a solid header added on the bottom.*

At the bottom of panel siding, a special type of drip cap is used. It is called a water table. It does the same thing, but has a slightly different shape. The water table may also be used with or without metal flashing.

Fig. 10-38 *Gable siding is done before wall siding.* (Fox and Jacobs)

Fig. 10-39 *Installing gable siding.* (Fox and Jacobs)

INSTALLING THE SIDING

Several shapes of siding are used. Boards, panels, shakes, and shingles are the most common. These may be made of many different materials. However, many of the techniques for installing different types of siding are similar. Shape determines how the siding is installed. For example, wood or asbestos shingles are installed in about the same way. Boards made of any

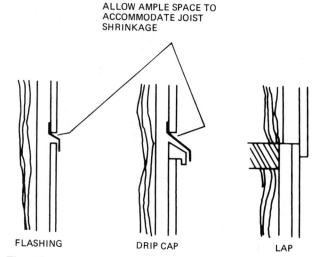

ALLOW AMPLE SPACE TO
ACCOMMODATE JOIST
SHRINKAGE

FLASHING   DRIP CAP   LAP

Fig. 10-40 *Drainage joints between gable and siding.* (Boise-Cascade)

material are installed alike. Panels made of plywood, hardboard, or fiber are installed in a similar manner. Special methods are used for siding made from vinyl or metal.

Putting Up Board Siding

There are three major types of board siding. These are plain boards, drop siding, and beveled siding. Each type is put up in a different manner.

Board siding may be of wood, plywood, or composition material. This material is a type of fiberboard made from wood fibers. Generally, plain boards and drop siding are made from real wood. Composition siding is usually made in the plain beveled shape.

Plain boards Plain boards are applied vertically. This means that the length runs up and down. There are no grooves or special edges to make the boards fit together. There are three major board patterns used. The *board and batten* pattern is shown in Fig. 10-41. In the board and batten style, the board is next to the wall. The board is usually not nailed in place except

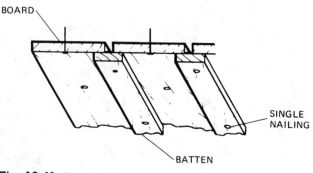

BOARD

SINGLE
NAILING

BATTEN

Fig. 10-41 *Board and batten vertical siding.* (Forest Products Laboratory)

sometimes at the top end. It is sometimes nailed at the top to hold it in place. A narrow opening is left between the boards. This opening is covered with a board called a *batten*. The batten serves as the weather seal. The nails are driven through the batten but not the boards, as shown in Fig. 10-41.

The *batten and board* style is the second pattern. The batten is nailed next to the wall. A typical nailing pattern is shown in Fig. 10-42. The wide board is then fastened to the outside as shown.

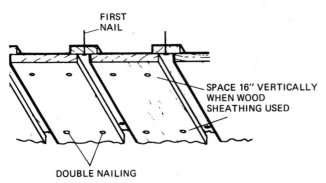

Fig. 10-42 *Batten and board vertical siding.* (Forest Products Laboratory)

Another style is the Santa Rosa style, shown in Fig. 10-43. All the boards are about the same width in this style. A typical nail pattern is shown in the figure. The inner board can be thinner than the outer board.

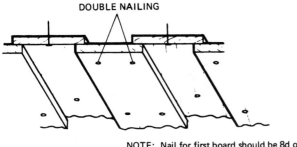

NOTE: Nail for first board should be 8d or 9d. Nail for second board should be 12d.

Fig. 10-43 *The Santa Rosa or board and board style of vertical siding.* (Forest Products Laboratory)

Plain board siding can be used only vertically. This is because it will not shed water well if it is used flat or horizontally. The surface appearance of the boards may vary. The boards may be rough or smooth. For the rough effect, the boards are taken directly from the saw mill. In order to make a smooth board finish, the rough-sawn boards must be surfaced.

Drop siding Drop siding differs from plain boards. Drop siding has a special groove or edge cut into it.

This edge lets each board fit into the next board. This makes the boards fit together and resist moisture and weather. Figure 10-44 shows some types of drop siding that are used. Nailing patterns for each type are also shown.

Beveled siding *Beveled* siding is made with boards that are thicker at the bottom. Figure 10-45 shows the major types of beveled siding. Common beveled siding is also called *lap* siding. The nailing pattern for lap siding is shown in Fig. 10-46.

The minimum amount of lap for lap siding is about 1 inch. For 10-inch widths, which are common, about 1½ inches of lap is suggested. Most lap siding today is made of wood fibers. However, in many areas wood siding is still used. When wood siding is used, the standard nailing pattern shown in Fig. 10-46 should be used. Some authorities suggest nailing the nails through both boards. However, this is not recommended. All wood tends to expand and contract, and few pieces do so evenly. Nailing the two boards together causes them to bend and bow. When this occurs, air and moisture can easily enter.

Another type of beveled siding is called *rabbeted beveled siding*. This is also called *Dolly Varden siding*. The Dolly Varden siding has a groove cut into the lower end of the board. This groove is called a rabbet. When installed from the bottom up, each successive board should be rested firmly on the one beneath it. The nails are then driven into the top of the boards.

Siding Layout

For all types of board siding, two or three factors are important to remember. Boards should be spaced to lay even with windows and doors. One board should fit against the bottom of a window. Another should rest on the top of a window. This way there is no cutting for an opening. Cutting should also be avoided for vertical siding. The spacing should be adjusted so that a board rests against the window on either side.

When siding is installed, guide marks are laid out on the wall. These are used to align the boards properly. To do this, a pattern board, called a story pole, is used just as in wall framing. The procedure for laying out horizontal lap siding will be given. The procedures for other types of siding are similar. This procedure may be adapted accordingly.

Procedure Choose a straight 1-inch board. Cut it to be exactly the height of the area to be sided. Many siding styles are allowed to overlap the foundation 1 to 2 inches. Find the width of the siding and the overlap to

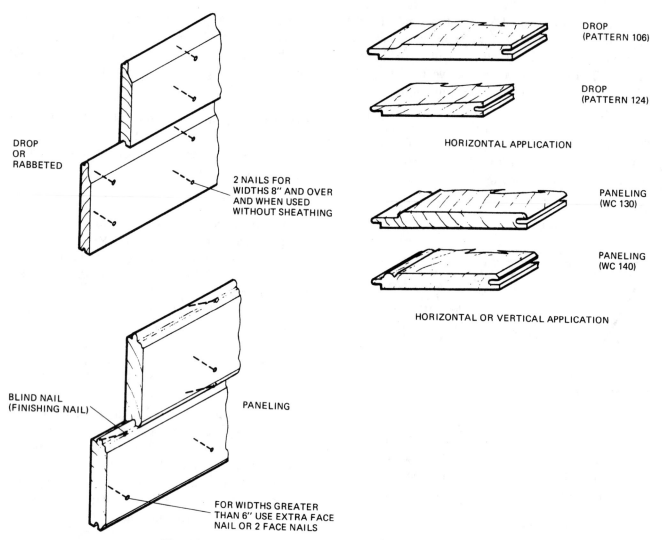

DROP OR RABBETED

2 NAILS FOR WIDTHS 8" AND OVER AND WHEN USED WITHOUT SHEATHING

DROP (PATTERN 106)

DROP (PATTERN 124)

HORIZONTAL APPLICATION

PANELING (WC 130)

PANELING (WC 140)

HORIZONTAL OR VERTICAL APPLICATION

BLIND NAIL (FINISHING NAIL)

PANELING

FOR WIDTHS GREATER THAN 6" USE EXTRA FACE NAIL OR 2 FACE NAILS

Fig. 10-44 *Shapes and nailing for drop siding.* (Forest Products Laboratory)

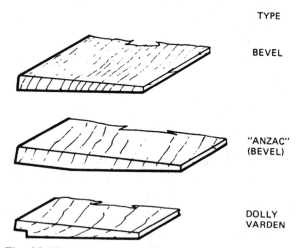

TYPE

BEVEL

"ANZAC" (BEVEL)

DOLLY VARDEN

Fig. 10-45 *Major types of bevel siding.* (Forest Products Laboratory)

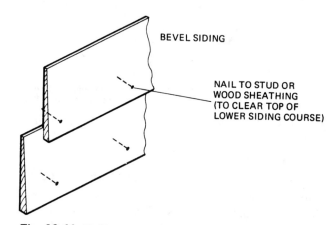

BEVEL SIDING

NAIL TO STUD OR WOOD SHEATHING (TO CLEAR TOP OF LOWER SIDING COURSE)

Fig. 10-46 *Nailing pattern for lap siding.* (Forest Products Laboratory)

be used. Subtract the overlap from the siding width. This determines the spacing between the bottom ends of the siding. The bottom end of lap siding is called the *butt* of the board. For example, siding 10 inches wide is used. The overlap is 1½ inches. Thus, the distance between the board butts is 8½ inches. On the story pole, lay out spaces 8½ inches apart.

Place the story pole beside a window. Check to see if the lines indicating the spacing between the boards line up even with the top and bottom of the windows. The amount of overlap over the foundation wall may be varied slightly. The amount of lap on the siding may also be changed. Small changes will make the butts even with the tops of the windows. They will also make the tops even with the window bottoms. See Fig. 10-47. Story pole markings should be changed to show adjustments. New marks are then made on the pole.

Use the story pole to make marks on the foundation. Also make marks around the walls at appropriate places. Siding marks should be made at edges of windows, corners, and doors. See Fig. 10-48. If the wall is long, marks may also be made at intervals along the wall. In some cases, a chalk line may be used to snap guide lines. A chalk line is often used to snap a line on the foundation. This shows where the bottom board is nailed.

Nailing

Normally siding is installed from the bottom to the top, with the first board overlapping the foundation at least 1 inch. The first board is tacked in place and checked for level. After leveling, the first board is then nailed firm. Usually bottom boards will be placed for the whole wall first. Then the other boards are put up, bottom to top. The level of the siding is checked after each few boards.

Siding can also be installed from the top down. This is done when scaffolding is used. The layout is the same. The lines should be used to guide the butts of the boards. However, the first board is nailed at the top. Two sets of nails are used on the top board. The first set of nails is nailed near the tops of the board. These may be nailed firm. Then, the bottom nail is driven 1½ inches from the butt of the board. This nail is not nailed down firm. About ¾ inch should be left sticking out. The butt of the board is pried up and away from the wall. A nail bar or the claw of a hammer is used. The board is left this way. Then, the next board is pushed against the nails of the first board. The second board is checked for level. Hold the level on the bottom of the board and check the second level. When the board is level, it is nailed at a place 1½ inches from the butt. Again, the nail in the second board is left out. About ¾ inch should be left sticking out. The board is nailed down at the butt for its length. Then the first board is nailed down firm. The second board is then pried up for the third board. This process is repeated until the siding has been applied.

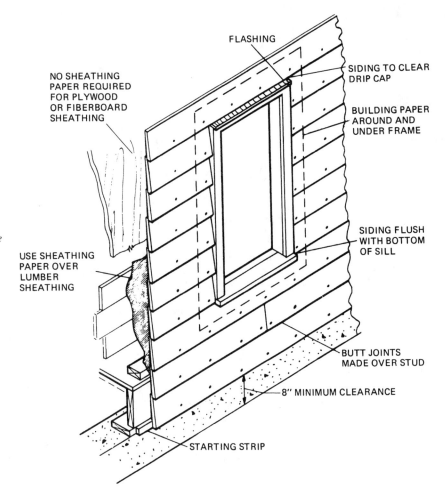

Fig. 10-47 *Bevel siding application. Note that pieces are even with top and bottom of window opening.* (Forest Products Laboratory)

FLASHING

NO SHEATHING PAPER REQUIRED FOR PLYWOOD OR FIBERBOARD SHEATHING

SIDING TO CLEAR DRIP CAP

BUILDING PAPER AROUND AND UNDER FRAME

USE SHEATHING PAPER OVER LUMBER SHEATHING

SIDING FLUSH WITH BOTTOM OF SILL

BUTT JOINTS MADE OVER STUD

8″ MINIMUM CLEARANCE

STARTING STRIP

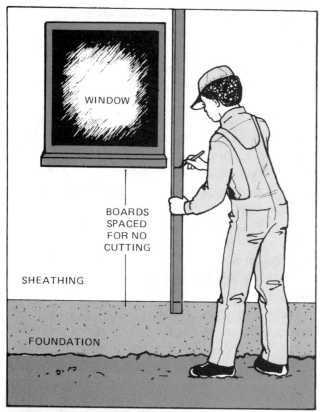

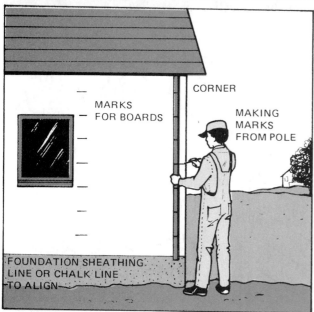

Fig. 10-48 *Marks are made to keep siding aligned on long walls.*

Corner Finishing

There are three ways of finishing corners for siding. The most common ways are shown in Fig. 10-49. Corner boards can be used for all types of siding. In one style of corner board, the siding is butted next to the boards. In this way, the ends of the siding fit snugly and the corner is the same thickness as the siding. See Fig. 10-49C.

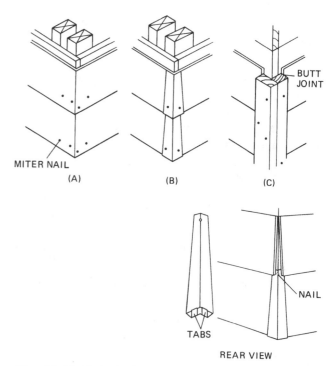

Fig. 10-49 *Methods of finishing corners.* (Forest Products Laboratory)

In Fig. 10-49B special metal corner strips are used. These are separate pieces for each width of board. The carpenter must be careful to select and use the right-size corner piece for the board siding used.

In other methods, the siding is butted at the corners. The corner boards are then nailed over the siding. See Fig. 10-50. This type of corner board is best used

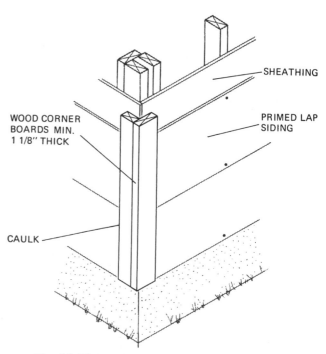

Fig. 10-50 *Wood outside corner detail.* (Boise-Cascade)

for plywood or panel siding. However, it is also common for board siding.

Another common method uses metal corners for lap siding. Figure 10-49B shows metal corner installation. These corners have small tabs at the bottom. They fit around the butt of each board. At the top is a small tab for a nail. The corners are installed after the siding is up. The bottom tab is put on first. Tabs are then put on other boards from bottom to top. The corners can be put on last. This is because the boards will easily spread apart at the bottoms. A slight spread will expose the tab for nailing.

Another method of finishing corners is to miter the boards. Generally, this method is used on more expensive homes. It provides a very neat and finished appearance. However, it is not as weatherproof. In addition, it requires more time and is thus more expensive. To miter the corners, a miter box should be used. Lap siding fits on the wall at an angle. It must be held at this angle when cut. A small strip of wood as wide as the siding is used for a brace. It is put at the base of the miter box. See Fig. 10-51. This positions the siding at the proper angle for cutting. If this is not handy, an estimate may be made. It is generally accurate enough. This method is shown in Fig. 10-52. A distance equal to the thickness of the siding is laid off at the top. The cut is then made on the line as shown.

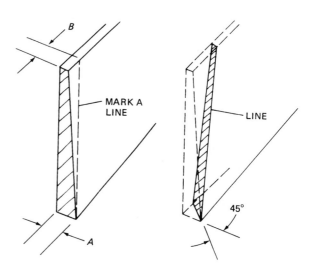

Fig. 10-52 *To cut miters without a miter box, make the top distance B equal to the thickness of A. Then cut back at about 45°.*

Inside corners are treated in two ways. Both metal flashing and wooden strips are used. The most common method uses wooden strips. See Fig. 10-53. A ¾-inch-square strip is nailed in the corner. Then each board of the siding is butted against the corner strip. After the siding is applied, the corner is caulked. This makes the corner weathertight.

Metal flashing for inside corners is similar. The metal has been bent to resemble a wooden square. The strips are nailed in the corner as shown in Fig. 10-54. As before, each board is butted against the metal strip.

PANEL SIDING

Panel siding is now widely used. It has several advantages for the builder. It provides a wider range of appearance. Panels may look like flat panels, boards, or battens and boards. Panels can also be grooved, and they can look like shingles. The surface textures range from rough lumber to smooth, flat board panels. Panels can show diagonal lines, vertical lines, or horizontal lines. Also, panels may be used to give the appearance of stucco or other textures.

Panel siding is easier to lay out. There is little waste and the installation is generally much faster. It is often stapled in place, rather than nailed. See Fig. 10-55. As a rule, corners are made faster and easier with panel siding.

Panel siding can be made from a variety of materials. Plywood is commonly used. Hardboard and various other fiber composition materials are also used. A variety of finishes is also available on panel siding. These are prefinished panels. Panels may be prefinished with paint, chemicals, metal, or vinyl plastics.

Fig. 10-51 *A strip must be used to cant the siding for mitering.*

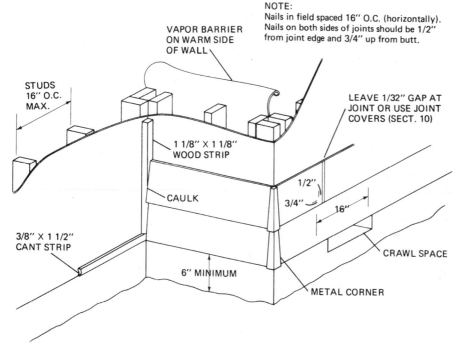

NOTE:
Nails in field spaced 16" O.C. (horizontally).
Nails on both sides of joints should be 1/2"
from joint edge and 3/4" up from butt.

VAPOR BARRIER
ON WARM SIDE
OF WALL

STUDS
16" O.C.
MAX.

1 1/8" X 1 1/8"
WOOD STRIP

LEAVE 1/32" GAP AT
JOINT OR USE JOINT
COVERS (SECT. 10)

CAULK

3/8" X 1 1/2"
CANT STRIP

1/2"
3/4"
16"

6" MINIMUM

CRAWL SPACE

METAL CORNER

Fig. 10-53 *Inside corners may be finished with square wooden strips.* (Boise-Cascade)

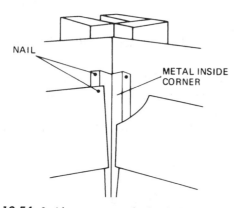

NAIL

METAL INSIDE
CORNER

Fig. 10-54 *Inside corners may be finished with metal strips.*

The standard panel size is 4 feet wide and 8 feet long. The most common panel thickness is ⅜ inch. However, other panel sizes are also used. Lengths up to 14 feet are also available. Thickness ranges from 5/16- to ¾-inch thick in 1/16-inch intervals.

The edges of panel siding may be flat. They may also be grooved in a variety of ways. Grooved edges form tight seams between plywood panels. Seams may be covered by battens. They may also be covered by special edge pieces. Outside corners may be treated in much the same manner as corners of regular board siding. Corners may be lapped and covered with corner boards. They may be covered with metal. They may also be mitered. Inside corners may be butted together. The edge of one panel is butted against the solid face of the first.

As a rule, panels are applied from the bottom up. However, they may be applied from top to bottom.

Fig. 10-55 *Panel siding is often stapled in place.* (Fox and Jacobs)

NAILS AND NAILING

Correct nails and nailing practices are essential in the proper application of wood siding. In general, siding and box nails are used for face nailing, and casing nails are used for blind nailing. Nails must be corrosion-resistant, and preferably rust proof. Avoid using staples. Stainless steel nails are the best choice. High tensile strength aluminum are economical and corrosion-resistant, and they will not discolor or cause deterioration of the wood siding. However, aluminum nails will react with galvanized metal, causing corrosion. Do not use aluminum nails on galvanized flashing (nor galvanized nails on aluminum flashing). The hot-dipped galvanized nail is the least expensive, but it might cause discoloration if precautions are not taken.

In some instances, the use of hot-dipped galvanized nails along with clear finishes on Western Red Cedar has resulted in stains around the nails. While this occurrence seems to be limited to the northeastern and north central regions of the country, the combination of hot-dipped galvanized nails with clear finishes on Western Red Cedar is not recommended.

Plastic hammer-head covers can be used when driving hot-dipped galvanized nails. This will reduce the potential for chipping and the subsequent potential for corrosion.

On siding you should not use staples or electroplated nails. These fasteners often result in black iron stains which can be permanent. Copper nails are not suitable for Western Red Cedar as cedar's natural extractives will react with the copper causing the nails to corrode, resulting in stains on the siding.

High-quality nails for solid wood siding are a wise investment. The discoloration, streaking, or staining that can occur with inappropriate nails ruins the appearance of the project, and is very difficult to remove.

Nail Shanks

Many nails have smooth shanks and will loosen as the siding expands and contracts under the extremes of seasonal changes in temperature and humidity. Ring or spiral-threaded nail shanks will increase holding power. Both types of shanks are readily available.

Nail Points

The most commonly used nail points include: blunt, diamond, and needle, as shown in Fig. 10-56. Blunt points reduce splitting while diamond points are the most commonly used. However, needle points should be avoided because they tend to cause splitting.

Recommended penetration into studs or blocking, or into a combination of wood sheathing and these members, is 1.5 inches. Penetration is 1.25 inches with ring shank nails.

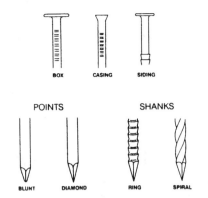

Fig. 10-56 *Nail types.* (Western Wood Products Association)

Vertical siding, when applied over wood bed sheathing, should be nailed to horizontal blocking or other wood framing members not exceeding 35 inches on center when face-nailed, or 32 inches on center when blind-nailed. Vertical siding, when installed without sheathing, should be nailed to wood framing or blocking members at 24 inches on center. Some building codes require 24 inches on center with or without sheathing. Check your local code for the requirements. Horizontal and diagonal siding should be nailed to studs at least 24 inches on center maximum when applied over wood-based, solid sheathing, and 16 inches on center maximum when applied without sheathing.

The siding pattern will determine the exact nail size, placement, and number of nails required. As a general rule, each piece of siding is nailed independently of its neighboring pieces. Do not nail through two overlapping pieces of siding with the same nail as this practice will restrict the natural movement of the siding and might cause unnecessary problems. Nail joints into the studs or blocking members—nailing into sheathing alone is not adequate.

Drive nails carefully. Hand nailing is preferred over pneumatic nailing because there is less control of placement and driving force with pneumatic nailers. Nails should be snug, but not overdriven. Nails that are overdriven can distort the wood and cause excessive splitting. Predrilling near the ends helps reduce any splitting that can occur with the thinner patterns. Some modern siding patterns with their recommended nailing procedures are shown in Fig. 10-57.

SHINGLE AND SHAKE SIDING

Shingles and shakes are often used for exterior siding. They are very similar in appearance. However, shakes have been split from a log. They have a rougher surface texture. Shingles have been sawn and are smoother in

SIDING PATTERNS		NOMINAL SIZES* Thickness & Width	NAILING 6" & Narrower	NAILING 8" & Wider
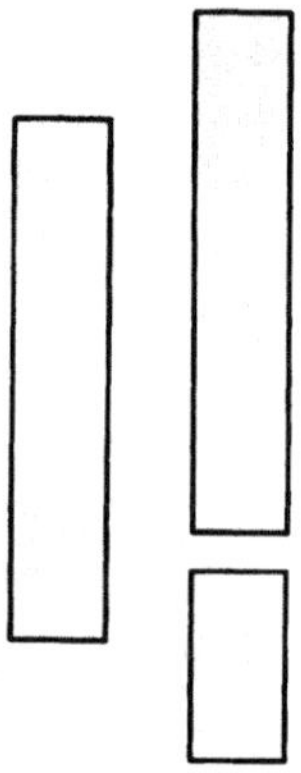	**TRIM BOARD-ON-BOARD BOARD-AND-BATTEN** Boards are surfaced smooth, rough or saw-textured. Rustic ranch-style appearance. Provide horizontal nailing members. Do not nail through overlapping pieces. Vertical applications only.	1 x 2 1 x 4 1 x 6 1 x 8 1 x 10 1 x 12 1¼ x 6 1¼ x 8 1¼ x 10 1¼ x 12	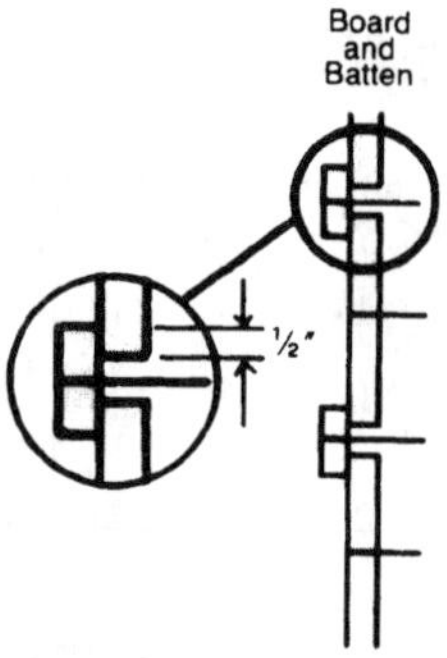Board and Batten ½" Recommend ½" overlap. One siding or box nail per bearing.	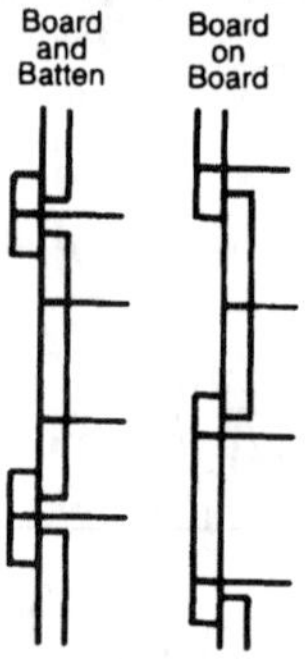Board and Batten Board on Board Increase overlap proportionately. Use two siding or box nails, 3-4" apart.
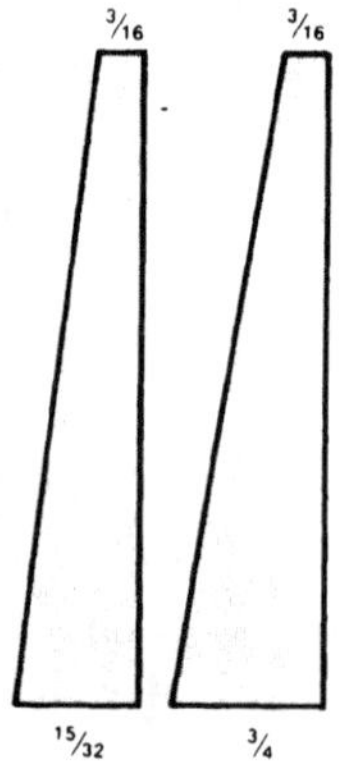3/16 3/16 15/32 3/4	**BEVEL OR BUNGALOW** Bungalow ("Colonial") is slightly thicker than Bevel. Either can be used with the smooth or saw-faced surface exposed. Patterns provide a traditional-style appearance. Recommend a 1" overlap. Do not nail through overlapping pieces. Horizontal applications only. Cedar Bevel is also available in ⅞ x 10,12.	½ x 4 ½ x 5 ½ x 6 ⅝ x 8 ⅝ x 10 ¾ x 6 ¾ x 8 ¾ x 10	Plain Recommend 1" overlap. One siding or box nail per bearing, just above the 1" overlap.	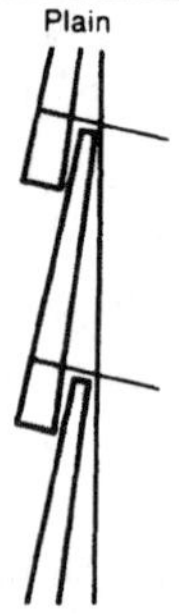Plain Recommend 1" overlap. One siding or box nail per bearing, just above the 1" overlap.
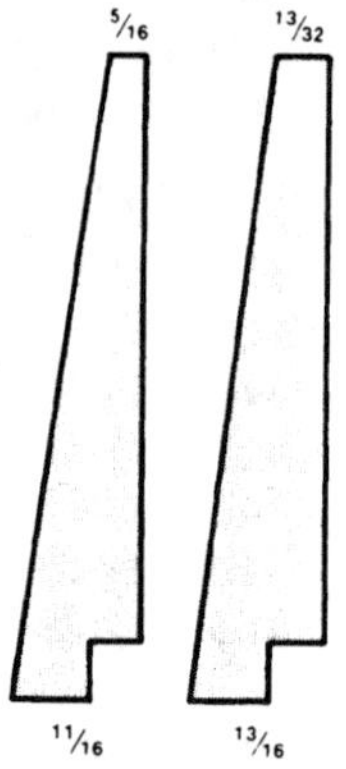5/16 13/32 11/16 13/16	**DOLLY VARDEN** Dolly Varden is thicker than bevel and has a rabbeted edge. Surfaced smooth or saw textured. Provides traditional-style appearance. Allows for ½" overlap, including an approximate ⅛" gap. Do not nail through overlapping pieces. Horizontal applications only. Cedar Dolly Varden is also available ⅞ x 10,12.	Standard Dolly Varden ¾ x 6 ¾ x 8 ¾ x 10 Thick Dolly Varden 1 x 6 1 x 8 1 x 10 1 x 12	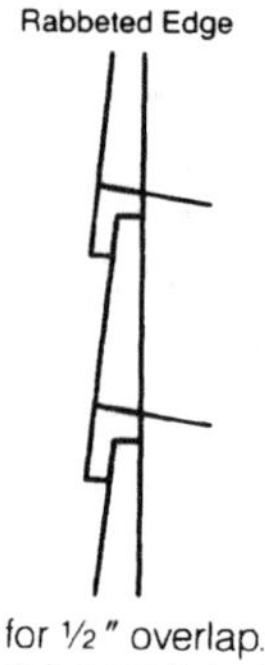Rabbeted Edge Allows for ½" overlap. One siding or box nail per bearing, 1" up from bottom edge.	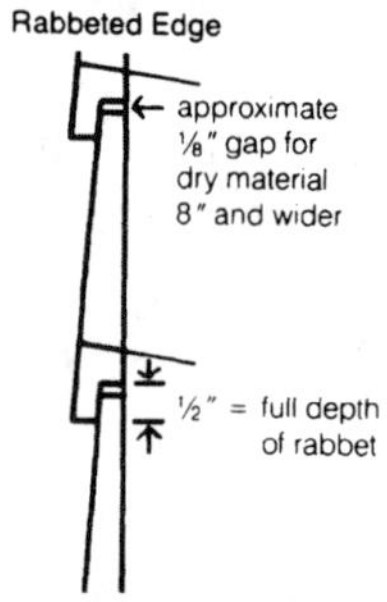Rabbeted Edge approximate ⅛" gap for dry material 8" and wider ½" = full depth of rabbet Allows for ½" overlap. One siding or box nail per bearing, 1" up from bottom edge.
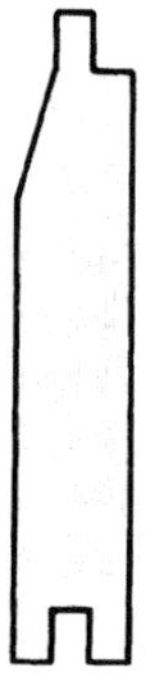	**DROP** Drop siding is available in 13 patterns, in smooth, rough and saw textured surfaces. Some are T&G, others shiplapped. Refer to "Standard Patterns" for dimensional pattern profiles. A variety of looks can be achieved with the different patterns. Do not nail through overlapping pieces. Horizontal or vertical applications. Tongue edge up in horizontal applications.	¾ x 6 ¾ x 8 ¾ x 10	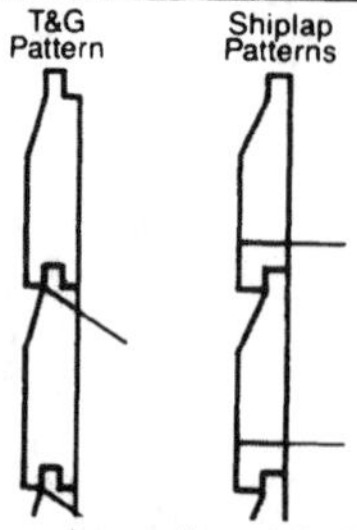T&G Pattern Shiplap Patterns Use casing nails to blind nail T&G patterns, one nail per bearing. Use siding or box nails to face nail shiplap patterns, one inch up from bottom edge.	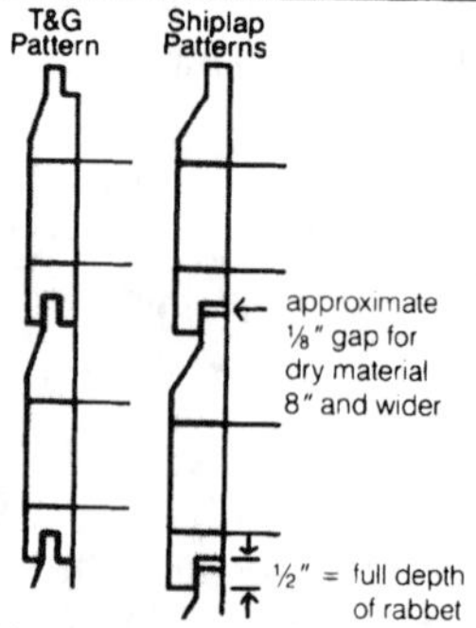T&G Pattern Shiplap Patterns approximate ⅛" gap for dry material 8" and wider ½" = full depth of rabbet Use two siding or box nails, 3-4" apart to face nail, 1" up from bottom edge.

Fig. 10-57 *Siding patterns, nominal sizes, and recommended nailing.* (Western Wood Products Association)

SIDING PATTERNS	NOMINAL SIZES* Thickness & Width	NAILING 6" & Narrower	NAILING 8" & Wider
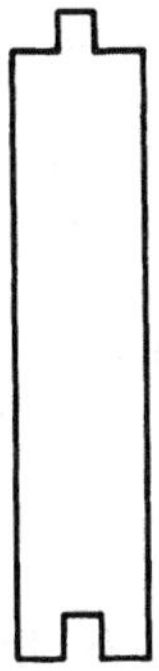 **TONGUE & GROOVE** Tongue & groove siding is available in a variety of patterns. T&G lends itself to different effects aesthetically. Refer to WWPA "Standard Patterns" (G-16) for pattern profiles. Sizes given here are for Plain Tongue & Groove. Do not nail through overlapping pieces. Vertical or horizontal applications. Tongue edge up in horizontal applications.	1 x 4 1 x 6 1 x 8 1 x 10 Note: T&G patterns may be ordered with ¼, ⅜ or ⁷⁄₁₆" tongues. For wider widths, specify the longer tongue and pattern.	Plain 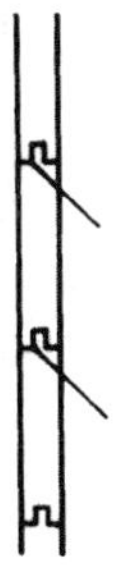 Use one casing nail per bearing to blind nail.	Plain 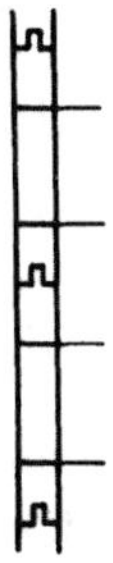 Use two siding or box nails 3-4" apart to face nail.
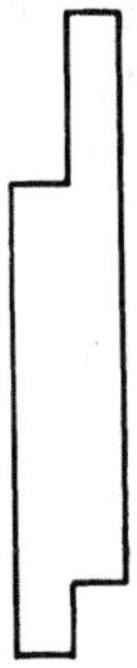 **CHANNEL RUSTIC** Channel Rustic has ½" overlap (including an approximate ⅛" gap) and a 1" to 1¼" channel when installed. The profile allows for maximum dimensional change without adversely affecting appearance in climates of highly variable moisture levels between seasons. Available smooth, rough or saw textured. Do not nail through overlapping pieces. Horizontal or vertical applications.	¾ x 6 ¾ x 8 ¾ x 10	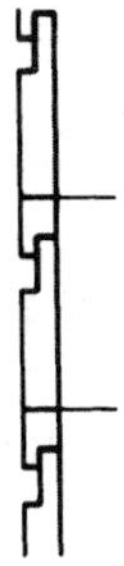 Use one siding or box nail to face nail once per bearing, 1" up from bottom edge.	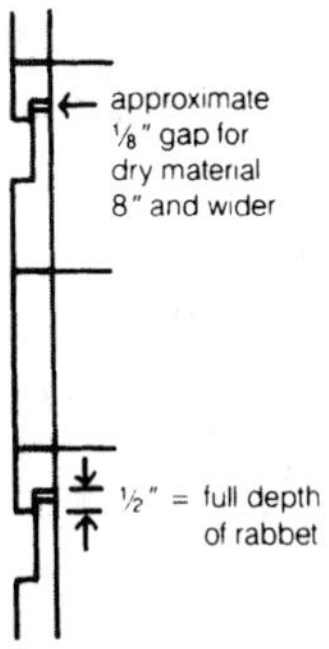 approximate ⅛" gap for dry material 8" and wider ½" = full depth of rabbet Use two siding or box nails 3-4" apart per bearing.
 **LOG CABIN** Log Cabin siding is 1½" thick at the thickest point. Ideally suited to informal buildings in rustic settings. The pattern may be milled from appearance grades (Commons) or dimension grades (2x material). Allows for ½" overlap, including an approximately ⅛" gap. Do not nail through overlapping pieces. Horizontal or vertical applications.	1½ x 6 1½ x 8 1½ x 10 1½ x 12	 Use siding or box nail to face nail once per bearing, 1½" up from bottom edge.	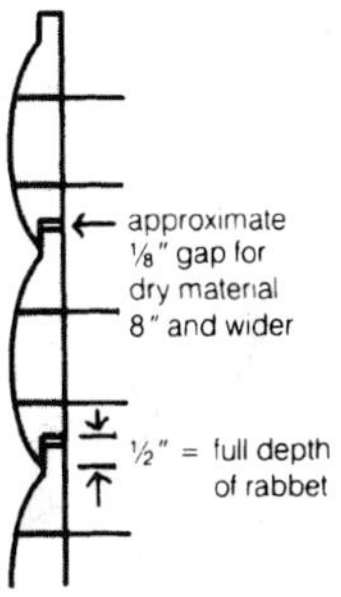 approximate ⅛" gap for dry material 8" and wider ½" = full depth of rabbet Use two siding or box nails, 3-4" apart, per bearing to face nail.

SIDING INSTALLATION TIPS

Do not nail through overlapping pieces. Use stainless steel, high tensile strength aluminum or hot-dipped galvanized nails with ring or spiral-threaded shanks. Use casing nails to blind nail; siding or box nails to face nail.

Horizontal applications only for Bevel, Bungalow and Dolly Varden.

Vertical applications only for Board-and-Board or Board-and-Batten; bevel cut ends of pieces and install so water is directed to outside.

Horizontal or vertical applications for Tongue & Groove, Channel Rustic, Log Cabin or Drop patterns. Tongue edge up in horizontal applications of Drop and T&G patterns

Read the section on Nail Penetration & Spacing to determine nail size.

Read the sections on Moisture Content and Prefinishing before installing siding.

Fig. 10-57 *Continued.*

appearance. The procedure is the same for either shingles or shakes.

Shingles are made from many different materials. Shakes are made only from wood. The standard lengths for wood shakes and shingles are 16, 18, and 24 inches.

Shingles

Shingles may be made of wood, flat composition, or mineral fiber composition. The last is a combination of asbestos fiber and portland cement. Shingles for roofs are lapped about two-thirds. About one-third of the shingle is exposed. When the shingles are laid, or coursed, there are three layers on the roof. However, when shingles are used for siding, the length exposed is greater. Slightly more than one-half is exposed. This makes a two-layer thickness instead of three as on roofs. For asbestos mineral sidings, sometimes less lapping is used.

Wooden shingles are made in random widths. Shingles will vary from 3 to 14 inches in width. The better grades will have more wide shingles than narrow shingles.

Nailing

All types of shingles may be nailed in either of two ways. In the first method, shown in Fig. 10-58, nailing strips are used. Note that a moisture barrier is almost always used directly under shingles. Builder's felt (tar paper) is the most common moisture barrier. Shingles should be spaced like lap siding. The butt line of the shingles should be even with the tops of wall openings. Top lines should be even with the bottoms. For utility buildings such as garages, no sheathing is needed. A moisture barrier can be put over the studs. Nailing strips are then nailed to the studs. However, for most residential buildings, a separate sheathing is suggested. Strips are often nailed over the sheathing. This is done over sheathing that is not a good nail base. Shingles may be nailed directly to board or plywood sheathing. This is shown in Fig. 10-59.

The bottom shingle course is always nailed in place first. Part of the bottom course can be laid for large wall sections. The course may reach for only a part of the wall. Then that part of the wall can be shingled. The shingled part will be a triangle from the bottom corner upward. As a rule, two layers of shingles are used on the bottom course. The first layer is nailed in place, and a second layer is nailed over it. The edges of the second shingle should not line up with the edges of the first. This makes the siding much more weather-resistant. This layout is shown in Fig. 10-59.

Another technique for shingles is called *double coursing*. This means that two thicknesses are applied. The first layer is often done with a cheaper grade of shingle. Again, each course is completed before the one above is begun. The edges should alternate for best weather-resistive qualities. Double coursing is shown in Fig. 10-60. The outside shingle covers the bottom of the inside shingle. This gives a dramatic and contrasting effect.

Shakes

The shake is similar to a shingle, but shakes may be made into panels. Figure 10-61 shows this type of siding. Shake panels are real wooden shakes glued to a plywood base. The strips are easier and quicker to nail than individual shakes. Also, shakes can be spaced more easily. A more consistent spacing is also attained. The individual panels may be finished at the factory. Color, spacing, appearance, and texture can be factory-made. Panels may also be combined with insulation and weatherproofing. The panels are applied in the same manner as shingles.

Corners

Corners are finished the same as for other types of siding. Three corner types are used: corner boards, metal

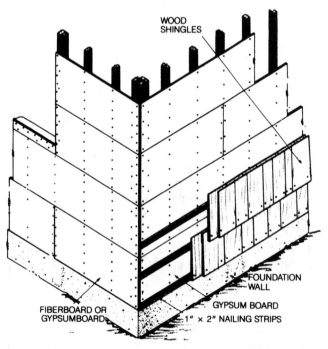

Fig. 10-58 *Nail strips are used over studs and sheathing. Shingles are then nailed to the strips. At least two nails are used for each shingle.*

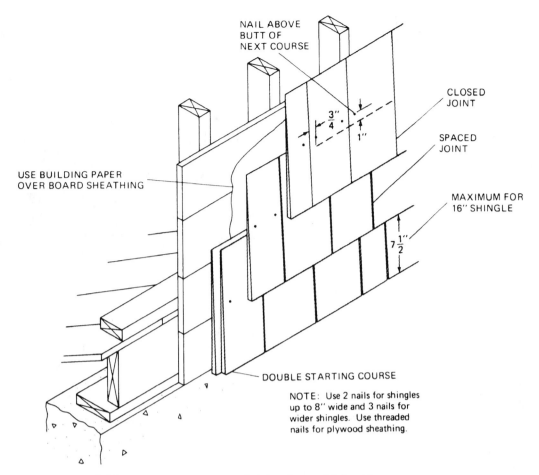

NAIL ABOVE
BUTT OF
NEXT COURSE

CLOSED
JOINT

SPACED
JOINT

$\frac{3}{4}''$

1''

MAXIMUM FOR
16'' SHINGLE

$7\frac{1}{2}''$

USE BUILDING PAPER
OVER BOARD SHEATHING

DOUBLE STARTING COURSE

NOTE: Use 2 nails for shingles
up to 8'' wide and 3 nails for
wider shingles. Use threaded
nails for plywood sheathing.

Fig. 10-59 *Single-course shingle nailed directly to solid sheathing.* (Forest Products Laboratory)

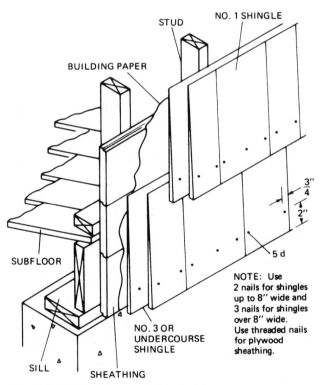

STUD

NO. 1 SHINGLE

BUILDING PAPER

$\frac{3}{4}''$

2''

5 d

SUBFLOOR

NO. 3 OR
UNDERCOURSE
SHINGLE

NOTE: Use
2 nails for shingles
up to 8'' wide and
3 nails for shingles
over 8'' wide.
Use threaded nails
for plywood
sheathing.

SILL

SHEATHING

Fig. 10-60 *Double-coursed shingle siding.* (Forest Products Laboratory)

Fig. 10-61 *Shakes are often made into long panels.* (Shakertown)

Shingle and Shake Siding   323

corners, and mitered corners. Again, mitering is generally used for more expensive buildings.

Corners may also be woven. Shingles may be woven as in Fig. 10-62. This is done on corners and edges of door and window openings.

Fig. 10-62 *Shingle and shake corners may be woven or lapped.*

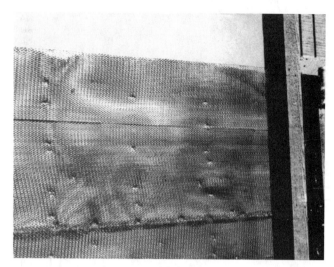

Fig. 10-63 *A wire mesh is nailed to the wall. Stucco will then be applied over the mesh.*

PREPARATION FOR OTHER WALL FINISHES

There are other common methods for finishing frame walls. These include stucco, brick, and stone. As a rule, the carpenter does not put up these wall finishes. However, the carpenter sometimes prepares the sheathing for these coverings. The preparation depends upon how well the carpenter understands the process.

Stucco Finish

Stucco is widely used in the South and Southwest. It is durable and less expensive than brick or stone. Like brick or stone, it is fireproof. It may be put over almost any type of wall. It can be prepared in several colors. Several textures can also be applied to give different appearances.

Wall preparation A vapor barrier is needed for the wall. The vapor barrier can be made of builder's felt or plastic film. The sheathing may also be used for the vapor barrier. It is a good idea to apply an extra vapor barrier over most types of walls. This includes insulating sheathing, plywood, foam, and gypsum. The vapor barrier is applied from the bottom up. Each top layer overlaps the bottom layer a minimum of 2 inches. Then a wire mesh is nailed over the wall. See Fig. 10-63. Staples are generally used rather than nails. A solid nail base is essential. The mesh should be stapled at 18- to 24-inch intervals. The intervals should be in all directions—across, up, and down. The wire mesh may

be of lightweight "chicken wire." However, the mesh should be of heavier material for large walls.

The mesh actually supports the weight of the stucco. The staples only hold the mesh upright, close to the wall.

Apply the stucco Usually two or three coats are applied. The first coat is not the finish color. The first coat is applied in a very rough manner. It is applied carefully but it may not be even in thickness or appearance. It is called the "scratch" coat. Its purpose is to provide a layer that sticks to the wire. The coat is spread over the mesh with a trowel. See Fig. 10-64.

This scratch coat should be rough. It may be scratched or marked to provide a rough surface. The rough surface is needed so that the next coat will stick. A trowel with ridges may be used. The ridges make grooves in the coat. These grooves may run in any direction. However, most should run horizontally. Another way to get a rough surface is to use a pointed tool. After the stucco sets, grooves are scratched in it with the tool.

One or more coats may be put up before the finish coat is applied. These coats are called "brown" coats. They are scratched so that the next coat sticks better. The last coat is called the finish coat. It is usually white or tan in color. However, dyes may be added to give other colors. The colors are usually light in shade.

Brick and Stone Coverings

Brick and stone are used to cover frame walls. The brick or stone is not part of the load-bearing wall. This means that they do not hold up any roof weight. The covering is called a brick or stone veneer. It adds beauty and weatherproofing. Such walls also increase the resistance of the building to fires.

Fig. 10-64 *Stucco is spread over mesh with a trowel.*

The preparation for either brick or stone is similar. First, a moisture barrier is used. If a standard wall sheathing has been used, no additional barrier is needed. However, it is not bad practice to add a separate moisture barrier.

The carpenter may then be asked to nail *ties* to the wall. These are small metal pieces, shown in Fig. 10-65. Ties are bent down and embedded in the mortar. After the mortar has set, these ties form a solid connection. They hold the brick wall to the frame wall. They also

Fig. 10-65 *Ties help hold brick or stone to the wall. They are nailed to studs and embedded in the mortar.*

help keep the space between the walls even. Small holes are usually left at the bottom. These are called *weep* holes. Moisture can soak through brick and stone. Moisture also collects at the bottom from condensation. The small weep holes allow the moisture to drain out. By draining the moisture, damage to the wood members is avoided. The joints are trimmed after the brick is laid (Fig. 10-66).

Fig. 10-66 *Joints are trimmed after brick is laid.* (Fox and Jacobs)

ALUMINUM SIDING

Aluminum siding is widely used on houses. Aluminum is used for new siding or can be applied over an old wall.

A variety of vertical and horizontal styles are used. Probably the most common type looks like lap siding. However, even this type may come as individual "boards" or as panels two or three "boards" wide. See Fig. 10-67. These types of siding may be hollow-backed or insulated. See Fig. 10-68.

Aluminum siding may also look like shingles or shakes, as in Fig. 10-69. For all types of aluminum siding, a variety of surface texture and colors are available.

Aluminum siding is put up using a special system. Note in Fig. 10-70 that the top edge has holes in it. All nails are driven in these holes. The bottoms or edges of the pieces interlock. In this way, the tops are nailed and the bottoms interlock. Each edge is then attached to a nailed portion for solid support.

To start, a special starter strip is nailed at the bottom. See Fig. 10-71. Then corner strips, as in Fig. 10-72, are added. Special shapes are placed around windows and other features as in Fig. 10-73. For these, the manufacturer's directions should be carefully followed.

Next, the first "board" is nailed in place. Note that it is placed at the bottom of the wall. The bottom of the

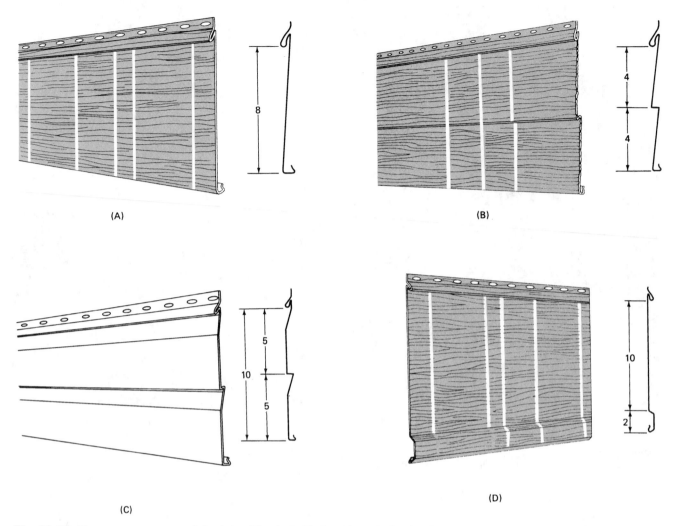

Fig. 10-67 *The most common type of aluminum siding locks like lap siding. (A) Single-"board" lap siding; (B) Two-"board" panel lap siding; (C) Double-"board" drop siding; (D) Vertical "board" siding.*

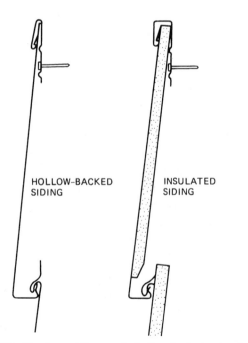

HOLLOW–BACKED SIDING

INSULATED SIDING

Fig. 10-68 *Aluminum siding may be hollow-backed or insulated.*
(ALCOA)

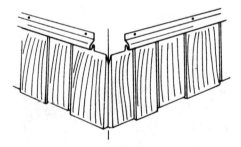

Fig. 10-69 *Aluminum "shake" siding.* (ALCOA)

first piece is interlocked with the starter strip and nailed. It is best to start at the rear of the house and work toward the front. This way, overlaps do not show as much. For the same reason, factory-cut ends should overlap ends cut on the site. See Fig. 10-74. Backer strips, as in Fig. 10-75, should be used at each overlap. These provide strength at the overlaps. Also, overlaps should be spaced evenly, as in Fig. 10-76. Grouping the overlaps together gives a poor appearance.

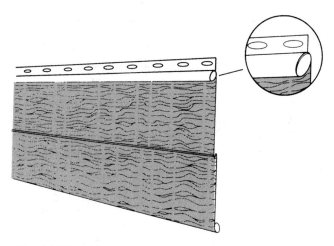

Fig. 10-70 *Panels are nailed at the top and interlock at the bottom.* (ALCOA)

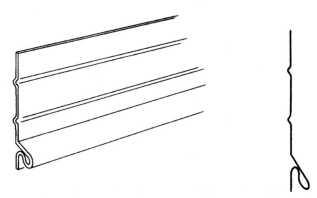

Fig. 10-71 *A starter strip is nailed at the bottom of the wall. The lower edge of the first board can interlock with it for support.* (ALCOA)

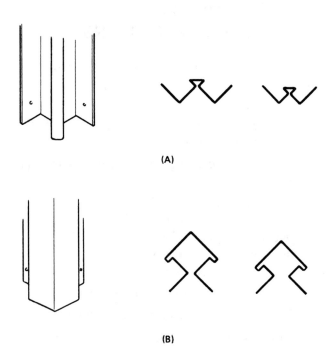

(A)

(B)

Fig. 10-72 *Corner strips for aluminum siding* (ALCOA). *(A) Inside; (B) Outside.*

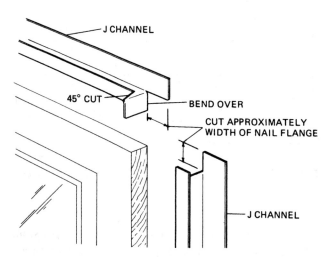

Fig. 10-73 *Special pieces are used around windows and gables.* (ALCOA)

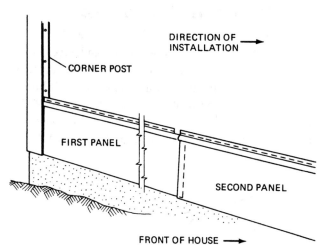

Fig. 10-74 *Siding is started at the back.* (ALCOA)

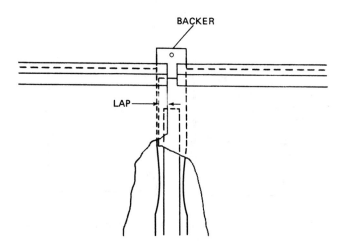

Fig. 10-75 *Backer strips support the ends at overlaps.* (ALCOA)

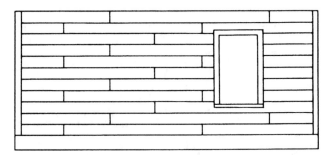

Fig. 10-76 *Overlaps should be equally spaced for best appearance.* (ALCOA)

Allowance must be made for heat expansion. Changes in temperature can cause the pieces to move. To allow movement, the nails are not driven up tightly. It is best to check the instructions that come with the siding.

Vertical Aluminum Siding

Most procedures for vertical siding are the same. Strips are put up at corners, windows, and eaves.

However, the starter strip is put near the center. A plumb line is dropped from the gable peak. A line is then located half the width of a panel to one side. The starter strip is nailed to this line. Panels are then installed from the center to each side. See Fig. 10-77.

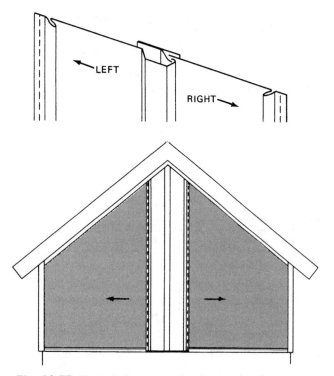

Fig. 10-77 *Vertical aluminum siding is started at the center.* (ALCOA)

SOLID VINYL SIDING

Solid vinyl siding is also used widely. It is available in both vertical and horizontal applications. A wide variety of colors and textures is also available.

As with aluminum siding, vinyl siding is nailed on one edge. The other edge interlocks with a nailed edge for support. See Fig. 10-78. Special pieces are again needed for joining windows, doors, gables, and so forth.

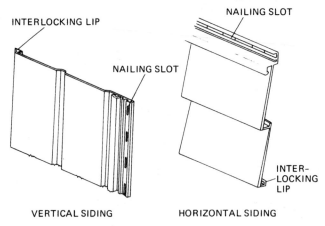

Fig. 10-78 *Solid vinyl siding.*

As with aluminum siding, the finish is a permanent part of the siding. No painting will ever be needed. It may be cleaned with a garden hose.

CHAPTER 10
STUDY QUESTIONS

1. What are the methods for enclosing eaves?
2. What sequence is used for finishing an exterior wall?
3. What can happen to painted wooden walls that have no vapor barrier?
4. How far should siding nails penetrate into the nail base?
5. How much siding is needed for:
   Wall size = 8 feet high × 30 feet long
   Openings = one, 3 × 4 feet
   Overlap = 1½ inches
   Siding width = 10 inches
6. Why is some siding started from the top?
7. What is the usual starting point for siding?
8. When are guard rails required on scaffolds?
9. What is a wall bracket?

10. What are the advantages of steel double-pole scaffolds?

11. How thick should platform planks be?

12. List six rules for ladder safety.

13. List six rules for scaffold safety.

14. Why are soffit vents used?

15. What types of board siding are used?

16. Why are boards not nailed together on lap siding?

17. Why is siding spaced even with openings?

18. What is the difference between shingles and shakes?

19. Why are shingles double-coursed on walls?

20. What is a "square" in shingling?

21. How are gable overhangs framed?

22. How is wooden gable and wall siding joined?

23. What holds brick veneer to a wall?

24. How are corners treated for wood panel siding?

25. What should be done to prepare a wall for stucco?

26. When can hot-dipped galvanized nails be used with Western Red Cedar?

27. Should vertical siding, when installed without sheathing, be nailed to wood framing or blocking members every 24 inches on center?

28. What does a blunt point nail do to wood?

29. Where can you use a needle pointed nail?

30. Why is hand nailing preferred over pneumatic nailing?

11
CHAPTER

Wiring
the House

ELECTRICITY IS NECESSARY FOR MODERN living. Many devices in the home rely on electricity to function and render the style and comfort expected of a new house. This electricity can be generated by fossil-fuel power generators, by atomic-powered generators, or by the use of falling water to drive generators. It is also possible to generate power by engine-driven generators. In Alaska, for instance, long distribution lines are impractical due to ice conditions that would result in line damage during storms. In Alaska, engine-driven generators are common.

Once electricity is generated in sufficient amounts for consumption in large quantities, the second necessary step is to get the energy to the consumer. Herein lies a distribution problem: that of stringing and maintaining long lines.

Figure 11-1 diagrams the process of getting power from the generating plant along high-voltage transmission lines, to voltage step-down substations, and then to office buildings, stores, apartments, and large factories. Further reductions are necessary in voltage to reduce the power to the proper voltages (120–240) for home use.

LOCAL DISTRIBUTION

Underground systems for distributing electrical energy are in the middle of a revolution that began about 20 years ago. A prime objective has been to reduce the cost of these systems so that they could be used in areas previously served only by overhead systems. Aluminum, long known as an economical electrical conductor, has contributed to this revolution.

Actually, underground distributing is as old as the electrical power industry itself. In 1882, Thomas A. Edison built the first central power station on Pearl Street in New York City. He installed feeders and mains under the city streets to distribute power generated in the station to its customers. These were pipe-type cables, with a pair of copper conductors that were separated from each other, and from the enclosing pipe, by specially designed washers. The pipe was then filled with insulating compound. This compound consisted of asphaltum boiled in oxidized linseed oil, with paraffin and a little beeswax added.

The development of lead-sheathed insulated cables that could be pulled into previously installed ducts greatly simplified construction of underground distribution systems. However, the high costs of such systems remained. As a result, for many years underground distribution was limited largely to the central portions of cities. Here load densities are high and the congestion that would result from overhead systems is highly undesirable.

Today, costs of underground distribution have been reduced to levels that are economically practical for service to light- and medium-density load areas. Underground systems now are being installed in most residential areas. Home buyers in these areas generally agree that the improved appearance and enhanced value of their property more than justify the added cost

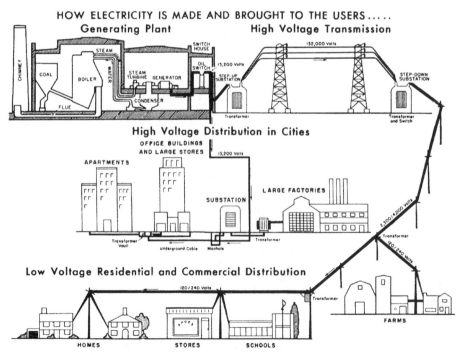

Fig. 11-1 *Generation and distribution of electricity.*

LINE-VOLTAGE SYMBOLS		LOW-VOLTAGE SYMBOLS	
Symbol	**Description**	—·—·—	Remote-Control Low-Voltage Wire
——————	2-Cond. 120V Wire or Cable	T	Low-Voltage Transformer
—///—	3-Cond. 120V Wire or Cable	⊣►—	Rectifier for Remote Control
◯	Ceiling Receptacle	B R	Box for Relays and Motor Master Controls
◯	Floodlight	R	Remote-Control Relay
⊏═◯	Valance Light	R P	Remote-Control Pilot-Light Relay
—Ⓒ	Clock Receptacle	P 11	Separate Pilot Light, R.C. Plate
Ⓛ	Keyless Lampholder	P 10	Separate Pilot Light, Inter. Plate
Ⓛ PS	Pull Chain Lampholder	MS	Master-Selector Switch
⊖	Double Receptacle, Split Wired	MM R	Motor Master Control for ON
⊖	Grounding Receptacle	MM B	Motor Master Control for OFF
⊖ WP	Weatherproof Grounding Receptacle	S M	Switch for Motor Master
⊖ R	Range Receptacle	S F6	R.C. Flush Switch
⊖ CD	Clothes Dryer Receptacle	S F7	R.C. Locator-Light Switch
S L	Lighted-Handle Mercury Switch	S F8	R.C. Pilot-Light Switch
S P	Push-Button, Pilot Switch	S K6	R.C. Key Switch
S D	Closet Door Switch	S K7	R.C. Locator Light Key Switch
		S K8	R.C. Pilot-Light Key Switch
		S T6	R.C. Trigger Switch
		S T7	R.C. Locator Light Trigger Switch
		S T8	R.C. Pilot-Light Trigger Switch
		S T4	Interchangeable Trigger Switch, Brown
		S T5	Interchangeable Trigger Switch, Ivory
		⊖ RO	Remote-Control Receptacle for Extension Switch

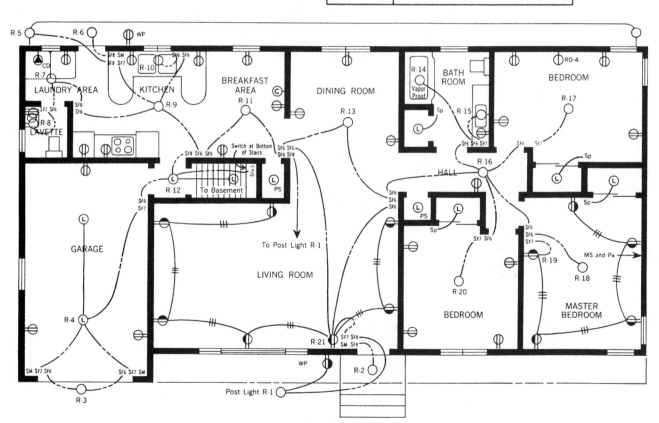

Fig. 11-1 Continued.

of underground service. However, in some places they locate the transformer in the front of the house and negate any advantage to the appearance value of underground feed. The transformers in the front yard are difficult to landscape and, if they are, the transformers are hard to gain access to in case of an emergency. Many communities either place the underground service completely underground with buried transformers or distribute the power from the rear of the house. Exposed upright telephone line terminals and cable television boxes also detract from the beauty of a residential area when they, too, are located in the front yard.

FARM ELECTRICITY

In 1935, Franklin D. Roosevelt created the Rural Electrification Administration as an emergency relief program. In May 1936, the Congress of the United States passed the Rural Electrification Act. This established REA as a leading agency in the federal government. It had the responsibility of developing a program for rural electrification. The act authorized and empowered REA to make self-liquidating loans to companies, cooperatives, municipalities, and public power districts. These loans were to finance the construction and operation of generating plants along with the transmission and distribution lines. See Fig. 11-2.

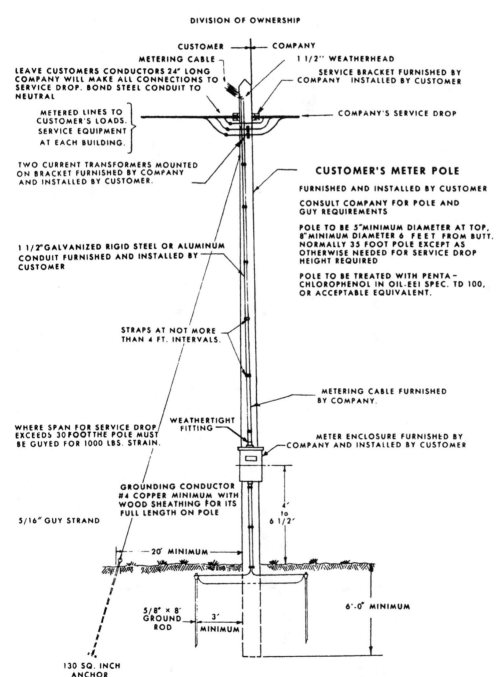

Fig. 11-2 *A farm meter pole. Single-phase, three-wire, 120/240 volts for loads exceeding 40 kilowatt demand. Some farmers purchase power from utility companies other than REA.*

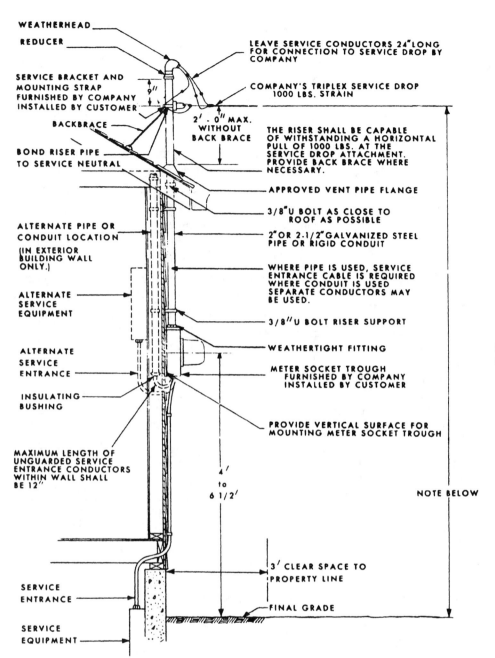

WEATHERHEAD
REDUCER

SERVICE BRACKET AND
MOUNTING STRAP
FURNISHED BY COMPANY
INSTALLED BY CUSTOMER

BACKBRACE

BOND RISER PIPE
TO SERVICE NEUTRAL

ALTERNATE PIPE OR
CONDUIT LOCATION
(IN EXTERIOR
BUILDING WALL
ONLY.)

ALTERNATE
SERVICE
EQUIPMENT

ALTERNATE
SERVICE
ENTRANCE

INSULATING
BUSHING

MAXIMUM LENGTH OF
UNGUARDED SERVICE
ENTRANCE CONDUCTORS
WITHIN WALL SHALL
BE 12"

SERVICE
ENTRANCE

SERVICE
EQUIPMENT

LEAVE SERVICE CONDUCTORS 24"LONG
FOR CONNECTION TO SERVICE DROP BY
COMPANY

COMPANY'S TRIPLEX SERVICE DROP
1000 LBS. STRAIN

2'-0" MAX.
WITHOUT
BACK BRACE

THE RISER SHALL BE CAPABLE
OF WITHSTANDING A HORIZONTAL
PULL OF 1000 LBS. AT THE
SERVICE DROP ATTACHMENT.
PROVIDE BACK BRACE WHERE
NECESSARY.

APPROVED VENT PIPE FLANGE

3/8"U BOLT AS CLOSE TO
ROOF AS POSSIBLE

2"OR 2.1/2"GALVANIZED STEEL
PIPE OR RIGID CONDUIT

WHERE PIPE IS USED, SERVICE
ENTRANCE CABLE IS REQUIRED
WHERE CONDUIT IS USED
SEPARATE CONDUCTORS MAY
BE USED.

3/8"U BOLT RISER SUPPORT

WEATHERTIGHT FITTING

METER SOCKET TROUGH
FURNISHED BY COMPANY
INSTALLED BY CUSTOMER

PROVIDE VERTICAL SURFACE FOR
MOUNTING METER SOCKET TROUGH

4'
to
6 1/2'

NOTE BELOW

3' CLEAR SPACE TO
PROPERTY LINE

FINAL GRADE

Fig. 11-3 Service entrance riser support on a low building.

WHERE THE BUILDING IS TOO LOW TO OBTAIN PROPER CLEARANCES IT IS
RECOMMENDED THAT CUSTOMER INSTALL AN UNDERGROUND SERVICE LATERAL.

Whether on the farm or in the suburbs, an electrician is usually required to place the meter socket trough. It is usually supplied by the power company and installed at the expense of the customer. Most meters are of the plug-in type. Once the service has been connected and the meter socket and house are properly wired, the meter is put into the socket completing the circuit for power into the house. See Fig. 11-3.

SAFETY AROUND ELECTRICITY

Most carpenters use extension cords to obtain power for their power tools. The proper size of cord makes a difference in the performance and life of the power equipment. A proper extension cord can make a difference in the safe operation of a piece of equipment. Make sure it has the proper capacity to handle the current needed. See Table 11-1.

Wire size and insulation of extension cords must be considered to ensure proper operation with a piece of equipment. Take a look at Table 11-2 for an explanation of the letters on a cord body. This table aids in determining what type of service for a specific cord is recommended.

Length of the extension cord is important in the sense that it must have a correct size wire to allow voltage to reach the consuming device. For instance, a 25-foot cord with No. 18 wire size is good for two amperes. If the distance is increased to 50 feet, you are still safe with No. 18 and two amperes. However, for a distance of 200 feet from the source, the size of the wire must be increased to No. 16. This is necessary to carry the two amperes without dropping the voltage along the cord and therefore producing a low voltage at the consuming device at the end of the line. See Table 11-3.

A visual inspection of the cord and harness should be made before each day's use. Check for loose or missing screws in the end and for loose or broken blades in the plug end. Check the continuity of each conductor with an ohmmeter at least once a week to make sure that there are no broken conductors. Check the cord when it is first used or taken from the box. Check again after it has been repaired or after an accident such as shown in Fig. 11-4. If the plug has been repaired, check to see if the clamp on the wire is securely attached.

Table 11-1 *Flexible Cord Ampacities***

Size	Type S, SO, ST	Type SJ, SJO, SJT
AWG	Amperes	
18	7, 10*	7, 10*
16	10, 13*	10, 13*
14	15, 18*	
12	20	
10	25	
8	35	
6	45	
4	60	

**Table 400.5. Reprinted by permission from NFPA 70-1981, National Electric Code, © 1980, National Fire Protection Association, Boston, Massachusetts.
*Where third conductor is used for equipment grounding only and does not carry load current.

Table 11-2 *Types of Flexible Cords*

Trade name	Type letter	Size AWG	No. of conductors	Insulation braid on	Braid on each conductor	Outer covering	Use
Junior Hard Service Cord	SJ SJO SJT	18, 14	2, 3 or 4	Rubber Thermoplastic or Rubber	None	Rubber Oil Resist Compound Thermoplastic	Pendant or Portable Damp Places Hard Usage
Hard Service Cord	S SO ST	6 18, to 10 incl.	2 or more	Rubber Thermoplastic or Rubber	None	Rubber Oil Resist Compound Thermoplastic	Pendant or Portable Damp Places Hard Usage

Table 11-3 *Extension Cord Sizes for Portable Electric Tools*

THIS TABLE FOR 115-VOLT TOOLS						
Full-load ampere rating of tool	0 to 2.0 A	2.1 to 3.4 A	3.5 to 5.0 A	5.1 to 7.0 A	7.1 to 12.0 A	12.1 to 16.0 A
Length of Cord	Wire size (AWG)					
25 ft	18	18	18	16	14	14
50 ft	18	18	18	16	14	12
75 ft	18	18	16	14	12	10
100 ft	18	16	14	12	10	8
200 ft	16	14	12	10	8	6
300 ft	14	12	10	8	6	4
400 ft	12	10	8	6	4	4
500 ft	12	10	8	6	4	2
600 ft	10	8	6	4	2	2
800 ft	10	8	6	4	2	1
1000 ft	8	6	4	2	1	0

NOTE—If voltage is already low at the source (outlet), have voltage increased to standard, or use a much larger cable than listed in order to prevent any further loss in voltage.

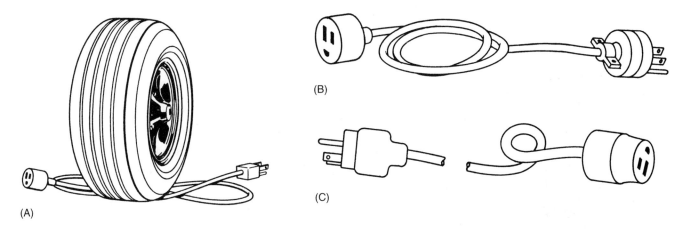

Fig. 11-4A *(A) After an accident with construction equipment, the cord should be checked for damage. (B) Check the extension cord before it is placed in service each day. (C) Check the wire clamp on the plug for proper fit.*

GROUNDED CONDUCTORS

A grounded conductor has a white-colored jacket in a two- or three-wire cable (neutral wire). It is terminated to the white or silver-colored terminal in a plug cap or connector, and it is terminated at the neutral bar in the distribution box.

An electrical fault might allow the hot line to contact the metal housing of electrical equipment (in a typical two-wire system) or some other ungrounded conductor. In these cases, any person who touches that equipment or conductor will be shocked. The person completes the circuit from the hot line to the ground, and current passes through the body. Because a human body is not a good conductor, the current is not high enough to blow the fuse. It continues to pass through the body as long as the body remains in contact with the equipment. See Fig. 11-5.

A *grounding conductor*, or equipment ground, is a wire attached to the housing or other conductive parts of electrical equipment that are not normally energized, to carry current from them to the ground. Thus, if a person touches a part that is accidentally energized, there will be no shock, because the grounding line furnishes a much lower resistance path to the ground. See Fig. 11-6. Moreover, the high current passing through the wire conductor blows the fuse and stops the current. In normal operation, a ground conductor does not carry current.

The grounding conductor in a three-wire conductor cable has a green jacket. It is always terminated at the green-colored hex head screw on the cap or connector. It utilizes either a green-colored conductor or a metallic conductor as its path to ground. In Canada, this conductor is referred to as the *earthing*

conductor. This term is somewhat more descriptive and helpful in distinguishing between grounding conductors and neutral wires or grounded conductors.

HOUSE SERVICE

For an electrical system to operate properly, it is necessary to design the system so that the proper amount of power is available. This requires the house to be wired so that wire of the proper size serves each plug in the house.

Ground Fault Circuit Interrupter (GFCI)

Some dangerous situations have been minimized by using ground fault interrupters (GFCI). See Fig. 11-7A. Since 1975, the National Electrical Code (NEC)® has required installation of GFCIs in outdoor outlets, outlets near kitchen sinks, and bathroom outlets in new construction, but most homes built before 1975 have no GFCI protection.

Retrofit GFCIs can protect one outlet or an entire circuit with multiple outlets. They can be installed in older homes to reduce the danger of electric shock. One of the simplest ways to achieve this protection in outdoor outlets or other outlets in which shock dangers are high is to use a plug-in type GFCI.

Two kinds of plug-ins are available. One has contact prongs attached to the housing, and it is simply plugged into a grounded outlet. The device to be used is plugged into a receptacle in the housing of the GFCI.

Another GFCI, more suitable for outdoor use, has a heavy-duty housing attached to a short extension cord. The extension cord type of GFCI is easily

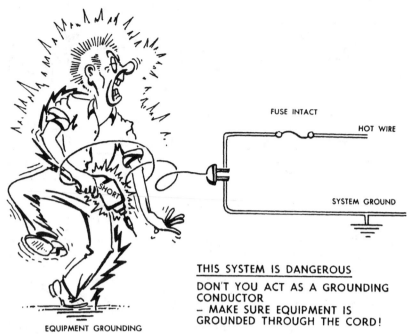

FUSE INTACT

HOT WIRE

SYSTEM GROUND

THIS SYSTEM IS DANGEROUS

DON'T YOU ACT AS A GROUNDING CONDUCTOR
— MAKE SURE EQUIPMENT IS GROUNDED THROUGH THE CORD!

EQUIPMENT GROUNDING

White-jacketed system grounds cannot conduct electricity from short circuits to the ground. Thus, they do not prevent the housing of faulty equipment from becoming charged. Therefore, a person who contacts the charged housing becomes the conductor in a short circuit to the ground. (National Safety Council)

(A)

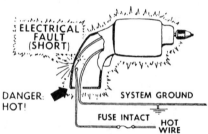

ELECTRICAL FAULT (SHORT)

DANGER: HOT!

SYSTEM GROUND

FUSE INTACT

HOT WIRE

Fig. 11-5 (A) Dangerous system. You become part of the circuit to ground and can be fatally injured. (B) Safe system uses the green wire to protect the power tool operator from shock.

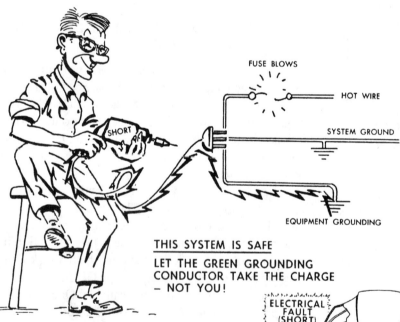

FUSE BLOWS

HOT WIRE

SYSTEM GROUND

EQUIPMENT GROUNDING

THIS SYSTEM IS SAFE

LET THE GREEN GROUNDING CONDUCTOR TAKE THE CHARGE
— NOT YOU!

Equipment stays at ground potential in spite of the short circuit, if the circuit has a grounding conductor. Internal electrical faults cause current to short-circuit harmlessly to equipment ground in systems with green-jacketed grounding conductors. (National Safety Council)

(B)

ELECTRICAL FAULT (SHORT)

SAFE

SYSTEM GROUND

FUSE BLOWS

HOT WIRE

EQUIPMENT GROUNDING

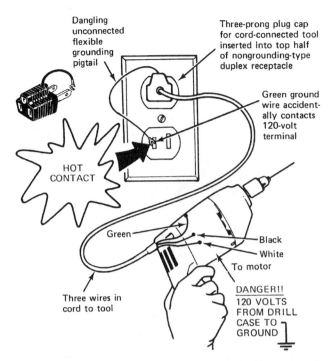

Fig. 11-6 *A pigtail adapter can cause problems if not properly attached to the screw.*

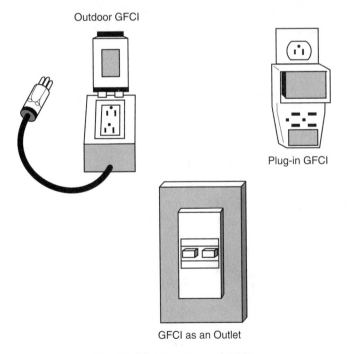

Fig. 11-7A *Three types of GFCIs.*

plugged into an outdoor outlet without its getting tangled with the outlet's lid or cover.

Keep in mind that GFCIs are not foolproof. They do, however, switch off the current in less than 0.025 second if a leakage is detected. They can detect as little as 2 milliamperes, although most operate at a threshold of 5 milliamperes—well below the level that would affect a person.

The GFCI should be tested about once a month to make sure it is working. Plug in something that consumes power and then push the test button. The red button should pop out and the power should be off. Then push in the red button to reset the device. This should turn the power on again. Some GFCIs that are built so that they fit outlet boxes have two buttons—black and red. These buttons are usually located side by side as in Fig. 11-7B, or they can be one above the other and a little longer than those in Fig. 11-7C.

Most power is distributed locally within a neighborhood by overhead wires. See Fig. 11-8. The wires are usually located in the rear of the house. In some communities, the wires are buried and come into the house near the basement or foundation wall. Power is brought from the pole or transformer into the rear of the house by one black, one red, and one white (uninsulated) wire. See Fig. 11-9. Once the cable is connected to the house, it is brought down to the meter by way of a sheathed cable with three wires. These are red, white, and black. The white wire is usually uninsulated and twisted as shown in Fig. 11-10. Figure 11-11 shows how the power is fed from the transformer to the distribution box.

Fig. 11-7B *This 15-amp GFCI is used to protect 120-volt receptacles in hotels and motels and homes where each bathroom is required to have GFCI protection. This is a Class A GFCI that trips when it senses 4 to 6 mA of ground fault leakage current. It is made in five colors to color-coordinate with bathroom interiors. Kitchens and outdoor outlets on homes must also be GFCI protected according to the National Electrical Code. (It takes over 7 mA for a human to feel the jolt of electricity.)*

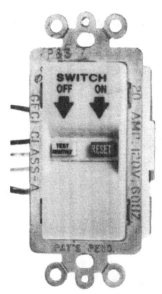

Fig. 11-7C *This GFCI provides protection for swimming pools, hot tubs, and industrial controls. It is designed for installation within 10 feet of a swimining pool. The reset button is red and will pop out when the circuit has been tripped and power turned off. Swimming pool wiring installed before the1965 Code may have wiring that can cause a GFCI to trip continuously.*

Service Entrance

All services should be three-wire. The capacity of service entrance conductors and the rating of service equipment should not be less than that shown in Table 11-4. This is a quick way to determine the type of service required. There are more accurate ways, which are explained later.

These capacities are sufficient to provide power for lighting, portable appliances, equipment for which

Fig. 11-9 *A transformer on a pole located at the rear of the house.*

individual appliance circuits are required, electric space heating of individual rooms, and air conditioning. A larger service might be required for larger houses. It also might be required if a central furnace or central hydronic-boiler is used for electric space heating.

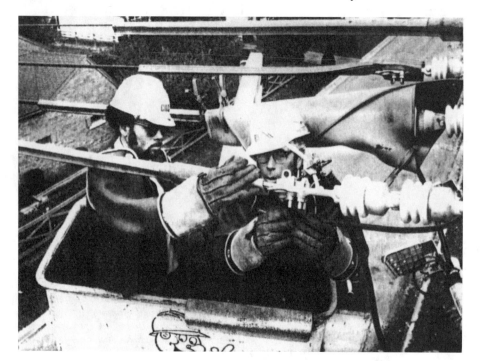

Fig. 11-8 *Stringing overhead lines for local service in a residential neighborhood.*

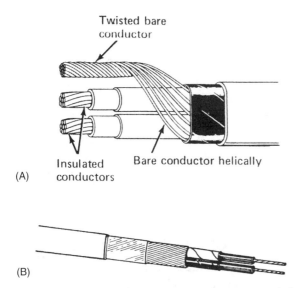

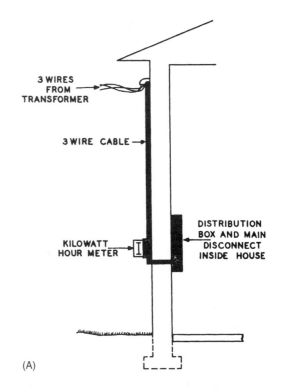

Twisted bare conductor

Insulated conductors

Bare conductor helically

(A)

(B)

Fig. 11-10 (A) Note how the twisted bare conductor is made from the outside coating of the other two wires. (B) Type SE cable. This cable has three wires: one red, one black, and an uninsulated wire that forms a protection for the other two. The stranded uninsulated wire is twisted at the end to make a connection, as in Fig. 11-9A.

3 WIRES FROM TRANSFORMER

3 WIRE CABLE

DISTRIBUTION BOX AND MAIN DISCONNECT INSIDE HOUSE

KILOWATT HOUR METER

(A)

Many major appliances in the kitchen require individual equipment circuits. Thus, where practical, electric service equipment should be located near or on a kitchen wall. This will minimize installation and wire costs. In addition, such a location is often convenient to the laundry. This minimizes circuit runs to laundry appliances.

Planning

The first step in wiring a house is planning. This means you should make accurate plans and know exactly how much material you need. Once you have learned about the boxes, wire, and tools, you can take a close look at the codes.

The local power company can advise you regarding the kind of service to use for your house. In most instances, the local power company will take the power right up to the house. The rest of the wiring is your responsibility.

The electrician works on a house at two of the construction phases. The electrician is needed to put in the wiring while the house is still in the roughed-in stage. This means a new house will have the wiring installed before the drywall goes onto the studs. Once the drywall workers and carpenters have finished their work, the electrician needs to return. At this time, the electrician installs the receptacles, receptacle face-plates, circuit breakers, chandeliers, and lights. Switches are connected and circuits tested for proper operation. In some cases the electrician has to install the furnace, the electric heating, or the electric range. Various states of construction make it obvious

(B)

Fig. 11-11 (A) Power from the transformer is fed through the wires and down the side of the house to the meter. Then it goes through the wall at a lower level to a distribution box in the basement. (B) Underground service feeds the meter on a slab-type house. Distribution box for the house is to the left of the meter housing.

Table 11-4A *Minimum Service Capacities for Various House Sizes*

Floor area		Minimum service capacity
m²	sq ft	
Up to 93	Up to 1000	125 amperes
93–186	1001–2000	150 amperes
186–279	2001–3000	200 amperes

Table 11-4B *Conductor Capacities for Household Equipment*

Item	Conductor capacity
Range (Up to 21-kW rating) *or*	50 A-3 W-115/230 V
⌈Built-In Oven	30 A-3 W-115/230 V
⌊Built-In Surface Units	30 A-3 W-115/230 V
Combination Washer-Dryer *or*	40 A-3 W-115/230 V
Electric Clothes Dryer	30 A-3 W-115/230 V
Fossil-Fuel-Fired Heating Equipment (if installed)	15 A or 20 A-2 W-115 V
Dishwasher and Waste Disposer (if necessary plumbing is installed)	20 A-3 W-115/230 V
Water Heater (if installed)	Consult Local Utility

that certain wiring should be done in an area before it is closed up. This is where planning is very important. It is much easier to wire a house while it is still unplastered than when it is walled in and ready for occupancy.

Permits

Before starting installation, check with your local power company to make sure what permits are necessary. Most permits specify that you will have someone inspect the installation after you have finished wiring it. This inspection will determine if the installation is safe and free from possible fire hazards.

Local Regulations

In some communities, local regulations supersede the National Electrical Code. Make sure you know what these local regulations are before starting the installation. Make sure the materials you use are approved by the local power company. If not, you might have trouble getting service once the job is complete.

Some regulations apply as a minimum. Those provided as guides by the National Electrical Code relative to the stringing of feeder wires are similar to those shown in Fig. 11-12. Here, a roof with a rise greater than 4 inches in 1 foot would be difficult to walk on easily. Therefore, one of the exceptions to the code

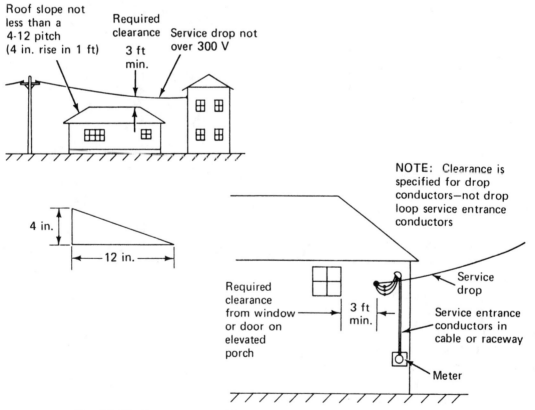

Fig. 11-12 *Clearance required for a 4/12 pitch roof in the path of a power line.*

will allow a clearance of 3 feet between the roof and an overhead power line if the line does not have over 300 volts. Figure 11-12 shows such an installation.

The electrician's main concern is the actual service drop and its entry into the house. The location of the entrance is important. The code does have something to say about where the service can be located. See Fig. 11-13. Note that the service is required to be at least 3 feet from a window.

The possibility of rot caused by water can be minimized by placing the service head as shown in Fig. 11-14.

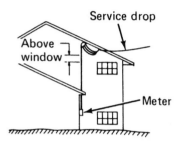

Fig. 11-13 *Minimum clearances for windows and doors for the installation of a service drop.*

The installation of this service head is usually the responsibility of the residential electrician.

SERVICE FROM HEAD TO BOX

The service wires are not connected to the power company's lines until the inside of the house is properly wired and inspected. Figure 11-15 shows how the service is connected and the meter socket is inserted in the line. The overhead connections can be made if the meter is left out of the socket. This causes the rest of the service to be inoperative until the meter is properly inserted in the socket by the power company.

Notice how the neutral wire is also connected to the cold water pipe inside the house. See Figs. 11-16 and 11-17. In some cases the house does not have a cold water system furnished from a local community water source. If a well is used for water, the neutral must be grounded, as shown in Fig. 11-18.

A conduit might be used for the installation of the service into the house. Here, three wires are used. One is black, one is white, and the other is red. The size of

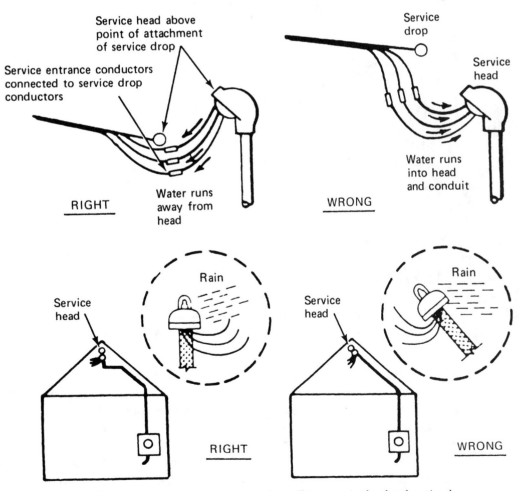

Fig. 11-14 *Proper and improper ways of installing a service head and service drop.*

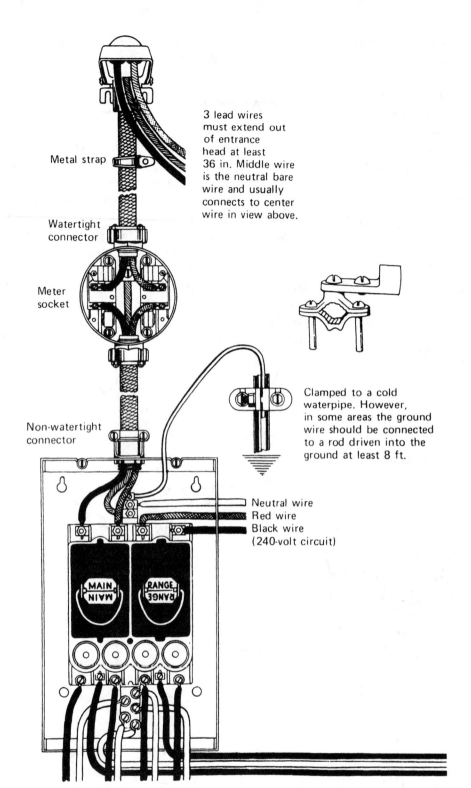

Metal strap

Watertight connector

Meter socket

Non-watertight connector

3 lead wires must extend out of entrance head at least 36 in. Middle wire is the neutral bare wire and usually connects to center wire in view above.

Clamped to a cold waterpipe. However, in some areas the ground wire should be connected to a rod driven into the ground at least 8 ft.

Neutral wire
Red wire
Black wire
(240-volt circuit)

MAIN

RANGE

Fig. 11-15 *Connected service using a fuse box and external ground.*

the wires will vary according to the amount of current needed.

Conduit is limited as to the size of wire it can handle. For instance, a 0.75-inch conduit can take three No. 8 wires. The 1.25-inch conduit can take three No. 2 wires, three No. 3 wires, three No. 4 wires, or three No. 6 wires. A 1.5-inch conduit can take three No. 1 wires. The 2-inch conduit will handle three No. 1/0 wires, three No.

2/0 wires, or three No. 3/0 wires. Therefore you can see the importance of knowing the requirements first. If you know the amount of current needed, you can determine the size of the wire. Then you can figure out which size conduit to use to hold the wires safely.

Wires from the utility company's pole to your building are called the *service drop*. These are usually furnished by the company. These wires must be high

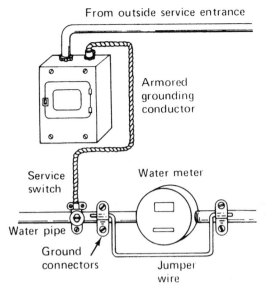

Fig. 11-16 *The usual method of grounding city and town systems.*

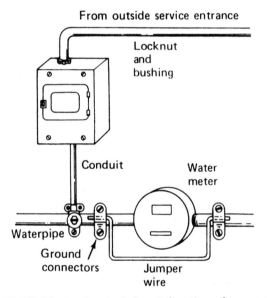

Fig. 11-17 *The usual method of grounding city and town systems using conduit.*

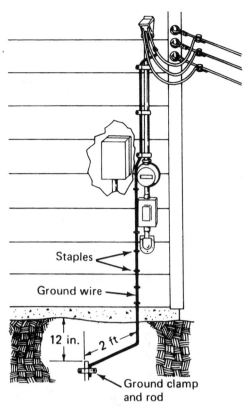

Fig. 11-18 *The approved REA method of grounding wire system with ground rod.*

enough to provide proper clearance above grade. They also must not come within 3 feet of doors, windows, fire escapes, or any opening. See Fig. 11-12. The structure to which the service drop wires are fastened must be sturdy to withstand the pull of ice and wind.

Installation of the Service

Figure 11-19 shows how the conduit for the service is connected to the head and to the meter socket. Note how it enters the house with a waterproof fitting and then goes unspliced into the distribution panel in the basement.

Circuits are run from the distribution panel to various parts of the house. Some examples of circuits are shown in Fig. 11-19. Notice that here there are 22 circuits available. Figure 11-15 shows how the fuse-type box is serviced by the entrance wires.

Conduit is fastened to the wall of the building by clamps placed every 4 feet. The conduit should use an entrance ell to turn the conduit into the house. Such an ell has two threaded openings corresponding to conduit size. Use an adapter to fasten the conduit into the threaded opening at the top of the ell. Into the lower opening, fasten a piece of conduit to run through the side of the house. Use a connector to attach the conduit to the electric service panel.

Inserting Wire Into a Conduit

After the conduit is installed, you can push the wires through the top hub of the meter, through the conduit, and out of the service head. All three wires must extend at least 36 inches out of the service head. This will allow enough wire for connecting to the power lines. Then the wires are brought down from the meter to the entrance ell. Remove the ell cover and pull the wires through to the service panel inside the house. Use a white-covered insulated wire for the neutral in the conduit.

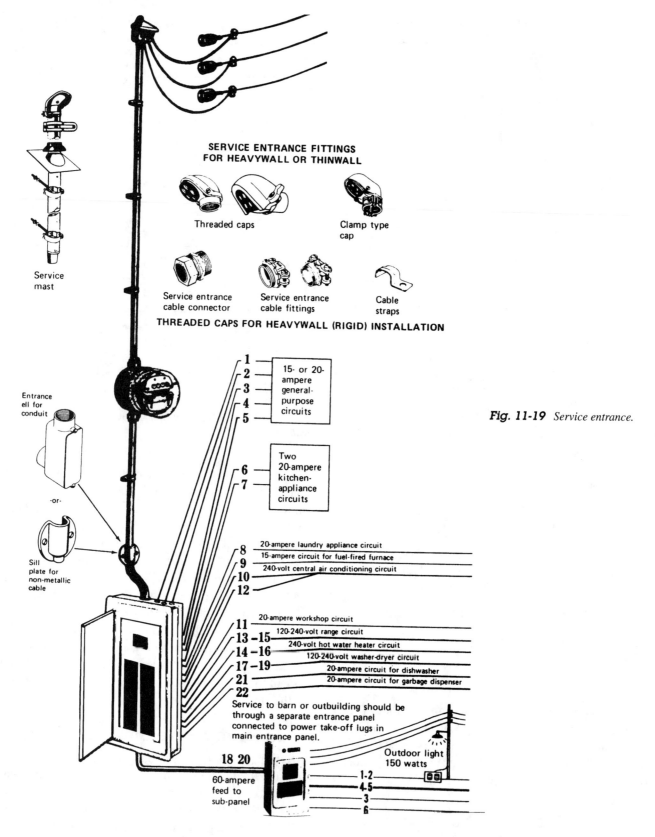

SERVICE ENTRANCE FITTINGS
FOR HEAVYWALL OR THINWALL

Threaded caps

Clamp type
cap

Service entrance
cable connector

Service entrance
cable fittings

Cable
straps

THREADED CAPS FOR HEAVYWALL (RIGID) INSTALLATION

Service
mast

Entrance
ell for
conduit

-or-

Sill
plate for
non-metallic
cable

1
2
3
4
5

15- or 20-
ampere
general-
purpose
circuits

6
7

Two
20-ampere
kitchen-
appliance
circuits

8   20-ampere laundry appliance circuit
9   15-ampere circuit for fuel-fired furnace
10   240-volt central air conditioning circuit
12

11   20-ampere workshop circuit
13 –15   120-240-volt range circuit
        240-volt hot water heater circuit
14 –16   120-240-volt washer-dryer circuit
17 –19   20-ampere circuit for dishwasher
21   20-ampere circuit for garbage dispenser
22

Service to barn or outbuilding should be
through a separate entrance panel
connected to power take-off lugs in
main entrance panel.

18 20

60-ampere
feed to
sub-panel

Outdoor light
150 watts

1-2
4-5
3
6

Fig. 11-19 *Service entrance.*

Distribution Panels

There are many types of distribution panels made for the home. One type has fuses that screw in. These fuses must be replaced when they are blown or open. See Fig. 11-20. Another type is the circuit breaker box. A circuit breaker can be reset if the device is tripped by an overload. This type is gaining in popularity for home use. Many people do not want to be bothered with looking for a new fuse every time one blows. "New builds" do not use fuses. The circuit

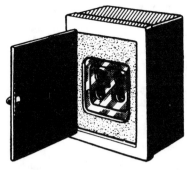

Fig. 11-20 A distribution box. This is a four-circuit box with screw-in fuses. It is used for an add-on in some older systems.

breaker can be reset by pushing it to the "off" position and then to the "on" setting. If it trips a second time, the circuit trouble is still present. It should be located and removed before resetting the breaker again. See Fig. 11-21.

For a closer look at the fuse replacement or circuit breaker, take a look at Fig. 11-22. This is a 100 ampere fuseless service entrance panel. (This size (100-amp) is obsolete today.) Most houses demand a larger current rating serving the home with so many electrical devices consuming power.

Fig. 11-21 Distribution box with circuit breakers.

Romex Cable

Romex cable is used to carry power from a distribution panel box to the individual outlets within the house. This nonmetallic sheathed cable has plastic insulation covering the wires to insulate it from the environment. Some types of cable can be buried underground. Figures 11-23 and 11-24 show the types most commonly used in house wiring.

Most new homes are wired with 12/2 (No. 12 wire with two conductors). One of the wires is white and the other black. The uninsulated conductor in the cable is usually the same size as the insulated black and white wires. In the past, the smallest wire size used in homes was No. 14, with two conductors (written on the cable as 14/2 with ground or 14/2WG). Single conductors can be used in conduits as stranded or solid conductors. See Fig. 11-25. Figure 11-26 shows how the wire is cut to remove the insulation without scoring the metal part of the wire. Scoring the copper makes it easy to break when bent around a screw for fastening to a switch or receptacle.

Wire Size

The larger the physical size of the wire, the smaller the number. For example, No. 14 is smaller than No. 12. Number 14 has been used for years to handle the 15-ampere home circuit. Today, the code calls for No. 12 for a 15-ampere circuit with aluminum or copper-clad aluminum wire. This is a safety factor. Aluminum wire cannot handle as much current as copper wire.

Figure 11-27 shows how size and numbers relate to determining wire diameter. Table 11-4 lists the capacities of wires needed for various household equipment. Table 11-5 shows conductor capacities for other household equipment. You should consider this information whenever you are planning an installation. Check the National Electrical Code and local codes for the latest changes and/or requirement.

PLANNING THE RIGHT SIZE SERVICE AND CIRCUITS

Most houses have either a 150-ampere or a 200-ampere service from the pole in the back of the house to the circuit breaker box in the basement. Number 1/0 or No. 3/0 (type RHW insulation) three-wire is usually used. This usually provides sufficient power for household needs. One way to find out how much power is needed is to list the devices that use power in a house, as well as how much power they use. Add up the requirements. Divide the wattage by 120 to find out how many amperes would be needed at any time.

EXAMPLE OF 100-AMPERE FUSELESS SERVICE ENTRANCE PANEL

100-ampere main breaker
(shuts off all power)

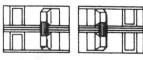

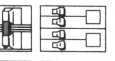

40-ampere circuit (120- to 240-
volt) for electric range.

30-ampere circuit (240-volt) for
dryer, hot-water heater, central
air conditioning, etc.

Four 15-ampere circuits for
general-purpose lighting, television, vacuum cleaner.

Four 20-ampere circuits for
kitchen and small appliances
and power tools.

Space for four 120-volt circuits
to be added for future loads
as needed.

Fig. 11-22 An example of a fuseless service entrance panel capable of handling a 100-ampere service.

Fig. 11-25 Single wire, stranded, with plastic coating. This can be used in conduits. It also comes in a solid copper wire.

Wrong way   **Right way**

Fig. 11-26 The right and wrong way to cut insulation from a piece of wire.

0   1   2   4   6   8   10 12 14

Fig. 11-27 Sizes of wires. Note the relationship between the size and diameter.

Fig. 11-23 Romex cable, 12/2, for use in home branch circuits.

In a house with an electric range, water heater, high-speed dryer or central air conditioning, together with lighting and the usual small appliances, there is at least a 150-ampere requirement. If the house also has electric heating, a 200-ampere line should be installed.

150-Ampere Service

A 150-ampere service provides sufficient electrical power for lighting and portable appliances. This would include, for example, a roaster, rotisserie, refrigerator, and clothes iron. If the dryer does not draw more than

Fig. 11-24 Romex that can be used for underground wiring.

SUNLIGHT RESISTANT — TYPE UF UNDERGROUND

Table 11-5 *Conductor Capacities for Other Household Equipment*

Item	Conductor capacity
Room Air Conditioners *or*	20 A-2 W-230 V
Central Air-Conditioning Unit *or*	40 A or 50 A-2 W-230 V
Attic Fan	20 A-2 W-115 V (Switched)
Food Freezer	20 A-2 W-115 or 230 V
Water Pump (where used)	20 A-2 W-115 or 230 V
Bathroom Heater	20 A-2 W-115 or 230 V
Workshop or Bench	20 A-3 W-115/230 V

8700 watts, the range no more than 12 kilowatts, and the air conditioner no more than 5000 watts, then the 150-ampere service is sufficient. Do not add any more than 5500 watts of appliances, however. Table 11-6 shows how much power is required to operate small devices.

Table 11-6 *Appliances and Their Wattages*

Device	Wattage	Voltage	Current
Small Appliances			
Blender	950	120	7.92
Clothes iron	1100	120	9.17
Electric fryer (small)	1200	120	10.00
Food processor	400	120	3.34
French fryer	900	120	7.50
Hair dryer	1200	120	10.00
Hand drill (⅛ HP)	324	120	2.70
Microwave oven	1680	120	14.00
Mixer	165	120	1.38
Toaster	1200	120	10.00
Vacuum cleaner (1 HP)	746	120	6.22
Waffle grill	1400	120	11.67
Large Appliances			
Clothes dryer	5000	240	20.83
Dishwasher	1200	120	10.00
Laundry circuit	3000	240	12.50
Range	12,000	240	50.00
Space heating	9000	240	37.50
Water heater	2500	240	10.40

Branch Circuits

General-purpose circuits should supply all lighting and all convenience outlets, except those in the kitchen, dining room (or dining areas of other rooms), and laundry rooms. General-purpose circuits should be provided on the basis of one 20-ampere circuit for not more than every 500 square feet or one 15-ampere circuit for not more than every 375 square feet of floor area. Outlets supplied by these circuits should be divided equally among the circuits. See Figs. 11-28 to 11-36.

These requirements for general-purpose branch circuits take into consideration the provision in the

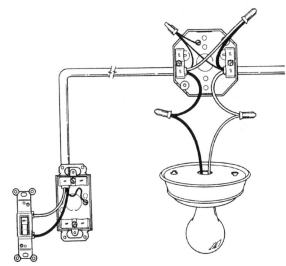

Fig. 11-28 *To add a wall switch for control of a ceiling light at the end of the run.*

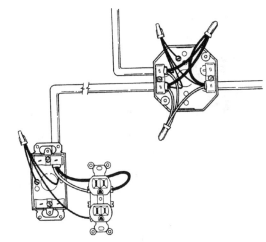

Fig. 11-29 *To add a new convenience outlet from an existing junction box.*

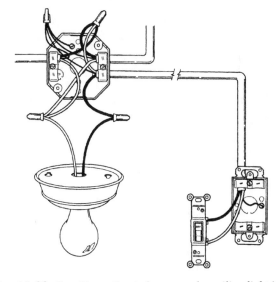

Fig. 11-30 *To add a wall switch to control a ceiling light in the middle of a run.*

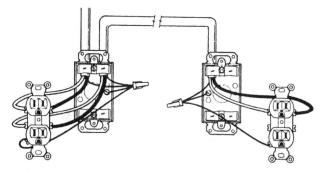

Fig. 11-31   *To add new convenience outlets beyond old outlets.*

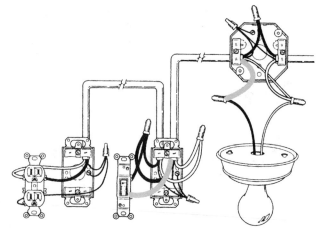

Fig. 11-34   *To add a switch and convenience outlet beyond the existing ceiling light.*

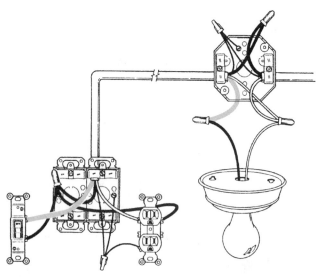

Fig. 11-32   *To add a switch and convenience outlet in one box, beyond an existing ceiling light.*

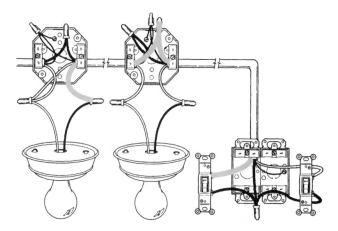

Fig. 11-35   *To install one ceiling outlet and two new switch outlets from an existing ceiling outlet.*

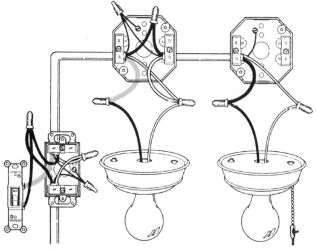

Fig. 11-33   *To install two ceiling lights on the same line; one controlled by a switch and the other with a pull chain.*

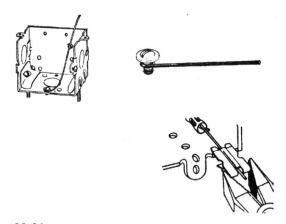

Fig. 11-36   *Two methods of attaching ground wire to a metal box.*

current edition of the National Electrical Code. Floor area designations are in keeping with present-day usage of such circuits. See Figs. 11-37 through 11-45. For a closer look at the internal wiring of a

house, before it is covered with drywall or plaster, see Figs. 11-46 through 11-52. Wiring of switches and receptacles, and locations within the circuits, are shown in Figs. 11-53 through 11-63. It is recommended that separate branch circuits be provided for

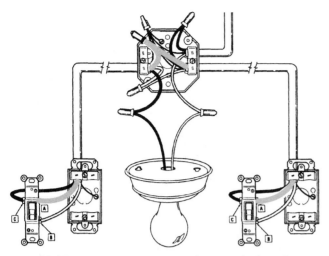

Fig. 11-37 *Three-way switches wired to control a lamp from two locations.*

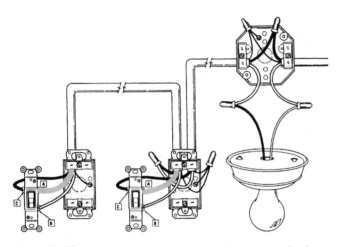

Fig. 11-38 *Another way to connect switches to control a lamp from two locations.*

both lighting and convenience outlets in living rooms and bedrooms. It is also recommended that the branch circuits servicing convenience outlets in these rooms be of the three-wire type, equipment with split-wired receptacles.

ELECTRIC SPACE HEATING

The capacity required for electric heating should be determined from Table 11-7. The table shows maximum winter heat loss, based on the total square feet of living space in the home. If electric space heating is installed initially, wiring should be as follows:

- For a central furnace, boiler, or heat pump: A three-wire, 120/240-volt feeder, sized to the installation.
- For individual room units: Ceiling cable, or panels, sufficient 15-, 20-, or 30-ampere two-wire 230-volt circuits to supply the heating units in groups, or individually.

AIR CONDITIONING

The capacity required for electric air conditioning is determined from the chart of maximum allowable summer heat gain, based on the total square feet of living space in the home. See Table 11-8. If electric air conditioning is installed initially, wiring should be provided for the following:

- A heat pump, providing both winter heat and summer air conditioning: A three-wire 120–240-volt feeder, sized to the installation.
- Individual room air-conditioning units: Sufficient 15-, 20-, or 30-ampere, 240-volt circuits to supply all units and a 20-ampere, 240-volt, three-wire outlet in each room, on an outside wall and convenient to a window.

In some cases, neither electric space-heating nor electric air-conditioning equipment is installed initially. In such an instance, the service entrance conductors and service equipment should have the capacity required, in accordance with the appropriate chart. Spare feeder or circuit equipment should be provided, or provision for these should be made in panel board bus space and capacity.

The plan should allow bus space and capacity for a feeder position that can serve a central electric heating or air-conditioning plant directly, or can supply a separate panel board to be installed later. The panel board would control circuits to individual room heating or air-conditioning units. Space heating and air conditioning are considered dissimilar and non-coincidental loads. Service and feeder capacity is provided only for the larger load, not for both.

SPACE HEATING AND AIR-CONDITIONING OUTLETS

Because many different systems and types of equipment are available for both space heating and air conditioning, it is impractical to show outlet requirements for those uses in each room. Electrical plans and specifications for the house should indicate the type of system to be supplied and the option of each outlet. Today, there are very few new homes built with individual window units for air conditioners. Most are central air-conditioning units which also serve as part of the heating plant.

ENTRANCE SIGNALS

Entrance push buttons should be installed at each commonly used entrance door and connected to the door

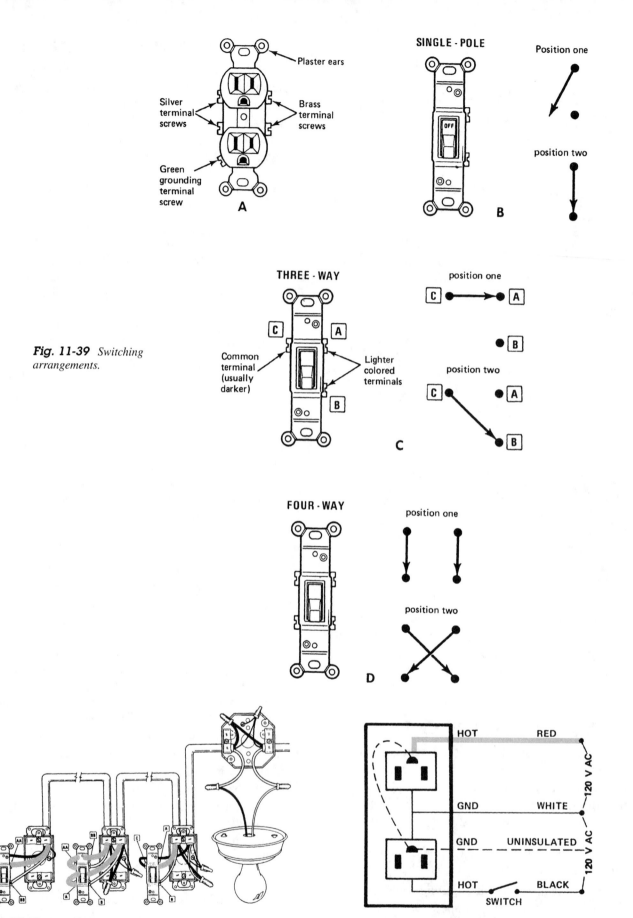

Fig. 11-39 Switching arrangements.

Fig. 11-40 Two three-way switches and a four-way for three-location control of a lamp.

Fig. 11-41 Three wires feeding a single receptacle. Switched lower outlet—top outlet hot all the time.

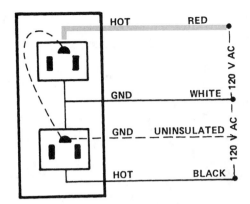

Fig. 11-42 *Three wires feeding a single receptacle.*

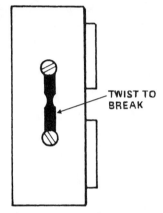

Fig. 11-43 *Breakaway connection between two outlets allows the red and black hot wires to be connected to the same duplex outlet.*

chime. They should give a distinctive signal for both front and rear entrances. Electrical supply for entrance signals should be obtained from an adequate bell-ringing or chime transformer.

A voice intercommunications system permits the resident and a caller to converse without opening the door. This is convenient and is an added protection. It

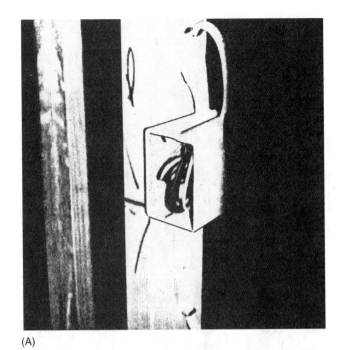

(A)

(B)

Fig. 11-46 *(A) The cable in this switch box is fed from the top. (B) Receptacle box is fed from the top.*

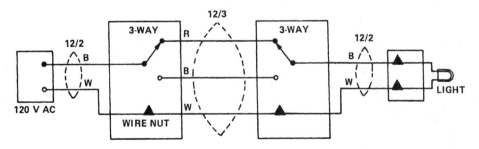

Fig. 11-44 *Schematic drawing of two three-way switches controlling a single lamp.*

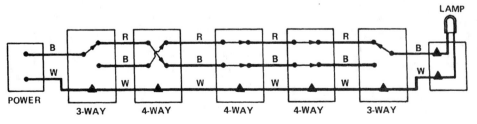

Fig. 11-45 *Schematic diagram of five locations to control a single lamp. Note there are never more than two three-way switches. Each time another location for control is added, a four-way switch is used.*

Fig. 11-47 *Location of an outlet where there is no insulation.*

Fig. 11-48 *Three-ganged switch boxes. Note how the wires are stapled to anchor them, and how each box has a cable entering and leaving. This will be a three-switch control center.*

Fig. 11-49 *Note that the Romex feeds the box. BX cable is the metallic armored one.*

Fig. 11-50 *Note that the studs are drilled and the wires are fed to the switch boxes.*

might be designed for this purpose alone, or it might be part of an overall intercommunications system for an entire house.

In a smaller home, the door chime is often installed in the kitchen, providing it will be heard throughout the home. If not, the chime should be installed at a more central location, usually in the entrance hallway. In a larger home, a second chime is often needed. This ensures that the signal is heard throughout the living quarters. Entrance signal conductors should be no smaller than the equivalent of a No. 18 AWG copper wire. (AWG is the abbreviation for American Wire Gage.)

(A)

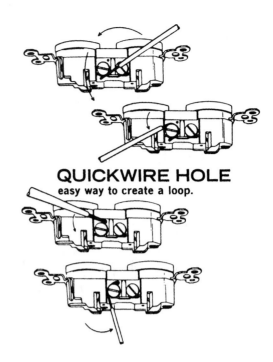

QUICKWIRE HOLE
easy way to create a loop.

(B)

Fig. 11-51 *Wires through the siding of a new house, to be attached to the outlet boxes later.*

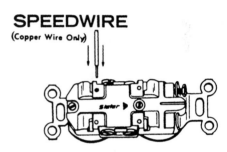

SPEEDWIRE
(Copper Wire Only)

Fig. 11-52 *Duplex receptacles being wired. A speedwire connection means that the wire is pushed into the hole to make contact.*

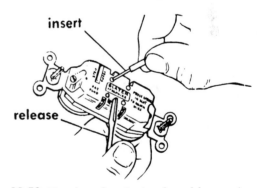

insert

release

Fig. 11-53 *Note how the wire is released by pressing with a screwdriver blade into the slot.*

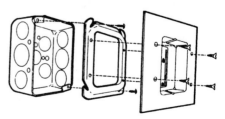

Fig. 11-54 *Recessed outlet mounted in a 4-inch square box.*

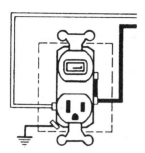

COMMON FEED ONLY

When plug is inserted into receptacle, pilot light automatically lights up.

Fig. 11-55 *Receptacle with pilot light.*

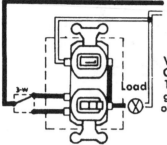

WIRED AS PILOT LIGHT

When 3-Way switch is ON, pilot light glows. This circuit requires a ground wire for this operation.

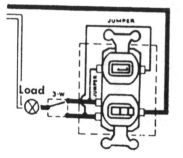

WIRED AS NIGHT LIGHT (Breakoff Tab Removed)

When 3-Way switch is OFF, indicator light glows.

Fig. 11-56 *Switch and pilot light.*

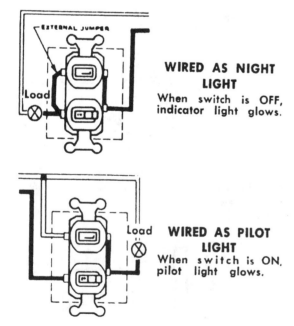

WIRED AS NIGHT LIGHT

When switch is OFF, indicator light glows.

WIRED AS PILOT LIGHT

When switch is ON, pilot light glows.

Fig. 11-57 *Switch and pilot light device wired with a three-way switch.*

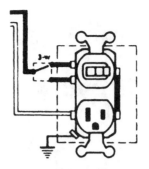

Grounded power outlet controlled from 3-Way switch and 3-Way switch at another location.

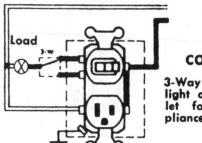

COMMON FEED

3-Way switch controls light only. Power outlet for grounded appliances.

Fig. 11-58 *Power outlet and switch can be used in a number of configurations for circuit control.*

The installation of door bells and chimes is an easy job. The wire used is No. 18. The connections are simple. Usually the wire comes in two- or three-conductor cable. This is taken from the transformer to the switch and the chime; as is shown in Fig. 11-64. The other side of the transformer goes to the 120-volt line. Usually the transformer is mounted on the side of the circuit breaker box. The primary wires are fed through a hole in the box. The transformer has a screw that makes it easily secured to the box. See Fig. 11-64.

CABLE TELEVISION

All new housing is made with cable television available in at least three locations. More than three locations requires an expensive broad-band amplifier and splitter. These cables are pulled into place using the same technique as for Romex wiring. They are terminated in a receptacle box just like the one used for the 120-volt wall outlet. The proper termination for the cable is usually placed on the end of the coaxial wire after the wiring job is finished and the plugs and switches are installed. The source of the entry cable varies with the areas. Some come in underground and others might be brought to the house from the street or pole through the air to be anchored near a termination determined by the local cable company. A special tool is used to attach the plug to the cable. The cable company usually has its installer do the job and check it out before energizing in house circuitry.

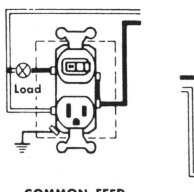

COMMON FEED

S. P. switch controls light only. Power outlet for grounded appliances.

Grounded power outlet controlled by S.P. switch.

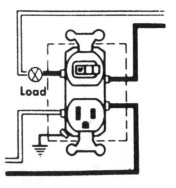

SEPARATED FEED (Breakoff Tab Removed)

S. P. switch and grounded power outlet on separate circuits.

Fig. 11-59 *Switch and outlet with a number of arrangements.*

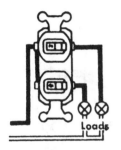

COMMON FEED

Two S.P. switches on same circuit. Each switch controls an independent light.

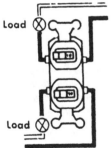

SEPARATED FEED (Breakoff Tab Removed)

Two S.P. switches on separate circuit. Each switch controls an independent light.

Fig. 11-60 *Two switches mounted in one unit with possibilities for controlling two loads in different ways.*

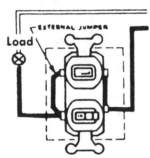

WIRED AS NIGHT LIGHT

When switch is OFF, indicator light glows.

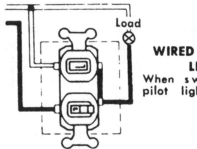

WIRED AS PILOT LIGHT

When switch is ON pilot light glows.

Fig. 11-61 *Switch and night-light or pilot light with the addition of an external jumper.*

INSTALLING ROMEX

Romex cable of the No. 14 or No. 12 size is usually employed in the wiring of homes, light industries, and businesses. To safely utilize the nonmetallic sheathed cable in a manner consistent with accepted wiring practices, it is necessary to use the switch and outlet box for terminations and connections. Splices are not allowed along the run of the cable unless the splice is housed in an appropriate box and totally enclosed with a cover plate.

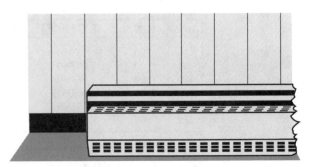

Fig. 11-62 *Baseboard electric heater.*

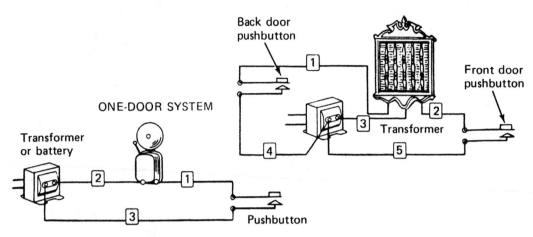

TWO-DOOR SYSTEM

Back door
pushbutton

Front door
pushbutton

ONE-DOOR SYSTEM

Transformer

Transformer
or battery

Pushbutton

Fig. 11-63 *Installation of door bells and chimes.*

Table 11-7 *Maximum Heat Loss Values**
(Based on infiltration rate of ¾ air change per hour.)

Degree days	Watts		Btuh	
	m²	sq ft	m²	sq ft
Over 8000	0.98	10.6	3.3	36
7001 to 8000	0.93	10.0	3.2	34
6001 to 7000	0.88	9.5	3.0	32
4501 to 6000	0.85	9.2	2.9	31
3001 to 4500	0.83	8.9	2.8	30
3000 and under	0.78	8.4	2.7	29

*For new homes. May be exceeded in converting existing homes.

When insulating material is used to make boxes and other equipment for wiring a house, it is not necessary to use a clamp or connector to hold the wire in place. Such a box is permitted when the wire or cable is supported within 8 inches of the box. See Fig. 11-65. Cable enters the box by way of a knock-out. All knock-outs should be closed if not used for cable.

Fig. 11-64 *A door bell or chime transformer. Note how the transformer is attached to the entrance panel with screws in the base of its cover. The curled wires attach to the 120-volt source and the screws between the wires are used to make connections to the stepped down 16 volts for operation of the chimes.*

Table 11-8 *Maximum Allowable Heat Gain (BTUH)**

Area to be conditioned		Design, dry-bulb temperature**			
m²	sq ft	32°C [90°F.]	35°C [95°F.]	38°C [100°F.]	40°C [105°F.]
70	750	15,750	18,000	19,500	21,000
93	1000	20,500	23,000	24,750	26,500
139	1500	27,000	30,000	31,500	33,000
186	2000	36,000	40,000	42,000	44,000
232	2500	45,000	50,000	52,500	55,000
279	3000	54,000	60,000	63,000	66,000

*Based on FHA minimum property standards.
**Based on dry-bulb temperature of less than 32° C (90° F.); use 32° C (90° F.) values. For designed dry-bulb temperature exceeding 40° C (105° F.); use 40°C (105°F.) values. (To find °C, simply subtract 32 from °F and multiply by 5/9.)

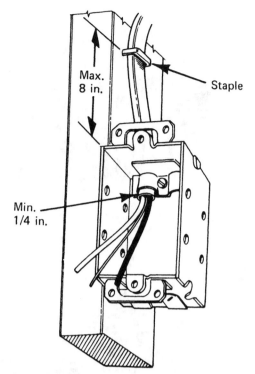

Fig. 11-65 *Installation of the nonmetallic (NM) wire known as Romex in a single-gang box.*

Box Volume

The volume of the box is the determining element in the number of conductors that can be allowed in the box. For instance, a No. 14 wire needs 2 cubic inches for each conductor. Therefore, if a two-wire Romex cable of No. 14 conductors is specified, it means that a ground wire (uninsulated) is also included, which will count as a conductor, and the three wires or conductors will require 6 cubic inches of space. A box used for this installation should have at least this amount. A 3 × 2 × 1.5-inch device box has only 9.0 cubic inches and is allowed to contain three of either No. 15, No. 12, or No. 10 conductors, but only two No. 8 conductors. Boxes manufactured recently have the cubic inch capacity stamped on them. You should check the National Electrical Code Handbook for the number of wires allowed until you are familiar with the standards for boxes. See Table 11-9.

Figures 11-66 through 11-79 show a representative sampling of devices made of insulating material that are acceptable in house wiring. Read the captions under each figure and identify the characteristics for future use in wiring buildings.

Duplex receptacles come in a variety of shapes and forms. For instance, Fig. 11-77B shows a metal cover plate rated at 10 amperes. This toggle switch comes with a gasket to go underneath, to seal the box from rain or snow. The 15-ampere covered receptacles shown

Table 11-9 *Volume Required for Each Wire Size**

Size of conductor	Free space within box for each conductor
No. 14	2.00 cubic inches
No. 12	2.25 cubic inches
No. 10	2.50 cubic inches
No. 8	3.00 cubic inches
No. 6	5.00 cubic inches

*Reprinted by permission from NFPA 70-1981, *National Electrical Code*, © 1980 National Fire Protection Association, Boston, Massachusetts.

Fig. 11-66 *An insulated metallic grounding bushing.*

Fig. 11-67 *A male insulating bushing.*

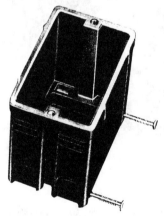

Fig. 11-68 *A nonmetallic switch box for a house.*

here come in both single and duplex types. One thing you should keep in mind is that the receptacles with a cap that stays in place when opened, such as those shown here, are for use in exterior locations but not in exposed areas. This type can be used on porches where they are sheltered from the weather. However, those

Fig. 11-69 *A nonmetallic switch box. It will hold nine conductors of No. 14 wire, eight conductors of No. 12 wire, or seven conductors of No. 10 wire.*

Fig. 11-71 *A nonmetallic switch box, designed with a bracket to mount to the wood or steel studs with an SST tool. The box will hold 15 No. 14 conductors, 13 No. 12, or 12 No. 10 conductors.*

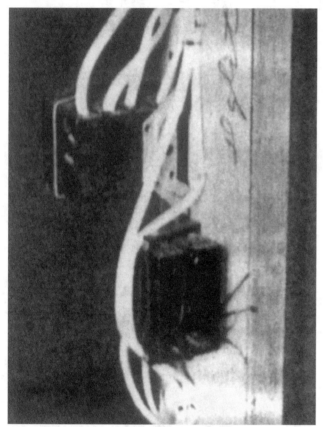

Fig. 11-70 *A nonmetallic switch box, with nails used to mount it. It will hold seven conductors of No. 14; six of No. 12 or No. 10.*

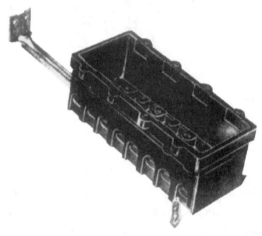

Fig. 11-72 *A nonmetallic switch box. It will hold four devices. The bracket will fit between two studs located on 16-inch centers. The box will hold 20 No. 14 conductors, 17 No. 12, or 15 No. 10 conductors. Remember, the ground wire counts as one conductor.*

Fig. 11-73 *A cover for an outlet box.*

designed for outside use in exposed-to-the-elements locations have a *spring-loaded cap* that closes automatically when you remove the plug from the socket. Be sure to use the correct type in each location.

Figure 11-80A shows how a connector is used to hold the Romex cable in place. Notice how the grounding clip is attached to the metal box. Also notice that part of the Romex insulation is extended through the connector into the box. Figure 11-80B shows how two or more ground wires are connected

(twisted) together and then grounded by a clip attached over the metal edge of the box.

Some Romex boxes are designed to be attached to 2×4 studs without nails. See Figs. 11-81 and 11-82. A switch box can be attached with nails as shown in Figs. 11-83A and B.

Fig. 11-74 *A nonmetallic cover for a toggle switch.*

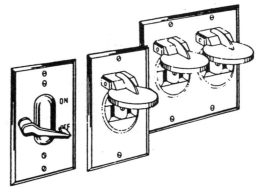

10-ampere toggle switch   15-ampere covered receptacles. Choice of single or duplex types.

Fig. 11-77B *Metal cover plates for exterior use.*

Fig. 11-75 *A nonmetallic surface box.*

A ground continuity tester for use on energized circuits only. It is equipped with a ground and a neutral blade. If the light comes on when it is plugged into a grounding-type receptacle, the ground continuity is complete. If the light does not come on, there is a fault in the circuit. It uses a penlite cell for power.

Fig. 11-78 *A ground continuity tester for use on energized circuits only.*

Fig. 11-76 *A nonmetallic conduit box, round, with four 0.5-inch threaded knockouts.*

A box finder and cable tracer used to locate boxes or cables that are hidden from view. To use, plug it into one of the receptacles on the covered box circuit wires. The transmitter will produce a signal. Tune in a small transistor radio to pick up the transmitted signal. Follow the cable by checking sound build-up due to more wire being folded back in the box. The box will be located at maximum signal strength. The unit is more efficient if the circuit ground wire is disconnected. It uses a 9-volt transistor battery.

A duplex receptacle cover (REC),with gasket and stainless steel screws. It may also be obtained with a single receptacle hole. Note the REC on the outside to identify the outlet.

Fig. 11-77A *A duplex receptacle cover.*

Fig. 11-79 *A box finder and cable tracer.*

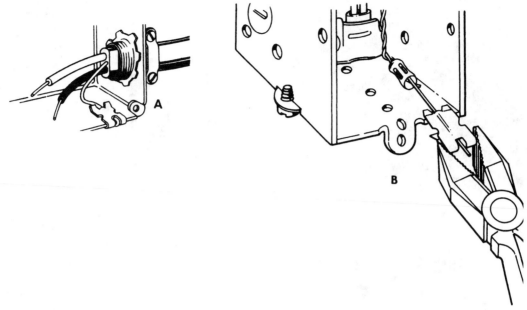

Fig. 11-80 *(A) Romex connector attached to a box. (B) Note how the ground clip is attached to the metal box.*

Fig. 11-81 *A metal switch box for Romex cable.*

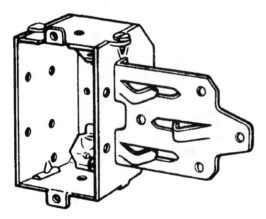

Fig. 11-82 *Beveled-corner switch box for Romex. Note mounting without nails.*

Lighting Fixtures

Lighting fixtures can be installed in several ways. There are hanger supports that thread onto a threaded stud, mounted in the box, as in Fig. 11-83. Straps and machine screws are used to mount the fixture in Fig. 11-84. If there is no stud, the metal strap can be used to hold the canopy of the light fixture in place, as in Fig. 11-85.

Glass-enclosed ceiling fixtures are easily attached to the ceiling box by using a threaded stud. See Fig. 11-86. Wall lights can be installed using the same method. Make sure the outlet in the wall fixture is wired for full-time service—not controlled by the switch that turns the light on and off. See Fig. 11-87.

Wires and Boxes

The box to be used is determined by its volume and the number of conductors you will be putting into the box. No matter how many ground wires that come into a box, a deduction of only one conductor must be made from the number of wires shown in Table 11-10. Table 11-10 shows the number of wires in a box. This assumes all the wires are the same size. If the wires are not the same size, you have to check the volume of the box and then take a look at Table 11-9. Here, the volume required for each wire is shown. Just add the number of wires of a particular size and multiply by the volume required by that conductor. Then do the same thing for the next size conductor. Once you have figured all the various sizes and their volume requirements, just add

(A)

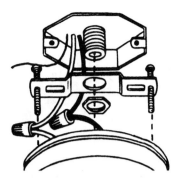

Outlet box has a stud in this case. Insert the machine screws in threaded holes of the metal strap shown. Slip the center hole of the strap over the stud in the outlet box. Hold the strap in position by a locknut. Connect wires with wire nuts and slip the canopy over the machine screws; fit flush and secure the fixture with two cap nuts. Don't forget to anchor the uninsulated ground wire to the box with a ground clip or to the box's threaded grounding screw.

Fig. 11-84 *Outlet box has a stud in this case for mounting of light fixture.*

(B)

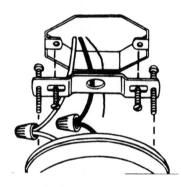

If there is no stud, insert the machine screws as shown here. Fasten the ears of the outlet box and the strap with screws. Then align the canopy onto the two screws pointing down and cap off with cap screws.

Fig. 11-85 *Use machine screws if there is no stud.*

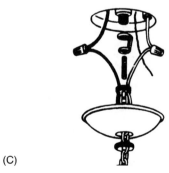

(C)

Mount large drop fixtures by simply using a screw hanger support onto the threaded stud in the outlet box. Use solderless connectors (wire nuts) to connect the electrical wires and a grounding clip for the extra uninsulated ground wire. Raise the canopy and anchor in position by means of a locknut.

Fig. 11-83 *(A) A switch box with nailing holes and hammer driving the nails into the stud. (B) A switch box with nails inserted for nailing. (C) Mounting of large drop fixture.*

Fig. 11-86 *Glass-enclosed fixtures installed by using a stud.*

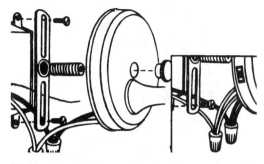

Wall brackets or lights are installed by strapping to the ears of the box, then using a nipple and cap to complete. Don't forget the ground wire and the ground clip. If there is an outlet which is "always on," wire according to the insert.

Fig. 11-87 *Wall brackets or lights are installed by strapping to the ears of the box.*

them. This will tell you whether or not you have too much for the volume of the box.

Any wire that runs unbroken through a box is counted as one wire. The ground wires can be in nonmetallic cable or ground wires that are run in metal or nonmetallic raceways. Wires that come into a box and are spliced are counted as one wire. It can be crimped or twisted in its connection. If a wire that comes into the box

is connected to a wiring device terminal, it is counted as one wire. Cable clamps, hickeys, and fixture studs count as one wire whether the box has one clamp, two clamps, or any combination of clamps, studs, or *hickeys*.

You can remove unused cable clamps from a box to provide more space in the box. If one clamp is left in the box, you must count the clamp as one connection. If both clamps are removed and you use a box connector, then the clamp is not deducted from the total allowed in the box.

If a jumper is used from the box screw to the receptacle grounding terminal, then the jumper is not counted as a conductor because it does not leave the box. If a switch or receptacle has a grounding strip on it, this must also be counted as a conductor in terms of the number of conductors used in a box.

One of the factors involved in mounting boxes and electrical equipment is the prevention of fire. Also, in case of fire, it must not be allowed to move from one level to another by way of a raceway or hole. The use of a hexagonal or square box in ceilings where there is sheetrock is not permitted. If they are used, they must have a "mud ring" installed to prevent fire from being allowed to contact wall or ceiling materials that may burn.

Table 11-10 *Boxes and Maximum Number of Wires.**

Box dimension, inches trade size or type	Min. cu. in. cap.	Maximum number of conductors			
		#14	#12	#10	#8
4 × 1¼ Round or Octagonal	12.5	6	5	5	4
4 × 1½ Round or Octagonal	15.5	7	6	6	5
4 × 2⅛ Round or Octagonal	21.5	10	9	8	7
4 × 1¼ Square	18.0	9	8	7	6
4 × 1½ Square	21.0	10	9	8	7
4 × 2⅛ Square	30.3	15	13	12	10
4¹¹⁄₁₆ × 1¼ Square	25.5	12	11	10	8
4¹¹⁄₁₆ × 1½ Square	29.5	14	13	11	9
4¹¹⁄₁₆ × 2⅛ Square	42.0	21	18	16	14
3 × 2 × 1½ Device	7.5	3	3	3	2
3 × 2 × 2 Device	10.0	5	4	4	3
3 × 2 × 2¼ Device	10.5	5	4	4	3
3 × 2 × 2½ Device	12.5	6	5	5	4
3 × 2 × 2¾ Device	14.0	7	6	5	4
3 × 2 × 3½ Device	18.0	9	8	7	6
4 × 2⅛ × 1½ Device	10.3	5	4	4	3
4 × 2⅛ × 1⅞ Device	13.0	6	5	5	4
4 × 2⅛ × 2⅛ Device	14.5	7	6	5	4
3¾ × 2 × 2½ Masonry Box/Gang	14.0	7	6	5	4
3¾ × 2 × 3½ Masonry Box/Gang	21.0	10	9	8	7
**FS-Minimum Internal Depth 1¾ Single Cover/Gang	13.5	6	6	5	4
**FD-Minimum Internal Depth 2⅜ Single Cover/Gang	18.0	9	8	7	6
**FS-Minimum Internal Depth 1¾ Multiple Cover/Gang	18.0	9	8	7	6
**FD-Minimum Internal Depth 2⅜ Multiple Cover/Gang	24.0	12	10	9	8

*Reprinted by permission from NFPA 70-1981, *National Electrical Code*, © 1980 National Fire Protection Association, Boston, Massachusetts.
**FS and FD are designations used for Unilets with conduit.

In single gang boxes, the nonmetallic cable does not have to be clamped to the box. It should be secured to a stud within 8 inches of the box, however. See Fig. 11-65. If the box is round or square two-gang or three-gang, the cable must be clamped to the box. See Fig. 11-88.

ELECTRIC RANGES

The largest appliance in the house is usually an electric range for cooking. It is wired with a permanent connection or with a plug and pigtails. The plug and receptacle must be capable of handling at least 50 amperes and provide connections for the three wires used in such circuits. In some instances a pigtail is used to connect the range to the power source. Figure 11-89 shows the pigtail used in ranges and in some instances for clothes dryers. However, the range and dryer pigtails are different because they are capable of supplying either 50 amperes for the range or 30 amperes for a dryer. The plugs, Fig. 11-90, are different also. Figure 11-91 shows how a pigtail is installed.

Connecting Ranges Permanently

The National Electrical Code specifies the size of wire that can be used for connecting electric ranges. Feeder capacity must be allowed for household cooking appliances according to Table 220-19 of the Code. This table applies if the appliance is rated over 1.75 kilowatts.

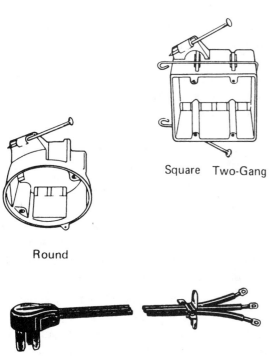

Round

Square Two-Gang

Fig. 11-88 *Boxes that must have a clamp or connector if used for wiring.*

Three-Gang

Fig. 11-89 *A pigtail and plug.*

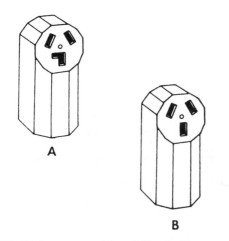

A

B

Fig. 11-90 *(A) L-type ground for a 250-volt, 30-ampere circuit. (B) 250-volt, 50-ampere range receptacle.*

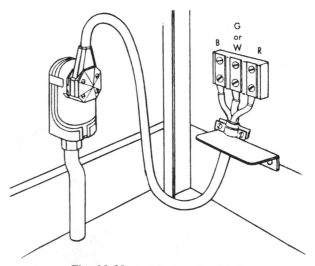

Fig. 11-91 *Installation of a pigtail.*

Section 210-19(b), Exception No. 1, refers to Table 220-19 for the sizing of a branch circuit to supply an electric range, a wall-mounted oven, or a counter-mounted cooking unit. Table 220-19 is also used for sizing the feeder wires that supply more than one electric range or cooking unit.

Figure 11-92 shows that the minimum size for a 12,000 watt range would be 35 amperes and the wire would have to be No. 8. The type of wire would be TW-40 A, THW-45A, XHHW, or THHN-50 A. Number 8 wire is the minimum size to be used for any range rated 8.75 kilowatts or more. The overload protection could be either a 40-ampere fuse or a 40-ampere circuit breaker.

Although the two hot legs of the wire must be No. 8, an exception to the Code says that the neutral conductor can be smaller. See Fig. 11-92. This is primarily because the heating elements of the range will be connected directly between the two hot wires and the maximum current will flow through these wires. The devices that operate on 120 volts will be the oven light and the lighting circuits over the top of the range. In some instances, the heating elements for the top portion will be wired to 120 volts at some heat levels.

The exception says that the smaller conductor can be not less than 70% of the capacity of the hot wires. In this case, the wire cannot be less than No. 10. The neutral size is figured as follows: 70% of 40 amperes is 28 amperes. Thus, 28 amperes calls for No. 10 wire. The neutral can be No. 10, but the other two wires must be No. 8.

If the range is over 8.75 kilowatts, then the minimum size of the neutral conductor must not be smaller than No. 10. Hot legs of the range must not be less than No. 8 wire and the neutral less than No. 10 wire.

The maximum demand for a range of 12 kilowatt rating according to the table in the code is 8 kilowatts. This 8,000-watt load can be converted to amperes by dividing by 230 volts (the midpoint between 220–240 volts). This, then, gives you 35 amperes. Look up the wire for handling 35 amperes and you find No. 8. Number 8 wire can handle up to 40 amperes.

On modern ranges, the heating elements of the surface units are controlled by five heat switches. The surface unit heating elements will not draw current from the neutral conductor unless the unit switch is in one of the low-heat settings.

Sizing a Range over 12,000 Watts

If you have a kitchen range rated at 16.6 kW, then the 230-volt supply must be capable of delivering at least 43 amperes. This is determined by the following:

1. Column A of the NEC Table for Demand Loads for Household Ranges, Wall-Mounted Ovens, Counter-Mounted Cooking Units, and Other Household Appliance Ovens 1.75 kW Rating states in Note 1 for ranges over 12 kW and up to 27 kW, the maximum demand in Column A must be increased by 5% for each additional Kw (or major fraction thereof) above 12 Kw.

2. This range is rated at 16.6 kW, which is 4.6 kW more than 12 kW. Column A says to use 5% of the 8,000 watts, or 400 watts.

3. The maximum demand for this 16.6 kW range must be increased above 8 kW by using the following: 400 watts × 5 (4 kW + 1 for the 0.6 kW more than 16). Therefore, 400 × 5 = 2,000. This is added to the 8,000 watts used as a base demand and produces 10,000 as the maximum demand load.

4. If you want the amount of current drawn at maximum demand, divide the 10,000 watts by 230 volts. The answer is 43.478 amperes. This is rounded to 43 because the fraction is less than 0.5. Look up the wire

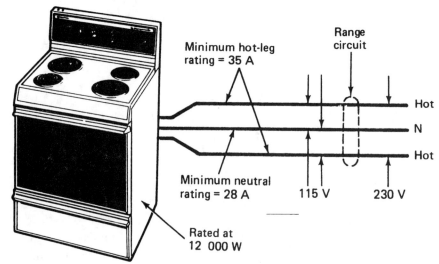

Fig. 11-92 *A 12-kW range with its minimum ampere rating.*

size for using a 60-degree C conductor and you will find the required UL wire for a branch circuit must have No. 6 TW conductors to handle the 43 amperes.

Tap Conductors

Section 210-19(b), Exception 1, of the NEC gives you permission to reduce the size of the neutral conductor for ranges or a three-wire range branch circuit to 70% of the current-carrying capacity of the ungrounded conductors. Keep in mind, however, that this does not apply to smaller taps connected to a 50-ampere circuit where the smaller taps must all be the same size. It does not apply when individual branch circuits supply each wall- or counter-mounted cooking unit, and all circuit conductors are of the same size and less than No. 10.

Exception No. 2 of the section previously mentioned allows tap conductors rated at not less than 20 amperes to be connected to a 50-ampere branch circuit that supplies ranges, wall-mounted ovens, and counter-mounted cooking units. These taps cannot be any longer than necessary for servicing the cooking top or oven. See Figs. 11-93 and 11-94.

Figure 11-95 shows two units treated as a single range load. Figure 11-96 shows how to determine branch circuit load for separate cooking appliances on a single circuit according to Section 210 of the Code.

Individual branch circuits can be used. There are some advantages to individual circuits. With this arrangement, smaller branch circuits supply each unit with no junction boxes required. Two additional fuses or circuit breaker poles are required in a panel board, however. Overall labor and material costs are less than those for the 50-ampere circuit shown in Fig. 11-93. There is a disadvantage, though. The smaller circuits will not handle larger units that you may wish to install later. See Fig. 11-97.

In some cases, a single 40-ampere circuit might supply the units. See Fig. 11-98. The NEC allows 40-ampere circuits in place of 50-ampere circuits where the total nameplate rating of the cook tops and ovens is less than 15.5 kilowatts. Because most ranges and combinations of cook tops and oven are less than 15.5 kilowatts, this can be a very popular arrangement.

CLOTHES DRYER

The electric clothes dryer uses 240 volts for the heating element. It also uses 120 volts for the motor and 120 volts for the light bulb. This means three wires are

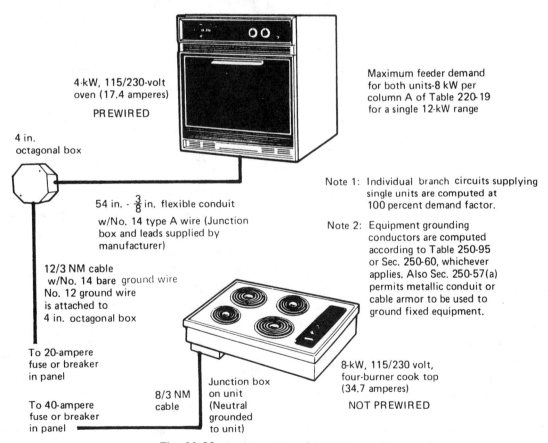

Fig. 11-93 *Cooking units with separate circuits.*

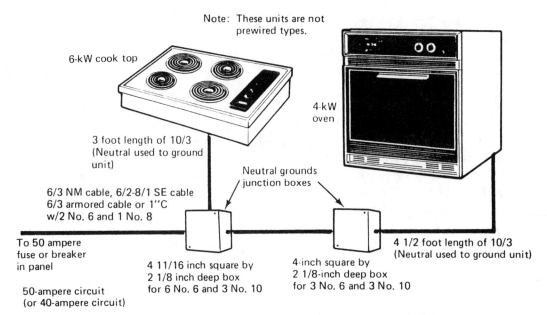

Note: These units are not prewired types.

6-kW cook top

4-kW oven

3 foot length of 10/3 (Neutral used to ground unit)

Neutral grounds junction boxes

6/3 NM cable, 6/2-8/1 SE cable 6/3 armored cable or 1''C w/2 No. 6 and 1 No. 8

To 50 ampere fuse or breaker in panel

50-ampere circuit (or 40-ampere circuit)

4 11/16 inch square by 2 1/8 inch deep box for 6 No. 6 and 3 No. 10

4-inch square by 2 1/8-inch deep box for 3 No. 6 and 3 No. 10

4 1/2 foot length of 10/3 (Neutral used to ground unit)

Fig. 11-94 *Using one branch circuit for cooking units.*

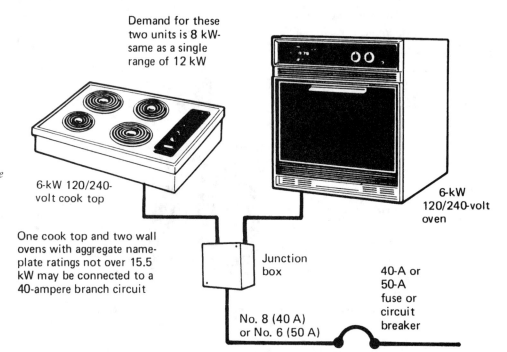

Demand for these two units is 8 kW-same as a single range of 12 kW

Fig. 11-95 *These two units are treated as one range load.*

6-kW 120/240-volt cook top

6-kW 120/240-volt oven

One cook top and two wall ovens with aggregate name-plate ratings not over 15.5 kW may be connected to a 40-ampere branch circuit

Junction box

40-A or 50-A fuse or circuit breaker

No. 8 (40 A) or No. 6 (50 A)

needed for proper operation. Figure 11-99 shows how the green and white wires are treated in the dryer. Because most home dryers draw about 4,200 watts, it is necessary to use a separate 30-ampere circuit with a pullout fuse for disconnecting it from the line. A circuit breaker in the main service panel can also be used to disconnect it from the line. The 30-ampere circuit uses No. 8 wire in most instances; however, it should be noted that some high-speed dryers use about 8,500 watts and need a 50-ampere circuit. That would call for a No. 6 wire. There is a difference in the configuration of the receptacle for a 50-ampere device and that

for a 40-ampere device. See Fig. 11-89. The surface-mounted receptacle with an L-shaped ground is used with dryers drawing up to 30 amperes. The surface-mounted receptacle capable of delivering 50 amperes can also be used, if needed. Note the difference in the slots in the receptacles. The pigtails that fit the two different plugs must be closely examined to make sure they will fit the L-shaped slot or the straight slots.

MICROWAVE OVENS

Microwave ovens are designed to operate on 120 volts. This means they can be plugged into the closest

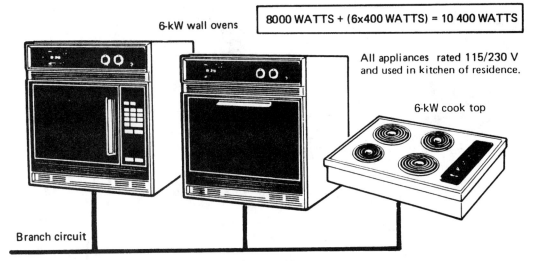

8000 WATTS + (6x400 WATTS) = 10 400 WATTS

6-kW wall ovens

All appliances rated 115/230 V and used in kitchen of residence.

6-kW cook top

Branch circuit

Note 4 of Table 220-19 says that the branch-circuit load for a counter-mounted cooking unit and not more than two wall-mounted ovens, all supplied from a single branch circuit and located in the same room, shall be computed by adding the nameplate ratings of the individual appliances and treating this total as a single range.

That means the three appliances shown may be considered to be a single range of 18-kW rating (6 kW + 6 kW + 6 kW).

From Note 1 of Table 220-19, such a range exceeds 12 kW by 6 kW and the 8-kW demand of Column A must be increased by 400 watts (5 percent of 8000 watts) for each of the 6 additional kilowatts above 12 kW.

Fig. 11-96 *Separate cooking units on a single circuit branch.*

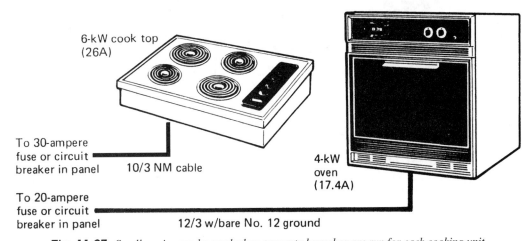

6-kW cook top
(26A)

To 30-ampere
fuse or circuit
breaker in panel   10/3 NM cable

4-kW
oven
(17.4A)

To 20-ampere
fuse or circuit
breaker in panel   12/3 w/bare No. 12 ground

Fig. 11-97 *Smaller wire can be used when separate branches are run for each cooking unit.*

convenient wall outlet. This might become an expensive mistake for many. The microwave oven usually pulls around 14 amperes. This means there is only 1 ampere of spare capacity left for the circuit. If there is another appliance on the circuit, a circuit breaker or fuse may go. Microwave ovens should have their own circuit. There should be nothing else on the line. They should have a direct connection to the distribution panel. This will also minimize the interference caused by the microwave's high-frequency energy radiation.

OVERHEAD GARAGE DOORS

Most overhead garage doors for residential use are furnished with a 0.33 horsepower, or 248.6-watt, electric motor. This means it draws about 2.07 amperes when running. It does, however, draw much more when starting, up to 35 amperes. This is what dims the lights when it starts. And, every time it turns off, it produces an inductive kickback up to about 300 or 400 volts. This spike on the line is what

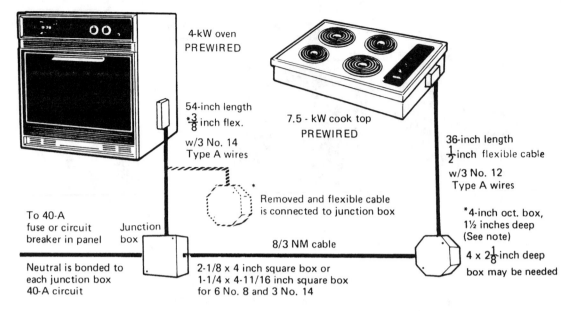

4-kW oven PREWIRED

54-inch length *$\frac{3}{8}$ inch flex.

7.5 - kW cook top PREWIRED

w/3 No. 14 Type A wires

36-inch length $\frac{1}{2}$ inch flexible cable

w/3 No. 12 Type A wires

Removed and flexible cable is connected to junction box

To 40-A fuse or circuit breaker in panel

Junction box

8/3 NM cable

*4-inch oct. box, 1½ inches deep (See note)

4 x 2$\frac{1}{8}$-inch deep box may be needed

Neutral is bonded to each junction box 40-A circuit

2-1/8 x 4 inch square box or 1-1/4 x 4-11/16 inch square box for 6 No. 8 and 3 No. 14

*Furnished with units

Note: Cubic inch capacity of 1½-by 4-inch octagonal box is 15.5; according to Table 370-6(b), 3 No. 8 and 3 No. 12 would require 15.75 cubic inches.

Fig. 11-98 *In some instances it is possible for a single 40-ampere circuit to supply both cooking units.*

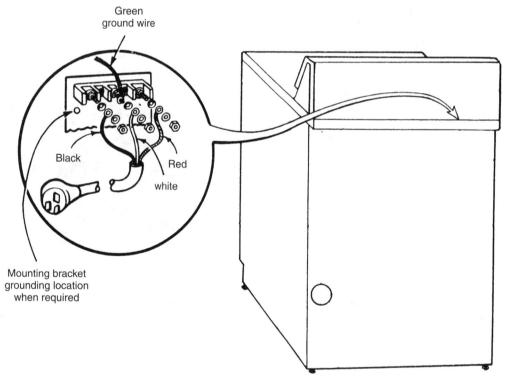

Green ground wire

Black

Red

white

Mounting bracket grounding location when required

Fig. 11-99 *Hookup for an electric dryer.*

reduces the life of the light bulbs in the door opener or anything else on the same circuit in the garage. See Fig. 11-100.

Garage Door Opener

The Homelink® Universal Transceiver replaces up to three remote controls (hand-held transmitters) that oper-

ate devices such as garage door openers, motorized gates, or home lighting. It triggers these devices at the push of a button, located on the overhead console. See Fig. 11-101. The universal transceiver operates off the car battery so no extra batteries are needed. For additional information on Homelink, call 1-800-355-3515, or check the internet at www.homelink.com.

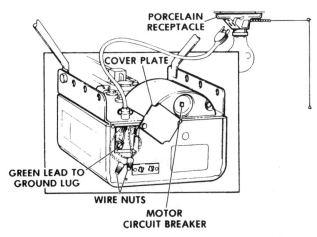

(A)

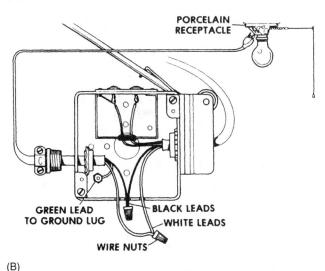

(B)

Fig. 11-100 *(A) Wiring of an overhead door opener. (B) Connection of a garage door opener to the power line.*

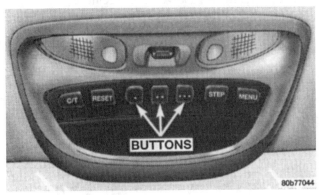

Fig. 11-101 *These three buttons on the overhead console of a Chrysler Concorde operate the garage door, the front gate, or the inside lighting system.* (Homelink®–Johnson Controls)

ELECTRIC WATER HEATERS

Electric water heaters draw current to heat the water. The larger the heater capacity, the higher the current rating of the heating element. They use 240 volts for heating the water.

A double-element water heater is probably better for larger families because it permits a more constant supply of hot water. Double-element heaters have two thermostats. The single-element type has only one thermostat. The sizes of the elements, the type of thermostats, and the method of wiring for heaters is usually specified by the local power company.

Figure 11-102 shows a typical installation of a hot water heater. Keep in mind that some electric companies offer a special rate for heating water. "Off-peak load" is the term that refers to this special rate. It simply means the power company furnishes electricity during its low load time. The time is already known by the electric company, so they can place a meter that measures the use of power for heating water. It also places a time switch on the line so that the usage of electricity is controlled or limited to the time when the power company has a very light load or demand for power. If you use power at any time other than off-peak time, you have to pay at a standard rate, instead of the reduced rate.

As far as the code is concerned, any fixed storage water heater with a capacity of 120 gallons or less must be treated as a continuous-duty load. This applies to about 90% of the residential water heaters that use electricity for heating water. The "continuous-duty load" means the ampere rating of the water heater must not exceed 80% of the ampere rating of the branch-circuit conductors.

The only case where the water heater current might load the circuit protective device (circuit breaker or fuse) to 100% of its rating is where the circuit protective

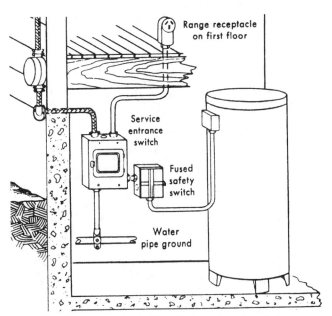

Fig. 11-102 *One method of installation of an electric hot water heater.*

device is listed for continuous operation at 100% of its rating. Presently, no standard protective device is rated in this way. Therefore, it is necessary to use the 80% rating as a guide to wire sizing.

GARBAGE DISPOSERS

The code emphasizes that the receptacle under the sink must be accessible and located to avoid physical damage to the flexible cord used to connect the motor to the power source. See Fig. 11-103.

The cord used for this type of installation should be type S, SO, ST, STO, SJO, SJT, SJTO, or SPT-3, which is a three-conductor terminating with a grounding-type plug. The cord must be at least 18 and not more than 36 inches long.

The hookup of trash compactors and dishwashers is the same as the hookup for garbage disposers. However, the cord must be from 3 to 4 feet long, instead of the 1.5 to 3 feet for the disposer. The ampere rating of an individual branch circuit to furnish power to these appliances must not be less than the marked rating of the appliance.

Each of the receptacles serving an appliance with a flexible cord must have some way of disconnecting the circuit so that it can be serviced or repaired by removing the plug from the receptacle. Most appliances need service from time to time. Thus it is a good idea to make sure the repair person is properly protected by the removal of the plug from the receptacle. It should have an easily accessible receptacle so the plug can be removed before work begins on the appliance.

The code permits electric ranges to be supplied by cord and plug connections to a range receptacle located at the rear base of the range. The rule allows such a

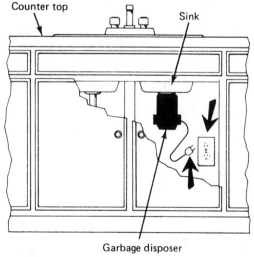

Fig. 11-103 *Installation of garbage disposers under the sink. Remember, double-insulated disposers do not require a ground.*

plug and receptacle to serve as the means for disconnecting the range if the connection is accessible from the front by the removal of a drawer.

In some cases, a gas range is used and the clock and oven lights are electric. Then, the receptacle for the lights and clock must be easily accessible without having to disassemble the range.

AIR CONDITIONERS

Some houses use a central air conditioner for cooling. This requires a unit with the compressor and condenser outside. The evaporator is usually located inside. A duct system carries the cool air throughout the house.

Electrical wiring in and to the units varies with the manufacturer. The extent to which the electrician must be concerned with the fuse and circuit breaker calculations depends on the manner in which the unit motors are fed. It also depends on the type of distribution system which the unit is to be connected. A packaged unit is treated as a group of motors. This is different from the approach used for the window units, which plug into an outlet on the wall nearby.

Some equipment will be delivered with the branch circuit selection current marked. This simplifies the situation. All the required controls and the size of the wire can be judged according to this information.

In sizing the wire for this type of unit, the selection of components for control must be the "rated-load current" marked on the equipment or the compressor. The disconnect for a hermetic motor (used in the compressor of the air conditioner) must be a motor-circuit switch, rated in horsepower, or a circuit breaker.

If a circuit breaker is used, it must have an ampere rating of not less than 115% of the nameplate "rated-load current" or the branch circuit selection current. The larger of the two ratings would be taken as a working base for the sizing of the wire.

NEWER WIRING SYSTEMS

The current method of wiring a house has not changed for decades. What has been and is usually installed is simple quad wiring supporting plain old telephone service and a low-performance coax cable for cable TV service. This cabling provides support for up to two basic, analog telephone lines and a limited number of CATV channels. This wiring situation does not allow for any home systems integration and does not include forethought for the future. The characteristics of a home wiring network need to be carefully defined. The products comprising a home network should be specifically designed for residential use. They should be easy to install in a new home construction as well as in existing homes. The system should also be modular and flexible to fit the homeowner's needs. A common in-home network provides great flexibility and lower overall cost than this separate dedicated wiring.

One of the Bell Telephone Companies' newer organizations has come up with a design for the modern home with computers, VCRs, television and fax machines, as well as the telephone. This system is especially useful when installed during the construction of a house. The system supports interactive voice, high-speed data, multimedia products, and communication services. It is a single network wiring system that provides instant "plug and play" access to ISDN services, Internet access, CATV programs, video on demand, digital satellite signals, and fax/modem plus controls for security systems and home automation from anywhere in the home.

HIGH-SPEED, HIGH-PERFORMANCE CABLE FOR VOICE AND DATA APPLICATIONS

Today's evolving technologies are requiring line speeds that are fast—up to 155 Mbps. The uses are diverse—not just voice communication, but also fax, electronic mail, video, data, and file transfer. Applications include everything from telecommuting to video conferencing to home-based businesses. To help the home keep pace with this changing communications mix, a high-quality coaxial cable must be installed in new homes. This high-speed, high-performance cable carries a full range of high-speed communications services up to 100 meters in home distribution systems. See Figs. 11-104 and 11-105. Two colors are used for the cables. Black is used for external video applications like cable television

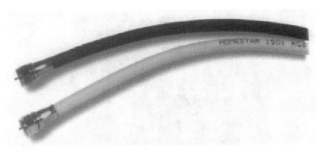

Fig. 11-105 *Coaxial cable.* (Lucent)

and white is used for internal video applications like security cameras.

Installation of Cable

The primary cable installed between the service center and outlets consists of one 4-pair Category 5 cable and two RG6 coaxial cables in a single jacket. See Fig. 11-106. All cable runs are direct from the service center to the outlet. One of the two RG6 coaxial cables in each homerun is external cable and carries output signal that drives TVs and VCRs. It is the electrical combination of the CATV input from the cable TV company and all input signals brought to the service center on the other RG6 internal cable. The RF modulated output of VCRs, CD players, and security cameras are sent to the service center on the internal coax cables. Up to 16 internal cable signals are combined together at the service center, and then are mixed with the CATV input to form the external signals.

Service Center

The Service Center (Fig. 11-107) is a centralized distribution point that connects communications devices to a single, uniform structure cabling system in the home. This service center contains distribution devices for

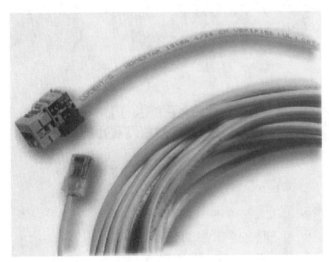

Fig. 11-104 *Twisted-pair cable shown after connectorization.* (Lucent)

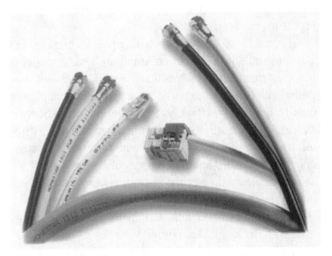

Fig. 11-106 *Hybrid cable for home wiring.* (Lucent)

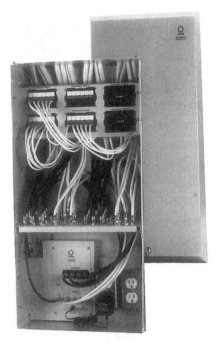

Fig. 11-107 *Service center.* (Lucent Technologies)

Fig. 11-108 *Mounting panel.* (Lucent)

Fig. 11-109 *Distribution amplifier and ac mounting panel.* (Lucent)

Fig. 11-110 *Service center box.* (Lucent)

voice data, and both baseband and broadband video. Up to 31 twisted pair connections can be made to the Service Center when it is fully configured. The unshielded twisted pair distribution can service telephone service, LAN transmission, or computer modems. The service center has many video applications: cable television, satellite dishes, or video cameras for in-home security. They can be connected through the Mounting Panels in the service center. This family of panels holds video splitters and combiners that service up to 16 dual video outlets when fully configured. A distribution amplifier is also located in the Service Center. It boosts the cable television (CATV) feed signal and combines it with in-home sources for distribution throughout the house. It is a necessary component when the home has more than four ports, or has cable runs exceeding 150 feet from the service center.

Figure 11-107 shows a mounting panel while Fig. 11-108 illustrates the Distribution Amplifier. The Amplifier is mounted on a panel, such as that in Fig. 11-109. The amplified video distribution is necessary for excellent picture quality. Figure 11-110 shows the mounting for the Service Center. It can be surface or flush mounted and the distribution modules are pre-assembled with front-facing fasteners. The enclosure and all components are grounded for safety. Cables can enter the enclosure from the top, bottom, or back for flexible, easy installation. Figure 11-111 shows a crimper and field terminator modulator plugs used with the HomeStar Wiring[1] system.

[1]Lucent Technologies' registered trademark.

CHAPTER 11
STUDY QUESTIONS

1. What are two good electrical conductors?

2. When were farms electrified?

3. Why do we have grounded conductors?

4. What color is the ground wire?

5. When does the electrician work on a house?

6. Who puts in the meter socket?

Fig. 11-111 *Crimper and field terminator for modular plugs used in HomeStar*® wiring.* (Lucent)

7. What is Romex cable? Where is it used?

8. How much current does a new house usually require?

9. Who installs the cable TV lines?

10. How are glass-enclosed ceiling fixtures attached to the ceiling box?

11. What voltage is needed for electric ovens?

12. How do you figure the size of line needed for a range drawing over 1200 watts?

13. What is the horsepower rating of most garage door openers?

14. What is meant by the term *code*?

15. What type of wiring is needed for a new house today?

16. What is CATV? Who is responsible for CATV wiring?

17. What voltages are usually available in a residential building?

18. What is another name for a neutral wire?

19. What is meant by service from head to box?

20. What is a conduit?

Plumbing the House

RESIDENTIAL BUILDINGS REQUIRE WATER. Water is used for laundry, bathing, drinking, and for moving human waste from the house. A modern home uses water in the kitchen for the sink, dishwasher, disposal, and refrigerator ice-maker. Water is needed in bathrooms for toilets, lavatories, bath tubs, and showers. Water is also used for the laundry, for outdoor faucets, and lawn sprinklers. Provision must be made for both hot and cold water service.

To provide for many uses for water in the home and for the disposal of human waste, a complex system of pipes is used to carry water to the building and its appliances. Pipes are also needed to carry away waste water.

In some cases, the term *plumbing* includes pipes used to move hot water from furnaces to various rooms of the house for heating. Other plumbing applications involve moving solar heated water into either the hot water system or a central heating system. In other instances, the black pipes used to carry natural gas (or propane in some locales) to a furnace is also considered part of the plumbing system. It is recommended that only steel pipe be used for fuel gas.

This chapter focuses on the design and installation of various pipes and drains that carry water to a building and waste water away. How each appliance or "fixture" is connected is not within the scope of this chapter.

The plumbing for a building is usually done in three stages. These are installation of: the main outside supply pipe and the main drain; inside pipes, vents, and drains; and water using fixtures. The first two steps are the focus of this chapter. The third step is usually done after the finished floor is done.

The first step, installing the main supply pipe and main drain pipe, is done at the time of the excavation

for the footings and foundation. Pipes for both water and waste are called *lines*. They also might be called the *water line* or the *sewage line*.

Because both water supply lines and drain lines carry water of some sort, both must be located sufficiently underground to prevent freezing during the winter. Refer to the depth of footings section in Chapter 3. This is why they are generally put in when the holes and trenches for the footings and foundation are dug. It is cheaper and more efficient to do all the digging at the same time.

The main supply line typically runs from the building to the water meter. See Fig. 12-1. In some areas, the water meter is located in the basement of the building; however, the trend today is to locate the water meter near the street. Here, it is much easier for the city to read and maintain. The city connects the water main to the meter, and the builder connects the meter to the building. The city will not turn the water on until the plumbing has been inspected and approved.

The main supply line and the main drain pipe that carries away waste water must be installed. See Fig. 12-1. Notice that both the supply and drain lines must be buried below the freeze line. Both pipes carry water and are subject to freezing outside the heated building space.

Water and sewage lines are placed into the area that will become the building and then they are *stubbed out*. See Fig. 12-2. This means that the pipes are located in the areas where they are to be used—the kitchen or bath. Some extra is left sticking up to make full installation a bit easier at a later time. The holes are filled in, and footings, foundations, floors, and such are built over the stubbed pipes.

The second stage of the plumbing is done after the floors and wall frames are started. The remaining

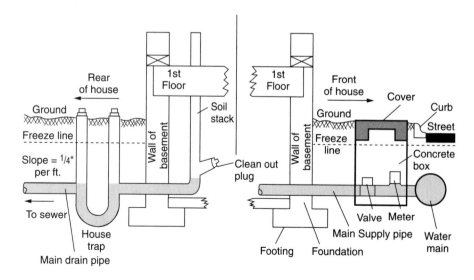

Fig. 12-1 *Initial supply and drain pipes.*

Fig. 12-2 *Supply lines and drains are "stubbed" off for later work. (A) Main line from street to house. (B) Plumbing vents and drains as well as soft copper lines encased in concrete slab. (C) Vents, drains, copper lines in concrete slab. (D) Vents and drains installed before slab is poured.*

pipes, drains, and vents are installed as the building is erected. Because the pipes and drains are hidden in the walls and under the floors, they should be installed before the inside wall board is applied.

For buildings erected on concrete slabs, much of the plumbing is placed under the slab. The pipes should be laid under the plastic vapor barrier and there should be some flexibility left in the pipe. This is because concrete has a high expansion and contraction rate with temperature changes. Leaving some slack in the lines, like wide curves in copper lines, allows the pipes to move with the expansion and contraction of the slab. Some builders prefer to locate all pipes outside the slab for ease in repairs and service, while others locate the pipes in the attics. Attic installations require careful attention to pipe insulation to prevent the pipes from freezing and bursting in the winter. This can cause serious water damage to the building.

PLUMBING SYSTEMS

A plumbing system consists of supply lines that bring water into the building, and drains that carry the water away from the building. In order for the drains to work efficiently, they must be vented to allow air to enter the system. Thus, the plumbing system has three major parts: a supply; a drain; and vents. The system is shown in Fig. 12-3.

Supply Lines

As mentioned earlier, one water line carries water from the main to the building. Common sizes for the first

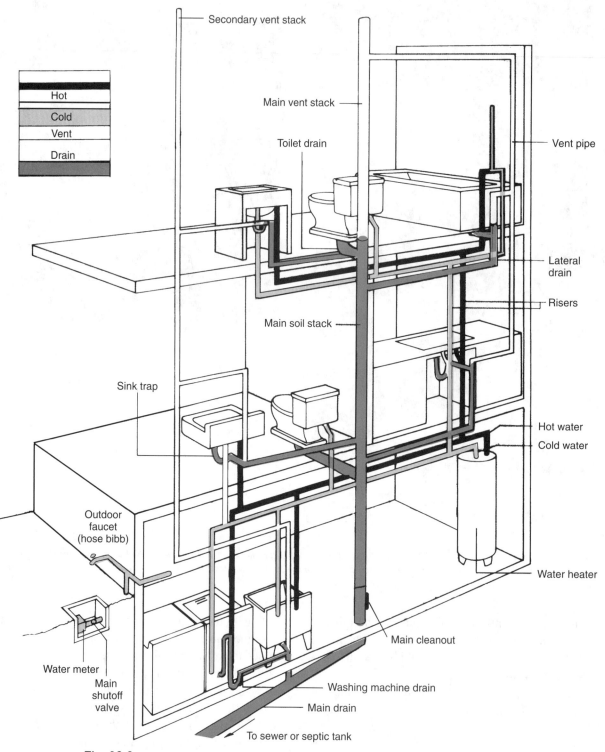

Fig. 12-3 *A typical plumbing system has three parts: (1) supply, (2) drains, and (3) vents.*

supply line are 1 inch, 1½ inches and 2 inches in diameter. Once the pipe is inside the building, smaller pipes are generally used to carry the water to various locations therein. Common sizes for this are 1-inch, ¾-inch, and ½-inch diameters.

Water lines can be assembled through holes in floors, wall studs, and so forth. Most water line assemblies are made through openings planned for them. It is a good idea to only run smaller lines through holes cut in wall studs or floor joists. There are two reasons for this. First, any hole or cut into a joist or stud can weaken it. Second, pipes in these holes might be struck by nails used for hanging wallboard. The nails could cause holes in the pipe that would not be detected until after the wall or floor is sealed and finished and the water turned on.

The supply pipes to most locations are capped with cut-off valves. See Fig. 12-4. This allows the water to be turned off to install the fixture or to make repairs without turning off the water supply for the entire building. Special flexible lines, such as the plastic line in Fig. 12-5, can be used to connect the supply valve to the fixture.

Air chambers must be installed near each supply outlet to prevent a loud noise from occurring each time the water is turned off. The noise is called *water hammer*. When the water is turned off, the full force of the moving water (remember that standard city water pressure is 80 pounds per square inch, which is a lot) slams against the valve. This force can make a loud banging noise, and the force can cause the pipes to physically jump.

The air in the chamber is momentarily compressed to cushion the force of the water stopping against the valve. Figure 12-6 shows this action and the proper location of the air chamber. It is important to remember that each supply outlet must have an air chamber. Air chambers are usually made of the same size pipe as the supply line.

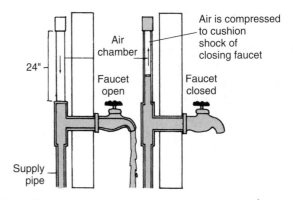

Fig. 12-6 *Typical air chamber action prevents water hammer.*

In addition to using air chambers, the pipes should be anchored to something solid, like a stud or joist, at appropriate intervals. The movement of the water, particularly when it is stopped or started, can cause the pipes to move. The movement can be enough to cause the pipes to "bang" against the floor or wall. Anchoring them reduces both the movement and noise.

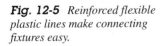
Fig. 12-4 *Cut-off valves on the pipes that supply fixtures.*

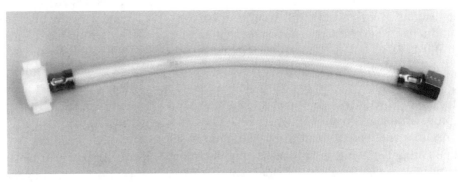

Fig. 12-5 *Reinforced flexible plastic lines make connecting fixtures easy.*

Plumbing Systems 381

Drains

Drain systems (often called *DWV* for drain, waste, and vent) are used to drain away waste water. Waste waters are divided into two categories. The first is called *gray* water and consists of the waste water from sinks, laundries, and bathing units. The second type, *black* water, contains both liquid and solid wastes from toilets. In most communities, both types of waste water are drained into one sewer system.

Pipes that carry waste water from a fixture (like a sink) are called *drains*. The drains may empty into another larger pipe to be carried across the building to the main drain. These parts that carry the water across the building are called *laterals*. The drain pipes that run vertically are called *stacks*, or *soil stacks*.

Because waste water is not under pressure, the drain systems must be angled down to allow gravity to move the water. There is typically about ¼ inch of downward slope per foot of horizontal run. This is usually expressed as "3 inches per 12 feet." This slight slope allows the water to carry the solid waste. If the drain is too steep, the water might run off before the solid wastes are moved and eventually cause clogs, which stop up the drains. Clogs can force the waste to back up into toilets and other drains preventing their effective use. These back-ups can have a terrible odor and might spill over onto a floor causing considerable damage.

The same slope for the drains must be used from each drain to the main drain line. All drains must slope down.

There is usually only one main drain connecting the building system to the sewer. This means that the several drains in the building must be planned so that they can be connected without eventually causing problems. Figure 12-7 shows typical installation factors.

Drains that connect sinks and lavatories, and tubs and showers to a system are typically 1.25, 1.5, or 2 inches in diameter. Size would be determined mainly by available space, building codes, and anticipated drainage flow. The cost of plastic drains is not much different regardless of size.

"Laterals" can be 2, 3, or 4 inches in diameter. Main drains should be larger in comparison and are typically 4 or 6 inches in diameter. Laterals can be suspended using *plumbers tape*. This is a perforated metal strip that allows the support to be installed at a length that will hold the lateral at the correct height to keep the proper drainage slope. See Fig. 12-8.

Another vital part of drains keeps the sewer gases from entering the living space. This is very important because sewer gases are very noxious, can be toxic,

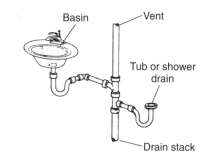

(A). Direct or "wet" vent drains are connected directly to the stack and vent

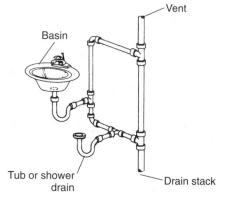

(B). Reventing (back venting) two drains connected to a stack. Note the loop connecting both units to the vent.

Fig. 12-7 *Typical multiple drain connections.*

Fig. 12-8 *Plumbers tape holding a lateral drain at the correct slope.*

and are explosive. A very simple device called a *trap* is used to block out these gases. Older systems might have an "S" trap, but most modern systems use a "P" trap, as in Fig. 12-9. The name is taken because the trap is shaped like the letter "P." This keeps a "plug" of standing water between the living space and the sewer. Every time the fixture is used, the water in the trap is replaced, keeping the trap fresh.

Sometimes small objects, such as rings, are accidentally dropped into a sink. The trap will hold them and keep them from tumbling into the main drain, provided that the water is not allowed to continue to flow through the drain.

Traps can be installed with clean-out plugs in them (Fig. 12-9A), or they can be plain, as in Fig. 12-9B. It

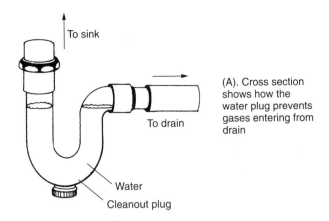

To sink

To drain

(A). Cross section shows how the water plug prevents gases entering from drain

Water

Cleanout plug

(B). Typical installation without cleanout plug

Fig. 12-9 *Typical "P" trap installation. (A) Cross section shows how the water plug prevents gases entering from the drain. (B) Typical installation without clean-out plug.*

is not a very difficult job to remove a trap. Both traps allow for the retrieval of lost objects or the removal of solid objects that are impeding the flow of waste water.

Installing traps is typically not a difficult job. However, a factor called *critical distance* must be considered. Simply expressed, it means that the outlet into the stack must never be completely below the water level of the trap. If it is, a pipe full of water would also siphon out the water in the trap and expose the living space to the odors and gases of the sewer. It is easy to figure the maximum amount of run you can use on a drain. First, determine the diameter of the drain, such as 1.5 inches. Next, divide by 0.25 inch—the amount of slope per foot. This would yield the number 6, or 6 feet. However, 6 feet would be the full 1.5 inches, so

the maximum run would be anything less than 6 feet. A good rule of thumb would be to use only whole numbers for the distance—which would then be 5 feet.

P traps are used on the drains for every single fixture in the building except water closets. The reason is that water closets have built-in P traps and are usually connected to a fitting called a *closet elbow*.

In many areas, a "house trap" is also required in the main drain outside the building. Check Fig. 12-1. This house trap protects the entire drain system from the odors and gases of the main sewer.

Drain systems are often made entirely of PVC plastic pipe of appropriate sizes. However, drain pipes also might be made of copper, cast iron, and heat-resistant glass. Glass drains are fragile and are only used when corrosive materials, such as acids, must be drained from the building to an appropriate collection point.

Vents

As mentioned earlier, vents allow air to enter the drain system so that the water can flow properly. Without the access to air, a vacuum could be formed in parts of the drain system that would prevent the waste water from flowing. Vents are simply a vertical pipe that rises from a drain up through the roof and allows air to enter the drain as in Fig. 12-10.

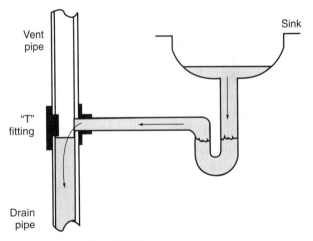

Vent pipe

Sink

"T" fitting

Drain pipe

Fig. 12-10 *Basic drain vent.*

As a rule, a vent is needed for every drain. However, this is often impractical and would require three vents for the typical bathroom, three for a kitchen, one for a laundry, and so forth. The vents can be combined as in Fig. 12-11 so that usually there is only one per bathroom, one per kitchen, etc.

Vents are often made from the same type of pipe as the drain system. However, it is common to use the

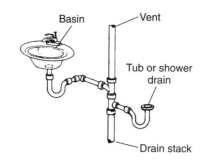

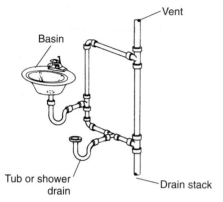

(A). Direct or "wet" vent drains are connected directly to the stack and vent

(B). Reventing (back venting) two drains connected to a stack. Note the loop connecting both units to the vent.

Fig. 12-11 *Multiple vent connections.*

least expensive material for vents because they do not actually carry water.

Clean-Out Plugs

Clogs are fairly common in any drain system, and for that reason, special "clean-out" plugs are advisable. These can be located either inside or outside the building, but should be easily accessible. See Fig. 12-12. This allows a clean-out tool called a *snake* or an *auger* to be inserted and used to unclog the drain. If no clean-out plug is used, when a clog occurs, a hole must be dug to reach the drain. Then an opening must be cut into the drain. Needless to say, cleaning out a drain is a dirty job and can be difficult and time-consuming.

It is advisable to install several clean-out plugs. It is even advisable to install a clean-out plug for each drain as well as the main drain. A typical plug is made by inserting a "wye" fitting into the line as in Fig. 12-13. Note that the slanted outlet angles into the drain line to make it easier to insert the clean-out tool. The wye must be the same diameter as the drain, and must be assembled to have the least interference with water flow.

PRELIMINARY PLANNING

A number of things must be considered before starting to plumb a building. These include: who does the work, the type of pipe used for supply and drains, the

Fig. 12-12 *A kitchen clean-out plug located for convenient access.*

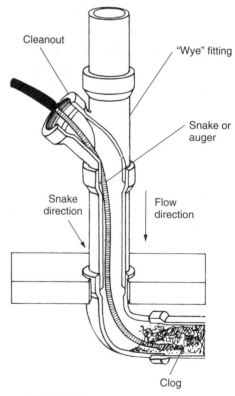

Fig. 12-13 *Clean-out plug detail.*

locations of the pipes, the best slope for the drains, and the location of clean out plugs for drains.

Who Does the Work

Normally, a contractor (or builder) arranges for professional plumbers, acting as subcontractors, to do the necessary work when it is needed. Developers of large tracts or subdivisions might employ a general contractor to make out all the building schedules. Sometimes, capable individuals who wish to build their own homes either do the work, or act as the contractor.

In the event that an individual wishes to do the work, the first step is to check with the local building inspection and permit office (see your phone book under city or county government), and find out what the rules are. In some areas work can only be done by professional plumbers certified in that area, while in other areas, the work can be done by anyone, but must be checked by a certified plumber. Other areas allow anyone to do the work provided it follows the appropriate rules, called building (or plumbing) *codes*. These codes can be very complex and specify who can do the work, what type and size pipes are to be used, and many other factors.

Some plumbers will provide advice and check the work of individuals and issue the proper certification for a fee. In some areas, the inspectors will check the work of individuals when called. It is very important, however, to find out what the rules are before starting.

Pipe Type

There are several types of pipe (copper, galvanized steel, and plastic) that can be used for both supply and drain pipes. Some types might be required for certain things in the local building codes, while others can be used for various reasons including durability, cost, ease of working, or special resistance to acids or chemicals.

It is also possible to use several different types of pipe in a single building. The major concern when using different types of pipe is whether or not any *electrolytic action* might result. Electrolytic action is what happens when the impurities in water (salt, minerals, alkalis, acids, etc.) react with two or more different metals. If, for example, you connect a copper water line directly to a steel water line, the reaction of the water on the two metals would cause a mineral build up at the joint that would clog up the water line. A simple "di-electric" union can be used to prevent this. It can be a union with a rubber or plastic washer to separate the two metals, or it can be a simple plastic coupling to separate the two.

Most pipe sizes are measured by the stream of water that will emerge from the pipe under standard city water pressures. Thus, the actual hole in the pipe is bigger than the "size" of the pipe. This is true even in the metric system, and inch sizes are used in the metric system for water flow planning.

Galvanized water pipe Galvanized water pipe is made of steel and coated with zinc to make it resistant to rust and mineral deposits. It is not used for drains. Steel pipe that is not galvanized (called *black* pipe) should never be used for water, but may be used for natural gas.

Galvanized pipe is available in most areas, it is tough, strong, heat resistant, and the cost is moderate. It is durable, and might last many decades. Its greatest disadvantage is in the amount of work required to install it across and around the many angles and curves in a typical building. The pieces of pipe must be laboriously assembled by screwing numerous fittings into place. When repairs must be made, in most instances, the whole unit must be taken apart piece by piece.

Working with galvanized pipe means the various pieces are screwed together in appropriate lengths and directions. It is important to use a sealing compound (pipe dope) or teflon tape on the threads before assembling the pipe. Most professionals use more teflon tape than pipe dope. It is wrapped on the end of the threads in the same direction the next piece is to be turned. See Fig. 12-14. About three or four wraps is enough, and the tape is simply torn off and the next piece screwed on.

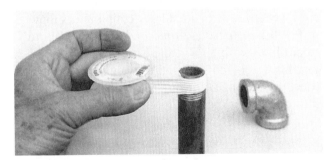

Fig. 12-14 *Using teflon tape for sealing threads.*

Pipe dope might either be in paste form in a squeeze tube (similar to toothpaste) or it might be a soft putty-like substance in a stick. In either case, the first three or four threads should be coated. Either dope or tape makes it easier to assemble the pieces and aids in making the joints water-tight.

A whole piece of pipe is called a *joint* and is usually 21 feet long. Obviously, this is often too long to work with. Shorter pieces are called *nipples* and are common in lengths of 4 feet, 3 feet, 2 feet, and 1 foot, and in inch sizes of 1, 2, 3, 4, 5, 6, and 9. A 4-inch and a 3-inch nipple can be combined for a 7-inch length.

Pieces are joined with either a *coupling* or a *union* as shown in Fig. 12-15. A union allows a person to disconnect the water line without taking the whole system apart. However, a union is much more expensive than a coupling.

Fig. 12-17 *A reducer coupling joins different pipe sizes.*

Fig. 12-18 *"T" coupling used to split lines.*

Fig. 12-15 *Couplings (on left) and unions (on right) are used to join pipe.*

Other lengths can be cut from a joint, but the ends must be threaded. Most professional plumbers are well equipped to do their own threading. However, the threading dies used to do this are expensive. Individuals who wish to do their own plumbing might wish to consider renting the threading dies. It is not cost-effective to rent the threading dies for just a few pieces, but larger jobs are fine. For smaller jobs, it is easier for the individual to build up the right lengths using nipples.

Direction is changed with *elbows*. Elbows are made in either 90° or 45° turns, as in Fig. 12-16. Size can be changed with a reducing coupling or elbow as in Fig. 12-17. A single line can be divided into different lines with a "T" connection. See Fig. 12-18. Provision for adding another water line at a later time can be made by using a "T" and putting a plug into one of the holes. Figure 12-19 shows several typical fittings.

Fig. 12-19 *Typical pipe fittings. The most common fittings are (left to right): 45° elbow; 90° elbow; nipple; coupling and union; "T" with a plug above it; and an end cap. A reducer coupling is at the bottom left.*

When assembling or disassembling threaded pipe, special wrenches, called *pipe wrenches* are used. The feature that makes these wrenches unique are the teeth in the jaws. These teeth are sharp and will dig into the surface to keep the piece tightly gripped. A professional-quality wrench has jaws that can be replaced as they dull or wear.

The wrenches are adjustable so that they can be used on several size pipes, and are available in several lengths to provide more torque to tighten or loosen pipes and fittings. When working on threaded pipe systems, two wrenches are usually used, as in Fig. 12-20. One wrench is used to hold the base, while the other is used to turn the piece being worked on.

Copper water pipe Copper pipe is very suited for water lines. It does not corrode easily, and it is easily cut, worked, and repaired. Copper is perhaps the most durable of the pipe materials, but it is also the most expensive. It might cost four or five times that of

Fig. 12-16 *Direction is changed with elbows. A 45° elbow is depicted on the left and a 90° elbow on the right.*

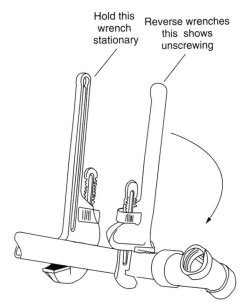

Fig. 12-20 *Proper use of two pipe wrenches.*

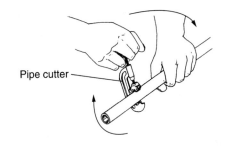

(A). Cutting galvanized, copper, or plastic pipe with a pipe cutter

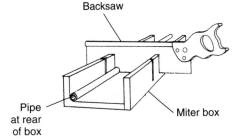

(B). Use a backsaw and miter box to cut plastic pipe

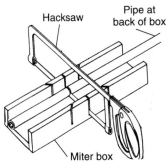

(C). Use a hacksaw box to cut copper pipe

Fig. 12-21 *Cutting pipe and tubing. (A) Cutting galvanized, copper, or plastic pipe with a pipe cutter. (B) Using a backsaw and miter box to cut plastic pipe. (C) Using a hacksaw and a miter box to cut copper pipe.*

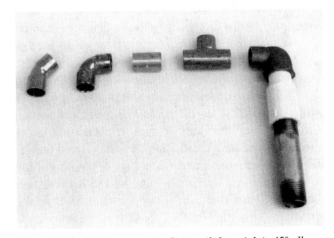

Fig. 12-22 *Rigid copper pipe fittings (left to right): 45° elbow; 90° elbow; coupling; T; 90° adapter elbow to threaded pipe; threaded plastic adaptor as a dielectric; a galvanized steel nipple.*

galvanized pipe. The pieces must either be soldered together, or joined using special fittings. Soldering takes a lot of time and must be carefully done to prevent leaks. The fittings are expensive.

There are two types of copper pipe commonly used. One is called *rigid* and the other is *soft* copper pipe. The sizes are slightly different as the hole in rigid pipe is measured like galvanized pipe, where the size of soft pipe is determined by the outside diameter. Because of this, soft copper is not considered pipe, but is correctly termed *tube* or *tubing*.

Both can be joined with appropriate and easily available fittings. Copper pipe is usually joined by soldering and tubing is usually joined with either compression or flared fittings. However, solder fittings for tubing can be obtained.

Joining a ¾-inch rigid pipe to a ¾-inch soft pipe does not make much difference in water flow. There are advantages to each.

It is very common for soft copper pipe to be used to connect a steel or plastic water line to a sink, refrigerator ice maker, toilet, or other fixture. This is because the soft copper can be easily bent to change direction or location. Refer back to Fig. 12-4.

Both types of pipe can easily be cut to almost any length. Either a tubing cutter or a hacksaw can be used, as in Fig. 12-21. When the pipe or tubing is cut with a saw, the edges must be dressed so that they are smooth both inside and outside.

Rigid copper pipe typically comes in 10- or 20-foot lengths. The most common diameters are ½ inch, ¾ inch and 1 inch. There is no need to buy nipples, because it is so readily cut to any length. However,

rigid pipe must be joined by soldering it into flanged fittings. See Fig. 12-22.

For best results, it is important to coat both the outside of the pipe and the inside of the fitting with a very thin coat of solder before the pieces are joined.

This is called *tinning* the joint. If the tinning is thick and lumpy, you cannot slide the pieces together. When the tinning is thick or lumpy, the excess can be wiped off using a cloth while the solder is still molten. This wiping with a rag leaves a very thin coating of solder on the pipe.

Perhaps the handiest method is to use a special soldering compound, which contains powdered solder in a grease-like mixture of *flux*. Flux is a fluid or paste used to clean the metal when it is being soldered. It is important to use the right type of flux. The flux is simply smeared on both pieces instead of tinning them, and the pieces are then assembled. After assembling either a tinned joint or a compounded joint, the unit must be heated to melt and fuse the solder. More solder can be added as shown in Fig. 12-23.

Copper pipe also can be soldered without tinning the joints. The solder joints are just coated with flux and assembled. The solder is added to the outside and flows into the joint. This is the easiest method and it works well.

It does take practice and skill to "tin" joints for soldering, but it is the best method. When starting a new job, it is a good idea to practice with a few short pieces before moving to the actual pipes.

The old systems were soldered using regular solder, which is a mixture of lead and tin. However, lead is a toxic material and could lead to cases of lead poisoning in extreme cases. To prevent lead poisoning, many systems now require a lead-free solder.

Both types of solder are easily obtainable at building supply stores. Either can be used, but the user should be sure to match the correct flux with the solder used.

Soft copper tubing is available in rolls of 10, 15, 50, and 100 feet. It is available in diameters from ⅛ through 1 inch in ⅛-inch intervals. It is soft enough to be bent with your hands in any direction. This means that special fittings such as elbows and nipples are not needed. However, it does require either compression or flared fittings (Fig. 12-24) to join it to appliances or other parts of the water system. Compression fittings are easier to install, but they are not as reliable under pressure. Flares on the end of the tubing are for the flare-fitting. The flared ends are easily made with a flaring tool, as in Fig. 12-25.

Plastic water pipe The majority of plastic water pipes are made of PVC or polyvinyl chloride. However, care should be taken to make sure that the correct pipe is used for hot and cold water applications. The most common type of PVC is not suitable for hot water applications, but a special PVC pipe, CPVC, is made for hot water. CPVC can be used for either hot or cold applications. Cold water PVC is white, while CPVC is a tan color. The outside dimensions of CPVC are a little bigger than the PVC so that the fittings used to join the pipe cannot be mixed up. It would be unwise to use cold water fittings with hot water pipe.

There is not much difference in the cost of either PVC or CPVC. PVC pipe is the least expensive, most durable, easiest, and quickest type of pipe to install. Its cost is about ¹⁄₁₀ of copper and about half of galvanized. It is not very strong nor heat resistant, but it is very light in weight. Like copper, it can be easily cut, which makes the purchase of nipples unnecessary. It can be cut with ordinary wood cutting tools. PVC is limber and can be bent slightly, but sharp turns require a 45° or 90° elbow fitting just like metal pipe. See Fig. 12-26. Size can be changed by using reducer couplings, and other couplings can be obtained to connect to threaded systems, as shown in Fig. 12-26.

"Welding" pieces of PVC pipe into the various fittings is rather easy. Simply coat the inside of the fitting and the outside of the pipe with a special solvent, using the swab in the can, and then push the two pieces together with a slight twisting motion. See Fig. 12-27. The solvent is usually available anywhere PVC pipe is sold and comes in various size containers for big or small jobs. There are separate solvents for both PVC and CPVC. However, there is a solvent that will work with either type of pipe. Some plumbers use a two-stage solvent, where the first application is a cleaning solution. The solvent is applied directly over the first coat, and both are applied in the same manner.

Plastic pipe also can be easily repaired. When trouble occurs, you simply cut out the bad part, and glue in new pieces. Neither the tools nor the processes require any special procedures. Compression unions also can be easily used for either initial installation or repairs, as in Fig. 12-28. They also can be used to join metal pipe to plastic pipe.

Location of Pipes

Inside the building, the various pipes and drains are worked between or through joists and studs. This takes a direct route and minimizes the amount of pipe and work needed for installation. However, once the area where the fixture is to be located is reached, care must be taken to locate the pipes or drains so that the fixtures can be installed. As in most systems, locations are fairly standardized for this purpose. This is important so that a fixture, such as a toilet (water closet), can be correctly installed. The pipe outlets must allow the fixture to be the correct distance from the wall, and

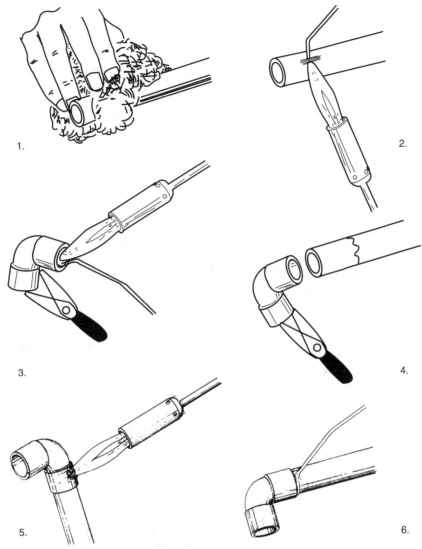

(A). Tinning copper pipe joints
1. Clean end of pipe with steel wool
2. Heat end of pipe and coat with solder
3. Wipe off excess
4. Grip fitting with pliers. Heat and apply solder to joint. Tap out excess
5. Heat fitting and push it onto pipe
6. Heat and apply more solder if needed. Finished joint should be smooth all the way around.

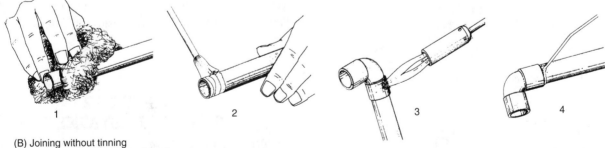

(B) Joining without tinning
1. Clean end of pipe with steel wool
2. Apply flux with brush or squirt tube
3. Push fitting onto pipe and heat
4. Apply solder, melted solder will flow up into joint.

Fig. 12-23 *Two methods of soldering copper pipe: (A) Tinning copper pipe joints. (1) Clean end of pipe with steel wool. (2) Heat end of pipe and coat with solder. (3) Wipe off excess. (4) Grip fitting with pliers. Heat and apply solder to joint. Tap out excess. (5) Heat fitting and push it onto pipe. (6) Heat and apply more solder if needed. Finished joint should be smooth all the way around. (B) Joining without tinning: (1) Clean end of pipe with steel wool. (2) Apply flux with brush or squirt tube. (3) Push fitting onto pipe and heat. (4) Apply solder. Melted solder will flow up into the joint.*

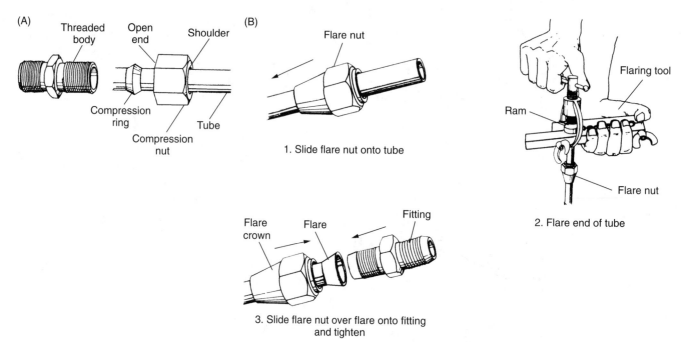

Fig. 12-24 *Compression or flare fittings: (A) Compression fitting (B) Flare fitting (1) Slide flare nut onto tube. (2) Flare end of tube. (3) Slide flare nut over flare onto fitting and tighten.*

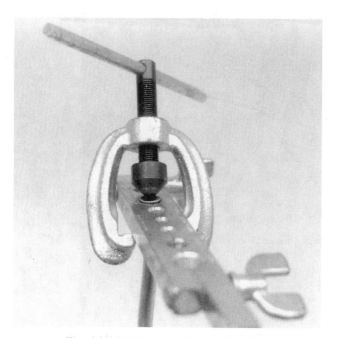

Fig. 12-25 *Flaring tool for copper tubing.*

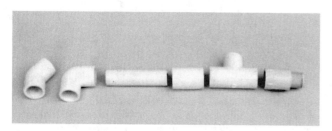

Fig. 12-26 *Typical plastic pipe fittings (left to right): 45° elbow; 90° elbow; nipple; coupling; T; threaded adaptor.*

have the water supply available without being very visible and yet remain accessible for repairs. Equally important is the location of the drain, so that the bottom portion can be installed over the drain in a correct manner.

There is more flexibility in the location of supply outlets that will be located inside cabinets, than for the outlets for tubs and toilets. Typical locating dimensions for various outlets and drains are shown in Fig. 12-29. Note that the hot water line is always installed on the left side.

Once the pipes are installed and either capped off or have cut-off valves installed, and all the drains are connected, the system is ready for inspection. If the system passes inspection, then the city will allow the water to be turned on, and the rest of the interior can be completed.

CHAPTER 12
STUDY QUESTIONS

1. What is the definition of "fixture" in plumbing terms?
2. Plumbing is done in three stages. Name them.
3. Why should water supply lines be buried underground?
4. When is the second stage of plumbing done when building a house?

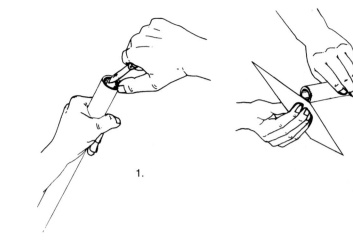

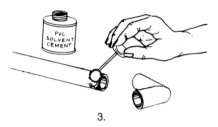

Fig. 12-27 *Welding plastic pipe: (1) Clear inside and outside edges of burrs. (2) Sand glaze off joint area. (3) Coat pipe with solvent. (4) Coat inside of fitting with solvent. (5) Push fitting onto pipe with a slight twist.*

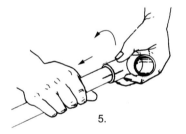

Welding plastic pipe
1. Clear inside and outside edges of burrs
2. Sand glaze off joint area
3. Coat pipe with solvent
4. Coat inside of fitting with solvent
5. Push fitting onto pipe with a slight twist

Fig. 12-28 *A plastic compression union: used to join plastic pipes, or plastic to metal.*

5. Name three elements of a plumbing system.

6. What is an air chamber?

7. How do you prevent air hammer?

8. What does DWV mean?

9. What is a lateral?

10. What is meant by soil stack?

11. How much slope should a drain system have?

12. What size pipe is used for drains that connect sinks and lavatories, tubs and showers?

13. What purpose does a trap serve?

14. Where are P traps used?

15. What purpose does a house trap serve?

16. What is the purpose of a vent?

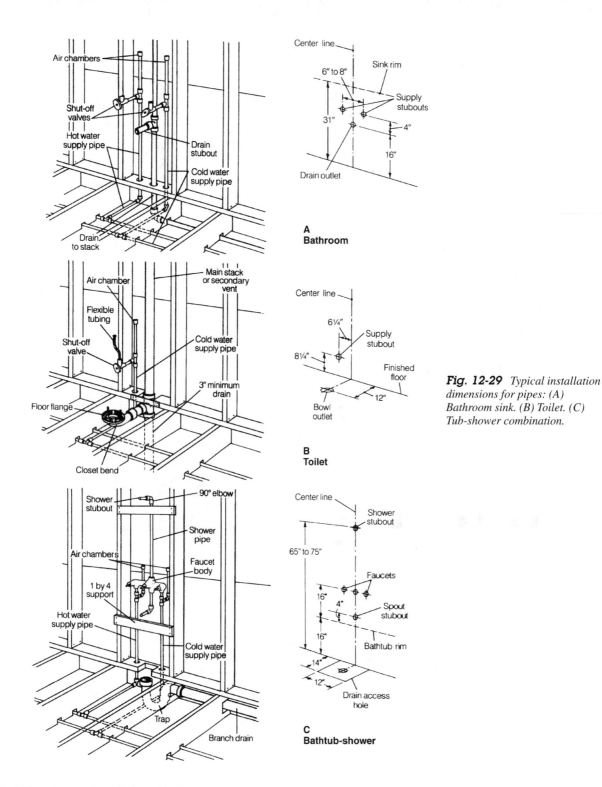

A
Bathroom

Center line
6" to 8"
Sink rim
Supply stubouts
31"
4"
16"
Drain outlet

Air chambers
Shut-off valves
Hot water supply pipe
Drain stubout
Cold water supply pipe
Drain to stack

Main stack or secondary vent
Air chamber
Flexible tubing
Shut-off valve
Cold water supply pipe
3" minimum drain
Floor flange
Closet bend

B
Toilet

Center line
6¼"
Supply stubout
8¼"
Finished floor
Bowl outlet
12"

Fig. 12-29 *Typical installation dimensions for pipes: (A) Bathroom sink. (B) Toilet. (C) Tub-shower combination.*

Shower stubout
90° elbow
Shower pipe
Air chambers
Faucet body
1 by 4 support
Hot water supply pipe
Cold water supply pipe
Trap
Branch drain

C
Bathtub-shower

Center line
Shower stubout
65" to 75"
Faucets
16"
4"
Spout stubout
16"
Bathtub rim
14"
12"
Drain access hole

17. What does a "wye" describe?

18. Name three types of pipe used for plumbing systems.

19. What is the difference between rigid and soft copper?

20. What does CPVC means? What color is it?

13
CHAPTER

Insulating for Thermal Efficiency

INSULATION MAKES A DIFFERENCE IN THE amount of energy used to heat or cool a house. Previously, houses were not insulated because of the low cost of energy. Today, things have changed. It is no longer economical to build without insulating.

According to a study conducted by the U.S. Department of Energy, 70 percent of the energy consumed in a house is used for heating and cooling. Another 20 percent is used to heat water. Only 10 percent is used for lighting, cooking, and appliances. See Fig. 13-1.

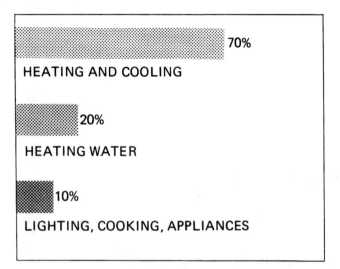

Fig. 13-1 *Consumption of energy in the home.* (U.S. Dept. of Energy)

Insulation can help conserve as much as 30 percent of the energy lost in a home. Note the air leakage test results shown in Fig. 13-2. Wall outlets are a source of 20 percent of this leakage. This can be stopped by using socket sealers. See Fig. 13-3. These are foam insulation pads that are placed between the face plate and the socket. Windows and doors have a leakage problem, too. They can be treated in a number of ways to cut down on leakage. The soleplate is an important place to check for proper insulation. To conserve energy, new codes call for leakage seals over corners and soleplates (Fig. 13-4). The ductwork in the heating or cooling system can be insulated with tape. Special tape is made for the job. See Fig. 13-5.

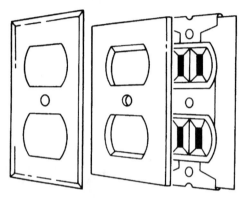

Fig. 13-3 *Socket sealers.* (Manco Tape)

There are a number of skills to be learned in this chapter. They are:

• How to check for insulation requirements

• How to apply insulation

• How to caulk around windows and doors

• How to apply various insulation materials

AIR LEAKAGE TEST RESULTS

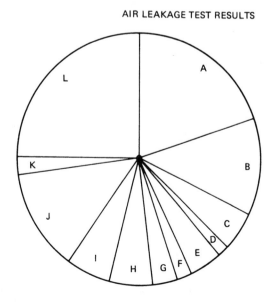

A – 20% WALL OUTLETS

B – 12% EXTERIOR WINDOWS

C – 5% RECESSED SPOT LIGHTS

D – 1% BATH VENT

E – 5% EXTERIOR DOORS

F – 2% SLIDING GLASS DOOR

G – 3% DRYER VENT

H – 5% FIREPLACE

I – 5% RANGE VENT

J – 14% DUCT SYSTEM

K – 3% OTHER

L – 25% SOLEPLATE

Fig. 13-2 *Air leakage test results.* (Manco Tape)

Fig. 13-4 *New codes call for leakage seals over corners and soleplates.* (Fox and Jacobs)

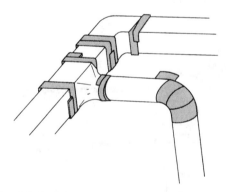

Fig. 13-5 *Insulating tape for ductwork.* (Manco Tape)

TYPES OF INSULATION

Batts and blankets are typically made of Fiberglas. They may also be made of cellulose fibers. Batts are best for use by carpenters and do-it-yourselfers. They can be easily applied in attics and between exposed ceiling joists.

Loose fill comes in bags. It can be poured or blown into the joist space in open attics. It can be blown by special machines into closed spaces such as finished walls.

Rigid types of plastic insulation come in boards. Foams can be applied with a spray applicator. Plastic has high insulation quality. Thinner layers of plastic may provide the same protection as thicker amounts of other materials. Thinner sections may provide higher *R values*. R values indicate the relative insulating qualities of a material.

All insulation material should have a vapor barrier. This barrier provides resistance to the passage of water vapor.

Another type of Fiberglas insulation is the tongue-and-groove sheathing that is available for exterior walls.

Batts and blankets (except friction-fit) have an asphalt-impregnated paper vapor barrier. Or, in some cases, they have a foil backing. All other types should have a 2-mil (0.002-inch) plastic sheet installed as a vapor barrier.

HOW MUCH IS ENOUGH?

For years the recommended amount of insulation was R-19 for ceilings. R-11 was suggested for walls almost anywhere in the country.

Recently there have been changes in these recommendations. The map in Fig. 13-6 shows how much is needed in various locations. The first number is for the ceiling. The second number is for the wall. Insulate the floor if the basement is not heated. These are the values recommended by Owens-Corning. Owens-Corning is a manufacturer of insulation.

In some parts of the country, utilities and cities have separate standards. Be sure to check the local requirements for your area. In most instances local codes follow Fig. 13-7. At present it appears that *more is better*. This has not been proven to everyone's satisfaction. More research needs to be done in this area. Different types of insulation are being developed. Foams appear to be the best bet for insulating an older house—that is, if it has no previous wall insulation.

WHERE TO INSULATE

Batts are made in specific thicknesses. Choose the proper R value for the space. Allow the material to expand. Fit it closely at the edges. Don't overlook the space between closely fitted timbers.

Vapor barrier protection is important. It should always be installed toward the warm side of the area being insulated. Vapor barriers must not be torn or broken. Repair any breaks with a durable tape.

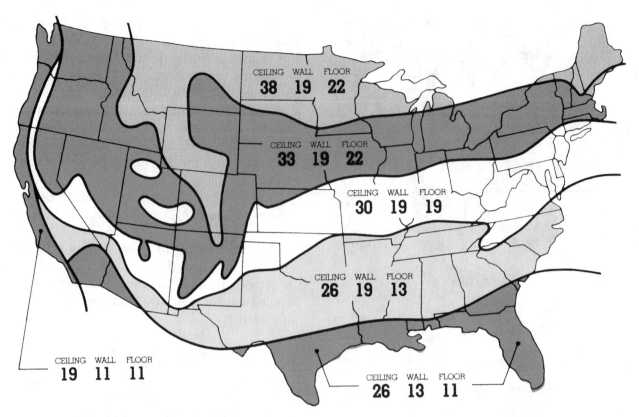

Fig. 13-6 *Owens-Corning Fiberglas minimum insulation recommendations of energy-efficient homes.*

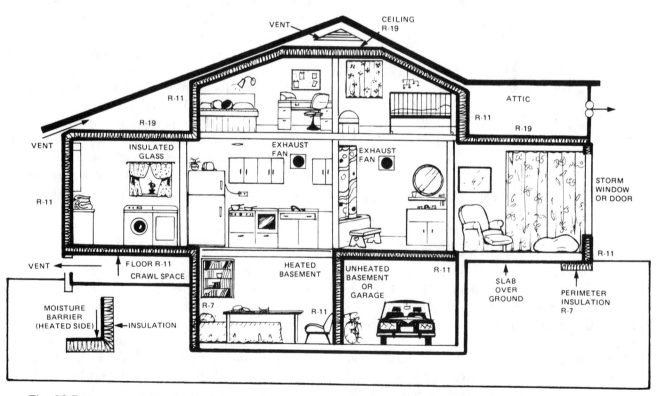

Fig. 13-7 *Recommendations for insulating your home. Note the R values and where they are assigned.* (New York State Electric and Gas)

Provide proper ventilation. Natural venting is important. It controls moisture and relieves heat buildup in the summer. See Fig. 13-8. In some cases there isn't enough natural ventilation. A fan is usually needed to exhaust the attic during the summer. This is when heat builds up and causes the air conditioner to work harder. See Fig. 13-9. The air space of the attic is often

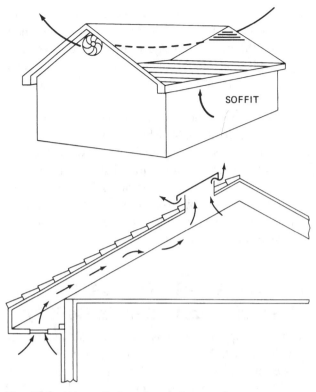

Fig. 13-9 *Attic ventilation—forced air removal.* (New York State Electric and Gas)

60 degrees F hotter than outside. Removing the heated air reduces the heat transfer to the home. Keep in mind that power venting is not an effective solution in conjunction with ridge venting. Power venting isn't always effective with vents located close to the unit.

Crawl space Crawl spaces need ventilation. See Fig. 13-10. There should be vents (2 with moisture seal) in

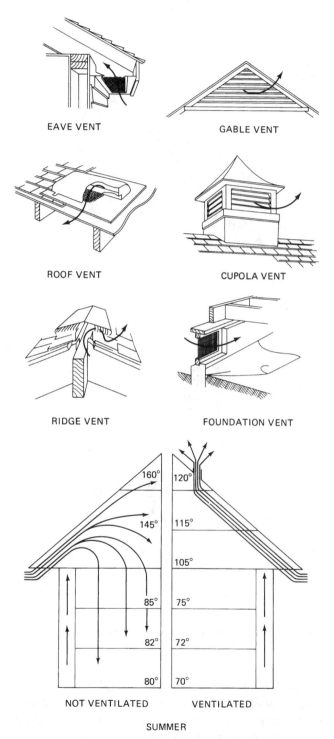

EAVE VENT

GABLE VENT

ROOF VENT

CUPOLA VENT

RIDGE VENT

FOUNDATION VENT

NOT VENTILATED

VENTILATED

SUMMER

Fig. 13-8 *Location of vents for adequate ventilation.* (New York State Electric and Gas)

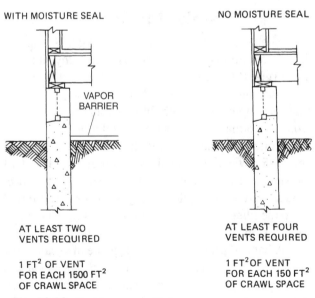

WITH MOISTURE SEAL

NO MOISTURE SEAL

VAPOR BARRIER

AT LEAST TWO VENTS REQUIRED

1 FT² OF VENT FOR EACH 1500 FT² OF CRAWL SPACE

AT LEAST FOUR VENTS REQUIRED

1 FT² OF VENT FOR EACH 150 FT² OF CRAWL SPACE

Fig. 13-10 *Crawl space ventilation.* (New York State Electric and Gas)

crawl spaces. Sealed and unsealed crawl spaces are treated differently. Vents in the basement wall help circulate the outside air and draw out moisture. This moisture is generated through the exposed earth in the crawl space. Four vents are needed for no moisture seal.

Figure 13-11 shows some of the points that may need special attention. Note how the vapor barrier is installed at these points. Figure 13-12 shows the best insulation for desired results. Note that some houses are now built with 2-×-6-inch instead of 2-×-4-inch studs in the walls. This allows for more insulation. Note how 1-inch foam sheathing board is used on the outside. This gives the added insulation needed for 2-×-4 studs.

INSTALLING INSULATION
Installing Insulation in Ceilings

Push batts up through joist spaces. Pull them back down even with bottom of the joists. See Fig. 13-13. This ensures full R-value performance because it prevents compression of the insulation. When high-R batts are used, it allows the batts to expand over the top of the joist. Compressing the batt reduces its insulating value.

Overlap the top plate. Do not block eave vents. Make sure the ends butt tightly against each other.

Obstructions (such as electrical boxes for light fixtures) should have insulation installed over them.

CAUTION: Insulation must be kept at least 3 inches away from recessed light fixtures.

Installation Safety

You should keep in mind some of the health hazards involved with installing Fiberglas. It has tiny fibers of resin (glass) that can be breathed in and cause trouble in your lungs. Just be careful and take the proper precautions. If you work with Fiberglas all day, make sure you wear a mask over your nose and mouth.

Notice that the person in the photos on pages 394 and 395 is wearing long sleeves and gloves. This will prevent the itching that this type of material can cause. You should wear goggles—not just glasses. These will keep the fibers out of your eyes and prevent damage to your vision.

A hat and proper long pant legs are also called for in this job. Proper shoes are needed, since you are working around a construction area. The nails are still exposed in some types of work, so make sure the shoes can protect your feet from the nails.

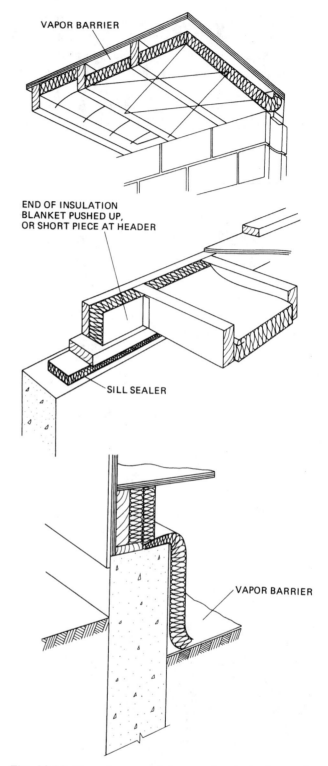

Fig. 13-11 *Note placement of the vapor barrier in the installation of insulation.* (New York State Electric and Gas)

The usual safety procedures should be followed in using a ladder and climbing around an attic or any place that may be elevated from floor level. A hard hat helps to keep you from getting roofing nails in your scalp when you are installing insulation near the underside of the roof.

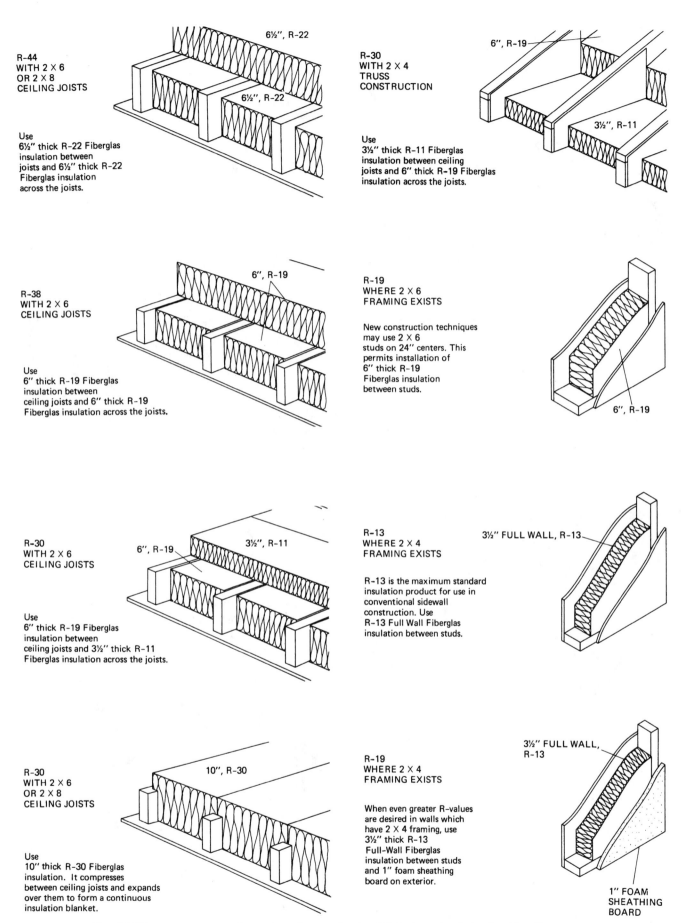

Fig. 13-12 (A) Optimum insulation for ceilings and attics. (B) Optimum insulation for sidewalls. (Certain-Teed)

Fig. 13-13 *Installing insulation in ceilings. Note the long-sleeved shirt, gloves, hat, respirator, and goggles.* (Owens-Corning)

Installing Insulation in Unfloored Attics

Install faced building insulation in attics that have no existing insulation. See Fig. 13-14. The vapor barrier should face down toward the warm-in-winter side of the structure.

Fig. 13-14 *Install insulation with a vapor barrier facing in attics with no existing insulation. The vapor barrier should face the warm-in-winter side of the structure.* (Owens-Corning)

Use unfaced insulation when you are adding to existing insulation. See Fig. 13-15. Batts or blankets can be laid either at right angles or parallel to existing insulation.

Start from the outside and work toward the center of the attic. See Fig. 13-16. Lay insulation in long runs

Fig. 13-15 *Use insulation without a vapor barrier facing when adding to existing insulation. Batts or blankets can be laid at right angles or parallel to existing insulation.* (Owens-Corning)

Fig. 13-16 *Work toward the center of the attic, laying long runs of insulation first. Fill in with short pieces. Be sure to butt insulation tightly at all joists. Insulation should cover top plate but must not block eave vents. Insulation must be kept at least 3 inches away from recessed light fixtures.* (Owens-Corning)

first and use leftovers for shorter spaces. Be sure to butt insulation tightly at joints for a complete barrier.

Insulation should extend far enough to cover the top plate. But it should not block the flow of air from the eave vents. See Fig. 13-17.

Fig. 13-17 *If the attic is used for storage, the floorboards must be removed. The flooring will limit the R value to the amount of uncompressed insulation which can fit between the joists.* (Owens-Corning)

Insulation should be pushed under wiring unless the wiring will compress the insulation. The ends of batts or blankets should be cut to lie snugly around cross-bracing and wiring. Fill spaces between the chimney and the wood framing with unfaced Fiberglas insulation.

Scuttle holes, pulldown stairways, and other attic accesses should also be insulated. Insulation can be glued directly to scuttle holes. Pulldown stairways may need a built-up framework to lay batts on and around.

Installing Insulation in Floored Attics

If the attic is not used for storage, unfaced batts or blankets may be laid directly on top of the floor. See Fig. 13-18.

Fig. 13-18 *If the attic is not used for storage, unfaced batts or blankets may be laid directly on top of the floor.* (Owens-Corning)

Floorboards must be removed before insulation is installed if the attic is to be used for storage. Since the flooring will be replaced, the R value will be limited. It is limited to that of the amount of uncompressed insulation that can be fitted between the joists.

Remove the flooring in the attic. It should then be insulated as shown in Fig. 13-19. Replace the floorboards and stairsteps.

Installing Insulation in Floors

Floors over unheated areas should be insulated. The recommended insulation R values should be used. See Fig. 13-20. With faced insulation, the vapor barrier should be installed face up. This is toward the warm-in-winter side of the house. Insulation should overlap the bottom plate. It should also overlap the band joist. See Fig. 13-21.

Fig. 13-19 *Once the flooring has been removed, the attic should be insulated in this manner.* (Owens-Corning)

Fig. 13-20 *Floors over heated areas should be insulated to the R values recommended.* (Owens-Corning)

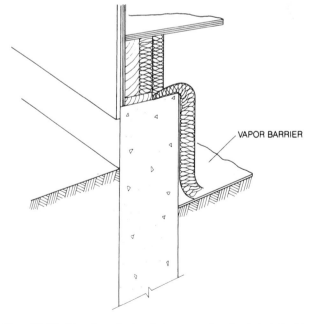

Fig. 13-21 *Note how the insulation vapor barrier is installed in the crawl space.*

Bow supports Fiberglas insulation can be supported by heavy-gage wires. The wire may be bowed or wedged into place under the insulation.

Crisscross wire support Insulation may also be supported by lacing wires around nails located at intervals along the joists.

Chicken-wire support Chicken wire nailed to the bottom of floor joists will also support the insulation.

Insulating Basement Walls

Masonry walls in conditioned areas may be insulated. Install a furring strip or 2-×-4 framework. Nail the top and bottom strips in place. Then, install vertical strips 16 or 24 inches O.C. Next cut small pieces of insulation. Push the pieces in place around the band joist. Be sure to get insulation between each floor joist and against the band joist.

Install insulation either faced or unfaced. Unfaced insulation will require a separate vapor barrier. Do not leave faced insulation exposed because the facing is flammable. It should be covered with a wall covering. Cover the insulation as soon as it has been installed.

Insulating Crawl Spaces

Measure and cut small pieces of insulation to fit snugly against the band joist. This is done so that there is no loss of heat through this area. Push pieces into place at the end of each joist space.

> CAUTION: Do not use faced insulation; the facing is flammable.

Use long furring strips. Nail longer lengths of insulation to the sill. Extend it 2 feet along the ground into the crawl space. Trim the insulation to fit snugly around the joists.

On walls that run parallel to the joists, it is not necessary to cut separate header strips. Simply use longer lengths of insulation and nail (with furring strips) directly to the band joist. Install the insulation. Lay the polyethylene film under the insulation and the entire floor area. This prevents ground moisture from migrating to the crawl space. Use 2-×-4-inch studs or rocks to help hold the insulation in place.

Installing Insulation in Walls

Insulating walls can pay dividends. Figure 13-22 shows that the sidewalls lose 15.2 percent of the heat lost in a ranch house. In the two-story house, where

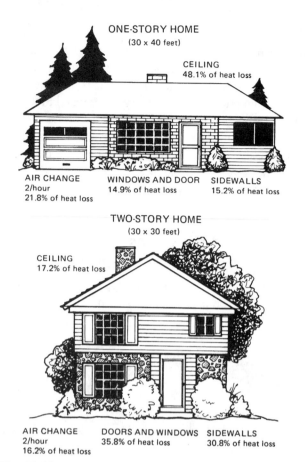

Fig. 13-22 *Heat loss in a one-story and a two-story house.* (New York State Electric and Gas)

there is more wall area, the loss is 30.8 percent. This means it is very important to properly insulate walls.

Figure 13-23 shows how insulation fits into the wall area. It is placed between the studs. The outside wood is also an insulation material.

Insulation for walls can be R-19. Use a combination of R-11 insulation with 1-inch (R-8) *high-R sheathing* or R-13 insulation with ¾-inch (R-8) *high-R sheathing* on the interior or exterior. See Fig. 13-24. R-13 insulation may also be used with ⅝-inch *high-R sheathing* only on the exterior.

In placing R-30, use a single-layer blanket. Or you can use R-19 and R-11 blankets. This combination is shown in Fig. 13-25.

High-R batts (R-26, R-30, and R-38) are made in full 16- and 24-inch widths. They expand over the top of the joists. This presents a continuous barrier to heat flow. See Fig. 13-26. Floors can have R values. Choose the right single layer of blanket insulation. See Fig. 13-27.

Cutting fiberglas insulation Measure the length of Fiberglas insulation needed. Place the insulation on a piece of scrap plywood or wallboard. Compress the

NO DIAGONAL WALL BRACING REQUIRED WITH PLYWOOD PANEL SIDING

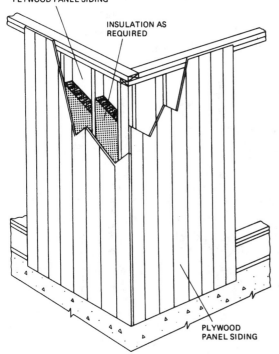
INSULATION AS REQUIRED

PLYWOOD PANEL SIDING

Fig. 13-23 *Placement of insulation inside a wall between studs.*
(American Plywood Association)

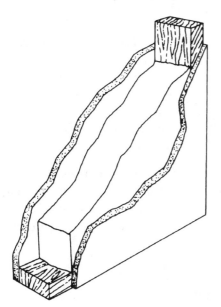

Fig. 13-24 *R-19 insulation can be obtained by combining R-11 insulation with 1-inch (R-8) high-R sheathing or combining R-13 insulation with ¾-inch (R-8) sheathing on the exterior. R-13 insulation may also be used with ⅝" high-R sheathing only on the exterior.* (Owens-Corning)

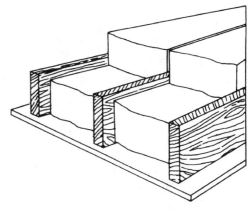

Fig. 13-25 *R-30 can be obtained by using a single R-30 blanket or R-19 and R-11 Fiberglas blankets in combination.* (Owens-Corning)

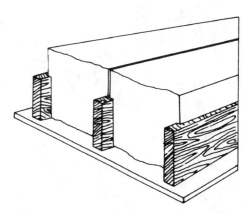

Fig. 13-26 *Single-layer Fiberglas high-R batts (R-26, R-30, R-38) are made in full 16-inch and 24-inch widths so that they expand over the top of the joists for a continuous barrier to heat flow.* (Owens-Corning)

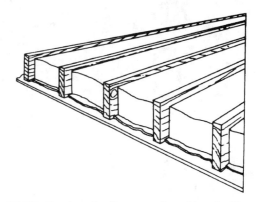

Fig. 13-27 *R values for floors can be achieved with a single layer of Fiberglas blanket insulation.*

material with one hand. Cut the material with a sharp knife. See Fig. 13-28.

Installing faced insulation Install faced insulation using either the insert or faced stapling method. Cut lengths 1 inch longer than needed. This is so that facing may be peeled back at the top and bottom. This makes it possible to staple to plates. Make sure the insulation fills the entire cavity. See Fig. 13-29.

Insert stapling Insulation is pushed into the stud or joist cavity. Vapor barrier flanges are stapled to the sides of the studs. Make sure that flanges do not gap. This will allow vapor to penetrate. See Fig. 13-30.

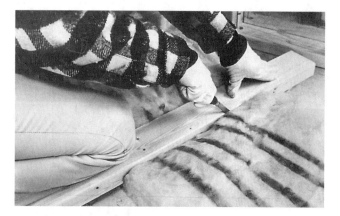
Fig. 13-28 *Cutting Fiberglas insulation.* (Owens-Corning)

Fig. 13-29 *Installing faced insulation.*

Faced stapling Vapor barrier flanges are overlapped. They are then stapled to the edge of the stud. Flanges must be kept smooth. This is necessary for proper application of the wall finish. See Fig. 13-31.

Fig. 13-31 *Faced stapling.* (Owens-Corning)

Installing unfaced insulation Wedge unfaced batts into the cavity. Make sure they fit snugly against the studs. Also fit the top and bottom plates snugly. See Fig. 13-32.

Filling small gaps Hand-stuff small gaps around window and door framing. This can be done with scrap pieces. Cover the stuffed areas with a vapor-resistant material. See Fig. 13-33. Insulation must fit snugly against both the top and bottom plates. It should also fit against the sides of the stud opening. See Fig. 13-34. Improperly installed, it will permit heat to escape.

Insulating behind wiring and pipes Insulation should be fitted behind or around heat ducts. It should also fit closely around pipes and electrical boxes. This will help keep pipes from freezing. It also prevents unnec-

Fig. 13-30 *Insert stapling.*
(Owens-Corning)

Fig. 13-32 *Installing unfaced insulation.* (Owens-Corning)

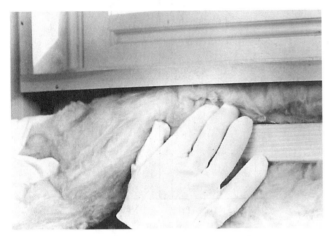

Fig. 13-33 *Filling small gaps.* (Owens-Corning)

Fig. 13-34 *Close doesn't count.* (Owens-Corning)

essary heat loss. Repair any tears in the vapor barrier of faced insulation. This prevents condensation problems. See Fig. 13-35.

Filling narrow stud spaces With unfaced insulation you can cut to stud width. Wedge the insulation into place. This area must be covered with a vapor barrier.

Fig. 13-35 *Insulating behind wiring and pipes.* (Owens-Corning)

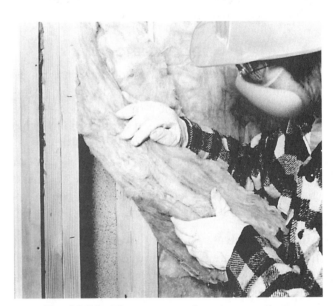

Fig. 13-36 *Filling narrow stud spaces.* (Owens-Corning)

With faced insulation, make sure the vapor barrier is properly stapled to the studs. See Fig. 13-36.

Insulating around wiring Split unfaced insulation and install it on both sides of the wiring. Install faced insulation behind the wiring. Pull it through to permit stapling. See Fig. 13-37.

VAPOR BARRIERS AND MOISTURE CONTROL

Excess moisture may be a concern in a well-insulated house because excess air changes have been eliminated. Therefore, some consideration must be given to the tight house. The insulation has been properly located. It keeps the inside air inside. This means it doesn't require

Fig. 13-37 *Insulation around wiring.* (Owens-Corning)

too much energy to reheat it. It doesn't have to be re-cooled too much during the summer.

Condensation

A house must *breathe*. Condensation that is not caused by insulation can be controlled. Condensation is simply water vapor. Water in the form of a fog comes from a variety of sources in a home. It is produced in baths, in the kitchen, and in the laundry. It will distribute itself throughout the house. Unless it is controlled, it can condense. Just as frost forms on the outside of an iced glass on a warm day, vapor condenses on house surfaces. Slowly, but surely, it can cause damage.

Moisture control There are two ways to control moisture. They are elimination and ventilation. Elimination means to get rid of the source of moisture. This can be leaks, wet ground in crawl spaces, or small leaks in the structure.

Ventilation means simply getting rid of moisture. It carries excess humidity outside. This is done with vent fans or hoods in the kitchen. The bathroom may be vented to the outside. Automatic clothes dryers must be vented to the outside. They are a major source of moisture.

Vent fans at high-moisture-producing locations are helpful; for example, the fan over a kitchen range. The bathroom may have a fan to remove moisture. Any of these fans can be equipped with a humidistat that will turn it on when the humidity gets to a preset level.

Attic ventilation Attic ventilation is very important for many reasons. It helps eliminate moisture in the attic. That moisture can leak into the living quarters and ruin the ceiling. Or, it can rot studs or other parts of the house. One way to provide ventilation is to place vents in the end of the roof structure. See Fig. 13-38. They

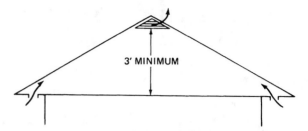

NOTE: There is 1 ft² inlet and 1 ft² outlet for each 600 ft² of ceiling area, with at least half of the vent area at the top of the gables and the balance at the eave.

(A)

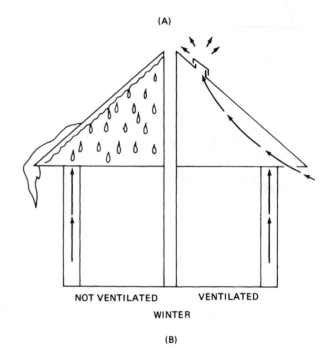

(B)

Fig. 13-38 *(A) Spacing of attic vents. (B) Ventilation can control condensation.*

can be various shapes. Some use cupolas to allow air to move upward and out. Some use a vented ridge section to allow hot air to escape in the summer and a constant flow of air all during the year.

These openings must be covered to prevent insects, birds, snow, and rain from entering. There are a number of materials used to cover the openings. Table 13-1 shows some of them.

Table 13-1 *Attic Ventilation Coverings*

Type of Covering	Size of Covering
¼-inch hardware cloth and rain louvers	2 × net vent area
8-mesh screen	1¼ × net vent area
¼-inch hardware cloth	1 × net vent area
8-mesh screen and rain louvers	2¼ × net vent area
16-mesh screen	2 × net vent area
16-mesh screen and rain louvers	3 × net vent area

Keep in mind that it takes 1 square foot of inlet area and 1 square foot of outlet area for each 600 square feet of ceiling area. At least half of the vent should be at the top of the gables and the balance should be at the eave.

THERMAL CEILINGS

Thermal ceiling panels have insulating values up to R-12. They are the only types of insulation that can be used for some types of homes. Take for instance the A frame. The only way the ceiling can be insulated is with thermal ceiling panels. See Figs. 13-39 and 13-40. These types of ceiling panels have a finish that is

pleasing. They also are sound-absorbing. They can be used in attics being turned into living space or attics opened up to the rest of the home. If the attic is already finished, this may be the only way to insulate.

A variety of sizes is available, with different types of finishes on the face. There are sizes and faces to suit almost every installation requirement. They come in sizes from 2 × 4 feet up to 4 × 16 feet. Figure 13-41 shows some of the applications for these panels.

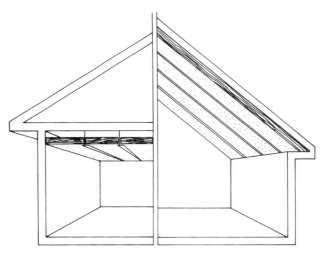

Fig. 13-41 *Thermal ceiling panels for cathedral ceilings and regular ceilings.* (Owens-Corning)

Installing Thermal Ceiling Panels

There are a number of ways to install thermal ceiling panels.

Grid system This uses the suspended interlocking metal grid system. It can be installed on the 2-×-4-foot or 4-×-4-foot system. A larger-faced grid system is needed for 3-inch R-12 panels. See Fig. 13-42A.

Solid wood beams Solid beams as in Fig. 13-42B are used in many homes. The beam becomes a support system for the insulation panels. The panels can be removed and replaced if they are damaged.

False-bottom beams A system of 1-×-6-inch false-bottom or box beams provides the beauty of beams

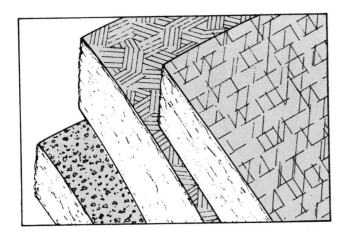

Fig. 13-39 *Thermal ceiling panels are easy to install.* (Owens-Corning)

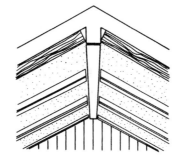

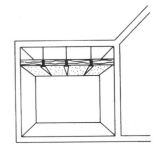

Fig. 13-40 *Typical installations of thermal ceiling panels.* (Owens-Corning)

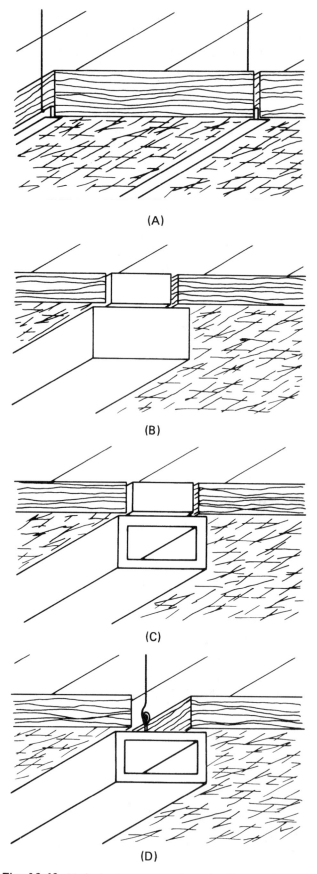

(A)

(B)

(C)

(D)

Fig. 13-42 *Methods of supporting thermal ceiling panels (Owens-Corning); (A) Grid system; (B) Solid wood beams; (C) False-bottom beams; (D) Suspended beams.*

without the expense. When wood beam support is used, wood block spacers must be installed between the existing ceiling or joists and the beam. This is so that panels are not compressed when the beams are nailed up.

Suspended beams False beams may also be used to lower a ceiling like a grid system. Beams are suspended. They use screw eyes attached to the existing ceiling joists.

Decorative Beams

The suspended-beams method of putting up ceiling insulation is easy.

In Fig. 13-43 note how the plastic beams are cut to length with a knife.

Fig. 13-43 *Cutting a plastic beam.*

Make a sketch on paper showing the pattern you want for your beams. Now it is simple to cut the beams to size. Use a knife or sharp tool with a razor blade.

Measure the exact length you need. Cut the beams to size. See Fig. 13-44. Use a ruler or chalk line to mark where you want to install the beam. Beams can be glued to other material. They can be glued to the ceiling with an adhesive. You might want to glue a piece of furring strip down the groove (Fig. 13-45) of the beam. This will allow you to attach screw hooks. They will be wired to the old ceiling later. See Fig. 13-42D. Angles can be cut in the beam with a miter box.

Fig. 13-44 *A ceiling using plastic foam beam.*

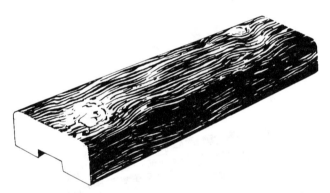

Fig. 13-45 *Note the groove in the beam. You may want to glue a furring strip in it.*

If you want to glue the beam up, use a brace such as that shown in Fig. 13-46. This can be done if you want to make a ceiling resemble the one in Fig. 13-42B. Nicks and scratches in this type of plastic foam beam

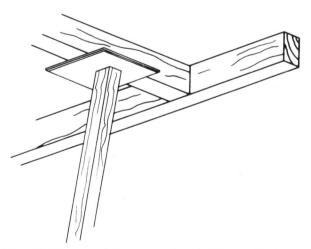

Fig. 13-46 *Method of supporting beam while the adhesive sets.*

can be touched up with shoe polish. Since the beam is made of foam, it also is very good insulation.

STORM WINDOWS

A single pane of glass is a very efficient heat transmitter. Window and door areas approach 20 percent of the sidewall area of the average home. It is important that heat loss from this source be minimized.

Properly fitted storm windows are a must for the well-insulated home. They come in styles suitable to the design of any home.

Older windows have storms fitted on the outside. This can cause some reduction in heat loss. In fact, an old window with a triple-deck outside storm window reduces heat loss by 67 percent. Adding another layer of insulating glass or plastic, such as in Fig. 13-47, makes it possible to reduce heat loss by up to 93 percent. The inside storm window is easily added to any window. Figures 13-48 through 13-54 show how the inside storm window is made and placed into position. Also check Table 13-2.

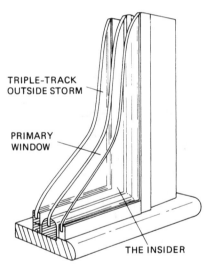

Fig. 13-47 *Standard window with storm window on exterior and the "insider" on the inside.* (Plaskolite)

Weather-stripping a window There are three basic types of weatherstripping for double hung windows. The foam-rubber type is probably the best; it is the easiest to work with. See Fig. 13-55. Each of the weatherstripping materials comes in coils. It often comes in complete cut-to-size kits. Metal-frame windows (such as casement, awning, or jalousie styles) use invisible vinyl tape, joint-sealing self-adhesive foam rubber, or casement aluminum strip.

Installing weatherstripping on a window Metal strip insulation should be nailed at three positions. The

Fig. 13-48 *Measuring for the inside storm window.* (Plaskolite)

Fig. 13-49 *Cutting the inside window.* (Plaskolite)

Fig. 13-50 *Cutting the trim for the window.* (Plaskolite)

Fig. 13-51 *Placing the trim around the plastic.* (Plaskolite)

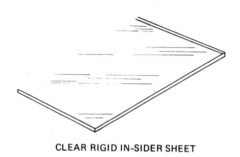

CLEAR RIGID IN-SIDER SHEET

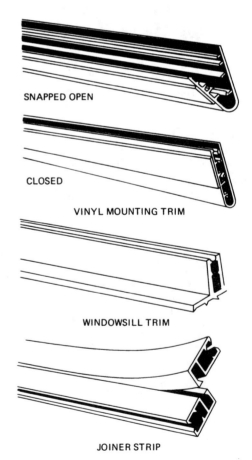

SNAPPED OPEN

CLOSED

VINYL MOUNTING TRIM

WINDOWSILL TRIM

JOINER STRIP

Fig. 13-52 *Pieces that go into the making of a storm window for the inside.* (Plaskolite)

Fig. 13-53 *Installing the window.*

Fig. 13-54 *Placing the plastic inside the window.*

sash channels inside the frame (making sure you leave the pulleys near the top uncovered) are one location. Underneath the upper sash on the inside is another location. Figure 13-56 shows how the spring metal type of window is insulated with weatherstripping.

Vinyl and adhesive types of weatherstripping can be installed as shown in Fig. 13-57. Vinyl strips should be nailed in place at three locations (see Fig. 13-58):

Table 13-2 *Storm Windows*

Window (Glass Area Only)	Approximate R of Unit*
Single glazing	0.88
Double glazing with 1/4-inch air space	1.64
Double glazing with 1/2-inch air space	1.73
Single glazing plus storm window	1.89
Double glazing plus storm window	2.67

Courtesy of New York State Electric and Gas Corp.
*R value for vertical air space is for air space from 3/4 to 4 inches thick. R values for glass are actual for type of window listed. R (resistance) indicates the amount of heat a material will prevent from passing through it in a given time. The higher the R value, the more heat the material will hold back, and hence, the better the insulation of that material.

1. On the outer part of the bottom rail of the upper sash.

2. On the same location on the lower sash.

3. On the outer surface of the parting strips.

There are other methods that can help. Invisible polyethylene tape replaces the old type of caulk. This fits around the windows to seal them inside. See Fig. 13-57.

STORM DOORS

In most of the country it is necessary to install storm doors. These fit on the outside and serve as screen doors during the summer. The glass and screen can be taken out and exchanged as needed. See Fig. 13-59. Of course, the doors aren't too effective unless they have door closers to make sure they close. See Fig. 13-60. Storm doors usually come in a prehung package. They are easily installed by simply placing screws through the holes and checking for plumb. They are designed for the do it-yourselfer. In most cases homes do not come with storm doors. The first owner has to install them. This is also true of the mailbox.

Take a look at Table 13-3. It will show you just how important the storm door really is. It can make quite a difference in heat loss.

SEALANTS

There are a number of materials on the market for sealing air leaks in a house. Leaks appear around doors, windows, and chimneys. There are a number of caulking compounds made to fill these gaps. Some caulks

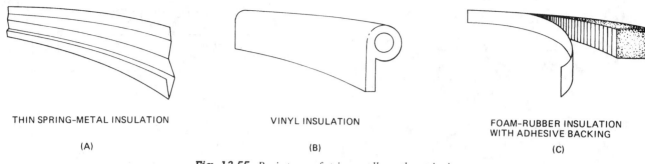

THIN SPRING-METAL INSULATION

(A)

VINYL INSULATION

(B)

FOAM-RUBBER INSULATION
WITH ADHESIVE BACKING

(C)

Fig. 13-55 *Basic types of strip or roll weatherstripping.*

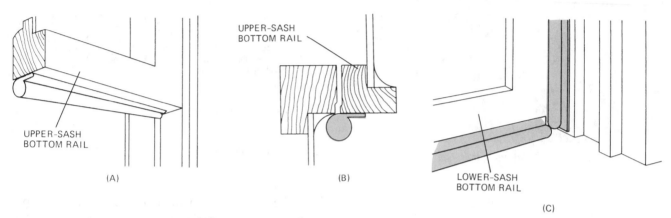

UPPER-SASH
BOTTOM RAIL

(A)

UPPER-SASH
BOTTOM RAIL

(B)

LOWER-SASH
BOTTOM RAIL

(C)

Fig. 13-56 *Vinyl tape and self-adhesive foam-rubber strip installation.*

will last longer than others. However, one of the best appears to be a paintable silicone sealant made by Dow-Corning. See Fig. 13-61. These sealants are usually packaged in 13-ounce tubes suitable for use in a caulking gun. Some, of course, are in tubes for use inside the house.

Acrylic latex sealant This sealant cures to a rubbery seal. Used inside and outside, it can seal any joint up to $\frac{1}{2} \times \frac{1}{2}$ inch in size. It is also used between common construction materials where movement is expected. It can be cleaned from tools or hands with water. Do not use under standing water or when rain is expected. It has an expected life of 8 to 10 years and is paintable in 30 to 60 minutes after application.

Vinyl latex caulk This is a middle-priced sealant. It has the same advantages and restrictions as the acrylic latex sealant. Life expectancy is 3 to 5 years.

Butyloid rubber caulk Especially useful in narrow openings up to $\frac{3}{8}$ inch, this is used where movement is expected. It can be used in glass-to-metal joints. It is also used between overlapping metal. It can be

used to seal gutter leaks or under metal or wooden thresholds. This type of caulk is useful under shower door frames. Life expectancy of the caulk is 7 to 10 years.

Roof cement Roof cement may be used on wet or dry surfaces. Cold weather application is possible. It stops leaks around vent pipes, spouts, valleys, skylights, gutters, and chimneys. This cement can be used to tack down shingles. It is also useful for sealing cracks in chimneys or foundations. Roof cement comes in black. It is not paintable.

Concrete patch This is a ready-made concrete and mortar patching compound. It contains portland cement. This patching compound is quick and easy. It is used to repair cracks in concrete sidewalks, steps, sills, or culverts. It is also recommended for cracked foundations, tuck pointing, and stucco repair.

WINTERIZING A HOME

There are a number of ways to winterize a home. They all deal with insulation and stopping air leaks. Check

Fig. 13-57 *Applying clear tape around windows.* (Manco Tape)

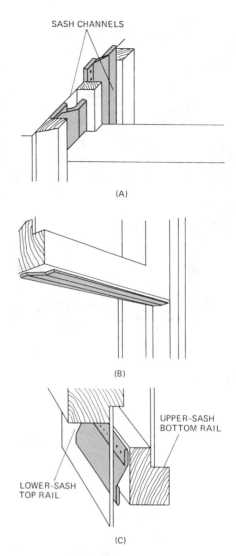

Fig. 13-58 *Metal strip insulation nailed at three positions; (A) sash channels inside the frame; (B) underneath the upper sash; and (C) on the lower sash bottom rail.*

the 33 ways listed in Table 13-4. Note how effective they are at saving energy.

INSULATING FOAM SEALANT

Energy conservation is not the only use for sealants. They can be used for a number of purposes in any construction job. The low expansion, rigid foam can be obtained in a pressurized can and applied to foundations, electrical outlets, switches, and in the attic, as well as everywhere an electrical cable goes through a stud or rafter. See Fig. 13-62. The foam can be used to prevent rodents, insects, and dust from entering a building. Once applied it expands and fills in a space about 5 to 6 times its original bead. When applied around the electrical receptacle and switch boxes, it expands to fill in the area between the box and the dry wall to seal out dust, ants, and other insects. When using foam, be sure to wear vinyl gloves; the foam is hard to remove from your

skin. The application releases a flammable gas. Do not smoke while applying the foam. Use in a well-ventilated area with proper respiratory protection. The product is extremely sticky and difficult to remove from the skin. Use acetone to remove. Acetone can be bought in liquid form, or you can use nail polish remover that contains acetone for a satisfactory job of removal.

The foam, which takes about an hour to cure, is waterproof, airtight, bonds to most materials, cures rigid, and trims easily. It is sandable, paintable, and can be used for interior or exterior areas. It fills, seals, and insulates. Figure 13-62 shows how the expanded foam is used to seal the hole that was drilled for Romex cable.

Fig. 13-59 *Metal storm door with full glass panel.*

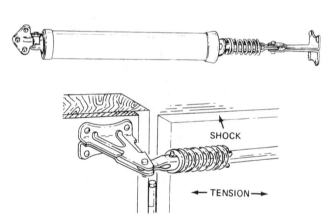

SHOCK

← TENSION →

Fig. 13-60 *Storm door closer (top) and spring for keeping the door from being blown off its hinges. (Stanley Tools)*

Table 13-3 *Storm Doors*

Door Type	Approximate R of Unit*
Solid wood, 1-inch	1.56
Solid wood, 2-inch	2.33
Solid wood, 1-inch plus metal/glass storm door	2.56
Solid wood, 2-inch plus metal glass/storm door	3.44
Solid wood, in-inch plus wood/glass (50%) storm door	3.33
Solid wood, 2-inch plus wood/glass (50%) storm door	4.17
Doors with rigid insulation core	up to 7

Courtesy of New York State Electric and Gas Corp.
*R (resistance) indicates the amount of heat a material will prevent from passing through it in a given time. The higher the R value, the more heat the material will hold back, hence the better the insulation. To find the R value of a building material, multiply above R value by actual thickness of the material.

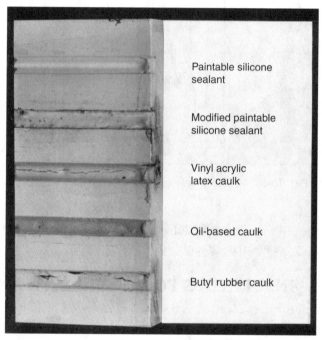

Paintable silicone sealant

Modified paintable silicone sealant

Vinyl acrylic latex caulk

Oil-based caulk

Butyl rubber caulk

Fig. 13-61 *Sealant samples ¼ inch wide by ¼ inch deep after 600 hours in an accelerated weathering machine. This is the equivalent of more than one year weathering outdoors in Florida. Only the paintable silicone sealant seems to have held up. (Dow-Corning)*

CHAPTER 13
STUDY QUESTIONS

1. Why weren't older homes insulated?

2. What does insulation do?

3. What is a vapor barrier?

4. What is meant by moisture control?

5. What does R value mean?

6. What is an insulation batt?

7. Why do crawl spaces have to be insulated?

8. What is the purpose of ventilation?

9. Which way should the vapor barrier face in an insulated home?

10. What are floorboards?

11. What does faced insulation mean?

12. What is the meaning of unfaced insulation?

13. Why can't you place faced insulation close to a chimney or source of heat?

Table 13-4 *Some Energy-Saving Ideas*

33 Ways
To Winterize Your Home

KEEPING THE COLD AIR OUT

1. Seal all outside doors, including basement doors, with weather-stripping material. In some cases, you can use old carpet strips.
2. Put masking tape around moving parts of windows. Caulk around window and door frames, including those in the basement. You can stop drafts under doors by placing rugs at the bottom.
3. Check pipes entering your home. You can keep the cold air out by packing rags around them.
4. Check light bulb fixtures for air leaks.
5. Put tape over unused keyholes.
6. Make sure unused flue or chimney covers fit tightly.
7. Keep fireplace dampers closed tightly when not in use.
8. Seal your foundation and sill plate with caulking material, insulation or rags.

KEEPING THE WARM AIR IN

9. Install storm windows and storm doors. If you don't have storm windows, you can substitute plastic sheeting but make sure it's tacked tightly all around the edges.
10. Install insulation between warm and cold areas. Begin by insulating the attic floor.
11. Wherever possible, carpet floors. If your attic floor can't be insulated, lay down a carpet.
12. Close off rooms you don't use, particularly those with the biggest windows and the largest outside walls.

MAINTAINING YOUR HEATING SYSTEM

13. Keep your heating system clean.
14. Make sure there is nothing blocking your registers, radiators or baseboard heaters.

15. Keep cold air returns clear.
16. Change or clean furnace filters monthly.
17. Have a qualified person adjust your heating system.

ADJUSTING YOUR LIVING HABITS

18. Keep the thermostat set at 68 degrees during the day. If this seems too cold for you, try wearing a sweater.
19. Set the temperature back at least 3 degrees at night.
20. If you're going away for a few days, set the thermostat at 60 degrees before you leave.
21. Try lowering the temperature in those rooms you don't spend much time in by adjusting registers, radiators or thermostats.
22. Keep humidifiers at the 30 percent mark or place pans of water on warm air registers or radiators. You'll feel more comfortable at relatively lower temperatures simply by maintaining the right humidity in your home.
23. Cover windows with drapes or curtains.
24. Open your drapes during the day to let the sunshine in and close them at night to keep the cold air out.
25. Try locating your furniture away from cold outside walls and windows.

SOME DO'S

26. Do fix leaky faucets, especially hot water taps.
27. Do use cold water for clothes washing.
28. Do turn off the lights, TV, radio or record player when not needed.

SOME DON'TS

29. Don't permanently fasten windows and doors shut—they may be needed for an emergency.
30. Don't use kitchen appliances to heat your home.
31. Don't use portable heaters as the main source of heat—be particularly cautious with oil or gas space heaters not vented or vented to your chimney.
32. Don't seal off attic ventilation.
33. Don't put insulation over recessed light fixtures.

Fig. 13-62 *Expandable foam used to fill the hole in an electrical wiring installation. It can serve to insulate, seal out insects, and prevent fire movement as well.*

14. Why should floors over unheated basements be insulated?

15. How do you cut Fiberglas insulation?

16. What is insert stapling?

17. How do you fill narrow stud spaces with insulation?

18. What is condensation?

19. How do you prevent condensation?

20. If you have a house with 1200 square feet of ceiling space and need to ventilate it, how much attic ventilation space would be needed if you used 8-mesh hardware cloth to cover the vent areas?

14
CHAPTER

Preparing Interior Walls & Ceilings

THE INSIDE WALLS AND CEILINGS OF A building are finished after the outside walls are finished and after doors and windows have been installed. This makes the building weathertight and protects the interior from damage. This is important because most interior materials are not weatherproof. Then the interior can be finished on bad days, when rain or snow makes work on the exterior impractical.

As a rule, ceilings are finished first; then the walls are finished. Floors are completed last. Keep in mind that workers will drop tools and materials as they work; it does no damage to drop tools on unfinished floors.

Both walls and ceilings can be finished the same way. This is because the same materials can be used for both walls and ceilings. Several steps can be involved. It is common for carpenters to apply vapor barriers. Other items to be installed include insulation, plumbing, and electrical wiring. The carpenter usually does not do the plumbing and wiring. However, the carpenter does install the insulation.

The most common interior material is gypsum board. Gypsum board is made from a chalk-like paste. This chalk is the gypsum. It is spread evenly between two layers of heavy paper. When the gypsum hardens, it forms a rigid board. The paper cover adds to its strength and durability. Finish may be applied directly to the paper. Wood panels or other materials may be applied over the gypsum board. Gypsum board is also called by other names such as *sheetrock* and *drywall*. Other interior wall panels are made of wood, plaster, and hardboard. Wooden boards are also used for interior walls.

Carpenters use special skills in covering walls and ceilings. These skills are related to working the various materials. Plaster is held up by strips called *lath*. Lath can be made of several things. Wood boards, special drywall, or metal mesh may be used. Carpenters often nail the wood edge guides and lath. However, the plasterer, not the carpenter, puts up the plaster. Skills carpenters need to prepare internal walls and ceilings include:

• Measuring and cutting wall materials

• Installing insulation

• Installing gypsum board

• Putting up paneling

• Preparing walls for plaster

• Putting up board panels

• Installing ceiling materials

SEQUENCE

As a rule, ceilings are done before either walls or floors. The carpenter usually does not select the interior materials. However, the carpenter must plan the way in which the work is done. To do this, the carpenter must know the various processes. The carpenter must also use the right process. The sequence for the carpenter to use is

1. Make sure the material and the fasteners are available.

2. Plan the correct work sequence.

3. Put insulation in the walls.

4. Put in the wall material.

5. Put in trim and molding.

PUTTING INSULATION IN WALLS

Today nearly all exterior walls are insulated. Insulation is also commonly used in interior walls. Interior wall insulation reduces sound transmission between rooms or apartments. Insulation must be put in before the walls are covered.

Most insulation today is made of loose woven fibers. The material comes in standard-width rolls or strips. These are made to fit between the wall studs. There are two commonly used standard widths. One is for walls with studs that are 16 inches O.C. The other is for walls with studs that are 24 inches O.C. The insulation is usually made from glass or mineral fibers that are woven into a thick mat. The mat is then glued to a heavy paper backing a little wider than the mat. This way, the edges can be used to nail the insulation to the studs. The insulation may come in long rolls, or it may come in precut lengths. The precut lengths are called batts.

The paper side of the material is placed toward the living space. See Fig. 14-1. The paper backing may also be coated with metal foil. The foil helps reduce the energy loss. See Fig. 14-2.

To put up insulation, the material must first be cut. Batts are cut to the proper length for the walls. The batts are then placed between the studs. The top of the batts should firmly touch the top plate in the walls. The batts should reach fully from top to the bottom. They should reach without being stretched. Either nails or staples may be used as fasteners. The fasteners may be driven on the inside of the studs. The edges of the paper are sometimes lapped on the edge of the studs. Fasteners are started at the top of the material. The nails

Fig. 14-1 *The paper back on the insulation should face the living space. It also acts as a vapor barrier.*

Fig. 14-3 *Insulation is stapled between the studs.*

Fig. 14-2 *Installing foil-backed insulation between the studs. Note that the foil is toward the living space. It is also the vapor barrier.*

should be placed about 12 inches apart. Fasteners are alternated from side to side. See Fig. 14-3.

INSTALLING A MOISTURE BARRIER

After the insulation has been installed, the carpenter should check the plans carefully. Vapor barriers are of-

ten put up on the inside. Plastic film from 2 to 4 mils thick may be used. It is fastened across the interior studs as in Fig. 14-4. Moisture barriers are essential in cold climates. They keep cold air and moisture from entering the building. Some sheathing can act as a vapor barrier. Also, the paper or foil back on the insulation acts as a vapor barrier. However, vapor barriers are needed on all outside walls and may also be used on interior walls. Vapor barriers may be made of plastic film, insulation paper, sheathing, or builder's felt.

PUTTING UP GYPSUM BOARD

As mentioned earlier, gypsum board is also called sheetrock or drywall. It is probably the most common interior wall material today. There are several advantages of using drywall: it is quick and easy to apply, reducing labor costs; it is not an expensive material; and there is no need to wait for it to dry.

Drywall comes in several sizes and thicknesses. Widths of 16, 24, and 48 inches are common. Lengths of 48 and 96 inches are used. One of the most often-used sizes is the standard 4-×-8-foot sheet. This large size means there are fewer seams to finish.

The edges of the sheets are finished or shaped in various ways. For most houses, the tapered edge is used. See Fig. 14-5. The edge is tapered to make a sunken bed for the seam. This way, the edges may be covered and hidden. The edges are first coated with

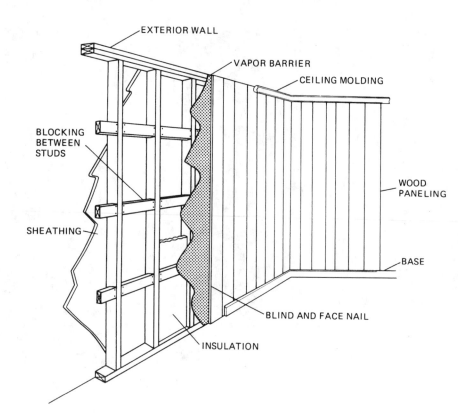

EXTERIOR WALL

VAPOR BARRIER

CEILING MOLDING

BLOCKING
BETWEEN
STUDS

SHEATHING

WOOD
PANELING

BASE

BLIND AND FACE NAIL

INSULATION

Fig. 14-4 *Plastic film may be applied over the inside of the studs and insulation for a vapor barrier.* (Forest Products Laboratory)

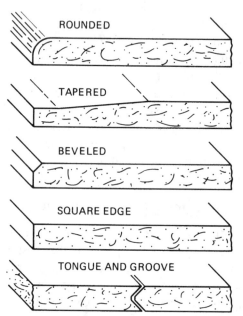

ROUNDED

TAPERED

BEVELED

SQUARE EDGE

TONGUE AND GROOVE

Fig. 14-5 *Standard edges for drywall.* (Gypsum Association)

Fig. 14-6 *After the sheetrock is nailed in place, the seams and nail dimples are covered.*

plaster and tape. Later coats are added to make a flat, smooth surface. See Fig. 14-6. This flat surface can then be covered, papered, or painted without seams showing through.

Drywall is held up by nails, screws, or adhesives. A special hole called a *dimple* is made for nails and screws. This hole is made by the hammer on the last stroke. The hole, or dimple, holds the plaster. The plaster helps cover the nail and screw to give a smooth surface. Figure 14-7 shows the dimple. On ceilings, nails or screws are also used with adhesives; however, not as many fasteners are needed. Adhesives are not used alone on ceilings.

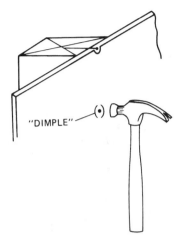

Fig. 14-7 *A dimple, or depression, is made around the nailhead. It is later covered with plaster.* (Forest Products Laboratory)

It is quicker to use adhesives rather than nails. However, some building codes prohibit the use of adhesives without nails. In such cases, the work is still fast because fewer fasteners are used. Panels can be put up very quickly with the adhesives. Then the nails may be added. No special holding is needed for nailing.

The thickness of drywall sheets can vary. Drywall is very fire-resistant. Layers can be used to increase the resistance to fire. Several layers can increase the strength. They will also reduce the transmission of noise between rooms. However, most construction is done with one thickness (single ply). There are three common thicknesses of drywall sheets. These are ⅜, ½, and ⅝ inch. In some locations, ceilings are made of ⅜-inch thickness. This reduces the weight being held. However, local building codes often detail what thickness can be used.

Drywall sheets can be used as a base for other types of walls. Drywall sheets are used for plaster lath. The sheets may be solid, or they may have holes (perforations) to help hold the plaster. See Fig. 14-8. Drywall sheets can also be used for backing thin finished panels. Drywall itself is also available with several types of surface finish.

Drywall sheets are applied in two ways. These are called horizontal (or parallel) and perpendicular applications. However, this refers to the direction of the long side with respect to the studs or joists. It does not refer to the direction with respect to the floor. For example, in perpendicular applications the long side is perpendicular, or at right angles, to the stud or joist.

Fig. 14-8 *Drywall sheets are often used for plaster lath.*

Horizontal application means that the long side runs in the same direction as the stud or joist. See Fig. 14-9.

Putting Up the Ceiling

Drywall is usually applied to ceilings first. Then the walls are covered. Drywall should be put on a ceiling in a horizontal application. Most manufacturers recommend that the first piece be placed in the middle of the ceiling. Then additional sheets are nailed up. A circular pattern from the center toward the walls is used. This procedure makes sure that the joists are properly spaced. The drywall acts as a brace to hold the joists in place. Joist spacing is important. Edges and ends of the drywall sheets can be joined only on well-spaced joists. Nonsupported joints will move. Movement can ruin the taped seams. This will give a bad appearance.

Measure and locate first piece Before the first sheet is applied, two base walls are selected. One wall should be parallel to the joists. See Fig. 14-10. The distance to the approximate center of the room is measured. Intervals of 4 feet are marked from this base wall. Next, the distance from the second wall to the center is found. Intervals of 4 feet are marked off, because the standard sheet size is 4 × 8 feet. The first sheet is located so that whole sheets can be applied from the center to that one wall. Cutting is done only on one side of the room. This saves time in trimming and piecing.

Before the first panel is applied, measure the distances between the joists at the center. The distance should be the proper 16 inches O.C. or 24 inches O.C. If the joists are spaced correctly, the first panel may be

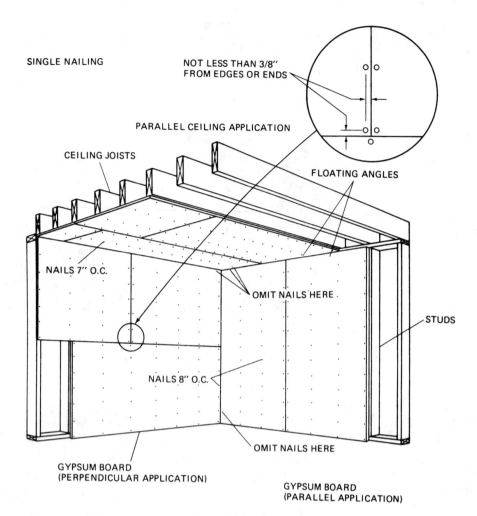

SINGLE NAILING

NOT LESS THAN 3/8"
FROM EDGES OR ENDS

PARALLEL CEILING APPLICATION

CEILING JOISTS

FLOATING ANGLES

Fig. 14-9 *Drywall sheets can be applied either parallel or perpendicular.* (Gypsum Association)

NAILS 7" O.C.

OMIT NAILS HERE

STUDS

NAILS 8" O.C.

OMIT NAILS HERE

GYPSUM BOARD
(PERPENDICULAR APPLICATION)

GYPSUM BOARD
(PARALLEL APPLICATION)

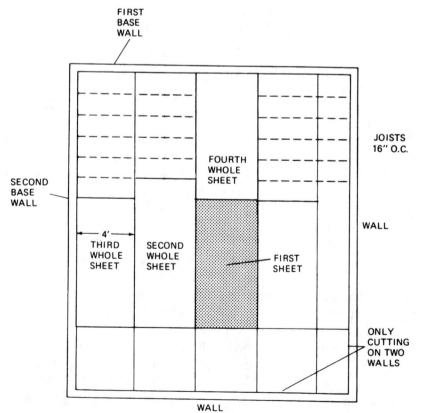

FIRST
BASE
WALL

JOISTS
16" O.C.

Fig. 14-10 *For ceilings, the first sheet is near the center. Measure so that whole sheets can be used toward the two base walls.*

SECOND
BASE
WALL

WALL

4'
THIRD
WHOLE
SHEET

SECOND
WHOLE
SHEET

FOURTH
WHOLE
SHEET

FIRST
SHEET

ONLY
CUTTING
ON TWO
WALLS

WALL

applied. The center nails should be driven first. Then joists should be spaced properly by slight pressure. Edges of the panels should rest on the centers of the joists. This provides a nail base for both panels.

Strongbacks Strongbacks are braces used across ceiling joists. They are made for two reasons. The first reason is to help space the joists properly. To space the joists, a flat board is nailed across the tops of them. The second reason is to help even up the bottom edges of the ceiling joists. One of the disadvantages of drywall is that studs and joists must be very even. Drywall cracks easily when it is nailed over uneven joists or studs.

To make a strongback, two boards are used. The bottom board is laid flat across the tops of the joists. A 2 × 4 should be used. The end of the board is nailed to the joist nearest the wall. Two 16d common nails should be used. The proper distance (16- or 24-inch marks) is measured. Pressure is applied to move the joist to the proper spacing. Then the joist is nailed at the proper spacing. This is done to each joist across the distance involved. This braces the joists into the proper spacing on centers.

The strongback is completed with the second brace. See Fig. 14-11. The second piece is made from a 2-×-4- or 2-×-6-inch board turned on edge. The edge provides a brace to even the joists to the same height. One end of the second brace is nailed to the end of the first brace. The carpenter should then stand on the flat piece over each joist. The carpenter's weight is used to force the joists to an even line. When the joists are even, the piece is nailed in place.

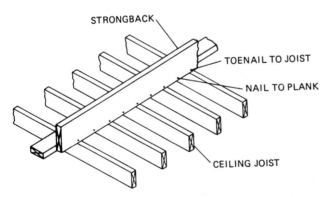

Fig. 14-11 *Strongbacks help even up and space joists. The flat piece is nailed first.* (Forest Products Laboratory)

Applying Ceiling Sheets

The ceiling sheets must be held up for nailing. Special braces are used in the nailing process. These temporary braces are called *cradles*. Figure 14-12 shows a cradle being used. Cradle braces are shaped like a large

Fig. 14-12 *A T-shaped "cradle" can be used to hold up pieces for nailing.* (Gold Bond)

T. The leg of the cradle is about ⅛ to ¼ inch longer than the ceiling height. This way, the cradle can be wedged in place to hold the panel. Once the sheet has been nailed, the brace can be removed and used again.

Sometimes, several workers are involved; some workers will hold materials in place and others will nail. Scaffolds can also be used to let the carpenters stand at a good working height. Stilts, as in Fig. 14-13, are also used. However, workers on stilts should not try to lift materials from the floor. A worker on stilts cannot bend or lean over. That person can only brace, nail, or plaster seams and joints.

Fasteners Sheets are fastened at the center of the panel first. Then fasteners are placed toward the edges. Fasteners are driven at intervals on ceilings. The spacing depends on the type of sheetrock used. Nails, screws, and staples are all used.

Special nails are used for sheetrock. Some are smooth and some have ridges on the shank of the nail. Special screws are also used, driven by an attachment that releases the screw when it is in place. Nails and screws for sheetrock are shown in Fig. 14-14A. As a rule, the nail or screw should penetrate into solid wood about 1 inch. Thus, for ⅜-inch drywall, a nail 1⅜ inch long would be recommended. When staples are used, the crown of the staple should be perpendicular to the joist or stud.

Edge spacing The seams of the sheets should be staggered. The edges of one junction should not align with

Fig. 14-14B *A magnetic sleeve to guide screws. Match the drive bit and screw.*

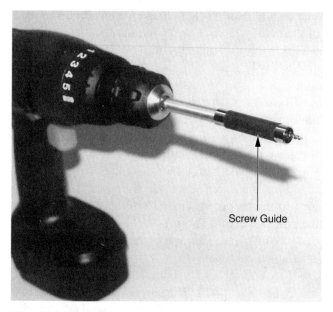

Screw Guide

Fig. 14-14C *Pull the sleeve over the screw. The screw can now be driven easily.*

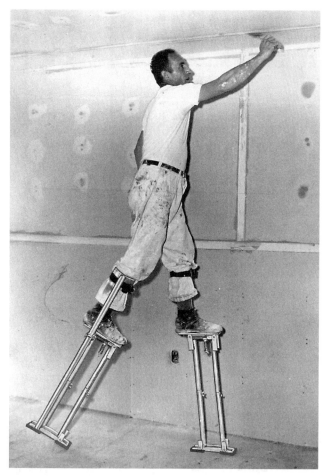

Fig. 14-13 *Stilts are often used for nailing and finishing ceilings.*
(Goldblatt Tools)

the edges of another junction. This is part of a process called *floating*. Figure 14-14B shows the use of a magnetic sleeve screw guide. Figure 14-15 shows how edges are floated or staggered. This allows each sheet to reinforce the next sheet. It makes a stronger, more rigid wall or ceiling. Floating reduces expansion and contraction of the walls and ceilings. With this controlled, the finished seams are less likely to crack. By eliminating the nails, the sheets are allowed to move or "float."

Cutting Gypsum Board

Gypsum board is easily cut. Careful measurements should be taken before cutting. It is a good idea to take all measurements at least twice to be sure. Sometimes, measurements are checked by measuring from two different places.

Cutting straight lines Straight lines are cut after being carefully marked. A pencil line is made on the

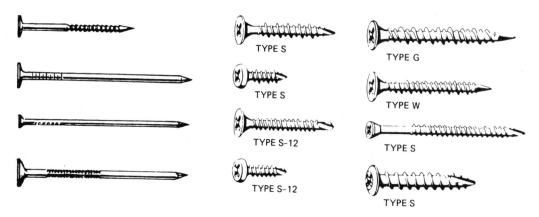

TYPE S

TYPE S

TYPE S-12

TYPE S-12

TYPE G

TYPE W

TYPE S

TYPE S

Fig. 14-14A *Typical nails and screws for sheetrock. Each is used for a certain application.* (Gypsum Association)

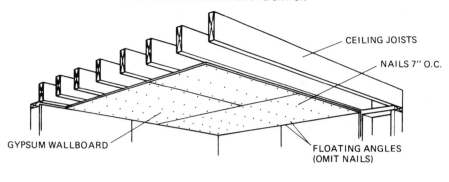

PERPENDICULAR CEILING APPLICATION

CEILING JOISTS

NAILS 7" O.C.

GYPSUM WALLBOARD

FLOATING ANGLES
(OMIT NAILS)

Fig. 14-15 *Floating angle construction helps eliminate nail popping and corner cracking. Fasteners at the intersection of walls or ceiling are omitted. Note that seams are not matched.*
(Gypsum Association)

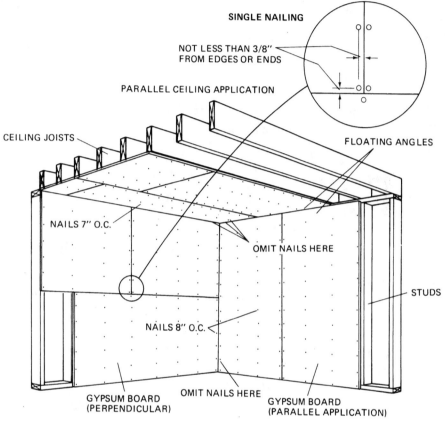

SINGLE NAILING

NOT LESS THAN 3/8"
FROM EDGES OR ENDS

PARALLEL CEILING APPLICATION

CEILING JOISTS

FLOATING ANGLES

NAILS 7" O.C.

OMIT NAILS HERE

STUDS

NAILS 8" O.C.

GYPSUM BOARD
(PERPENDICULAR)

OMIT NAILS HERE

GYPSUM BOARD
(PARALLEL APPLICATION)

good, finished side of the sheet. See Fig. 14-16. This line is then scored with a sharp knife. The cut should go through the paper and slightly into the gypsum core.

The board or sheet is then placed over a board or a rigid back. The carpenter can sometimes use a knee for a back. See Fig. 14-17. The board is then snapped or broken on the scored line. The paper on the back side is then cut with a knife. This finishes the cut. See Fig. 14-18.

Cutting openings Openings must be carefully measured and marked. Then the lines are heavily scored. Next the piece may be strongly tapped with a hammer. If a sheetrock hammer is used, the hatchet end may be used. The cutting edge is placed into the score. The blade is then pushed through with even pressure. The back side is cut with a knife and the section falls free.

Another method is to use a special device to cut the holes. This device resembles a box with teeth on the edges. The teeth are placed in the outlet box in the wall. The approximate center is marked on the panel. The panel is put in place. The handle of the special tool is forced into the marked center. Prongs are engaged into a slot in the teeth. The teeth quickly cut through the gypsum core. The tooth plate and core are simply pulled out. Thin saws, knives, and punches are also commonly used. Care should be taken with any method. Edges should be as smooth and even as possible.

Applying Wall Sheets

Most wall panels are applied horizontally to the studs. This means the long edge is parallel to the studs. It also means that the long edge will be vertical to the floor.

Note that one corner of the first sheet is not nailed. It is held in place when the second piece at that corner is applied. This technique is also used on ceilings. See Fig. 14-19. This is part of the process of floating. Floating allows for expansion and contraction of the walls. It helps keep seams from cracking.

Fig. 14-16 *Sheetrock is marked and scored with a knife for cutting.* (Gold Bond)

Fig. 14-17 *After scoring, sheetrock can be "cut" by breaking it over the knee or a piece of lumber.* (Gold Bond)

Fig. 14-18 *After scoring and breaking, the paper back may be cut. This separates the pieces.*

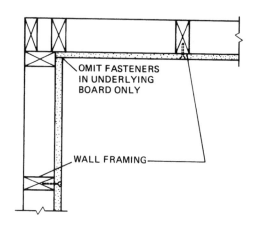

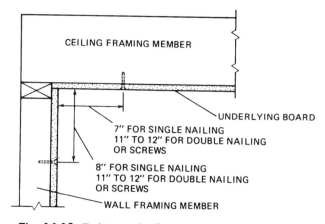

Fig. 14-19 *Techniques for floating corners.* (Gypsum Association)

Fig. 14-20 *A kick lever.* (Goldblatt Tools)

Fig. 14-21 *A kick lever being used to raise the panel in place.*
(Goldblatt Tools)

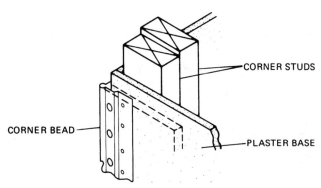

CORNER STUDS

CORNER BEAD

PLASTER BASE

Fig. 14-22 *Corners are butted together. A metal bead is then applied to protect the corner.* (Forest Products Laboratory)

Most ceilings are slightly more than 8 feet from the subfloor. This slight distance gives some working space and clearance. With this extra distance, standard 8-foot lengths will not catch. The drywall should be butted against the ceiling. It should be raised off the floor when applied. To start, one edge of the panel is laid on the floor. The top is laid against the wall studs. The sheet of drywall is then raised off the floor. Then it is nailed in place. A kick lever as in Fig. 14-20 is used. Kick levers may be purchased or may be made from lumber. A carpenter can step on the pedal to raise the panel. By using a kick lever, the carpenter keeps both hands free for nailing and holding (Fig. 14-21).

The first piece is usually put at a corner of the room. The work is done from corner to corner. The carpenter does not start in the middle of a wall. Nails, staples, or screws are driven at 8-inch intervals on the studs.

Corner treatments Outside edges and corners are reinforced with metal strips. Outside edges and corners are easily damaged and broken. The metal strips help prevent damage to these exposed edges. The metal strips are called *beads*. The most commonly used bead is the corner bead as shown in Figs. 14-22 and 14-23. Corners are butted together. The bead is then used to cover the corner. The reinforcement bead protects the corner or edge. The bead helps to prevent ugly damage to the edge or corner.

As a rule, wood is not covered with plaster. Plaster and wood expand and contract at different rates. The difference in movement of these two materials would cause cracking. The wood is covered with drywall. This reduces cracking.

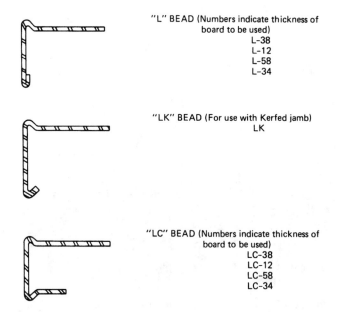

CORNERBEAD (Numbers indicate width of flanges,
i.e.—118 is 1 1/8-inches wide flange)
CB—100 X 100
CB—118 X 118
CB—114 X 114
CB—100 X 114
CB—PF (Paper flange,
steel corner,
combination
bead)

"L" BEAD (Numbers indicate thickness of
board to be used)
L-38
L-12
L-58
L-34

"LK" BEAD (For use with Kerfed jamb)
LK

"U" BEAD (Numbers indicate thickness of
board to be used, i.e.—38 is 3/8 inches)
U-38
U-12
U-58
U-34

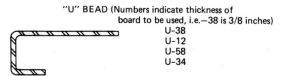

"LC" BEAD (Numbers indicate thickness of
board to be used)
LC-38
LC-12
LC-58
LC-34

Fig. 14-23 *Metal trim and casing.* (Gypsum Association)

Inside corners are normally not reinforced with beads. They are taped just as are other seams. The taping and beading process will be explained later.

Double-Ply Construction

Two layers of drywall are often used. The second layer increases the ability of the wall or ceiling to resist fire. It also helps reduce the amount of noise transmitted from one room to another.

Double-ply ceilings are glued and then nailed. However, the nails are driven from 16 to 24 inches apart. The length of the nail is also increased. A longer nail will penetrate through the second thickness of the drywall.

The joints of each layer should overlap. The joint of the top layer should occur over a solid sheet. Figure 14-24 shows this. Adhesives are applied, as in Fig. 14-25. The sheets of drywall should be firmly pressed in place. It is best to nail the glued sheets immediately. If they must be left, a temporary brace should be used. A brace will hold the panels firmly until the glue dries. Nails are driven when the brace is removed.

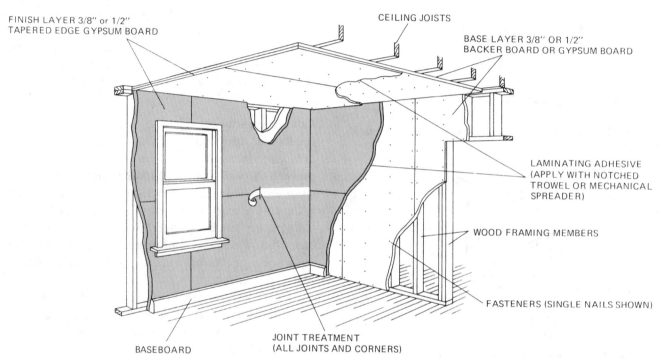

FINISH LAYER 3/8" or 1/2" TAPERED EDGE GYPSUM BOARD

CEILING JOISTS

BASE LAYER 3/8" OR 1/2" BACKER BOARD OR GYPSUM BOARD

LAMINATING ADHESIVE (APPLY WITH NOTCHED TROWEL OR MECHANICAL SPREADER)

WOOD FRAMING MEMBERS

FASTENERS (SINGLE NAILS SHOWN)

BASEBOARD

JOINT TREATMENT (ALL JOINTS AND CORNERS)

Fig. 14-24 *Double-ply application. Note that joints are not in the same place for each layer.* (Gypsum Association)

Fig. 14-25 *Notched spreaders are used to apply adhesive for ceiling sheet lamination.*

Finishing Joints and Seams

Drywall is often used as a base. Both panels and plaster may be applied over it. However, in most cases, it is finished directly. It may be painted or papered. Before gypsum board is finished, the seams should be smoothly covered. Some building codes require seams to be covered even when the drywall is merely a base. The covered seams increase the fire-resistive properties.

Seams are covered with successive coats of plaster and special tape. The tape may be either paper or Fiberglas mesh (Fig. 14-26). Figure 14-27A through F shows how the layers are applied. Covering seams are also called *finishing joints*. The process is often called "taping and bedding" (Fig. 14-28). The seams on all corners and edges are taped and bedded with plaster. However, when paneling or plaster is applied, it is not always necessary.

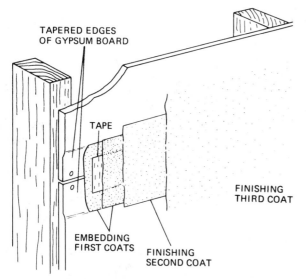

Fig. 14-26 *Seams are finished in layers.* (Gypsum Association)

Usually, the carpenter does not finish drywall seams. This is normally done by workers in the trowel trades. Plasterers or sheetrock® workers do most of this.

Ceiling Panels

One of the biggest and most expensive concerns of drywallers is callbacks and what they can do to their reputation. But with the new Sheetrock-brand interior ceiling panel, there is less worry about those callbacks brought on by sagging ceilings, and panel and joint cracking. The typical ceiling panel of ⅝-inch-thick Sheetrock weighs 70 pounds. The ½-inch panel weighs 50 pounds. That's almost a 30% reduction in weight. This means less to ship and handle. All this can be translated into savings if the ½-inch panel doesn't sag noticeably. Today's Fiberglas-reinforced panels allow the use of the ½-inch panel because it has less sag associated with it than the ⅝-inch size.

Sag-resistant panels are designed for parallel or perpendicular application to framing components spaced up to 24 inches on center with a maximum of 2.2 lb./sq. ft. insulation loading and wet texturing for ceiling application. For single-layer wood-framed ceilings, nails are spaced 7 inches on center and 1.25-inch type W screws are spaced 12 inches on center. Adhesive and/or nail-on fastening improves bond strength and reduces face nailing.

In new construction or renovation applications, steel furring channels can be used if spaced a maximum of 24 inches on center. Pay careful attention to framing construction and alignment. Problems will "telegraph" through the board if the framing is not true. Excessively long drying times might also result in problems with the ceiling finish, such as joint banding

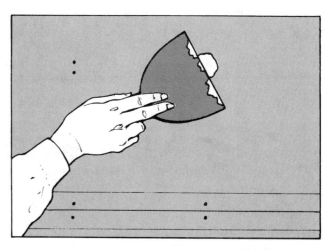

Fig. 14-27A *Spot nailheads with a first coat of compound either as a separate operation before joint treatment or after applying tape.* (Gold Bond)

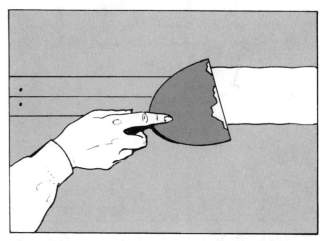

Fig. 14-27B *The first coat of joint compound fills the channel formed by the tapered edges of the wallboard.* (Gold Bond)

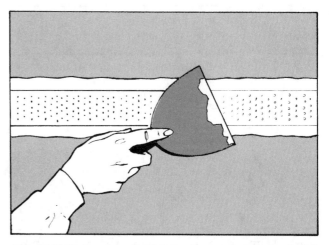

Fig. 14-27C *Tape is embedded directly over the joint for the full length of the wall. Smooth joint compound around and over the tape to level the surface.* (Gold Bond)

Fig. 14-27D *The first finishing coat is applied after the first coat has dried. Apply it thinly and feather out 3 to 4 inches on each side of the joint. Apply a second coat to the nailheads if needed.* (Gold Bond)

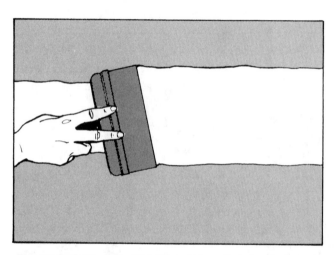

Fig. 14-27E *The second finishing coat is applied when the last coat has dried. Spread it thinly, feathering out 6 to 7 inches on each side of the joint. Finish nail spotting may be done at this time.* (Gold Bond)

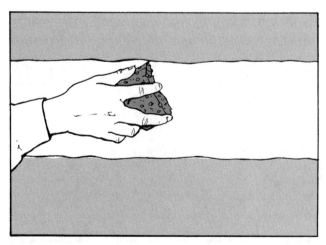

Fig. 14-27F *After 24 hours, smooth the finished joints with a damp sponge. If light sanding is required, use a respirator to avoid inhaling the dust.* (Gold Bond)

Fig. 14-28 *Taping and bedding sheetrock seams.* (Fox and Jacobs)

and staining. You should ensure proper ventilation to remove excess moisture during and after finishing. Supplemental heat or dehumidification might be required.

When panels are used as a base for water-based spray-applied finish, the weight of overlaid insulation should not exceed 2.2 psf. Thorough ventilation should be ensured to dry texture finish.

PREPARING A WALL FOR TUBS AND SHOWERS

Walls around tubs and showers must be carefully prepared. This prevents harmful effects from water and water vapor. Walls are finished with tile, panels, and other coverings. Water-resistant gypsum board can be used as a wall base. Special water-resistive adhesives should also be used. Edges and openings around pipes and fixtures should be caulked. A waterproof, nonhardening caulking compound should be used. The caulking should be flush with the gypsum. The wall finish should be applied at least 6 inches above tubs. It should be at least 6 feet above shower bases. See Fig. 14-29. Figure

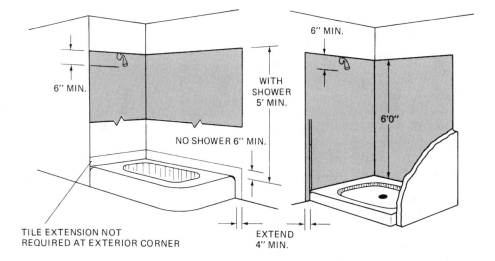

Fig. 14-29 *Water resistive wall finishes should be applied around bathtubs and showers.* (Gypsum Association)

6" MIN.

6" MIN.

WITH SHOWER 5' MIN.

NO SHOWER 6" MIN.

6'0"

TILE EXTENSION NOT REQUIRED AT EXTERIOR CORNER

EXTEND 4" MIN.

14-30 shows a tub support board. It is made from 2- × 4-inch lumber, and is nailed over the gypsum wallboard.

Another technique for preparing walls is shown in Fig. 14-31. In this situation, gypsum wallboard is not used. Instead, a vapor barrier is applied directly over the studs. It can be made of water-resistant sheathing paper or plastic film. A metal spacer is used around the edge of the tub, as shown in the illustration. Next, metal lath is applied over the vapor barrier. The wall is then finished by applying plaster. A water-resistive plaster is used for the tub or shower area.

Special tub and shower enclosures are also used. These may be made of metal or Fiberglas. See Fig. 14-32.

These are framed and braced to manufacturer's recommendations. No base wall or vapor barrier is used.

PANELING WALLS

Walls are often finished with standard-size panels made of wood, fiberboard, hardboard, or gypsum board. Panels made from hardboard, fiberboard, or gypsum are usually prefinished. The surfaces of these panels can be made to resemble wood or tile. Paint, stain, or varnish is not needed. This gives a fast finish with little labor. They can also be covered with wallpaper or plastic laminates. Paint and other materials lend an extremely wide range of appearances. Wood panels are also commonly used. They provide a wide range of wood grains and finishes. Panels may be prefinished. However, unfinished panels are also used. Stains may be applied to customize the appearance of the interior.

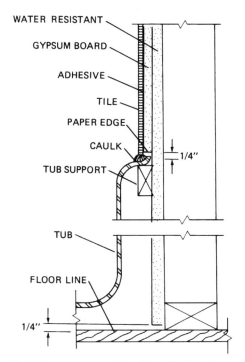

WATER RESISTANT

GYPSUM BOARD

ADHESIVE

TILE

PAPER EDGE

CAULK

TUB SUPPORT

1/4"

TUB

FLOOR LINE

1/4"

Fig. 14-30 *Tubs are supported on the walls. Note the two layers of gypsum board and the gap used.* (Gypsum Association)

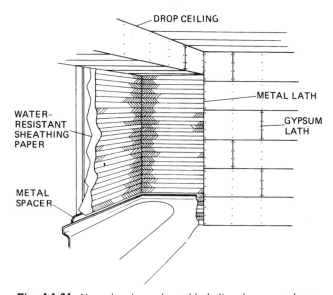

DROP CEILING

WATER-RESISTANT SHEATHING PAPER

METAL SPACER

METAL LATH

GYPSUM LATH

Fig. 14-31 *Vapor barrier and metal lath directly over studs.* (Forest Products Laboratory)

(A)

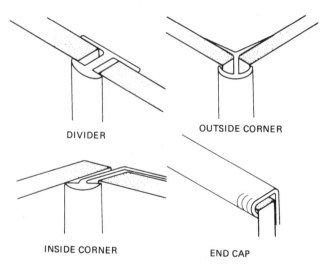

(B)

Fig. 14-32 *(A) Special tub and shower units are framed and braced by carpenters. (B) This shower stall is made of Fiberglas. No base wall is needed. (Cori Industries)*

All panels come in standard 4-×-8-foot sizes. They are put up with nails, screws, or glue. Special nails and screws are available, colored to match the surface finish. Plain nails are also used. Casing or finish-head nails are recommended. These should be set below the surface and filled. This filling should match the color and texture of the surface.

Adhesives are applied in patterns, as shown in Fig. 14-25. On studs where panels join, two lines of adhesives are used.

Panels are available in various materials and range from ⅛ to 1 inch in thickness. As a rule, wood paneling ¼ inch thick needs no base wall. However, the wall is more substantial if a base wall is used, especially with thinner panels. The wall is stronger and more fire-resistant. Many building codes require gypsum base walls. Gypsum drywall is the most common base wall for panels. Some codes require it to have the joints covered. The ⅜-inch thickness is commonly used.

Edges and corners Panel edges and corners are finished in several ways. Figure 14-33 shows several methods. Divider clips are used between panels. To install the panel, a special trick is often used. The edge trim is nailed down first. Then one edge of the panel is inserted into one clip. The second edge is moved away from the wall. Pressure is applied against the edge. This causes the panel to bow slightly. The second edge of the panel can then be slipped into the second clip. When the pressure is released, the edge will spring into the clip. The panel will be held flat against the wall. The panel can then be pressed against the adhesives or nailed in place.

Spacing and nailing Panels are often fastened directly to the base walls. In this situation, panels should be carefully spaced. A line may be snapped at 4-foot

DIVIDER

OUTSIDE CORNER

INSIDE CORNER

END CAP

Fig. 14-33 *Joining edges and corners of paneling.*

intervals to show where panels should join. These lines may be used for guides. They show where the adhesives are applied. They also show where the panels are applied. Chalk lines may be snapped at 4-foot intervals on vertical or horizontal references.

Wood paneling is also nailed in place. Nails are placed approximately 4 inches O.C. around the edges. Nails should be placed at 8-inch intervals on the inside studs. Panels made of laminates, hardboards, and other materials may require special methods. The carpenter should always read the manufacturer's suggestions; they usually describe special methods and materials that should be used.

Panels over studs First, crooked or uneven studs should be straightened. They can be planed or sawed to make a flat base. Sometimes panels cannot be joined over studs; special boards must be added between the studs. This is often true for vertical applications of gypsum board. Fire stops and special studs may be added.

Panels over masonry Panels are often used over concrete walls. Basements are prime examples. They are paneled in both new and remodeling jobs.

Special nail bases must be provided. They are made from wood boards called *furring strips*. Furring strips are fastened to the walls first. Top and bottom plates are also needed. Masonry nails or screw anchors are used. They may be added horizontally or vertically.

A vapor barrier should also be used. It may be applied over the masonry or over the furring strips. In either case, insulation should also be added, placed between the furring strips. See Figs. 14-34 through 14-37. For concrete blocks, masonry screws are better than nails. Nails driven into blocks can crack or break

the thin walls of the block. To use masonry screws, match the drill bit size with the screw size. Then select the correct length. For example, if you are attaching $3/4$-inch lath, use flat head screws that are at least 1-inch longer than the lath. The screw should be at least $1^3/_4$-inch long. Hold the lath in place and drill the hole through the lath and into the concrete as in Fig. 14-35B. It is easier if you use two electric drills for the job. The carbide-tipped drill is used for the holes in the first one. The second one holds the drive-bit that matches the slot on the screw, (Fig. 14-35C). Then drive the screw through the lath and into the concrete, (Fig. 14-35D). After insulation and vapor barriers have been installed, the wall cover can be applied. Either adhesives or nails can be used.

Fig. 14-35A *Furring strips are attached to the concrete wall. They are aligned with the level. They form a nail base for the panels. (Masonite)*

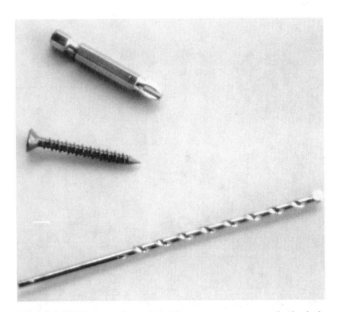

Fig. 14-35 B *A matching drive bit, masonry screw, and pilot hole bit. Used correctly, this 3/16-inch screw grips with 360 pounds of tension.*

Fig. 14-34 *A vapor barrier is applied over a concrete wall. This is a must for best paneling. (Masonite)*

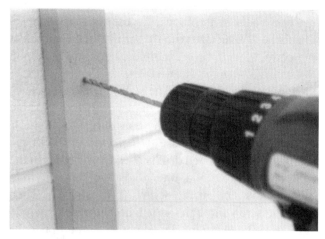

Fig. 14-35C *Hold the lath in place. Drill the pilot hole through the wood and into the concrete.*

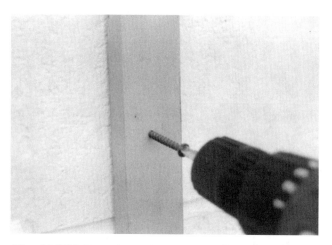

Fig. 14-35D *Drive the screw into the concrete.*

Fig. 14-36 *Insulation is added between the furring strips.*

Fig. 14-37 *Finally, the panels are nailed in place. A level may be used to align the panels.* (Masonite)

Board Walls

Walls can also be finished with boards. Figure 14-38 shows a boarded interior. The boards may be wood, plywood, or composition. The boards may be prefinished or unfinished. Most are shaped on the edges for joining. Tongue-and-groove joints and rabbeted joints are common.

Boards laid horizontally (parallel to the floor) are braced in several places by the studs. For vertical-board walls, the boards are nailed to the top and bottom plates. Sometimes more nail base is needed. Then special headers or fire stops can be used. See Fig. 14-39.

Fig. 14-38 *Boards may be used for attractive interior walls.* (Weyerhauser)

TOP AND BOTTOM PLATES
ARE NAIL BASE

FIRE STOP
MAY BE
ADDED AS
EXTRA NAIL
BASE

NOTE: Nails are driven in
edge of board. They will
not show.

Fig. 14-39 *Nailing vertical board walls.*

Fig. 14-41 *Vertical board walls accent height.* (California Redwood Association)

The tongue-and-groove joints are strong. Also, nails can be hidden in the tongues of the boards. This way, the nails are covered by the next board. Boards are tapped into place with a block and hammer. See Fig. 14-40. Boards may be applied horizontally, vertically, or diagonally. See Figs. 14-41, 14-42, and 14-43. However, as in Fig. 14-39, special bracing may be needed for vertical or diagonal boards.

Fig. 14-42 *Horizontal wood boards provide a natural look that goes well with modern designs.* (Potlatch)

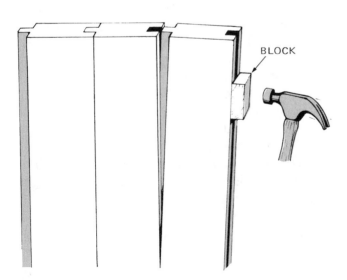

BLOCK

Fig. 14-40 *Boards may be forced into place using a scrap block and hammer. The scrap blocks protect the edges.*

Fig. 14-43 *Diagonal boards can give dramatic wall effects.* (California Redwood Association)

PREPARING A WALL FOR PLASTER

As a rule, the carpenter does not apply plaster to walls. However, the carpenter sometimes prepares the wall for the plaster. This includes nailing edge strips around the walls, doors, and windows. These strips are called *grounds*. The grounds may be made of wood or metal. They are used to help judge the thickness of the plaster. They also guide the application of the plaster.

The carpenter will often apply the lath that holds up the plaster. The most common lath is made from drywall. Both solid and perforated drywall is used. Figure 14-44 shows perforated drywall used for lath. Metal mesh, as in Fig. 14-45, is also used. Wooden lath is no longer used much because it costs more and takes longer to install.

Nailing Plaster Grounds

Plaster grounds are strips of wood or metal. They are nailed around the edges of a wall. Grounds are also nailed around openings, corners, and floors. The frame of the door or window is usually put up first. Grounds are nailed next to the frames. Sometimes the edges of the window or door casing serve as the grounds.

Grounds on interior doors and openings Plaster is usually applied before the doors are installed. A temporary ground is nailed in place. Figure 14-46 shows grounds around an interior opening. There are two

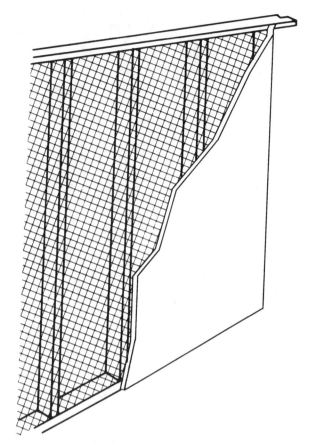

Fig. 14-45 *Wire lath for plaster.*

methods that are used. The standard width of the plastered wall is 5¼ inches. A piece of lumber 5¼ inches wide may be used. Note that it is centered on the stud and nailed in place. Thus the outside edges of the board form the grounds. These help guide the application of the plaster. The other method is shown in Fig. 14-46B. This method uses less material. Two strips are nailed in place as shown. Here, the carpenter must be careful with the measurements.

Plaster is normally applied to a certain thickness for both the lath and the plaster together. Two thicknesses are often used, either ¾ or ⅞ inch. Local building codes and practices determine which is used. If ⅜-inch drywall lath is used for a ⅞-inch wall, then ½ (⁴⁄₈) inch of plaster must be applied.

The plaster is applied in two or three coats. The first coat is called the "scratch" coat. The scratch coat is the thickest coat. It is put on and allowed to harden slightly. It is then scratched to make the next coat hold better. The next coat may be either "brown" coat or the finish coat. The finish coat is a thin coat. In most residential buildings it is ⅛ inch thick or less. For this application, the guides become like screeds in concrete work. The final coat is put on as a flat, even surface.

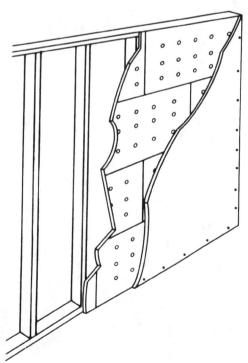

Fig. 14-44 *Most plaster lath for houses is made of perforated gypsum board.*

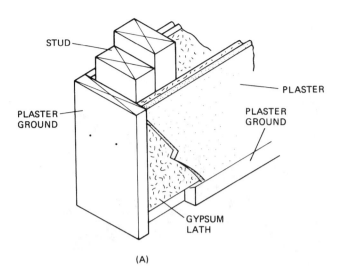

(A)

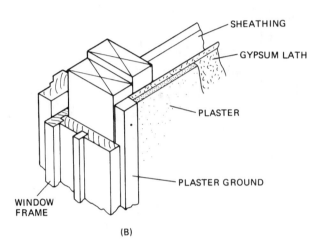

(B)

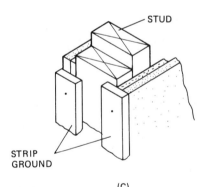

(C)

Fig. 14-46 *Plaster grounds around openings. (A) Recommended for doors. (B) Recommended for windows. (C) Temporary.* (Forest Products Laboratory)

The plaster ground strips are left in place after the plaster is applied. They become the nail base for the molding and trim. Trim is applied around ceilings, floors, and door and window openings. The grounds on interior door openings are usually removed. Then, the jambs for the interior doors are built or installed.

Plastering is normally done by people in the trowel trades. Two types of finish coats may be put on by these workers. The *sand float* finish gives a textured finish. This can then be painted. The *putty* finish is a smoother finish. It is commonly used in kitchens and bathrooms. In these locations, the plaster is often painted with a gloss enamel. This hard finish makes the wall more water-resistive. In addition, an insulating plaster can be used. It is made of vermiculite, perlite, or some other insulating aggregate. It may also be used for wall and ceiling finishes. Figure 14-47 shows plaster being applied with a trowel. Figure 14-48 shows a worker applying plaster with a machine.

FINISHING MASONRY WALLS

Masonry walls (brick, stone, or concrete block) are not comfortable walls. Although common in basements, they are cold and they sometimes sweat. This can give a room a cold, clammy feeling. To make them comfortable, they are finished with care. A vapor barrier and insulation should be installed, covered with paneling or drywall. This process was explained earlier.

INSTALLING CEILING TILE

Ceiling tiles can be installed over gypsum board, plaster, joists, or furring strips. Ceiling tile is available in a wide variety of surface textures and appearance. It is also available in a variety of sizes. A standard tile is

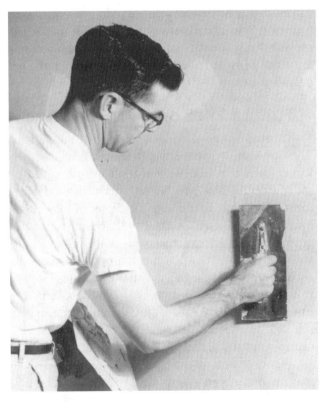

Fig. 14-47 *Plaster being applied with a trowel.* (Gold Bond)

Fig. 14-48 *Plaster being applied by machine.* (Gypsum Association)

Fig. 14-49 *Special decorative trim can be used with ceiling tile.* (Armstrong Cork)

12 × 12 inches. However, sizes such as 24 × 24, 24 × 48, and 16 × 32 inches are also available. Appearance, texture, light reflection, and insulation effect should be considered. Also, tiles can help absorb and deaden sounds. This ability is called the *acoustical quality.* Other factors in choosing tile include fire resistance, cost, and ease of installation.

Ceiling tile can be used for new or old ceilings. It can add a new decorative element or improve an old appearance. Special decorative trim, as in Fig. 14-49, can be added. These artificial beams and corner supports are made of rigid foam. They greatly enhance the appearance. Figure 14-50 shows the overall appearance.

Putting Tiles over Flat Ceilings

This method is good for both new and old ceilings. Tiles are usually applied with a special cement. Interlocking tiles may also be used in this situation. To install the ceiling tiles, first brush the loose dirt and grime from the ceiling. Then locate the center of the ceiling. Snap chalk lines on the surface of the ceiling to provide guides. Then check the squareness of the lines with the walls. Make sure that the tiles on opposite walls are equally spaced. If the space is not even, tiles on each wall should be equally trimmed. Tiles should not be trimmed on only one side of the ceiling. The tile on both sides must be trimmed an equal amount. This makes the center symmetrical and pleasing in appearance.

Apply cement to the back of the tile. The cement may be spread evenly or spotted. Use the chalk lines as guides. Then press each tile firmly in place. The work should progress from the center out in a spiral.

Using Furring Strips to Install Ceiling Tile

Small tiles may be attached to ceiling joists. However, furring strips are also needed to provide a nailing base. Figure 14-51 shows how the furring strips provide the nailing base. Small strips of wood 1 × 2 or 1 × 3 inches are used. These are nailed with one nail at each joist. They are perpendicular to the joists as shown. Insulation may be added to conserve energy or deaden sound. The furring strips are spaced to the length of the tile. Each edge of the tile should rest in the center of a strip.

The tile may be fastened with staples, adhesives, or nails. Nails or staples are applied to the inside of the tongue. This way, the next tile covers the nail so that it cannot be seen.

Installing Suspended Ceilings

Most suspended ceilings use larger panels. The 24-×-24- or 24-×-48-inch panels are common. The panels are suspended into a metal grid. This grid is composed

Fig. 14-50 *A basement area using ceiling tile and artificial beams and posts.* (Armstrong Cork)

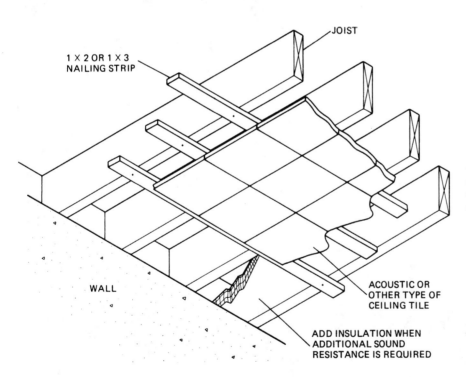

JOIST

1 × 2 OR 1 × 3
NAILING STRIP

WALL

ACOUSTIC OR
OTHER TYPE OF
CEILING TILE

ADD INSULATION WHEN
ADDITIONAL SOUND
RESISTANCE IS REQUIRED

Fig. 14-51 *Furring, or nailing, strips are used as a nail base for ceiling tile. Nails are driven in edges so that they will not show.* (Forest Products Laboratory)

of special T-shaped braces. These are called *cross T's* and *runners*. The grid system is suspended on wires from the ceiling or the joists.

Suspended ceilings are popular for remodeling. They also give a pleasing appearance in new situations.

These ceilings are economical, quick, and easy to install. They provide good sound-deadening qualities. They also give access to pipes and utilities. Lighting fixtures may be built into the grid system. This way no special brackets are used.

A suspended ceiling is installed in five steps:

1. First, nail the molding to the wall at the proper ceiling height. This supports the panels around the edge of the room.

2. Attach hanger wires to the joists. These are usually nailed in place at 4-foot intervals.

3. Fasten the main runners of the metal grid frame to the hanger wires. Check to make sure that the main runners are suspended at the proper distance. Or the main runners may be nailed directly to the joists. See Figs. 14-52 and 14-53.

4. Snap the cross T into place between the runners. The cross T and runners form a grid. See Fig. 14-54.

5. Lay the ceiling panels into the grid. See Fig. 14-55.

Concealed Suspended Ceilings

In concealed suspended ceilings, a special technique is used to hide the metal grid. Panels with a grooved edge are used. See Figs. 14-56 and 14-57. The metal then holds the panels from the groove. In this system, panels are installed in much the same manner as regular suspended ceilings; however, the metal grids fit into the grooves. Once the tiles are in place, no grid system is visible.

CHAPTER 14
STUDY QUESTIONS

1. What materials are used for interior walls?

2. What are the advantages of gypsum wallboard?

3. Why is gypsum wallboard used under other panels?

Fig. 14-52 *Main runners can be suspended from the joists.* (Armstrong Cork)

Fig. 14-53 *Main runners can also be nailed directly to joists.* (Armstrong Cork)

Fig. 14-54 *Cross T pieces are then attached to the runner to form a grid.* (Armstrong Cork)

Fig. 14-55 *Last, the panels are laid in place. The pattern of the tile helps to disguise the grid.* (Armstrong Cork)

Fig. 14-56 *Panels with grooved edges may be used for a system where no grid shows.* (Armstrong Cork)

Fig. 14-57 *No metal grid shows when grooved-edge tile is used.* (Armstrong Cork)

4. How are vapor barriers installed?

5. What is the sequence for covering a wall?

6. What are the three methods of applying wall panels?

7. Why should a carpenter know how plaster is applied?

8. What are some other names for gypsum board?

9. Why are strong backs used?

10. Why are ceiling panels started near the center of the room?

11. What is floating?

12. How is gypsum board cut?

13. How are ceiling panels held for nailing?

14. How are wall panels held for nailing?

15. What are the nailing intervals for walls and ceilings?

16. What is done for bath or kitchen walls?

17. What is done to edges and corners?

18. What ways are used to hang ceiling tile?

19. How are boards for walls positioned in place?

20. Why should ceiling tiles be started in the exact center?

15
CHAPTER

Finishing the Interior

THE INTERIOR OF A BUILDING IS THE LAST part finished. The frame has been built and covered. Exterior walls and roof are complete. Exterior doors and windows are done. The interior walls and ceilings have also been built; however, they have not been finished. Also, windows and doors on interior sections have not been installed.

Finishing the interior consists of several things. Interior doors are installed, and the molding and trim are applied to the inside of the windows and around the interior doors. Cabinets are then built or installed. Cabinets are needed in bathrooms and kitchens. Other special cabinets or shelves (called built-ins) are also built.

Then plumbing is installed. The woodwork is finished with paint or stain, and the walls and ceilings are finished. Then wiring is connected to the electrical outlets and to built-in units. Appliances such as dishwashers, ovens, lights, and other things are installed and connected. Finally, the final floor layer is finished to complete the interior.

Skills needed to finish the interior include:

- Installing cabinets in kitchens and baths
- Building shelves and cabinets
- Applying interior trim and molding
- Applying finishes
- Painting or papering walls
- Installing floor materials

SEQUENCE

As a rule, the sequence can be varied. However, a few factors must be considered in planning the sequence. The first factor is the type of floor involved. Some buildings are constructed on a slab. Here, the interior may be completed before the finish flooring is applied. However, with a wooden frame, two layers of flooring are laid. Here the second layer of flooring may be applied before the interior is finished. The second layer, however, is not finished until later. For a frame building on a wooden (frame) floor, the general sequence should be:

1. Install cabinets in kitchens and baths. ⎫ In any
2. Install interior doors. ⎬ sequence
3. Trim out interior doors and windows. ⎭
4. Paint or stain wood trim and cabinets.
5. Finish walls with paint, texture paint, or paper.
6. Install electrical appliances.
7. Lay the finish floor.
8. Finish floors by sanding, staining, varnishing, or laying linoleum or carpet.

INTERIOR DOORS AND WINDOW FRAMES

Interior door units are installed to finish the separation of rooms and areas. Then the insides of the windows must be cased or trimmed. Trim is also applied to all interior door units. The trim is needed to cover framing members. It also seals these areas from drafts and airflow.

Installing Interior Doors

Openings for interior doors are larger than the door. They are higher and wider than the actual door width. Frames are usually 3 inches higher than the door height. They are usually 2½ inches wider than the door width. This provides space for the members of the door. Interior door frames are made up of two side pieces and a headpiece. These cover the rough opening frame. The pieces that cover these areas are called *jambs*. They consist of side jambs and head jambs. Other strips are installed on the jamb. These are called *stops*. The stops, also called door stops, form a seat for the door. It can latch securely against the stops. Most jambs are made in one piece as in Fig. 15-1A. However, two- and three-piece adjustable jambs are also made. Adjustable jambs are used because they can be adapted to a variety of wall thicknesses. See Fig. 15-1B.

Today, interior door frames can be purchased with the door prehung. These are ready for installing.

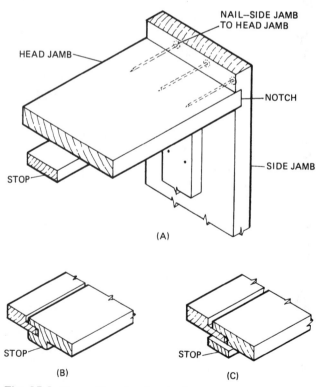

Fig. 15-1 *Types of interior door jambs.* (Forest Products Laboratory)

Sometimes, however, the carpenter must cut off the door at the bottom. This is done to give the proper clearance over the finished floor material.

Door frames The door unit is assembled first. Then it is adjusted vertical and square. This is called being *plumb*. The adjustment is made with small wedges, usually made from shingles, as shown in Fig. 15-2. The jamb is plumbed vertical and square. Then nails are driven through the wedges into the studs as shown. Any necessary hinging and adjusting of the door height is made at this point. For instructions on installing door hinges, see the chapter on exterior doors and windows.

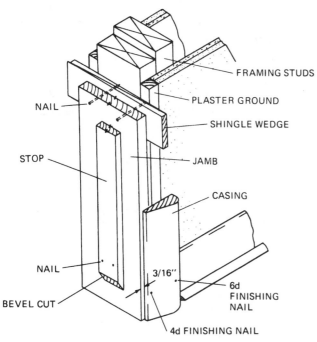

Fig. 15-2 *Door jambs are shimmed in place with wedges, usually made from shingles.* (Forest Products Laboratory)

Trim is attached after the door frame is in place. The door is first properly hung. Then trim, called *casing*, may be applied. Casing is the trim around the edge of the door opening. Casing is also applied over interior door and window frames. Casing is nailed to the

jamb on one side. On the other side it is nailed to the plaster ground or the framing stud. It should be installed to run from the bottom of the finished floor. Note, in Fig. 15-3, that two procedures may be used for the top casing. The top casing piece can be elaborately shaped. Other pieces of molding are used to enhance the line. This is often done to fit historical styles. A slight edge of the jamb is exposed as shown in Fig. 15-3C.

Next, the door stop may be nailed in place. The door is closed. The stop should be butted close to the hinged door. However, a clearance of perhaps 1/16 inch should be allowed. The door stop is usually 1½ or 3 inches wide. It is wide enough for two nails to be used. The finishing nails may be driven as shown. When a stop is spliced, a beveled cut in either direction is made. See Fig. 15-4. Casing may be shaped in several ways, as shown in Fig. 15-5.

Door details Two interior door styles are the flush and the panel door. Other types of doors are also commonly

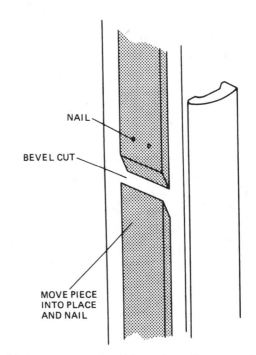

Fig. 15-4 *Miter joints should be used to splice stop and casing pieces where necessary.*

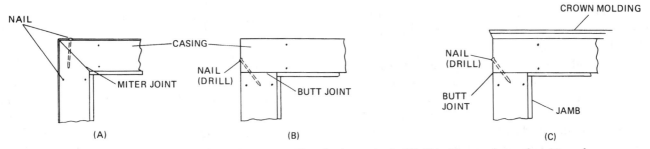

Fig. 15-3 *Two types of joints for door casing. Note the decoration in (C). This does not change the joint used.*

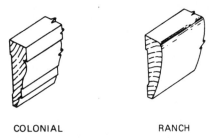

Fig. 15-5 *Two common casing shapes: (A) Colonial; (B) Ranch.* (Forest Products Laboratory)

COLONIAL   RANCH

used. They include folding doors and louvered doors. These are known as novelty doors.

Most standard interior doors are 1⅜ inches thick. They are used in common widths. Doors for bedrooms and other living areas are 2 feet 6 inches. Bathrooms usually have doors with a minimum width of 2 feet 4 inches. Small closets and linen closets have doors with a minimum width of 2 feet. Novelty doors come in varied widths for special closets and wardrobes. They may be 6 feet or more in width. In most cases, regardless of the door style, the jamb, stop, and casing are finished in the same manner.

The standard height for interior and exterior doors is 6 feet 8 inches. However, for upper stories, 6 feet 6 inch doors are sometimes used.

The flush interior door is usually a hollow-core door. This means that a framework of some type is covered with a thin outside layer. The inside is hollow, but may be braced and stiffened with cores. These cores are usually made of cardboard or some similar material. They are laid in a zigzag or circular pattern. This gives strength and stiffness but little weight. Cover layers for hollow-core doors are most often hardwood veneers. The most commonly used wood veneers are birch, mahogany, gum, and oak. Other woods are occasionally used as well. Doors may also be faced with hardboard. The hardboard may be natural, or it may be finished in a variety of patterns. These patterns include a variety of printed wood-grain patterns.

Hinging doors Doors should be hinged to provide the easiest and most natural use. Doors should open or swing in the direction of natural entry. If possible, the door should open and rest against a blank wall. It should not obstruct or cover furniture or cabinets. It should not bar access to or from other doorways. Doors should never be hinged to swing into hallways. Doors should not strike light fixtures or other fixtures as they are opened.

Installing door hardware Hardware for doors includes hinges, handles or locksets, and strike plates.

Hinges are sold separately. Door sets include locks, handles, and strike plates. They are available in a variety of shapes and classes. Special locks are used for exterior doors. Bathroom door sets have inside locks with safety slots. The safety slots allow the door to be opened from the outside in an emergency. Bedroom locks and passage locks are other classes. Bedroom locks are sometimes keyed. Often, they have a system similar to the bathroom lock. They are also available without keyed or emergency opening access. Passage "locks" cannot be locked.

To install door sets, two or more holes must be drilled. Most locksets today feature a bored set. This is perhaps the simplest to install. A large hole is drilled first. It allows the passage of the handle assembly. The hole is bored in the face of the door at the proper spacing. The second hole is bored into the edge of the door. The rectangular area for the face plate is routed or chiseled to size. Variations occur from manufacturer to manufacturer. The carpenter should refer to specifications with the locks before drilling holes. The door handle should be 36 to 38 inches from the floor. Other dimensions and clearances are shown in Fig. 15-6. A machine used to cut mortises and bores for locks is shown in Fig. 15-7.

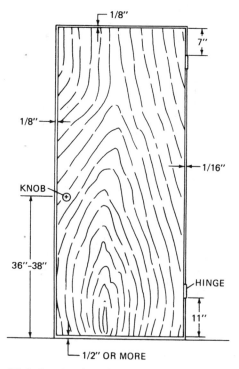

Fig. 15-6 *Interior door dimensions.* (Forest Products Laboratory)

A second type of lockset is called a mortise lock. Its installation is shown in Fig. 15-8. For the mortise lock, two holes must be drilled in the face of the door. One is for the spindle. The other is for the key. Also,

Fig. 15-7 *A machine used for cutting mortises and bores for locks.* (Rockwell International, Power Tool Division)

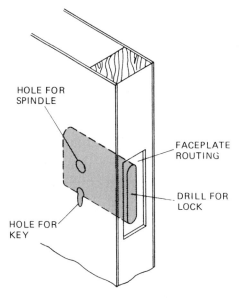

HOLE FOR
SPINDLE

FACEPLATE
ROUTING

DRILL FOR
LOCK

HOLE FOR
KEY

Fig. 15-8 *Mortise lock preparation.* (Forest Products Laboratory)

the area inlet into the edge of the door is larger. This requires more work. As shown in the figure, setting the face plate also requires more work than the other style.

Installing hinges Interior doors need two hinges, but exterior doors need three. The third hinge offsets the tendency of doors to warp. They tend to do this because the weather on the outside is very different from the "weather" on the inside.

Special door hinges should be used in all cases. Loose-pin *butt* hinges should be used. For 1⅜-inch-thick exterior doors, 4-×-4-inch butt hinges should be used. For 1¾-inch-thick interior doors, 3½-×-3½-inch butt hinges may be used.

First, the door is fitted to the framed opening. The proper clearances shown in Fig. 15-6 are used. Hinge halves are then laid on the edge of the door. A pencil is used to mark the location. The portion of the edge of the door is inlet or removed. This is called cutting a "gain." A chisel is often used. But, for many doors, gains are cut using a router with a special attachment.

The hinge should be placed square on the door. Both top and bottom hinges should be placed on the door. The door is then blocked in place. The hinge location on the jamb is marked. The door is removed, and the hinges are taken apart. The hinge halves are screwed in place. Next, the gain for the remaining hinge half is cut on the door jamb. The door is then positioned in the opening. The hinge pins are inserted into the hinge guides. The door can then be opened and swung. If the hinges are installed properly, the door will swing freely. Also, the door will close and touch the door stops gently. If the door tends to bind, it should be removed and planed or trimmed for proper fit. Also, note that a slight bevel is recommended on the lock side of the door. This is shown in Fig. 15-9.

Installing the strike plate The strike plate holds the door in place. It makes contact with the latch and holds it securely. This holds the door in a closed position. See Fig. 15-10. To install the strike plate, close the door with the lock and latch in place. Mark the location of the latch on the door jamb. This locates the proper position of the strike plate. Outline the portion to be removed with a pencil. Use a router or a chisel to remove the portion required. Note that two depths are required for the strike plate area. The deeper portion receives the lock and latch. It may be drilled the same diameter used for the lockset.

Finishing door stops Stops are first nailed in place with nail heads out. This way their positions can be changed after the door is hung. Once the door has been hung, the stops are positioned. Then they are securely nailed in place. Use 6d finish or casing nails. The stops at the lock side should be nailed first. The stop is pressed gently against the face of the door. The pressure is applied to the entire height. Space the nails 16 to 18 inches apart in pairs as in Fig. 15-10.

Next, nail the stop behind the hinge side. However, a very slight gap should be left. This gap should be approximately the thickness of a piece of cardboard (¹⁄₃₂ inch). A matchbook cover makes a good gage. After the lock and the hinge door stops have been nailed in place properly, the door is checked. When the clearances are good, the head jamb stop is nailed.

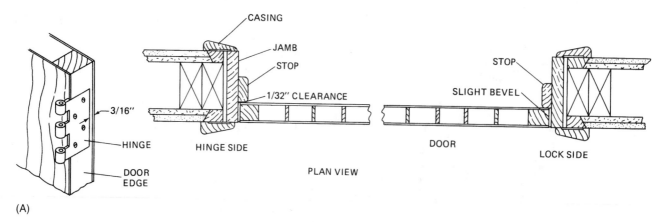

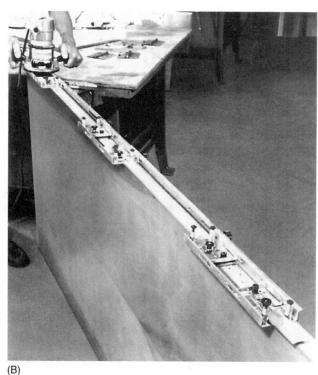

Fig. 15-9 *(A) Door details* (Forest Products Laboratory)*; (B) Using templates and fixtures to cut hinge gains.* (Rockwell International, Rockwell Power Tools Division)

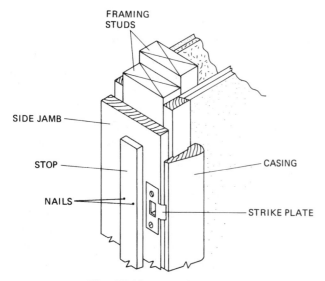

Fig. 15-10 *Strike plate details.*

WINDOW TRIM

There are two major ways of finishing window interiors. The window interior should be finished to seal against drafts and air currents. Trim is applied to block off the openings. The trim also gives the window a better appearance. The trim completely covers the rough opening.

Finishing Wooden Frame Windows

The traditional frame building has wooden double-hung windows. The frame of the window consists of a jamb around the sides and top. The bottom piece is a sill. The bottom sill is sloped. Water on the outside will drain to the outside. The frame covers the inside of the rough opening. However, there is still a gap between the window frame and the interior wall. This must be covered to seal off the window from air currents. Two ways are commonly used to trim, or case, around the window.

Making a window stool A separate ledge can be made at the bottom on the inside. See Fig. 15-11. The bottom piece extends like a ledge into the room. This bottom piece is called a *stool*. However, a finishing piece is applied directly beneath it. This is called the *apron*. See Fig. 15-11. The sides and top of the window are finished with casing.

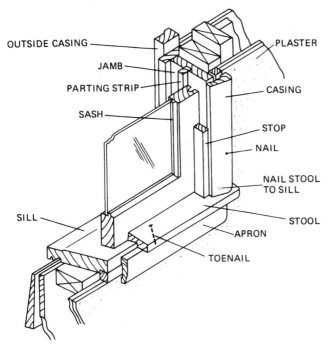

Fig. 15-11 *Stool and apron window trim.* (Forest Products Laboratory)

Casing a window The second method of finishing windows also uses trim. However, trim or casing is applied completely around the opening. Shown in Fig. 15-12, this method requires fewer pieces, takes less time to install, and is faster.

Sometimes, the carpenter also nails the stops in place around the window jamb. The stop in the window provides a guide for moving the window. It also forms a seal against air currents. To install the window stop, the window is fully closed. Next, the bottom stop is placed against the window sash. If the window is unpainted, 1/32 inch (about the thickness of a stiff piece of cardboard) is left between the stop and the sash. The stop is nailed in place with paired nails. This was also done for the doors. Next, the side stops are placed on the side jamb. The same amount of clearance is allowed. The stops on each side are nailed in place. The nail heads are left protruding. Then, the window is opened

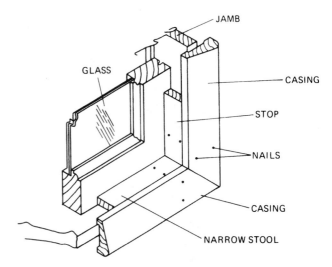

Fig. 15-12 *Finishing a window with casing.* (Forest Products Laboratory)

and shut. If the window slides evenly and smoothly, the stops are nailed down in place. If the window binds, not enough clearance was used. If the window wobbles, too much clearance was used. In these cases, pull the nails and reposition them.

Finishing Metal Window Frames

The use of windows with metal frames is increasing. See Fig. 15-13. The metal frame is nailed to the frame

Fig. 15-13 *Metal window frames have different details.*

members of the rough opening. This is done after the wall has been erected. Later, the window is sealed with insulation, caulk, or plastic foam. The window is finished in two steps. The inside of the rough opening is framed with plain boards. These boards are trimmed flush with the finish wall. Then, casing is applied to cover the gap between the wall and the window framing. See Fig. 15-14.

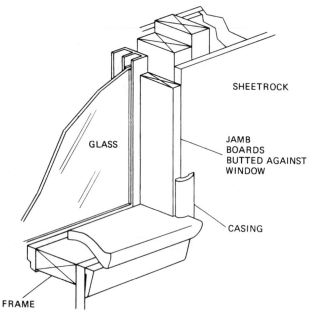

Fig. 15-14 *For metal windows, the jamb is butted next to the window unit. The jamb is then nailed to the rough opening frame and cased.*

For a different appearance, the window may be cased with a stool and apron. See Fig. 15-15. A flush-width board is used on the sides and tops. However, for the bottom, a wider board is used. The ends are notched to allow the side projections as in the figure. For a more finished appearance, the edges can be trimmed with a router. The apron is then added to finish out the window.

CABINETS AND MILLWORK

Millwork is a term used for materials made at a special factory, or mill. It includes both single pieces of trim and big assemblies. Interior trim, doors, kitchen cabinets, fireplace mantels, china cabinets, and other units are all millwork. Most of these items are sent to the building ready to install. So the carpenter must know how to install units.

However, not all cabinets and trim items are made at a mill. The carpenter must know how to both construct and install special units. This is called *custom* work. Custom units are usually made with a combination of

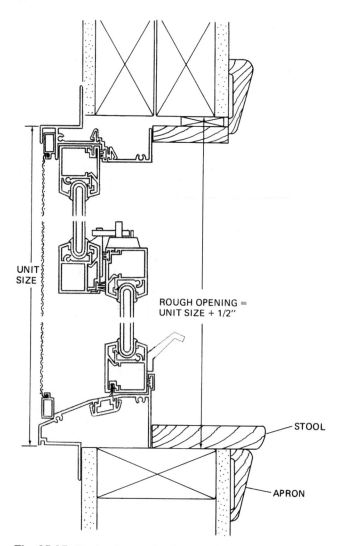

Fig. 15-15 *Stool and apron detail for an aluminum window.* (ALCOA)

standard dimension lumber and molding or trim pieces. Also, many items that are considered millwork do not require a highly finished appearance. For example, shelves in closets are considered millwork. But they generally do not require a high degree of finish. On the other hand, cabinets in kitchens or bathrooms do require better work.

Various types of wood are used for making trim and millwork. If the millwork is to be painted, pine or other soft woods are used most often. However, if the natural wood finish is applied, hardwood species are generally preferred. The most common woods include birch, mahogany, and ash. Other woods such as poplar and boxwood are also used. These woods need little filling and can be stained for a variety of finishes.

Installing Ready-Built Cabinets

Ready-built cabinets are used most in the kitchens and bathrooms. They may be made of metal, wood, or

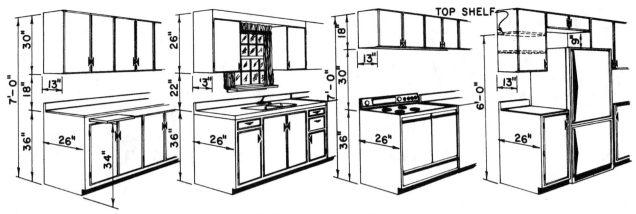

Fig. 15-16 *Typical cabinet dimensions.*

wood products. The carpenter should remember that overhead cabinets may be used to hold heavy dishes and appliances. Therefore, they must be solidly attached. Counters and lower cabinets must be strong enough to support heavy weights. However, they need not be fastened to the wall as rigidly as the upper units. Ready-builts are obtained in widths from 12 to 48 inches. The increments are 3 inches. Thus cabinets of 12, 15, 18, 21 inches, and so forth, are standard. They may be easily obtained. Figure 15-16 shows some typical kitchen cabinet dimensions.

Ordering cabinets takes careful planning. There are many factors for the builder or carpenter to consider. The finish, the size and shape of the kitchen, and the dimensions of built-in appliances must be considered. For example, special dishwashing units and ovens are often built-in. The cabinets ordered should be wide enough for these to be installed. Also, appliances may be installed in many ways. Manufacturer's data for both the cabinet and the appliance should be checked. Special framing may be built around the appliance. The frame can then be covered with plastic laminate materials. This provides surfaces that are resistant to heat, moisture, and scratching or marring. See Fig. 15-17A and B.

Sinks, dishwashers, ranges, and refrigerators should be carefully located in a kitchen. These locations are important in planning the installation of cabinets. Plumbing and electrical connections must also be considered. Also, natural and artificial lighting can be combined in these areas.

Cabinets and wall units should have the same standard height and depth. It would be poor planning to have wall cabinets with different widths in the same area.

Five basic layouts are commonly used in the design of a kitchen. These include the sidewall type as shown in Fig. 15-18A. This type is recommended for small kitchens. All elements are located along one wall.

The next type is the parallel or pullman kitchen. See Fig. 15-18B. This is used for narrow kitchens and can be quite efficient. Arrangement of sink, refrigerator, and range is critical for efficiency. This type of kitchen is not recommended for large homes or families. Movement in the kitchen is restricted, but it is efficient.

The third type is the L shape. See Fig. 15-18C. Usually the sink and the range are on the short leg. The refrigerator is located on the other. This type of arrangement allows for an eating space on the open end.

The U arrangement usually has a sink at the bottom of the U. The range and refrigerator should be located on opposite sides for best efficiency. See Fig. 15-18D.

The final type is the island kitchen. This type is becoming more and more popular. It promotes better utilization and has better appearance. This arrangement makes a wide kitchen more efficient. The island is the central work area. From it, the appliances and other work areas are within easy reach. See Fig. 15-18E.

Screws should be used to hang cabinet units. The screws should be at least No. 10 three-inch screws. The screws reach through the hanging strips of the cabinet. They should penetrate into the studs of the wall frame. Toggle bolts could be used when studs are inaccessible. However, the walls must be made of rigid materials rather than plasterboard.

To install wall units, one corner of the cabinet is fastened. The mounting screw is driven firm; the other end is then plumbed level. While someone holds the cabinet, a screw is driven through at the second end. Next, screws are driven at each stud interval.

When installing counter units, care must be taken in leveling. Floors and walls are not often exactly

(A)

(B)

Fig. 15-17 *(A) Appliance is built-in to extend. (B) Appliance is built-in flush.*

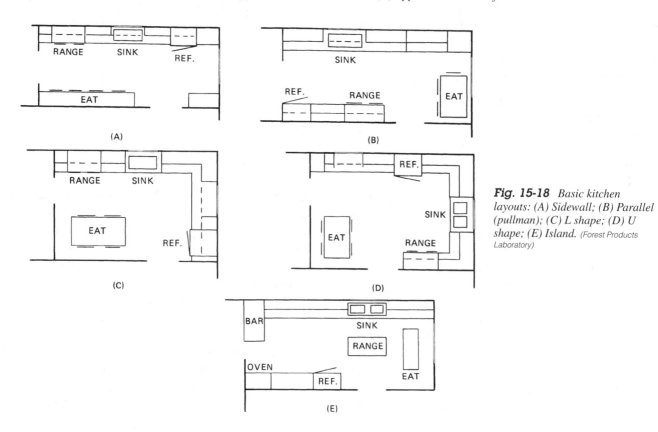

Fig. 15-18 *Basic kitchen layouts: (A) Sidewall; (B) Parallel (pullman); (C) L shape; (D) U shape; (E) Island.* (Forest Products Laboratory)

square or plumb. Therefore, care must be taken to install the unit plumb and level. What happens when a unit is not installed plumb and level? The doors will not open properly and the shelves will stick and bind.

Shims and blocking should be used to level the cabinets. Shingles or planed blocks are inserted beneath the cabinet bases.

The base unit and the wall unit are installed first. Then the countertop is placed. Countertops may be supplied as prefabricated units. These include dashboard and laminated tops. After the countertops are applied, they should be protected. Cardboard or packing should be put over the top. It can be taped in place with masking tape and removed after the building is completed. Sometimes the cabinets are hung but not finished. The countertop is installed after the cabinets are properly finished.

Ready-built units may be purchased three ways. First, they may be purchased assembled and prefinished. Counters are usually not attached. The carpenter must install them and protect the finished surfaces.

Second, cabinets may be purchased assembled but unfinished. These are sometimes called *in-the-white* and are very common. They allow all the woodwork and interior to be finished in the same style. Such cabinets may be made of a variety of woods. Birch and ash are among the more common hardwoods used. The top of the counter is usually provided but not attached.

Also, the plastic laminate for the countertop is not provided. A contractor or carpenter must purchase and apply this separately.

In the third way, the cabinets are purchased unassembled. The parts are precut and sanded. However, the unit is shipped in pieces. These are put together and finished on the job.

Cabinets are sometimes a combination of special and ready-built units. Many builders use special crews that do only this type of finish work. The combination of counter types gives a specially built look with the least cost.

Kitchen Planning

The kitchen is the focal point for family life in most homes. See Fig. 15-19. Some sources estimate that family members spend up to 50 percent of their time in the kitchen. For this reason, this is generally the most used and the most remodeled room of a home. Remodeling calls for a carpenter to be able to rearrange according to the desires of the owners.

The kitchen is a physically complex area in that it sustains heavy traffic flow from people using it and passing through it. It contains hot and cold water sources, drains, plumbing, and electrical outlets and fixtures, and is exposed to high moisture, splashing water and other liquids, items that are intensely hot and items that are very cold. In addition, activities involve

Fig. 15-19 *Efficient arrangement of the kitchen cooking area. (Jenn-Air)*

sharp instruments and considerable forces may be exerted in blending, rolling, pressing, cutting, and shaping. Other activities include mixing, washing, transferring, and storing.

Four major functions must be considered in kitchen planning: cooking, storing, eating, and entertaining. While most people consider cleaning up as part of cooking, it should be considered separately. The kitchen may also be the base of operations for several other functions done by one or more family members: studying, using a computer, home office work, sewing, and laundry. A larger kitchen may sometimes include home entertainment equipment or be used as an area for children.

The work area in a kitchen must be considered in laying out the cabinets, refrigerator, range, and sink. The work triangle is very important since that is where most of the food preparation is done. See Fig. 15-20.

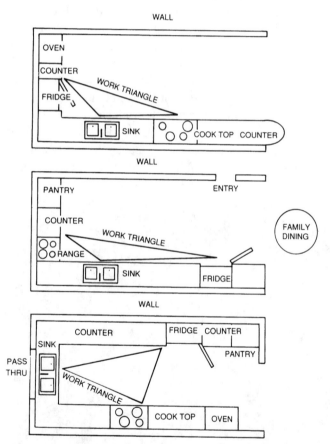

Fig. 15-20 *Location of the work triangle for three different kitchen arrangements.*

Cabinets may be finished in a variety of ways. They should be coordinated with the kitchen fixtures. Wooden cabinets with white plastic laminate covers can be made or purchased. These lend a European style that contrasts the white and the warm wood tones. See Fig. 15-21 for the rough layout before the tile is placed on the counter-

top. Figure 15-22 shows how the finished countertop looks. As you can see from these illustrations, the cabinets can be made on the job or purchased and the finishing touches made by the carpenter on the job.

Fig. 15-21 *Plywood base installed for the application of the countertop tile.* (American Olean Tile)

Countertops Countertops may be built as part of the cabinet and then covered with almost any counter material or they may be purchased as slabs cut to the right length. Openings in either type may be cut out. These slabs may have a splashboard molded in as part of the slab or the splashboard may be added later. Some building codes require all sinks and basins to have splashboards; others do not. See Fig. 15-23.

Kitchen light Kitchens need lots of light to be most useful and enjoyable. It is a good idea to have a good general light source combined with several additional lights for specific areas. Areas in which lights are necessary include the sink, cooktop, under cabinets, and in areas where recipes or references are kept. Figure 15-24 is a good example of a kitchen with a skylight and plenty of cooking and work areas.

Kitchen safety Unfortunately, kitchens are the scene of many accidents. Sharp objects, wet floors, boiling pots, and open flames are all potential hazards.

Cooktops should never be installed under windows because most people hang curtains, shades, or blinds in them. These can catch fire from the flames or heating elements. In addition, people also lean over to look out a window or to open or shut it. This puts them directly over the cooktop where they could be burned.

Cooktops should have 12 to 15 inches of counter around all sides. This keeps the handles of the pots from

Fig. 15-22 *Finished countertop.*
(American Olean Tile)

Fig. 15-23 *Note the splashboards on these countertops.*

Fig. 15-24 *Ceramic tile is used on the floors and the walls as well as the countertop to provide a durable kitchen with plenty of light.* (American Olean Tile)

extending beyond the edges where they can be bumped by passing adults or grabbed by curious children. Either situation can cause bad spills and nasty burns.

Using revolving shelves or tiered and layered swingout shelves can reduce falls caused by trying to reach the back of top shelves.

Heavy objects should be stored at or near the level at which they are used. Above all, they should not be stored high overhead or above a work area.

Making Custom Cabinets

Custom cabinets are special cabinets that are made on the job. Many carpenters refer to these and any type of millwork as *built-ins*. See Fig. 15-25. These jobs include building cabinets, shelves, bookcases, china closets, special counters, and other items.

A general sequence can be used for building cabinets. The base is constructed first. Then a frame is made. Drawers are built and fitted next. Finally the top is built and laminated.

Pattern layout Before beginning the cabinet, a layout of the cabinet should be made. This may be done

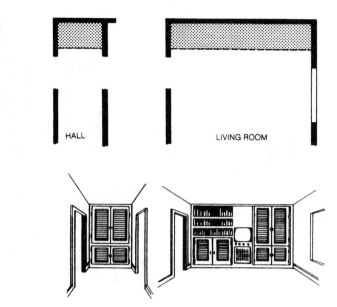

Fig. 15-25 *Typical built-ins.*

on plywood or cardboard. However, the layout should be done to full size if possible. The layout should show sizes and construction methods involved. Figure 15-26 shows a typical layout.

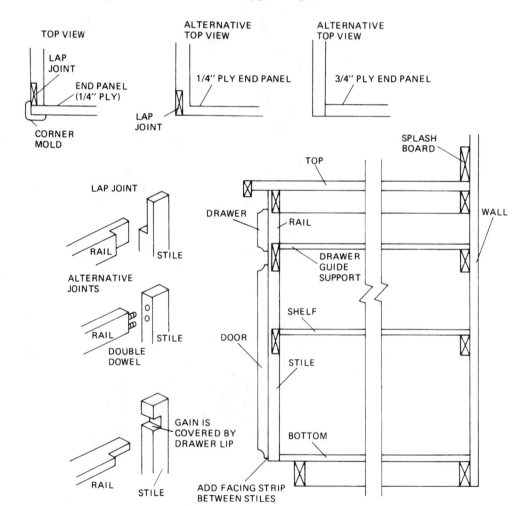

Fig. 15-26 *A cabinet layout. Dimensions and joints would also be added.*

Custom cabinets can be made in either of two ways. The first involves cutting the parts and assembling them *in place*. Each piece is cut and attached to the next piece. When the last piece is done, the cabinet is in place. The cabinet is not moved or positioned.

For the second procedure, all parts are cut first. The cabinet unit is assembled in a convenient place. Then, it is moved into position. The cabinet is leveled and plumbed and then attached.

Several steps are common to cabinet making. A bottom frame is covered with end and bottom panels. Then partitions are built and the back top strip is added. Facing strips are added to brace the front. Drawer guides and drawers are next. Shelves and doors complete the base.

Making the base The base for the cabinets is made first. Either 2-×-4- or 1-×-4-inch boards may be used for this. No special joints are used. See Fig. 15-27. Then the end panel is cut and nailed to the base. The toeboard, or front, of the base covers the ends. The end panel covers the end of the toeboard. A temporary brace is nailed across the tops of the end panels. This braces the end panels at the correct spacing and angles.

Bottom panels are cut next. The bottom panels serve as a floor over the base. The partition panels are placed next. They should be notched on the back top. This allows them to be positioned with the back top strip.

The back top strip is nailed between the end panels as shown. The temporary brace may then be removed. The partition panels are placed into position and nailed to the top strip. For a cabinet built in position, the partitions are toenailed to the back strip. The process is different if the cabinet is not built in place. The nails are driven from the back into the edges of the partitions. The temporary braces may be removed.

Cutting facing strips Facing strips give the unit a finished appearance. They cover the edges of the panels, which are usually made of plywood. This gives a better and more pleasing appearance. The facing strips also brace and support the cabinet. And, the facing strips support the drawers and doors. Because they are supports, special notches and grooves are used to join them.

The vertical facing strip is called a *stile*. The horizontal piece is called the *rail*. They are notched and joined as shown in Fig. 15-28. Note that two types of rail joints are used. The flat or horizontal type uses a notched joint. The vertical rail type uses a notched lap joint in both the rail and the stile.

As a rule, stiles are nailed to the end and partition panels first. The rails are then inserted from the rear.

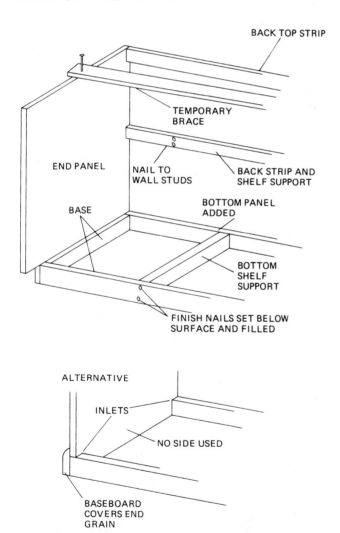

Fig. 15-27 *The base is built first. Then end panels are attached to it.*

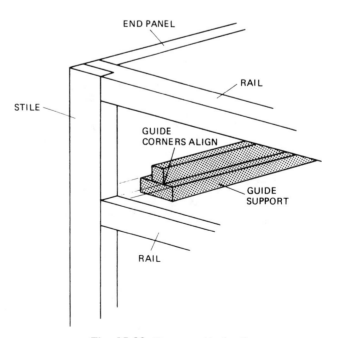

Fig. 15-28 *Drawer guide detail.*

Glue may be used, but nails are preferred. Nail holes are drilled for nails near the board ends. The hole is drilled slightly smaller than the nail diameter. Finishing nails should be driven in place. Then the head is set below the surface. The hole is later filled.

Drawer guides After the stiles and rails, the drawer assemblies are made. The first element of the drawer assembly is the drawer guide. This is the portion into which the drawer is inserted. The drawer guides act as support for the drawers. They also guide the drawers as they move in and out. The drawer guide is usually made from two pieces of wood. It provides a groove for the side of the drawer. The side of the guide is made from the top piece of wood. It prevents the drawer from slipping sideways.

A strip of wood is nailed to the wall at the back of the cabinet. It becomes the back support for the drawer guide. The drawer guide is then made by cutting a bottom strip. This fits between the rail and the wall. The top strip of the guide is added next. See Fig. 15-28. It becomes the support for the drawer guide at the front of the cabinet. Glue and nails are used to assemble this unit. The glue is applied in a weaving strip. Finish nails are then driven on alternate sides about 6 inches apart to complete the assembly. Each drawer should have two guides. One is on each side of the drawer.

There are three common types of drawer guides that are made by carpenters. These are side guides, corner guides, and center guides. Special guide rails may also be purchased and installed. These include special wheels, end stops, and other types of hardware. A typical set of purchased drawer guides is shown in Fig. 15-29.

The corner guide has two boards that form a corner. The bottom edge of the drawer rests in this corner. See Fig. 15-30A. The side guide is a single piece of wood nailed to a cabinet frame. As in Fig. 15-30B, the single piece of wood fits into a groove cut on the side of the drawer. The guides serve both as a support and as guides.

For the center guide, the weight of the drawer is supported by the end rails. However, the drawer is kept in alignment by a runner and guide. See Fig. 15-30C. The carpenter may make the guides by nailing two strips of wood on the bottom of the drawer. The runner is a single piece of wood nailed to the rail.

In addition to the guides, a piece should be installed near the top. This keeps the drawer from falling as it is opened. This piece is called a *kicker*. See Fig. 15-31. Kickers may be installed over the guides. Or, a single kicker may be installed.

Drawer guides and drawer rails should be sanded smooth. Then they should be coated with sanding sealer. The sealer should be sanded lightly and a coat of wax applied. In some cases, the wood may be sanded and wax applied directly to the wood. However, in either case, the wax will make the drawer slide better.

Making a drawer Most drawers made by carpenters are called *lip drawers*. This type of drawer has a lip

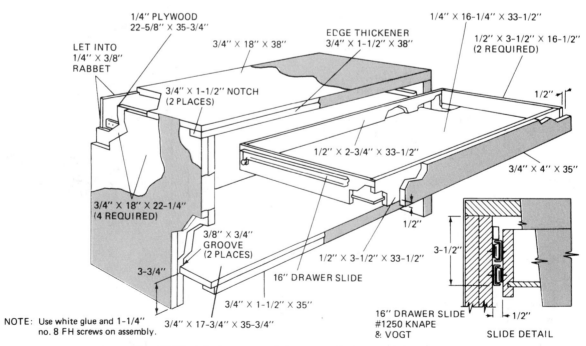

Fig. 15-29 A built-in unit with factory-made drawer guides. (Formica)

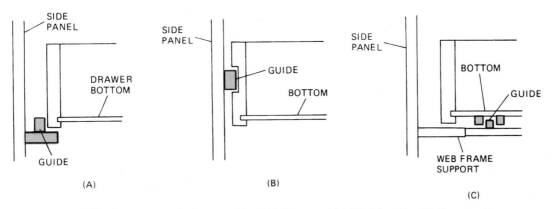

Fig. 15-30 *Carpenter-made drawer guides; (A) Corner guide; (B) Sideguide; (C) Center guide.*

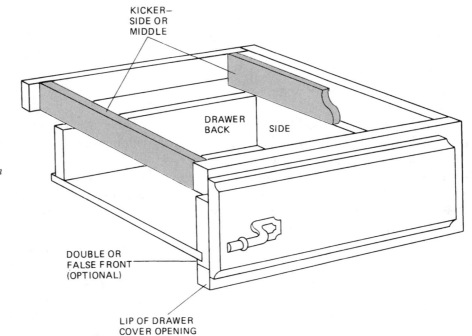

Fig. 15-31 *Kickers keep drawers from tilting out when opened.*

around the front. See Fig. 15-32. The lip fits over the drawer opening and hides it. This gives a better appearance. It also lets the cabinet and opening be less accurate.

The other type of drawer is called a *flush drawer*. Flush drawers fit into the opening. For this reason, they must be made very carefully. If the drawer front is not accurate, it will not fit into the opening. Binding or large cracks will be the result. It is far easier to make a lip drawer.

Drawers may be made in several ways. As a rule, the procedure is to cut the right and left sides as in Fig. 15-33. The back ends of these have dado joints cut into them. Note that the back of the drawer rests on the bottom piece. The bottom is grooved into the sides and front piece as shown.

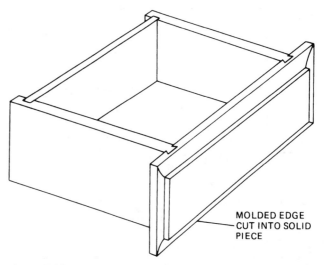

Fig. 15-32 *Lip drawers have a lip that covers the opening in the frame. Drawer fronts can be molded for appearance.*

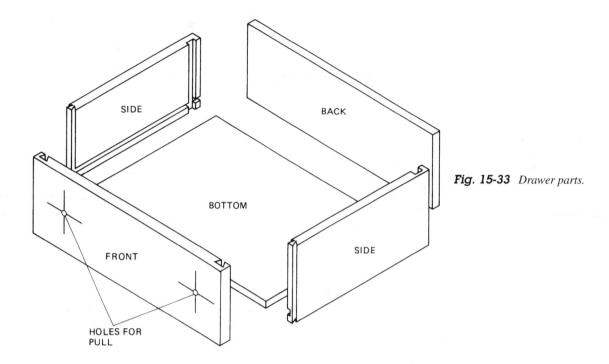

Fig. 15-33 *Drawer parts.*

The front of the drawer should be made carefully. It takes the greatest strain from opening and closing. The front should be fitted to the sides with a special joint. Several types of joint may be used. These are shown in Fig. 15-34.

The highest quality work can feature a special joint called the *dovetail joint*. However, this joint is ex-pensive to make and cut. As a rule, other types of dove-tail joints are used. The dovetailed dado joint is often used.

Drawers are made after the cabinet frame has been assembled. Fronts of drawers may be of several shapes. The front may be paneled as in Fig. 15-35A. Or it can be molded, as in Fig. 15-35B. Both styles can be made thicker by adding extra boards.

The wood used for the drawer fronts is selected and cut to size. Joints are marked and cut for assembly. Next, the groove for the bottom is cut.

Then the stock is selected and cut for the sides. As a rule, the sides and backs are made from different wood than the front. This is to reduce cost. The drawer front may be made from expensive woods finished as required. However, the interiors are made from less ex-pensive materials. They are selected for straightness and sturdiness. Plywood is not satisfactory for drawer sides and backs.

The joints for the back are cut into the sides. The joints should be cut on the correct sides. Next, the grooves for the drawers are cut in each side. Be sure that they align properly with the front groove.

Next, the back piece is cut to the correct size. Again, note that the back rests on the bottom. No groove is cut. All pieces are then sanded smooth.

A piece of hardboard or plywood is chosen for the bottom. The front and sides of the drawer are assembled. The final measurements for the bottom are made. Then, the bottom piece is cut to the correct size. It is lightly sanded around the edges. Then the bottom is inserted

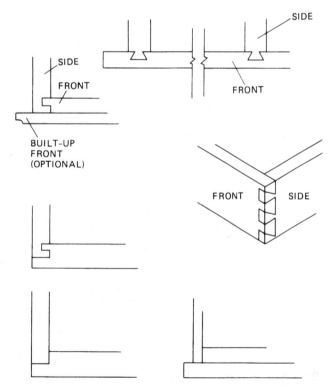

Fig. 15-34 *Joints for drawer fronts.*

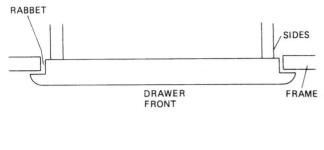

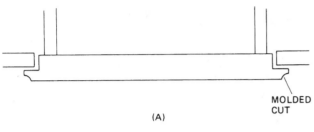

(A)

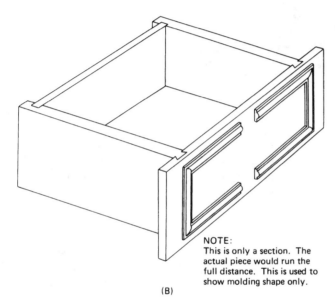

NOTE:
This is only a section. The actual piece would run the full distance. This is used to show molding shape only.

(B)

Fig. 15-35 *(A) These drawer types cover the opening. The accuracy needed for fitting is less. (B) Molding can be used to give a paneled effect.*

into the grooves of the bottom and front. This is a trial assembly to check the parts for fit. Next, the drawer is taken apart again. Grooves are cut in the sides for the

drawer guides. Also, any final adjustments for assembly are made.

The final assembly is made when everything is ready. Several processes may be used. However, it is best to use glue on the front and back pieces. Sides should not be glued. This allows for expansion and contraction of the materials.

The drawer is checked for squareness. Then one or two small finish nails are driven through the bottom into the back. One nail should be driven into each side as well. This will hold the drawer square. The bottom of the drawer is numbered to show its location. A like number is marked in the cabinet. This matches the drawer with its opening.

Cabinet door construction Cabinet doors are made in three basic patterns. These are shown in Fig. 15-36. A common style is the rabbeted style. A rabbet about half the thickness of the wood is cut into the edge of the door. This allows the door to fit neatly into the opening with a small clearance. It also keeps the cabinet door from appearing bulky and thick. The outside edges are then rounded slightly. The door is attached to the frame with a special offset hinge. See Figs. 15-37 and 15-38.

The molded door is much like the lip door. See Fig. 15-36B. As can be seen in the figure, the back side of the door fits over the opening. However, no rabbet is cut into the edge. This way, no part of the door fits inside the opening. The edge is molded to reduce the apparent thickness. This can be done with a router. Or, it may be done at a factory where the parts are made. This method is becoming more widely used by carpenters. Its advantages are that it does not fit in the opening. That means no special fitting is needed. Also, special hinges are not required. The appearance gives extra depth and molding effects. These are not available for other types of cabinets.

Both styles of door are commonly paneled. Paneled doors give the appearance of depth and contour. Doors may be made from solid wood. However, when

Fig. 15-36 *Cabinet door styles; (A) Rabbeted; (B) Molded; (C) Flush.*

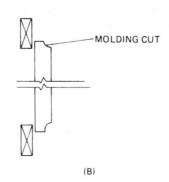

(A)

(B)

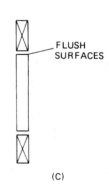

(C)

Fig. 15-37 *A colorful kitchen with rabbeted lip drawers and cabinets. Note the lack of pulls.* (Armstrong Cork)

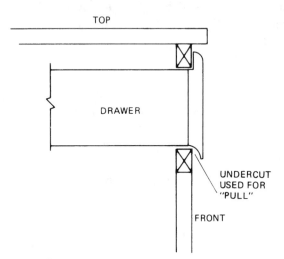

Fig. 15-38 *How drawers and doors are undercut so that pulls need not be used.*

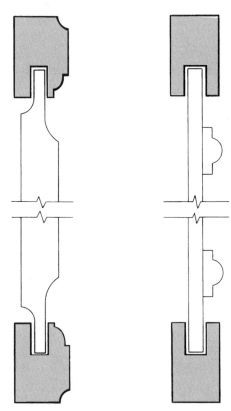

Fig. 15-39 *(A) Cross section of a solid door panel. (B) Cross section of a built-up door panel.*

cabinets are purchased, the panel door is very common. The panel door has a frame much like that of a screen door. Grooves are cut in the edges of the frame pieces. The panel and door edges are then assembled and glued solidly together. Cross sections of solid and built-up panels are shown in Fig. 15-39.

Flush doors fit inside the cabinet. These appear to be the easiest to make. However, they must be cut very carefully. They must be cut in the same shape as the opening. If the cut is not carefully made, the door will not fit properly. Wide or uneven spaces around the door will detract from its appearance.

Sliding doors Sliding doors are also widely used. These doors are made of wood, hardboard, or glass. They fit into grooves or guides in the cabinet. See Fig.

15-40. The top groove is cut deeper than the bottom groove. The door is installed by pushing it all the way to the top. Then the bottom of the door is moved into the bottom groove. When the door is allowed to rest at the bottom of the lower groove, the lip at the top will still provide a guide. Also, special devices may be purchased for sliding doors.

Fig. 15-41 *Countertops may be flat or may include splashboards.*

NOTE: Gap is included so that door can be lifted up.

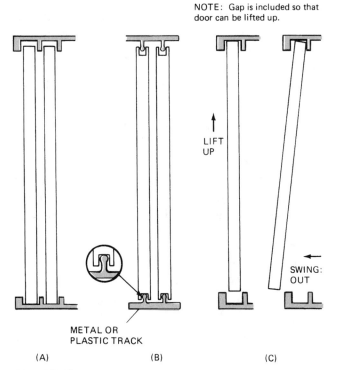

METAL OR PLASTIC TRACK

LIFT UP

SWING: OUT

(A)   (B)   (C)

Fig. 15-40 *Sliding doors. (A) Doors may slide in grooves. (B) Doors may slide on a metal or plastic track. (C) To remove, lift up, and swing out.*

Making the countertop Today, most cabinet tops are made from laminated plastics. There are many trademarks for these. These materials are usually ¹⁄₁₆ to ⅛ inch thick. The material is not hurt by hot objects, and does not stain or peel. The laminate material is very hard and durable. However, in a thin sheet it is not strong. Most cabinets today have a base top made from plywood or chipboard. Usually, a ½-inch thickness, or more, is used for countertops made on the site.

Also, specially formed counters made from wood products may be purchased. These tops have the plastic laminates and the mold board permanently formed into a one-piece top. See Fig. 15-41. This material may be purchased in any desired length. It is then cut to shape and installed on the job.

The top pieces are cut to the desired length. They may be nailed to the partitions or to the drawer kickers

on the counter. The top should extend over the counter approximately ⅜ inch. Next, sides or rails are put in place around the top. These pieces can be butted or rabbeted as shown in Fig. 15-42. These are nailed to the plywood top. They may also be nailed to the frame

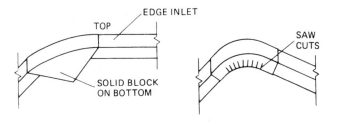

EDGE INLET
TOP
SOLID BLOCK ON BOTTOM
SAW CUTS

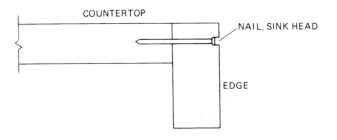

COUNTERTOP
NAIL, SINK HEAD
EDGE

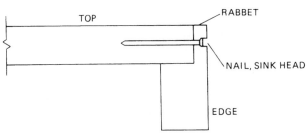

TOP
RABBET
NAIL, SINK HEAD
EDGE

Fig. 15-42 *Blocking up counter edges.*

of the counter. Next, they should be sanded smooth. Uneven spots or low spots are filled.

Particle board is commonly used as a base for the countertop. It is inexpensive and it does not have any grain. Grain patterns can show through on the finished surface. Also, the grain structure of plywood may form pockets. Glue in pockets does not bond to the laminate. Particle board provides a smooth, even surface that bonds easily.

Once the counter has been built, the top is checked. Also, any openings should be cut. Openings can be cut for sinks or appliances. Next, the plastic laminate is cut to rough size. Rough size should be ⅛ to ¼ inch larger in each dimension. A saw is used to cut the laminate as in Fig. 15-43.

Next, contact cement is applied to both the laminate and the top. Contact cement should be applied with a brush or a notched spreader. Allow both surfaces to dry completely. If a brush is used, solvent should be kept handy. Some types of contact cement are water-soluble. This means that soap and water can be used to wash the brush and to clean up.

When the glue has dried, the surface is shiny. Dull spots mean that the glue was too thin. Apply more glue over these areas.

It usually takes about 15 or 20 minutes for contact cement to dry. As a rule, the pieces should be joined within a few minutes. If they are not, a thin coat of contact cement is put on each of the surfaces again.

To glue the laminate in place, two procedures may be used. First, if the piece is small, a guide edge is put in place. The straight piece is held over the area and the guide edge lowered until it contacts. The entire piece is lowered into place. Pressure is applied from the center of the piece to the outside edges. The hands may be used, but a roller is better.

For larger pieces, a sheet of paper is used. Wax paper may be used, but almost any type of paper is acceptable. The glue is allowed to dry first. Then the paper is placed on the top. The laminate is placed over the paper. Then, the laminate is positioned carefully. The paper is gently pulled about 1 inch from beneath the laminate. The position of the laminate is checked. If it is in place, pressure may be applied to the exposed edge. If it is not in place, the laminate is moved until it is in place. Then, the exposed edge is pressed until a bond is made. Then the paper is removed from the entire surface. Pressure is applied from the middle toward the edges. See Fig. 15-44A to M.

Trim for laminated surfaces The edges should be trimmed. The pieces were cut slightly oversize to allow for trimming. The tops should extend over the sides slightly. The tops and corners should be trimmed so that a slight bevel is exposed. This may be done with a special router bit as in Fig. 15-45A. It may also be done with a sharp and smooth file as shown in Fig. 15-45B.

The back of most countertops has a raised portion called a splashboard. Splashboards and countertops may be molded as one piece. However, splashboards are also made as two pieces, in which case metal cove and cap strips are applied at the corners. Building codes may set a minimum height for these splashboards. The FHA requires a minimum height of 4 inches for kitchen counters.

Installing hardware Hardware for doors and drawers means the knobs and handles. These are frequently

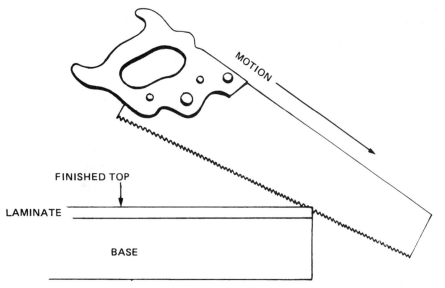

Fig. 15-43 *Cut into the laminate to avoid chipping.*

MOTION

FINISHED TOP

LAMINATE

BASE

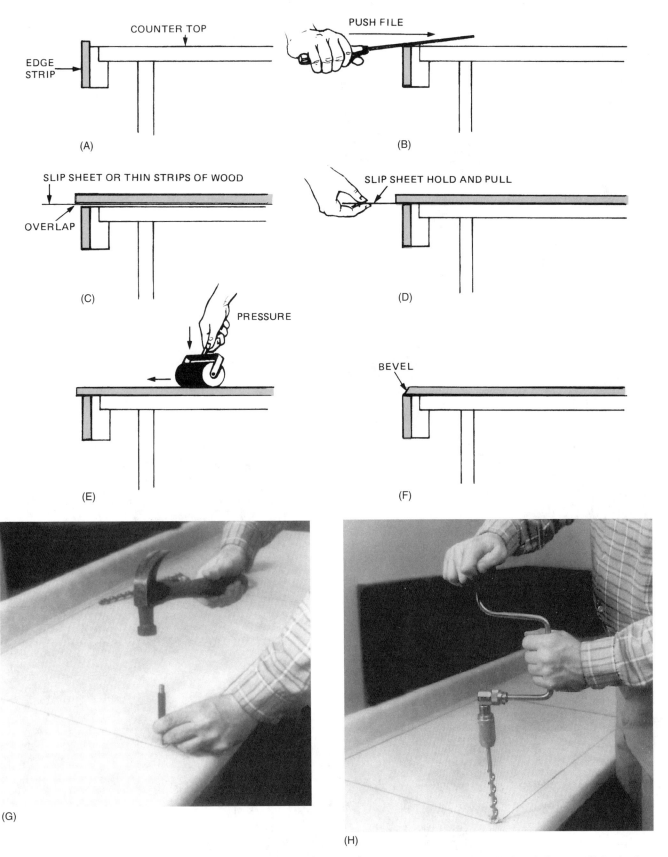

Fig. 15-44 *Applying plastic laminate to countertops. (A) Apply the edge strip. (B) Trim the edge strip flush with the top. (C) Apply glue. Lay a slip sheet or sticks in place when the glue is dry. (D) Position the top. Remove the slip sheet or sticks. (E) Apply pressure from center to edge. (F) Trim the edge at slight bevel. (G) To cut holes for sinks, first center punch for drilling. (H) Next drill holes at corners. (Formica)*

(I)

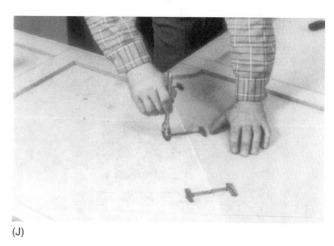

(J)

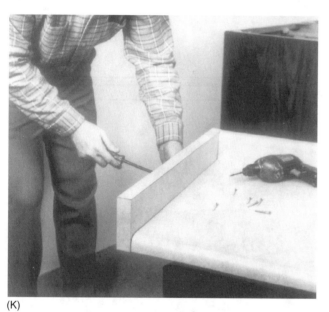

(K)

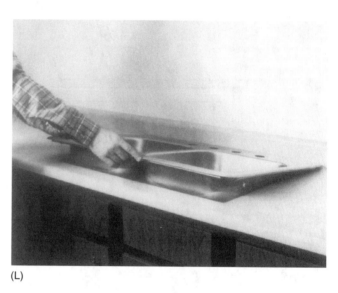

(L)

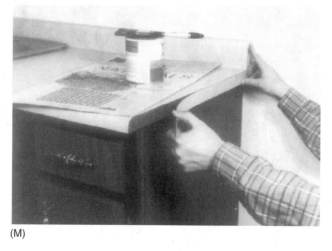

(M)

Fig. 15-44 *Continued. (I) Then cut out opening* (Rockwell International, Power Tool Division). *(J) Mitered corners can be held with special clamps placed in cut-outs on the bottom* (Formica). *(K) Ends may also be covered by splashboards* (Formica). *(L) Lay sink in opening.* (Formica). *(M) Ends may be covered by strips.* (Formica).

(A)

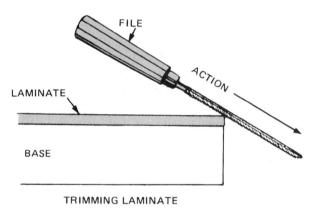

FILE

ACTION

LAMINATE

BASE

TRIMMING LAMINATE

Fig. 15-45 *(A) Edges may be trimmed with edge trimmers or routers* (Rockwell International, Power Tool Division). *(B) Edges may also be filed. Note direction of force.*

(A)

(B)

Fig. 15-46 *(A) Pulls may be located in the center for dramatic effect. (B) Regular pull location.*

called *pulls*. A variety of styles are available. As a rule, drawers and cabinets are put into place for finishing. However, pulls and handles are not installed. Hinges are applied in many cases. In others, the doors are finished separately. However, pulls are left off until the finish is completed.

Drawer pulls are placed slightly above center. Wall cabinet pulls are placed in the bottom third of the doors. Door pulls are best put near the opening edge. For cabinet doors in bottom units, the position is different. The pull is located in the top third of the door. It is best to put it near the swinging edge. Some types of hardware, however, may be installed in other places for special effects. See Fig. 15-46A and B. These pulls are installed in the center of the door panel.

To install pulls, the location is first determined. Sometimes a template can be made and used. Whatever method is used, the locations of the holes are found. They are marked with the point of a sharp pencil. Next, the holes are drilled from front to back. It is

a good idea to hold a block of wood behind the area. This reduces splintering.

There are other types of hardware. These include door catches, locks, and hinges. The carpenter should always check the manufacturer's instructions on each.

As a rule, it is easier to attach hinges to the cabinet first. Types of hinges are shown in Fig. 15-47. Types of door catches are shown in Fig. 15-48.

Shelves

Most kitchen and bathroom cabinets have shelves. Also, shelves are widely used in room dividers, bookshelves, and closets. There are several methods of shelf construction that may be used. Figure 15-49 shows some types of shelf construction. Note that each of these allows the shelf location to be changed.

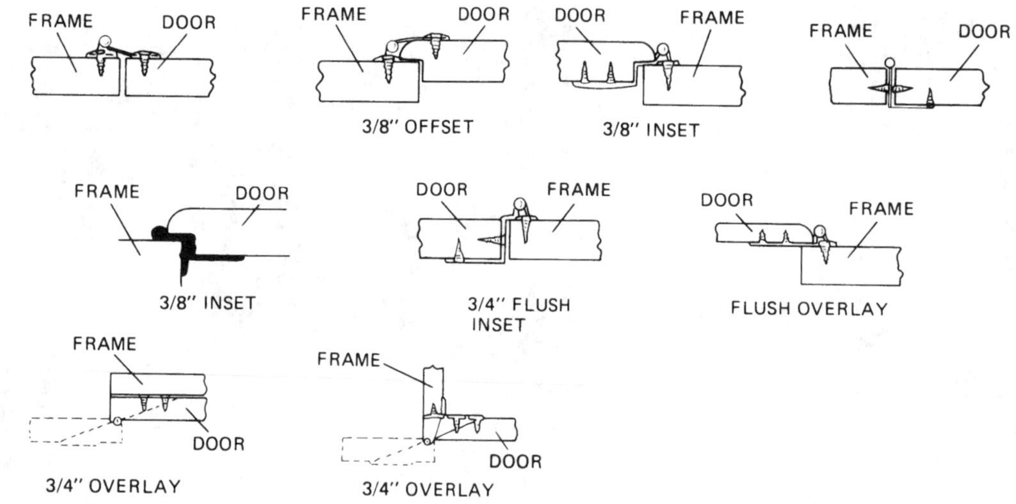

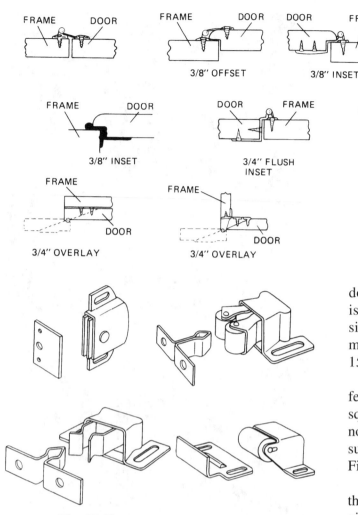

FRAME DOOR

FRAME DOOR
3/8" OFFSET

DOOR FRAME
3/8" INSET

FRAME DOOR

FRAME DOOR
3/8" INSET

DOOR FRAME
3/4" FLUSH INSET

DOOR FRAME
FLUSH OVERLAY

Fig. 15-47 *Hinge types.*

FRAME
DOOR
3/4" OVERLAY

FRAME
DOOR
3/4" OVERLAY

Fig. 15-48 *Types of cabinet door catches.*

For work that is not seen, shelves are held up by ledgers. Figure 15-50 shows the ledger method of shelf construction.

Special joints may also be cut in the sides of solid pieces. These are types of dados and rabbet joints. Figure 15-51 shows this type of construction.

As a rule, a ledger-type shelf is used for shelves in closets, lower cabinets, and so forth. However, for exposed shelves a different type of shelf arrangement is used. Adjustable or jointed shelves look better on bookcases, for instance.

Facing pieces are used to hide dado joints in shelves. See Fig. 15-52. The facing may also be used to make the cabinet flush with the wall. The facing can be inlet into the shelf surface.

Applying Finish Trim

Finish trim pieces are used on the base of walls. They cover floor seams and edges where carpet has been laid. Also, trim is used around ceilings, win-

dows, and other areas. As a rule, a certain procedure is followed for cutting and fitting trim pieces. Outside corners, as in Fig. 15-53, are cut and fit with miter joints. These are cut with a miter box. See Fig. 15-54.

However, trim for inside corners is cut with a different joint. This is done because most corners are not square. Miter joints do not fit well into corners that are not square. Unsightly gaps and cracks will be the result of a poor fit. Instead, a coped joint is used. See Fig. 15-53.

For a coped joint, the first piece is butted against the corner. Then the outline is traced on the second piece. A scrap piece is used for a guide. The outline is then cut using a coping saw. See Fig. 15-55. The coped joint may be effectively used on any size or shape of molding.

APPLYING FINISH MATERIALS

As a rule, the millwork is finished before the wall surfaces. Many wood surfaces are stained rather than painted. This enhances natural wood effects. Woods such as birch are commonly stained to resemble darker woods such as walnut, dark oak, and pecan. These stains and varnishes are easily absorbed by the wall. They are put on first so that they do not ruin the wall finish.

Paint for wood trim is usually a gloss or semigloss paint. These paints are more washable, durable, and costly. Flat or nonglare paints are widely used on walls.

Paints, stains, and varnishes are often applied by spraying. This is much faster than rolling. When spray equipment is used, there is always an overspray. This overspray would badly mar a wall finish. However, the

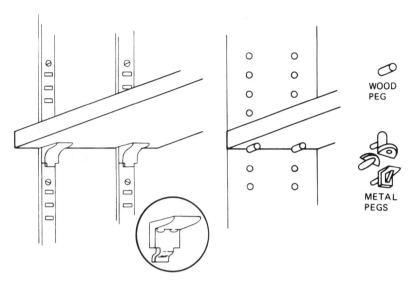

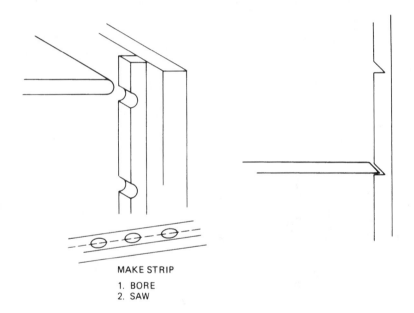

Slotted bookshelf standards and clips are ideal if you want to adjust bookcase shelves, but they add to total cost.

If you use wood or metal pegs set into holes, you must be sure to drill holes at the same level and 3/8 inch deep.

WOOD PEG

METAL PEGS

Fig. 15-49 *Methods of making adjustable shelves.*

MAKE STRIP
1. BORE
2. SAW

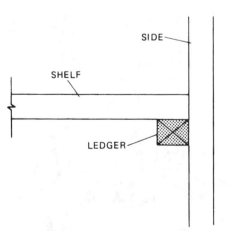

SIDE

SHELF

LEDGER

Fig. 15-50 *Ledgers or cleats are used for shelves and steps where appearance is not important.*

wall finish can be put on easily over the overspray. See Fig. 15-56.

To prepare wood for stain or paint, first sand it smooth. Mill and other marks should be sanded until they cannot be seen. Hand or power sanding equipment may be used.

Applying Stain

Stain is much like a dye. It is clear and lets the wood grain show through. It simply colors the surface of the wood to the desired shade. It is a good idea to make a test piece. The stain can be tested on it first. A scrap piece of the same wood to be finished is used. It is sanded smooth and the stain is applied. This lets a

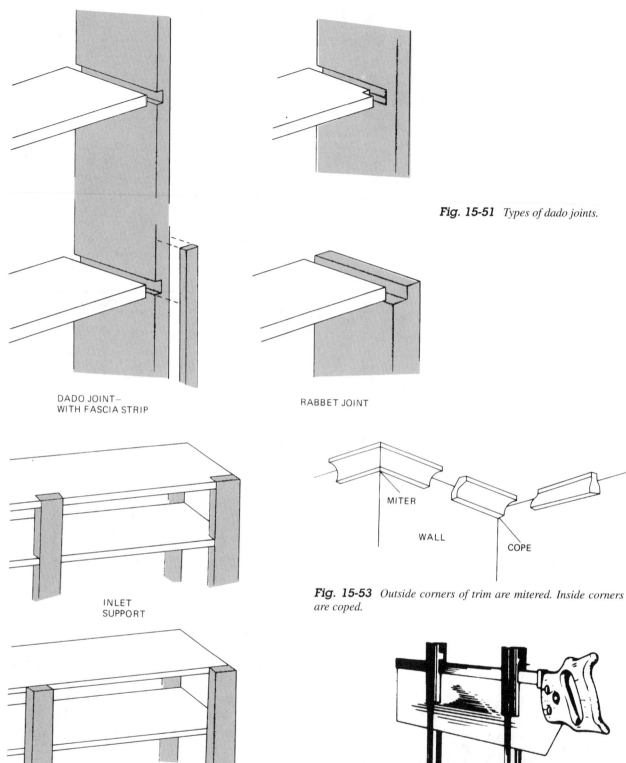

DADO JOINT—
WITH FASCIA STRIP

RABBET JOINT

Fig. 15-51 *Types of dado joints.*

INLET
SUPPORT

EXTERNAL
SUPPORT

Fig. 15-52 *Facing supports for shelves and cabinets.*

MITER

WALL

COPE

Fig. 15-53 *Outside corners of trim are mitered. Inside corners are coped.*

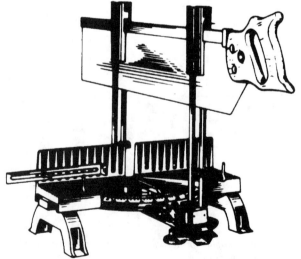

Fig. 15-54 *A miter box used for cutting miters.*

worker check to see if the stain will give the desired appearance.

Stain is applied with a brush, a spray unit, or a soft cloth. It should be applied evenly with long, firm

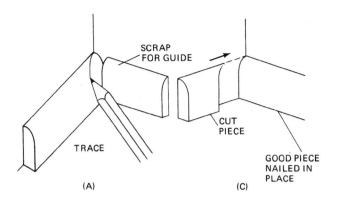

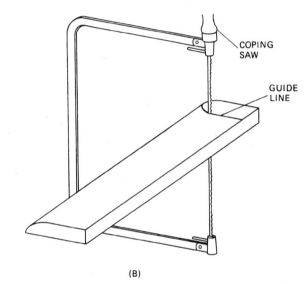

Fig. 15-55 *Making a coped joint. (A) Tracing the outline. (B) Cutting the outline. (C) Nailing the molding in place.*

strokes. The stain is allowed to sit and penetrate for a few minutes. Then it is wiped with a soft cloth. It is also a good idea to check the manufacturer's application instructions. There are many different types of stains. Thus there are several ways of applying stains.

The stain should dry for a recommended period of time. Then it should be smoothed lightly with steel wool. Very little pressure is applied. Otherwise, the color will be rubbed off. If this happens, the place should be retouched with stain.

The stain is evenly smoothed. Then, varnish or lacquer may be applied with a brush or with a spray gun. Today, most interior finishes are sprayed. See Fig. 15-56.

Applying the Wall Finish

Two types of wall finishes are commonly used. The first is wall paint. The other is wallpaper.

Painting a wall The wood trim for doors and so forth is finished. Then, the walls may be painted. Most wall

Fig. 15-56 *Cabinets are stained before walls are painted or papered. Note the overspray around the edges.*

paint is called *flat* paint. This means that it does not shine. It makes a soft, nonglare surface. Most wall paints are light-colored to reflect light. Before painting, all nail holes or other marks are filled. Patching paste should be used for this.

There are several methods of applying paint. It may be brushed, sprayed, or rolled. Also, special types of paint may be applied. Plain paint is a liquid or gel that gives a smooth, flat finish. However, special "texture" paints are also used. Sand-textured paints leave a slightly roughened surface. This is caused by particles of sand in the paint. Also, thick mixtures of paints are used. These may be rolled on to give a heavy-textured surface. Thicker paints can be used for a shadowed effect.

Putting wallpaper on walls Wallpaper is widely used in buildings today. However, the "paper" may not be paper. It may be various types of vinyl plastic films. Also, a mixture of plastic and paper is common. In either case, the wall surface should be prepared. It should be smooth and free of holes or dents. As a rule, a single roll of wallpaper will cover about 30 square feet of wall area. A special wall sealer coat called size

Applying Finish Materials   471

or sizing is used. This may be purchased premixed and ready to use. Powder types may be purchased and mixed by the worker. In either case, the sizing is painted on the walls and allowed to dry. As it dries, it seals the pores in the wall surface. This lets the paste or glue adhere properly to the paper.

A corner is chosen for a starting point. It should be close to a window or door. The width of the wallpaper is measured out from the corner. One inch is taken from the width of the wallpaper. Make a small mark at this distance. The mark should be made near the top of the wall. For a 27-inch roll, 26 inches is measured from the corner. A nail is driven near the ceiling for a chalk line. The chalk line is tied to this nail at the mark. The end of the chalk line is weighted near the floor. The line is allowed to hang free until it comes to a stop. When still, the line is held against the wall. The line is snapped against the wall. This will leave a vertical mark on the wall. See Fig. 15-57A.

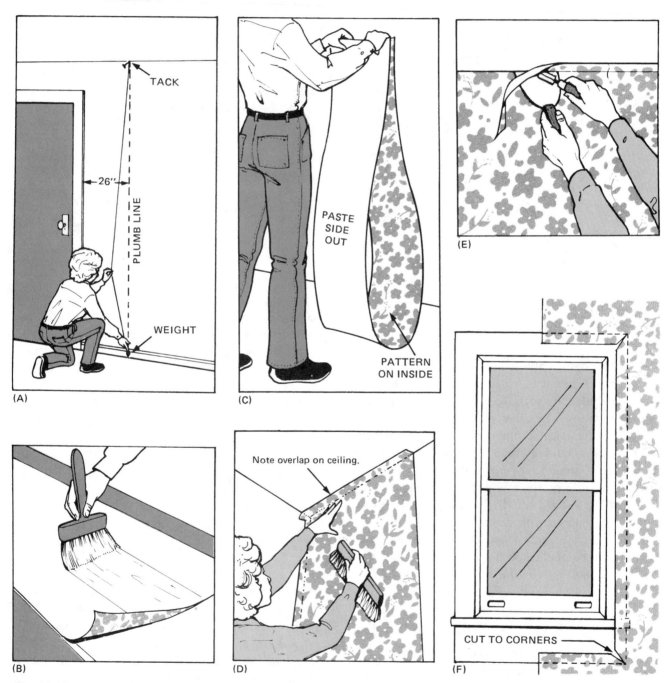

Fig. 15-57 *Putting wallpaper on walls. (A) A chalk line and plumb are used to mark the starting point. (B) After the paper has been measured and cut, paste is brushed on the back. (C) The bottom and top edges are used to carry the pasted wall covering. (D) Strips are lapped at the ceiling and brushed down. (E) Use a putty knife or straightedge as a guide to trim overlaps. (F) Cut corners diagonally at windows and doors.*

Next, several strips of paper are cut for use. The distance from the floor to the ceiling is measured. The wallpaper is unrolled on the floor or a table. The pattern side is left showing. The distance from the floor to the ceiling is laid off and 4 inches is added. This strip is cut, and several more are cut.

The first precut strip is laid on a flat surface. The pattern should be face up. The strip is checked for appearance and cuts or damage. Next, the strip is turned over so that the pattern is face down. The paste is applied with the brush. See Fig. 15-57B. The entire surface of the paper is covered. Next, the strip is folded in the middle. The pattern surface is on the inside of the fold. This allows the worker to hold both the top and the bottom edges. See Fig. 15-57C.

The plumb line is used as a starting point. The first strip is applied at the ceiling. About 1 inch overlaps the ceiling. This will be cut off later. A stiff bristle brush is moved down the strip. See Fig. 15-57D. Take care that the edges are aligned with the chalk line. About 1 inch will extend around the corner. This will be trimmed away later. This is an allowance for an uneven corner.

The entire strip is then brushed to remove the air bubbles. The brush is moved from the center toward the edge. The second strip is prepared in the same manner. However, care is taken to be sure that the pattern is matched. The paper is moved up or down to match the pattern. The edge of the second strip should exactly touch the edge of the first piece.

Next, the second piece is folded down. The edge is exactly matched with the edge of the first piece. The edge should exactly touch the edge of the previous piece. Then, any bubbles are smoothed out. The process is repeated for each strip until the wall is finished. Then the wall is trimmed.

The extra overlap at the ceiling and at the base are cut. A razor blade or knife is used with a straightedge as a guide. See Fig. 15-57E.

A new line is plumbed at each new corner. The first strip on each corner is started even with the plumb line. This way each strip is properly aligned. The corners will not have unsightly gaps or spaces.

A diagonal cut is made at each corner of a window. About ½ inch is allowed around each opening. The diagonal cut forms a flap over the molding. This allows the opening to be cut for an exact fit. See Fig. 15-57F.

FLOOR PREPARATION AND FINISH

The floor is finished last. This really makes sense if you think about it. After all, people will be working in the building. They will be using paint, varnish, stain, plaster, and many other things. All of these could damage a finished floor if they spilled on it.

Also, workers will be often entering and leaving the building. Mud, dirt, dust, and trash will be tracked into the building. It is very difficult to paint or plaster without spilling. A freshly varnished floor or a newly carpeted one could be easily ruined. So the floor is finished last to avoid damage to the finish floor.

Special methods are sometimes needed to lay flooring over concrete. Figure 15-58 shows some steps in this. Also, this was discussed in an earlier chapter.

(A)

(B)

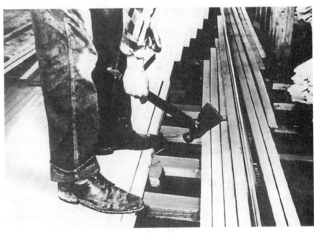
(C)

Fig. 15-58 *Laying wood floors on concrete. (A) Tar or asphalt is poured on. It acts as a vapor barrier. (B) Wood strips are positioned on the asphalt. (C) Flooring is nailed to the strips. (National Oak Flooring Manufacturers Association)*

For all types of floors, the first step is to clean them. See Fig. 15-59. The surface is scraped to remove all plaster, mud, and other lumps. Then the floor is swept with a broom.

Fig. 15-59 *Subflooring must be scraped and cleaned.* (National Oak Flooring Manufacturers Association)

Laying Wooden Flooring

Wooden flooring most often comes in three shapes. The first is called *strip* flooring (Fig. 15-60). The second is called *plank* flooring as in Fig. 15-61. The third type is *block* or *parquet* flooring (Fig. 15-62). The blocks may be solid as in Fig. 15-63A. Here the grain runs in one direction and the blocks are cut with tongues and grooves. The type shown in Fig. 15-63B may be straight-sided. They are made of strips with the grain running in the same direction. Such blocks may have a spline at the bottom or a bottom layer.

The type of block shown in Fig. 15-63C is made of several smaller pieces glued to other layers. The bottom layer is usually waterproof.

Most flooring has cut tongue-and-groove joints. Hidden nailing methods are used so that the nails do not

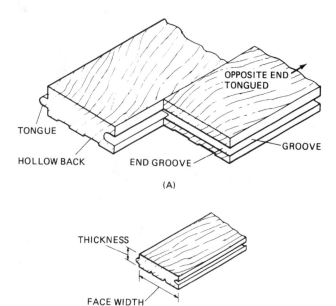

Fig. 15-60 *Strip flooring.* (Forest Products Laboratory)

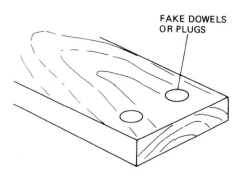

Fig. 15-61 *Plank flooring.*

show when the floor is done. Strip and plank flooring will also have an undercut area on the bottom. This undercut is shown in Fig. 15-60. The undercut or *hollow* helps provide a stable surface for the flooring. Small bumps will not make the piece shift. Strip flooring has narrow, even widths with tongue-and-groove joints on the ends. The strips may be random end matched. Plank flooring comes in both random widths and lengths. It may be drilled and pegged at the ends. However, today most plank flooring has fake pegs that are applied at the factory. The planks are then nailed in much the same manner as strip flooring. Block flooring may be either nailed or glued, but glue is used most often.

Most flooring is made from oak. Several grades and sizes are available. However, the width or size is largely determined by the type of flooring.

Carpeting can cost much more than a finished wood floor. Also, carpeting may last only a few years. As a rule, wood floors last for the life of the building. It is a good idea to have hardwood floors underneath carpet. Then the carpet can be removed without greatly reducing the resale value of the building.

Preparation for Laying Flooring

Manufacturers recommend that flooring be placed inside for several days before it is laid. The bundles should be opened and the pieces scattered around the room. This lets the wood reach a moisture content similar to that of the room. This will help stabilize the flooring. If the flooring is stabilized, the expansion and contraction will be even.

Check the subfloor for appearance and evenness. The subfloor should be cleaned and scraped of all deposits and swept clean. Nails or nail heads should be removed. All uneven features should be planed or sanded smooth. See Fig. 15-72.

A vapor barrier should be laid over the subfloor. It can be made of either builder's felt or plastic film. Seams should be overlapped 2 to 4 inches. Then chalk lines should be snapped to show the centers of the floor joists. See Fig. 15-64.

Fig. 15-62 *A parquet floor of treated wood blocks.* (Perma Grain Products)

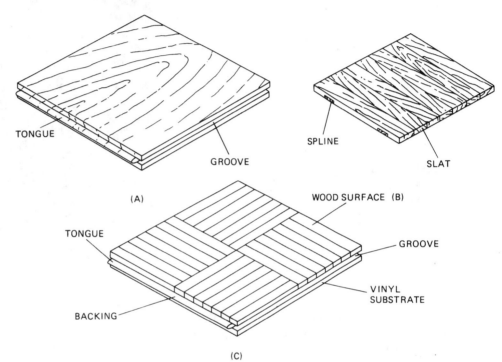

Fig. 15-63 *Wood block (parquet) flooring. (A) Solid.* (Forest Products Laboratory) *(B) Splined.* (Forest Products Laboratory). *(C) Substrated.*

TONGUE

GROOVE

(A)

SPLINE

SLAT

WOOD SURFACE (B)

TONGUE

GROOVE

BACKING

VINYL SUBSTRATE

(C)

Fig. 15-64 *Vapor barriers are needed over board subfloors.* (National Oak Flooring Manufacturers Association)

Installing wooden strip flooring Wooden strip flooring should be applied perpendicular to the floor joists. See Fig. 15-65. The first strip is laid with the grooved edge next to the wall. At least ½ inch is left between the wall and the flooring. This space controls expansion. Wooden flooring will expand and contract. The space next to the wall keeps the floor from buckling or warping. Warps and buckles can cause air gaps beneath the floor. They also ruin the looks of the floor. The space next to the wall will be covered later by molding and trim.

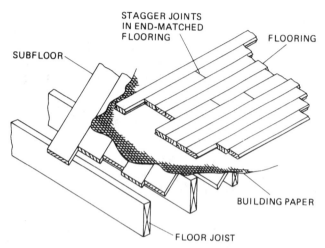

Fig. 15-65 *Strip flooring is laid perpendicular to the joists.* (Forest Products Laboratory)

Fig. 15-66 *Nailing the first strip.* (National Oak Flooring Manufacturers Association)

The first row of strips is nailed using one method. The following rows are nailed differently. In the first row, nails are driven into the face. See Figs. 15-66 and 15-67. Later, the nail is set into the wood and covered.

Hardwood flooring will split easily. Most splits occur when a nail is driven close to the end of a board. To prevent this, drill the nail hole first. It should be slightly smaller than the nail.

The next strip is laid in place as shown in Fig. 15-68. The nail is driven blind at a 45° or 50° angle as shown. Note that the nail is driven into the top corner of the tongue. The second strip should fit firmly against the first layer. Sometimes, force must be used to make it fit firmly. A scrap piece of wood is placed over the tongue as shown in Fig. 15-69. The ends of the second layer should not match the ends of the first layer. The ends should be staggered for better strength and appearance. The end joints of one layer should be at least 6 inches from the ends on the previous layer. A nail set should be used to set the nail in place. Either the vertical position or the position shown in Fig. 15-69 may be used. Be careful not to damage the edges of the boards with the hammer.

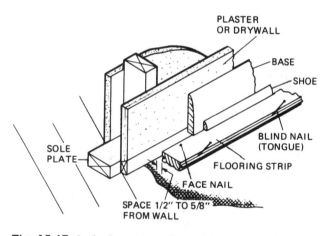

Fig. 15-67 *In the first strip, nails are driven into the face.* (Forest Products Laboratory)

The same amount of space (½ inch) is left at the ends of the rows. End pieces may be driven into place with a wedge. Pieces cut from the ends can become the first piece on the next row. This saves material and helps stagger the joints. Figures 15-70 and 15-71 show other steps.

Wooden plank flooring Wooden plank flooring is installed in much the same manner. Today, both types

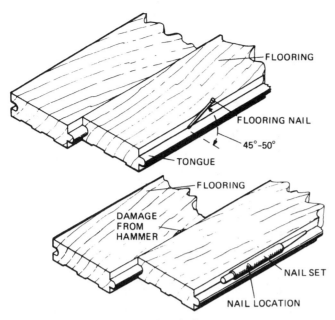

Fig. 15-68 *Nailing strips after the first.* (Forest Products Laboratory)

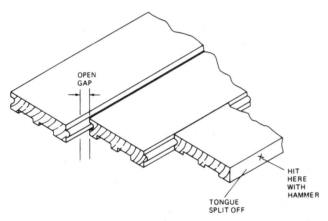

Fig. 15-69 *A scrap block is used to make flooring fit firmly. This prevents damage to either tongues or grooves.*

Fig. 15-70 *Boards are laid in position and nailed in place.* (National Oak Flooring Manufacturers Association)

Fig. 15-71 *The last piece is positioned with a pry bar. Nails are driven in the face.* (National Oak Flooring Manufacturers Association)

are generally made with tongue-and-groove joints. The joints are on both the edges and ends. The same allowance of ½ inch is made between plank flooring and the walls. The same general nailing procedures are used.

Wood block floors There are two types of wood block floors. The first type is like a wide piece of board. The second type is called parquet. Parquet flooring is made from small strips arranged in patterns. Parquet must be laminated to a base piece. Today, both types of block floors are often laminated. When laminated, they are plywood squares with a thick veneer of flooring on the top. Usually, three layers of material make up each block.

Most blocks have tongue-and-groove edges. Common sizes are 3-, 5-, and 7-inch squares. With tongue-

and-groove joints, the blocks can be nailed. However, today most blocks are glued. When they are glued, the preparation is different than for plank or strip flooring. No layer of builder's felt or tar paper is used. The blocks are glued directly to the subfloor or to a base floor. As a rule, plywood or chipboard is used for a base under the block flooring.

Blocks may be laid from a wall or from a center point. Some patterns are centered in a room. Then, blocks should be laid from the center toward the edges. To lay blocks from the center, first the center point is found. Then a block is centered over this point. The outline of the block is drawn on the floor. Then a chalk line is snapped for each course of blocks. Care is taken that the lines are parallel to the walls.

Sometimes blocks are laid on the floor, proceeding from a wall. The chalk lines should be snapped for each course from the base wall.

Blocks are often glued directly to concrete floors. These floors must be properly made. Moisture barriers and proper drainage are essential. When there is any doubt, a layer process is used. First, a layer of mastic cement is applied. Then a vapor barrier of plastic or felt is applied over the mastic. Then a second layer of mastic is applied over the moisture barrier. The blocks are then laid over the mastic surface.

Wooden blocks may also be glued over diagonal wood subflooring. The same layer process as above is used.

FINISHING FLOORS

In most cases, floors are finished after walls and trim. When resilient flooring is applied, no finish step is needed. However, floors of this type should be cleaned carefully.

Finishing Wood Floors

The first step in finishing wood floors is sanding. A special sanding machine is used, such as the one in Fig. 15-72. When floors are rough, they are sanded twice. For oak flooring, it is a good idea to use a sealer coat next. The sealer coat has small particles in it. These help fill the open pores in the oak. One of two types of sealer or filler should be used. The first type of sealer is a mixture of small particles and oil. This is rubbed into the floor and allowed to sit a few minutes. Then, a rotary sander or a polisher is used to wipe off the filler. The filler on the surface is wiped off. The filler material in the wood pores is left.

Floors may be stained a darker color. The stain should be applied before any type of varnish or lacquer

Fig. 15-72 *After the floor is laid, it is sanded smooth using a special sanding machine.* (National Oak Flooring Manufacturers Association)

is used. It is a good idea to stain the molding at the same time. The stain is applied directly after the sanding. If a particle-oil type filler is used, the stain should be applied after the filler.

The second filling method uses a special filler varnish or lacquer. This also contains small particles which help to fill the pores. It is brushed or sprayed onto the floor and allowed to dry. The floor is then buffed lightly with an abrasive pad.

After filling, the floor should be varnished. A hard, durable varnish is best. Other coatings are generally not satisfactory. Floors need durable finishes to avoid showing early signs of wear.

Molding can be applied after the floors are completely finished. The base shoe and baseboard are installed as in Fig. 15-73. Many workers also finish the molding when the floor is finished.

Base Flooring for Carpet

Often two floor layers are used but neither is the "finished" floor surface. The first layer is the subfloor and the next layer is a base floor. Both base and subflooring may be made from underlayment. Underlayment may be a special grade of plywood or chipboard. In many cases, nailing patterns are printed on the top side of the underlayment.

It is a good idea to bring underlayment into the room to be floored and allow it to sit for several days exposed to the air. This allows it to reach the same moisture content as the rest of the building components.

Base flooring is used when added strength and thickness are required. It is also used to separate resilient flooring or other flooring materials from concrete or other types of floors. It provides a smooth, even base for carpet. It is much cheaper to use a base

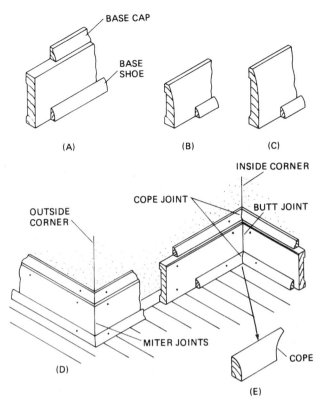

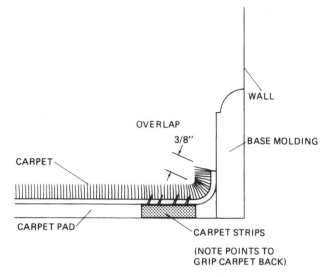

Fig. 15-74 *Carpet strips hold the carpet in place.*

Fig. 15-73 *Base molding* (Forest Products Laboratory)*: (A) Square-edged base; (B) Narrow ranch base; (C) Wide ranch base; (D) Installation; (E) A cope joint.* (Forest Products Laboratory)

floor than a hardwood floor under carpets. The costs of nailing small strips and of sanding are saved.

INSTALLING CARPET

Carpet is becoming more popular for many reasons. It makes the floor a more resilient and softer place for people to stand. Also, carpeted floors are warmer in winter. Carpeting also helps reduce noise, particularly in multistory buildings,

To install carpet, several factors must be considered. First, most carpet is installed with a pad beneath it. When carpet is installed over a concrete floor, a plastic film should also be laid. This acts as an additional moisture seal. The carpet padding may then be laid over the film.

The first step in laying carpet is to attach special carpet strips. These are nailed to the floor. They are laid around the walls of a room to be carpeted. These carpet strips are narrow, thin pieces of wood with long tacks driven through them. They are nailed to the floor approximately ¼ inch from the wall. See Fig. 15-74. The carpet padding is then unrolled and cut. The padding extends only to the strips.

Next, the carpet is unrolled. If the carpet is large enough, it may be cut to exactly fit. However, carpet should be cut about 1 inch smaller than the room size.

The carpet is wedged between the carpet strip and one wall. A carpet wedge is used as in Fig. 15-75. The carpet is then smoothed toward the opposite wall. All wrinkles and gaps are smoothed and removed. Next, the carpet is stretched to the opposite wall. The person installing the carpet will generally walk around the edges pressing the carpet into the tacks of the carpet

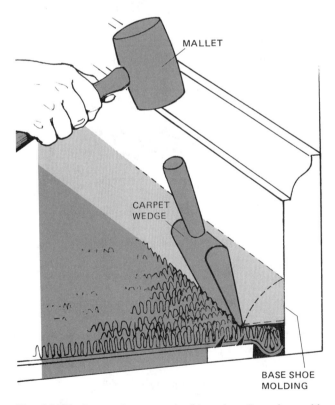

Fig. 15-75 *Carpet edges are wedged into place. Base shoe molding may then be added.*

strips. This is done after each side is wedged. After the ends are wedged, the first side is attached. The same process is repeated. After the first side is wedged the opposite side is attached.

To seam a carpet, special tape is used. This tape is a wide strip of durable cloth. It has an adhesive on its upper surface. It is rolled out over the area or edge to be joined. Half of the tape is placed underneath the carpet already in place. The carpet is then firmly pressed onto the adhesive. Some adhesive tapes use special heating tools for best adhesion. The second piece of carpet is carefully butted next to the first. Be sure that no great pressure is used to force the two edges together. No gaps should be wider than 1/16 inch. The edges should not be jammed forcefully together, either. If edges are jammed together, lumps will occur. If wide gaps are left, holes will occur. However, the nap or shag of the carpet will cover most small irregularities.

To cut the carpet for a joint, first unroll the carpet. Carefully size and trim the carpet edges as straight as possible. The joint will not be even if the edges are ragged. Various tools may be used. A heavy knife or a pair of snips may be used effectively.

Metal end strips are used where carpet ends over linoleum or tile. The open end strips are nailed to the floor. The carpet is stretched into place over the points on the strip. Then the metal strip is closed. A board is laid over the strip. See Fig. 15-76. The board is struck sharply with a hammer to close the strip. Do not strike the metal strip directly with the hammer. Doing so will leave unsightly hammer marks on the metal.

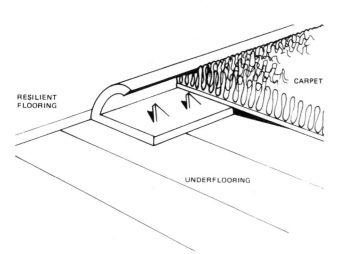

Fig. 15-76 *A metal binder bar protects and hides carpet edges where carpet ends over linoleum or tile.*

RESILIENT FLOORING

Resilient flooring is made from chemicals rather than from wood products. Resilient flooring includes com-

positions such as linoleum and asphalt tile. It may be laid directly over concrete or over base flooring. Resilient flooring comes in both sheets and square tiles. Frequently, resilient flooring sheets will be called linoleum "carpets."

Installing Resilient Flooring Sheets

The first step is to determine the size of the floor to be covered. It is a good idea to sketch the shape on a piece of paper. Careful measurements are made on the floor to be covered. Corners, cabinet bases, and other features of the floor are included. It is a good idea to take a series of measurements. They are made on each wall every 2 or 3 feet. This is because most rooms are not square. Thus, measurements will vary slightly from place to place. The measurements should be marked on the paper.

Next, the floor is cleaned. Loose debris is removed. A scraper is used to remove plaster, paint, or other materials. See Fig. 15-59. Then, the area is swept. If necessary, a damp mop is used to clean the area. Neither flooring nor cement will stick to areas that are dirty. Next, the surface is checked for holes, pits, nail heads, or obstructions. Holes larger than the diameter of a nail are patched or filled. Nails or obstructions are removed.

Most rooms will be wider than the roll of linoleum carpet. If not, the outline of the floor may be transferred directly to the flooring sheet. The sheet may then be cut to shape. The shaped flooring sheet is then brought into the room. It is positioned and unrolled. A check is made for the proper fit and shape. Any adjustments or corrections are made. Then, the sheet is rolled up approximately halfway. The mastic cement is spread evenly (about 3/32 inch thick) with a toothed trowel. The unrolled portion is rolled back into place over the cemented area. Then, the other end is rolled up to expose the bare floors. Next, the mastic is spread over the remaining part of the floor. The flooring is then rolled back into place over the cement. The sheet is smoothed from the center toward the edge.

However, most rooms are wider than the sheets of flooring. This means that two or more pieces must be joined. It is best to use a factory edge for the joint line. Select a line along the longest dimension of the room as shown in Fig. 15-77. On the base floor, measure equal distances from a reference wall as shown. These are the same width as the sheet. Then snap a chalk line for this line. Often more than one joint or seam must be used.

The same measuring process is used. Measurements are taken from the edge of the first sheet. How-

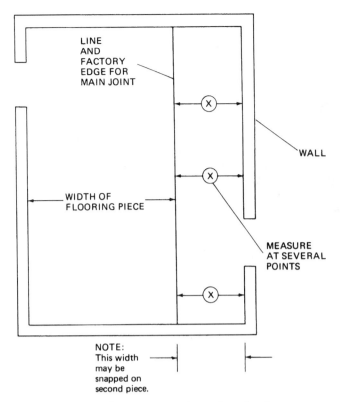

LINE
AND
FACTORY
EDGE FOR
MAIN JOINT

WALL

WIDTH OF
FLOORING PIECE

MEASURE
AT SEVERAL
POINTS

NOTE:
This width
may be
snapped on
second piece.

Fig. 15-77 *Floor layout for resilient flooring.*

ever, for smaller pieces, a different process is used. The center of the last line is found. A line is snapped at right angles. This shows the pattern for the pieces of flooring that must be cut. The second line should run the entire width of the room. A carpenter's square may be used to check the squareness of the lines.

The two chalk lines are now the reference lines. These are used for measuring the flooring and the room area. Measurements to the cabinets or other features are made from these lines. Take several measurements from the paper layout. Walls are seldom square or straight. Frequent measurements will help catch these irregularities. The fit will be more accurate and better.

Next, in a different area unroll the first sheet of flooring to be used. Find the corresponding wall and reference line. The factory edge is aligned on the first line. The dimensions are marked on the resilient flooring. The necessary marks show the floor outline on the flooring sheet. A straight blade or linoleum knife is used to make these cuts. For best results, a guide is used. A heavy metal straightedge is used for a guide when cutting. A check is made to be sure that there is nothing underneath the flooring. Anything beneath it could be damaged by the knife used to cut the flooring.

Next, the cut flooring piece is carried into the area. It is unrolled over the area, and the fit is checked. Any

adjustments necessary are made at this point. Next, the material is rolled toward the center of the room. The area is spread with mastic. The flooring is rolled back into place over the cement. The sheet is smoothed as before. Do not force the flooring material under offsets or cabinets. Make sure that the proper cuts are made.

The adhesive should not be allowed to dry more than a few minutes. No more than 10 or 15 minutes should pass before the material is placed. A heavy roller is recommended to smooth the flooring. It should be smoothed from the center toward the edges. This removes air pockets and bubbles.

Where seams are made, a special procedure is recommended. Unroll the two pieces of flooring in the same preparation area. The two edges to be joined are slightly overlapped. See Fig. 15-78. Next, the heavy metal guide is laid over the doubled layer. Then both layers are cut with a single motion. The straightedge is used as a guide to make the straightest possible cut. Edges cut this way will match, even if they are not perfectly straight or square. Figure 15-79 shows how edges are trimmed.

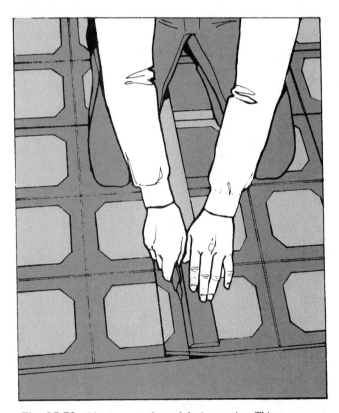

Fig. 15-78 *Edges are overlapped during cutting. This way seams will match even if the cut is not perfectly straight.* (Armstrong Cork)

To Install Resilient Block Flooring

Resilient block flooring is often called *tile*. This term includes floor tile, asphalt tile, linoleum tile, and others.

Fig. 15-79 *Trimming the edges.
A metal straightedge or
carpenter's square is used to
guide a utility knife.* (Armstrong Cork)

As a rule, the procedure for all of these is the same. The sizes of these tiles range from 6 to 12 inches square. To lay tile, the center of each of the end walls is found. See Fig. 15-80. A chalk line is tied to each of these points. Lines are then snapped down the middle of the floor. Next, the center of the first line is located. A square or another tile is used as a square guide. A second line is snapped square to the first line.

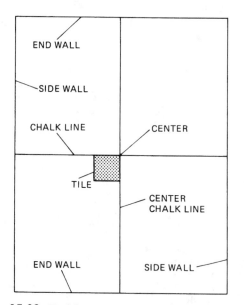

Fig. 15-80 *Find the intersecting midpoints.* (Armstrong Cork)

Next, a row of tiles is laid along the perpendicular chalk line as shown. They are not cemented. Then the distance between the wall and the last tile is measured. If the space is less than half the width of a tile, a new line is snapped. It is placed half the width of a tile away from the center line. See Fig. 15-81. A second

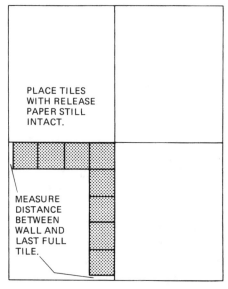

Fig. 15-81 *Lay tiles in place without cement to check the spacing.* (Armstrong Cork)

line is snapped half the width of a tile from the perpendicular line. The first tile is then aligned on the second snapped line. The tile can now be cemented. The first tile becomes the center tile of the room.

Another method is used if the distance at the side walls is greater than one-half the width of a tile. Then the tile is laid along the first line. This way, no single tile is the center.

The first two courses are laid as a guide. Then the mastic is spread over one-quarter of the room area. See Fig. 15-82. Lay the tiles at the center first. The first tiles should be laid to follow the snapped chalk lines. Tiles should not be slid into place. Instead, they are pressed firmly into position as they are installed. It is

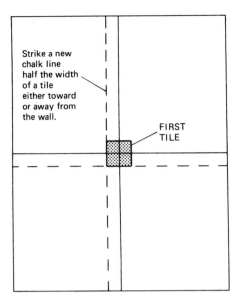

Strike a new chalk line half the width of a tile either toward or away from the wall.

FIRST TILE

Fig. 15-82 *The next chalk line is used as a guide for laying tile. This method places one tile in the center.*

best to hold them in the air slightly. Then the edges are touched together and pressed down.

The first quarter of the room is covered. Only the area around the wall is open. Here, tiles must be cut to fit.

Cutting the tiles A loose tile is placed exactly on top of the last tile in the row. See Fig. 15-83A. Then a third tile is laid on top of this stack. It is moved over until it

touches the wall. The edge of the top of the tile becomes the guide. See Fig. 15-83B. Then, the middle tile is cut along the pencil line as shown. This tile will then have the proper spacing.

A pattern is made to fit tile around pipes or other shapes. The pattern is made in the proper shape from paper. This shape is traced on the tile. The tile is then cut to shape.

LAYING CERAMIC TILE

Several types of ceramic tile are commonly used: ceramic tile, quarry tile, and brick. Tiles are desirable where water will be present, such as in bathrooms and wash areas. However, ceramic tiles also make pleasing entry areas and lobbies. Two methods are used for installing ceramic floor tile.

The first, and more difficult, method uses a cement-plaster combination. A special concrete is used as the bed for the tile. This bed should be carefully mixed, poured, leveled, and troweled smooth. It should be allowed to sit for a few minutes. Just before it hardens, it is still slightly plastic. Then the tiles are embedded in place. The tiles should be thoroughly soaked in water if this method is used. They should be taken out of the water one at a time and allowed to drain slightly. The tiles are then pressed into place in the cement base. All the tiles are installed. Then special grout is

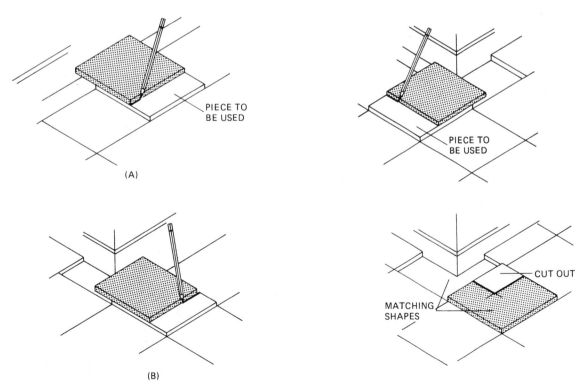

PIECE TO BE USED

(A)

PIECE TO BE USED

CUT OUT

MATCHING SHAPES

(B)

Fig. 15-83 *Cutting and trimming tile to fit. (A) Marking tile for edges. (B) Marking tile for corners.*

pressed into the cracks between the tiles. This completely fills the joints between the tiles. The grouted joints should be cleaned and tooled within a very few hours after installation.

Adhesives are most widely used today. Adhesives are much easier to use. The adhesives are spread evenly on the tiles. This is much like laying resilient tile. The adhesive should be one that is recommended for use with ceramic tiles. After the adhesive has been spread, the tiles are placed. Then a common grout is forced between the tiles. It is wiped and allowed to dry.

Tiles are available as large individual pieces. They are also available as assortments. Small tiles may be preattached to a cloth mesh. This mesh keeps the tiles arranged in the pattern. This also makes the tile easy to space evenly.

CHAPTER 15
STUDY QUESTIONS

1. How is expansion controlled on strip flooring?
2. When is a vapor barrier used under finish flooring?
3. How are board ends kept from splitting?
4. How are floor tiles laid out?
5. Why are floor tiles started in the center?
6. When is strip flooring laid? When is it finished?
7. How are seams in linoleum cut?
8. When is door trim applied?
9. What is the easiest way to lay ceramic tile?
10. How is interior trim used on metal window frames?
11. What are built-ins?
12. What is the sequence for hanging doors?
13. What are three ways of making shelves?
14. List the steps in building a cabinet counter.
15. How are ready-built cabinets hung?
16. How is plastic laminate applied?
17. Why are coped joints used?
18. Why are coatings smoothed from the center toward the edges?
19. When should floors be stained?
20. What are the advantages of lip drawers?

16
CHAPTER

Special Construction Methods

N BUILDING, CARPENTERS HAVE MANY JOBS to do. They build the frames and cover them; however, the carpenter must often do other jobs. The regular methods are not always used for the frames. Special jobs include fireplaces, chimneys, and stairs. The post-and-beam method of building is also included.

Carpenters need special skills to do the special jobs. Almost every building will involve one or more of these special jobs. Specific skills that can be learned in this unit are:

- Laying out, cutting, and installing stair parts
- Designing, cutting, and making fireplace frames according to the manufacturer's or builder's specifications
- Planning and building basic post-and-beam frames

STAIRS

Carpenters should know how to make several types of stairs. Main stairs should be pleasing and attractive. This is because they are visible in the living or working areas. They should also be safe, sturdy, and easy to climb. It is a good idea to make them as wide as possible. Service stairs are not as visible. They are mainly used to give occasional access or entry. They are used in basements, attics, and other such areas. Service stairs are not as wide or attractive as main stairs. Service stairs can be steeper and harder to climb. However, all stairs should be sturdy and safe to use.

Stair Parts

The carpenter will most often build stairs around a notched frame. See Fig. 16-1. This frame is called a *carriage*. The carriage is also called a *stringer* by some carpenters. The carriage should be made of 2-×-10-inch or 2-×-12-inch boards. This way one solid board extends from the top to the bottom. The other stair parts are then attached to the carriage.

The part on which people step is called the *tread*. The vertical part at the edge is called a *riser*. A stair unit is made of one tread and one riser. The unit run is the width of the tread. The unit rise is the height of the riser. The tread is usually rounded on the front edge. The rounded edge also extends over the riser. The part that hangs over is called the *nose or nosing*. See Fig. 16-2A.

Stairs must also have a handrail. Where a stair is open, a special railing is used. The fence-like supports are called *balusters*. The handrail or banister rests on the balusters. The end posts are called *newels*. See Fig. 16-2B.

Fig. 16-1 *A notched frame called a carriage is the first step in building stairs.*

Some stairs rise continuously from one level to the next. Other stairs rise only part way to a platform. The platform is called a *landing*. Often landings are used so that the stairs can change direction. See Fig. 16-3.

Stair Shapes

Stairs are made in several shapes. See Fig. 16-4A through E. Stairs may be in open areas or they may have walls on one or both sides. Stairs with no wall or one wall on the side are called *open* stairs. Stairs that have walls on both sides are called *closed* stairs.

Straight-run stairs The straight run is the simplest stair shape. This type of stair rises in a straight line from one level to the next. The stairs may be open or closed. See Fig. 16-4A and B.

L-shaped stairs Figure 16-4C shows L-shaped stairs. These stairs rise in two sections around a corner. To make these, the carpenter builds two sets of straight runs. The first is to the platform or landing. The second is from the landing to the next level.

U-shaped stairs The stairs in Fig. 16-4D are U-shaped stairs. They are much like L-shaped stairs. However, two corners are turned instead of one. As a rule, a platform or landing is used at each corner. Again the carpenter makes straight-run stairs from one platform to the next.

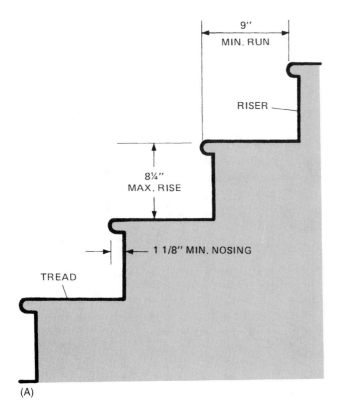

(A)

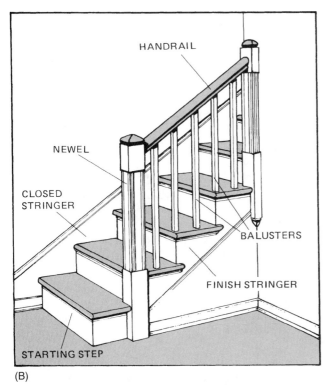

(B)

Fig. 16-2 *Stair parts. (A) Risers and treads* (Forest Products Laboratory). *(B) Railing. The newels, balusters, and rails together are called a balustrade.*

Winders Some stairs have steps that "fan" or turn as in Fig. 16-4E. No landing is used to turn a corner. These turning steps are called *winders* or winding stairs. These are the hardest to make and they take spe-

Fig. 16-3 *A landing is used to change direction.*

cial framing. As a rule, the stairs are pie-shaped only on corners. However, some special jobs may feature the true winding staircase.

Winders are considered attractive by many people. However, carpenters make winders only when there is not enough room for a landing. The winder is not as safe as a regular landing.

Stair Design

Most stairs are made with 2-×-12-inch carriages rising through a framed opening. As a rule, the framed opening is made when the floors are framed. The carpenter may check for proper support and framing before making the carriage attachments.

Of course, carpenters build stairs as the designer or architect specifies. However, carpenters should know the fundamentals of stair design. Sometimes, architects or designers will not specify all the details. In these cases, carpenters must use good practices to complete the job.

Stairs should rise at an angle of 30 to 35°. The minimum width of a tread is about 9 inches. There is no real maximum limit for tread width. However, a long step results when the tread is very wide. As a rule, treads 9 inches wide are used only for basements and service stairs. For main stairs, the tread width is normally 10 to 12 inches.

The riser for most main stairs should not be very high. If the riser is too high, the stairs become steep and hard to climb. As a rule, 8¼ inches is considered the maximum riser height. Less than that is desirable. For main stairs, 7 to 7½ inches is both more comfortable and safer.

Find the number of steps The amount of rise per step determines the number of steps in stairs. If a 7-inch rise is desirable, then the total rise is divided by 7 inches. The total rise is the total distance from the floor level of one landing to the next. A two-story building

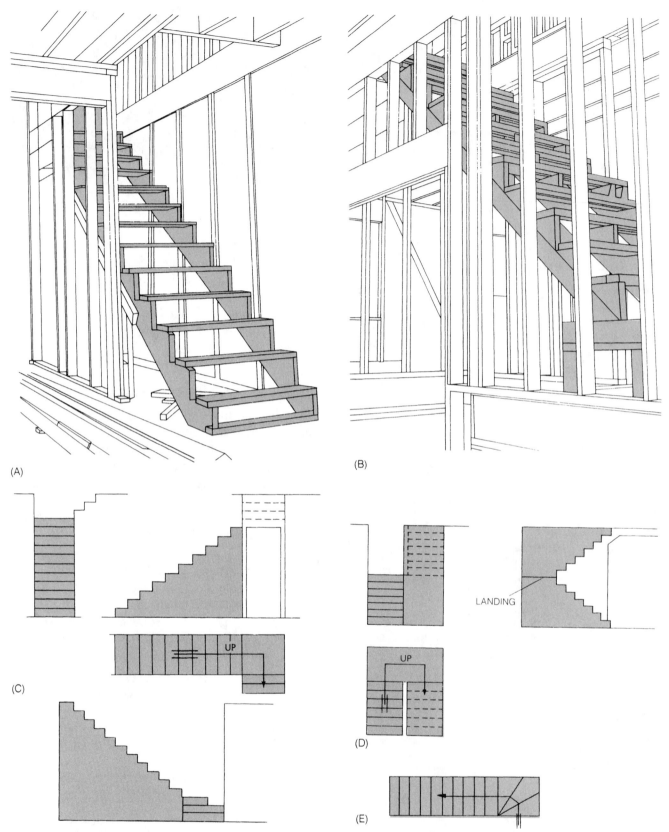

Fig. 16-4 *(A) Straight-run stairs do not change direction. This stair is part open. (B) This stair will be closed. It will have walls on both sides. (C) A long L-shaped stair* (Forest Products Laboratory)*. (D) U-shaped stairs* (Forest Products Laboratory)*. (E) A winder saves space. It does not use a landing to change direction.* (Forest Products Laboratory)

with an 8-foot ceiling is an example. The width of the floor joists and second-story finished flooring is added. The second-floor joists are 2×10 inches. The total rise would be about 8 feet 10 inches. For 7-inch risers, this distance (8'10") is divided by 7:

$$8'10" = 106"$$

$$\text{Height/rise} = \text{number of steps}$$

$$\frac{106}{7} = 15.14$$

15 would be the number of steps

In practice, most stairs in houses have 13, 14, or 15 steps. Fourteen steps (a riser of about 7½ inches) is probably the most common.

Using ratios To find the step width, a riser-to-tread ratio is used. There are three rules for this ratio. Remember, one tread and one riser is considered one unit. See Fig. 16-5.

1. Unit rise + unit run = 17 to 18.
2. Two unit rises + one unit run = 24 to 25.
3. Unit rise × unit run = 72 to 75.

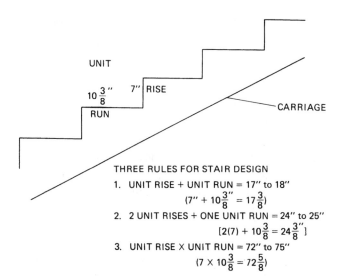

Fig. 16-5 *A unit is one run and one rise.*

THREE RULES FOR STAIR DESIGN
1. UNIT RISE + UNIT RUN = 17" to 18"
$$(7" + 10\tfrac{3}{8} = 17\tfrac{3}{8})$$
2. 2 UNIT RISES + ONE UNIT RUN = 24" to 25"
$$[2(7) + 10\tfrac{3}{8} = 24\tfrac{3}{8}"]$$
3. UNIT RISE × UNIT RUN = 72" to 75"
$$(7 \times 10\tfrac{3}{8} = 72\tfrac{5}{8})$$

Now, the total length of run is divided by the number of steps. If this length is 13'0", then the step width would be 156 inches divided by 15. This would be approximately 10⅜ inches.

These two numbers (7 inches for the riser and 10⅜ inches for the tread) are used to check the stair shape. To apply the rules:

1. Allowable unit rise + unit run 17 to 18. Actual sum is
$$7 \text{ inches} + 10\tfrac{3}{8} \text{ inches} = 17\tfrac{3}{8}$$
This figure is OK.

2. Allowable length of two unit rises + one unit run = 24 to 25. Actual length is
$$
\begin{aligned}
2 \times 7 &= 14 \text{ inches} \\
1 \times 10\tfrac{3}{8} &= \underline{+10\tfrac{3}{8}} \text{ inches} \\
&\quad\; 24\tfrac{3}{8}
\end{aligned}
$$
This is OK.

3. Allowable value of unit rise × unit run = 72 to 75. Actual value is
$$7 \times 10\tfrac{3}{8} = 72\tfrac{5}{8}.$$
This is also OK.

Sometimes, two or three combinations have to be tried to get a good design. Experience has shown that these ratios give a safe, comfortable, and well-designed stair.

Headroom Headroom is the distance between the stairs and the lowest point over the opening. The minimum distance should be above 6 feet 4 inches. It is better to have 6 feet 8 inches. The headroom can be quickly checked. The carpenter can lay a board at the desired angle. Or a line can be strung. The distance between the board or line and the lowest point can be measured.

Stair width The width of the stair is the distance between the rail and the wall or baluster. This is the *walking area*. This area between the rail and the wall is called the *clear* distance. Narrow stairs are not as safe as wide stairs. Also, it is hard to move furniture and appliances in and out through narrow stairs. The width of the stair is a major factor.

The minimum clear distance allowed by the FHA is 2 feet 8 inches on main stairs. Sometimes 2 feet 6 inches is used on service stairs. However, added width is always desirable.

Tread width on winders should be equal to the regular width. Winder treads are checked at the midpoint. Or, the tread width can be checked at the regular walking distance from the handrail. For example, tread width on a stair is 10⅜ inches. Then, the tread width on the middle of the winder should also be 10⅜ inches. See Fig. 16-6.

Sequence in Stair Construction

Carpenters should follow a given sequence in building the stairs. The stair is laid out on the carriage piece. The step (riser and tread) notches are cut. Two carriage pieces are used for stairs up to 3 feet in width. Three or more carriage boards are used for stairs that are wider than 3'0". One is on each side, and one is in the center.

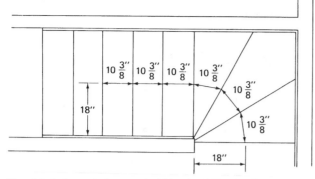

Fig. 16-6 *In a winder, tread widths should be equal on the main path.* (Forest Products Laboratory)

Carriage boards are often positioned as walls and floors are framed. However, the stairs are not finished until the walls have been finished. Often, 2-inch lumber is used for temporary steps during the construction. See Fig. 16-7. Stairs that rest on concrete floors are not built until the concrete floor is finished.

Fig. 16-7 *Two-inch lumber is used for temporary steps during construction. It can be left as a base for finish treads.*

Three methods of making stairs are used. The first two methods require the carpenter to make the carriage. The treads are finished later. In the first method, regular lumber is used for a tread. These treads are later covered with finished hardwood flooring.

In the second method, hardwood treads are nailed directly to the carriage. These boards are usually oak 1⅛ or 1½6 inches thick.

The third method is called a *housed carriage*. The carriage board has grooves to receive the treads and risers. As a rule, these pieces are precut at a factory or mill. The treads are inserted from the top to the bottom. Glue and wedges are used to wedge the stairs tightly in place. See Fig. 16-8. This type of stair is very sturdy and normally does not squeak.

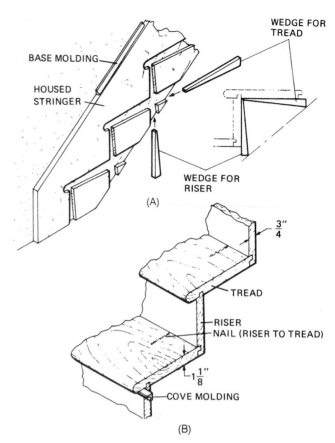

Fig. 16-8 *Housed carriage stairs: (A) Housing* (Forest Products Laboratory); *(B) Tread detail.* (Forest Products Laboratory)

Carriage Layout

Riser height and tread width are found from the ratios. The example given used a 7-inch riser with a 10⅜-inch tread. A 2-×-12-inch board of the proper length is selected. The board is laid across the stair area to see whether the board is long enough. If so, the board is placed on a sawhorse or work table.

A framing square is used as in Fig. 16-9. The square is placed at the top end of the carriage as shown. The short blade of the square is held at the corner. The outside edge is used for these measurements. The height of the riser is measured down from the corner on the outside tongue. The blade is swung until the

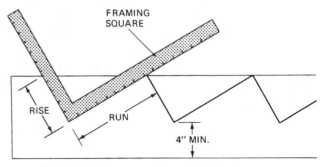

Fig. 16-9 *Lay out carriage with a framing square.*

tread width is shown on the scale. This is shown when the tread width (10⅜ inches) meets the edge of the board. Lines are drawn along the edges of the squares.

This process is repeated for all the steps. On the last step, two things are done. First, a riser is marked for the first step height. See Fig. 16-10. Then the height is adjusted to allow for the tread thickness. This must be done. When the tread is nailed to the carriage, it will raise the riser distance. For example, suppose a carriage is cut so that the full first step rises 7 inches. Then a board 1⅛ inches thick is added for the tread. This would make the total riser height 8⅛ inches. So the thickness of the tread is subtracted as in Fig. 16-10.

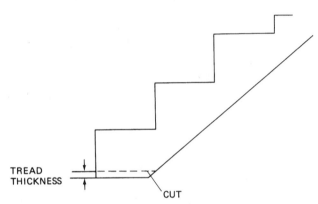

Fig. 16-10 *The carriage bottom must be trimmed to make the first rise correct.*

Notches are cut. A bayonet saw or a handsaw should be used. It is not a good idea to use a circular saw. The blade does not cut vertically through the wood. See Fig. 16-11.

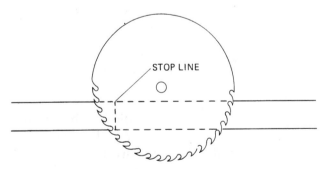

Fig. 16-11 *A circular saw should not be used, since it will cut past the line.*

For some outside stairs, the carriage is cut differently. The layout is the same. However, the steps are dadoed into place as in Fig. 16-12. To do this, the dado is cut on each side of the line. Cut the dado one-third to one-half the thickness of the carriage. Standard 2-inch lumber is only 1½ inches thick. The dado should

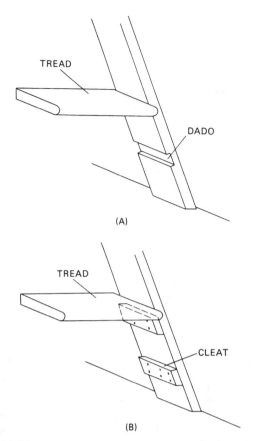

Fig. 16-12 *Exterior stair construction. (A) Dadoed carriage. (B) Cleated carriage.*

be cut ½ inch (one-third the thickness) to ¾ inch (one-half the thickness) deep.

Frame the Stairs

First, the ledger for the top is cut. Then it is nailed in place at the top of the landing or second story. The ledger is made from 2-×-4-inch lumber. It is nailed with 16d common nails. See Fig. 16-13. Next, a kicker is cut from 2-×-4-inch lumber. It is nailed (16d common nails) at the bottom as in Fig. 16-13. Then, notches are cut in the carriage for the kicker and ledger. Next, a backing stringer is nailed on the walls. This should be a clear piece of 1-inch lumber. It gives a finished appearance to the stairs. Use 8d common nails. See Fig. 16-14. Next, the carriage is nailed to the ledgers, kickers, wall, and backing strips. Heavy spikes are recommended. Next, the outside carriage is nailed to the ledger and kicker.

The outside finish board is nailed to the outside of the carriage. If the finish board is 2 inches thick, 16d finish nails should be used. If the outside board is 1-inch lumber, an 8d nail should be used. After the wall is complete, the newel is nailed in place. A finish stringer can be used as in Fig. 16-15. The finish stringer is cut from a 1-×-12-inch board. A finish stringer is blocked and nailed in place as shown. Finish nails should be used.

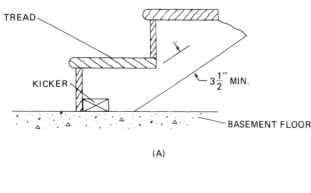

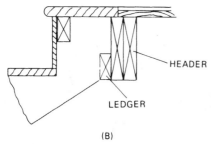

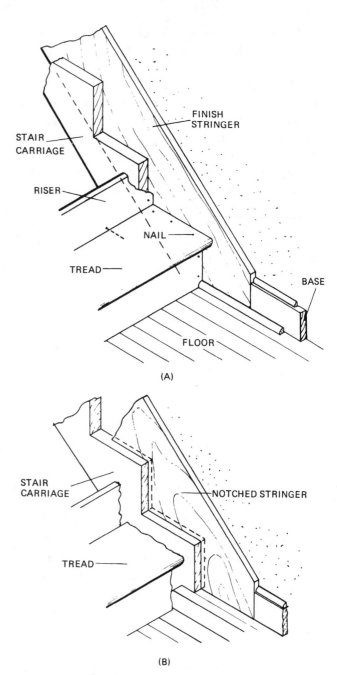

Fig. 16-13 (A) A kicker anchors the bottom of the carriage. (B) The ledger anchors the top. (Forest Products Laboratory)

Temporary steps are now removed from the carriage. As a rule, cutting is avoided on stair pieces. Tread overhang less than 1½ inches is ignored. The overhang becomes the nosing. However, excess width is cut from the back side.

In hardwoods it's a good idea to drill holes for nails. Doing this keeps the nails from splitting the board. The drilled hole should be slightly smaller than the nail. The nails should be set and filled after they are driven. Of course, only finish nails are used.

First, the bottom riser is nailed in place. It must be flush with the top of the carriage riser. It is normal to have a small space at the bottom of the riser. The space at the bottom can be covered by molding. The riser is always laid first. A convenient number of steps is worked at one time. This includes all the risers that can be easily reached. Probably two or three steps can be worked at the same time. Next, the bottom tread is placed in position. It is moved firmly against the riser. The tread is drilled and nailed as needed. Treads are laid until one is on each placed riser. Then, the worker moves up the steps. Again, a convenient number of risers and treads are laid. This process is repeated until the stair is finished.

The baluster is started when all the treads are installed. The ends of balusters are round like dowels. Remember, holes are located by the center of the hole. The centers of the holes for the balusters are located. They are spaced and located carefully. The holes are drilled to receive the baluster ends. The top

Fig. 16-14 Backing or finish stringers on a wall. (Forest Products Laboratory)

newel post is attached as specified by the manufacturer. Next, the bottom newel post is fastened in place. Large spikes should be used. It is best to attach the newel post to the subflooring. Sometimes, carpenters must first chisel an area in the finished floor. This receives the bottom of the newel post. Next, glue is put on the round bottoms of the balusters. These are inserted in the holes drilled in the treads. Glue is never put in the hole. All the bottom pieces are glued and placed. Glue is then put on the top stems of the balusters. Then the top rail, or banister, is laid over the tops of the balusters. It is tapped

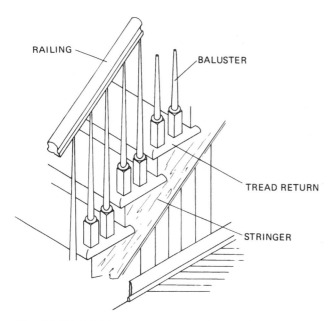

RAILING
BALUSTER
TREAD RETURN
STRINGER

Fig. 16-15 *A finish stringer on open stairs. (Forest Products Laboratory)*

firmly into position and allowed to dry. Then, finish caps, molding, handrails, and trim are installed.

Install Housed Carriages

To install housed carriages, it is best to work from the back. Here, the carpenter should start at the top and work toward the bottom. The first riser is applied and lightly wedged. Then the tread is attached. The wedge is coated with glue and tapped firmly into place. Then the riser wedge is coated with glue. It is then wedged firmly into place. Refer to Fig. 16-8A.

Wedged and glued stairs squeak less than other types. Stair squeaks are usually caused by loose nails. Since the closed carriage has fewer nails, it squeaks less.

Although this type of stair is commonly precut, carpenters sometimes cut these on the site. Special template guides and router bits should be used. See Fig. 16-16

Fig. 16-16 *Templates are used to rout the housed stringer on the site. (Rockwell International, Power Tool Division)*

STAIR SYSTEMS

There is a faster way of making a stair system. You might be able to locate a lumber dealer with all the supplies you need to produce a top-quality stairway. Such manufacturers as L.J. Smith Company in Ohio make it possible for the do-it-yourselfer, or the carpenter who wants to improve his skills, to quickly produce a stairway that meets all standard requirements.

Two types of systems are available: the post-to-post system and the over-the-post system. Figure 16-17 shows the post-to-post system with the necessary parts identified. Check Fig. 16-18 for the guidelines when ordering a system. As you can see in this type of system, there are no curved railings to bother with, just use the correct upeasing or rail drop to take care of the changes in direction for the railing. Notice in this type of system, the posts have the knob or ball on top exposed as part of the decorative features of the stairway.

If railings must be bent, bending rails and bending molds make for a curved balustrade system that is very practical and inexpensive. The bending rails can be successfully glued to a 30-inch radius on the rake and a 36-inch radius on the level, provided installation instructions are accurately followed. Installation sheets come with the wooden pieces. The rail is made up on the job by gluing either seven or nine pieces and using a polyvinyl bending mold because the glue does not stick to the Enduromold surface of the mold. See Fig. 16-19. The molds are available in 8-foot lengths and fit the contours of the rail. By applying the glue and aligning the pieces of wood and then placing the bending mold over the outside while clamping the rail to the stairs near the steps, the rail will then be bent according to the curve in the stairway. Figure 16-20 shows the curved staircase and the starting step made of solid oak. The system can be made of beech, oak, poplar, or cherry.

The L.J. Smith system uses a Conect-A-Kit™ method of assembling and installing handrail fittings. No bolts are required. All of the Conect-A-Kit fittings have a base with machined pockets and a removable top lid (or bottom filler) for easy installation. See Fig. 16-21. All assembly hardware is concealed within the base of each fitting. Most of these fittings are manufactured with a 45 degree precision miter cut on the end or ends so they can be used in combination with other fitting components to build a landing fitting assembly. The mitered ends are easily square cut for attachment to a straight handrail. See Fig. 16-22.

There are a number of advantages to the kit assembly method. You can make left or right turns with the same fitting. Fittings can be used to make up a variety

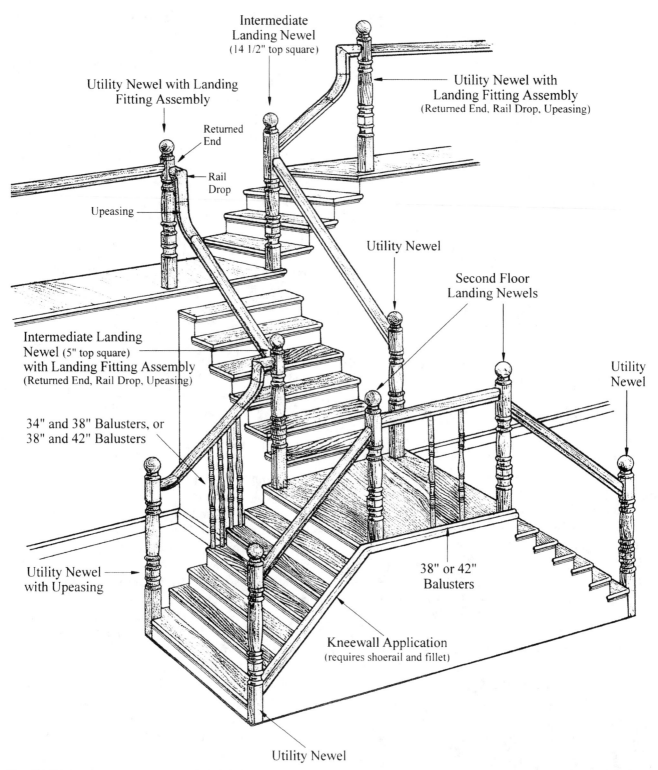

Intermediate
Landing Newel
(14 1/2" top square)

Utility Newel with Landing
Fitting Assembly

Returned
End

Rail
Drop

Upeasing

Utility Newel with
Landing Fitting Assembly
(Returned End, Rail Drop, Upeasing)

Utility Newel

Second Floor
Landing Newels

Intermediate Landing
Newel (5" top square)
with Landing Fitting Assembly
(Returned End, Rail Drop, Upeasing)

Utility
Newel

34" and 38" Balusters, or
38" and 42" Balusters

Utility Newel
with Upeasing

38" or 42"
Balusters

Kneewall Application
(requires shoerail and fillet)

Utility Newel

Fig. 16-17 *Post-to-post system. Note the labeled parts. This is a combination stairway to show the possible number of parts needed and where they fit to make a complete system.* (L.J. Smith)

of fitting combinations. See Fig. 16-23. Most joint connections are made on top of the rail system for better access. No rail bolts are required. The hardware included with each fitting provides better strength and tighter connections.

Upeasings, Caps, and Quarterturns

The upeasings and over easings are used when rise changes. The level quarter turn is used without a newel. The quarter turn with cap is used with a newel. The tandem cap allows for additional newels to be inserted into

	ITEM	GUIDELINES
		Support System
1	TREADS	Select one **tread** for each step. For a stair open on one side order miter-returned (MRT) and add 1 1/4" to the stringer to stringer measurement. For a stair open both sides order MRT both and use the finished stringer to stringer measurement (measured outside to outside).
2	RISERS	Select one **riser** for each step. Select one more **riser** than treads per each flight because of landing tread (see #3). (Landing tread replaces the nosing over the last riser).
3	LANDING TREAD	Select sufficient lineal footage for the entire balcony and width of stairs at each landing (8090-5 is used with 3 1/2" or wider newels - check the systems catalog for newel dimensions).
4	COVE MOULD	Select sufficient lineal footage to go under all tread nosing (including miter-returns) and under all landing tread.
		Balustrade
5	UTILITY NEWEL	For a 30" - 34" rake rail height, use the shortest available **utility newel**. For a 34" - 38" rake rail height, use the longest available **utility newel**. Select one for each side with handrail.
6	INTERMEDIATE LANDING NEWEL (Rake-to-Rake)	Select an **intermediate landing newel** with 14 1/2" face when not using a landing fitting assembly. If a landing fitting assembly is used, select the 5" face **intermediate landing newel** (see pages 9 - 19 for newel length requirements at level landings, 2-winder landings, and 3-winder landings).
7	SECOND FLOOR LANDING NEWEL (Rake-to-Level)	Select the **2nd floor landing newel** with 11" face when not using a landing fitting assembly (36" balcony height). 42" balcony height requires the use of a landing fitting assembly. Use the shortest available **utility newel** for surface mount when using a landing fitting assembly. Use the longest available **utility newel** if the newel extends below the floor surface (with landing fitting assembly).
8	BALCONY NEWEL (On the Level Run)	Match the **balcony newel(s)** to the **2nd floor landing newel(s)**. For a run of 10 feet or more, use a newel at the midpoint or every 5 to 6 feet. Place a newel at every corner.
9	HALF NEWEL OR	Select the **half-newel** of the same style as the other *full* **newels** on the balcony. (i.e. 11" face or 5")
10	ROSETTES	Select the **round rosette** for all level run rail connections into a wall. Select the **oval rosette** for all angled rail connections into a wall (when the rail meets the wall on a rake).
11	NEWEL MOUNTING HARDWARE	Select one of the **newel mounting kits** for each newel post (if desired).
12	RAKE BALUSTERS (Open Stairs)	For 30" - 34" handrail height compliance, use the **34" baluster** for the 1st baluster on the tread and use **38"** for the 2nd and 3rd balusters on the tread. If using 3 balusters per tread, and a fitting, substitute a **42"** for the **3rd baluster** under each landing fitting assembly. For 34" - 38" handrail height compliance, use the **38" baluster** for the 1st baluster on the tread and use the **42" baluster** for the 2nd baluster on the tread. If using 3 balusters per tread, use the **38" baluster** for the 1st and 2nd balusters on the tread, and use the **42" baluster** for the 3rd baluster on the tread.*
13	RAKE BALUSTERS (Kneewall Stairs)	Select the shortest **baluster** available at a rate of 2 per tread. Place on 4" to 6" centers.
14	LEVEL RUN BALUSTERS	Use the **38" baluster** for all 36" level runs/balconies. Use the **42" baluster** for all 42" level runs/balconies. Place on 4" to 6" centers. Subtract one **baluster** from the calculated total to account for the end of the run. Subtract one **baluster** for each newel post on the level run. Do not however, subtract one for the newel post beneath the landing fitting at the 2nd floor landing.
15	HANDRAIL	Select **handrail** at a rate of 13" per each tread and include enough for all level runs.
16	SHOERAIL (Kneewall-Rake)	Select **shoerail** at a rate of 13" per each tread.
17	SHOERAIL (Level Runs)	Select **shoerail** to cover all balcony landing tread (if desired).
18	FILLET	Select enough **fillet** to fill all plowed handrail and all shoerail.
19	PLUGS	Select two **wood plugs** for every newel (depending on installation requirements).
20	DOUBLE-END SCREW	Select one **Dowel-Fast™** double-end wood screw for each baluster. This is optional but highly recommended. Double-end wood screws are not needed for balusters installed within shoerail.
21	BRACKETS (Open Stairs)	Select one **bracket** for each tread (if desired).
22	HANDRAIL MOUNTING HDWE	Select one **Flush Mount Kit** or **Rail & Post Fastener** to fasten the handrail to newels.

* Note: when using 3 balusters per tread for 34" - 38" rail height, the **42" baluster** may not be long enough for use under a landing fitting assembly.

Fig. 16-18 *Post-to-post system guidelines for ordering a system.* (L.J. Smith)

a long level stretch of rail for strength. Figure 16-24 shows over-the-post upeasings, caps, and quarter turns.

Figure 16-25 illustrates the applications of an Over-The-Post System. Guidelines for ordering the system are given in Fig. 16-26.

A variety of post styles are shown in Fig. 16-27. Notice the differences in the two systems. The over-the-post system has a dowel rod sticking up to fit into the rail, whereas the post-to-post styles have a knob or ball on top with the exception of the square top. The

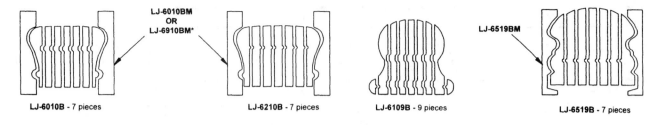

LJ-6010B - 7 pieces   LJ-6210B - 7 pieces   LJ-6109B - 9 pieces   LJ-6519B - 7 pieces

*LJ-6910BM is a polyvinyl bending mold in a reusable form. Glue does not stick to Enduromold surface. Available in 8' lengths.

Fig. 16-19 *Profiled bending rails, shoerails and fillets, and over-the-post volute, turnout and starting easing with cap.* (L.J. Smith)

(A)

(B)

Fig. 16-20 *(A) A curved stairway with an over-the-post system and closed-in stairs. (B) Note the oak starting step.*

Climbing Volutes

The Climbing Volutes are an addition to our line of code compliance products to provide a more attractive alternative to the extra long newel post on the starting step. The Climbing Volutes are also available in the following <u>non-plowed</u>, handrail profiles. Each volute requires 4 or 5 balusters. Climbing Volutes are a one-piece fitting, and are <u>not</u> part of the Conect-A-Kit fitting line.

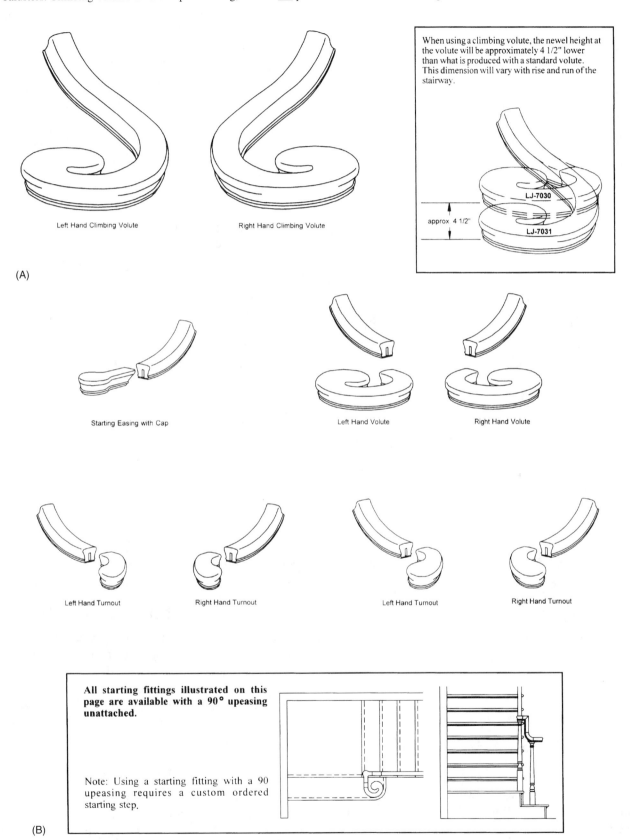

When using a climbing volute, the newel height at the volute will be approximately 4 1/2" lower than what is produced with a standard volute. This dimension will vary with rise and run of the stairway.

LJ-7030

approx 4 1/2"

LJ-7031

(A)

Left Hand Climbing Volute

Right Hand Climbing Volute

Starting Easing with Cap

Left Hand Volute

Right Hand Volute

Left Hand Turnout

Right Hand Turnout

Left Hand Turnout

Right Hand Turnout

All starting fittings illustrated on this page are available with a 90° upeasing unattached.

Note: Using a starting fitting with a 90 upeasing requires a custom ordered starting step.

(B)

Fig. 16-21 *(A) Climbing volutes for an over-the-post system.* (L.J. Smith) *(B) Starting fittings.* (L.J. Smith)

Quarterturn With Cap

Upeasing

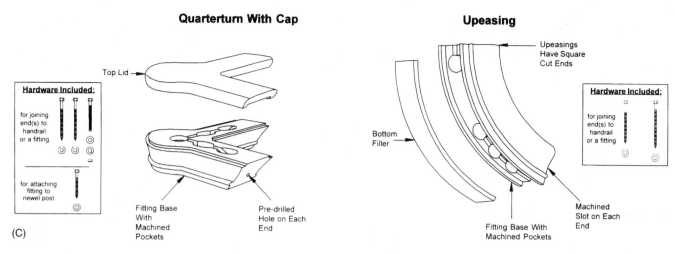

Top Lid

Hardware Included:

for joining
end(s) to
handrail
or a fitting

for attaching
fitting to
newel post

(C)

Fitting Base
With
Machined
Pockets

Pre-drilled
Hole on Each
End

Upeasings
Have Square
Cut Ends

Bottom
Filler

Hardware Included:

for joining
end(s) to
handrail
or a fitting

Machined
Slot on Each
End

Fitting Base With
Machined Pockets

Fig. 16-21 *Continued. (C) Exploded view of two Conect-A-Kit™ fittings. (L.J. Smith)*

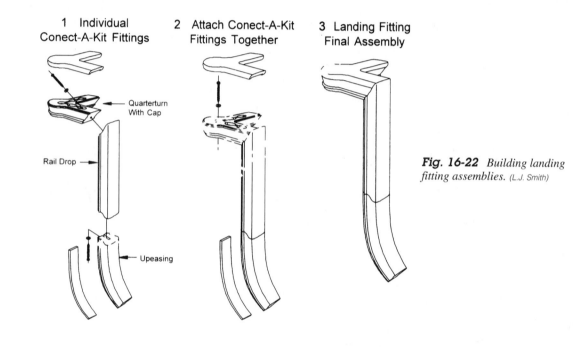

1 Individual
Conect-A-Kit Fittings

2 Attach Conect-A-Kit
Fittings Together

3 Landing Fitting
Final Assembly

Quarterturn
With Cap

Rail Drop

Upeasing

Fig. 16-22 *Building landing fitting assemblies. (L.J. Smith)*

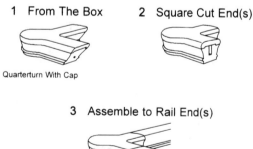

1 From The Box

2 Square Cut End(s)

Quarterturn With Cap

3 Assemble to Rail End(s)

Fig. 16-23 *Traditional stairway applications. (L.J. Smith)*

rail fits against the post or newel in this style and terminates against the wall in a rosette (see Fig. 16-31). Wall rails and wall rail brackets are shown in Fig. 16-31, as is the quarter turn and starting easing. The handrail is also required, in some housing codes, on the wall side of the staircase. See Fig. 16-28.

Starting Steps

All starting steps are shipped with necessary cove and shoe molding. All starting steps are available in longer lengths, 48 inches being the standard. Single bullnose starting steps are reversible and can be job cut for shorter lengths. Tread is $1\frac{1}{32}$ inches $\times$ 11.5 inches. Total rise is $8\frac{1}{32}$ inches. Riser measures $\frac{3}{4}$ inch $\times$ 7 inches. See Fig. 16-29.

Treads and Risers

The false tread kit has three possible applications: open treads-left hand, open treads-right hand, and closed treads. Various treads and risers are shown in Fig. 16-30.

Handrail Profile	LJ-6010	LJ-6010P	LJ-6210	LJ-6210P	LJ-6400	LJ-6400P	LJ-6109	LJ-6109P0	LJ-6109P1	LJ-6519	LJ-6519P

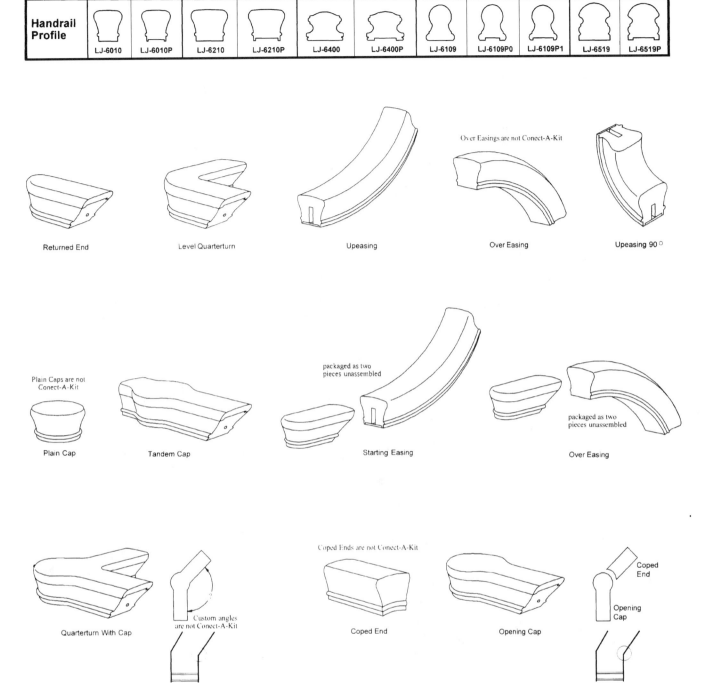

Returned End   Level Quarterturn   Upeasing   Over Easings are not Conect-A-Kit   Over Easing   Upeasing 90°

Plain Caps are not Conect-A-Kit   Plain Cap   Tandem Cap   packaged as two pieces unassembled   Starting Easing   Over Easing   packaged as two pieces unassembled

Quarterturn With Cap   Custom angles are not Conect-A-Kit   Coped Ends are not Conect-A-Kit   Coped End   Opening Cap   Coped End   Opening Cap

Fig. 16-24 *Upeasings and over easings used where rise changes.* (L.J. Smith)

Rosettes and Brackets

Rosettes are the round or oval pieces used against the wall and terminates the rail in a fitting arrangement. See Fig. 16-31.

In some cases wall rails are needed. Going down basement steps is a common location. Some communities have housing codes for new houses to include a handrail against the wall opposite the handrail on the "open" side of the staircase. The rails are available in 8-foot lengths with various quarter turns and starting easing as well as over easing. Three shapes are shown here that represent the typical wall rail.

Figure 16-32A is an example of over-the-post style with the handrail terminating in a rosette. However, Fig. 16-32B illustrates the post-to-post style with a turn.

A well chosen combination of materials properly assembled and installed can make a difference in the way a house is valued. The stairway can become the focal point of the interior and as much a part of the decorations as furniture, drapes, and carpets.

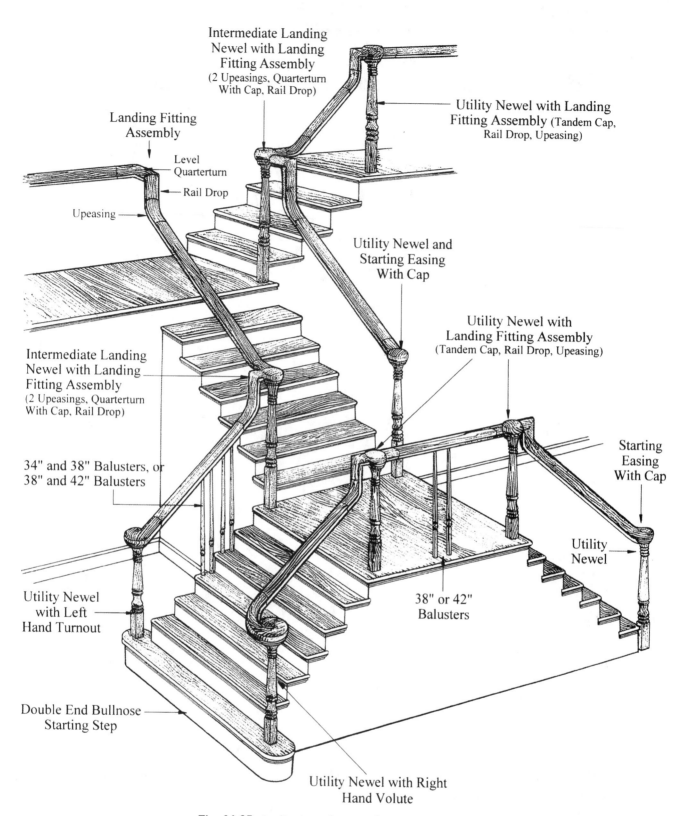

Intermediate Landing
Newel with Landing
Fitting Assembly
(2 Upeasings, Quarterturn
With Cap, Rail Drop)

Utility Newel with Landing
Fitting Assembly (Tandem Cap,
Rail Drop, Upeasing)

Landing Fitting
Assembly

Level
Quarterturn

Rail Drop

Upeasing

Utility Newel and
Starting Easing
With Cap

Utility Newel with
Landing Fitting Assembly
(Tandem Cap, Rail Drop, Upeasing)

Intermediate Landing
Newel with Landing
Fitting Assembly
(2 Upeasings, Quarterturn
With Cap, Rail Drop)

Starting
Easing
With Cap

34" and 38" Balusters, or
38" and 42" Balusters

Utility
Newel

Utility Newel
with Left
Hand Turnout

38" or 42"
Balusters

Double End Bullnose
Starting Step

Utility Newel with Right
Hand Volute

Fig. 16-25 *Applications of an over-the-post system.* (L.J. Smith)

	ITEM	GUIDELINES
		Support System
1	**STARTING STEP**	For use with volutes and turnouts. Select for bullnose on one or both ends (match the floor plan on p. 38 of the Systems Catalog). Measure finished stringers from outside to outside.
2	**TREADS**	Select one **tread** for each step (except the starting step). For a stair open one side order miter-returned (MRT) and add 1 1/4" to stringer measurement, then refer to the next longer standard length available. For a stair open both sides order MRT both and refer to stringer to stringer measurement (measured outside to outside).
3	**RISERS**	Select one **riser** for each step (except the starting step). Select one more **riser** than treads per each flight because of landing tread (see #4). Landing tread replaces nosing over last riser.
4	**LANDING TREAD**	Select sufficient lineal footage for the entire balcony and width of stairs at each landing. (8090-5 is used with 3 1/2" or wider newels - check the systems catalog for newel dimensions).
5	**COVE MOULD**	Select sufficient lineal footage to go under all tread nosing (including miter-returns) and under all landing tread.
		Balustrade
6	**STARTING FITTING**	Select either a **volute, turnout, or starting easing w/cap**. Choose a **climbing volute** to eliminate the need for an unusually long starting newel.
7	**UTILITY NEWEL**	Use everywhere except at the intermediate landing corner of an L-shaped stair.
8	**UTILITY NEWEL** (50")	Use for **balcony newel(s)** that will extend below the floor surface. Also use under a starting easing w/cap when a starting step is _not_ used and the rake handrail height is 34" or higher.
9	**INTERMEDIATE LANDING NEWEL**	Use the **57"** intermediate landing newel at the intermediate landing corner of an L-shaped stair. Use the **65"** intermediate landing newel in 2-winder landing situations. Use the **73"** intermediate landing newel in 3-winder situations.
10	**LEVEL RUN NEWEL** (Utility Newel)	If the balcony is 10 feet or longer, use the **utility newel** every 5 or 6 ft. under a tandem cap. Place a newel at every corner under a quarterturn with cap. Use the **50"** utility newel if the newel is to extend below the 2nd floor surface.
11	**HALF NEWEL OR**	Select the **half-newel** of the same style as the other _full_ newels on the balcony. (i.e. utility newels)
12	**ROSETTES**	Select the **round rosette** for all level run rail connections into a wall. Select the **oval rosette** for all angled rail connections into a wall (when the rail meets the wall on a rake).
13	**NEWEL MOUNTING HARDWARE**	Select one of the **newel mounting kits** for each newel post (if desired). The Shortest and Longest Utility Newels are packaged with newel mounting hardware.
14	**BALUSTERS FOR VOLUTES AND TURNOUTS**	Volutes require **4 or 6 - 1 1/4"** or **4 - 1 3/4" balusters**. Turnouts require **2 - 1 1/4"** or **1 - 1 3/4" baluster(s)**. For 30" - 34" rake rail height, use **38" balusters** under all volutes and use **42" baluster(s)** under all turnouts. For 34" - 38" rake rail height, use **42" balusters** under all volutes and turnouts. Climbing Volute Requirements: For 30" rake rail height, use **3 - 34"** and 1 or 2 - 38" baluster(s); For 34" rake rail height, use **3 - 38"** and 1 or 2 - 42" baluster(s); For 36" rake rail height, use **4 - 42" balusters.**
15	**BALUSTERS FOR STARTING EASING WITH CAP** (Open Stairs)	Use **1 - 38" baluster** for 30" - 34" handrail height compliance. Use **1 - 42" baluster** for 34" - 38" handrail height compliance.
16	**RAKE BALUSTERS** (Open Stairs)	For 30" - 34" handrail height compliance, use the **34" baluster** for the 1st baluster on the tread and use **38"** for the 2nd and 3rd balusters on the tread. If using 3 balusters per tread, substitute a **42"** for the **3rd baluster** under each landing fitting assembly. For 34" - 38" handrail height compliance, use the **38" baluster** for the 1st baluster on the tread and use the **42" baluster** for the 2nd baluster on the tread. If using 3 balusters per tread, use the **38" baluster** for the 1st and 2nd balusters on the tread, and use the **42" baluster** for the 3rd baluster on the tread.*
17	**RAKE BALUSTERS** (Kneewall Stairs)	Use the **34" baluster** at a rate of 2 per tread. Place on 4" to 6" centers. Subtract one baluster from the calculated total as the starting newel replaces the first baluster.
18	**LEVEL RUN BALUSTERS**	Use the **38" baluster** for all 36" level runs/balconies. Use the **42"** for all 42" level runs/balconies.** Place on 4" to 6" centers. Subtract one **baluster** from the calculated total to account for the end of the run. Subtract one **baluster** for each newel post on the level run. Do not however, subtract one for the newel post beneath the landing fitting at the 2nd floor landing.
19	**HANDRAIL**	Select **handrail** at a rate of 13" per each tread and include enough for all level runs.
20	**HANDRAIL FITTINGS** (Landing Fitting Components)	Match each corner of the floor plan to a corresponding plan in the systems catalog. Specify each **landing fitting component** needed to construct the landing fitting assembly. Assembled goosenecks have been replaced with components.
21	**HANDRAIL FITTINGS** (Over half-newel)	Select the **opening cap** which corresponds to the rail to cover each half-newel (i.e. LJ-6010 requires LJ-7019). This will be cut on the job.
22	**HANDRAIL FITTINGS** (Miscellaneous)	Each newel or half-newel must be covered with a **fitting**. Select the fittings from the drawings in the systems catalog.
23	**SHOERAIL** (Kneewall-Rake)	Select **shoerail** at a rate of 13" per each tread.
24	**SHOERAIL** (Level Runs)	Select **shoerail** to cover all balcony landing tread (if desired).
25	**FILLET**	Select enough **fillet** to fill all plowed handrail and all shoerail.
26	**PLUGS**	Select two **wood plugs** for every newel (depending on installation requirements).
27	**DOUBLE END SCREW**	Select one Dowel-Fast™ double-end wood screw for each baluster. This is optional but highly recommended. (Not for use with shoerail)
28	**BRACKETS** (Open Stairs)	Select one **bracket** for each tread (if desired).

* Note: when using 3 balusters per tread for 34" - 38" rail height, the **42" baluster** may not be long enough for use under a landing fitting assembly. ** An Over The Post rake rail height of 34" - 38" requires **42" balusters** for 36" and 42" level balconies.

Fig. 16-26 _Guidelines for ordering an over-the-post system._ (L.J. Smith)

Post to Post Styles

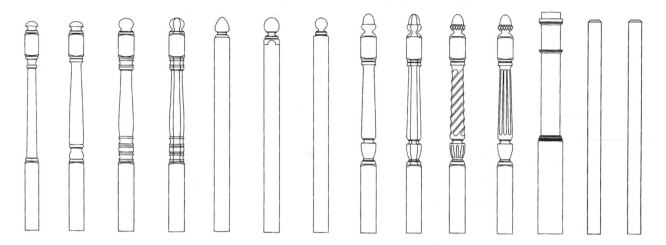

Shortest Utility Newel (5" top square) - Use this newel for the starting newel on all stairs with a 30" - 34" rake rail height. May use this newel as a balcony newel that is surface mounted.

Longest Utility Newel (5" top square) - Use this newel as a starting newel for stairs with a 34" - 38" rake rail height and for balcony newels that will extend below the floor surface.

2nd Floor Landing Newel (11" top square) - Use this newel for the 2nd floor landing newel when not using a landing fitting assembly. Will achieve a 36" balcony railing height.

Intermediate Landing Newel (14 1/2" top square) - Use this newel for level intermediate landings with no landing fitting assembly.

Intermediate Landing Newel (29" bottom square) - Use this newel for level landings with a landing fitting assembly.

Intermediate Landing Newel (37" bottom square) - Use this newel for intermediate landings with 2-winder treads and a landing fitting assembly.

Intermediate Landing Newel (45" bottom square) - Use this newel for intermediate landings with 3-winder treads and a landing fitting assembly.

Over the Post Styles

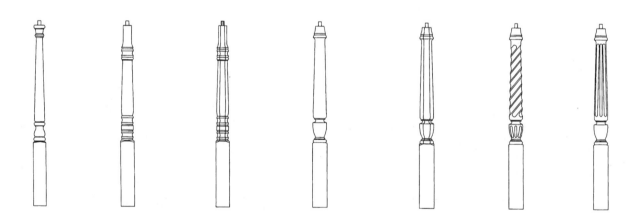

Shortest Utility Newel (43") - Use this newel under all starting fittings and as a balcony newel that is surface mounted. (See "Longest Utility Newel" for exception on a starting easing with cap).

Longest Utility Newel (50") - Use this newel under a starting easing with cap when a starting step is not used and the rake handrail height is 34" or higher. May also be used as a balcony newel that will extend below the floor surface.

Intermediate Landing Newel (57") - Use this newel for level intermediate landings.

Intermediate Landing Newel (65") - Use this newel for intermediate landings with 2-winder treads.

Intermediate Landing Newel (73") - Use this newel for intermediate landings with 3-winder treads.

Fig. 16-27 *Posts for post-to-post style and over-the-post style stairway.* (L.J. Smith)

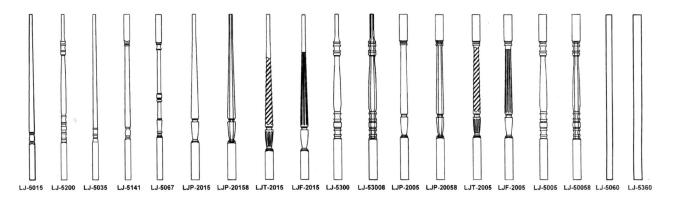

LJ-5015 LJ-5200 LJ-5035 LJ-5141 LJ-5067 LJP-2015 LJP-20158 LJT-2015 LJF-2015 LJ-5300 LJ-53008 LJP-2005 LJP-20058 LJT-2005 LJF-2005 LJ-5005 LJ-50058 LJ-5060 LJ-5360

All balusters are available in 3 lengths: 34", 38", 42"
The following guidelines will achieve a 32" minimum/36" maximum rake rail height

Rake Balusters:
For 30" - 34" rake rail height, use 34" for the 1st baluster on the tread and 38" for the 2nd and 3rd balusters on the tread. If using 3 balusters per tread, use a 42" baluster under each landing fitting assembly (formerly called gooseneck fittings).

For 34" - 38" rake rail height, use 38" for the 1st baluster on the tread and 42" for 2nd baluster on the tread. If using 3 balusters per tread, use 38" for the 1st and 2nd balusters on the tread and use 42" for the 3rd baluster on the tread. Note: when using 3 balusters per tread, the 42" baluster may not be long enough for use under a landing fitting assembly (formerly called gooseneck fittings).

Use the 34" baluster for <u>all</u> kneewall stairs.

Balcony Balusters:
Use 38" balusters for all 36" level balconies and use 42" balusters for all 42" level balcony rails (exception: an Over The Post rake rail height of 34" - 38" requires 42" balusters for all 36" <u>and</u> 42" level balconies).

Starting Fitting Balusters:
For 30" - 34" rake rail height, use 38" balusters under each volute and starting easing with cap, and use 42" balusters under each turnout.

For 34" - 38" rake rail height, use 42" balusters under <u>all</u> starting fittings.

Climbing Volute requirements: for 30" rake rail height, use 34" and 38" balusters; for 34" rake rail height, use 38" and 42" balusters; for 36" rake rail height, use 42" balusters.

Select a handrail that corresponds to the baluster choice.

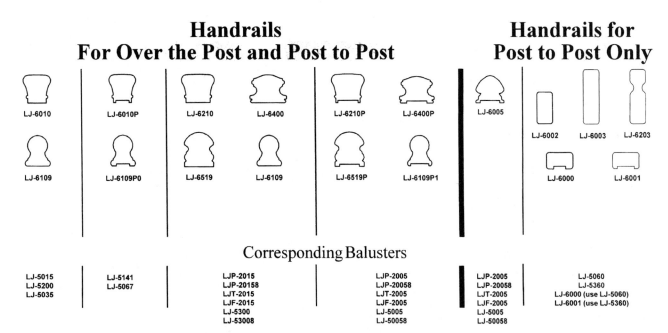

Fig. 16-28 *Balusters and handrails.* (L.J. Smith)

For Use with Volutes and Turnouts

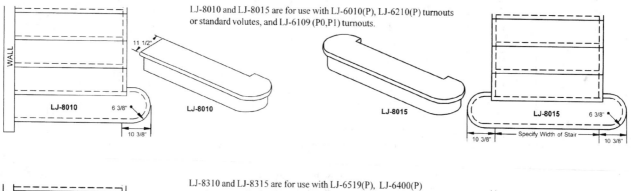

LJ-8010 and LJ-8015 are for use with LJ-6010(P), LJ-6210(P) turnouts or standard volutes, and LJ-6109 (P0,P1) turnouts.

LJ-8310 and LJ-8315 are for use with LJ-6519(P), LJ-6400(P) turnouts or standard volutes, LJ-6109(P0,P1) standard volutes, and all climbing volutes.

For Use with Square Top & Box Newels, and 90° Starting Fittings

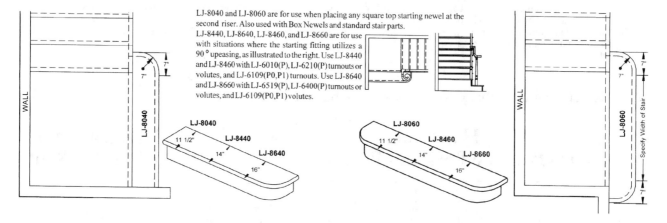

LJ-8040 and LJ-8060 are for use when placing any square top starting newel at the second riser. Also used with Box Newels and standard stair parts.
LJ-8440, LJ-8640, LJ-8460, and LJ-8660 are for use with situations where the starting fitting utilizes a 90° upeasing, as illustrated to the right. Use LJ-8440 and LJ-8460 with LJ-6010(P), LJ-6210(P) turnouts or volutes, and LJ-6109(P0,P1) turnouts. Use LJ-8640 and LJ-8660 with LJ-6519(P), LJ-6400(P) turnouts or volutes, and LJ-6109(P0,P1) volutes.

For Use with Box Newels

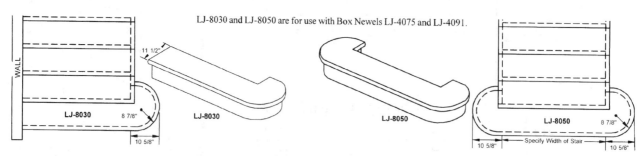

LJ-8030 and LJ-8050 are for use with Box Newels LJ-4075 and LJ-4091.

Fig. 16-29 *Starting steps.* (L.J. Smith)

False Ends

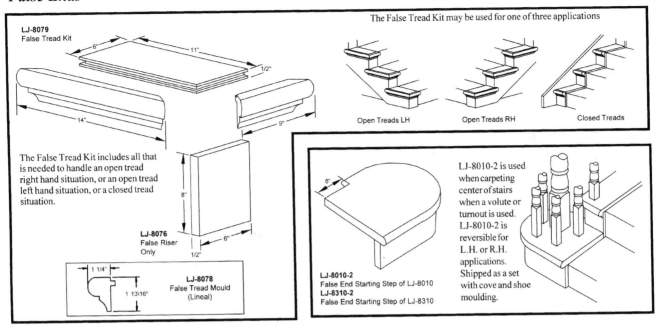

LJ-8079
False Tread Kit

6"

11"

1/2"

14"

9"

The False Tread Kit includes all that is needed to handle an open tread right hand situation, or an open tread left hand situation, or a closed tread situation.

LJ-8076
False Riser Only

8"

6"

1/2"

1 1/4"

1 13\16"

LJ-8078
False Tread Mould
(Lineal)

The False Tread Kit may be used for one of three applications

Open Treads LH

Open Treads RH

Closed Treads

6"

LJ-8010-2
False End Starting Step of LJ-8010
LJ-8310-2
False End Starting Step of LJ-8310

LJ-8010-2 is used when carpeting center of stairs when a volute or turnout is used. LJ-8010-2 is reversible for L.H. or R.H. applications. Shipped as a set with cove and shoe moulding.

Treads, Risers and Mouldings

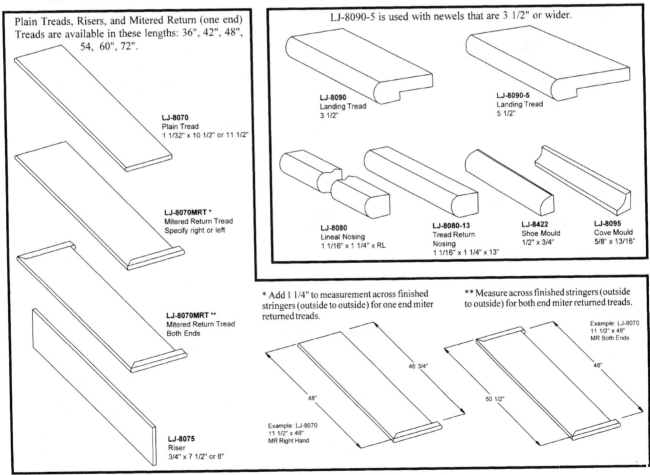

Plain Treads, Risers, and Mitered Return (one end) Treads are available in these lengths: 36", 42", 48", 54, 60", 72".

LJ-8070
Plain Tread
1 1/32" x 10 1/2" or 11 1/2"

LJ-8070MRT *
Mitered Return Tread
Specify right or left

LJ-8070MRT **
Mitered Return Tread
Both Ends

LJ-8075
Riser
3/4" x 7 1/2" or 8"

LJ-8090-5 is used with newels that are 3 1/2" or wider.

LJ-8090
Landing Tread
3 1/2"

LJ-8090-5
Landing Tread
5 1/2"

LJ-8080
Lineal Nosing
1 1/16" x 1 1/4" x RL

LJ-8080-13
Tread Return
Nosing
1 1/16" x 1 1/4" x 13"

LJ-8422
Shoe Mould
1/2" x 3/4"

LJ-8095
Cove Mould
5/8" x 13/16"

* Add 1 1/4" to measurement across finished stringers (outside to outside) for one end miter returned treads.

** Measure across finished stringers (outside to outside) for both end miter returned treads.

46 3/4"

48"

Example: LJ-8070
11 1/2" x 48"
MR Right Hand

50 1/2"

48"

Example: LJ-8070
11 1/2" x 48"
MR Both Ends

Fig. 16-30 *Treads and risers.* (L.J. Smith)

Rosettes and Brackets

Use LJ-7026 and LJ-7027 with LJ-6005, LJ-6010(P), LJ-6210(P), LJ-6109(P0,P1), LJ-6519(P), LJ-6400(P).

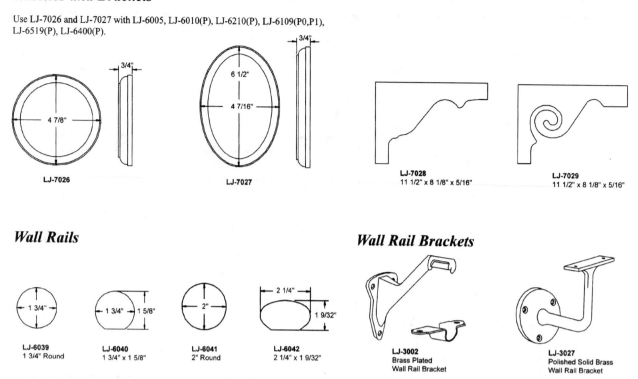

Wall Rails

Wall Rail Brackets

Wall Rail Fittings

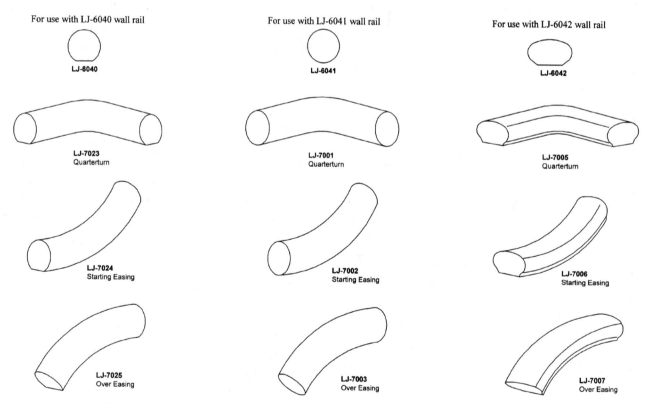

Fig. 16-31 *Rosettes, brackets, wall rail brackets, wall rails, and wall rail fittings.* (L.J. Smith)

FOLDING STAIRS

Folding stairs have a number of uses. They are used to reach the attic, but fold up and out of the way when not in use. They are also handy whenever it is necessary to have access to an area but there is not enough room for a permanent type of stairway. See Fig. 16-33.

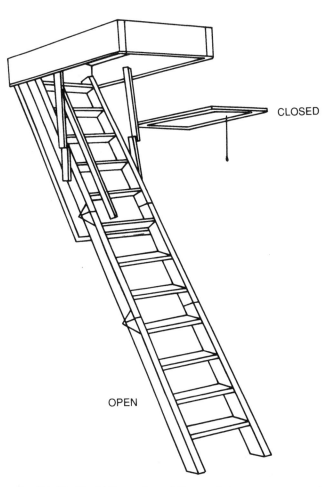

CLOSED

OPEN

Fig. 16-33 *The folding stairway fully extended.* (Memphis Folding Stairs)

Locating the Stairway

Allow sufficient space for a safe landing area at the bottom of the stairway. Be sure there is enough clearance for the swing of the stair as it is being unfolded to its full length. See Fig. 16-34.

Making the Rough Opening

Cut and frame the rough opening through the plaster or ceiling material the same size as shown on the carton. Generally, the rough opening size of the stair as listed on the carton will be ½ inch wider and ⅝ inch longer than the *actual net size* of the stairway. This allows for shimming and squaring the stair in the opening.

In most cases, stairways are installed parallel to ceiling joists. See Fig. 16-35. However, in some cases,

(A)

(B)

Fig. 16-32 *(A) A complete over-the-post system showing the handrail, utility newel, balusters, starting easing with cap, shoerail, fillet, and rosette.* (L.J. Smith) *(B) A complete post-to-post system with handrail, intermediate landing newel, second-floor landing newels, balusters, shoerail, fillet, and rosette.* (L.J. Smith)

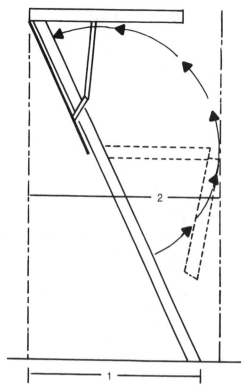

ROUGH OPENING	CEILING HEIGHT	LANDING SPACE (1)
22 × 48″	8′5″	59″
22 × 54″	8′9″	67″
22 × 54″	10′	73″
*22 × 60″	8′9″	67″
*22 × 60″	10′	73″
25 1/2 × 48″	8′5″	59″
25 1/2 × 54″	8′9″	67″
25 1/2 × 54″		73″
*25 1/2 × 60″	8′9″	67″
*25 1/2 × 60″	10′	73″
**30 × 54″	8′9″	67″
**30 × 54″	10′	73″
**30 × 60″	8′9″	67″
**30 × 60″	10′	73″

Fig. 16-34 *Space needed for the stairway.* (Memphis Folding Stairs)

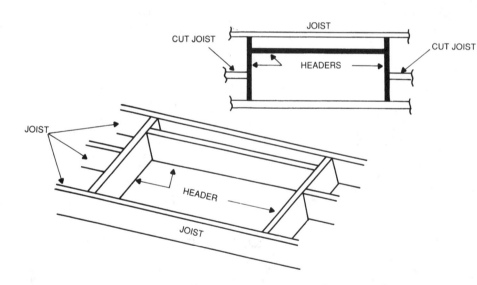

Fig. 16-35 *Framing the support for the stairway.* (Memphis Folding Stairs)

the stair must be installed perpendicular to the ceiling joist. See Fig. 16-36.

CAUTION: If the house uses roof trusses, do not cut ceiling joists without engineering consultation and approval. If it is necessary to cut the ceiling joists or truss cord, watch out for electrical wiring and be sure to tie the cut members to other joist or trusses with 2-x-6 or 2-x-8 headers forming a four-sided frame or stairwell to install the stair. Keep the corners square to simplify

the installation. Figures 16-35 and 16-36 show the frame that has to be built before installing the stair if the stair is put in after the house is constructed. The dark area shows the frame needed before installing the stair.

Figures 16-35 and 16-36 also show how to frame the rough opening for the stair. Installation parallel to existing joists requires only single joists and headers. Installation perpendicular to the joists requires double headers and joists. Make the ceiling joists and header

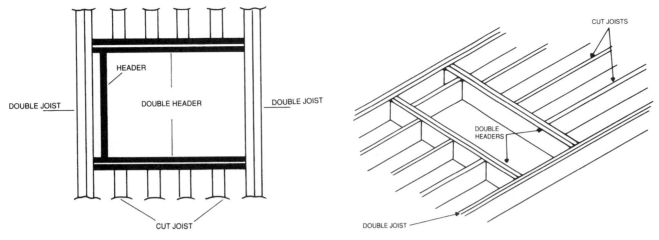

Fig. 16-36 *Rough opening for the stairway.* (Memphis Folding Stairs)

sections from the same size lumber as the existing joists. When making double headers, fasten members together with 10d common nails. The double joist sections shown in Fig. 16-36 must be long enough to be supported by a load-bearing wall at both ends.

Temporary Support for the Stairway

It is necessary to hold the stairway in the prepared rough opening by use of temporary boards which extend across the width of the rough opening and form a ledge of ½ inch at the main hinge end and a ⅞-inch ledge at the pull cord end. These boards should be nailed securely enough to hold the weight in the rough opening.

CAUTION: Do not place any weight on the stair at this time. See Fig. 16-37.

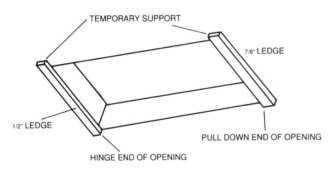

Fig. 16-37 *Placement of temporary supports.* (Memphis Folding Stairs)

Placing Stairway in the Rough Opening

Raise the stairway into the attic by turning it sideways. Pull it up through the rough opening and then lower it carefully until the hinge end rests on the ½-inch ledge and the pull cord end rests on the ⅞-inch ledge. This can be done from above in the attic while a helper on the floor is needed to lower the stair ladder sections out of the way for nailing the stair frame to the rough opening frame. As an additional safeguard against dropping the stair through the rough opening, it is best to secure it with several 8d common nails driven through the stair frame into the rough opening frame to help hold the stair in place. Do not drive these nails home as they can be removed after permanent nailing is completed.

Be sure the stair is square and level in the rough opening. Blocks of wood can be used for shims to straighten the stair frame in the event it has become bowed in inventory. This is normal since these wood parts are subjected to strong spring tension sometimes several months before installation, but can be straightened by the use of nails and shims.

The next step involves lowering the stairway and unfolding the stairway sections. Do not stand on the stair at this time. Use a step ladder or extension ladder.

Nail the sides (jambs) of the stairwell to the rough opening frame using 16d nails or 3-inch wood screws. Holes are provided in the pivot plate and the main header. Also nail through the main hinge into the rough opening header, then complete the permanent nailing by nailing sufficient 16d nails to securely fasten the stair to all four sides of the opening. Remove the temporary support slats and the 8d finish nails used for temporary support.

Adjustments

Pull the stairway down. Open the stair sections, folding the bottom section under the middle section so that the top and middle sections form a straight line.

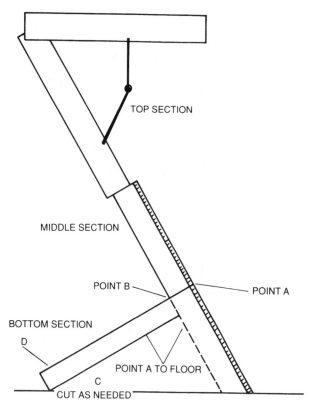

Fig. 16-38 Measuring the bottom section for fitting the floor.
(Memphis Folding Stairs)

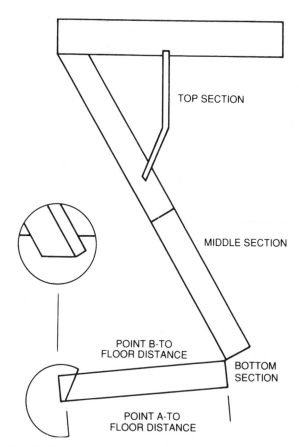

Fig. 16-39 Bottom section measurements for cutting. (Memphis Folding Stairs)

See Fig. 16-38. Apply pressure so that the hardware arms are fully extended. Maintain this pressure and use a straightedge placed on top of the middle section (see Fig. 16-38). Slide the straightedge down until it contacts the floor. Measure from point A to the floor. Record on the top side of bottom section C. Using the same procedure, measure bottom side B to the floor. Record on the bottom side D. Cut from C to D (Fig. 16-38). It is possible for your landing area to be uneven due to a floor drain, unlevel floor, or other reasons. Be sure to measure both sides of the bottom section using these procedures. The bottom section should fit flush to the floor on both sides after cutting. See Figs. 16-39 and 16-40. All joints should be tight at each section with weight on the stair (Fig. 16-40).

When the stair is correctly installed, stand on the second step of the bottom section. Be sure the stair is slanted from the ceiling to the floor and all sections of the stair are in a completely straight line as shown in Fig. 16-41. This should occur whenever the stair is used.

The feet of the stairway that rests upon the floor must always be trimmed so that each part of the foot or bottom section of the stair always fits flush to the floor and rests firmly and snugly on the floor. See Fig. 16-42. Failure to do so may produce undue stress on the com-

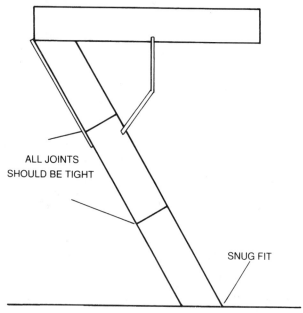

Fig. 16-40 All joints should fit tightly and a snug fit should be seen at the floor. (Memphis Folding Stairs)

ponents of the stairway and cause a break in the stairway. It could possibly cause bodily injury. Figure 16-43 shows the completed unit.

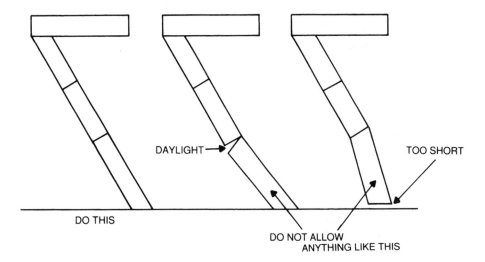

Fig. 16-41 *Proper and improper fits.* (Memphis Folding Stairs)

DAYLIGHT

TOO SHORT

DO THIS

DO NOT ALLOW
ANYTHING LIKE THIS

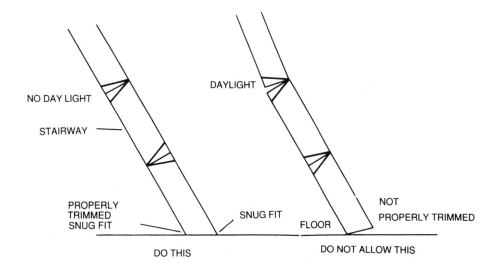

Fig. 16-42 *Right and wrong way floor fit.* (Memphis Folding Stairs)

NO DAY LIGHT

DAYLIGHT

STAIRWAY

PROPERLY
TRIMMED
SNUG FIT

SNUG FIT

FLOOR

NOT
PROPERLY TRIMMED

DO THIS

DO NOT ALLOW THIS

FIREPLACE FRAMES

As a rule, the carpenter does not make a fireplace. Fireplaces are usually made of brick, stone, or concrete block. These are constructed by people in the trowel trades. However, the carpenter must make framed openings for the fireplace and flues (or chimney).

Also, today many builders use metal fireplace units. The chimneys for these are made of metal too. These fireplaces and chimneys are installed in framed openings. See Fig. 16-44.

Fireplaces make attractive areas. They give the appearance and feeling of comfort and security. Decorative fireplaces are low in heating efficiency. However, there are popular devices which help offset their inefficiencies. Fireplaces can be made with various types of ventilators and reflector devices. These allow the heat to be directed and radiated into the living area for greater efficiency. Chimneys are also used to vent furnaces, water heaters, and other fuel-burning appliances. Often these vents are brought through the roof inside the chimney unit. This makes the roofline more attractive. Fewer unsightly vents pierce the roof structure.

Rough Openings in Ceilings and Roofs

The rough openings in ceilings and roofs are framed in by the carpenter. The carpenter makes these openings from the plans. These should be made as the roof and walls are constructed. They should be built before the fireplace is started. Special care must be taken to maintain proper clearances. A minimum distance of 2 inches is required between all frame members and the chimney. This allows for uneven settling of the chimney foundation. It also gives additional safety from fire.

The sequence used for constructing stair openings is also used here. First, an inner joist is nailed in place. This becomes the side of the opening. Then a first header is nailed in place. This boxes in the sides of the rough opening. Then tail joists or rafters are nailed in place. After this step, second headers are nailed on the

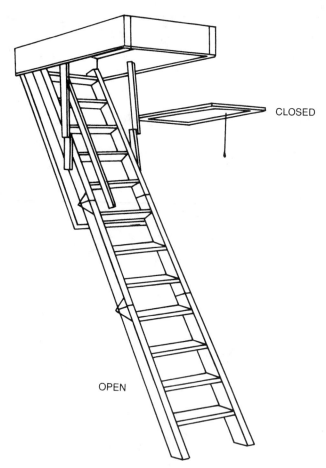

CLOSED

OPEN

Fig. 16-43 *Fully extended stairway.* (Memphis Folding Stairs)

insides. Next, the outer joist or rafter is added to complete the rough opening. After the rough opening is framed, the deck for the roof is applied.

At this point the fireplace, flue, and chimney should be built. Proper flashing techniques should be used where the chimney goes through the roof. After the flashing is installed, the roof is applied. See Fig. 16-45A through C.

Fireplace Types

Several types and shapes of fireplaces are common. Others, such as the one in Fig. 16-46B, are used occasionally. These special units add beauty, warmth, and utility to a living area.

A shape that is commonly used is the standard rectangular shape. Another shape is the double opening, with open areas on the face and one side. "See-through" fireplaces have openings on both faces, as in Fig. 16-46. This fireplace serves two rooms as shown.

Freestanding fireplaces are also popular. These are usually metal units located over a hearth area on the floor. See Fig. 16-47. Another version of the freestanding fireplace is shown in Fig. 16-48. This type of fire-

place features an open fire without walls. A large hood is placed over the unit for the chimney.

Fireplaces are made of many materials. Brick (Fig. 16-49) and natural stone are the most common. However, these are metal units faced with brick and stone. The insides must be lined with firebrick. Most bricks will absorb moisture. Fire changes the moisture to steam. This expands and cracks the brick. Firebrick is a special brick. It has been glazed so that it will not absorb moisture. Using firebrick, which does not have moisture, reduces cracking.

Many fireplaces are made from masonry materials. They are constructed piece by piece in a proper shape. However, getting the correct shape is very critical. A poorly designed fireplace will not draw the smoke. This will make the smoke enter the house or building and cause discomfort.

To make the design of the fireplace easier, builders often use metal units. There are two types of these units. One is a plain unit made of sheet metal. The sheet-metal unit is placed on a reinforced concrete hearth or bottom. A properly designed frame is built around it. The outside may be faced with brick or stone as desired.

Another type of prefabricated metal fireplace has hollow chambers or tubes. These tubes pass through the heated portion. They allow air to be drawn in and heated. The heated air is circulated into the living area. These openings or tubes are vented to the front of the fireplace. Since heated air rises naturally, natural ventilation occurs. See Fig. 16-50. These tubes can be connected to the regular furnace air system. Special blowers are also used to increase their heating effect. These systems make the fireplace more efficient.

General Design Factors

The design of a fireplace and chimney is extremely important. Certain practices must be followed. The heat from the burning fire will cause the smoke to rise. It must pass through the chimney to the outside. As a rule, the chimney and fireplace are built as a single unit. A special foundation is required for this combined unit. The chimney is a vertical shaft. The smoke from the burning fire passes through it to the outside. The chimney may also be used for vents or flues from furnaces and water heaters. Footings for the fireplace should extend beneath the frostline. Also they should extend 6 inches or more beyond the sides of the fireplace. Chimney walls should be at least 4 inches thick. Manufacturers' specifications should always be followed closely.

Fig. 16-44 *Metal fireplace units must be enclosed by frames.* (Martin Industries)

The lower part of the fireplace is called the hearth. See Fig. 16-51. The hearth has two sections. The first is the outer hearth. This is the floor area directly in front of the fireplace. The second part is the bottom of the chamber where the fire is built. The front hearth should be floored with fire-resistive material. Tile, brick, and stone are all used. These contrast with the regular floor. The contrast makes the hearth decorative as well as practical. Some fireplaces have a raised hearth. See Fig. 16-52. The raised hearth is also an advantage in some cases. As in Fig. 16-53, a special ash pit may be built at floor

level. This is an advantage when houses are built on concrete slabs. The raised hearth allows the clean-out unit to be installed.

The back hearth, the back wall, and the sides of the fireplace are lined with firebrick. The back wall should be approximately 14 inches high. The depth from the front lintel to the back wall should be 16 to 18 inches. See Fig. 16-51. Sidewalls and the top portion of the back are built at an angle. This angle helps the brick to radiate and reflect heat into the room. The back wall should slope forward. The amount of slope is almost one-half the height of the opening. This slope reflects

(A)

(B)

(C)

Fig. 16-45 *(A) Openings in roofs must be framed. Metal units may be added after a house is finished* (Martin Industries)*. (B) A metal unit being set in place on a new home. (C) The same unit after framing and finishing.*

heat to the room. It also directs the smoke into the throat area of the fireplace.

Fireplace throat The throat is the narrow opening in the upper part of the fireplace. See Fig. 16-51. The throat makes a ledge called a *smoke shelf*. The smoke shelf does three things for the fireplace. First, it pro-

vides a barrier for downdrafts. The smoke and heated air rise rapidly in the chimney. This movement causes an opposite movement of cold air from the outside. This is called a *downdraft* (Fig. 16-53). The smoke shelf blocks the downdraft as shown. It also keeps the downdraft from disrupting the movement of the smoke. The throat narrows the area in the chimney.

Fig. 16-46 *"See-through" fireplace serves two rooms.* (Potlatch)

Fig. 16-47 *A freestanding metal fireplace.* (Majestic)

This increases the speed of the smoke. The faster speed helps create a good draft. Without proper throat design, the chimney will not draw well. The smoke will reenter the living area.

Second, the smoke shelf provides a rain ledge. The ledge shields the burning fire from direct rainfall. Third, the throat and smoke shelf provide a fastening ledge for the damper.

The area of the throat should not be less than that of the flue. For best results, the throat area should be slightly bigger than the flue. For this reason, the throat opening should be long. Its length should be the full width of the fireplace. The throat width is approximately one-half the width of the flue. To have the same area, the throat must be longer than the flue.

Fireplace damper The narrowest part of the throat has a door. The door is called a *damper*. Special mechanisms keep the damper open in normal use. The

Fig. 16-48 *A freestanding fireplace that is open on all sides.* (Weyerhauser)

Fig. 16-49 *This metal fireplace is faced with brick and trimmed with a formal wooden mantel.* (Martin Industries)

damper should open as in Fig. 16-51. This way, the door helps form a wall at the bottom of the smoke shelf. When the fireplace is not being used, the damper may be closed. This keeps heated or cooled air from being lost through the flue.

Smoke chamber The *smoke chamber* is the area above the smoke shelf. The throat and the smoke shelf are the full width of the fireplace. The smoke chamber narrows from the width of the fireplace to that of the flue. This also helps to reduce the velocity of downdrafts. The effects of wind gusts are lessened here. A smoke chamber is necessary for proper draft. Smoke chambers should be plastered smooth with fire clay.

Chimney flue The chimney flue is the passage through the chimney. Its size is determined by the area of the fireplace opening. The flue area should be one-tenth the opening area. Lower chimney heights are often used in modern buildings. Here, flues should be about 10 percent larger than this ratio.

A rule of thumb concerning flue areas is as follows: Fourteen square inches of flue are allowed per

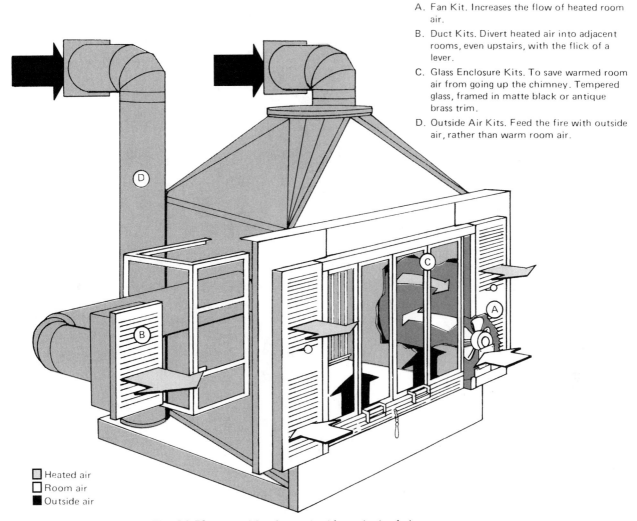

A. Fan Kit. Increases the flow of heated room air.
B. Duct Kits. Divert heated air into adjacent rooms, even upstairs, with the flick of a lever.
C. Glass Enclosure Kits. To save warmed room air from going up the chimney. Tempered glass, framed in matte black or antique brass trim.
D. Outside Air Kits. Feed the fire with outside air, rather than warm room air.

☐ Heated air
☐ Room air
■ Outside air

Fig. 16-50 *A metal fireplace unit with an air circulation system.* (Majestic)

square foot of the opening of the main fireplace. For example, a fireplace opening has a height of 30 inches (2½ feet). The width is 36 inches (3 feet). The area of the fireplace opening would be 7.5 square feet. The chimney flue area should be 14 square inches for every square foot. For the opening of 7.5 square feet, the flue size should be 7.5 × 14. This would be a total of 105 square inches. A good flue size would be 10½ × 10 inches.

Flues should be as smooth as possible on the inside to prevent buildup of soot, oily smoke, or tar. Fire clay or mortar is most often used to line flues. Industrial chimneys are exposed to chemical action. Special porcelain liners may be needed for these. For most buildings today, tile flue liners are used.

The drawing capability of a chimney is affected by the air differential. The air rising from the chimney is hot. The outside air is cold. The difference in the temperature of these two is the air differential. The greater the difference, the better the chimney will draw. Locating chimneys on outside walls does not help this. Locating them on inside sections conserves the heat of the air rising in the chimney. Thus, chimneys located on the inside are usually more efficient.

Flues and chimneys Chimneys are made of several different materials. For standard masonry chimneys, brick, stone, and concrete block are all used. Special metal chimneys are also used. In many cases, the metal chimneys are encased with special coverings. See Fig. 16-54. These covers must be built according to manufacturer's specifications. However, in most cases metal standoffs must be used to hold the metal chimney away from the wood members. The chimney cover is framed from wood and siding materials. Some chimney covers have a fake brick or masonry appearance.

Many chimneys have special covers over them. These keep out rain or snow. Several methods are used

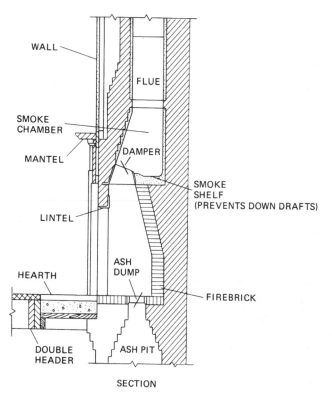

Fig. 16-51 Fireplace parts.

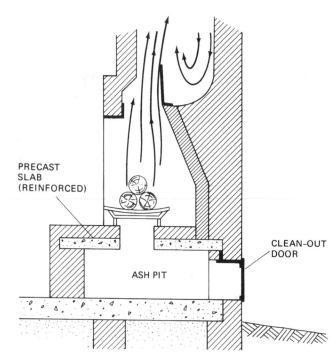

Fig. 16-53 Proper construction blocks downdrafts that push smoke into a room.

to fasten these covers to the chimney. Some types slip inside the flue opening. Other types are fastened around the outside areas of the chimney by pressure bands. These are simply clamps that are tightened against the masonry of the chimney. Other types must be fastened by special expanding screw anchors. For these, holes are drilled into the brick or stone of the chimney. The special screw anchors are inserted and expanded firmly against the brick. This provides a firm anchor for regular bolts. These screw anchors should

Fig. 16-52 A fireplace may have a raised hearth. (Martin Industries)

Fig. 16-54 *The sequence for installing special chimney covers.* (Martin Industries)

always be placed in the solid brick or stone. Placing them into the mortar will cause the joint to crack. This will loosen the masonry from its mortar bed. This weakens the chimney.

GAS VENTS

If you are building the house yourself, there are a number of things you must be aware of and capable of doing in order to complete the building and make it habitable. One of these is the installation of gas vents for hot water heaters, fireplaces, and other devices used in modern day living. There are some do's and don'ts associated with the venting of exhaust gases in various appliances.

Round Type B gas vents are for gravity draft venting of "listed" L.P. or natural gas-burning appliances. Appliances that might be so vented are: Those equipped with draft hood, those designated as "Category I," and those for which the installation instructions call for a permit venting with Type B gas vents. These appliances include, but are not limited to, the following types: furnaces, boilers, water heaters, room heaters, unit heaters, duct furnaces, floor furnaces, and decorative appliances.

Do not use Type B gas vents for wall furnaces "listed" for use only with Type B gas vents. Do not use Type B gas vents for Category II, III, or IV appliances, or for any gas-burning appliance that requires either a pressure-tight or liquid-tight venting system. Do not use on any appliance for which the installation instructions call for taping or sealing the joints to prevent leakage.

All sizes of Type B gas vents made by Eljer Manufacturing can be used in single- and multi-story buildings. All Type B gas vents can be used for both individual and multiple appliance venting. Eljer Manufacturing's Type B gas vents are to be installed and used in accordance with the National Fuel Gas Code NFPA No. 54 or standard for Chimneys, Fireplace and Venting Systems, NFPA No. 211. Type B gas vents are also suitable for use in existing, otherwise unused, masonry chimneys to protect the chimney from the damaging effects of moist combustion products from the appliances listed previously. In general, all Type B gas vent systems must include draft hoods at the appliance. However, certain advanced technology gas appliances have flue outlet characteristics acceptable for venting with Type B gas vents without draft hoods. Never use Type B gas vents on any appliance that is not listed and approved for venting with Type B gas vents.

A notice should be posted near the point where the gas vent is connected to the appliance, with the following wording: Connect this gas vent only to gas burning appliances as indicated in Installation Instructions. Do not connect incinerators, or liquid or solid fuel-burning appliances.

Round Gas Vent

All joints in gas vents must be secured using the appropriate method. Draft hood connectors must be attached to the appliance outlet with screws. Single connectors, if used, must be secured to the appliance, to the gas vent, and at all joints with three sheet metal screws per joint. See Table 16-1 for the method of joining various size round vents. Figures 16-55 A and B show the fitting of gas vents. Diagram A shows the 3- to 8-inch type of vent connected with six sheet metal screws per joint while diagram B shows the 10- to 48-inch vents that must be joined with eight sheet metal screws.

Table 16-1 *Method of Joining and Clearance to Combustion*

Size	Shape	Method of Joining	Min. Clear. to Combust.
3–8" incl.	Round	Integral Coupler	1"
10–24" incl.	Round	6 sheet metal screws/joint	1"
26–48" incl.	Round	6 sheet metal screws/joint	2"

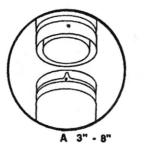 

A 3" - 8"   **B 10" - 48"**

Fig. 16-55 *Round Type B gas vent.* (Eljer)

Air Supply

Gas appliances must have an adequate supply of air for combustion, vent operation, and ventilation. See Fig. 16-56. Special provisions for bringing in outside air might be necessary in tight buildings or when appliances are in small rooms. Consideration must be given to climate in choice of the air supply method. Check the local building code or National Fuel Gas Code for methods of air supply. Figure 16-56 shows typical air supply requirements.

Vent Connector Type and Size

In Fig. 16-57, examine the vent pipe connections that meet all building code and safety standards for use as gas appliance vent connectors and breechings.

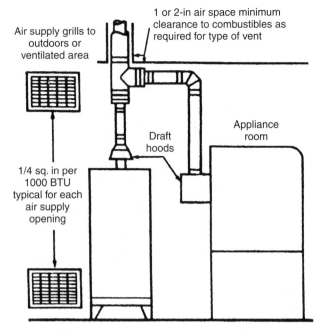

Fig. 16-56 *Typical air supply requirements.* (Eljer)

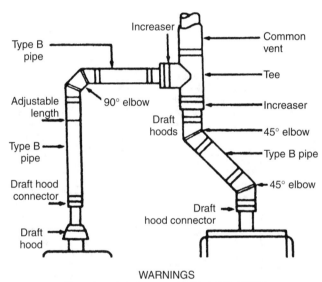

WARNINGS
Do not apply tape or sealant to joints of Eljer manufacturing type B gas vents. Do not use type B gas vent for: incinerators, or for appliances burning liquid or solid fuels, or for any appliance which needs a chimney such as "Factory-Built" or masonry.

Fig. 16-57 *Combined vent system using Type B parts starting with draft hood connections.* (Eljer)

Vent Location

Metal vents are suitable for indoors or outdoors. See Figs. 16-59 and 16-61. Outdoor gas vents should be designed as close to maximum capacity as possible. Appliances served by an outside gas vent must have an air supply to the appliance room adequate to balance indoor and outdoor pressures. Otherwise, "stack action" of the heated building can cause reverse venting

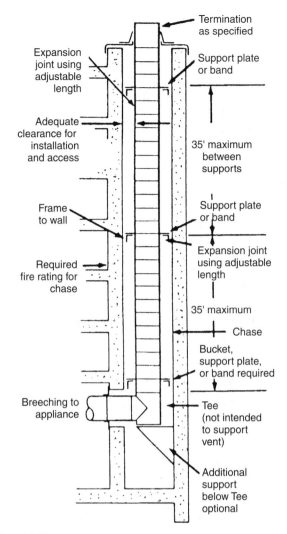

Fig. 16-58 *Chase installation of large Type B gas vent in multi-story building.* (Eljer)

action when the appliance is off, or operating only on its pilot.

In multi-family residential, high rise, and many other types of buildings, vents are not permitted to penetrate the floor structure. They must be located in noncombustible shafts or chases with no openings except for inspection access. See Fig. 16-58. Building code requirements in such cases must be carefully followed with respect to wall construction, access, clearance, support, initial penetration of breeching, and method of termination. See the NFPA Standards No. 54 and No. 211 for such situations in which gas vents are installed.

Clearances and Enclosures

Type B gas vents that are from three inches to 24 inches in diameter can be installed with one inch minimum air space clearance to combustibles. The 26- to 48-inch sizes require two inches minimum air space

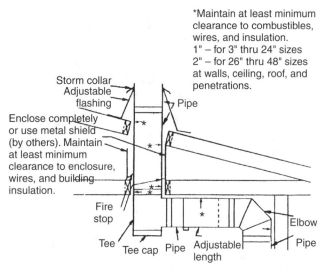

*Maintain at least minimum clearance to combustibles, wires, and insulation.
1" – for 3" thru 24" sizes
2" – for 26" thru 48" sizes at walls, ceiling, roof, and penetrations.

Fig. 16-59 *Typical ceiling and roof penetrations.* (Eljer)

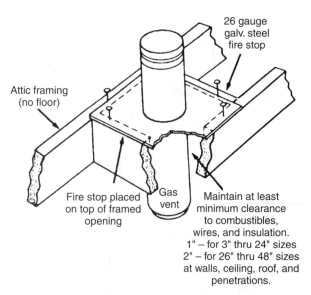

26 gauge galv. steel fire stop

Attic framing (no floor)

Fire stop placed on top of framed opening

Gas vent

Maintain at least minimum clearance to combustibles, wires, and insulation.
1" – for 3" thru 24" sizes
2" – for 26" thru 48" sizes at walls, ceiling, roof, and penetrations.

Fig. 16-60 *Fire stopping required for all ceiling and floor penetrations.* (Eljer)

clearance to combustibles. These clearances are marked on all gas-carrying items. They apply to indoor or outdoor vents, whether they are open, enclosed, horizontal or vertical, or pass through floors, walls, roofs, or framed spaces. The appropriate clearance should be observed to joists, studs, subfloors, plywood, drywall, plaster enclosures, insulating sheathing, rafters, roofing, and any other materials classed as combustible. The required minimum air space clearances apply also to electrical wires and any kind of building insulation. Keep electrical wires and building insulation away from the gas vents and out of their required air space clearance.

Reduced clearances must be used with 4-inch oval gas vents in 2 × 4 stud walls. Gas vents can be fully enclosed or installed in the open. Single wall materials used as vent connectors are not permitted to be enclosed. There is no need to enclose gas vents when used as connectors under floors, in crawl spaces and basements, or in normally unoccupied or in inaccessible attics. Enclosing the vertical portions of vents is recommended where they pass through rooms, closets, halls, or other occupied spaces.

Fire Stopping

All Type B gas vents passing through floors, ceilings, or in framed walls must be fire stopped at floors or ceilings using 26-gage or heavier galvanized steel. Figure 16-60 illustrates how a fire stop is built. The fire stop must close the area between the outer wall of the pipe and the opening in the structure. In attics the fire stop should be placed on top of the framed ceiling opening. Keep wires and insulation out of required air space clearance around gas vents.

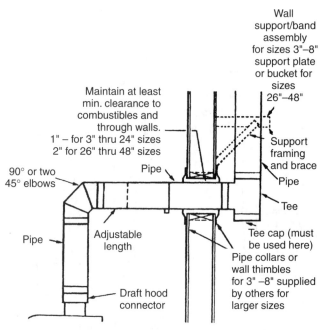

Maintain at least min. clearance to combustibles and through walls.
1" – for 3" thru 24" sizes
2" for 26" thru 48" sizes

Wall support/band assembly for sizes 3"–8" support plate or bucket for sizes 26"–48"

90° or two 45° elbows

Pipe

Support framing and brace

Pipe

Tee

Pipe

Adjustable length

Draft hood connector

Tee cap (must be used here)

Pipe collars or wall thimbles for 3" –8" supplied by others for larger sizes

Fig. 16-61 *Wall penetrations for Type B.* (Eljer)

Fire stops can be used as vent pipe supports provided that the aluminum inner wall of the gas vent is not penetrated or damaged by the method of attachment. See gas vent piping later on to see how the vents must be secured. For gas vents within a shaft or chase, fire stopping is provided by the vertical walls of the shaft.

Use of Gas Vent Fittings

Figure 16-61 shows adjustable lengths that can be telescoped over a fixed length, to make up in between lengths of vent or connector. An adjustable length suspended below a support serves as an expansion joint between two fixed points of properly supported gas vent. Ordinarily, the adjustable length must be secured, but for expansion joints it should just maintain good contact and a minimum 1.5-inch overlap. Do not use adjustable lengths to suspend any weight of pipe below. Elbows for 3 through 8 inches are fully adjustable. Elbows for sizes 10 through 48 inches are 45° fixed only.

Most building codes require the use of full-size fittings for vent interconnections. Tees, elbows, increasers, and short lengths are especially designed to facilitate interconnections.

Tees used as fittings to start vertical vents must be tightly capped to prevent air leakage. Gas vents and vent connectors having more than 90° turns (an elbow and a tee) might require additional height or a size increase to compensate for added flow resistance. All unused openings in a gas vent must be sealed to prevent loss of effective vent action.

Minimum Gas Vent Height

A minimum gas vent height of 5 feet above the appliance draft hood is required. See Fig. 16-62. Where the vent has a lateral, or offset, or serves multi-appliances, greater heights might be required for proper venting. Special care must be taken that short gas vents on duct furnaces, unit heaters, and furnaces in attics have sufficient vent height to ensure complete venting. Check appliance manufacturer instructions and local codes for the required minimum heights. Along with these suggestions, minimum heights of gas vents must also comply with the rules mentioned in *Vent Termination.*

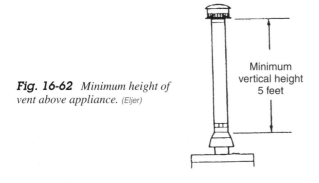

Fig. 16-62 *Minimum height of vent above appliance.* (Eljer)

Minimum vertical height 5 feet

Support

Gas vent piping must be securely supported. Lateral runs are to be supported at least every 5 feet; when offsets are necessary, adequate support above and below the offset is required. In addition, securing the offset

elbow with three maximum 0.25-inch long sheet metal screws is recommended. Vertical runs fire-stopped at 8- to 10-foot intervals need only be supported near the bottom. Tees used as vent inlets can be supported by sheet metal plates or brackets. Plumbers tape can be used to space both horizontal and vertical piping. Short vents with less than 6 feet of vertical pipe below the flashing can be suspended with the flashing. The pipe can be supported by the storm collar resting on the top of the flashing. Use 0.29-inch-long sheet metal screws to attach the storm collar to the pipe at the appropriate place. Gas vents supported only by the flashing must be guyed above or below the roof to withstand snow and wind loads. All gas vents extending above the roof more than 5 feet must be securely guyed or braced.

Indoors, gas vent sizes 3 through 8 inches can be supported, as shown in Fig. 16-63, and fire-stopped as well, using the support plate as illustrated. Cut away all floors and ceilings to 1-inch minimum air space clearance. The support plate will support 35 feet of vent.

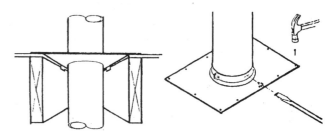

Fig. 16-63 *Storm collar on top of flashing.* (Eljer)

Vent sizes 10 to 24 inches can be supported within or adjacent to fireproof or noncombustible structures or shafts. Check Fig. 16-58. Use a split plate support (10 through 24 inches) which fits around the groove near the ends of the pipe and rests on masonry or a metal frame. See the A section of Fig. 16-64. Use bucket support for 26 inches and larger sizes; or suitable structural iron bands can be constructed using 16-gage steel, or heavier, for the same purpose.

Flashing

The roof opening should be located and sized such that the vent is vertical and has the required air space clearance. The tall cone flashing is for flat roofs only. See Fig. 16-65A. It is nailed in place through all four sides of the base flange. The adjustable roof flashing shown in Fig. 16-65B is positioned with the lower portion of the base flange over roofing material. Nail through only the upper portion and sides of the base

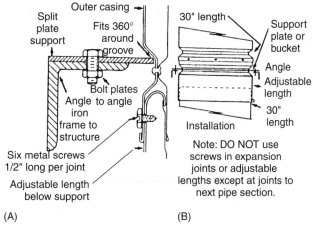

Fig. 16-64 *(A) Split plate suggestion support detail.* (Eljer) *(B) Typical expansion joint below each support.* (Eljer)

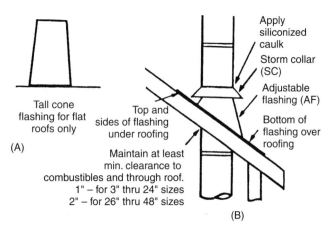

Fig. 16-65 *(A) Tall cone flashing for flat roofs only.* (Eljer) *(B) Elements for proper roof flashing installation.* (Eljer)

flange. Do not nail through the lower flange. Use nails with a neoprene washer, or cover the nail heads with siliconized caulk. Finish roofing around the flashing, covering the sides and upper areas of the flange with roofing material.

Gas Termination

Gas vent piping must extend through the flashing to a height above the roof determined by Rule I or Rule II shown below. A storm collar is installed on the vent pipe over the opening between pipe and flashing. Siliconized caulk is used over the joint between pipe and storm collar. The top is securely attached to the gas vent using the proper method for the model of pipe. See Figs. 16-67 and 16-68.

Termination height for gas vent sizes 12 inches and less can be in accordance with Rule I height table. (See Table 16-2.)

Tops 14-inch size and over and roof assemblies must be located in accordance with Rule II.

Table 16-2 *Minimum Height From Roof to Lowest Discharge*

Roof Pitch	Opening, Foot
Flat to 7/12	1.0
Over 7/12 to 8/12	1.5
Over 8/12 to 9/12	2.0
Over 9/12 to 10/12	2.5
Over 10/12 to 11/12	3.25
Over 11/12 to 12/12	4.0
Over 12/12 to 14/12	5.0
Over 14/12 to 16/12	6.0
Over 16/12 to 18/12	7.0
Over 18/12 to 20/12	7.5
Over 20/12 to 21/12	8.0

RULE 1: Tops for gas vents sizes 12 inches and smaller.

The cap is suitable for installation on listed gas vents terminating a sufficient distance from the roof so that no discharge opening is less than 2 feet horizontally from the roof surface, and the lowest discharge opening will be no closer than the minimum height specified in Figure 16-66B. These minimum heights may be used provided that the vent is no less than 8 feet from any vertical wall.

RULE II: Tops for gas vent sizes 14 inches and larger.

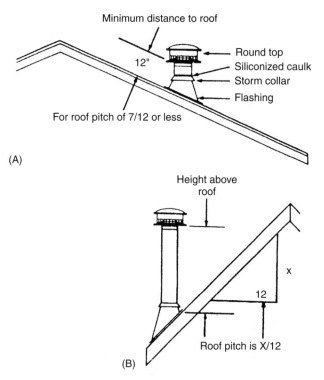

Fig. 16-66 *(A) For roof pitches of 7/12 or less.* (Eljer) *(B) Location rules for flat to 21/12 pitch.* (Eljer)

For installations other than covered by Table 16-2 in Fig. 16-66B, or closer than 8 feet to any vertical wall, the cap shall be not less than 2 feet above the highest point where the vent passes through the roof and at least 2 feet higher than any portion of a building within 10 feet. Vent caps 14-inch size and larger and chimney style roof housings must comply with Rule II regardless of roof pitch.

These rules were established on the basis of tests conducted in accordance with American National Standard ANSI/UL 441.

Top Installation

Round tops for pipes sizes 3 to 8 inches have a spring clip that locks into the upper end of the vent. The clip engages automatically when the top is pushed into the pipe. See Fig. 16-67A. The top will also fit any sheet metal pipe with full nominal inch dimensions. To attach securely, bend a one-inch length of the pipe under end inward about 0.125 inch. The spring clip will lock under this bent edge. See Fig. 16-67B. To remove the top from any pipe, pull up evenly on opposite sides of the skirt of the top.

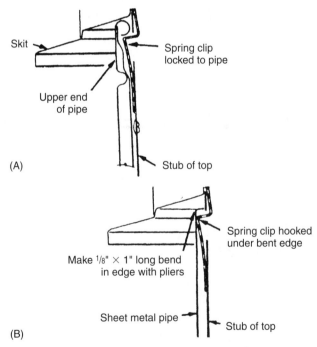

Fig. 16-67 *(A) Round top installation on Eljer Manufacturing's gas vent. (Eljer) (B) Round top installation on single wall pipe. (Eljer)*

Top Installation of 10- to 24-inch Vents

The round tops for this size pipe feature an expanding collar that clamps to the inside of the gas vent on any

other sheet metal pipe of similar nominal size. To install, loosen the screw on top of the collar and squeeze the bottom so that it will enter the pipe easily as shown in Fig. 16-68A. Press down evenly on the skirt until it contacts the upper end of the pipe. See Fig. 16-68B. Tighten the screw to expand the collar against the inner pipe. See Fig. 16-68C. Don't overtighten. Now, at-

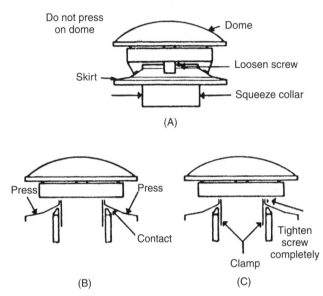

Fig. 16-68 *(A) Dome mounted on top of squeeze collar. (Eljer) (B) Dome being mounted. (Eljer) (C) Location of screws to hold dome in place. (Eljer)*

tempt to lift the skirt to be sure the top is secure.

Checking Vent Operation

Complete all gas piping, electrical, and vent connections. After adjusting the appliance and lighting the main burner, allow a couple of minutes for warm-up. Hold a lighted match just under the rim of the draft hood relief opening. Proper venting will draw the match flame toward or into the draft hood. Improper venting is indicated by escape or spillage of the burned gas, and will cause the match to flicker or go out. Smoke from a cigarette should also be pulled into the draft hood if the vent is drawing correctly. See Fig. 16-69.

Painting

To prolong the life and appearance of the galvanized steel outer casing and other parts of the gas vents located outdoors, use proper painting procedures at the time of installation. Remove oil and dirt with a solvent. Paint first with a good-quality zinc primer or other primer recommended for use on galvanized steel. Next, apply an appropriate finish coat. Ordinary

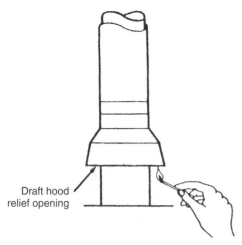

Draft hood
relief opening

Fig. 16-69 *Match test for spillage. (Eljer)*

house paints might not adhere well and do not prevent underfilm corrosion, which leads to paint loosening and peeling.

POST-AND-BEAM CONSTRUCTION

Post-and-beam construction has long been used. Warehouses, barns, and other large buildings are all examples of post-and-beam construction. Designers are now using post-and-beam construction for schools, homes, and office buildings as well. Post-and-beam construction can have wider, longer rooms and higher ceilings. Exposed beams, as in Fig. 16-70, add beauty, line, and depth to the building. Beams can be laminated or boxed in several shapes. This makes unusually shaped buildings possible.

Wall supports are widely spaced. Wide spacing makes possible many very large windows. Using many windows makes a building interior light and cheerful.

Wooden beams give a natural appearance. These and the natural wood finishes can add greatly to a feel of natural beauty. The actual construction methods are simple. Floors and roofs are constructed in a similar manner. Several different materials and techniques may be used with pleasing effect.

Both roof decking and floors are done in much the same manner. Wooden planks, plywood sheets, and special panels may all be used on either floors or roofs. In many construction processes, the roof and ceiling are the same. This cuts down the complexity and cost. Only one layer is used instead of two or three. The walls may combine a variety of glass, stone, brick, wood, and metal materials. There are other names for the post-and-beam type of construction. It is sometimes called *plank and beam* when used in floors and roofs. For ceilings, the term *cathedral ceiling* is often used.

General Procedures

Regular footing and foundation methods are used. The floor beams or joists are laid as in Fig. 16-71. Decking is applied directly over the floor beams as shown. Grooved-plank decking or plywood sheets are common. As a rule, any piece of decking should span at least two openings. Figure 16-72 shows spans. End-joined boards may be used so that joints need not occur over beams. However, if end-joined pieces are not used, joints should be made over floor beams.

Wall stud posts In regular framing, wall pieces are nailed together while they are flat on the subfloor. This is called the conventional system. Studs are spaced 12 to 24 inches apart.

Post-and-beam stud posts may also be nailed to sole and top plates. See Fig. 16-73. However, these

Fig. 16-70 *Post-and-beam construction allows wider, longer open areas.*

Fig. 16-71 *Widely spaced floor or roof joists are decked with long planks.* (Weyerhauser)

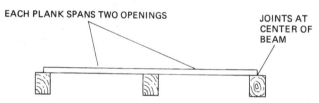

Fig. 16-72 *Any piece of decking should span at least two openings.*

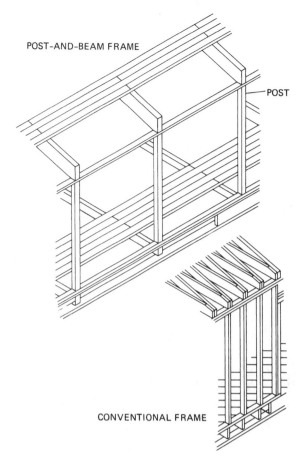

Fig. 16-73 *Posts may be nailed to top and soleplates. However, note the greater space between posts than for conventional frames.*

posts should not be less than 4 inches square. They may be spaced 4 to 12 feet apart. This gives a much wider area between supports. This allows larger glass areas without special headers to help support the roof.

Stud posts may also be applied directly to floors. See Fig. 16-74. In either method, various metal straps and cleats are used. They anchor the stud posts to the floor or soleplate. The posts should be attached directly over the floor joists or beam. Where soleplates are used, the stud posts may be attached at any point.

Stud posts should not be attached directly to concrete floors exposed to the outside weather. A special metal bracket should be used. This bracket raises the post off the concrete as in Fig. 16-75. These brackets must be used to help prevent rot and deterioration of the post caused by moisture. Outside concrete retains moisture. This moisture leads to wood decay. Any wood piece that directly touches the concrete can decay. The special metal brackets make an air space so that the wood does not touch the concrete. The circulating air keeps the wood dry and reduces decay.

Beams Beams are supports for roof decking. Two systems are used. Beams may run the length of the building as in Fig. 16-76A. They may also run across the building (Fig. 16-76B). Both straight gable and shed roof shapes are common. Other beams may be the frame of a roof and wall combined. Beams may combine the wall studs and roof beams. The special beam shapes shown in Fig. 16-77 are typical. These beams combine interior beauty with structural integrity. They are common in churches, schools, and similar buildings.

Beams are made in four basic types. These are solid beams, laminated beams, bent laminated beams, and plywood box beams. Solid beams are made of solid pieces of wood. Obviously, for larger beams, this becomes impractical. To make larger beams, pieces of wood are laminated and glued together. Laminated beams may be used horizontally or vertically. To provide graceful curved shapes, beams are both laminated and curved to shape. These are built up of many layers of wood. Each layer is bent to the shape desired. They are extremely strong and fire-resistant and provide graceful beauty and open space. Laminated beams

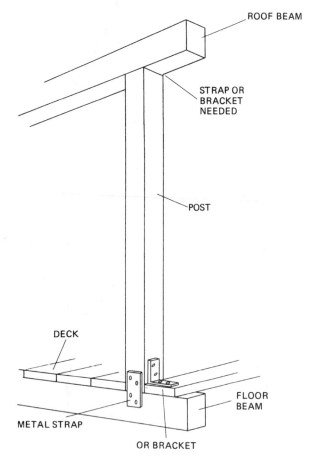

Fig. 16-74 *Posts may also be anchored directly to floors. They must be anchored solidly.*

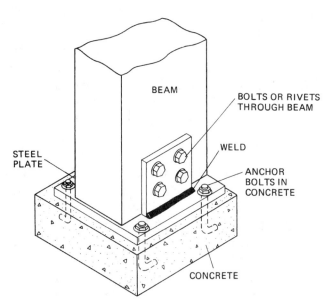

Fig. 16-75 *Foot plates help to reduce moisture damage.*

have a more uniform moisture content. This means that they expand and contract less. Plywood box beams provide a heavy appearance, great strength, and comparatively light weight. They are made as in Fig. 16-78. They have one or more wooden beams at top

and bottom. These beams are stiffened and spaced by lumber spacers as shown. However, the interiors are hollow. The entire framework is then covered with a plywood skin as shown. Although these beams appear to be solid beams, they are much lighter.

Beam systems may also be used. These use small beams between larger beams. This allows large beams to be spaced wider apart. The secondary beams are called *purlins*. See Fig. 16-79.

Beams must be anchored to either posts or top plates. There are two ways to anchor beams. The first is to use special wooden joints. These are cut into the beams, posts, and other members. To prevent side slippage and other shear forces, special pegs, dowels, and splines are used. Many of the old buildings in Pennsylvania and other areas were built in this manner. However, today, this method is time-consuming and costly. Today, the second method is used more often. In this method, special metal or wooden brackets are used. See Fig. 16-80. There are many of these bracket devices. They hold two or more pieces together and prevent slipping and twisting. They should be bolted rather than nailed. Nails will not hold the heavy pieces. Although lag bolts may be used, through bolts are both stronger and more permanent.

Ridge beams are used for gable construction using post-and-beam methods. Where beams are joined to the ridge beams, special straps are used. See Fig. 16-81. These straps may resemble gussets and are applied to the sides. However, an open appearance without brackets can be made. The straps are inlet and applied across the top as shown.

Purlins are held by hangers and brackets like joist hangers. They are most often made of heavy metal. Most metal fasteners are made specifically for each job to the designer's recommendations.

Bearing plates Where several beams join, special allowances must be made for expansion and contraction. The surface of this joining area should also be enlarged. All beams may then have enough area to support them. Special steel plates are used for these types of situations. These are called *bearing plates*. Bearing plates are used at tops of posts, bottoms of angular braces, and other similar situations.

Decking Common decking materials include plywood boards and 2-inch-thick lumber. Decking is often exposed on the interior for a natural look. Both lumber and plywood should be grooved for best effect. The grooves should be on both the edges and ends. Great strength is needed because of the longer spans between supports.

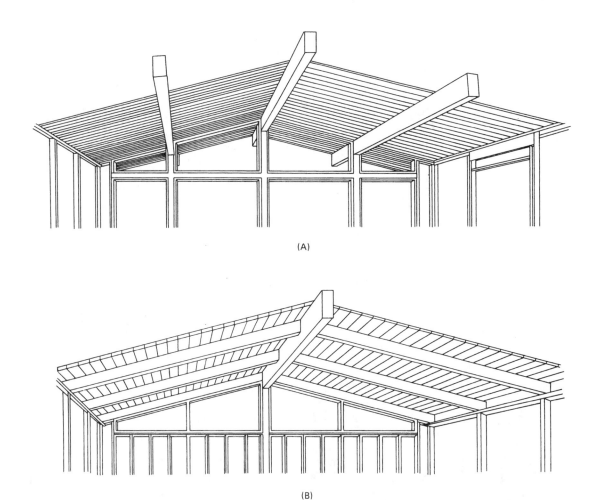

(A)

(B)

Fig. 16-76 (A) Longitudinal beams run the length of a building. (B) Cross or transverse beams run across the width.

Fig. 16-77 Special beam shapes may combine wall and roof framing. (Weyerhauser)

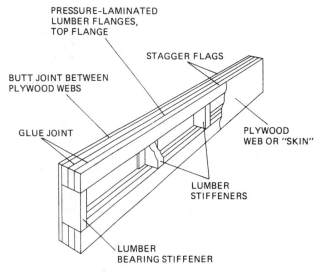

PRESSURE-LAMINATED
LUMBER FLANGES,
TOP FLANGE

STAGGER FLAGS

BUTT JOINT BETWEEN
PLYWOOD WEBS

GLUE JOINT

PLYWOOD
WEB OR "SKIN"

LUMBER
STIFFENERS

LUMBER
BEARING STIFFENER

Fig. 16-78 Box beams look massive, but they are made of plywood over board frames.

Other materials used for roof coverings include concrete and special insulated gypsum materials. Common post-and-beam construction has insulation

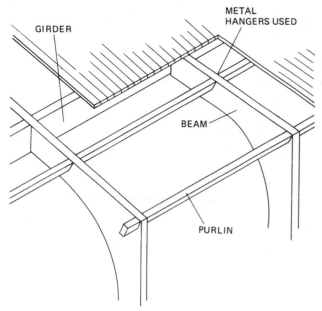

Fig. 16-79 *Post-and-beam roof frame.*

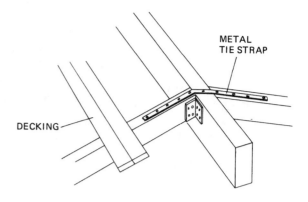

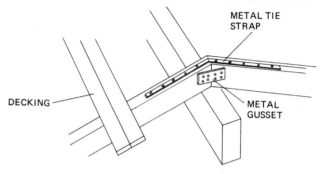

Fig. 16-81 *Beams are tied to girders with gussets and straps.*

Fig. 16-80 *Special brackets are often used to hold beams and other parts together. (Weyerhauser)*

applied on top of roof decking. See Fig. 16-82. A moisture barrier should be used on top of the decking and beneath the insulation. A rigid insulation should be used on top of the decking. It should be strong enough to support the weight of workers without damage. As a rule, a built-up roof is applied over this type of construction. Roofs of this type generally are shed roofs with very little slope. When gables are used, a low pitch is deemed better. Built-up roofs may not be used on steep-pitched roofs.

However, on special shapes, several different roofing techniques may be used. Decking or plywood may be used to sheath the beams or bents. Metal or wooden roofing can be applied over the decking. Commonly sheet copper, sheet steel, and wood shingles are all used.

Stressed-skin panels Stressed-skin panels are large prebuilt panels. They are used for walls, floors, and roof decks. They can be made to span long distances and still give strong support. They are built in a factory and hauled to the building site. They cost more to make, but they save construction time.

The panels are formed around a rigid frame. Interior and exterior surfaces are included. The panels can also include ducts, insulation, and wiring. One surface can be roof decking. The interior side can be insulating

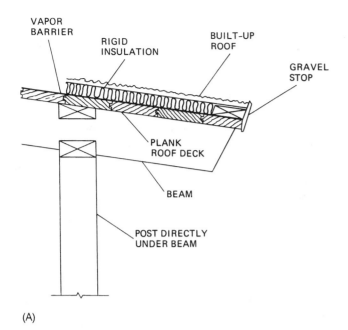

VAPOR
BARRIER

RIGID
INSULATION

BUILT-UP
ROOF

GRAVEL
STOP

PLANK
ROOF DECK

BEAM

POST DIRECTLY
UNDER BEAM

(A)

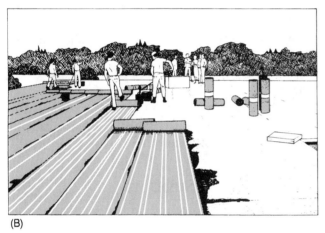

(B)

Fig. 16-82 (A) Insulation can be applied on top of either beams or roof-decking. (B) A composition roof is then laid. (Gold Bond)

Fig. 16-83 Stressed-skin panels are used on floors and roof decks.

sound board to make the finished ceiling. The exterior would be heavy plywood for the roof decking. A standard composition roof could be applied over the panels. The same techniques can be used for wall and floor sections. See Fig. 16-83. The panels are usually assembled with glue and fasteners such as nails or staples.

DECKS

Patios can crack and become unsightly in time. They have to be removed and new slabs poured. This can be an expensive job. Most of the damage is done when the fill against the footings or basement walls begins to settle after a few years of snow and rain. One way to renew the usable space that was once the patio is to make a deck. And, of course, if you are building, the best bet is to attach a deck onto the house while it is being constructed. See Fig. 16-84.

There are a few things you should know before replacing the concrete patio slab with a wooden deck. Working drawings for new houses or for remodeling seldom show much in the way of details. The proposed location of any deck should be shown on the floor plan. Its exact size, shape, and construction, however, are usually determined on-site after a study of the site conditions, the contours of the land, orientation of the deck, and the locations of nearby trees. See Fig. 16-85. Decks must be built to the same code requirements as floor framing systems inside the house. Every deck has four parts: a platform, a floor frame, vertical supports, and guard rails.

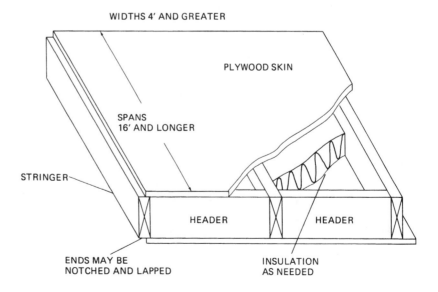

WIDTHS 4' AND GREATER

PLYWOOD SKIN

SPANS
16' AND LONGER

STRINGER

HEADER

HEADER

ENDS MAY BE
NOTCHED AND LAPPED

INSULATION
AS NEEDED

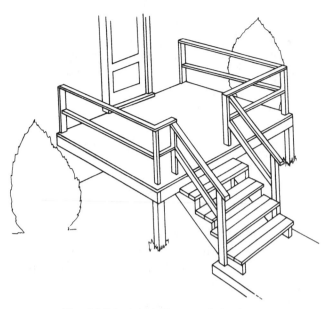

Fig. 16-84 *A deck above grade level.* (Teco)

Platform

The deck itself is built of water-resistant 2 × 4s or 2 × 6s laid flat. All lumber should be of such species as to assure long duration. That means redwood, cedar, cypress, or pressure-treated wood certified as being suitable for exterior use.

If the decking has vertical graining, the grain runs up and down when the lumber is flat. That means either side may be placed face up. If the decking has flat graining (the grain curves from side to side), each piece must be laid with the annular rings pointing down at the edges. Annular rings are produced as the tree grows. They cause wood to have hard and soft sections. Parts of these rings can be observed at the end of a piece of lumber.

Alternative platform layouts are shown in Figs. 16-86 through 16-90. These decks have the lumber and the necessary connectors for making the deck

that particular size. Decking may run parallel to the wall of the house, at right angles to it, or in a decorative pattern, such as a parquet or herringbone. See Fig. 16-91.

Whatever the pattern, each piece of decking must be supported at both ends and long lengths must be given intermediate support. Decking must run at right angles to the framing beneath it. To prevent water buildup from rain or snow, the boards are spaced about ¼ inch apart before they are nailed down. A quick way to do the spacing without having to guess is to put a nail between the decking pieces, so they cannot be pushed together while being nailed.

Frame

The deck can be supported by its perimeter on posts or piers. This means the frame will be a simple box-shaped arrangement. When the decking runs parallel to the wall of the house, supporting joists are attached with joist hangers to an edge joist bolted through sheathing into a header. Or, joist hangers can be used to anchor the joist of the deck to the header of the house. See Fig. 16-92. Refer to Figs. 16-87 through 16-90 for alternate ideas of making a deck frame. Single joists can be spaced up to 2 feet 6 inches apart with the joist span no more than 6 feet.

Support

The other edges of a deck frame are usually supported on 4-×-4 wood posts. Figure 16-93 shows the post joint assembly detail. Figure 16-94 shows exterior joist-to-beam connection while beam-to-post connection is shown in Fig. 16-95. The posts rest on 2-inch wood caps anchored to the tops of concrete piers. Figure 16-96 shows one method used in anchoring. Caps may be omitted if posts are less than 12 inches long.

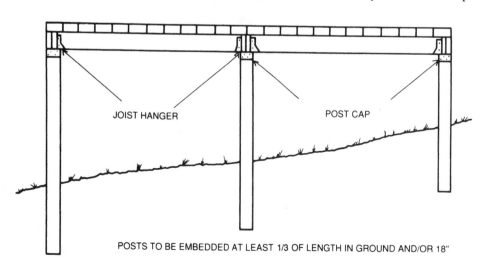

JOIST HANGER   POST CAP

POSTS TO BE EMBEDDED AT LEAST 1/3 OF LENGTH IN GROUND AND/OR 18"

Fig. 16-85 *Wood deck elevation.* (Teco)

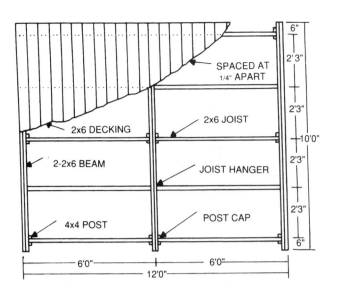

LUMBER		CONNECTORS	
QTY.	TYPE	QTY.	TYPE
9	4x4 POSTS X LENGTH	9	POST CAPS
6	2x6 10/0 BEAMS	20	JOIST HANGERS
10	2x6 6/0 JOISTS		
25☆	2x6 10/0 DECKING	10	FRAMING ANGLES
*	2x6 STAIR TREAD CLEATS	*	3/8" x 4" BOLT
*	2x6 STAIR CARRIAGES	9	5- 1/2" x 6 1/2" PLYWOOD FILLER SHIMS
*	2x6 STAIR TREADS		
*	DETERMINED BY WIDTH & RISE OF STAIR		

☆ IF 2x4 DECKING IS USED QUANTITY REQUIRED IS 39

BASIC WOOD DECK PLAN

Fig. 16-86 *(A) Basic wood deck plan. (Teco) (B) Adding to the basic plan can produce a porch. Notice the roof.*

The tops of the piers must be at least 8 inches above grade. If the side is not level, the lengths of posts must be varied so that their tops are at the same level to support the deck frame. The beams that span from post-to-post may be 4-inch timbers or a pair of 2-inch joists. The sizes of the beams and the spacing of post have to be considered in the basic design at the beginning. Maximum spans of typical decks are shown in Table 16-3.

As you can see from the table, it is usually less expensive to use larger beams and fewer posts. The posts need more work since they call for a hole to be dug and

6'0" MODULE (ALL MEMBERS 2x6)

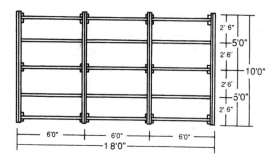

LUMBER		CONNECTORS	
QTY.	TYPE	QTY.	TYPE
12	4x4 POSTS X LENGTH	12	POST CAPS
8	2 × 6 10/0 BEAMS	30	JOIST HANGERS
15	2 × 6 6/0 JOISTS	12	5- 1/2" x 6 1/2" PLYWOOD FILLER SHIMS
37☆	2 × 6 10/0 DECKING		

☆ IF 2x4 DECKING IS USED QUANTITY REQUIRED IS 58

Fig. 16-87 *A deck plan using 6-foot modules.* (Teco)

9'0" x 5'0" MODULE (ALL MEMBERS 2x6)

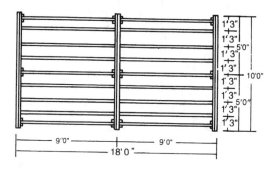

LUMBER		CONNECTORS	
QTY.	TYPE	QTY.	TYPE
9	4x4 POSTS X LENGTH	9	POST CAPS
6	2x6 10/0 BEAMS	36	JOIST HANGERS
18	2x6 9/0 JOISTS	9	5- 1/2" x 6 1/2" PLY- WOOD FILLER SHIMS
37☆	2x6 10/0 DECKING		

☆ IF 2x4 DECKING IS USED QUANTITY REQUIRED IS 58

Fig. 16-89 *A deck plan using 9-foot by 5-foot modules.* (Teco)

ADDITION OF 5'0" MODULE
(ALL MEMBERS 2x6)

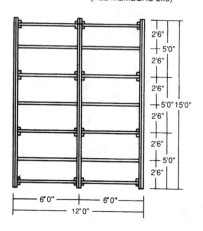

LUMBER		CONNECTORS	
QTY.	TYPE	QTY.	TYPE
12	4x4 POSTS X LENGTH	12	POST CAPS
6	2x6 15/0 BEAMS	28	JOIST HANGERS
14	2x6 6/0 JOISTS	12	5- 1/2" x 6 1/2" PLYWOOD FILLER SHIMS
41☆	2x6 10/0 DECKING		

☆ IF 2x4 DECKING IS USED QUANTITY REQUIRED IS 38

Fig. 16-88 *A deck plan using 5-foot modules.* (Teco)

8'0" x 4'0" MODULE (ALL MEMBERS 2 x 6)

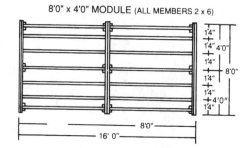

LUMBER		CONNECTORS	
QTY.	TYPE	QTY.	TYPE
9	4x4 POSTS X LENGTH	9	POST CAPS
6	2x6 8/0 BEAMS	28	JOIST HANGERS
14	2x6 8/0 JOISTS	9	5- 1/2" x 6 1/2" PLY - WOOD FILLER SHIMS
34☆	2x6 8/0 DECKING		

☆ IF 2x4 DECKING IS USED QUANTITY REQUIRED IS 52

Fig. 16-90 *A deck plan using 8-foot by 4-foot modules.* (Teco)

concrete to be mixed and poured. Plan for the spacing of posts before you begin the job. They have to be placed in concrete and allowed to set for at least 48 hours before you nail to them.

If you space the joists as shown in Fig. 16-86, make sure they correspond to the local code, which should be checked before you start the job. The layouts in Figs. 16-86 through 16-90 are suggestions only and do not indicate compliance with any specific structural, code, service, or safety requirements.

Guard Rails

A deck more than 24 inches above grade must be surrounded with a railing. Supporting posts must be part of the deck structure. The railing cannot be toenailed to the deck. The posts may extend upward through the deck, or they may be bolted to joists below deck level. See Fig. 16-97. Spacing of support every 4 feet provides a sturdy railing. Support on 6-foot centers is acceptable. See Fig. 16-98.

Making a Hexagonal Deck

A hexagonal deck can add beauty to a house or make it the center of attraction for an area located slightly away from the house. Figure 16-99 shows the basics of the foundation and layout of stringers. Figure 16-100

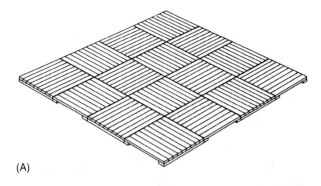

(A)

(B)

Fig. 16-91 (A) Parquet deck pattern. (Western Wood) (B) Nailing treated southern pine 2 × 4s to a deck attached to the house. (Southern Forest Products Association)

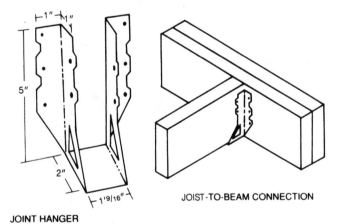

JOINT HANGER

JOIST-TO-BEAM CONNECTION

Fig. 16-92 Hanger used to connect joists to beam for a deck. (Teco)

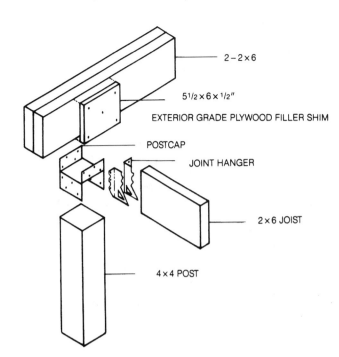

2 – 2 × 6

5 1/2 × 6 × 1/2"

EXTERIOR GRADE PLYWOOD FILLER SHIM

POSTCAP

JOINT HANGER

2 × 6 JOIST

4 × 4 POST

Fig. 16-93 Post joint detail assembly. (Teco)

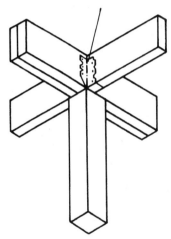

Fig. 16-94 Beam connection. (Teco)

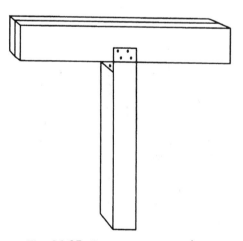

Fig. 16-95 Beam-to-post connection. (Teco)

shows the methods needed to support the deck. To make a hexagonal deck:

1. Lay out the deck dimensions according to the plan and locate the pier positions.

Excavate the pier holes to firm soil. Level the bottom of the holes and fill with gravel to raise the piers to the desired height. To check the pier height, lay a 12-foot stringer between piers and check with a level.

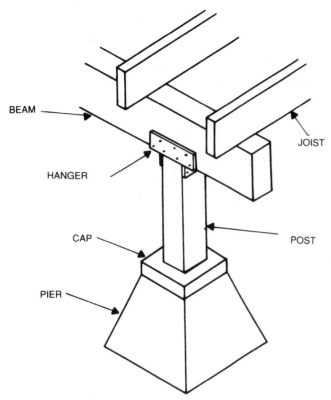

Fig. 16-96 Wood posts that support the deck rest on caps anchored to concrete piers. Piers may be bought precast or you can pour your own.

Table 16-3 *Deck Beam Spacing*

Deck Width (feet)	Distance Between Posts		
	4 × 6 Beam	4 × 8 Beam	4 × 10 Beam
6	6'9"	9'0"	11'3"
8	6'0"	8'0"	10'0"
10	5'3"	7'0"	8'9"
12	4'6"	6'0"	7'6"

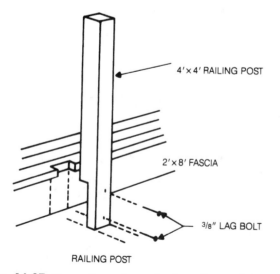

Fig. 16-97 Connections of post for the railing and the fascia board. Note that lag bolts are used and not nails. (Western Wood)

2. When all piers are equal height, place a 12-foot stringer in position. Cut and fit 6-foot stringers and toenail them with 10d nails to the 12-foot stringer at the center pier. Use temporary bracing to position the stringers in the correct alignment. Cut and apply the fascia. Cut and apply the stringers labeled *B* in Fig. 16-99. Note the details in Fig. 16-100.

3. Drive the bracing stakes into the ground and nail them to the stringers to anchor the deck firmly into position. See Fig. 16-101.

4. Apply the decking. Begin at the center. See Fig. 16-101. Center edge of the first deck member is over the 12-foot stringer. Use 10d nails for spacing guides between deck members. Apply the remaining decking. Nail the decking to each stringer with two 10d nails. Countersink the nails. Check the alignment every five or six boards. Adjust the alignment by increasing or decreasing the width between deck members.

5. Tack the trim guide in place and trim the edges, allowing a 2-inch overhang. See Fig. 16-102. Smooth the edges with a wood rasp or file. See Fig. 16-103.

Raised Deck

A raised deck (Fig. 16-98) uses a different technique for anchoring the frame to the building. If the deck is attached to a building, it must be inspected by the local building inspector and local codes must be checked for proper dimensions and proper spacing of dimensional lumber. The illustrations here are general and will work in most instances to support a light load and the wear caused by human habitation.

The deck shown in Fig. 16-106 should be laid out according to the sketch in Fig. 16-104. Figure 16-105 shows how the deck is attached to the existing house. Figure 16-106 shows how the decking is installed. Use 10d nails for spacing guides between the deck members. Nail the deck member to each stringer with two 10d nails. Countersink the nails. Check the alignment every five or six boards. Adjust the alignment by increasing or decreasing the width between the deck members.

The railing posts are predrilled and then the fascia is marked and drilled for insertion of the ⅜-×-3-inch lag bolts per post. The railing cap can be installed with two 10d nails per post. This 12-×-12-foot deck takes about 5 gallons of Penta or some other type of wood preservative—that is, of course, if you didn't start with pressure-treated wood.

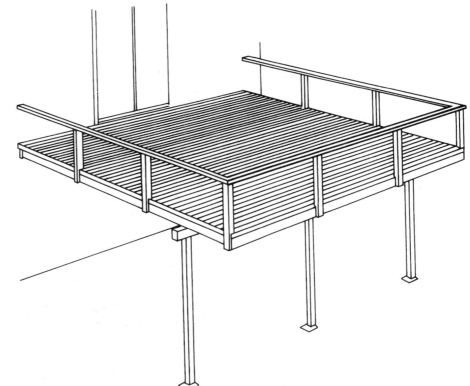

Fig. 16-98 *A deck located well above-grade. (Western Wood)*

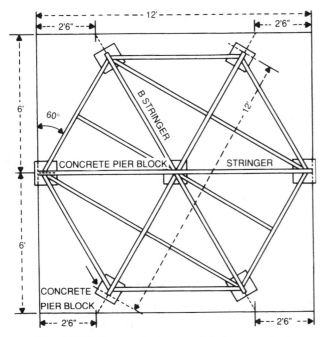

Fig. 16-99 *Layout of a hexagonal deck. (Western Wood)*

Steps

If the deck is located at least 12 inches above the grade you will have to install some type of step arrangement to make it easy to get onto the deck from grade level. Steps will also make it easier to get off the deck.

By using the Teco framing angles it is possible to attach the steps to the deck rather easily. See Fig. 16-106.

These galvanized metal angles help secure the steps to the deck without the sides of the steps being cut and thus weakened. Another method can also be used to attach steps to the deck. Figure 16-107 shows how step brackets (galvanized steel) are used to make a step without having to cut the wood stringers. They come in a number of sizes to fit almost any conceivable application.

CONCRETE PATIOS

Before you pour a patio a number of things must be taken into consideration: size, location in relation to the house, shape, and last but not least, cost. Cost is a function of size and location. Size determines the amount of excavation, fill, and concrete. Location is important in terms of getting the construction materials to the site. If materials must be hauled for a distance, it will naturally cost more in time and energy and therefore the price will increase accordingly.

Sand and Gravel Base

If the soil is porous and has good drainage, it is possible to pour the concrete directly on the ground, if it is well tamped. If the soil has a lot of clay and drainage is poor, it is best to put down a thin layer of sand or gravel before pouring the concrete. See Fig. 16-108. Just before pouring the concrete, give the soil a light

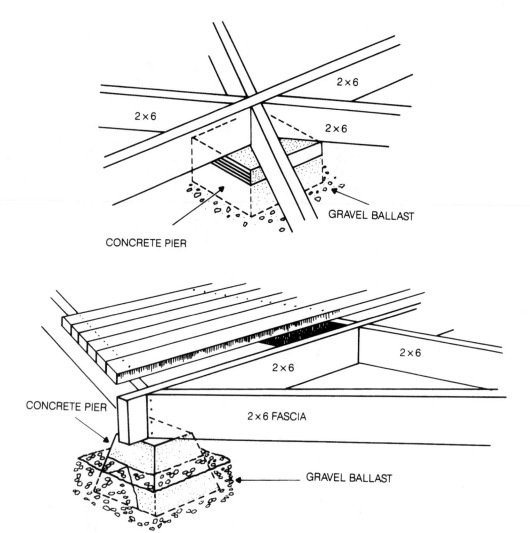

Fig. 16-100 *Supports for the deck.* (Western Wood)

water sprinkling. Avoid developing puddles. When the earth is clear of excess water you can begin to pour the concrete. In some instances voids need to be filled before the concrete is poured. When these additional areas need attention, bring them up to the proper grade with granular material thoroughly compacted in a maximum of 4-inch layers.

If you have a subgrade that is water-soaked most of the time, you should use sand, gravel, or crushed stone for the top 6 inches of fill. This will provide the proper drainage and prevent the concrete slab from cracking when water gets under it and freezes. When you have well-drained and compacted subgrades, you do not need to take these extra precautions.

Expansion Joints

When you have a new concrete patio abutting an existing walk, driveway, or building, a premolded material, usually black and ½-inch thick, should be placed at the joints. See Fig. 16-109. These expansion joints are placed on all sides of the square formed by the intersection of the basement wall or floor slab and the patio slab. Whenever a great expanse is covered by the patio it is best to include expansion joints in the layout before pouring the concrete.

The Mix

If you are going to mix your own, which is a time-consuming job, keep in mind that the concrete should contain enough water to produce a concrete that has relatively stiff consistency, works readily, and does not separate. Concrete should have a slump of about 3 inches when tested with a standard *slump cone.* Adding more mixing water to produce a higher slump than specified lessens the durability and reduces the strength of the concrete.

The slump test performed on concrete mix measures the consistency of the wet material and indicates

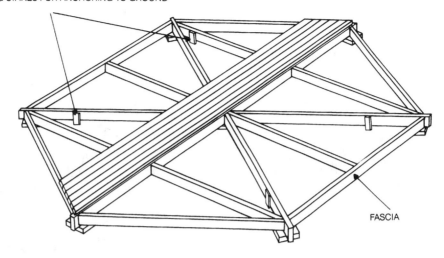

BRACING STAKES FOR ANCHORING TO GROUND

FASCIA

(A)

(B)

Fig. 16-101 *(A) Anchoring the deck to the ground.* (Western Wood) *(B) Finished project.* (Southern Forest Products Association)

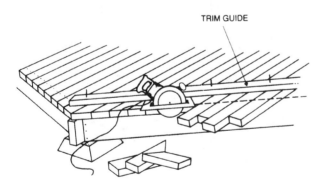

TRIM GUIDE

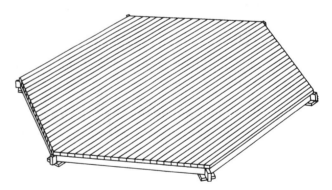

Fig. 16-103 *The finished hexagonal deck.* (Western Wood)

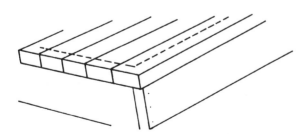

Fig. 16-102 *Trimming the overhang.* (Western Wood)

if the mix is too wet or dry. The test is performed by filling a bucket with concrete and letting it dry (Fig. 16-109). Then take another wet sample and dump it alongside. Test for the amount of slump the wet mix displays. It is obvious if the mixture is too wet: the cone will slump down or wind up in a mess as shown in Fig. 16-110.

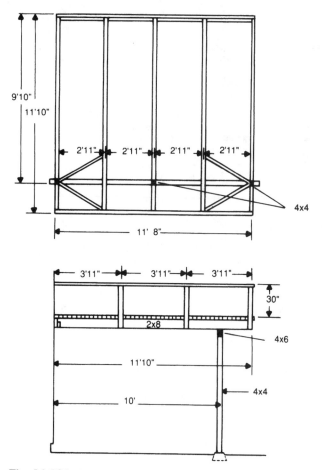

Portland cement or by adding an air-entrained agent during mixing.

Before the concrete is poured, the subgrade should be thoroughly dampened. Make sure it is moist throughout, but without puddles of water.

Keep in mind that concrete should be placed between forms or screeds as near to its final position as practicable. Do not overwork the concrete while it is still plastic because an excess of water and fine materials will be brought to the surface. This may lead to scaling or dusting later when it is dry. Concrete should be properly spaced along the forms or screeds to eliminate voids or honeycombs at the edges.

Forms

Forms for the patio can be either metal or wood. See Fig. 16-111. Dimensional lumber (2×4s or 2×6s) are usually enough to hold the material while it is being worked and before it sets up. Contractors use metal forms that are used over and over. If you are going to use the 2×4s for more than one job, you may want to oil them so that the concrete will not stick to them and make a mess of the edges of the next job. You can use old crankcase oil and brush it on the wooden forms.

Forms should be placed carefully since their tops are the guides for the screeds. Make sure the distances apart are measured accurately. Use a spirit level to assure that they are horizontal. If the forms are used on an inclined slab, they must follow the incline. Forms or curved patios or driveways are made from ½-inch redwood or plywood. If you want to bend the redwood,

Fig. 16-104 *Supporting the raised 12-×-12-foot deck. (Western Wood)*

In northern climates where flat concrete surfaces are subjected to freezing and thawing, *air-entrained concrete* is necessary. It is made by using an air-entraining

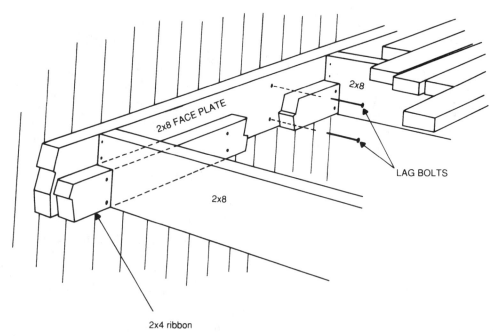

Fig. 16-105 *Notched stringers are used to attach to the nailing ribbon and to the beam. (Western Wood)*

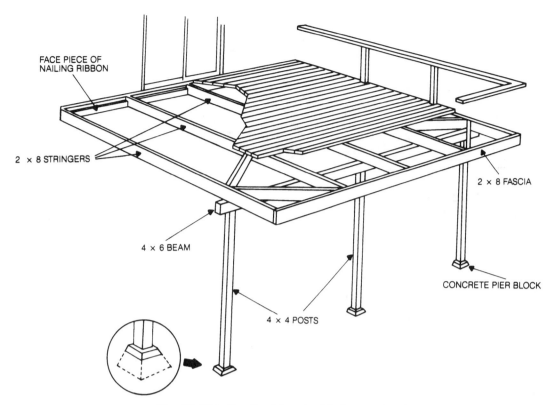

Fig. 16-106 *Details of the raised deck.* (Western Wood)

Labels in figure:
FACE PIECE OF NAILING RIBBON
2 × 8 STRINGERS
2 × 8 FASCIA
4 × 6 BEAM
CONCRETE PIER BLOCK
4 × 4 POSTS

soak it in water for about 20 minutes before trying to bend it.

Place stakes at intervals along the outside of the forms and drive them into the ground. Then nail the stakes to the forms to hold them securely in place. The tops of the stakes must be slightly below the edge of the forms so they will not interfere with the use of the strike-off board for purposes of screeding later.

Placing the Joints

Joints are placed in concrete to allow for expansion, contraction, and shrinkage. It is best not to allow the slab to bond to the walls of the house, but allow it to move freely with the earth. To prevent bonding, use a strip of rigid waterproof insulation, building paper, polyethylene, or something similar. Also use the expansion joint material where the patio butts against a walk or other flat surface. See Fig. 16-112.

Wide areas such as a patio slab should be paved in 10- to 15-foot-wide alternate strips. A construction joint is made by placing a beveled piece of wood on the side forms. See Fig. 16-113. This creates a groove in the slab edges. As the intermediate strips are paved, concrete fills the groove and the two slabs are keyed together. This type of joint keeps the slab surfaces even and transfers the load from one slab to the other when heavy loads are placed on the slabs.

You have seen contraction joints, often called *dummy joints,* cut across a slab (Fig. 16-114). They are cut to a depth of one-fifth to one-fourth the thickness of the slab. This makes the slab weaker at this point. If the concrete cracks due to shrinkage or thermal contraction, the crack usually occurs at this weakened section. In most instances, the dummy joint is placed in the concrete after it is finished off. A tool is drawn through the concrete before it sets up. It cuts a groove in the surface and drops down into the concrete about ½ inch. In case the concrete cracks later after drying, it will usually follow these grooves. These grooves are usually placed 10 to 15 feet apart on floor slabs or patios.

Pouring the Concrete

Most concrete is ordered from a ready-mix company and delivered with the proper consistency. If you mix your own, some precautions should be taken to assure its proper placement. Concrete should be poured within 45 minutes of the time it is mixed. Some curing begins to take place after that and the concrete may become too thick to handle easily. If you are pouring a large area, mix only as much as you can handle within 45 minutes. The first batch should be a small one that you can use for trial purposes. After the first batch, you can determine whether the succeeding batches require

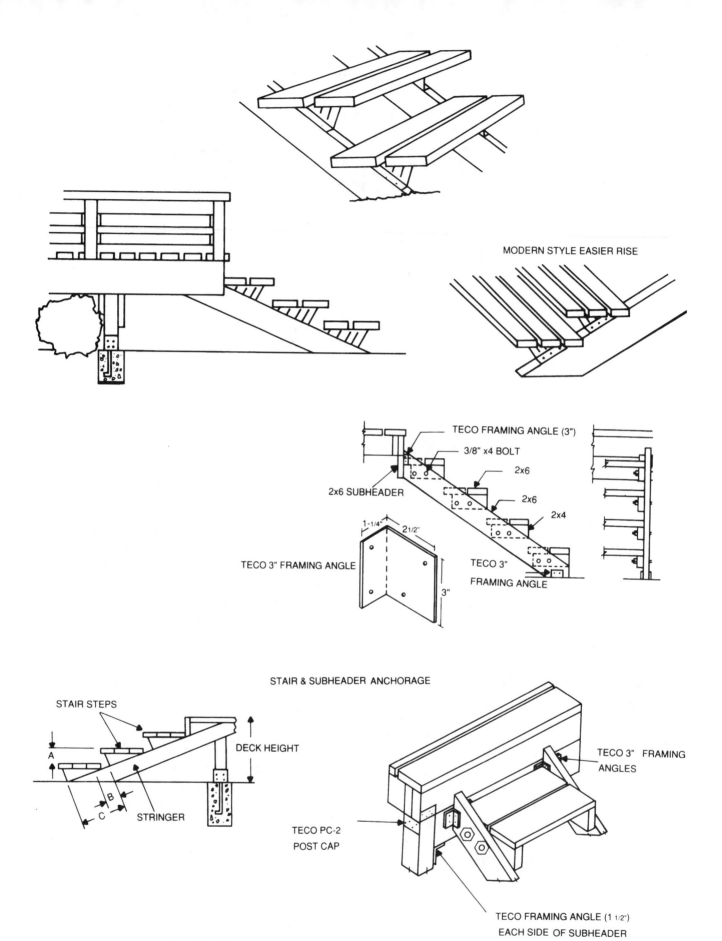

MODERN STYLE EASIER RISE

TECO FRAMING ANGLE (3")

3/8" x4 BOLT

2x6

2x6 SUBHEADER

2x6

2x4

1-1/4"

2 1/2"

TECO 3" FRAMING ANGLE

TECO 3"
FRAMING ANGLE

3"

STAIR & SUBHEADER ANCHORAGE

STAIR STEPS

DECK HEIGHT

A

B

STRINGER

C

TECO 3" FRAMING
ANGLES

TECO PC-2
POST CAP

TECO FRAMING ANGLE (1 1/2")
EACH SIDE OF SUBHEADER

Fig. 16-107 *Methods of mounting steps to a deck.* (United Steel Products)

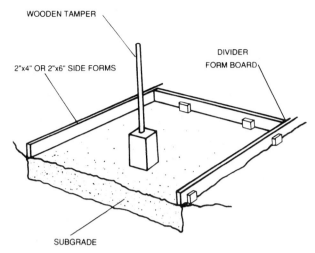

NOTE: SUBGRADE MAY CONSIST OF CINDER, GRAVEL, OR OTHER SUITABLE MATERIAL WHERE CONDITIONS REQUIRE. THE SUBGRADE SHOULD BE WELL-TAMPED BEFORE PLACING CONCRETE

Fig. 16-108 *Using wooden tamper to compact the sand or crushed stone before pouring concrete.*

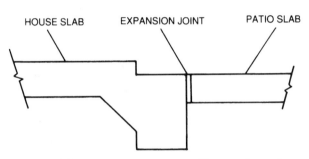

Fig. 16-109 *Expansion joints are used between large pieces.*

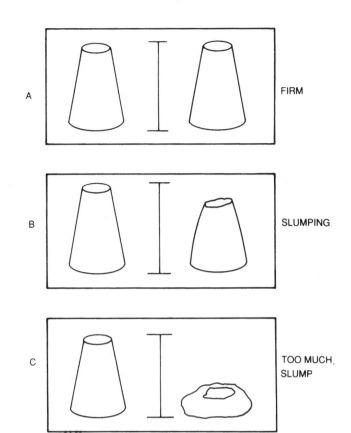

Fig. 16-110 *(A) Slump of the mix is too little. It stands alongside the test cone. (B) The slump is acceptable. (C) The slump is too much.*

more or less sand. Remember, do not vary the water for a thicker or thinner concrete—only the amount of sand. Once you have established the workability you like, stay with that formula.

Pour the concrete into the forms so it is level with the form edges. Then immediately after the first batch is poured into place, spade the concrete with an old garden rake or hoe. This will even out the wet concrete so it is level with the form boards. Make sure there are no air voids in the mix.

Use a striker board to level off the concrete in the form. It usually requires two people to do this job—one at each end of the board. The idea is to get a level and even top surface on the poured concrete by using the board forms as the guide.

Draw the striker board across the concrete while using the form edges as the guides. Then you can see-saw the board as you move it across. Now it is obvious why the stakes for the forms had to be below the edges of the form boards: they should not interfere with the movement of the strike board. The strike board takes off the high spots and levels the concrete. If the board

skips over places where it is low and lacks concrete, fill in the low spots and go over them again to level off the new material.

Once you have finished striking the concrete and leveling it off, you may notice a coat of water or a shiny surface. This may not be evenly distributed across the top, but wait until this sheen disappears before you do any other work on the concrete. This may take an hour or two; the temperature determines how quickly the concrete starts to set up. The humidity of the air and the wind are also factors. The concrete will begin to harden and cure. See Fig. 16-115.

Finishing

When the water sheen and bleed water have left the surface of the concrete, you can start to finish the surface of the slab. This may be done in one or more ways, depending on the type of surface you want. Keep in mind that you can overdo the finishing process.

You can bring water under the surface up to the top to cause a thin layer of cement to be deposited near or on the surface. Later, after curing, it becomes a scale that will powder off with use.

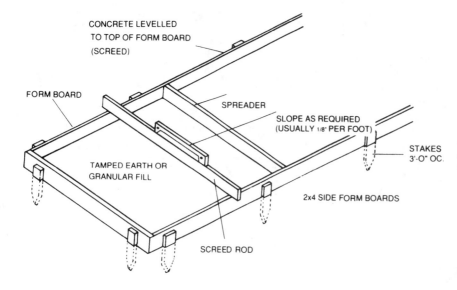

CONCRETE LEVELLED
TO TOP OF FORM BOARD
(SCREED)

FORM BOARD

SPREADER

SLOPE AS REQUIRED
(USUALLY 1/8" PER FOOT)

STAKES
3'-0" OC.

TAMPED EARTH OR
GRANULAR FILL

2x4 SIDE FORM BOARDS

SCREED ROD

Fig. 16-111 *Forms are made of 2 × 4s or 2 × 6s and staked to prevent movement when the concrete is added and worked.*

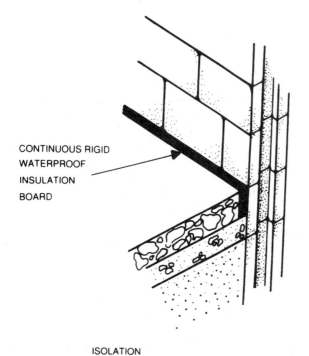

CONTINUOUS RIGID
WATERPROOF
INSULATION
BOARD

ISOLATION

Fig. 16-112 *Isolation joint prevents slab from cracking wall as it expands.*

CONSTRUCTION JOINT

2" LUMBER WITH
BEVELED 1"x2" STRIP

Fig. 16-113 *Construction joint.*

Finishing can be done by hand or by rotating power-driven trowels or floats. The size of the job will determine which to use. In most cases, you will have to rent the power-driven trowel or float. Therefore, economy of operation becomes a factor in finishing. Can you do the job by yourself or will you need help to get it done before the concrete sets up too hard to work? You should make up your mind before you start so you can have additional help handy or have the power-driven machines around to speed up the job.

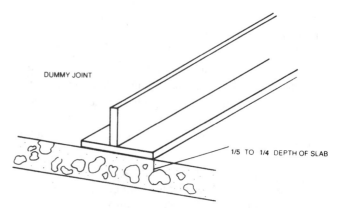

DUMMY JOINT

1/5 TO 1/4 DEPTH OF SLAB

Fig. 16-114 *Dummy or contraction joint.*

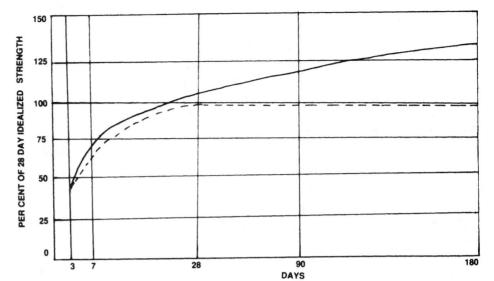

Fig. 16-115 *Relative concrete strength versus curing method.*

The type of tool used determines the type of finish on the surface of the patio slab. A wood float puts a slightly rough surface on the concrete. A metal trowel or float produces a smooth finish. Extra-rough surfaces are produced by using a stiff-bristled broom across the top.

Floating

A float made of a piece of wood with a handle for use by hand is used to cause the concrete surface to be worked. In some cases it has a long handle so the concrete can be worked by a person standing up and away from the forms. See Fig. 16-116. A piece of plywood or other board can be used to kneel on while floating the surface. The concrete should be set up sufficiently to support the person doing the work. Floating has some advantages. It embeds the large aggregate (gravel) beneath the surface and removes slight imperfections such as bumps and voids. It also consolidates the cement near the surface in preparation for smoother finishes. Floating can be done before or after

edging and grooving. If the line left by the edger and groover is to be removed, floating should follow the edging and grooving operation. If the lines are to be left for decorative purposes or to provide a crack line for later movement of the slab, edging and grooving will have to follow the floating operation.

Troweling

Troweling produces a smooth, hard surface. It is done right after floating. For the first troweling, whether by hand or power, the trowel blade must be kept as flat against the surface as possible. If the trowel blade is tilted or pitched at too great an angle, an objectionable washboard or chatter surface will be produced. For first troweling, do not use a new trowel. An older one that has been broken in can be worked quite flat without the edges digging into the concrete. The smoothness of the surface can be improved by a number of trowelings. There should be a lapse of time between successive trowelings to allow the concrete to increase its set or become harder. As the surface stiffens, each successive troweling should be made by a smaller trowel. This gives you sufficient pressure for proper finishing.

Brooming

If you want a rough-textured surface you can score the surface after it is trowelled. This can be done by using a broom. Broom lines should be straight lines, or they can be swirled, curved, or scalloped for decorative purposes. If you want deep scoring of the surface use a wire-bristled broom. For a finer texture you may want to use a finer-bristled broom. Draw the broom toward

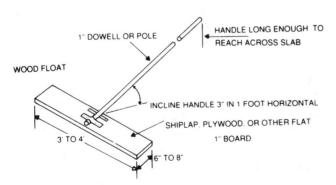

Fig. 16-116 *Construction details for a long-handled float.*

you one stroke at a time with a slight overlap between the edges of each stroke. The broom should be wet when it is first drawn across the surface. You can use a pail of water to wash off the excess concrete. Each time a completed stroke is accomplished, dip the broom into the water and shake off the excess water before drawing the broom across the surface again. Most patios, however, look better with a smooth silky surface.

Grooving

To avoid random cracking due to heaving of the concrete slab after it has cured, it is best to put in grooves after the troweling is done. Then cracks that form will follow the grooves instead of marring the surface of an otherwise smooth finish.

FENCES

Installing a fence calls for the surveyor's markers to be located in the corners of the property. Lot markers are made of ¾-inch pipe or they could be a corner concrete monument. Many times they are buried as much as 2 feet under the surface. The plot plan may be of assistance in locating the surveyor's markers. Locating proper limits will make sure the fence is on the land it is intended for and not on the property next door.

The local zoning board usually has limitations on the height of the fence and just where it can be placed on a property. If ordinances allow it to be placed exactly on the property line, be sure to check with the abutting neighbors and obtain their consent in order to prevent any future law suits.

In most states it is generally understood that if the posts are on the inside facing your property and the fence on the other side, except for post-and-rail fencing, the fence belongs to you. Local officials will be more than happy to help you in regard to the details.

Installation

Most fences are variations of a simple post, rail, and board design. The post and rail support structure is often made of standard dimension lumber while the fenceboards come in different shapes and sizes, giving the fence its individual style.

On a corner of the lot place a stake parallel to the surveyor's marker and attach a chalk line to the stake. Pull the chalk line taut and secure it at the opposite stake. Repeat until all boundaries are covered. See Fig. 16-117. Repeat the procedure to measure the required setback from the original boundary and restake accordingly. You are now ready to install the fence.

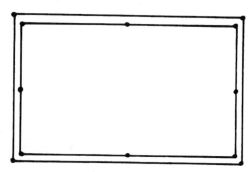

Fig. 16-117 *Layout of the lot and location of posts for the fence.*

To get the fencing started correctly, start from one corner of the lot and place the first post in that corner.

Setting Posts

Setting posts requires patience and is the most critical aspect of fence building. Posts must be sturdy, straight, and evenly spaced for the fence to look and perform properly. Redwood, western cedar, or chemically treated pine can be used for the posts. Posts are commonly placed 8 feet on center. Mark the post locations with stakes.

Dig holes about 10 inches in diameter with a post hole digger. Holes dug with a shovel will be too wide at the mouth to provide proper support. Auger-type diggers work well in rock-free earth. If you are likely to encounter stones, use a clam-shell type digger.

Set corner posts first. String a line between corner posts to mark the fenceline and align the inside posts. For a 5- or 6-foot fence, post holes should be at least 2 feet deep. A 3-foot hole is required for an 8-foot fence. See Fig. 16-118.

Proper drainage in post holes eliminates moisture and extends the life of posts. Fill the bottom of holes with gravel. Large flat base stones also aid drainage. Fill in with more gravel, 3 or 4 inches up the post.

For the strongest fence, set the posts in concrete. You may figure your concrete requirements at roughly an 80-pound bag of premixed concrete per post. If you extend the concrete by filling with rocks or masonry rubble, be sure to tap it down. Cleats or metal hardware can be used to strengthen fenceposts. Cleats are 2-×-4 wood scraps attached horizontally near the base of the post that provide lateral stability. Large lag screws or spikes partially driven into the post can be used to do the same job when posts are set in concrete.

Make sure the concrete completely surrounds the post to make a collar. It doesn't have to be a full shell. This reduces the tendency to hold water and promote

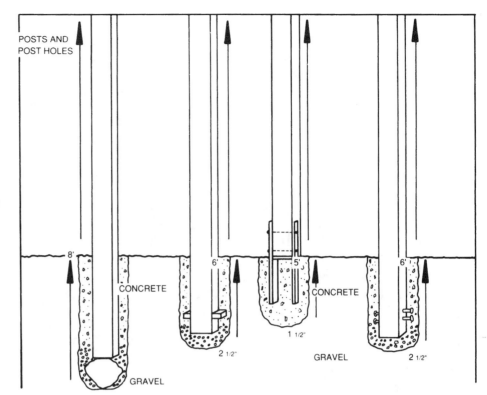

Fig. 16-118 *Depths needed for mounting fence posts. (California Redwood Association)*

POSTS AND POST HOLES

CONCRETE

CONCRETE

GRAVEL

GRAVEL

8'

6'

5'

6'

2 1/2"

1 1/2"

2 1/2"

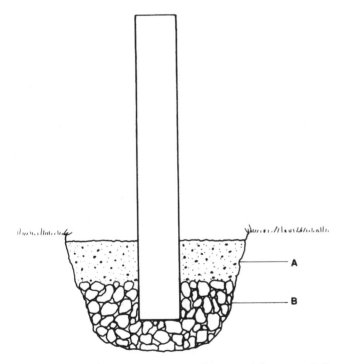

Fig. 16-119 *Location of concrete collar around the post at A. B is the gravel fill.*

A

B

decay. See Fig. 16-119. When you have placed the posts to the height desired, pour the concrete collar around the post. Be sure the posts are perpendicular to the surface and allow the concrete to cure at least 48 hours before attaching the fencing.

Attaching the Rails

Two or three horizontal rails run between the posts. The number of rails depends on the fence height. Rails 8 feet long are common because this length of 2 × 4 is readily available and provides enough support for most styles of fenceboards. Upper rails should rest on the posts. Rails can be butt-joined and mitered at corners. Bottom rails can be toenailed into place, but the preferred method is to place a block underneath the joint for extra support. Metal hardware such as L brackets can be used to secure rails, but make sure all metal fasteners including nails are noncorrosive. See Fig. 16-120.

Attaching Fenceboards

Nailing the fenceboard in place is the easiest and most satisfying part of the project. Boards 1 inch in thickness and 4, 6, 8, or 10 inches in width are common. The dimensions of the fenceboards and the way they are applied will give the fence its character.

There are several creative ways to do this. For fenceboards of widths 4 inches and less, use one nail per bearing. For wider fenceboards use two nails per bearing. Do not overnail.

Picket fences leave plenty of room between narrow fenceboards. They are often used to mark boundaries and provide a minimal barrier. Just enough to keep small children or small dogs in the yard, they are typically 3 or 4 feet high. See Fig. 16-121.

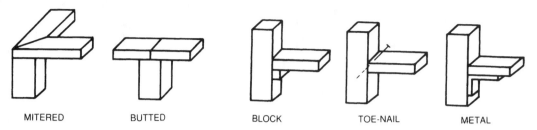

Fig. 16-120 *Attaching rails to the posts.* (California Redwood Association)

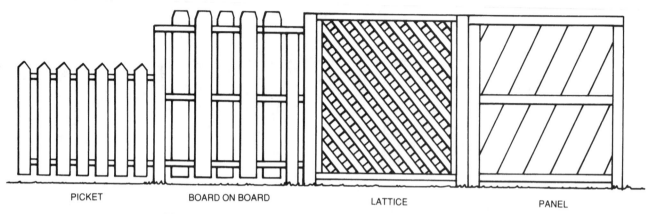

Fig. 16-121 *Various types of fence designs.* (California Redwood Association)

Board on board fences are taller and provide more of a barrier. Distinguished by the fenceboards alternating pattern, board and board fences look the same from either side. This allows great flexibility in design and function. Depending on the placement of fenceboards, this fence will block the wind and the view.

Lattice panels made from redwood lath or ½ × 2's can be used to create a lighter, more delicate fence. Lattice panels can be prefabricated within a 2 × 4 frame, then nailed to the posts.

Panel fences create a solid barrier. Panels are formed by nailing boards over rails and posts. By alternating the side that the fenceboards are nailed to at each post, you can build a fence that looks the same from both sides.

Stockade fencing is built level. See Fig. 16-122. However, not all ground is flat. You can adjust for the slope of the ground working downhill if possible. Fasten the section into a post that is plumb, then exert downward pressure on the end of the section before attaching it to the post. See Fig. 16-123. This is called *racking*. When finished, all posts should follow the slope of the ground.

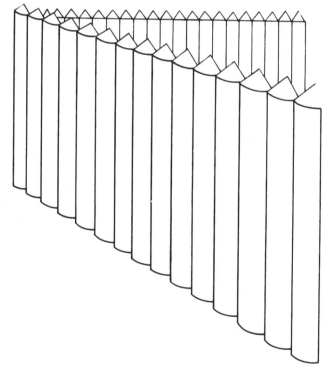

Fig. 16-122 *Stockade fence.*

Nails and Fasteners

Use noncorrosive nails with redwood outdoors. Stainless steel, aluminum and top-quality, hot dipped galvanized nails will perform without staining. Inferior hardware, including cheap or electroplated galvanized nails, will corrode and cause stains when in contact with moisture.

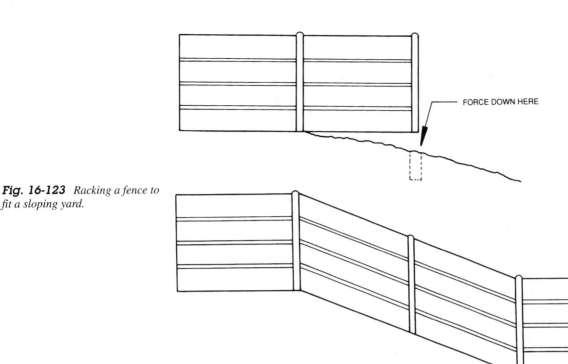

Fig. 16-123 *Racking a fence to fit a sloping yard.*

If you use redwood do not finish off with varnishes and clear film finishes, oil treatments, or shake and shingle type paints.

Gates

Getting the gate to operate properly is another problem with fencing. By using the proper hardware and making sure the gate posts are plumb and sturdy, it is possible to install a gate that will work for years. See Fig. 16-124. To apply the hinges, position the straps approximately 4 inches from both the top and bottom of the gate on the side to be hinged. See Fig. 16-125. Position the gate in the opening and allow adequate clearance between the bottom of the gate and the ground. You can use a wooden block under the gate to ensure proper clearance and aid in the installation. Mark and secure the vertical leaves to the fence posts. Drill pilot

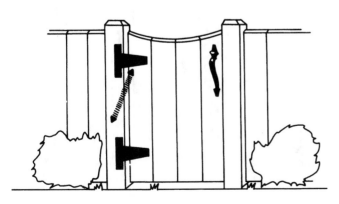

Fig. 16-124 *Fence gate spring closer.*

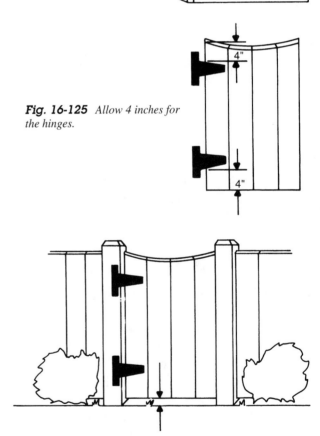

Fig. 16-125 *Allow 4 inches for the hinges.*

Fig. 16-126 *Place a spacer block of wood under the gate while attaching the hinges.*

holes for the screws. See Fig. 16-126. For flat gates to be mounted on the fence post use an ornamental screw hook and strap hinge. See Fig. 16-127. For gates

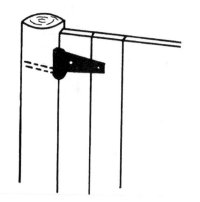

Fig. 16-127 *Attaching another type of hinge.*

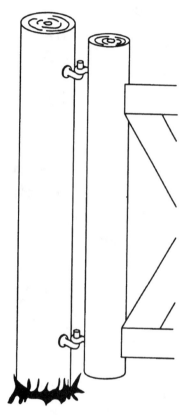

Fig. 16-128 *The hook and eye hinge mounted to a round post.*

framed with round posts use the screw hook and eye hinge shown in Fig. 16-128.

To keep the gate closed you can use a spring that is mounted such as the one shown in Fig. 16-129. Note that the spring is always inclined toward the right. Place the spring as near vertical as conditions permit. Place the adjusting end at the top.

There are at least three ways to latch a gate. The cane bolt drops by gravity. It can be held in raised position and the bolt cannot be moved while mounted. This drops down into a piece of pipe that is driven into the ground. Then it becomes very difficult to

open the gate. This type is usually used when there are two gates and one is kept closed except when large loads are to pass through. See Fig. 16-130. The middle latch is self-latching. It can be locked with a padlock. The ornamental thumb latch is a combination pull and self-latching latch with padlock ability. See Fig. 16-131.

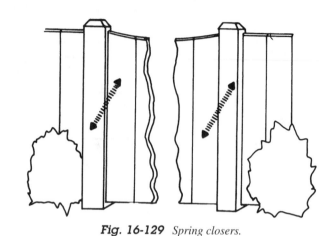

Fig. 16-129 *Spring closers.*

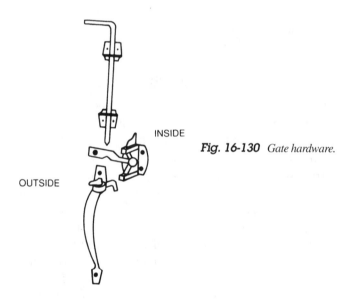

INSIDE

OUTSIDE

Fig. 16-130 *Gate hardware.*

Fig. 16-131 *Completed installation.*

CHAPTER 16
STUDY QUESTIONS

1. Why is firebrick used in fireplaces?

2. How many steps would be cut for stairs with:

   Rise of 8'2"   **Run** of 14"0"   **Riser:** 7 inches

3. For the steps above, what would the tread width be?

4. What is the opening-to-flue ratio?

5. What would be a good flue size for a fireplace with an opening 30 inches high and 48 inches long?

6. What are the advantages of metal fireplaces?

7. What are the advantages of post-and-beam construction?

8. How are posts and beams connected?

9. Would you embed an exterior beam into concrete? Why or why not?

10. Why are laminated beams better than solid wooden beams?

11. How can a stressed-skin panel be used for both roof decking and finished ceilings?

12. What kind of insulation is used over lumber roof decking?

13. What type of roof is not used on steep-pitched roofs?

14. What is a *housed carriage*?

15. What causes stairs to squeak?

16. Can stud posts be fastened at any point on a subfloor?

17. What are two stairway post systems?

18. What is a *rosette*?

19. What is *reverse venting*?

20. How much clearance must you have in Type B gas vents from 3 to 24 inches in diameter?

17
CHAPTER

Maintenance & Remodeling

PLANNING IS AN IMPORTANT PART OF ANY job. The job of remodeling is no exception. It takes a plan to get the job done correctly. The plan must have all the details worked out. This will save money, time, and effort. The work will go smoothly if the bugs have been worked out before the job is started.

You will learn:

• How to diagnose problems
• How to identify needed maintenance jobs
• How to make minor repairs
• How to do some minor remodeling jobs

PLANNING THE JOB

Maintenance means keeping something operating properly. It means taking time to make sure a piece of equipment will operate tomorrow. It means doing certain things to keep a house in good repair. Many types of jobs present themselves when it comes to maintenance. The carpenter is the person most commonly called upon to do maintenance. This may range from the replacement of a lock to complete replacement of a window or door.

Remodeling means just what the word says. It means changing the looks and the function of a house. It might mean you have to put in new kitchen cabinets. The windows might need a different type of opening. Perhaps the floor is old and needs a new covering. The basement might require new paneling or tiled floors.

Working on a house when it is new calls for a carpenter who can saw, measure, and nail things in the proper place. Working on a house after it is built calls for many types of operations. The carpenter may be called upon to do a number of different things related to the trade.

Diagnosing Problems

A person who works in maintenance or remodeling needs to know what the problem actually is. That person must be able to find out what causes a problem. The next step is, of course, to decide what to do to correct the problem.

If you know how a house is built, you should be able to repair it. This means you know what has to be done to properly construct a wall or repair it if damaged. If the roof leaks, you need to know where and how to fix it.

In other words, you need to be able to diagnose problems in any building. Since we are concerned with the residential types here, it is important to know how

things go wrong in a well-built home. In addition, it is important to be able to repair them.

Before you can remodel a house, or add on to it, you need to know how the original was built. You need to know what type of foundation was used. Can it support another story, or can the soil support what you have in mind? Are you prepared for the electrical loads? What about the sewage? How does all this fit into the addition plans? What type of consideration have you given the plumbing, drainage, and other problems?

Identifying Needed Operations

If peeling paint needs to be removed and the wall repainted, can you identify what caused the problem? This will be important later when you choose another paint. Figure 17-1 shows what can happen with paint.

Paint problems by and large are caused by the presence of moisture. The moisture may be in the wood when it is painted. Or it may have seeped in later. Take a look at Fig. 17-1 to see just what causes cracking and alligatoring. Also notice the causes of peeling, flaking, nailhead stains, and blistering.

Normal daily activity in the home of a family of four can put as much as 50 pints of water vapor in the air in one day. Since moisture vapor always seeks an area of lesser pressure, the moisture inside the house tries to become the same as the outside pressure. This equalization process results in moisture passing through walls and ceilings, sheathing, door and window casings, and the roof. The eventual result is paint damage. This occurs as the moisture passes through the exterior paint film.

Sequencing Work to Be Done

In making repairs or doing remodeling, it is necessary to schedule the work properly. It is necessary to make sure things are done in order. For example, it is difficult to paint if there is no wall to paint. This might sound ridiculous, but it is no more so than some other problems associated with getting a job done.

It is hard to nail boards onto a house if there are no nails. Somewhere in the planning you need to make sure nails are available when you need them. If things are not properly planned—in sequence—you could be ready to place a roof on the house and not have the proper size nails to do the job. You wouldn't try to place a carpet on bare joists. There must be a floor or subflooring first. This means there is some sequence that must be followed before you do a job or even get started with it.

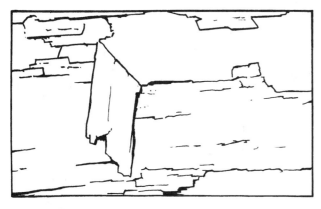

FLAKING

CONDITION:

Siding alternatively swells and shrinks as moisture behind it is absorbed and then dries out. Paint film cracks from swelling and shrinking and flakes away from surface.

CORRECTION:

* All moisture problems must be corrected before surface is re-painted.
* Scrape and sand all peeling paint to bare wood including several inches around damaged areas. Feather edges.
* Apply primer according to label directions.
* Apply topcoat according to label directions.

BLISTERING

CONDITION:

Blistering is actually the first stage of the peeling process. It is caused by moisture attempting to escape through the existing paint film, lifting the paint away from the surface.

CORRECTION:

* All moisture problems must be corrected before repainting.
* Scrape and sand all blistering paint to bare wood several inches around blistered area.
* Feather or smooth the rough edges of the old paint by sanding.
* Apply primer according to label directions.
* Apply topcoat according to label directions.

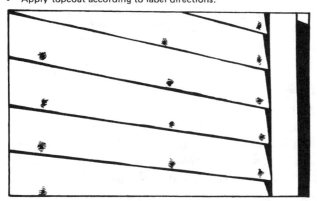

CRACKING AND ALLIGATORING

CONDITION:

Cracking and "alligatoring" are caused by (a) paint that is applied too thick, (b) too many coats of paint, (c) paint applied over a paint coat which is not completely dry, or (d) an improper primer.

CORRECTION:

* Removal of entire checked or alligatored surface may be necessary.
* Scrape and sand down the surface until smooth. Feather edges.
* Apply primer. Follow label directions.
* Apply topcoat. Follow label directions.

PEELING

CONDITION:

Peeling is caused by moisture being pulled through the paint (by the sun's heat), lifting paint away from the surface.

CORRECTION:

* All moisture problems must be corrected before surface is repainted.
* Remove all peeling and flaking paint.
* Scrape and sand all peeling paint to bare wood including several inches around damaged areas. Feather edges.
* Apply primer according to directions on label.
* Apply topcoat according to directions.

NAILHEAD STAINS

CONDITION:

Nailhead stains are caused by moisture rusting old or uncoated nails.

CORRECTION:

* All moisture problems must be corrected before repainting.
* Sand or wire brush stained paint and remove rust down to bright metal of nailhead.
* Countersink nail if necessary.
* Apply primer to nailheads. Allow to dry.
* Caulk nail holes. Allow to dry. Sand smooth.
* Apply primer to surface, following label directions.
* Apply topcoat according to label directions.

Fig. 17-1 *Causes of paint problems on houses.* (Grossman Lumber)

Make a checklist to be sure you have all the materials you need to do a job *before* you get started. If there is the possibility that something won't arrive when needed, try to schedule something else so the operation can go on. Then you can pick up the missing part later when it becomes available. This means that sequencing has to take into consideration the problems of supply and delivery of materials. The person who coordinates this is very important. This person can make the difference between the job being a profitable one or a money loser.

MINOR REPAIRS AND REMODELING

When doors are installed, they should fit properly. That means the closed door should fit tightly against the door stop. Figure 17-2 shows how the door should fit. If it doesn't fit properly, adjust the strike jamb side of the frame in or out. Do this until the door meets the weather-stripping evenly from top to bottom.

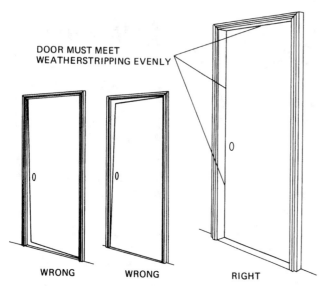

Fig. 17-2 *Right and wrong ways of nailing the strike jamb.* (General Products)

Adjusting Doors

The strike jamb can be shimmed like the hinge jamb. Place one set of shims behind the strike plate mounting location. Renail the jamb so that it fits properly. If it is an interior door, there will be no weather-stripping. Figure 17-3 shows how to shim the hinges to make sure the door fits snugly. In some cases you might have to remove some of the wood on the door where the hinge is attached. See Fig. 17-4. This shows how the wood is removed with a chisel. You must be careful not to remove the back part of the door along the outside of the door.

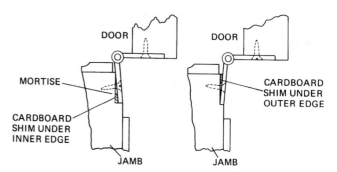

Fig. 17-3 *Hinge adjustment for incorrectly fitting doors. Note the shim placement.*

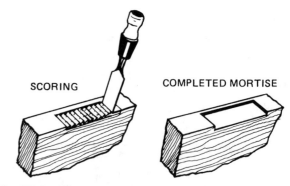

Fig. 17-4 *Removing extra wood from a door. Be careful not to remove too much.* (Grossman Lumber)

If the door binds or does not fit properly, you might have to remove some of the lock edge of the door. Bevel it as shown in Fig. 17-5 so that it fits. An old carpenter's trick is to make sure the thickness of an 8d nail is allowed all around the door. This usually allows enough space for the door to swell some in humid weather and not too much space when winter heat in the house dries out the wood in the door and causes it to shrink slightly.

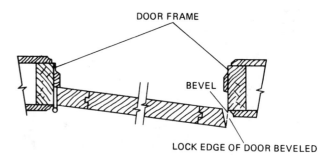

Fig. 17-5 *Beveling the lock stile of a door.* (Grossman Lumber)

Occasionally it is necessary to remove the door from its hinges by removing the hinge pins. Take the door to a vise or workbench so that the edge of the door can be planed down to fit the opening. If the amount of wood to be removed from the door is more than ¼ inch, it will be necessary to trim both edges of the door equally to one-half the width of the wood to be removed. Trim off the wood with a smooth or jack plane as shown in Fig. 17-6.

Fig. 17-6 *Trimming the width of a door.* (Grossman Lumber)

Sometimes it is necessary to remove the doors from their hinges and cut off the bottom because the doors were not cut to fit a room where there is carpeting. If this occurs, remove the door carefully from its hinges. Mark off the amount of wood that must be removed from the bottom of the door. Place a piece of masking tape over the area to be cut. Redraw your line on the masking tape, in the middle of the tape. Tape the other side of the door at the same distance from the bottom. Set the saw to cut the thickness of the door. Cut the door with a power saw or handsaw. Cutting through the tape will hold the finish on the wood. You will not have a door with splinters all along the cut edge. If the door is cut without tape, it may have splinters. This can look very bad if the door has been prefinished.

If the top of the door binds, it should be beveled slightly toward the stop. This can let it open and close more easily.

In some cases the outside door does not meet the threshold properly. It may be necessary to obtain a thicker threshold or a piece of plastic to fit on the bottom of the door as in Fig. 17-7.

Adjusting Locks

Installing a lock can be as easy as following instructions. Each manufacturer furnishes instructions with

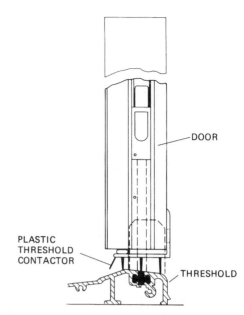

Fig. 17-7 *Adding a piece of plastic to a threshold to make sure the bottom of the door fits snugly. This keeps rain from entering the room as well as keeping out cold air in the winter.* (General Products)

each new lock. However, in some cases you have to replace one that is around the house and no instructions can be found. Take a look at Fig. 17-8 for a step-by-step method of placing a lock into a door. Note that the lockset is typical. It can be replaced with just a screwdriver. There are 18 other brands that can be replaced by this particular lockset.

Various locksets are available to fit the holes that already exist in a door. Strikes come in a variety of shapes too. You should choose one that fits the already-grooved door jamb.

A deadbolt type of lock is called for in some areas. This is where the crime rate is such that a more secure door is needed.

Figure 17-9 shows a lockset that is added to the existing lock in the door. This one is key-operated. This means you must have two keys to enter the door. It is very difficult for a burglar to cause this type of lock bolt to retract.

In some cases it is desired that once the door is closed, it is locked. This can be both an advantage and a disadvantage. If you go out without your key, you are in trouble. This is especially true if no one else is home. However, it is nice for those who are a bit absent-minded and forget to lock the door once it is closed. The door locks automatically once the door is pushed closed.

Figure 17-10 shows how to insert a different type of lock, designed to fit into the existing drilled holes. All you need is a screwdriver to install it. Figure 17-11

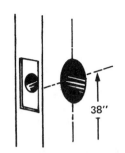

STEP 1: Prepare door; drill for bolt and lock mechanism.

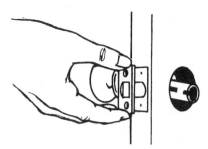

STEP 2: Insert bolt and lock mechanism.

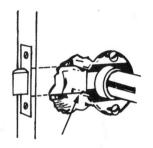

STEP 3: Engage bolt and lock mechanism; fasten face plate.

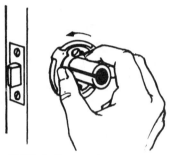

STEP 4: Put on clamp plate; Tighten screws.

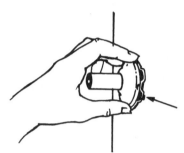

STEP 5: "Snap-On" rose.

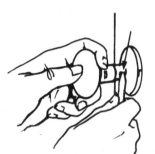

STEP 6: Apply knob on spindle by depressing spring retainer.

STEP 7: Mortise for latch bolt; fasten strike with screws.

Fig. 17-8 Fitting a lockset into a door. (National Lock)

Fig. 17-9 Key-operated auxiliary lock. (Weiser Lock)

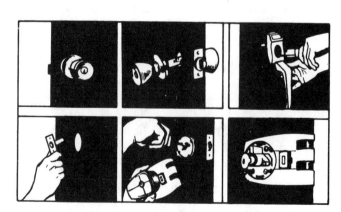

Fig. 17-10 Putting in a lockset. (Weiser Lock)

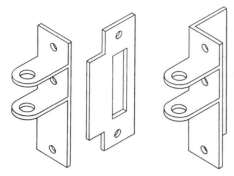

Fig. 17-11 *Strikes for single- and double-cylinder locks.*

shows how secure the lock can be with double cylinders to fit through the holes on the door jamb.

About the only trouble with a lockset is the loosening of the doorknob. It can be tightened with a screwdriver in most cases. If the lockset has two screws near the knob on the inside of the door, simply align the lockset and tighten the screws. If it is another type—with no screws visible—just release the small tab that sticks up inside the lockset; it can be seen through the brass plate around the knob close to the wooden part of the door. This will allow the brass plate or ring to be rotated and removed. Then it is a matter of tightening the screws found inside the lockset. Tighten the screws and replace the cover plate.

The strike plate may work loose or the door may settle slightly. This means the striker will not align with the strike plate. Adjust the plate or strike screws if necessary. In some cases it might be easier to remove the strike plate and file out a small portion to allow the bolt to fit into the hole in the door jamb.

If the doorknob becomes green or off-color, remove it. Polish the brass and recoat the knob with a covering of lacquer and replace. In some cases the brass is plated and will be removed with the buffing. Here you may want to add a favorite shade of good metal enamel to the knob and replace it. This discoloring does occur in bathrooms where the moisture attacks the doorknob lacquer coating.

Installing Drapery Hardware

One often-overlooked area of house building is the drapery hardware. People who buy a new house are faced with the question, "What do I do to make these windows attractive?" Installing window hardware can be a job in itself. Installing conventional adjustable traverse rods is a task that can be easily done by the carpenter, in some cases, or the homeowner. There are specialists who do these things. However, it is usually an extra service that the carpenter gets paid for after the house is finished and turned over to the home owner.

Decorative traverse rods are preferred by those who like period furniture. See Fig. 17-12. Installation of a traverse rod is shown step by step in Figs. 17-13 through 17-17.

Now that you have been studying some of the different types of windows in Chapter 9, have you wondered how the draperies would be fitted?

Figure 17-18 shows some of the regular-duty types of corner and bay window drapery rods. Note the various shapes of bay windows that are made. The rods are made to fit the bay windows.

Figure 17-19 shows more variations of bay windows. These call for some interesting rods. Of course, in some cases the person looks at the cost of the rods

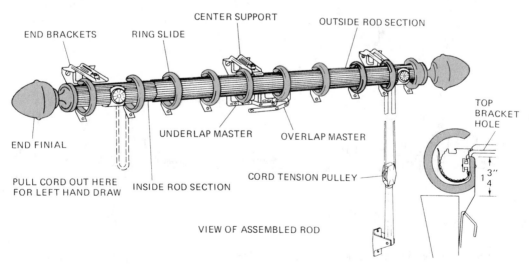

Fig. 17-12 *An adjustable decorative traverse rod.*

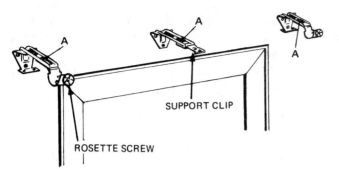

Install end brackets above and 6" to 18" to the sides of casing. When installing on plaster or wallboard and screws do not anchor to studding, plastic anchors or other installation aids may be needed to hold brackets securely. Place center supports (provided with longer size rods only) equidistant between end brackets. Adjust bracket and support projection at screws. "A." *NOTE: Attach rosette screws to end brackets if not already assembled.*

Fig. 17-13 *Install end brackets and center support.*

Place end finials on rod. Finial with smaller diameter shaft fits in "inside" rod section, larger one fits in "outside" section. Place last ring between bracket and end finial.

Extend rod to fit brackets. Place rod in brackets so tongue on bottom of bracket socket fits into hole near end of rod. Tighten rosette screw.

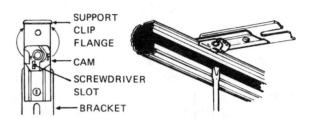

When center supports are used, insert top of rod in support clip flange. If outside rod section, turn cam with screwdriver counterclockwise to lock in rod. If inside rod section, turn cam clockwise.

Fig. 17-14 *Place rod in bracket.*

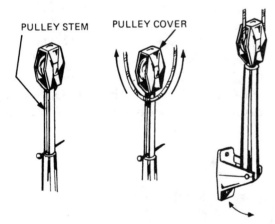

Right-hand draw is standard. If left hand draw is desired, simply pull cord loop down from pulley wheels on left side of rod.

Fasten tension pulley to sill, wall or floor at point where cord loop falls from traverse rod. Lift pulley stem and slip nail through hole in stem. Pull cord up through bottom of pulley cover. Take up excess slack by pulling out knotted cord from back of overlap master slide. Re-knot and cut off extra cord. Remove nail. Pulley head may be rotated to eliminate twisted cords.

Fig. 17-15 *Install cord tension pulleys.*

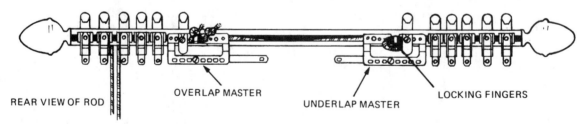

Pull draw cord to move overlap master to end of rod. Holding cords taut, slide underlap master to opposite end of rod. Then fasten cord around locking fingers on underlap master to stop underlap master from slipping. Then center both masters.

Fig. 17-16 *Adjusting master slides.*

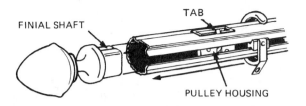

FINIAL SHAFT

TAB

PULLEY HOUSING

Slip off end finial. Remove extra ring slides by pulling back on tab and sliding rings off end of rod. Be sure to leave the last ring between pulley housing and end finial.

To train draperies, "break" fabric between pleats by folding the material toward the window. Fold pleats together and smooth out folds down the entire length of the draperies. Tie draperies with light cord or cloth. Leave tied two or three days before operating.

Fig. 17-17 (A) Remove unused ring slides; (B) Training draperies.

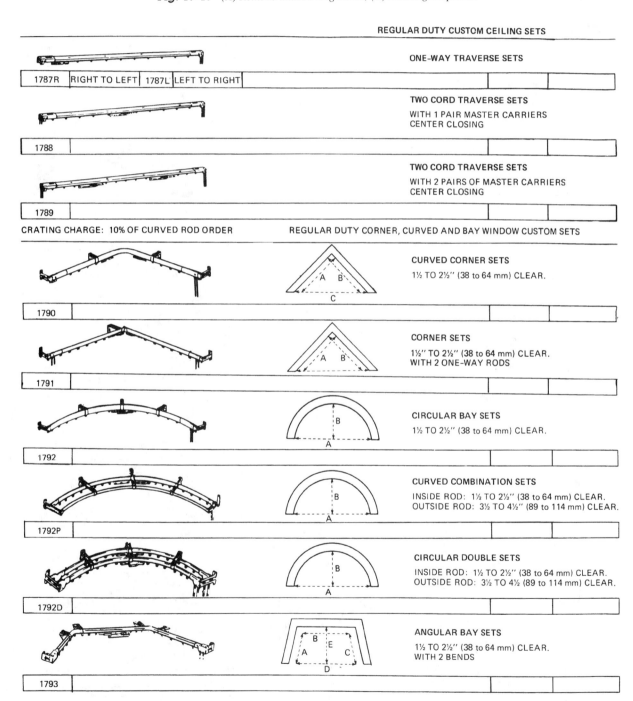

REGULAR DUTY CUSTOM CEILING SETS

ONE-WAY TRAVERSE SETS

| 1787R | RIGHT TO LEFT | 1787L | LEFT TO RIGHT | | | |

TWO CORD TRAVERSE SETS
WITH 1 PAIR MASTER CARRIERS CENTER CLOSING

| 1788 | | | |

TWO CORD TRAVERSE SETS
WITH 2 PAIRS OF MASTER CARRIERS CENTER CLOSING

| 1789 | | | |

CRATING CHARGE: 10% OF CURVED ROD ORDER REGULAR DUTY CORNER, CURVED AND BAY WINDOW CUSTOM SETS

CURVED CORNER SETS
1½ TO 2½" (38 to 64 mm) CLEAR.

| 1790 | | | |

CORNER SETS
1½" TO 2½" (38 to 64 mm) CLEAR.
WITH 2 ONE-WAY RODS

| 1791 | | | |

CIRCULAR BAY SETS
1½ TO 2½" (38 to 64 mm) CLEAR.

| 1792 | | | |

CURVED COMBINATION SETS
INSIDE ROD: 1½ TO 2½" (38 to 64 mm) CLEAR.
OUTSIDE ROD: 3½ TO 4½" (89 to 114 mm) CLEAR.

| 1792P | | | |

CIRCULAR DOUBLE SETS
INSIDE ROD: 1½ TO 2½" (38 to 64 mm) CLEAR.
OUTSIDE ROD: 3½ TO 4½ (89 to 114 mm) CLEAR.

| 1792D | | | |

ANGULAR BAY SETS
1½ TO 2½" (38 to 64 mm) CLEAR.
WITH 2 BENDS

| 1793 | | | |

Fig. 17-18 Regular-duty custom ceiling sets of traverse rods; regular-duty corner, curved, and bay window custom sets. (Kenny)

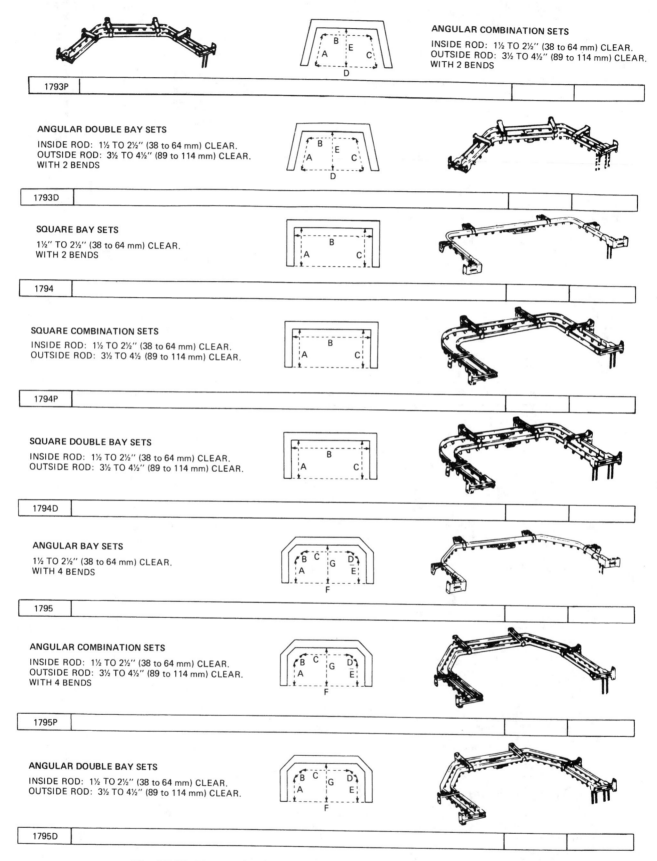

ANGULAR COMBINATION SETS

INSIDE ROD: 1½ TO 2½'' (38 to 64 mm) CLEAR.
OUTSIDE ROD: 3½ TO 4½'' (89 to 114 mm) CLEAR.
WITH 2 BENDS

1793P

ANGULAR DOUBLE BAY SETS

INSIDE ROD: 1½ TO 2½'' (38 to 64 mm) CLEAR.
OUTSIDE ROD: 3½ TO 4½'' (89 to 114 mm) CLEAR.
WITH 2 BENDS

1793D

SQUARE BAY SETS

1½'' TO 2½'' (38 to 64 mm) CLEAR.
WITH 2 BENDS

1794

SQUARE COMBINATION SETS

INSIDE ROD: 1½ TO 2½'' (38 to 64 mm) CLEAR.
OUTSIDE ROD: 3½ TO 4½ (89 to 114 mm) CLEAR.

1794P

SQUARE DOUBLE BAY SETS

INSIDE ROD: 1½ TO 2½'' (38 to 64 mm) CLEAR.
OUTSIDE ROD: 3½ TO 4½'' (89 to 114 mm) CLEAR.

1794D

ANGULAR BAY SETS

1½ TO 2½'' (38 to 64 mm) CLEAR.
WITH 4 BENDS

1795

ANGULAR COMBINATION SETS

INSIDE ROD: 1½ TO 2½'' (38 to 64 mm) CLEAR.
OUTSIDE ROD: 3½ TO 4½'' (89 to 114 mm) CLEAR.
WITH 4 BENDS

1795P

ANGULAR DOUBLE BAY SETS

INSIDE ROD: 1½ TO 2½'' (38 to 64 mm) CLEAR.
OUTSIDE ROD: 3½ TO 4½'' (89 to 114 mm) CLEAR.

1795D

Fig. 17-19 *More regular-duty corner, curved, and bay window custom rods.* (Kenny)

and hardware and decides to put one straight rod across the bay and not follow the shape of the windows. This can be the least expensive, but it negates the effect of having a bay window in the first place.

Repairing Damaged Sheetrock Walls (Drywall)

In drywall construction the first areas to show problems are over joints or fastener heads. Improper application of either the board or the joint treatment may be at fault. Other conditions existing on the job can also be responsible for reducing the quality of the finished gypsum board surface. A discussion of some of these conditions follows on pages 563–565.

Panels improperly fitted

Cause Forcibly wedging an oversize panel into place. This bows the panel and builds in stresses. The stress keeps it from contacting the framing. See Fig. 17-20.

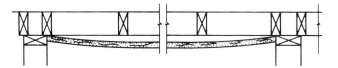

Fig. 17-20 *Forcibly fitted piece of gypsum board.* (U.S. Gypsum)

Result After nailing, a high percentage of the nails on the central studs probably will puncture the paper. This may also cause joint deformation.

Remedy Remove the panel. Cut it to fit properly. Replace it. Fasten from the center of the panel toward the ends and edges. Apply pressure to hold the panel tightly against the framing while driving the fasteners.

Panels with damaged edges

Cause Paper-bound edges have been damaged or abused. This may result in ply separation along the edge. Or it may loosen the paper from the gypsum core. Or, it may fracture or powder the core itself. Damaged edges are more susceptible to ridging after joint treatment.

Remedy Cut back any severely damaged edges to the sound board before application.

Prevention Avoid using board with damaged edges that may easily compress. Damaged edges can take on moisture and swell. Handle sheetrock with care.

Panels loosely fastened

Cause Framing members are uneven because of misalignment or warping. If there is lack of hand

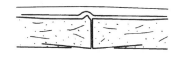

Fig. 17-21 *Damaged edges and their effect on a joint.* (U.S. Gypsum)

pressure on the panel during fastening, loosely fitting panels can result. See Fig. 17-21.

Remedy When panels are fastened with nails, during final blows of the hammer use your hand to apply additional pressure to the panel adjacent to the nail. See Fig. 17-22.

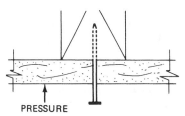

Fig. 17-22 *Apply hand pressure (see arrow) while nailing the board to the stud.* (U.S. Gypsum)

Prevention Correct framing imperfections before applying the panels. Use screws or adhesive method instead of nails.

Surface fractured after application

Cause Heavy blows or other abuse have fractured finished wall surface. If the break is too large to repair with joint compound, do the following.

Remedy In the shape of an equilateral triangle around the damaged area, remove a plug of gypsum. Use a keyhole saw. Slope the edges 45°. Cut a corresponding plug from a sound piece of gypsum. Sand the edges to an exact fit. If necessary, cement an extra slat of gypsum panel to the back of the face layer to serve as a brace. Butter the edges and finish as a butt joint with joint compound.

Framing members out of alignment

Cause Because of misaligned top plate and stud, hammering at points X in Fig. 17-23 as panels are applied on both sides of the partition will probably result in nail heads puncturing the paper or cracking the board. If framing members are more than ¼ inch out of alignment with adjacent members, it is difficult to bring panels into firm contact with all nailing surfaces.

Remedy Remove or drive in problem fasteners and drive new fasteners only into members in solid contact with the board.

Prevention Check the alignment of studs, joists, headers, blocking, and plates before applying panels.

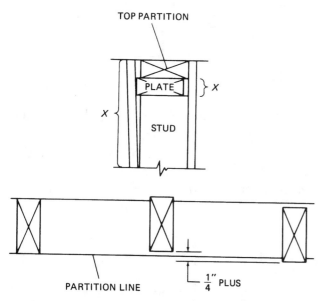

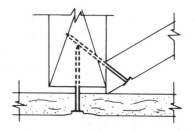

Fig. 17-25 *Nailhead goes through but doesn't pull the board up tightly against the stud. The stud is prevented from meeting the nailing surface by a piece of bridging out of place.* (U.S. Gypsum)

Fig. 17-23 *If framing members are bowed or misaligned, shims are needed if the wall board is to fit properly.* (U.S. Gypsum)

Correct before proceeding. Straighten badly bowed or crowned members. Shim out flush with adjoining surfaces. Use adhesive attachment.

Members twisted

Cause Framing members have not been properly squared with the plates. This gives an angular nailing surface. See Fig. 17-24. When panels are applied, there is a danger of fastener heads puncturing the paper or of reverse twisting of a member as it dries out. This loosens the board and can cause fastener pops.

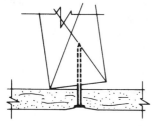

Fig. 17-24 *Framing members improperly squared.* (U.S. Gypsum)

Remedy Allow the moisture content in the framing to stabilize. Remove the problem fasteners. Refasten with carefully driven screws.

Prevention Align all the twisted framing members before you apply the board.

Framing protrusions

Cause Bridging, headers, fire stops, or mechanical lines have been installed improperly. See Fig. 17-25. They may project out past the face of the framing member. This prevents the board or drywall surface

from meeting the nailing surface. The result can be a loose board. Fasteners driven in this area of protrusion will probably puncture the face paper.

Remedy Allow the moisture content in the framing to stabilize. Remove the problem fasteners. Realign the bridging or whatever is out of alignment. Refasten with carefully driven screws.

Puncturing of face paper

Cause Poorly formed nailheads, careless nailing, excessively dry face paper, or a soft core can cause the face paper to puncture. Nailheads that puncture the paper and shatter the core of the panel are shown in Fig. 17-26. They have little grip on the board.

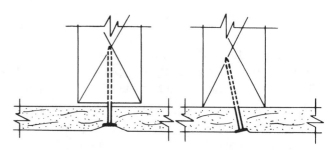

Fig. 17-26 *Puncturing of the face paper.* (U.S. Gypsum)

Remedy Remove the improperly driven fastener. Properly drive a new fastener.

Prevention Correcting faulty framing and driving nails properly produce a tight attachment. There should be a slight uniform dimple. See Fig. 17-27 for the proper installation of the fastener. A nailhead bears on the paper. It holds the panel securely against the framing member. If the face paper becomes dry and brittle, its low moisture content may aggravate the nail cutting. Raise the moisture content of the board and the humidity in the work area.

Nail pops from lumber shrinkage

Cause Improper application, lumber shrinkage, or a combination of the two. With panels held reason-

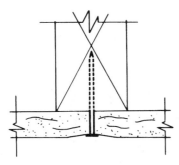

Fig. 17-27 *Proper dimple made with the nailhead into the drywall board.* (U.S. Gypsum)

ably tight against the framing member and with proper-length nails, normally only severe shrinkage of the lumber will cause nail pops. But if panels are nailed loosely, any inward pressure on the panel will push the nailhead through its thin covering pad of compound. Pops resulting from "nail creep" occur when shrinkage of the wood framing exposes nail shanks and consequently loosens the panel. See Fig. 17-28.

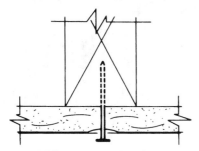

Fig. 17-28 *Popped nailhead.* (U.S. Gypsum)

Remedy Repairs usually are necessary only for pops that protrude 0.005 inch or more from the face of the board. See Fig. 17-28. Smaller protrusions may need to be repaired if they occur in a smooth gloss surface or flat-painted surface under extreme lighting conditions. Those that appear before or during decoration should be repaired immediately. Pops that occur after one month's heating or more are usually caused wholly or partly by wood shrinkage. They should not be repaired until near the end of the heating season. Drive the proper nail or screw about 1½ inches from the popped nail while applying sufficient pressure adjacent to the nailhead to bring the panel in firm contact with the framing. Strike the popped nail lightly to seat it below the surface of the board. Remove the loose compound. Apply finish coats of compound and paint.

These are but a few of the possible problems with gypsum drywall or sheetrock. Others are cracking, surface defects, water damage, and discoloration. All

can be repaired with the proper tools and equipment. A little skill can be developed over a period of time. However, in most instances it is necessary to redecorate a wall or ceiling. This can become a problem of greater proportions. It is best to make sure the job is done correctly the first time. This can be done by looking at some of the suggestions given under *Prevention*.

Installing New Countertops

Countertops are usually covered by a plastic laminate. Formica is usually applied to protect the wooden surface and make it easier to clean. (However, Formica is only one of the trademarks for plastic laminates.) Plastic laminates are easy to work with since they can be cut with either a power saw or a handsaw. Just be sure to cut the plastic laminate face down when using a portable electric saw. This minimizes chipping. See Fig. 17-29.

Fig. 17-29 *Cutting plastic laminate with a saber saw.* (U.S. Gypsum)

If you use a handsaw as in Fig. 17-30, make sure you use a low angle and cut only on the downward stroke.

Before applying the laminate to the surface of the counter, you have to coat both the laminate and the

Fig. 17-30 *Cutting plastic laminate with a handsaw.* (Grossman Lumber)

counter with adhesive. See Fig. 17-31A. This is usually a contact cement. That means both surfaces must be dry to the touch before they are placed in contact with one another. See Fig. 17-31B and C. Make sure the cement is given at least 15 minutes and no more than 1 hour for drying time. See Fig. 17-32. If the cement sinks into the

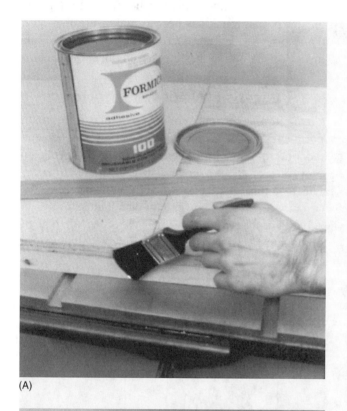

(A)

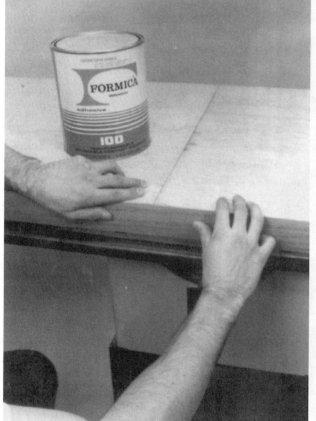

(B)

(C)

Fig. 17-31 *Applying the strip of laminate to the edge. (A) Flip the top right side up. Apply contact cement to laminate and core self-edge* (Formica). *(B) Bond the strip of laminate to the edge using your fingertips to keep the surfaces apart as you go* (Formica). *(C) Apply pressure immediately after bonding by using a hammer and a clean block of hardwood.* *(Formica)*

Fig. 17-32 *Placing the glue-coated plastic laminate on the bottom over a piece of paper.* (Grossman Lumber)

work surface, it may be best to apply a second coat. See Fig. 17-33A and B. Make sure it is dry before you apply the laminate to the work surface of the counter.

In order to get a perfect fit, you might have to place a piece of brown paper on the work surface. Then slide the plastic over the paper. Slip the paper from under the plastic laminate slowly. As the paper is removed, the two adhesive surfaces will contact and stick. Don't let the cemented surfaces touch until they are in the proper location. Figure 17-33C and D shows another method of applying the laminate.

Once the paper has been removed and the surfaces are aligned, apply pressure over the entire area.

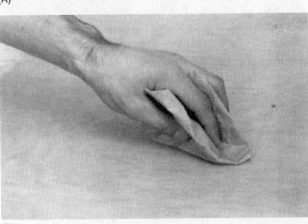

(A)

(B)

(C)

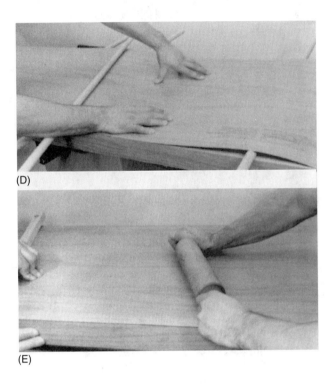

(D)

(E)

Fig. 17-33 *(A) Use a paint roller and work from a tray to apply the contact cement. Lining the tray with aluminum foil speeds cleanup later* (Formica). *(B) Surfaces are ready for bonding when the glue does not adhere to clean kraft paper* (Formica). *(C) Align the laminate over the core. As you remove the ¼-inch sticks, the laminate is bonded to the wood* (Formica). *(D) Dowel rods can also be used to prevent bonding until they are removed* (Formica). *(E) Use an ordinary rolling pin to apply pressure to create a good bond.* (Formica)

Carefully pound a block of wood with a hammer on the laminated top. This bonds the laminate to the surface below. See Fig. 17-33E for another method.

Trimming edges Use a router as shown in Fig. 17-34 to finish up the edges. A saber saw or hand plane can be used if you do not have a router. The edge of the plywood that shows can be covered with plastic laminate, wood molding, or metal strips made for the job. Figure 17-35 shows the cleanup operation after the laminate has been applied.

Fig. 17-34 *Using a router to edge the plastic laminate.* (Formica)

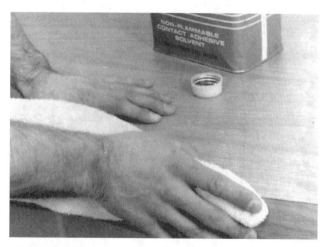

Fig. 17-35 *Wipe the surfaces clean with a rag and thinner. Use the thinner very sparingly, since it can cause delamination at the edges.* (Formica)

Fitting the sink in the countertop is another step in making a finished kitchen. You will need a clamp-type sink rim installation kit. Apply the kit to your sink to be sure it is the correct size. Then, cut the countertop hole about ⅛ inch larger all around than the rim. Make the cutout with a keyhole saw, a router, or a saber saw. See Fig. 17-36.

Fig. 17-36 *Marking around the sink rim for the cutout in the countertop.* (Grossman Lumber)

Start the cutout by drilling a row of holes and leaving a ¹⁄₁₆-inch to accommodate the leg of the rim. Install the sink and rim. Or, you can go on to the backsplash.

The backsplash is a board made from ¾-inch plywood which is nailed to the wall. It is perpendicular to the countertop. Finish off the backsplash with the same plastic laminate before mounting it to the wall permanently. In some cases it may be easier to mount the plywood to the wall with an adhesive, since the drywall is already in place at this step in the construction. Use a router to trim the edges of the backsplash. It should fit flush against the countertop. You may want to finish it off with the end grain of the plywood covered with the same plastic laminate or with metal trim.

Figure 17-37 shows how the sink is installed in the countertop. Also note the repair method used to remove a bubble in the plastic laminate.

Repairing a Leaking Roof

In most areas when a reroofing job is under consideration, a choice must be made between removing the old roofing or permitting it to remain. It is generally not necessary to remove old wood shingles, old asphalt shingles, or old roll roofing before applying a new asphalt roof—that is, if a competent inspection indicates that:

1. The existing deck framing is strong enough to support the weight of workers and the additional new roofing. This means it should also be able to support the usual snow and wind loads.

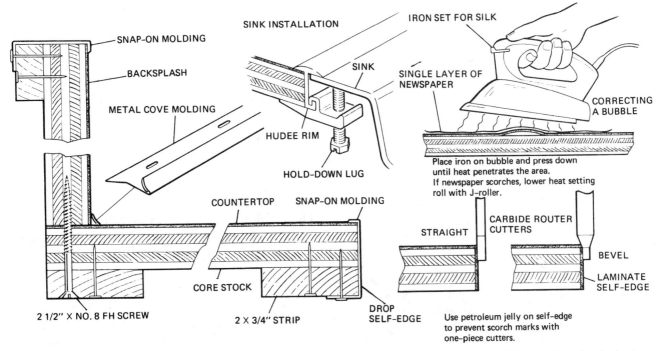

Fig. 17-37 *Details of installing a sink in a laminated countertop. Note the method used to remove a bubble caught under the laminate.*
(Formica)

2. The existing deck is sound and will provide good anchorage for the nails used in applying the new roofing.

Old roofing to stay in place If the inspection indicates that the old wood shingles may remain, the surface of the roof should be carefully prepared to receive the new roofing.

This may be done as follows:

1. Remove all loose or protruding nails, and renail the shingles in a new location.

2. Nail down all loose shingles.

3. Split all badly curled or warped old shingles and nail down the segments.

4. Replace missing shingles with new ones.

5. When shingles and trim at the eaves and rakes are badly weathered, and when the work is being done in a location subject to the impact of unusually high winds, the shingles at the eaves and rakes should be cut back far enough to allow for the application, at these points, of 4- to 6-inch wood strips, nominally 1 inch thick. Nail the strips firmly in place, with their outside edges projecting beyond the edges of the deck the same distance as did the wood shingles. See Fig. 17-38.

6. To provide a smooth deck to receive the asphalt roofing, it is recommended that beveled wood

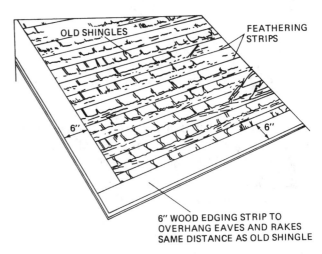

Fig. 17-38 *Treatment of rakes and eaves when reroofing in windy location.* (Bird and Son)

"feathering" strips be used along the butts of each course of old shingles.

Old roofing (asphalt) shingles to remain in place If the old asphalt shingles are to remain in place, nail down or cut away all loose, curled, or lifted shingles. Remove all loose and protruding nails. Remove all badly worn edging strips and replace with new. Just before applying the new roofing, sweep the surface clean of all loose debris.

Square-butt strip shingles to be recovered with self-sealing square-butt strip shingles The following application procedure is suggested to minimize uneven appearance of the new roof. All dimensions are given assuming that the existing roof has been installed with the customary 5-inch shingle exposure.

Starter Course Cut off the tabs of the new shingle using the head portion equal in width to the exposure of the old shingles. This is normally 5 inches for the starter shingle. See Fig. 17-39.

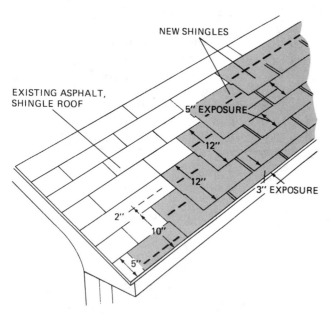

Fig. 17-39 *Exposure of new shingles when reroofing.* (Bird and Son)

First Course Cut 2 inches from the top edge of a full-width new shingle. Align this cut edge with the butt edge of the old shingle.

Second Course Use a full-width shingle. Align the top edge with the butt edge of the old shingle in the next course. Although this will reduce the exposure of the first course, the appearance should not be objectionable, as this area is usually concealed by the gutter.

Third Course and All Others Use full-width shingles. Align the top edges with the butts of the old shingles. Exposure will be automatic and will coincide with that of the old roof.

Old lock-down or staple-down shingles These shingles should be removed before reroofing. They have an uneven surface, and the new shingles will tend to conform to it. If a smoother-surface base is desired, the deck should be prepared as described in *Old Roofing to Be Removed*, below.

New shingles over old roll roofing When new asphalt roofing is to be laid over old roll roofing without

removing the latter, proceed as follows to prepare the deck:

1. Slit all buckles, and nail segments down smoothly.
2. Remove all loose and protruding nails.
3. If some of the old roofing has been torn away, leaving pitchy knots and excessively resinous areas exposed, cover these defects with sheet metal patches made from galvanized iron, painted tin, zinc, or copper having a thickness approximately equal to 26 gauge.

Old roofing to be removed When the framing supporting the existing deck is not strong enough to support the additional weight of roofing and workers during application, or when the decking material is so far gone that it will not furnish adequate anchorage for the new roofing nails, the old roofing, regardless of type, must be removed before new roofing is applied. The deck should then be prepared for the new roofing as follows:

1. Repair the existing roof framing where required to level and true it up and to provide adequate strength.
2. Remove all rotted or warped old sheathing and replace it with new sheathing of the same kind.
3. Fill in all spaces between boards with securely nailed wood strips of the same thickness as the old deck. Or, move existing sheathing together and sheath the remainder of the deck.
4. Pull out all protruding nails and renail sheathing firmly at new nail locations.
5. Cover all large cracks, slivers, knot holes, loose knots, pitchy knots, and excessively resinous areas with sheet metal securely nailed to the sheathing.
6. Just before applying the new roofing, sweep the deck thoroughly to clean off all loose debris.

Old built-up roofs

If the deck has adequate support for nails When the pitch of the deck is below 4 inches per foot but not less than 2 inches per foot, and if the deck material is sound and can be expected to provide good nail-holding power, any old slag, gravel, or other coarse surfacing materials should first be removed. This should leave the surface of the underlying felts smooth and clean. Apply the new asphalt shingles directly over the felts according to the manufacturer's recommendations for low-slope application.

If the deck material is defective and cannot provide adequate security All old material down to the upper surface of the deck should be removed. The existing deck material should be repaired. Make it secure to the underlying supporting members. Sweep it clean before applying the new roofing.

Patching a roof In some cases it is not necessary to replace the entire roof to plug a leak. In most instances you can visually locate the place where the leak is occurring. There are a number of roof cements that can be used to plug the hole or cement the shingles down. In some cases it is merely a case of backed-up water. To keep this from happening, heating cables may have to be placed on the roof to melt the ice. See Chapter 8 for more details.

Replacing Guttering

Guttering comes in both 4- and 5-inch widths. The newer aluminum gutter is usually available in 5-inch widths. The aluminum has a white baked-on finish. It does not require soldering, painting, or priming. It is easily handled by one person. The light weight can have some disadvantages. Its ability to hold ice or icicles in colder climates is limited. It can be damaged by the weight of ice buildup. However, its advantages usually outweigh its disadvantages. Not only is it lighter, prefinished on its exterior, and less expensive, but it is also very easy to put up.

The aluminum type of guttering is used primarily as a replacement gutter. The old galvanized type requires a primer before it is painted. In most cases, it does not hold paint even when primed. It becomes an unsightly mess easily with extremes in weather. It is necessary to solder the galvanized guttering. These soldered points do not hold if the weather is such that it heats up and cools down quickly. The solder joints are subject to breaking or developing hairline cracks which leak. These leaks form large icicles in northern climates and put an excessive load on the nails that support the gutter. Once the nails have been worked loose by the expansion and contraction, it is only a matter of time before the guttering begins to sag. If water gets behind the gutter and against the fascia board, it can cause the board to rot. This further weakens the drainage system. The fascia board is used to support the whole system.

Figure 17-40A shows that the drainage system should be lowered at one end so that the water will run down the gutter to the downspout. About ¼ inch for every 10 feet is sufficient for proper drainage. Figure 17-40B shows a hidden bracket hanger. It is used to support the gutter.

In Fig. 17-40C you can see the method used to mount this concealed bracket. The rest of the illustrations in Fig. 17-40 show how the system is put together to drain water from the roof completely. Each of these pieces has a name. The names are shown in Fig. 17-41. If you are planning a system, its cost is easily estimated by using the form provided in Fig. 17-41. Note the tools needed for installing the system. The caulking gun comes in handy to caulk places that were left unprotected by removal of the previous system. The holes for the support of the previous system should be caulked. Newer types of cements can be used to make sure each connection of the inside and outside corners and the end caps are watertight.

Once the water leaves the downspout, it is spread onto the lawn or it is conducted through plastic pipes to the storm sewer at the curb. Some locations in the country will not allow the water to be emptied onto the lawn. The drains to the storm sewer are placed in operation when the house is built. This helps control the seepage of water back into the basement once it has been pumped out with a sump pump. The sump pump also empties into the drainage system and dumps water into the storm sewer in the street.

Replacing a Floor

Many types of floor coverings are available. You may want to check with a local dealer before deciding just which type of flooring you want. There are continuous rolls of linoleum or there are 9-×-9-inch squares of tile. There are 12-×-12-inch squares of carpet that can be placed down with their own adhesive. The type of flooring chosen determines the type of installation method to use.

Staple-down floor This type of floor is rather new. It can be placed over an old floor. A staple gun is used to fasten down the edges.

Figures 17-42 through 17-44 show the procedure required to install this type of floor. This type of flooring is so flexible it can be folded and placed in the trunk of even the smallest car. Unroll the flooring in the room and move it into position (Fig. 17-42). In the 12-foot width (it also comes in 6-foot width) it covers most rooms without a seam. In the illustration it is being laid over an existing vinyl floor. It can also be installed over plywood, particle board, concrete, and most other subfloor materials.

To cut away excess material (see Fig. 17-43) use a metal straightedge or carpenter's square to guide the utility knife. Install the flooring with a staple every 3

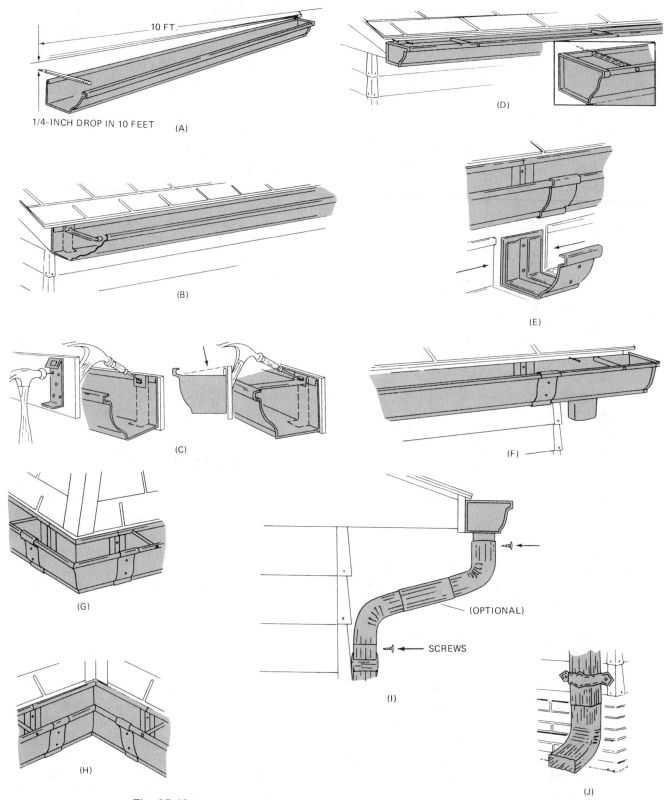

Fig. 17-40 *Replacing an existing gutter and downspout system.* (Sears, Roebuck and Co.)

inches close to the trim. (See Fig. 17-44.) This way the quarter-round trim can be installed over the staples and they will be out of sight. Cement is used in places where a staple can't penetrate. For a concrete floor, use

a special adhesive around the edges. Any dealer who sells this flooring has the adhesive.

The finished job looks professional even when it is done by a do-it-yourselfer. This flooring has a built-in

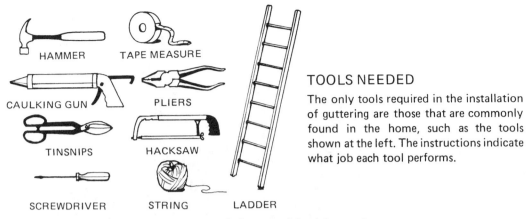

TOOLS NEEDED

The only tools required in the installation of guttering are those that are commonly found in the home, such as the tools shown at the left. The instructions indicate what job each tool performs.

HAMMER TAPE MEASURE
CAULKING GUN PLIERS
TINSNIPS HACKSAW
SCREWDRIVER STRING LADDER

Fig. 17-41 *How to select the right fittings and the right quantity.* (Sears, Roebuck and Co.)

Fig. 17-42 *Unroll the flooring and allow it to sit face up overnight before installing it.* (Armstrong Cork)

Fig. 17-43 *Use a utility knife and carpenter's square to make sure the cut is straight.* (Armstrong Cork)

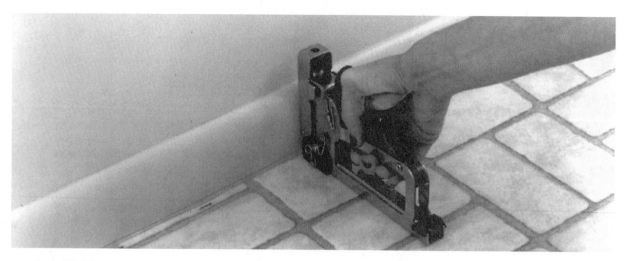

Fig. 17-44 *Place a staple every 3 inches along the kickboard. These staples will be covered by molding.* (Armstrong Cork)

memory. When it was rolled face-side-out at the factory for shipment, the outer circumference of the roll was stretched. After it is installed in the home, the floor gently contracts, trying to return to the dimensions it had before it was rolled up. This causes any slack or wrinkles that might have been left in the flooring to gradually be taken up by the memory action.

No-wax floor One of the first rooms that comes in for improvement is the kitchen. Remodeling may be a major project, but renewing a floor is fairly simple.

No-wax flooring comes in the standard widths—6 and 12 feet. In some instances it doesn't need to be tacked or glued down. However, in most cases it should be cemented down. It fits directly over most floors, provided they are clean, smooth, and well bonded. Make sure any holes in the existing flooring are filled and smoothed over. In the case of concrete

basement floors, just vacuum them or wash them thoroughly and allow them to dry.

The tools needed to install a new floor are a carpenter's square, chalk line, adhesive, trowel, and knife or scissors.

The key to a perfect fit is taking accurate room measurements. See Fig. 17-45. Diagram the floor plan on a chart, noting the positions of cabinets, closets, and doorways.

After transferring the measurements from the chart to the flooring material, cut along the chalk lines, using a sharp knife and a straightedge. Transfer the measurements and cut the material in a room where the material can lie flat. Cardboard under the cut lines will protect the knife blade.

Return the material to the room where it is to be installed. Put it in place. Roll back one-half of the

Fig. 17-45 *Room measurements are essential to a good job.* (Armstrong Cork)

Fig. 17-46 *Spread the adhesive on half of the floor and place the flooring material down. Then do the other half.* (Armstrong Cork)

material. Spread the adhesive. Unroll the material onto the adhesive while it is still wet. Repeat the same steps with the rest of the material to finish the job. See Fig. 17-46.

The finished job makes any kitchen look new. All that is needed in the way of maintenance is a sponge mop with detergent.

Floor tiles Three of the most popular kinds of self-adhering tiles are vinyl-asbestos, no-wax, and vinyl tiles that contain no asbestos filler.

The benefit of no-wax tiles is obvious from the name. They have a tough, shiny, no-wax wear surface. You pay a premium price for no-wax tiles.

Vinyl tiles are not no-wax, but they are easier to clean than vinyl-asbestos tiles. Their maintenance benefits are the result of a nonporous vinyl wear surface that resists dirt, grease, and stains better than vinyl-asbestos.

Any old tile or linoleum floor can be covered with self-adhering tiles provided the old material is smooth and well bonded to the subfloor. Just make sure the surface is clean and old wax is removed.

Putting down the tiles Square off the room with a chalk line. Open the carton, and peel the protective paper off the back of the tile. See Fig. 17-47.

Do one section of the room at a time. The tiles are simply maneuvered into position and pressed into place. See Fig. 17-48.

Border tiles for the edges of the room can be easily cut to size with a pair of ordinary household shears. See Fig. 17-49.

It doesn't take long for the room to take shape. There is no smell from the adhesive. The floor can be

Fig. 17-47 *Peel the paper off the back of a floor tile and place the tile in a predetermined spot.* (Armstrong Cork)

used as soon as it is finished. This type of floor replacement or repair is commonplace today.

Paneling a Room

The room to be paneled may have cracked walls. This means you'll need furring strips to cover the old walls and provide good nailing and shimming for a smooth wall. See Fig. 17-50. Apply furring strips (1 × 1s or 1 × 3s) vertically at 16-inch intervals for full-size 4-×-8-foot panels. Apply the furring strips horizontally at 16-inch intervals for random-width paneling. Start at the end of the wall farthest from the main entrance to the room. The first panel should be plumbed from the

Fig. 17-48 *Do one section of the room at a time. The tiles can be maneuvered into position and pressed into place. (Armstrong Cork)*

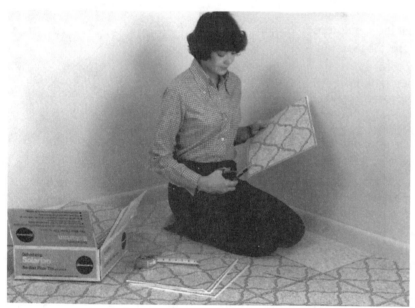

Fig. 17-49 *Border tiles for the edges of the room can be easily cut to size with a pair of household shears. (Armstrong Cork)*

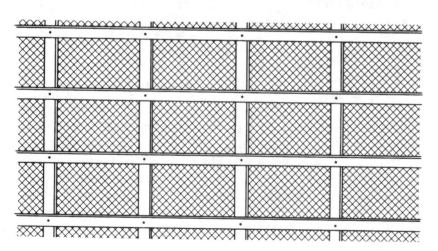

Fig. 17-50 *Applying vertical and horizontal furring strips on an existing wall. (Valu)*

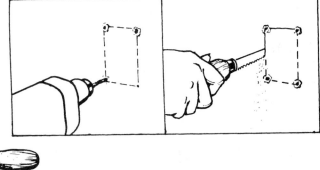

Fig. 17-51 *Cutting holes in the paneling for electrical switches and outlets.* (Valu)

WALLBOARD SAW

corner by striking a line 48 inches out. Trim the panel in the corner (a rasp works well) so that the panel aligns on the plumb line. Turn the corner, and butt the next panel to the first panel. Plumb in the same manner as the first panel.

To make holes in the paneling for switch boxes and outlets, trace the box's outline in the desired place on the panel. Then drill holes at the four corners of the area marked. Next, cut between the holes with a keyhole saw. Then rasp the edges smooth. See Fig. 17-51.

Before you apply the adhesive, make sure the panel is going to fit. If the panel is going to be stuck right onto the existing wall, apply the adhesive at 16-inch intervals, horizontally and vertically. See Fig. 17-52A.

Spacing   Avoid a tight fit. Above grade, leave a space approximately the thickness of a matchbook cover at the sides and a ⅛-inch space at top and bottom. Below grade, allow ¼ inch top and bottom. Allow not less than ¹⁄₁₆ inch space (the thickness of a dime) between panels in high humidity areas. See Fig. 17-52B.

Vapor barrier   A vapor barrier is needed if the panels are installed over masonry walls. It doesn't matter if the wall is above or below grade. See Fig. 17-52C. If the

(A)

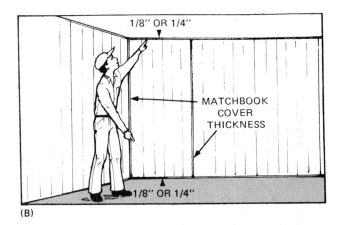

(B)

(C)

Fig. 17-52   *(A) Make sure the panel fits before you apply adhesive* (Valu). *(B) Allow space at top and bottom and at each side for the panel to expand* (Valu). *(C) Apply a vapor barrier to the walls before you place the furring strips onto the wall.* (Abitibi)

plaster is wet or the masonry is new, wait until it is thoroughly dry. Then condition the panels to the room.

Adhesives In Fig. 17-53 a caulking gun is being used to apply the adhesive to the furring strips. In some cases it is best to apply the adhesive to the back of the panel. Follow the directions on the tube. Each manufacturer has different instructions.

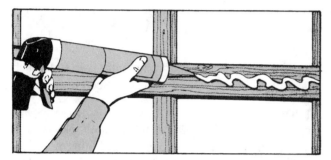

Fig. 17-53 *Use a caulking gun to apply the adhesive.* (Valu)

Hammer a small nail through the top of the panel. It should act as a hinge. Press the panel against the wall or frame to apply glue to both sides. See Fig. 17-54A.

Pull out the bottom of the panel and keep it about 8 to 10 inches away from the wall or framework. Use a wood block. See Fig. 17-54B. Let the adhesive become tacky. Leave it for about 8 to 10 minutes.

> CAUTION: Methods of application for some types of panel adhesives may differ. Always check the instructions on the tube.

Let the panel fall back into position. Make sure that it is correctly aligned. The first panel should butt one edge against the adjacent wall. The panel should be completely plumb. Trim the inner edge as needed. This is so that the outer edge falls on a stud.

Scribing Use a compass to mark panels so that they fit perfectly. It can be used to mark the variations in a butting surface. See Fig. 17-54C.

Nailing Use finishing nails (3d) or brads (1¼ inch) or annular hardboard nails (l-inch). Begin at the edge. Work toward the opposite side. Never nail opposite ends first, then the middle. With a hammer and a cloth-covered block, hammer gently to spread the adhesive evenly. See Fig. 17-54D. If you are using adhesive alone to hold the panels in place, don't remove the nails until the adhesive is thoroughly dry. Then, after they're out, fill the nail holes with a matching putty stick.

Moldings can be glued into place or nailed.

Fig. 17-54 *(A) Hammer a nail through the top of the panel. This will act as a hinge so that the panel can be pressed against the wall to apply the glue to both surafces* (Valu). *(B) Pull out the bottom of the panel and keep it about 8 to 10 inches away from the wall with a wood block until the adhesive becomes tacky* (Valu) . *(C) Use a compass to mark panels so that they fit perfectly against irregular surfaces* (Abitibi). *(D) Use a hammer and block of wood covered with a cloth to spread the adhesive evenly.* (Valu)

Installing a Ceiling

There are a number of methods used to install a new ceiling in a basement or a recreation room. In fact, some very interesting tiles are available for living rooms as well.

Replacing an old ceiling The first thing to do in replacing an old, cracked ceiling is to lay out the room. See Fig. 17-55. It should be laid out accurately to scale. Use ½" = 1'0" as a scale. Do the layout on graph paper.

Then, on tracing paper, draw ½-inch squares representing the 12-inch ceiling tile. Lay the tracing paper over the ceiling plan. Adjust the paper until the borders are even. Border pieces should never be less than half a tile wide. Use cove molding at the walls to cover the trimmed edges of the tile. See Fig. 17-56. Tile can be stapled or applied with mastic. This is especially true of fiber tile. It has flanges that will hold staples. See Fig. 17-57. You can staple into wallboard or into furring strips nailed to joists. When applying tile to furring strips, follow your original layout but start in one corner and work toward the opposite corner.

If you apply tile with mastic, snap a chalk line as shown in Fig. 17-58. This will find the exact center of

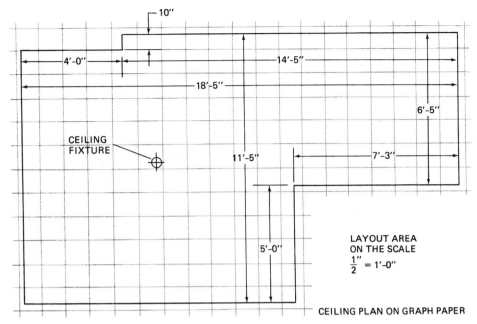

Fig. 17-55 *Lay out your room accurately to scale on graph paper.* (Grossman Lumber)

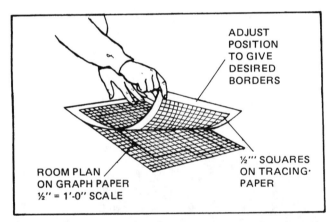

Fig. 17-56 *Draw ½-inch squares on tracing paper. Lay the tracing paper over the ceiling plan. Adjust the paper until the borders are even. Border pieces should never be less than half a tile wide.* (Grossman Lumber)

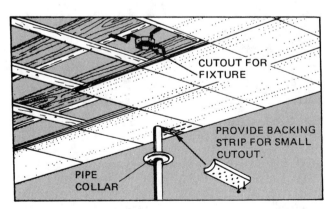

Fig. 17-57 *Fiber tile is stapled to furring strips applied to the joists.* (Grossman Lumber)

Fig. 17-58 *If you are going to glue the tiles to the ceiling, snap a chalk line to find the exact center of the ceiling.* (Grossman Lumber)

your ceiling. In applying the mastic to the ceiling tile, put a golf-ball-sized blob of mastic on each corner of the tile and one in the middle. See Fig. 17-59.

Apply the first tile where the lines cross. Then, work toward the edges. Set the tile just out of position, then slide it into place, pressing firmly to ensure a good bond and a level ceiling. See Fig. 17-60. Precut holes for lighting fixtures and pipes. Use a sharp utility knife. Always be sure to make the cutouts with the tile face up. See Fig. 17-61.

The drop ceiling If an old ceiling is too far damaged to be repaired inexpensively, it might be best to install a

Fig. 17-59 *Place a blob of mastic at each corner and in the middle of the tile before sliding it into place on the ceiling surface.* (Grossman Lumber)

Fig. 17-60 *Apply the first tile where the lines cross on the ceiling. Work toward the edges.* (Grossman Lumber)

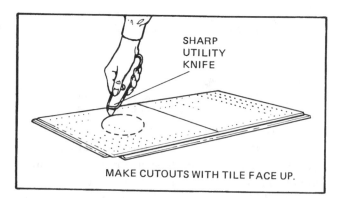

Fig. 17-61 *Precut holes for lighting fixtures and pipes. Use a sharp utility knife. Always be sure to cut the tiles with the face up.* (Grossman Lumber)

drop ceiling. This means the ceiling will be completely new and will not rely upon the old ceiling for support. This way the old ceiling does not have to be repaired.

Figure 17-62 shows the first operation of a drop ceiling. Nail the molding to all four walls at the desired ceiling height. Either metal or wood molding may be used. Figure 17-63 shows how a basement dropped ceiling is begun the same way, by nailing molding at the desired height.

Fig. 17-62 *For a new suspended ceiling, nail molding to all four walls.* (Armstrong Cork)

Fig. 17-63 *A suspended ceiling in a basement is begun the same way. Nail molding on all four walls.* (Armstrong Cork)

The next step is to install the main runners on hanger wires as shown in Fig. 17-64. The first runner is always located 26 inches out from the sidewall. The remaining units are placed 48 inches O.C., perpendicular to the direction of the joists. Unlike conventional suspended ceilings, this type requires no complicated

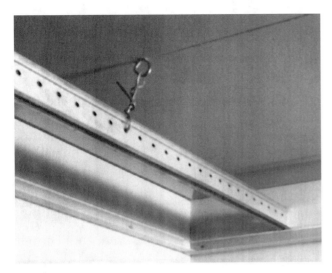

Fig. 17-64 *Main runners are installed on hanger wires.* (Armstrong Cork)

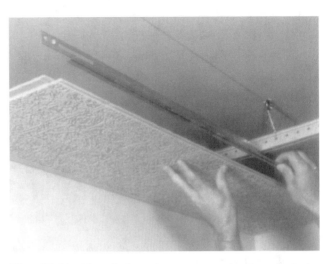

Fig. 17-66 *After all the main runners are in place, begin installing ceiling tile in a corner of the room. Note how the T slides into the tile and disappears.* (Armstrong Cork)

measuring or room layout. Figure 17-65 shows how the main runner is installed by fastening hanger wires at 4-foot intervals—the conventional method for a basement dropped ceiling.

Fig. 17-65 *In the basement installation, install the main runners by fastening them to hanger wires at 4-foot intervals.* (Armstrong Cork)

Fig. 17-67 *When the main runners are in place, install cross Ts between them. The T's have tabs that engage slots in the main runner to lock firmly in place. This type of suspended ceiling will have the metal showing when the job is finished.* (Armstrong Cork)

After all main runners are in place, begin installing ceiling tile in a corner of the room. Simply lay the first 4 feet of tile on the molding, snap a 4-foot cross T onto the main runner, and slide the T into a special concealed slot on the leading edge of the tile. See Fig. 17-66. This will provide a ceiling without the metal supports showing.

In the conventional method, as shown in the basement installation in Fig. 17-67, cross T's are installed between main runners. The T's have tabs that engage slots in the main runner to lock firmly in place. After

this is done, the tiles are slid into place from above the metal framework and allowed to drop into the squares provided for them.

In Fig. 17-68 tile setting is continuing. The tiles and cross T's are inserted as needed. Note how all metal suspension members are hidden from view in the finished portion of the ceiling.

The result of this type of ceiling is an uninterrupted surface. There is no beveled edge to produce a line across the ceiling. All supporting ceiling metal is concealed.

Fig. 17-68 *Continue across the room in this manner. Insert the tiles and cross Ts.* (Armstrong Cork)

One of the advantages of dropping a ceiling is the conservation of energy. Heat rises to the ceiling. If the ceiling is high, the heat is lost to the room. The lower the ceiling, the less the volume to be heated and the less energy needed to heat the room.

Replacing an Outside Basement Door

Many houses have basements that can be made into playrooms or activity areas. The workshop that is needed but can't be put anywhere else will probably wind up in the basement. Basements can be very difficult to get out of if a fire starts around the furnace or hot-water-heater area. One of the safety measures that can be taken is placing a doorway directly to the outside so you don't have to try to get up a flight of stairs and then through the house to escape a fire.

Figure 17-69 shows how an outside stairway can be helpful for any basement. This type of entrance or exit from the basement can be installed in any house. It takes some work, but it can be done.

Figure 17-70 shows how digging a hole and breaking through the foundation in stages will permit using tools to get the job done. A basement wall of concrete blocks is easier to break through than a poured wall.

Start by building the areaway. This is done by laying a 12- to 16-inch concrete footing. This footing forms a level base for the first course of concrete block. See Fig. 17-71. Allow the footing to set for two to three days before laying blocks for the walls.

The areaway for a size C door is shown in Fig. 17-72. Figure 17-72A shows the starting course. Figure 17-72B shows how the next course of blocks is laid. The top course should come slightly above ground level. See Fig. 17-73. It should be about 3 inches from

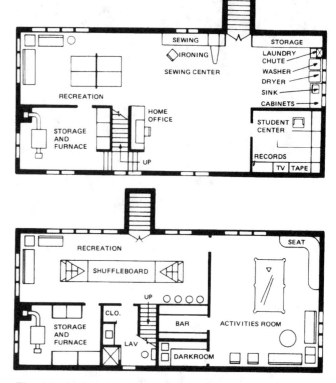

Fig. 17-69 *Note the location of the outside basement door or entry on these drawings.* (Bilco)

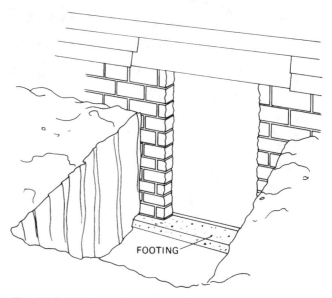

Fig. 17-70 *Excavate the area around the proposed entrance.* (Bilco)

the required areaway height as given in the construction guide provided with the door.

Build up the areaway to the right height in the manner shown. You may want to waterproof the outside of the new foundation. Use the material recommended by your local lumber dealer or mason yard.

Now build a form for capping the wall. See Fig. 17-74. Fill the cores of the top blocks halfway up with

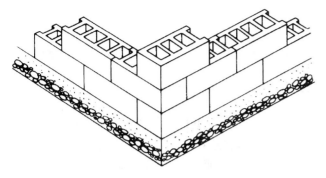

Fig. 17-71 *Start the blocks on top of the footings.* (Bilco)

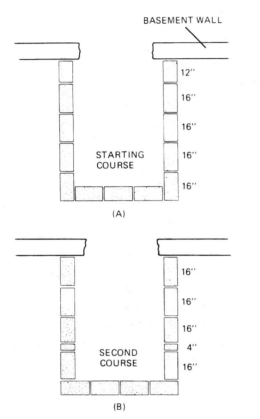

Fig. 17-72 *(A) The starting course, looking down into the hole* (Bilco). *(B) The second course, on top of the first, looks like this.* (Bilco)

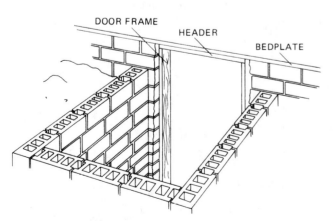

Fig. 17-73 *The finished block work ready for the cap of concrete.* (Bilco)

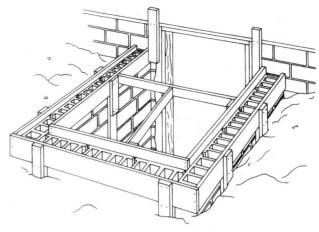

Fig. 17-74 *Forms for the concrete cap are built around the concrete block.* (Bilco)

crushed balls of newspaper or insulation. The cap to be poured on this course will bring the areaway up to the required height.

When the cap is an inch or so below the desired height, set the door back in position with the header flange between the siding and the sheathing underneath. Make sure the frame is square. Insert the mounting screws with the spring-steel nuts in the side pieces and the sill. Embed them in the wet concrete to hold the screws tightly. See Fig. 17-75. Continue pouring the cap. Bring the concrete flush with the bottom of the sill and side-piece flanges. Do not bring the capping below the bottom of the door. The door should rest on top of the foundation. With a little extra work, the cap outside the door can be chamfered downward as shown in Fig. 17-75. This ensures good drainage. Trowel the concrete smooth and level.

Fig. 17-75 *Anchor bolts are embedded into the concrete cap to allow attachment of the metal doors.* (Bilco)

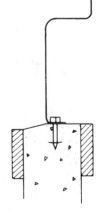

Install a prehung door in the wall of the basement. These come in standard sizes. The door should be selected to fit the hole made in the wall. Use the widest standard unit that fits the entryway you have built.

Figure 17-76 shows how the steps are installed in the opening. The stringers for the steps are attached to the walls of the opening. See Fig. 17-77.

The outside doors will resemble those shown in Fig. 17-78. There are a number of designs available for almost any use. Lumber for the steps is 2 × 10s cut to length and slipped into the steel stair stringers.

Seal around the door and foundation with caulk. Seal around the door in the basement and the wall. Allow the stairwell to *air out* during good weather by keeping the outside doors open. This will allow the moisture from the masonry to escape. After it has dried out, the whole unit will be dry.

CONVERTING EXISTING SPACES

There is never enough room in any house. All it takes is a few days after you unpack all your belongings to find out that there isn't enough room for everything.

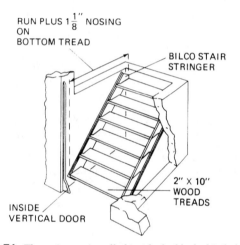

Fig. 17-76 *The stairs are installed inside the blocked-in hole.* (Bilco)

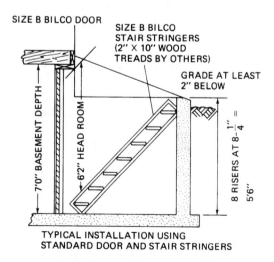

The next thing new homeowners do is look around at the existing building to see what space can be converted to other uses. It is usually too expensive to add on immediately, but it is possible to remodel the kitchen, porch, or garage.

Adding a Bathroom

As the family grows, the need for more bathrooms becomes very apparent. First you look around for a place to put the bathroom. Then you locate the plumbing and check to see what kind of a job it will be to hook up the new bathroom to the cold and hot water and to the drains. How much effort will be needed to hook up pipes and run drains? How much electrical work will be needed? These questions will have to be answered as you plan for the additional bathroom.

Start by getting the room measurements. Then make a plan of where you would place the various necessary items. Figure 17-79 shows some possibilities. Your plan can be as simple as a lavatory and water closet, or you can expand with a shower, a whirlpool bath, and a sauna. The amount of money available will usually determine the choice of fixtures.

Look around at books and magazines as well as literature of manufacturers of bathroom fixtures. Get some ideas as to how you would rearrange your own bathroom or make a new one.

If you have a larger room to remodel and turn into a bathroom, you might consider what was done in Fig. 17-80. This is a Japanese bathroom dedicated to the art of bathing. Note how the tub is fitted with a shower to allow you to soap and rinse on the bathing platform before soaking, as the Japanese do.

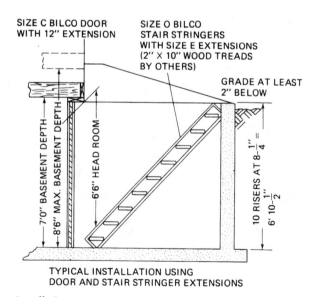

Fig. 17-77 *Typical door installations.* (Bilco)

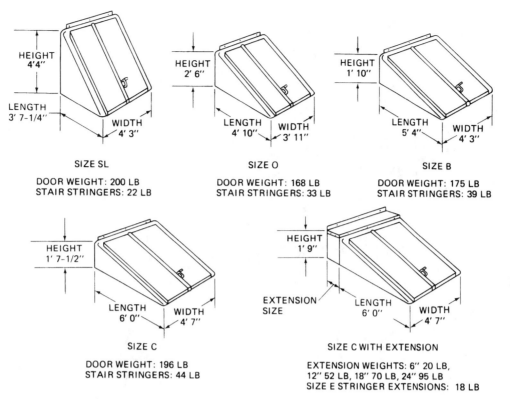

SIZE SL

DOOR WEIGHT: 200 LB
STAIR STRINGERS: 22 LB

SIZE O

DOOR WEIGHT: 168 LB
STAIR STRINGERS: 33 LB

SIZE B

DOOR WEIGHT: 175 LB
STAIR STRINGERS: 39 LB

SIZE C

DOOR WEIGHT: 196 LB
STAIR STRINGERS: 44 LB

SIZE C WITH EXTENSION

EXTENSION WEIGHTS: 6" 20 LB,
12" 52 LB, 18" 70 LB, 24" 95 LB
SIZE E STRINGER EXTENSIONS: 18 LB

Fig. 17-78 *Various shapes and sizes of outside basement doors.* (Bilco)

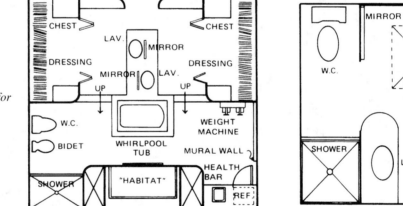

Fig. 17-79 *Floor plans for bathrooms.* (Kohler)

You may not want to become too elaborate with your new bathroom. All you have to do then is decide how and what you need. Then draw your ideas for arrangements. Check the plumbing to see if your idea will be feasible. Order the materials and fixtures and then get started.

Providing Additional Storage

Cedar-lined closet Aromatic red cedar closet lining is packed in a convenient, no-waste package which contains 20 square feet. It will cover 16⅓ square feet of wall space for lifetime protection from moths. In order to install the cedar, follow these steps. See Fig. 17-81.

1. Measure the wall, ceiling, floor, and door area of the closet. Figure the square footage. Use the length times width to produce square feet.

2. Cedar closet lining may be applied to the wall either vertically or horizontally. See Fig. 17-82. When applied to the rough studs, pieces of cedar must be applied horizontally.

3. When applying the lining to the wall, place the first piece flush against the floor with the grooved edge down. Use small finishing nails to apply the lining to the wall. See Fig. 17-83.

4. When cutting a piece of cedar to finish out a course or row of boards, always saw off the tongued end so that

Fig. 17-80 *Japanese bathroom design. The platform gives the illusion of a sunken bath. Many of these features can be incorporated into a remodeled room which can serve as a bathroom.* (Kohler)

Fig. 17-81 *Cedar-lined closet. The aromatic red cedar serves as a moth deterrent.* (Grossman Lumber)

Fig. 17-82 *Placing a piece of cut-to-fit red cedar lining vertically in a closet.* (Grossman Lumber)

Fig. 17-83 *Work from the bottom up when applying the cedar boards.* (Grossman Lumber)

the square sawed-off end will fit snugly into the opposite corner to start the new course. See Fig. 17-84.

5. Finish one wall before starting another. Line the ceiling and floor in a similar manner. Each piece of cedar is tongued and grooved for easy fit and application. See Fig. 17-85.

The cedar-lined closet will protect your woolens for years. However, in some instances you might not have a closet to line. You might need extra storage space. In this instance take a look at the next section.

Building extra storage space Any empty corner can provide the back and one side of a storage unit. This simple design requires a minimum of materials for a maximum of storage. Start with a 4-foot unit now and add 2 feet later. See Fig. 17-86.

Fig. 17-84 *To start a new course, saw off the tongued end. The square sawed end will fit snugly into the opposite corner.* (Grossman Lumber)

Fig. 17-85 *Finish one wall before starting another.* (Grossman Lumber)

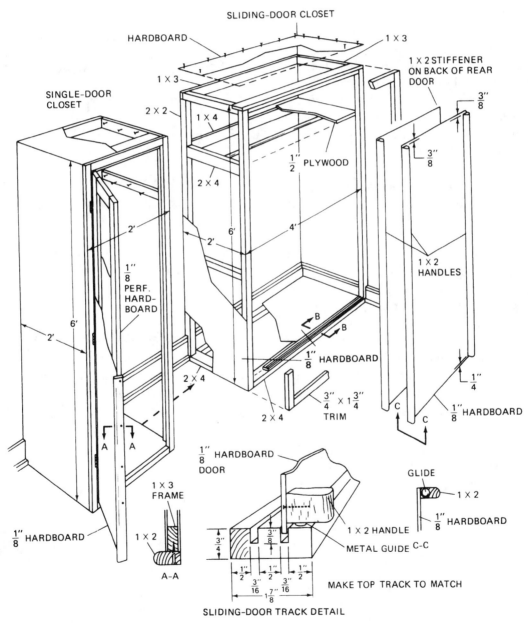

SLIDING-DOOR CLOSET

HARDBOARD

1 × 3

1 × 2 STIFFENER ON BACK OF REAR DOOR

$\frac{3''}{8}$

SINGLE-DOOR CLOSET

1 × 3

2 × 2

1 × 4

$\frac{3''}{8}$

2 × 4

$\frac{1}{2}''$ PLYWOOD

2'

2 × 4

6'

4'

2'

1 × 2 HANDLES

$\frac{1}{8}''$ PERF. HARD-BOARD

6'

$\frac{1}{4}''$

2'

$\frac{1}{8}''$ HARDBOARD

B B

$\frac{1}{8}''$ HARDBOARD

2 × 4

$\frac{3''}{4} × 1\frac{3''}{4}$ TRIM

C C

2 × 4

$\frac{1}{8}''$ HARDBOARD

$\frac{1}{8}''$ HARDBOARD DOOR

GLIDE

1 × 3 FRAME

1 × 2

$\frac{3''}{4}$

$\frac{3''}{8}$

1 × 2 HANDLE

METAL GUIDE C-C

1 × 2

$\frac{1}{8}''$ HARDBOARD

A–A

$\frac{1''}{2}$ $\frac{3''}{16}$ $\frac{1''}{2}$ $\frac{3''}{16}$ $\frac{1''}{2}$

$1\frac{7''}{8}$

MAKE TOP TRACK TO MATCH

SLIDING-DOOR TRACK DETAIL

Fig. 17-86 *Extra storage space. The sliding-door closet and the single-door closet.* (Grossman Lumber)

Use a carpenter's level and square to check a room's corners for any misalignment. Note where the irregularities occur. The basic frame is built from 1-inch and 2-inch stock lumber. It can be adjusted to fit any irregularities in the walls or floor. Nail the top and shelf cleats into the studs. Cut the shelf from ½-inch plywood. Slip it into place and nail it to the cleat. Attach the clothes rod with the conventional brackets. Fit, glue, and nail the floor, side, and top panels. Make them from ⅛-inch hardboard. You can buy metal or hardwood sliding doors and tracks. Or you can make the doors and purchase only the tracks. Doors are made of single sheets of ⅛-inch hardboard stiffened with full-length handles of 1-×-2 trim set on the edge. Metal glides on the ends of handles carry the weight of the door and prevent binding.

Figure 17-87 shows some of the details for making the door operational.

A swinging door unit is built almost like the double door unit. The framework should be fastened to the wall studs where possible. Make the door frame of 1 × 3s laid flat with ⅛-inch hardboard (plain) on the face and ⅛-inch perforated hardboard on the inside. Again, a full length 1 × 2 makes the handle.

Attach three hinges to the door. Make sure that the pins line up so that the door swings properly. Place the door in the opening and raise it slightly. Mark the frame and chisel out notches for the hinges.

Other types of storage space For adequate, well-arranged storage space, plan your closets first. Minimum depth of closets should be 24 inches. The width can vary to suit your needs. But provide closets with large doors for adequate access. See Fig. 17-88.

Figure 17-89 shows some of the arrangements for closets. Use 2-×-4 framing for dividers and walls so that shelves may be attached later. Carefully measure and fit ⅜-inch gypsum board panels in place. Finish the interior or outside of the closet as you would any type

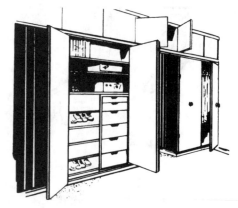

Fig. 17-88 *A variation of cabinet storage designs.* (Grossman Lumber)

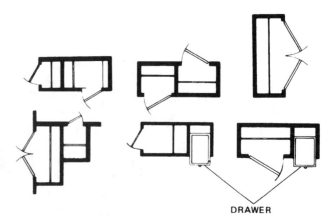

DRAWER

Fig. 17-89 *Different plans and door openings for closets.* (Grossman Lumber)

of drywall installation. Install the doors you planned for. These can be folding, bifold, or a regular prehung type.

Remodeling a Kitchen

In remodeling a kitchen the major problem is the kitchen cabinets. There are any number of these available already made. They can be purchased and installed to make any type of kitchen arrangement desired.

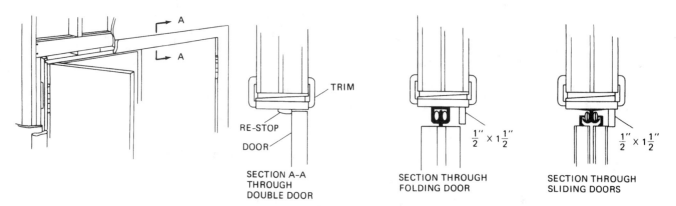

TRIM

RE-STOP

DOOR

SECTION A-A
THROUGH
DOUBLE DOOR

½" × 1½"

SECTION THROUGH
FOLDING DOOR

½" × 1½"

SECTION THROUGH
SLIDING DOORS

Fig. 17-87 *Details of the door installation for the closets in Fig. 17-86.* (Grossman Lumber)

The countertops have already been covered. The color of the countertop laminate must be chosen so that it will blend with the flooring and the walls. This is the job of the person who will spend a great deal of time in the kitchen.

Planning the kitchen The kitchen begins with a set of new cabinets for storage and work areas. The manufacturer of the cabinets will supply complete installation instructions.

Check the drawing and mark on the walls where each unit is to be installed. Mark the center of the stud lines. Mark the top and the bottom of the cabinet so that you can locate them easily at the time of installation.

As soon as the installation is completed, wipe the cabinets with a soft cloth dampened with water. Dry the cabinets immediately with another clean soft cloth. Follow this cleaning with a very light coat of high-quality liquid or paste wax. The wax helps keep out moisture and causes the cabinets to wear longer.

Finishing up the kitchen After the cabinets have been installed, it is time to do the plumbing. Have the sink installed and choose the proper faucet for the sink.

Kitchen floor Now it's time to put down the kitchen floor. In most cases the flooring preference today is carpet, although linoleum and tile are also used. It is easier to clean carpet—just vacuuming is sufficient. It is quieter and can be wiped clean easily if something spills. If a total remodeling job has been done, it may be a good idea to paint or wallpaper the walls before installing the flooring. A new range and oven are usually in order, too. The exhaust hood should be properly installed electrically and physically for exhausting cooking odors and steam. This should complete the kitchen remodeling. Other accents and touches here and there are left up to the user of the kitchen.

Enclosing a Porch

One of the first things to do is to establish the actual size you want the finished porch to be. In the example shown here, a patio (16 × 20 feet) is being enclosed. A quick sketch will show some of the possibilities (Fig. 17-90). This becomes the foundation plan. It is drawn ¼ inch to equal 1 foot. The plan can then be used to obtain a building permit from the local authorities.

The floor plan is next. See Fig. 17-91. It shows the location of the doors and windows and specifies their size.

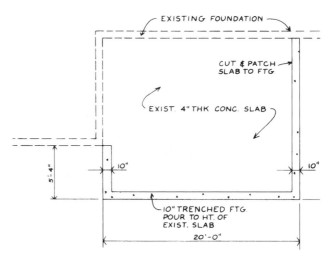

Fig. 17-90 *Foundation plan for an addition to an already existing building: enclosing a patio.*

A cross section of the addition or enclosure is next. See Fig. 17-92. Note the details given here dealing with the actual construction. The scale here is ½" = 1'0". Note that the roof pitch is to be determined with reference to the window location on the second floor of the existing building. Figure 17-95A shows what actually happened to the pitch as determined by the window on the second floor. As you can see from the picture, the pitch is not 5:12 as called for on the drawing. Because of the long run of 16 feet, the 2-×-6 ceiling joists, 16 O.C., had to be changed to 2 × 10s. This was required by the local code. It was a good requirement, since with the low slope on the completed roof, the pile-up of snow would have caused the roof to cave in. See Fig. 17-95A.

Elevations Once the floor plan and the foundation plan have been completed, you can begin to think about how the porch will look enclosed. This is where the elevation plans come in handy. They show you what the building will look like when it is finished. Figure 17-93A shows how part of the porch will look when it is extended past the existing house in the side elevation.

Figure 17-93B shows how the side elevation looks when finished. Note the storm door and the outside light for the steps.

The rear elevation is simple. It shows the five windows that allow a breeze through the porch on days when the windows can be opened. See Fig. 17-94 for a view of the enclosed porch viewed from the rear.

A side elevation is necessary to see how the other side of the enclosure will look. This shows the location of the five windows needed on this side to provide ventilation. Figure 15-95A illustrates the way the

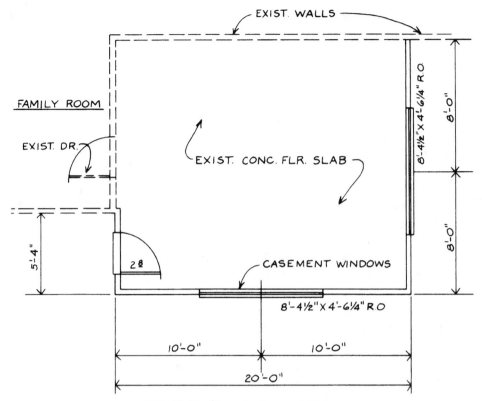

Fig. 17-91 *Floor plan for an addition.*

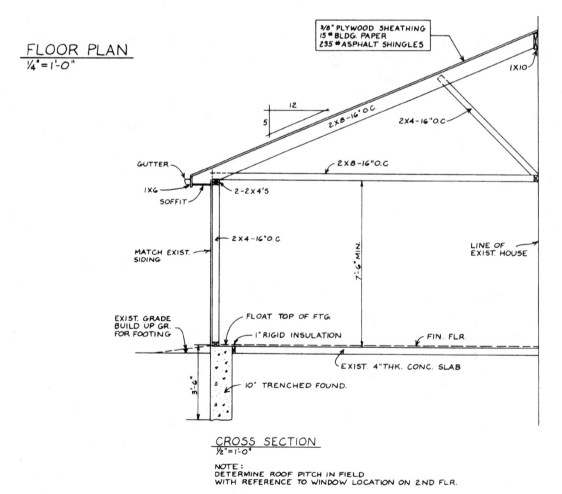

FLOOR PLAN
¼"=1'-0"

CROSS SECTION
½"=1'-0"

NOTE:
DETERMINE ROOF PITCH IN FIELD
WITH REFERENCE TO WINDOW LOCATION ON 2ND FLR.

Fig. 17-92 *Cross section view of the addition.*

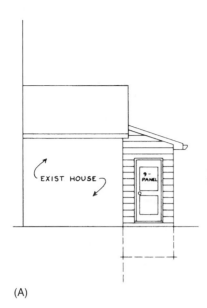

(A)

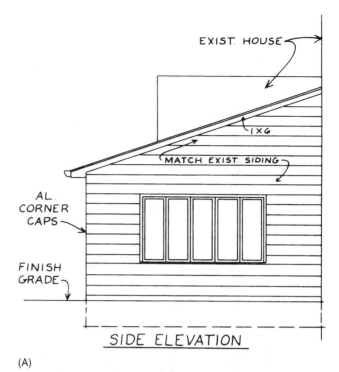

SIDE ELEVATION

(A)

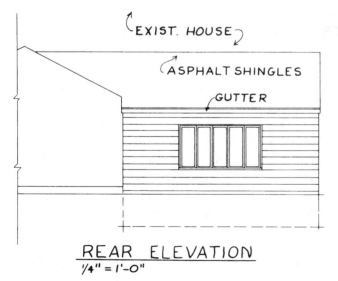

(B)

Fig. 17-93 *(A) Side elevation of the addition. (B) What the side will look like when finished.*

(B)

Fig. 17-95 *(A) Side elevation of the addition. (B) What the side elevation will look like when finished.*

REAR ELEVATION
1/4" = 1'-0"

Fig. 17-94 *Rear elevation of the addition.*

enclosure should look. Figure 17-95B shows how the finished product looks with landscaping and the actual roofline created by the second-floor window location. With this low-slope roof, heating cables must be installed on the roof overhang. This keeps ice jams from forming and causing leaks inside the enclosed porch.

Once your plans are ready and you have all the details worked out, it is time to get a building permit. You have to apply and wait for the local board's decision. If you comply with all the building codes, you can go ahead. This can become involved in some communities. The building permit shown in Fig. 17-96 shows some of the details and some of the people involved in

Fig. 17-96 *Application for a building permit.*

issuing a building permit. This building permit is for the porch enclosure shown in the previous series of drawings and pictures.

Starting the project Once you have the building permit, you can get started. The first step is to dig the hole for the concrete footings or for the trench-poured foundation shown here. Once the foundation concrete has set up, you can add concrete blocks to bring the foundation up to the existing level. After the existing grade has been established with the addition, you can proceed as usual for any type of building. Put down the insulation strip and the sole plate. Attach the supports to the existing wall and put in the flooring joists. Once the flooring is in, you can build the wall framing on it and push the framing up and into place. Put on the rafters and the roof. Get the whole structure enclosed with a nail base or plywood on the outside walls. Place the windows and doors in. Put on the roof and then the siding. Trim the outside around the door and windows. Check the overhang and place the trim where it belongs. Put in the gutters and connect them to the existing system. If there is to be electrical wiring, put that in next. In most porches there is no plumbing, so you can go on to the drywall. Finish the ceiling and walls.

After the interior is properly trimmed, it should be painted. The carpeting or tile can be placed on the floor and the electrical outlets covered with the proper plates. The room is ready for use.

Finishing the project Throughout the building process, the local building inspector will be making calls to see that the code is followed. This is for the benefit of both the builder and the owner.

There is another *benefit* (to the local government) of the inspections and the building permit. Once the addition is inhabited, the structure can go on the tax roles and the property tax can be adjusted accordingly.

ADDING SPACE TO EXISTING BUILDINGS

Adding space to an already-standing building requires some special considerations. You want the addition to look as if it "belongs." That means you should have the siding the same as the original or as close as possible. In some cases you may want it to look added-on so you can contrast the new with the old. However, in most in-

stances, the intention is to match up the addition as closely as possible.

First, the additional space must have a function. You may need it for a den, an office, or a bedroom. In any of these instances you don't need water or plumbing if you already have sufficient bathroom facilities on that floor. You will need electrical facilities. Plan the maximum possible use of the building and then put in the number of electrical outlets, switches, and lights that you think you can use. Remember, it is much cheaper to do it now than after the wallboard has been put up.

Planning an Addition

In the example used here, we will add a 15-×-22-foot room onto an existing, recently built house. The addition is to be used as an office or den. It is located off the dining room, so it is out of the way of through-the-house foot traffic. The outside wall will also serve to deaden the sound. It was insulated when the house was built. This means that only three walls will have to be constructed. Make a rough sketch of what you think will be needed. See Fig. 17-97 for an example. Note that the light switch has been added and that two lights over the bookcases will be added later. Note the outlets

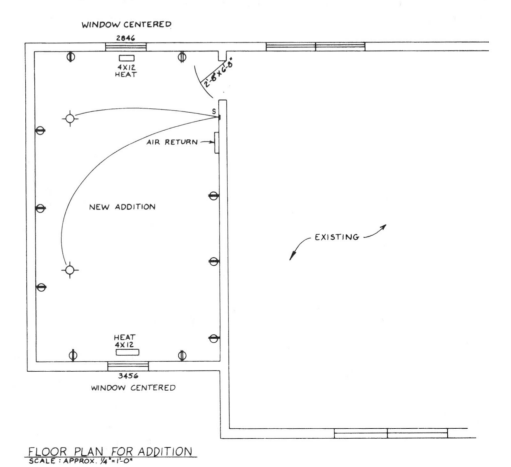

Fig. 17-97 *Floor plan for an addition.*

WINDOW CENTERED
2846
4X12 HEAT
2·8×6·0·
S
AIR RETURN →
NEW ADDITION
EXISTING
HEAT 4X12
3456
WINDOW CENTERED

FLOOR PLAN FOR ADDITION
SCALE : APPROX. ¼"=1'-0"

for electricity and heat. The windows are of two different sizes. The local code calls for a certain amount of square footage in the windows for ventilation. One window has been taken from the existing upstairs bedroom. It is too large and will interfere with the roofline of this addition. A smaller window is put in up there, and the one used in the bedroom is moved down to the addition. Only two windows need to be purchased. Only one door is needed. These should be matched up with the existing doors and windows so that the house looks complete from the inside, too.

Elevations must be drawn up to show to the building code inspectors. You must get a building permit; therefore the elevation drawings will definitely be needed. See Figs. 17-98, 17-99, and 17-100 for the front, rear, and side elevations. This information will help you obtain a building permit. If you use Andersen's window numbers, the town board will be able to see that these windows provide the right amount of ventilation.

To make the finished product fit your idea of what you wanted, it is best to write a list of specifications. Make sure you list everything you want done and how you want it done. For example, the basement, siding, overhang, flooring, windows, door, walls, roof, and even the rafters should be specified.

Specifications

Addition for 125 Briarhill in Town of Amherst. (See detailed sketches for rear, front, and side elevations.)

Basement

- Crawl space—skim coat of concrete over gravel. Drainage around footings and blocks to prevent moisture buildup.
- Blocks on concrete footings—42 inches deep.
- To be level with adjoining structure.
- To be waterproofed. Fill to be returned and leveled around the exterior.

Siding

- National Gypsum Woodrock Prefinished (as per existing).
- Size to match existing.
- Must match at corners and overlap in rear to look as if built with original structure.
- To be caulked around windows.

Overhang

- To match existing as per family room extension.
- Gutters front and rear to match existing. Downspouts (conduits) to match existing and to connect.

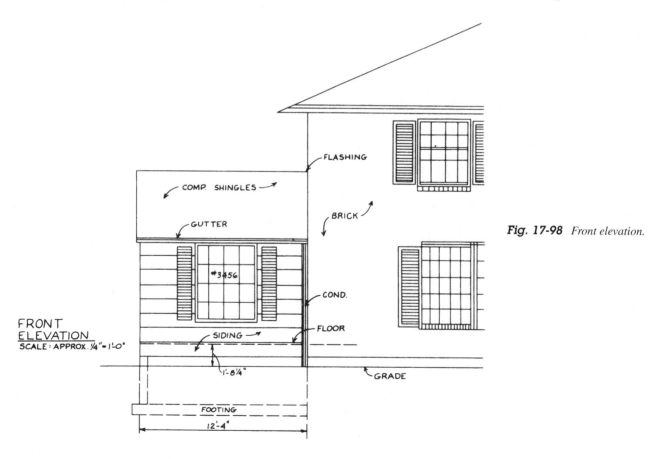

Fig. 17-98 *Front elevation.*

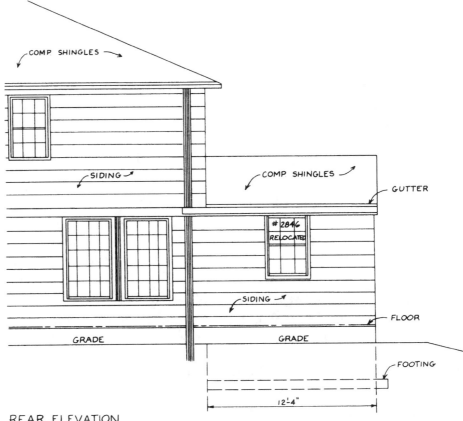

Fig. 17-99 *Rear elevation.*

REAR ELEVATION
SCALE : APPROX. ¼"=1'-0"

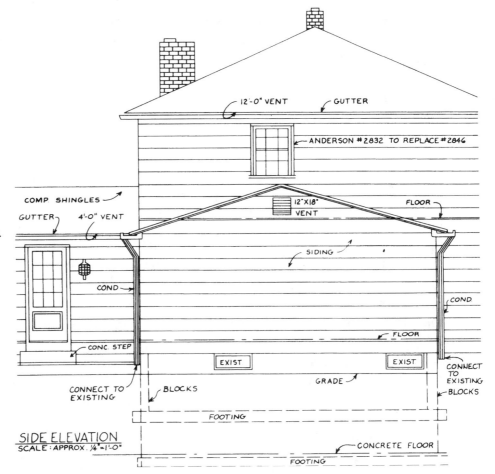

Fig. 17-100 *Side elevation.*

SIDE ELEVATION
SCALE : APPROX. ¼"=1'-0"

- Ventilation screen in Upson board overhang.
- Exterior fascia and molding to match existing.

Flooring

- Kiln-dried 2 × 10 (construction or better) on 16-inch centers.
- Plywood subfloor, ⅝-inch exterior, A—D.
- Plywood subfloor, ½-inch A—D, interior.
- Fiberglas insulation strip between sole plate and blocks.
- Must meet with existing room. Allowance made for carpeting.
- Must be level.

Windows

- Anderson No. 3456 for south wall.
- Replace existing window with Anderson No. 2832 and fir in. (See Fig. 17-100. Note: Size difference of windows.)
- Windows to be vinyl-coated and screened as per existing.
- Shutters on south window wall to match existing.

Door

- Interior with trim as per existing, mounted and operating.
- Size, 32" × 6'8"; solid, flush, walnut with brass lock and hardware.
- Dining room to be left in excellent condition.

Permits

- To be obtained by the contractor.

Walls

- Studs, 2-× -4 kiln-dried (construction or better), 16 inches O.C.
- Double or better headers at windows and door and double at corners where necessary.

Roof

- Composition shingles as per existing, black of same weight as existing.
- Felt paper under shingles and attached to ⅝-inch exterior plywood (A—D) sheathing.

Rafters

- Kiln-dried 2 × 4 truss type as per existing, 16 inches O.C.

- Roof type to be shown in attached drawings. Check with existing garage to determine type of construction if necessary.

In order to add any information you may have left out, you can make drawings illustrating what you need. Figure 17-101 shows a cornice detail that ensures there is no misunderstanding of what the overhang is. Other details are also present in this drawing. The scale of the cornice is 1½" = 1'0".

The contractor is protected when the details of work to be done are spelled out in this way. There is little or no room for argument if things are written out. A properly drawn contract between contractor and owner should also be executed to make sure both parties understand the financial arrangements. Figures 17-102 and 17-103 show the completed addition as it looks with landscaping added.

CREATING NEW STRUCTURES

A storage building can take the shape of any number of structures. In most instances you want to make it resemble some of the features of the home nearby.

Small storage sheds are available in precut packages from local lumber yards. Figure 17-104 shows three versions of the same package. It can be varied to meet the requirements of the buyer. Lawn mowers, bicycles, or almost anything can be stored in a structure of this type.

All you need is a slab to anchor the building permanently. In some instances you may not want to anchor it, so you just drive stakes in the ground and nail the sole plate to the stakes. Most of the features are simple to alter if you want to change the design.

Before you choose any storage facility, you should know just how you plan to use it. This will determine the type of structure. It will determine the size, and in some cases the shape. The greenhouse in Fig. 17-104 is nothing more than the rustic or contemporary shed with the siding left off. The frame is covered with glass, Plexiglas, or polyvinyl according to your taste or pocketbook.

Custom-Built Storage Shed

In some cases it is desirable to design your own storage shed. The example used here, shown in Fig. 17-105, was designed to hold lawn mowers and yard equipment. It was designed for easy access to the inside. Take a look at the overhead garage door in the rear of the shed. The 36-inch door was used along with the window to make it more appealing to the eye. It also

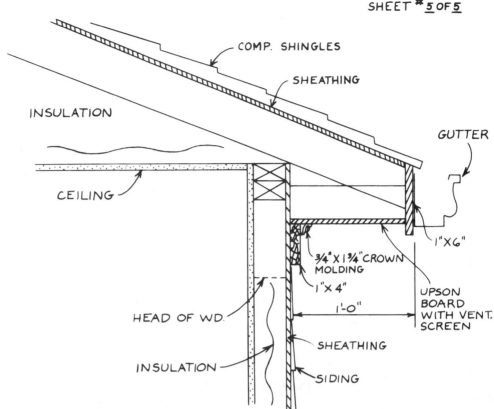

Fig. 17-101 *Cornice detail.*

COMP. SHINGLES

SHEATHING

INSULATION

GUTTER

CEILING

1"×6"

¾"×1¾"CROWN MOLDING

1"×4"

1'-0"

HEAD OF WD.

UPSON BOARD WITH VENT. SCREEN

INSULATION

SHEATHING

SIDING

SCALE: APPROX. 1½"=1'-0"

CORNICE DETAIL

Fig. 17-102 *Side elevation in finished form.*

has a hip roof to match the house to which it is a companion. See Fig. 17-106.

The design is 10 × 15 feet. See Fig. 17-107. That produces a 2-to-3 ratio which produces a pleasing ap-

pearance. A 3-to-4 ratio is also very common. The concrete slab was placed over a bed of crushed rocks and anchored by bolts embedded in the concrete slab. A 9-foot-wide and 7-foot-long slab is tapered down

Fig. 17-103 *Rear elevation in finished form.*

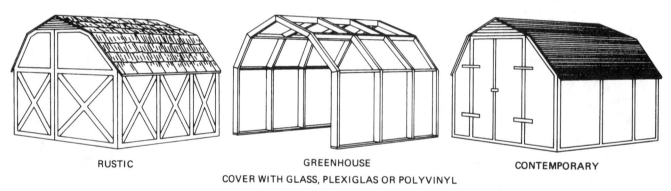

RUSTIC   GREENHOUSE   CONTEMPORARY

COVER WITH GLASS, PLEXIGLAS OR POLYVINYL

Fig. 17-104 *Various designs for storage facilities.* (TECO)

from the floor to the yard. This allows rider lawn mowers to be driven into the shed. The outside pad also serves as a service center.

Wires serving the structure are buried underground and brought up through a piece of conduit. They enter the building near the small door. There is only one window, so the wall space can be used to hang yard tools. The downspouts empty into the beds surrounding the structure to water the evergreens. An automatic light switch turns on both lights at sundown and off again at dawn.

The overhead door faces the rear of the property. This produced some interesting comments from the concrete installers, who thought the garage had been turned around by mistake. It does resemble a one-car garage, but it is specifically designed for the storage of yard work equipment.

Don't forget to get a building permit. In some areas, even a tool shed requires a building permit. This doesn't necessarily mean it goes on the tax rolls, but does call for a number of inspections by the building inspector, which helps protect both builder and owner.

Buildings for storage take all shapes. They may be garages or barns. The design of a new structure should be carefully chosen to harmonize with the rest of the buildings on the property.

Fig. 17-105 *View of one end of a storage shed.*

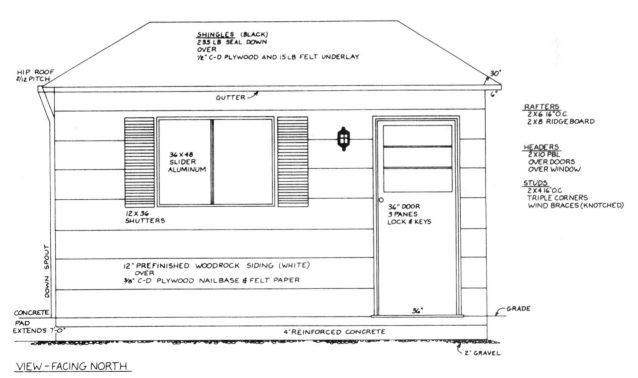

SHINGLES (BLACK)
2 35 LB SEAL DOWN
OVER
½" C-D PLYWOOD AND 15 LB FELT UNDERLAY

HIP ROOF
2/12 PITCH

30°

6°

GUTTER

RAFTERS
2 X 6 16" O.C.
2 X 8 RIDGE BOARD

HEADERS
2 X 10 PBL
OVER DOORS
OVER WINDOW

STUDS
2 X 4 16" O.C.
TRIPLE CORNERS
WIND BRACES (KNOTCHED)

36 X 48
SLIDER
ALUMINUM

12 X 36
SHUTTERS

36" DOOR
5 PANES
LOCK & KEYS

DOWN SPOUT

12" PREFINISHED WOODROCK SIDING (WHITE)
OVER
⅜" C-D PLYWOOD NAILBASE & FELT PAPER

CONCRETE
PAD
EXTENDS 7'-0"

36"

GRADE

4" REINFORCED CONCRETE

2" GRAVEL

VIEW - FACING NORTH

Fig. 17-106 *View of the finished storage shed.*

CHAPTER 17
STUDY QUESTIONS

1. What is the meaning of the word maintenance?

2. What do you need to know about a house before you start remodeling it?

3. How does moisture passing through the exterior paint cause problems with a house?

4. What is the importance of scheduling in getting a job done?

5. What is meant by shimming a strike jam of a door?

6. What is meant by side elevation?

7. What is a deadbolt lock? Where is it used?

8. What is a traverse rod? Where is it used?

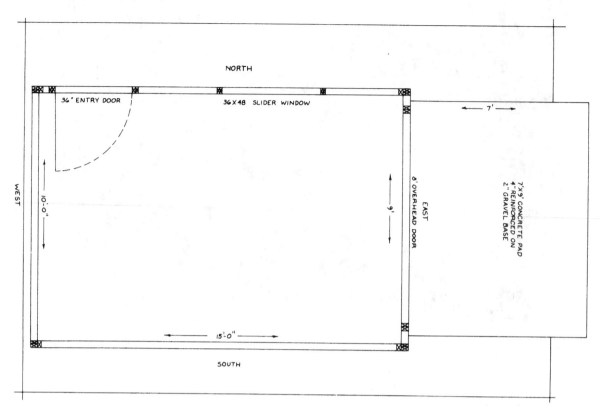

Fig. 17-107 *Outside dimension details.*

9. What is another name for gypsum board?

10. How do you eliminate popped nails and their unsightliness?

11. Why are nails used to install drywall pop?

12. What do you use to cut plastic laminates for counter tops?

13. How does contact cement work? Why do you have to wait for it to dry before you place the two pieces together?

14. What is meant by a course of shingles?

15. What are the two standard sizes of floor tiles?

16. What is the standard size of a piece of paneling?

17. What do you mean when you say the panel should be hung perfectly plumb?

18. What is a drop ceiling? Where would you use one?

19. What type of cedar is used to line a closet?

20. What is a bifold door? Where do you use them?

18
CHAPTER

The Carpenter
&
the Industry

THIS CHAPTER EXAMINES THE LATEST building methods and materials. Also discussed are building codes and zoning provisions, along with recent trends in manufactured housing. You will learn:

- How construction procedures are changing
- Why building codes are necessary and how to fill out a building permit
- How factory-built commercial buildings and homes are constructed

BROADENING HORIZONS IN CARPENTRY

Carpentry, like other trades, is constantly changing. New materials are being introduced to replace old time-honored ones that have become too expensive or scarce. New tools are available to work with new materials and plastics have taken over where glass once reigned supreme. Reading trade journals and other printed material is absolutely necessary. Manufacturers usually include a set of instructions with each prefab unit or a newer type of product using a different kind of material.

New Building Materials

Figure 18-1 shows a house made of plywood. Only a few years ago all floors and walls were made of pieces of wood that measured 1 × 6 inches. Plywood has some special features. It is stronger, it has a good nail surface, and it is good insulation. It is easily placed in position, and it can be bought cheaply compared with single pieces of siding or flooring.

New materials are used as a nail base for siding. Carpenters have to adapt to these types of material and be able to handle them. See Fig. 18-2.

Changing Construction Procedures

Construction procedures are changing. Note that in Fig. 18-3 the wall construction is different from the usual. This is a double-wall partition. It is used to separate rooms in apartments so that sound is not easily transmitted between rooms.

Figure 18-4 shows how adhesives can be used to apply panels. These panels are held in place with glue instead of nails. New clips are available for holding 2 × 4s and 2 × 10s. They are also designed to hold plywood sheets. Using these clips cuts installation time.

Fig. 18-1 *Building materials used in construction are changing. Note the extensive use of plywood in a modern house.* (Western Wood Products)

Fig. 18-2 *Carpenters must learn to adapt to new materials like this composition nail base used for siding.*

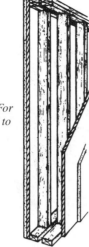

Fig. 18-3 *Construction procedures are also changing. For example, double walls are used to help keep the noise down in apartments.*

They also make buildings stronger. The carpenter then becomes a valuable person on the job.

Innovations in Building Design

Architects are constantly coming up with new designs. These new designs call for different ways of doing things. The carpenter has to be able to work with new woods and new combinations of materials. Figure 18-5 shows one of the newer designs for a "modern" home. A mixture of brick and wood is used for the outside covering. The inside also calls for some new methods, since the open ceiling is used here. Different window sizes call for a carpenter with ability to innovate on the job. Doors are different from the standard types. This calls for an up-to-date carpenter, or one who can adapt to the job to be done.

Figure 18-6 shows a modern design for a condominium building. This uses many carpentry skills. Note the different angles and the wood siding. Even the fence calls for close following of drawings.

Different types of home units can draw upon the carpenter's abilities. The skilled carpenter can adapt to the demands made by newer designs.

New materials and new ways of doing things are being developed. The carpenter today has to be able to adapt to the demands. The carpenter has to keep up-to-date on new materials and new techniques. New designs will demand an even more adaptable person in this trade in the future.

Fig. 18-4 *Adhesives are used extensively in the building industry today.* (U.S. Gypsum)

Fig. 18-5 *Modern construction using brick and wood. A number of various materials are used in house exteriors.*

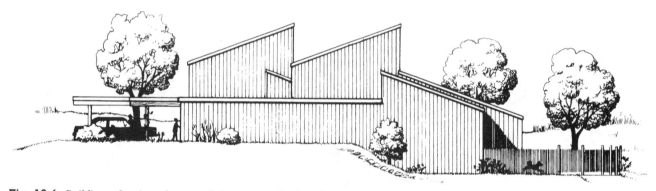

Fig. 18-6 *Buildings of various shapes and sizes use wood today. The carpenter is needed to apply acquired skill in a building of this nature.*

A person interested in doing something new and different can surely find it in carpentry.

BUILDING CODES AND ZONING PROVISIONS
Building Codes

Building codes are laws written to make sure buildings are properly constructed. They are for the benefit of the buyer. They also benefit all the people in a community. If an expensive house is built next to a very inexpensive one, it lowers the property value of the expensive house. Codes are rules which direct people who build homes. They say what can and cannot be done with a particular piece of land. Some land is hard to build on. It may have special surface problems. There may be mines underneath. There are all kinds of things which should be looked into before building.

If a single-family house is to be built, the building codes determine the location, materials, and type of construction. Codes are written for the protection of individuals and the community. In areas where there are no codes, there have been fires, collapsed roofs, and damage from storms.

Figure 18-7 shows inspections needed for a building permit. This is one way the community has to check on building. You have to obtain a permit to add on to a house. It is necessary to get a permit if you build a new house. Figure 18-8 shows the permit for a tool shed. One reason for requiring a building permit is to let the tax assessor know so that the property value can be changed.

Note in Fig. 18-8 how the application for a building permit is filled out. Note also the possibilities to be checked off. The town board is required to sign, since they are responsible to the community for what is built and where.

NOTICE
INSTRUCTIONS AND REQUIRED INSPECTIONS

A. Builder's name, phone number, property house number, and building permit number must be displayed on all construction or building projects.

B. A reasonable means of ingress must be provided to each structure and each floor. (Planks or other; and ladders or stairs from floor-to-floor.)

C. The following inspections are mandatory on all construction within the Town of Amherst:

1. FIRST WALL: When footing is ready to be poured, including trench pour.
2. PRE-LATH: Before insulating and after ALL electrical, telephone, plumbing, and heating rough work is complete, including metal gas vent and range-hood exhaust duct.
3. DRAIN TILE: Subfloor and/or footing drain tile, before pouring of concrete floor.
4. FINAL: Make application for Certificate of Occupancy, file copy of survey with application. Everything must be completed. All work must be finished, including sidewalks and drive aprons.) Each structure must have a house number displayed and visible from the street.

NOTE: An approved set of building plans and plot plans shall be made available to the Building Inspector at the above four inspections.

Call the Building Department 24 hours before you are ready for inspection.
Phone 555-6200, Ext. 42, 43, or 49

Fig. 18-7 *Notice of the inspections needed when a building permit is required for a house.*

In Fig. 18-7, look at the number of inspections required before the building is approved. Each inspection is made by the proper inspector. This way the person who buys the house or property is protected.

The certificate of occupancy is shown in Fig. 18-9. This is required before a person can move into the house. The certificate is given with the owner in mind. It means the house has been checked for safety hazards before anyone is allowed to live in it.

Community Planning and Zoning

Zoning laws or codes are designed to regulate areas to be used for building. Different types of buildings can be placed in different areas for the benefit of everyone. Shopping areas are needed near where people live. Working areas or zones are needed to supply work for people. In most cases, we like to keep living and working areas separated. This is because of the nature of the two types of building. Some people don't like to live near a factory with its smoke and noise.

Living near a sewage treatment plant can also be very unpleasant. Certain types of buildings should be near one another and away from living zones.

Some areas are designated as industrial. Others are designated as commercial. Still others are marked for use as residential areas. Residential means homes. Homes may be single houses or apartments. An industrial building cannot be built in a residential area. All over the United States there are regional planning boards. They decide which areas can be used for what. Master plans are made for communities. Master plans designate where various types of buildings are located.

Community plans also include maps and areas outlined as to types of buildings. Parks are also designated in a community plan. Streets are given names, and maps are drawn for developments. A development means land to be developed for housing or other use. Figure 18-10 shows a typical plan for a residential development. Note how the streets are laid out. They are not straight rows. This has a tendency to slow down traffic. The safety of children is important in a residential area. It is best to have residential areas off main traveled roads or streets.

Overbuilding

Where there has been residential, commercial, and industrial overbuilding, the utilities have not been able to keep up. The additional activity puts a strain on the existing facilities. That is another reason for having a community plan.

For example, there can be problems with sewage. Plants may not be big enough to handle the extra water

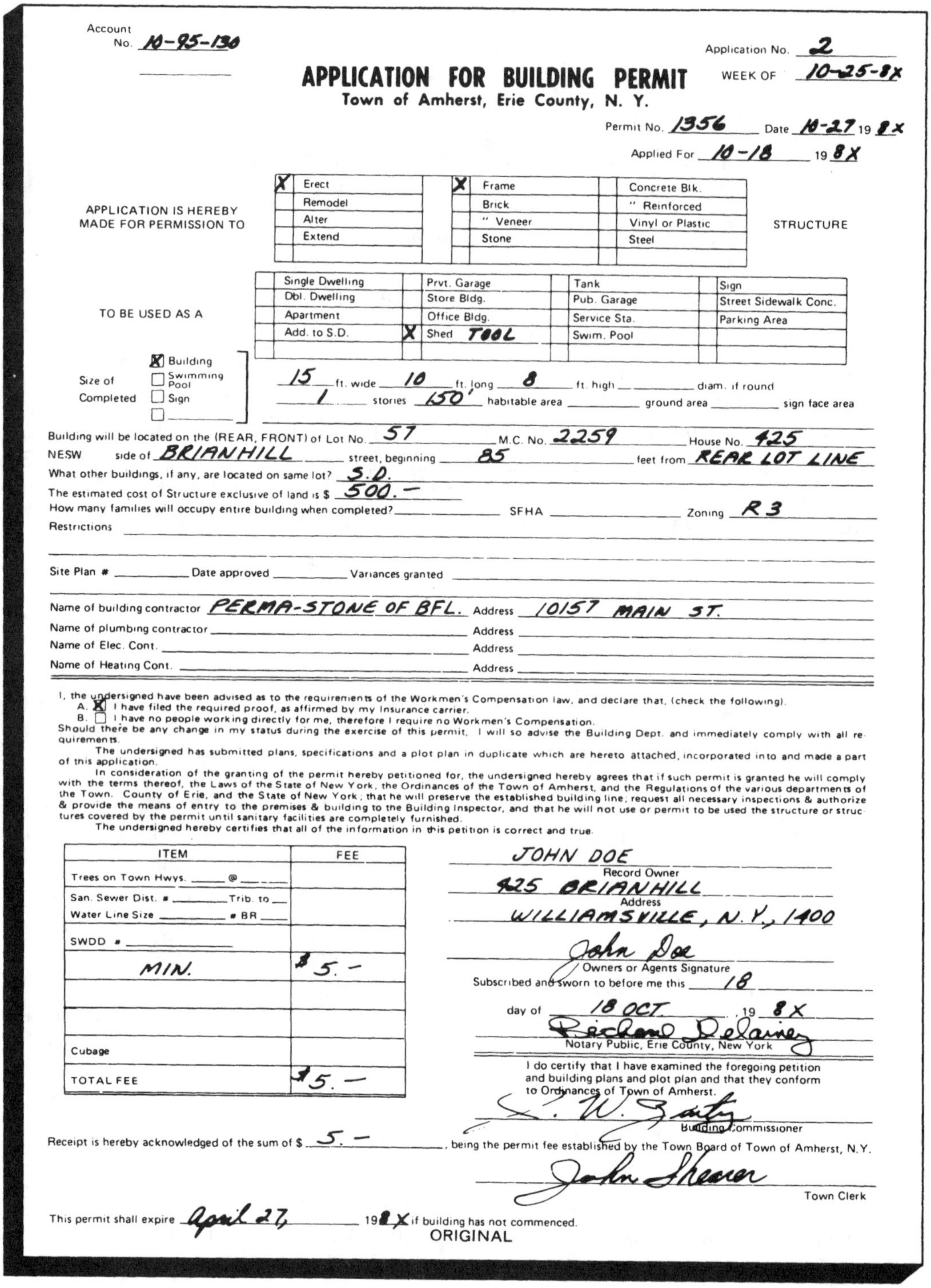

APPLICATION FOR BUILDING PERMIT
Town of Amherst, Erie County, N. Y.

Account No. *10-95-130*

Application No. *2*
WEEK OF *10-25-8X*

Permit No. *1356*   Date *10-27* 19 *8X*

Applied For *10-18*   19 *8X*

APPLICATION IS HEREBY
MADE FOR PERMISSION TO

X	Erect	X	Frame	Concrete Blk.
	Remodel		Brick	" Reinforced
	Alter		" Veneer	Vinyl or Plastic
	Extend		Stone	Steel

STRUCTURE

TO BE USED AS A

Single Dwelling	Prvt. Garage	Tank	Sign
Dbl. Dwelling	Store Bldg.	Pub. Garage	Street Sidewalk Conc.
Apartment	Office Bldg.	Service Sta.	Parking Area
Add. to S.D.	X Shed *TOOL*	Swim. Pool	

Size of
Completed

[X] Building
[ ] Swimming Pool
[ ] Sign
[ ]

15 ft. wide *10* ft. long *8* ft. high _____ diam. if round
1 stories *150'* habitable area _____ ground area _____ sign face area

Building will be located on the (REAR, FRONT) of Lot No. *57* M.C. No. *2259* House No. *425*
NESW side of *BRIANHILL* street, beginning *85* feet from *REAR LOT LINE*
What other buildings, if any, are located on same lot? *S.D.*
The estimated cost of Structure exclusive of land is $ *500.—*
How many families will occupy entire building when completed? _____ SFHA _____ Zoning *R3*
Restrictions _____

Site Plan # _____ Date approved _____ Variances granted _____

Name of building contractor *PERMA-STONE OF BFL.* Address *10157 MAIN ST.*
Name of plumbing contractor _____ Address _____
Name of Elec. Cont. _____ Address _____
Name of Heating Cont. _____ Address _____

I, the undersigned have been advised as to the requirements of the Workmen's Compensation law, and declare that, (check the following).
A. [X] I have filed the required proof, as affirmed by my Insurance carrier.
B. [ ] I have no people working directly for me, therefore I require no Workmen's Compensation.
Should there be any change in my status during the exercise of this permit, I will so advise the Building Dept. and immediately comply with all requirements.
The undersigned has submitted plans, specifications and a plot plan in duplicate which are hereto attached, incorporated into and made a part of this application.
In consideration of the granting of the permit hereby petitioned for, the undersigned hereby agrees that if such permit is granted he will comply with the terms thereof, the Laws of the State of New York, the Ordinances of the Town of Amherst, and the Regulations of the various departments of the Town, County of Erie, and the State of New York; that he will preserve the established building line, request all necessary inspections & authorize & provide the means of entry to the premises & building to the Building Inspector, and that he will not use or permit to be used the structure or structures covered by the permit until sanitary facilities are completely furnished.
The undersigned hereby certifies that all of the information in this petition is correct and true.

ITEM	FEE
Trees on Town Hwys. ____ @ ____	
San. Sewer Dist. # ____ Trib. to ____	
Water Line Size ____ # BR ____	
SWDD # ____	
MIN.	*$5.—*
Cubage	
TOTAL FEE	*$5.—*

JOHN DOE
Record Owner
425 BRIANHILL
Address
WILLIAMSVILLE, N.Y., 1400
John Doe
Owners or Agents Signature
Subscribed and sworn to before me this *18*
day of *18 OCT* 19 *8X*
Richard Delaney
Notary Public, Erie County, New York

I do certify that I have examined the foregoing petition and building plans and plot plan and that they conform to Ordinances of Town of Amherst.
S. W. Zaty
Building Commissioner

Receipt is hereby acknowledged of the sum of $ *5.—*, being the permit fee established by the Town Board of Town of Amherst, N.Y.
John Shearer
Town Clerk

This permit shall expire *April 27,* 19 *8X* if building has not commenced.
ORIGINAL

Fig. 18-8 *Application for building permit.*

and effluent. Storm sewers may not be available or may not be able to handle the extra water. Water can accumulate quickly if there is a large paved surface. This water has to be drained fast during a rainstorm. If it is not, flooding of streets and houses causes damage. Some areas are just too flat to drain properly. It takes a lot of money to build sewers.

Sewers are of two types. The sanitary sewer takes the fluid and solids from toilets and garbage disposals. This is processed through a plant before the water is returned to a nearby river or creek. The storm sewer is usually much larger in diameter than the sanitary sewer. It has to take large volumes of water. The water is dumped into a river or creek without processing.

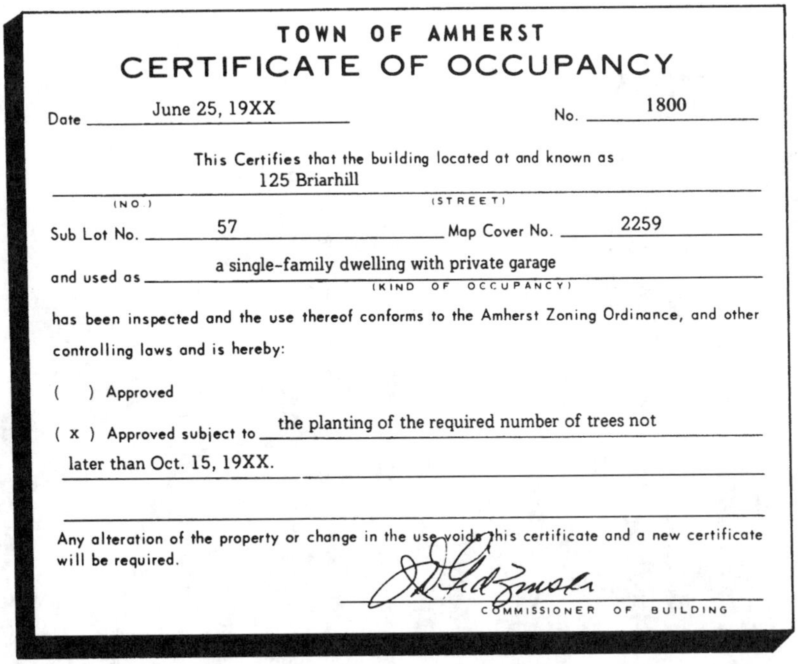

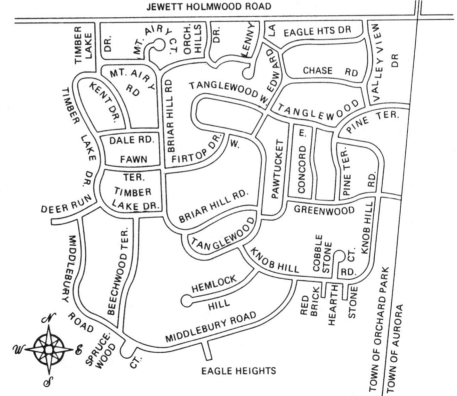

Fig. 18-9 *Certificate of occupancy.*
(Courtesy of Town of Amherst, NY)

TOWN OF AMHERST
CERTIFICATE OF OCCUPANCY

Date _____June 25, 19XX_____   No. _____1800_____

This Certifies that the building located at and known as
_____125 Briarhill_____
_(N O)_   _(STREET)_

Sub Lot No. _____57_____   Map Cover No. _____2259_____

and used as ____a single-family dwelling with private garage____
(KIND OF OCCUPANCY)

has been inspected and the use thereof conforms to the Amherst Zoning Ordinance, and other controlling laws and is hereby:

(   ) Approved

(x) Approved subject to ____the planting of the required number of trees not____
later than Oct. 15, 19XX. _____

Any alteration of the property or change in the use voids this certificate and a new certificate will be required.

_____
COMMISSIONER OF BUILDING

Fig. 18-10 *Map of a planned subdivision development.*

Local communities can have a hand in controlling their growth and problems. They can form community planning boards. These boards enforce zoning requirements, and the development of the community can progress smoothly.

TRENDS AND EFFECTS
Manufactured Housing

Some of the early attempts at manufacturing housing are shown in Fig. 18-11. Here are a number of houses

Fig. 18-11 Poured concrete houses. Example of an early type of manufactured house. (Universal Form Clamp)

that look alike. They have been made one after the other, as in a factory setup. They are made of poured concrete. The forms were moved from one place to the next and concrete was poured to make a complete house. As you can see, the housing looks rather dull. It would be hard to find your house if you didn't know the house number.

Factory-produced buildings are relatively recent. They are made in a number of sizes and shapes. Some of them are used as office buildings, as in Fig. 18-12. Large sections of the building are made in a factory. They are shipped to the construction site. Here they are bolted together. In this case, the parts make a very interesting building.

One of the advantages of this type of building is the minimum of waste and lost time in its manufacture. Construction workers do not have to move from one building to another, but can do the same job day after day. They can work inside. After a while the worker becomes very skilled at the job. Little material is wasted. Such things as plumbing, floors, walls, and electric facilities are included in the package.

Figure 18-13 shows another type of commercial building. This bank was built in sections inside a factory and then assembled on site. Everything is measured closely. This means little time is wasted on the job. The building can be put together in a short time if everything fits. It is very important that the foundation

and the water and plumbing lines are already in place in the floor.

Types of Factory-Produced Buildings

There are two types of factory-produced buildings. Modular buildings are constructed of modules. The module is completely made at the factory. A module is a part of a building, such as a wall or a room. The modular technique is very efficient. All it takes at the site is a crane to place the module where it belongs. Then the unit is bolted together. This type of construction can be used on hotels and motels. Housing of this type is very practical. It is used as a means of making dormitories for colleges.

In the other type of factory-produced building, only panels are constructed inside. The panels are assembled and erected at the site.

Some companies specialize in factory-produced buildings. Some are specialists in commercial and industrial buildings. Others are specialists in houses.

Some building codes will not allow factory-made housing. It is up to the local community whether or not this type of construction is allowed. In some cases the low-cost construction advantage is lost. The community can have code restrictions that make it expensive to put up such a house or building. In some cases this

Fig. 18-12 *Office building made in a factory and assembled at the site.* (Butler)

Fig. 18-13 *Commercial building made in a factory and assembled on site.* (Butler)

is good for the community. The type of construction should blend in with the community. If not, the manufactured house can become part of a very big slum in a short time.

A builder who does not adhere to the zoning and building codes in a locality may have to tear down the building. With zoning and building codes, a building permit system is usually used. The permit system enables the local government to monitor construction. This will make sure the type of building fits into the community plan.

Premanufactured Apartments

A combination of plant-built cores and precut woods is a key to volume production. This is especially true with apartment houses. Figure 18-14 shows a well-thought-out utility core production line. Components are fed into the final assembly, where they are integrated into serviceable units. The units get to the job site in a hurry. Production scheduling can then be easily controlled.

Figure 18-15 shows that wood floor joists, precut in the plant, go together quickly. The frame is equipped with wheels. It is then turned over. Sub and finish flooring is applied. Then the floor is rolled to the main production line.

In Fig. 18-16 you see wall panels framed with 2 × 4s. The bathroom wall uses 2 × 6s to accommodate the stand pipe. For extra strength on the main floor units, special 3 × 4 studs are used. This size replaces the standard 2 × 6. This dimension will carry the added weight yet still fit precisely with the upper floor cores.

After the drywall is applied to one side, the wall panels are stored. They are stored on wheeled carts. This means they are ready to move to the production line when they are needed. The utility core is beginning to take shape. Carpenters nail the walls in place. Plumbing fixtures are installed before the end wall goes up.

Ceiling panels complete the framing. Next, the plumbing and wiring is done and the furnace and water heater are installed. In the final step, the cabinets and appliances are installed. The interior walls are finished next. See Fig. 18-17. Each unit is hooked up and tested before it leaves the plant.

The units are wrapped and shipped to the job site. See Fig. 18-18. The cores are stacked on the foundation and hooked up to utilities. Local contractors add the interior and exterior walls. They use either panelized wood framing or conventional construction. This conventional framing is precut to fit the house or apartment unit.

Designs can vary. They can be adapted to meet the floor plan and outside requirements. See Fig. 18-19 for an example of manufactured apartments.

Manufactured Homes

One- and two-story homes can be constructed and completely finished in the plant. Interior finishing is completed on outdoor "stations," then the homes are moved onto their concrete foundations. It requires

Fig. 18-14 *Assembly line for premanufactured apartment house units.* (Western Wood Products)

Fig. 18-15 *Lifting a floor section and turning it over for sub- and finish flooring.* (Western Wood Products)

Fig. 18-16 *Framing a wall panel.* (Western Wood Products)

Fig. 18-17 *Ceiling panels are added to complete the framing.* (Western Wood Products)

Fig. 18-18 *Units are wrapped before being shipped to the job site.* (Western Wood Products)

Fig. 18-19 *Variations in design are possible in the premanufactured apartment house.* (Western Wood Products)

Fig. 18-20 *Floor joists being laid for a manufactured house.* (Western Wood Products)

unusually sturdy construction to move a home of this size. A web floor system of 2 × 4s and plywood is used for strength. See Fig. 18-20. Two homes are built simultaneously, with floors built over pits so that workers can install heating ducts, plumbing, and electrical facilities from below.

While the floor system is under construction, carpenters are building wall panels from precut wood. See Fig. 18-21. Wood-frame construction is used throughout the house.

At the building site, a giant machine places the foundation. See Fig. 18-22. Then, similar equipment moves the finished house from the factory onto the foundation. See Fig. 18-23. This is done in just 1 hour and 10 minutes.

Although in-plant production is standardized to cut costs, the homes avoid the manufactured look. The wood-frame construction makes it easy to adapt the designs.

The advantages of this type of building are lower cost of materials, less waste, and lower-cost worker skills. It is a great advantage when it comes to maintaining production schedules. The weather does not play too much of a role in production schedules when the house is built indoors.

Fig. 18-21 *Wall panels attached to the floor system.* (Western Wood Products)

Fig. 18-22 *A machine laying the foundation.* (Western Wood Products)

Fig. 18-23 *A machine moves the house over the foundation.* (Western Wood Products)

This type of construction is changing the way homes are being built. It requires a carpenter who can adapt to the changes.

CHAPTER 18
STUDY QUESTIONS

1. What does the future look like for a carpenter?
2. What are the opportunities for a carpenter?
3. When are carpenters usually nonunion?
4. What is a premanufactured house?
5. What is a building code?
6. How do you obtain a building permit?
7. What does the term *community planning* mean?
8. What is meant by overbuilding?
9. What happens if a builder does not follow local building codes?
10. What is the advantage of making a house in a plant or factory?

19
CHAPTER

Bathrooms

SEVERAL TRENDS INFLUENCE BATHROOM design and appearance. More bathrooms are being built per home and they include more open areas, more use of light, more outdoor views, and more features such as saunas, hot tubs, and spas. Many are expanded to provide extra space within the bath area or even a private terrace, garden patio, or deck. Decks and enclosed decks may include sauna, Japanese furo tub, or spa. See Figs. 19-1 through 19-4.

Unusual materials and combinations of materials are also used on walls. For example, wood, stone, tile, plastic laminates, and wainscotting. Color plays a more important role as well, with traditional antiseptic whites and grays being replaced by bright, chromatic colors sharply contrasting with sparkling whites or the natural textures of wood and stone. Walls, floors, and ceilings now contrast and supplement each other (Fig. 19-5) and often match the decor of nearby rooms.

Bathrooms are an essential part of any house. They are designed with various objectives in mind. Some are large; others small. Some have the minimum

Fig. 19-1 New trends in bathrooms include open space and comfort. (American Olean Tile)

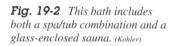

Fig. 19-2 This bath includes both a spa/tub combination and a glass-enclosed sauna. (Kohler)

Fig. 19-3 *A spacious bath with luxury and easy maintenance for compart shower, toilet, bidet, and twin vanities. The adjoining sunken bath overlooks a small garden.* (American Olean Tile)

Fig. 19-4 *The gleaming smooth tile contrasts with the wooden walls.* (American Olean Tile)

Fig. 19-5 *Contrasting tile colors.* (American Olean Tile)

of fixtures while others are very elaborate. In this chapter you will learn:

- How building codes influence bathroom design
- How plumbing, proper ventilation, and lighting are installed
- How to select the right toilet, tub, and shower
- How to space fittings
- How to select floors, fittings, countertops, and vanities

ROOM ARRANGEMENT

The current trend is toward large bathrooms, but many new homes still use the standard size of 5 × 8 feet. These smaller sizes are appropriate for extra bathrooms used mainly by guests or for smaller homes. Size is relatively unimportant if the desired features can be arranged within the space. When adding hot tubs, saunas, and so forth, it may be necessary to make projecting bay windows or decks. The key to maximum appearance, utility, and satisfaction is not size, but good arrangement.

In planning room arrangement, the typical pattern of movement in the room should be examined. Avoid major traffic around open doors or other projections. Also, consider the number of people who will be using the room at the same time. Many master bathrooms now have two lavatory basins because both members of the couple must rise, dress, and groom themselves at the same time. Having a slightly longer counter with two basins minimizes the frustrations of waiting or trying to use the same basin at the same time.

Entry to the bathroom should be made from a less public area of the house such as a hall. Doors should preferably open into a bathroom rather than out from it. Also, it is a good idea to consider what is seen when the bathroom door is open. The most desirable arrangement is one in which the first view is of the vanity or basin area. Next, a view of the tub or bathing area is appropriate. If possible, the toilet should not be the first item visible from an open door.

Placing new fixtures close to existing lines and pipes minimizes carpentry and plumbing costs. Kitchen plumbing can be located near bathroom plumbing, or plumbing for two bathrooms can be located from the same wall. See Fig. 19-6. Locating the plumbing core in a central area is a good idea.

Water hammer is a pounding noise produced in a water line when the water is turned off quickly. The noise can be reduced by placing short pieces of pipe, called air traps, above the most likely causes of quick turnoffs: the clothes washing machine and the dishwasher. Air traps are rather simple to install and can help quiet a noisy plumbing system.

If future changes are anticipated, rough in the pipes needed for the future and cap them off. This way not all of the walls will have to be opened later to make complicated connections. Upstairs and downstairs plumbing can be planned to run through the same wall. Again, this makes plumbing accessible and reduces the amount of carpentry and wall work required.

FUNCTION AND SIZE

Where the bathroom is large enough and where two or more people are frequently expected to use it at the same time, compartments provide required privacy and yet accessibility to common areas at all times. See Fig. 19-7. The bath and toilet are frequently set into separate compartments, which allows one to use either area in a degree of privacy while a lavatory or wash basin is used by someone else.

Sometimes, instead of one large bathroom, the space can be converted into two smaller ones, so that more family members can have access at the same time (Fig. 19-8).

There is nothing wrong with small. There are certain advantages in having small bathrooms. They are more economical to build, they can still accommodate several people, and they are much easier to keep clean. Small bathrooms can have all of the main features that larger bathrooms have by careful use of space through

Fig. 19-6 *Locating a second bath near the first minimizes the plumbing needs and expense.*

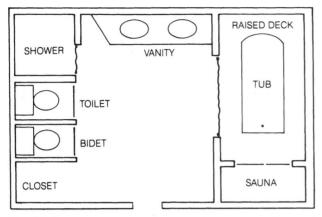

Fig. 19-7 *By putting in compartments for showers, toilets, and bidets, more privacy results as well as the use by more than one person at a time.*

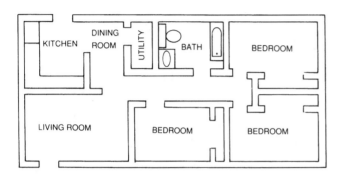

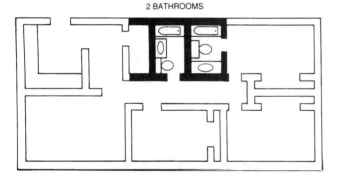

Fig. 19-8 *Two smaller baths can be placed where one larger one once existed.*

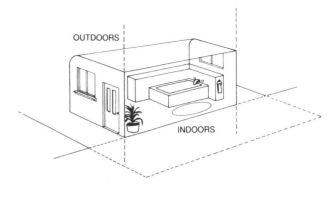

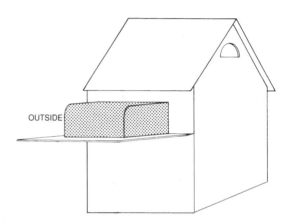

Fig. 19-9 *A cantilever can be used to extend the bathroom for a different result.*

built-in units, vanity counter cabinet space, and pocket doors for compartments. The comfort factors of ventilation, heating and air conditioning, and sound proofing, together with privacy and availability when needed, should be the main considerations.

It is relatively simple to make a small bathroom larger if it is located on an outside wall. Cantilever projections can be used to make windows into large bay windows (Fig. 19-9), or a deck to provide extra space. Also, a bathroom can be enlarged simply by taking other interior space to allow the construction of custom areas, the addition of privacy gardens or patios, or the addition of spas, saunas, and so forth.

BUILDING CODES

Because bathrooms are complex, building codes may be involved. Codes may designate the types of floors, the materials used, the way things are constructed, and where they are placed. There are usually good reasons for these regulations—even though they might not be obvious. Most cities require rigid inspections based on these codes.

Plumbing

The plumbing is perhaps the most obvious thing affected by codes. Rules apply to the size and type of pipes that can be used, placement of drains, and the placement of shut-off valves.

Shut-off valves allow the water to be turned off to repair or replace fixtures. They are used in two places. The first controls an area. It usually shuts off the cold water supply to different parts of the house. A modern three-bedroom house built over a full basement would typically have three area valves—one for the master bath, one for the main bath, and one for the kitchen. Outdoor faucets may be part of each subsystem based on locations, or they may be on a separate circuit. Second, each fixture, such as a hot-water heater, toilet, lavatory, and so forth, will have a cut-off valve located beneath it for both hot and cold water lines. Most building codes now require both kinds of valves.

Electrical

Many building codes specify three basic electrical requirements:

1. The main light switch must be located next to the door but outside the bathroom itself.
2. The main light switch must turn on both the light and a ventilation unit.
3. At least one electrical outlet must be located near the basin and it must be on a separate circuit from the lights. It should also have a Ground Fault Circuit Interrupter (GFCI).

Ventilation

Ventilation is often required for bathrooms. It is a good idea and has many practical implications. In the past years, doors and windows were the main sources of bathroom ventilation. They consumed no energy but allowed many fluctuations in room temperature. Forced ventilation is not required if the room has an outside window, but most codes require that all interior bathrooms (those without exterior walls or windows) have ventilation units connected with the lights.

Ventilation helps keep bathrooms dry to prevent the deterioration of structural members from moisture, rot, or bacterial action. It also reduces odors and the bacterial actions that take place in residual water and moisture.

Fans are vital in humid climates. They should discharge directly to the outdoors, either through a wall or through a roof, and not into an attic or wall space. Ventilation engineers suggest the capacity of the fan be enough to make 12 complete air changes each hour.

Spacing

Building codes may also affect the spacing of the fixtures such as the toilet, tub, and wash basin. Figure 19-10 shows the typical spaces required between these units. It is acceptable to have more space, but not less.

The purpose of these codes is to provide some minimum distance that allows comfortable use of the facilities and room to clean them. If there were no codes, some people might be tempted to locate facilities so

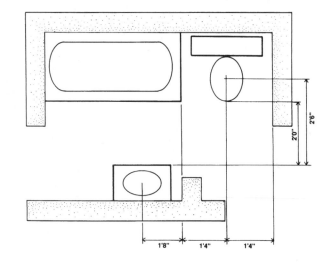

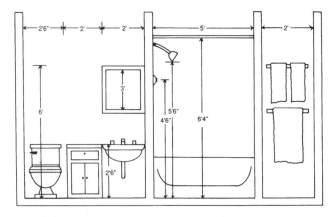

Fig. 19-10 *Typical spacing requirements for bathroom fixtures.*

close together that they could not be safely or conveniently used.

Other Requirements

Local codes might require the bathroom door to be at least two doors away from the kitchen. Some locations specify floors to be made of tile or marble, while others mandate tile or marble thresholds. Certain localities insist that a plastic film or vapor barrier be in place beneath all bathroom floors, and that all basin counters have splash backs or splashboards.

Some specifications might be strict, requiring rigid enclosures on showers or prohibiting the use of glass in shower enclosures. Others might be open, stating the minimum simply to be a rod on which to hang a shower curtain.

FURNISHINGS

Furnishings are the things that make a bathroom either pleasant or drab. They include the fixtures, fittings, and vanity area. The vanity area consists of a lavatory or basin, lights, mirror, and perhaps a counter.

Fixtures

The term *bathroom fixtures* refers to just about everything in the room that requires water or drain connections, such as lavatories (or basins), toilets, bidets, tubs, and showers. Features to consider for each include color, material, quality, cost, and style.

Generally, the better the quality, the higher the cost. Assuming three grades, the cheapest will not be made to withstand long, heavy use. The difference between medium and high quality will be the thickness of the plating and the quality of the exterior finish.

Toilets, bidets, and some lavatories are made of vitreous china, which is a ceramic material that has been molded, fired, and glazed, much like a dinner plate. This material is hard, waterproof, easy to clean, and resists stains. It is very long-lasting; in fact, some china fixtures are still working well after 100 years or more. White is the traditional color, but most manufacturers now provide up to 16 additional colors. Shopping around can give many insights into colors and features available.

Toilet Selection

Toilets, or water closets, come in several different mechanisms and styles. The styles include those that fit in corners (Fig. 19-11), rest on the floor, and have their weight supported entirely by a wall. Corner toilets are designed to save space and are particularly functional in very small rooms. Even the triangular tank fits into a corner to save space. The wall-hung toilet is expensive and requires sturdy mounts in the wall. The most common is the floor-mounted toilet (Fig. 19-12).

Floor-mounted toilets can be obtained in different heights. Older people generally find that an 18-inch-

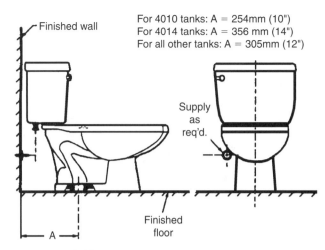

For 4010 tanks: A = 254mm (10")
For 4014 tanks: A = 356 mm (14")
For all other tanks: A = 305mm (12")

Fig. 19-12 *Roughing in dimensions.* (American Standard)

high toilet is easier to use than the 14-inch height of conventional units. Heights range from 12 to 20 inches and some can be purchased with handles and other accessories for the ill or handicapped.

The siphon jet is perhaps the most common type and is most recommended. It is quiet and efficient. Most of the bowl area is covered by water, making it easier to clean. The siphon action mechanism is an improvement over the siphon jet. It leaves no dry surface, thus making it easier to clean. It is efficient, attractive, and almost silent. It is also the most expensive. Most builders recommend the siphon jet because it costs less.

Up-flush toilets are used in basements when the main sewer line is above the level of the basement floor. These require special plumbing and must be carefully installed.

Toilet Installation

Installing the two-piece toilet requires some special attention to details to prevent leakage and ensure proper operation. The unit itself is fragile and should be handled with care to prevent cracking or breaking. Keep in mind that local codes have to be followed.

Roughing in Use Fig. 19-12 as a reference. Notice the distance from the wall to closet flange centerline. The distance varies according to the unit selected. For instance, American Standard's 4010 tanks require the distance *A* to be 254mm or 10 inches. Model 4014 needs 356mm or 14 inches for distance *A*. All other tanks require 12 inches or 305mm. The tank should not rest against the wall. Also notice the location of the water supply.

Install the closet bolts as shown in Fig. 19-13. Install the closet bolts in the flange channel and turn 90° and slide into place 6 inches apart and parallel to the wall.

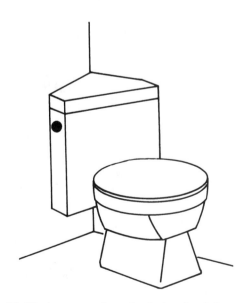

Fig. 19-11 *A space-saving toilet designed to fit into a corner.*

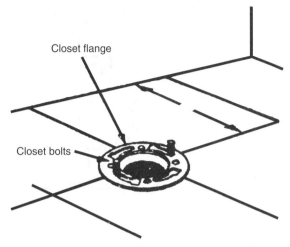

Fig. 19-13 *Closet flange and bolts.* (American Standard)

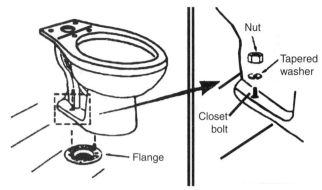

Fig. 19-15 *Positioning the toilet on the flange.* (American Standard)

Distance *A* shown here is the same as that in the previous figure. Next, install the wax seal; see Fig. 19-14. Invert the toilet on the floor (cushion to prevent damage). Install the wax ring evenly around the waste flange (horn), with the tapered end of the ring facing the toilet. Apply a thin bead of sealant around the base flange.

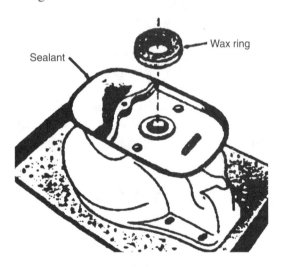

Fig. 19-14 *Installing the wax seal.* (American Standard)

Position the toilet on the flange as shown in Fig. 19-15. Unplug the floor waste opening and install the toilet on the closet flange so the bolts project through the mounting holes. Loosely install the retainer washers and nuts. The side of washers marked "This side up" *must* face up!

Install the toilet as per Fig. 19-16. Position the toilet squarely to the wall, and with a rocking motion press the bowl down fully on the wax ring and flange. Alternately tighten the nuts until the toilet is firmly seated on the floor. CAUTION! Do not overtighten the nuts or the base may be damaged. Install the caps on

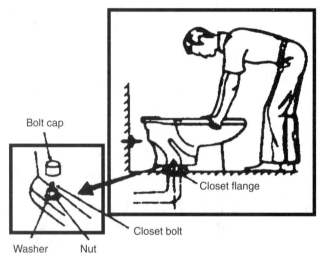

Fig. 19-16 *Installing the toilet.* (American Standard)

the washers, and if necessary, cut the bolt height to size before installing the caps. Smooth off the bead of sealant around the base. Remove any excess sealant. Next, install the tank. In some cases, where the tanks and bowls use the Speed Connect System, the tank mounting bolts are pre-installed. Install the large rubber gasket over the threaded outlet on the bottom of the tank and lower the tank onto the bowl so that the tapered end of the gasket fits evenly into the bowl water inlet opening (see Fig. 19-17) and the tank mounting bolts go through the mounting holes. Secure with metal washers and nuts. With the tank parallel to the wall, alternately tighten the nuts until the tank is pulled down evenly against the bowl surface. CAUTION! Do not overtighten the nuts more than required for a snug fit.

In those instances where the bolts are not pre-installed, start by installing large rubber gaskets over the threaded outlet on the bottom of the tank and then lower the tank onto the bowl so that the tapered end of the gasket fits evenly into the bowl water inlet opening. See Fig. 19-18. Insert the tank mounting bolts and

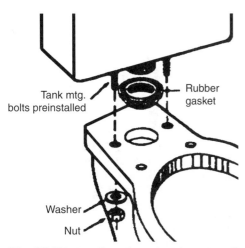

Fig. 19-17 *Installing the tank with pre-installed bolts.* (American Standard)

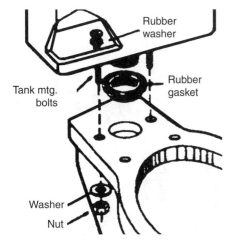

Fig. 19-18 *Installing the tank without pre-installed bolts.* (American Standard)

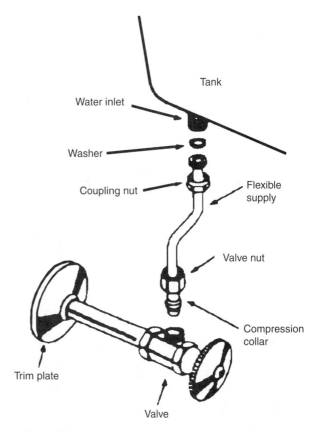

Fig. 19-19 *Connecting to the water supply.* (American Standard)

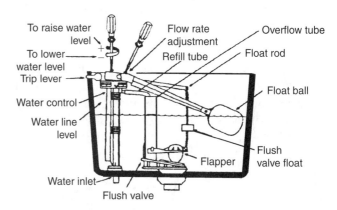

Fig. 19-20 *Making adjustments after installation.* (American Standard)

rubber washers from the inside of the tank, through the mounting holes, secure with metal washers and nuts. With the tank parallel to the wall, you can then alternately tighten the nuts until the tank is pulled down evenly against the bowl surface. Again, caution is needed to make sure the nuts are not overtightened. Install the toilet seat according the manufacturer's directions.

Connect the water supply line between the shut-off valve and tank water inlet fitting. See Fig. 19-19. Tighten the coupling nuts securely. Check that the refill tube is inserted into the overflow tube. Turn on the supply valve and allow the tank to fill until the float rises to the shut-off position. Check for leakage at the fittings; tighten or correct as needed.

Adjustments There are some adjustments that need to be made in most installations to ensure proper operation. See Fig. 19-20.

1. Flush the tank and check to see that the tank fills and shuts off within 30 to 60 seconds. The tank water level should be set as specified by the mark on the inside of the tank's rear wall.

2. To adjust the water level, turn the water level adjustment screw counter-clockwise to raise the level and clockwise to lower the level.

3. To adjust the flow rate (tank fill time), turn the flow rate adjustment screw clockwise to decrease the flow rate. This increases the fill time. Turn the adjustment screw counter-clockwise to increase the flow rate or decrease the fill time.

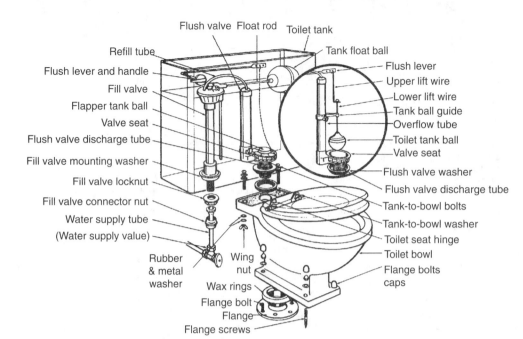

Flush valve Float rod Toilet tank
Refill tube
Flush lever and handle
Fill valve
Flapper tank ball
Valve seat
Flush valve discharge tube
Fill valve mounting washer
Fill valve locknut
Fill valve connector nut
Water supply tube
(Water supply value)
Rubber & metal washer
Wing nut
Wax rings
Flange bolt
Flange
Flange screws

Tank float ball
Flush lever
Upper lift wire
Lower lift wire
Tank ball guide
Overflow tube
Toilet tank ball
Valve seat
Flush valve washer
Flush valve discharge tube
Tank-to-bowl bolts
Tank-to-bowl washer
Toilet seat hinge
Toilet bowl
Flange bolts caps

Fig. 19-21 *All parts of the water closet or 1.6 gallon per flush, two-piece toilet.* (Plumbshop)

4. Carefully position the tank cover on the tank.

5. The flush valve float has been factory set and does not require adjustment. Repositioning the float will change the amount of water used, which might affect the toilet's performance.

Figure 19-21 identifies all the parts of the toilet.

Bidets

Bidets are common in Europe and are increasing in popularity in North America. They are used for sitz bath and are similar in shape and construction to a toilet (Fig. 19-12). The bidet is usually located outside the toilet. It is provided with both hot and cold water, and a spray or misting action is available as an extra component.

Vanity Areas

Vanity areas include the lavatory, lights, mirror, and often a shelf or counter. Lavatories, or basins, can be obtained in a wide variety of shapes sizes and colors. Two basic styles comprise the majority: counter and wall-mounted basins.

Basins and counters are typically located 31 to 34 inches above the floor, although they can be placed higher or lower. Most builders and designers suggest an 8-inch space between the top of the basin or counter and the bottom of any mirror or cabinet associated with it. Splash backs may or may not be required by local codes. They can be part of the basin or part of the counter.

Double basins should be widely separated. A minimum of 12 inches should separate the edges, but where space is available this space should be even

Fig. 19-22 *Toilet and bidet combinations are popular.* (Kohler)

greater. Basin should not be located closer than 6 to 8 inches from a wall. See Figs. 19-23 and 19-24.

Countertop Basins

Countertop basin units are popular. See Fig. 19-25. They have one or two basins and frequently incorporate storage areas beneath them. Most basins are designed for counter use and are made from steel coated with a porcelain finish. They can be obtained in traditional white or in a variety of colors that will match the colors of toilet, tub, and bidet.

Most basins used currently are self-rimming units that seal directly to the countertops for neater, quicker installations. Other styles are available that flush-mount with the countertop and are sealed by a metal or plastic

Fig. 19-23 *Double basins extend the vanity area so that it can be used by two people at one time—ideal for working couples.* (American Olean Tile)

Fig. 19-24 *A double-basin counter combines crisp lines and easy maintenance.* (Formica)

rim, and recessed units that mount below the surface of the counter. Recessed units require more care to install and are sometimes difficult to keep clean. Note that the self-rimming unit in Fig. 19-26 also has a spray unit for hair care.

Countertop materials Countertops are made from a variety of materials. The most common are probably ceramic tile and plastic laminates. In recent years, the plastic laminates have changed from marble-like patterns to bright, solid colors. While white remains a constant favorite, laminates are available in a wide variety of colors including black, earth tones, and pastels. They can be obtained in a variety of surface finishes and styles, including special countertops with molded splash back and front rims.

Ceramic tile is also an ideal material for counters. It is a hard, durable surface that is waterproof and easy

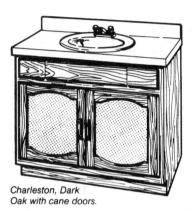

Charleston, Dark
Oak with cane doors.

Nova, Honey Oak
with planked doors.

Fig. 19-25 *Cabinets topped by one or two basin counters are economical and practical.* (NuTone)

to clean, and has a beauty that does not fade or wear out. Figure 19-23 features tile counters. Figure 19-24 shows good use of plastic laminates for a countertop.

Other materials commonly used for countertops include slate, marble, and sometimes wood. Butcher block construction is gaining in popularity and can be

Fig. 19-26 *Self-rimming basins may incorporate hair grooming features.* (Kohler)

Fig. 19-27 *Pedestal basins are obtainable in a wide variety of shapes and colors.* (Kohler)

finished with either natural oils and waxes or be heavily coated with special waterproof plastic finishes. Counters need not be waterproof, but if they are not, water should be wiped up immediately.

Integral tops and basins Counters and basin can be made as one solid piece. The advantage of one-piece construction is that there are no seams to discolor or leak. Both fiberglass and synthetic marble are used for these units. Some provide a complete enclosure for the vanity unit. This protects walls and underlying structures from water damage while being striking to look at.

Wall-Mounted Basins

Wall-mounted basins may be entirely supported by the wall or may be placed to the wall and supported by a pedestal. See Fig. 19-27. Other styles are supported by metal legs at the two front corners. Wall-hung basins placed in corners save space and allow easier movement in smaller bathrooms.

BATHING AREAS

A bathing area may be a tub, a shower, or a combination of both. There are many types and varieties of each, and custom units may be built for all type baths and combination baths.

Bathtubs

Many people like to soak and luxuriate in a tub. Tubs can be purchased in a variety of shapes and sizes (see Fig. 19-28) and can exactly match the color of the other fittings and fixtures in the bathroom. Tubs can also be

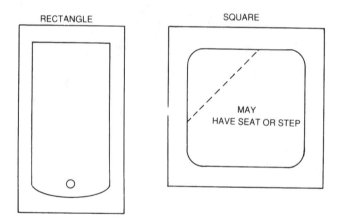

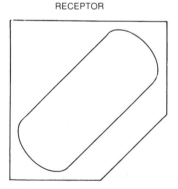

Fig. 19-28 *The three most common tub shapes are rectangular, made to fit into an alcove, square, and receptor. The square and receptor tubs have a longer tub area.*

purchased in shower-and-tub combinations that match the color of the other fixtures.

Whether or not to replace an old unit can be an important decision when remodeling. Old tubs can be refinished and built in to provide a newer and more modern appearance. Fittings can be changed and showers added, along with updated wall fixtures such as shelves and soap dishes. Sometimes the antique appearance is preferred, in which case reworking is more desirable than replacing.

Sunken tubs may be standard tubs with special framing to lower them below the surface of the floor. They may be custom-made or may incorporate specially manufactured tubs. Before installing a sunken tub, be sure there is room beneath the bathroom floor. When space below is not available, the alternative is to raise the level of the floor in the remainder of the room. Of course, this presents considerable complications with existing doors and floors. One compromise is to construct a wide pedestal around the lip of the tub (Fig. 19-29). This pedestal can then become a sitting area or a shelf for various articles, and can even have built-in storage.

Japanese tubs (*furos*) can be built-in and sunken (Fig. 19-30). The tub is simply a deep well that accommodates one or more people on a seat. The seat can be made of wood (the traditional Japanese style), tile, or other material. The soaker is immersed to the neck or shoulders for relaxation. The actual washing with soap or cleaners is traditionally done outside the tub area.

Fig. 19-30 *Furos, or Japanese style tubs, are becoming increasingly popular.* (American Olean Tile)

The standard rectangular tub (Fig. 19-31) is 60 inches long, 32 inches wide, and 16 inches high. It is enclosed or sided on one side, but open at both ends and the remaining side. This shape was designed to fit into an alcove as shown.

Receptor tubs (Fig. 19-31) are squarish, low tubs ranging in height from 12 to 16 inches. Rectangular shapes makes them ideal for corner replacement. They are approximately 36 inches long and 45 inches wide. Square tubs are similar to receptor tubs in that they can be recessed easily into corners and alcoves. Some have special shelves set into corners and some incorporate controls in these areas. Square tubs are approximately 4 to 5 feet square, increasing in three-inch increments. The receptor tub has a diagonal opening while the square tub may have a truly square basin. The disadvantage of a square tub is that it requires a larger volume

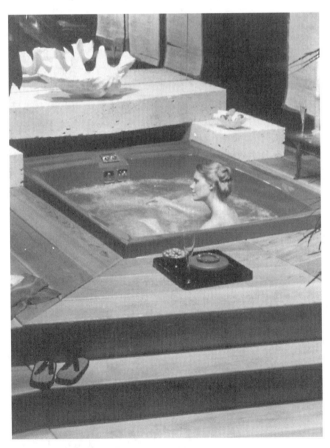

Fig. 19-29 *A pedestal can be built around a tub for many reasons. The pedestal can hide pumps, plumbing, and electrical support units.* (Jacuzzi)

Fig. 19-31 *This receptor tub provides a longer than normal bathing area but requires less space and water than a full square tub. It also incorporates a hydromassage unit.* (Kohler)

of water and getting in and out is sometimes difficult, particularly for elderly people.

Custom tubs (Fig. 19-32) often feature striking use of ceramic tiles. Tiles can be used in combination with metal, stone, and wood. Specially designed custom showers can have stone and glass walls with low tile sides and are used for sitting and storage.

Fig. 19-32 *Tubs can be custom-built to any size and shape. This tile tub features the same color and style of tile for the tub, walls, counter, and floor.* (American Olean Tile)

A spa, also called a hydromassage, whirlpool, water jet, or hot tub, requires extra space to house the jet pump mechanism and the special piping and plumbing. Also, wiring is needed to power the motors. These hydromassage units are available in a variety of sizes

and can be incorporated into standard tubs. If they are to be installed in addition to the tub, extra space is required. People with smaller bathrooms find that the combination tub and hydromassage unit (Fig. 19-33) is satisfactory.

Fig. 19-33 *The hydromassage action can be combined with more conservative settings to look like a conventional tub.* (Jacuzzi)

Specially shaped tubs can be built by making a frame (Fig. 19-34) that is lined or surfaced with a material such as plywood, which generally conforms to the size and shape desired. Make sure the frame and lining will hold the anticipated weight and movement. Next, tack or staple a layer of fiberglass cloth to the form. Pull the cloth to form the shapes around the corners that are desired. If additional support is desired during the shaping process, corner spaces can be filled in and rounded with materials such as fiber insulation. The tub will not need support in the corners, and the fiberglass material itself will be strong enough once completed.

After the cloth has been smoothed to the shape desired (a few seams are all right and will be sanded smooth later on), coat the cloth with a mixture of resin. Tint the resin the color desired for the tub and use the same color for all coats. Allow the first coat to harden and dry completely. This will stiffen the cloth and give the basic shape for the tub. Next, apply another coat of resin and lay the next layer of cloth onto it. Allow this to harden and repeat the process. At least three layers of fiberglass cloth or fiber will be needed. It is best to add several coats of resin after the last layer of cloth. Three layers of glass fiber are generally applied followed by three more coats of resin. The last three coats of resin are sanded carefully to provide a smooth, curving surface in the exact shape desired.

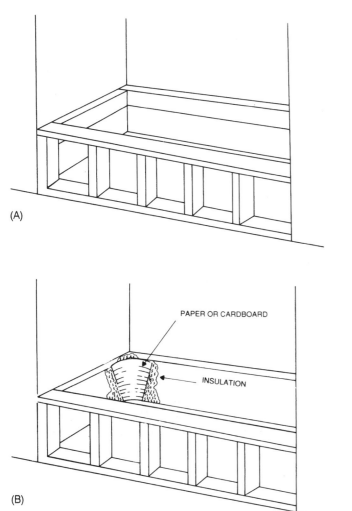

(A)

PAPER OR CARDBOARD

INSULATION

(B)

STAPLE OR TACK CLOTH
TO FRAME AT
TOP AND BOTTOM

LAY CLOTH PATCH
OVER CORNERS

STUFFING IN
CORNERS

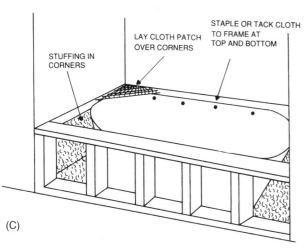

(C)

Fig. 19-34 *Fiberglass tubs can be custom-built to almost any shape. (A) Wooden frame gives dimension and support. (B) Insulation or cardboard defines the approximate shape and contour. (C) A layer of fiberglass is applied. It can be tinted any color to match decor.*

Showers

Showers may be combined with the tub (Fig. 19-35) or may be separate (Fig. 19-36). In some smaller bath-

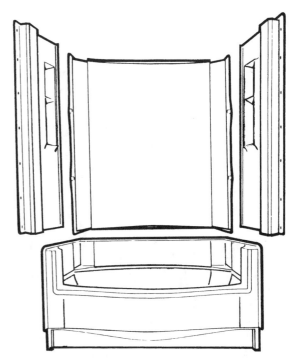

Fig. 19-35 *Combination tubs and showers are perhaps the most common unit. They can be made of almost any material or any combination of materials.* (Owens-Corning Fiberglas)

Fig. 19-36 *Showers are often separate compartments for increased privacy.*

rooms, showers are the only bathing facility. They can be custom-made to fit an existing space, or standard units made from metal or fiberglass may be purchased. Shower stalls are available with the floor, three walls, and sometimes molded ceiling as a single large unit. When made of fiberglass, they are molded as a single integral unit. When made of metal, they are joined by permanent joints or seams.

Standard showers are also available unassembled. This allows a unit to be brought in through halls and doors. The components consist of the drain mechanism, a floor unit, and wall panels.

Manufactured shower units often have handholds, rail ledges for shampoo, built-in soap dishes, and so forth, molded into the walls. Custom-built units can also have conveniences molded into the walls but use separate pieces (Fig. 19-37).

Fig. 19-37 *A wide variety of accessory fittings such as hand-holds, soap dishes, and storage are available for showers.* (NuTone)

Fiberglass units are usually more expensive than metal ones. Metal units have greater restraints on their design and appearance and are noisier than fiberglass units. Fiberglass should not be cleaned with abrasive cleaners.

Custom-built units are made from a variety of materials, including ceramic tile, wood, and laminated plastics. Tile is an ideal material but is relatively expensive. The grout between the tiles is subject to stains and is difficult to clean, but special grouts can be used to minimize these disadvantages. Floors for custom-made shower units must be carefully designed to include either a metal drain pan or special waterproofing membranes beneath the flooring.

Using laminated plastics for walls of showers provides several advantages. The material is almost impervious to stains and water, and a variety of special moldings allow the materials to be used. The large size of the panels makes installation quick and easy.

The bottom surface of a shower should have a special non-slip texture; that is, it should be rough enough to prevent skidding but smooth enough to be comfortable. Neither raised patterns on the bottom of a bathtub nor stick-ons are very good.

Shelves, recessed hand-holds, and other surfaces in a shower area should be self-draining so that they will not hold accumulated water. Shelves approximately 36 to 42 inches above the floor of the shower are convenient for the soap, shampoo and other items. Hand-holds, vertical grab-bars, and other devices used for support while entering or leaving the shower should be firmly anchored to wall studs.

FITTINGS

Fittings is the plumber's word for faucets, handles, and so forth. The available array of size, shape, and finish of fittings is almost endless. Both single and double faucets are obtainable with chrome, stainless, or gold-tone finishes. They can be operated by one or two handles that may be made of any material from metal to glass.

LIGHTING AND ELECTRICAL CONSIDERATIONS

Some of the work of the carpenter is influenced by the nature of the custom-made bathrooms. The following is shown to ensure proper installation of the bathroom.

Older bathrooms were usually lit with a single overhead light. Later, one or two lights were added near the mirrored medicine chest above the basin. Older bathrooms frequently have neither sufficient lighting nor electrical power outlets for hair dryers, electric razors, electric toothbrushes, and water jets for dental hygiene.

General lighting can be enhanced by using ceiling panels or by wall lamps. See Fig. 19-38. Another lighting idea is to use hanging swag lamps. See Fig. 19-39. A large, free-hanging swag lamp would be inappropriate for a small bathroom. General lighting and lighting for the basin areas may be combined for smaller bathrooms.

Basin, or vanity, lamps should be placed above or to the sides of mirrors. They can also be placed in both locations. Light should not shine directly into the eyes but should come from above or to the side. One good

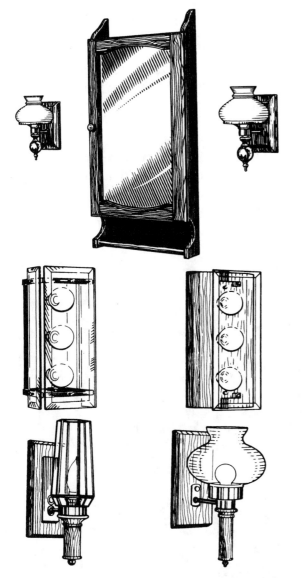

idea is to use the special "Hollywood" make-up lights around the mirrors (see Fig. 19-40). They eliminate glare and give good light for grooming. They can be a single string of lights above the mirror or surround the mirror on the sides and top. They are best controlled near the basin area. Separate controls may be desirable so that the user can adjust the lighting and reduce the number of bulbs lit. In addition, dimmer switches vary the intensity.

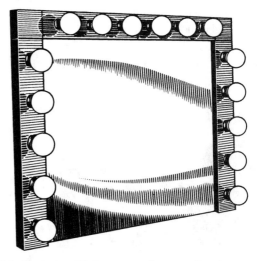

Fig. 19-40 *Strips of lights surrounding a vanity mirror are both useful and popular. This Hollywood-style can be controlled by a dimmer switch for increased flexibility.* (NuTone)

Fig. 19-38 *Specialized lighting for vanity areas can be mounted on walls.* (NuTone)

Special lights may also be desired for radiant heating and for keeping a suntan. The controls for these lights might be housed in several locations, or with the general lighting switch. The controls for special areas such as bathing or toilet compartments are controlled inside the bathroom.

Bathroom outlets should be protected by a GFCI. This is a term used for the *ground fault circuit interrupter*. If you are about to be shocked it will turn off the circuit.

Newer bathrooms use skylights to give natural light for grooming. They also make good use of picture windows opening onto a patio or deck. Stained glass is also used in some bathroom designs.

The main thing in planning is to avoid a single light source and to use special area lights where needed. Also plan enough outlets to power everything that will be used.

BATHROOM BUILT-INS

A variety of storage space is needed for a bathroom. Tissues and towels are stored for instant use. Also, if several people use a bathroom, more towel space is

Fig. 19-39 *Swag lamps are used in larger bathrooms for special lighting effects.* (NuTone)

needed. Clean towels and washcloths, soaps, shampoo, dental articles, grooming items, cleaning, equipment, dirty clothes hampers, and even the family linens are all potential storage problems.

Many bathrooms include deep shelves that are often underused because no one can reach the back of them, particularly the top ones. Often the items stacked in the front part of the shelf block the accessibility of the items in back. Sometimes built-ins are simply shelves or drawers that rotate, swing out, or move to allow better use of these back areas that cannot be reached easily. Figure 19-41 shows some ideas to improve efficiency. A closet or cabinet is a must for a bathroom. It should have shelves for a variety of sizes and may incorporate a laundry hamper.

Fig. 19-41 *Bathroom storage and convenience can be increased by using racks and shelves that swing out or roll out to save shelf space.* (Closet Maid by Clairson)

A laundry chute can be built into the space normally used for a hamper. It will take less space and perhaps save a lot of stair-climbing. If a home has a basement laundry, a hole can be cut through the floor deck (but not the frame) so that soiled laundry can be dropped directly into the basement. It is not necessary

for the laundry to drop directly into the laundry area, although that would be preferable.

Cabinets beneath the basins are extremely popular because they provide additional storage space housed in an attractive unit. In smaller bathrooms, this becomes more important because there is less space for closets or other built-ins.

People who have basins with wide counters may have to lean forward when using a mirror mounted on the wall. Medicine cabinets with mirrors (Fig. 19-42) hold the storage area out from the wall, which reduces the need to lean.

Fig. 19-42 *A cabinet that extends out from the wall minimizes the strain caused by leaning toward the mirror.* (NuTone)

Where large mirrors are used behind basins, medicine cabinets can be built-in on the ends of the counter. The doors may be either mirrors or wood. Mirrors provide more light and more three-dimensional view, while the wood finishes may fit in with the decor. See Fig. 19-43.

Building racks and shelves into walls is especially helpful when space is at a premium or when storage space is located in an area of high traffic flow. By using the space between the wall studs, items are recessed out of the way. Items such as toothbrushes, water jets, and electric razors can be kept readily accessible, hidden from view and protected from splashing water by building them into the walls. Electrical outlets can also

Fig. 19-43 *Some examples of the many types of cabinets and doors available.* (NuTone)

be built into these areas to power these items. Sliding or hinged doors will hide them from view and keep water and dust off them.

FLOORS AND WALLS

Bathroom floors and walls are subject to water spills and splashes, heat, and high humidity. They should be capable of withstanding the heaviest wear under the most extreme conditions. Good flooring materials for bathrooms include ceramic and quarry tile, stone and brick, wood, resilient flooring, and special carpeting. Good wall materials include stone, tile, wood, and plastic laminates. Walls can also be painted, but regular flat paint is not advised around splash areas such as basins, tubs, and showers. If paint is to be used in those areas, use the best waterproof gloss or semigloss paint.

CHAPTER 19 STUDY QUESTIONS

1. How do building codes influence bathroom design?
2. What are the new trends in bathrooms?
3. What is water hammer?
4. What is an air trap?
5. What are the advantages to small bathrooms?
6. What do building codes in local communities have to do with bathrooms? How much water does it take to flush a new toilet today?
7. How is ventilation obtained for bathrooms with no windows?
8. Where are up-flush toilets used?
9. Where are wax seals used?
10. What are Japanese tubs called?

20
CHAPTER

Construction for Solar Heating

SOLAR SYSTEMS FOR HEATING AND COOLING the house are often thought of as a means of getting something for nothing. After all, the sun is free. All you have to do is devise a system to collect all this energy and channel it where you want it to heat the inside of the house or to cool it in the summer. Sounds simple, doesn't it? It can be done, but it is not inexpensive.

There are two ways to classify solar heating systems. The *passive* type is the simplest. It relies entirely on the movement of a liquid or air by means of the sun's energy. You actually use the sun's energy as a method of heating when you open the curtains on the sunny side of the house during the cold months. In the summer you can use curtains to block the sun's rays and try to keep the room cool. The passive system has no moving parts. (In the example, the moving of the curtains back and forth was an exception.) The design of the house can have a lot to do with this type of heating and cooling. It takes into consideration if the climate requires the house to have an overhang to shade the windows or no overhang so the sun can reach the windows and inside the house during the winter months.

The other type of system is called active. It has moving parts to add to the circulation of the heat by way of pumps to push hot water around the system or fans to blow the heated air and cause it to circulate.

Solar heating has long captivated people who want a care-free system to heat and cool their residences. However, as you examine this chapter, you will find that there is no free source of energy. Some types of solar energy systems are rather expensive to install and maintain. In this chapter you will learn:

- How active and passive systems work
- How cooling and heating are accomplished
- Advantages and disadvantages of an underground house
- How various solar energy systems compare with the conventional methods used for heating and cooling

PASSIVE SOLAR HEATING

Three concepts are used in the passive heating systems: direct, indirect, and isolated gain. See Fig. 20-1. Each of these concepts involves the relationship between the sun, storage mass, and living space.

Indirect Gain

In indirect gain, a storage mass is used to collect and store heat. The storage mass intercedes between the sun and the living space. The three types of indirect

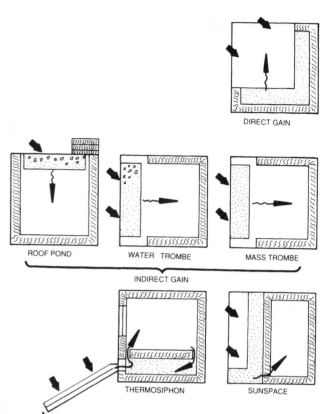

Fig. 20-1 *Three concepts for solar heating: direct gain, indirect gain, and isolated gain.*

gain solar buildings are mass trombe, water trombe, and roof pond. See Fig. 20-1.

Mass trombe buildings involve only a large glass collector area with a storage mass directly behind it. There may be a variety of interpretations of this concept. One of the disadvantages is the large 15-inch-thick concrete mass constructed to absorb the heat during the day. See Fig 20-2. The concrete has a black coating to aid the absorption of heat. Decorating around this gets to be a challenge. If it is a two-story house, most of the heat will go up the stairwell and remain in the upper bedrooms. Distribution of air by natural convection is viable with this system since the volume of air in the space between the glazing and storage mass is being heated to high temperatures and is constantly trying to move to other areas within the house. The mass can also be made of adobe, stone, or composites of brick, block, and sand.

Cooling is accomplished by allowing the 6-inch space between the mass and the glazed wall to be vented to the outside. Small fans may be necessary to move the hot air. Venting the hot air causes cooler air to be drawn through the house. This will produce some cooling during the summer. The massive wall and

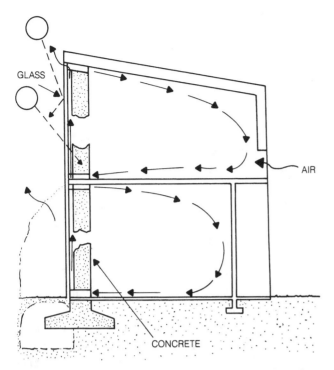

Fig. 20-2 *Indirect gain passive solar heating: mass trombe.*

ground floor slab also maintains cooler daytime temperatures. Trees can be used for shade during the summer and then will drop their leaves during the winter to allow for direct heating of the mass.

A hot-air furnace with ducts built into the wall is used for supplemental heat. Its performance evaluation is roughly 75 percent passive heating contribution. Performance is rated as excellent. For summer, larger vents are needed and in the winter too much heat rises up the open stairwell.

Water trombe buildings are another of the indirect-gain passive heating types. See Fig 20-3. The buildings have large glazed areas and an adjacent massive heat storage. The storage is in water or another liquid, held in a variety of containers, each with different heat exchange surfaces to storage mass ratios. Larger storage volumes provide greater and longer-term heat storage capacity. Smaller contained volumes provide greater heat exchange surfaces and faster distribution. The trade-off between heat exchange surface versus storage mass has not been fully developed. A number of different types of storage containers, such as tin cans, bottles, tubes, bins, barrels, drums, bags, and complete walls filled with water, have been used in experiments.

A gas-fired hot water heating system is used for backup purposes. That is primarily because this system has a 30 percent passive heating contribution. When fans are used to force air past the wall to improve the heat circulation, it is classified as a *hybrid system*.

The *roof pond* type of building is exactly what its name implies. The roof is flooded with water. See Fig. 20-4. It is protected and controlled by exterior movable insulation. The water is exposed to direct solar gain that causes it to absorb and store heat. Since the heat source is on the roof, it radiates heat from the ceiling to the living space below. Heat is by radiation only. The ceiling height makes a difference to the individual being warmed since radiation density drops off with distance. The storage mass should be uniformly spaced so it covers the entire living area. A hybrid of the passive type must be devised if it is to be more efficient. A movable insulation has to be utilized on sunless winter

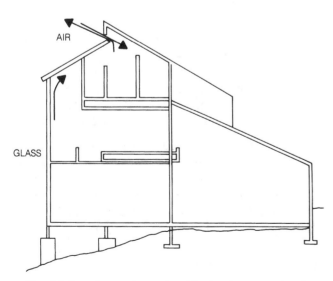

Fig. 20-3 *Indirect gain passive solar heating: water trombe.*

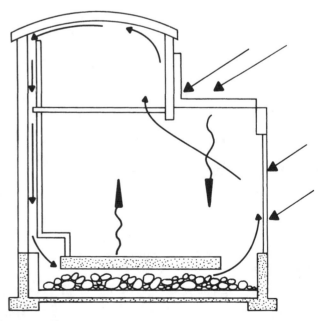

Fig. 20-4 *Indirect gain passive solar heating: roof pond.*

days and nights to prevent unwanted heat losses to the outside. It is also needed for unwanted heat gain in the summer.

This type of system does have some cooling advantages in the summer. It works well for cooling in parts of the country where significant day-to-night temperature swings take place. The water is cooled down on summer evenings by exposure to the night air. The ceiling water mass then draws unwanted heat from the living and working spaces during the day. This takes advantage of the temperature stratification to provide passive cooling.

This type of system has not been fully tested and no specifics are known at this time. It is still being tested in California.

Direct Gain

The direct gain heating method uses the sun directly to heat a room or living space. (See Fig. 20-5.) The area is open to the sun by using a large windowed space so the sun's rays can penetrate the living space. The areas should be exposed to the south, with the solar exposure working on massive walls and floor areas that can hold the heat. The massive walls and thick floors are necessary to hold the heat. That is why most houses cool off at night even though the sun has heated the room during the day. Insulation has to be utilized between the walls and floors and the outside or exterior space. The insulation is needed to prevent the heat loss that occurs at night when the outside cools down.

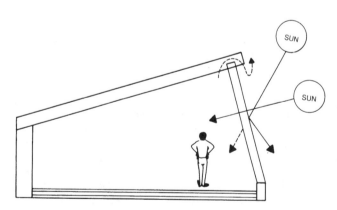

Fig. 20-5 *Direct gain passive solar heating.*

Woodstoves and fireplaces can be used for auxiliary heat sources with this type of heating system. Systems of this type can be designed with 95 percent passive heating contribution. Overhangs on the south side of the building can be designed to provide shading against unwanted solar gain.

Isolated Gain

In this type of solar heating the solar collection and storage are thermally isolated from the living spaces of the building.

The *sunspace* isolated gain passive building type collects solar radiation in a secondary space. The isolated space is separate from the living space. This design stores heat for later distribution.

This type of design has some advantages over the others. It offers separation of the collector-storage system from the living space. It is midway between the direct gain system where the living space is the collector of heat, and a mass or water trombe system that collects heat indirectly for the living space.

Part of the design may be an atrium, a sun porch, a greenhouse, and a sunroom. The southern exposure of the house is usually the location of the collector arrangement. A dark tile floor can be used to absorb some of the sun's heat. The northern exposure is protected by a berm and a minimum of window area. The concrete slab is 8 inches thick and will store the heat for use during the night. During the summer the atrium is shaded by deciduous trees. A fireplace and small central gas heater are used for auxiliary heat. Performance for the test run has reached the 75 to 90 percent level.

The *thermosiphon* isolated gain passive building type generates another type of solar heated building. See Fig. 20-6. In this type of solar heating system the collector space is between the direct sunshine and the living space. It is not part of the building. A thermosiphoning heat flow occurs when the cool air or liquid naturally falls to the lowest point—in this case, the collectors. Once heated by the sun, the heated air or liquid rises up into an appropriately placed living space, or it can be moved to a storage mass. This causes a somewhat cooler air or liquid to fall again. This movement causes a continuous circulation to begin. Since the collector space is completely separate from the building, the thermosiphon system resembles an active system. However, the advantage is that there are no external fans or blowers needed to move the heat transfer medium. The thermosiphon principle has been applied in numerous solar domestic hot water systems. It offers good potential for space heating applications.

Electric heaters and a fireplace with a heatilator serve as auxiliary sources. The south porch can be designed to shade the southern face with overhangs to protect the *clerestory* windows. Clerestory windows are those above the normal roofline such as in Fig. 20-6. Cross-ventilation is provided by these windows.

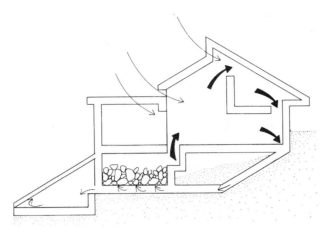

Fig. 20-6 *Isolated gain passive heating system: thermosiphon.*

Time Lag Heating

Time lag heating was used by the American Indians of the Southwest where it got extremely hot in the day and cooled down during the night. The diurnal (day-to-night) temperature conditions provided a clear opportunity for free or natural heating by delaying and holding daytime heat gain for use in the cool evening hours.

In those parts of the country where there are significant (20 to 35 degrees F) day-to-night temperature swings, a building with thermal mass can allow the home itself to delay and store external daytime heating in its walls. The captured heat is then radiated to the building interior during the cool night. Internal heat gains come from people, lights, and appliances. This heat can also be absorbed and stored in the building structure.

Massive or heavy construction of walls, floors, and ceilings are used for this type of solar heating. Because it is dense, concrete, stone, or adobe has the capacity to hold heat. As the outside temperature rises during the day, so does the temperature of the building surface. The entire wall section heats up and will gradually release the stored heat to the room by radiation and convection. Two controls can make time lag heating systems most effective for passive heating. Two, four, eight, even twelve hours of delay can be guaranteed by building walls of the right thickness and density. Choosing the right material and thickness can allow you to control what hour in the evening you begin heating. Exterior or sheathing insulation can prevent the heat storage wall from losing its carefully gained heat to the outside, offering more passive heat to the inside. For additional heat, winter sunshine can also be collected and stored in massive walls, provided adequate shading is given to these walls to prevent overheating in the summer.

Uninsulated massive walls can cause problems with the auxiliary heating system if improperly used. In climates where there is no day-to-night temperature swing, uninsulated massive walls can cause problems. Continually cold or continually hot temperatures outside will build up in heavy exterior walls and will draw heat from the house for hours until these walls have been completely heated from the inside.

This type of house has been built and tested in Denver, Colorado with a 65 percent passive heating contribution and 60 percent passive cooling contribution.

Underground Heating

The average temperature underground, below the frostline, remains stable at approximately 56 degrees. This can be used to provide effective natural heating as outside temperatures drop below freezing. The massiveness of the earth itself takes a long time to heat up and cool down. Its average annual temperature ranges between 55 and 65 degrees, with only slight increases at the end of summer and slight decreases at the end of winter. In climates with severe winter conditions or severe summer temperatures, underground construction provides considerably improved outside design temperatures. It also reduces wind exposure. The underground building method removes most of the heating load for maximum energy conservation.

One of the greatest disadvantages of the underground home is the humidity. If you live in a very humid region of the country, excessive humidity, moisture, and mildew can present problems. This is usually no problem during the winter season, but can become serious in the summer. Underground buildings cannot take maximum advantage of comfortable outside temperatures. Instead, they are continuously exposed to 56 degree ground temperatures. So, if you live in a comfortable climate, there is no need to build underground. You could, however, provide spring and fall living spaces outside the underground dwelling and use it only for summer cooling and winter heat conservation. Make sure you do not build on clay that swells and slides. And do not dig deep into slopes without shoring against erosion.

An example is a test building that was constructed in Minneapolis. The building was designed to be energy efficient. It was also designed to use the passive cooling effects of the earth. Net energy savings over a conventional building are expected to be 80 to 100 percent during the heating period and approximately 45 percent during the cooling period.

PASSIVE COOLING SYSTEMS

There are six passive cooling systems. These include natural and induced ventilation systems, desiccant systems, and evaporative cooling systems, as well as the passive cooling that can be provided by night sky temperature conditions, diurnal (day-to-night) temperature conditions, and underground temperature conditions.

Natural Ventilation

Natural ventilation is used in climates where there are significant summer winds and sufficient humidity (more than 20 percent) so that the air movement will not cause dehydration.

Induced Ventilation

Induced ventilation is used in climate regions that are sunny but experience little summer wind activity.

Desiccant Cooling

Another name for desiccant cooling is *dehumidification*. This type of cooling is used in climates where high humidities are the major cause of discomfort. Humidities greater than 70 or 80 percent RH (relative humidity) will prevent evaporative cooling. Thus, methods of drying out the air can provide effective summer cooling.

This type of cooling is accomplished by using two desiccant salt plates for absorbing water vapor and solar energy for drying out the salts. The two desiccant salt plates are placed alternately in the living space, where they absorb water vapor from the air, and in the sun, to evaporate this water vapor and return to solid form. The salt plates may be dried either on the roof or at the southern wall, or alternately at east and west walls responding to morning and afternoon sun positions. Mechanized wheels transporting wet salt plates to the outside and dry salt plates to the inside could also be used.

Evaporative Cooling

This type of cooling is used where there are low humidities. The addition of moisture by using pools, fountains, and plants will begin an evaporation process that increases the humidity but lowers the temperature of the air for cooling relief.

Spraying the roof with water can also cause a reduction in the ceiling temperature and cause air movement to the cooler surface. Keep in mind that evaporative cooling is effective only in drier climates. Water has to be available for make up of the evaporated moisture. The atrium and the mechanical coolers should be kept out of the sun since it is the air's heat you are trying to use for the evaporation process not solar heat. For the total system design, the pools of water, vegetation, and fountain court should be combined with the prevailing summer winds for efficient distribution of cool, humidified air.

Night Sky Radiation Cooling

Night sky radiation is dependent on clear nights in the summer. It involves the cooling of a massive body of water or masonry by exposure to a cool night sky. This type of cooling is most effective when there is a large day-to-night temperature swing. A clear night sky in any climate will act as a large heat sink to draw away the daytime heat that has accumulated in the building mass. A well-sized and exposed body of water or masonry, once cooled by radiation to the night sky, can be designed to act as a cold storage, draining heat away from the living space through the summer day and providing natural summer cooling.

The roof pond is one natural conditioning system that offers the potential for both passive heating and passive cooling. The requirements for this system involve the use of a contained body of water or masonry on the roof. This should be protected when necessary by moving insulation or by moving the water. The house has conventional ceiling heights for effective radiated heating and cooling. During the summer when it is too hot for comfort, the insulating panels are rolled away at night. This exposes the water mass to the clear night sky, which absorbs all the daytime heat from the water mass and leaves the chilled water behind. During the day the insulated panels are closed to protect the roof mass from the heat. The chilled storage mass below absorbs heat from the living spaces to provide natural cooling for most or all of the day.

This type of cooling offers up to 100 percent passive, nonmechanical air conditioning.

Time Lag Cooling

This type of cooling has already been described in the section on time lag heating. It is used primarily in the climates that have a large day-to-night temperature swing. The well-insulated walls and floor will maintain the night temperature well into the day, transmitting little of the outside heat into the house. If 20-inch eaves over all the windows are used, they can exclude most of the summer radiation thereby controlling the direct heat gain.

Underground Cooling

Underground cooling takes advantage of the fairly stable 56-degree temperature conditions of the earth below the frostline. The only control needed is the addition of perimeter insulation to keep the house temperatures above 56 degrees F. In climates of severe summer temperatures and moderate-to-low humidity levels, underground construction provides stable and cool outside design temperatures as well as reduced sun exposure to remove most of the cooling load for maximum energy conservation.

ACTIVE SOLAR HEATING SYSTEMS

Active solar systems are modified systems that use fans, blowers, and pumps to control the heating process and the distribution of the heat once it is collected.

Active systems currently are using the following six units to collect, control, and distribute solar heat.

Unit	Function
1. Solar collector	Intercepts solar radiation and converts it to heat for transfer to a thermal storage unit or to the heating load.
2. Thermal storage unit	Can be either an air or liquid unit. If more heat than needed is collected, it is stored in this unit for later use. Can be either liquid, rock, or a phase change unit.
3. Auxiliary heat source	Used as a backup unit when there is not enough solar heat to do the job.
4. Heat distribution system	Depending on the systems selected, these could be the same as those used for cooling or auxiliary heating.
5. Cooling distribution system	Usually a blower and duct distribution capable of using air or liquid directly from either the solar collector or the thermal storage unit.

Operation of Solar Heating Systems

It would take a book in itself to examine all the possibilities and maybe a couple of volumes more to present details of what has been done to date. Therefore, it is best to take a look at a system that is commercially available from a reputable firm that has been making heating and cooling systems for years. See Fig. 20-7.

Domestic Water Heating System

The domestic water heating system uses water heated with solar energy. It is more economically viable than whole-house space heating because hot water is required all year-round. The opportunity to obtain a return on the initial investment in the system every day of the year is a distinct economic advantage. Only moderate collector temperatures are required

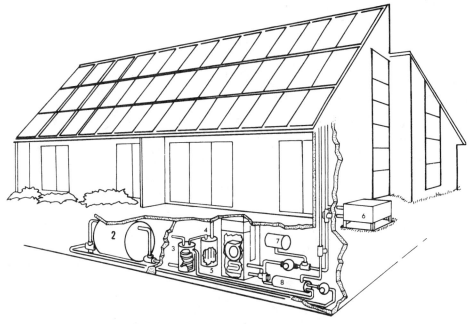

Fig. 20-7 *Components of a liquid to air solar system.* (Lennox Furnace Co.)

Key components in a liquid air solar system: 1. Solar collectors, 2. Storage tank, 3. Hot water heat exchanger, 4. Hot water holding tank, 5. Space heating coil, 6. Purge coil (releases excess solar heat), 7. Expansion tank, 8. Heat exchanger.

to cause the system to function effectively. Thus domestic water can be heated during less than ideal weather conditions.

Indirect Heating/Circulating Systems

Indirect heating systems circulate antifreeze solution or special heat transfer fluid through the collectors. This is primarily to overcome the problem of draining liquid collectors during periods of subfreezing weather. See Fig. 20-8. Air collectors can also be used. As a result there is no danger of freezing and no need to drain the system.

Circulating a solution of ethylene glycol and water through the collector and a heat exchanger is one means of eliminating the problem of freezing. See Fig. 20-9. Note that this system requires a heat exchanger and an additional pump. The heat exchanger permits the heat in the liquid circulating through the collector to be transferred to the water in the storage tank. The extra pump is needed to circulate water from the storage tank through the heat exchanger.

The extra pump can be eliminated if:

• The heat exchanger is located below the storage tank.

• The pipe sizes and heat exchanger design permit thermosiphon action to circulate the water.

• A heat exchanger is used that actually wraps around and contacts the storage tank and transfers heat directly through the tank wall.

Safety is another consideration in the operation of this type of system. Two major problems might develop with liquid solar water heaters:

1. Excessive water may enter the domestic water service line.
2. High temperature–high pressure may damage collectors and storage unit. See Fig. 20-10.

If you want to prevent the first problem, you can add a mixing valve between the solar storage tank and the conventional water heater. See Fig. 20-11.

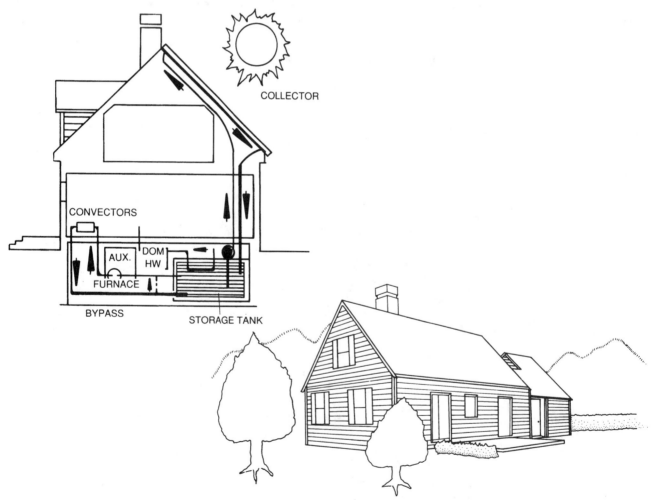

Fig. 20-8 *Collectors on the roof heat water that is circulated throughout the house.*

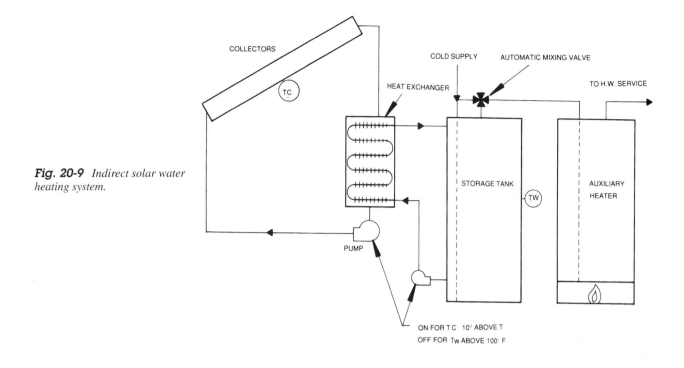

Fig. 20-9 *Indirect solar water heating system.*

Within Fig. 20-9, the following labels appear:

COLLECTORS

COLD SUPPLY   AUTOMATIC MIXING VALVE

HEAT EXCHANGER

TO H.W. SERVICE

TC

STORAGE TANK

TW

AUXILIARY HEATER

PUMP

ON FOR T C   10° ABOVE T
OFF FOR Tw ABOVE 100° F

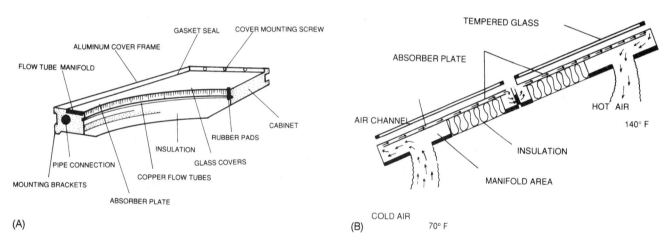

Fig. 20-10 *(A) Liquid collector. (B) Air collector.*

Within Fig. 20-10 (A), the following labels appear:

GASKET SEAL   COVER MOUNTING SCREW

ALUMINUM COVER FRAME

FLOW TUBE MANIFOLD

CABINET

INSULATION   RUBBER PADS

PIPE CONNECTION   GLASS COVERS

MOUNTING BRACKETS   COPPER FLOW TUBES

ABSORBER PLATE

(A)

Within Fig. 20-10 (B), the following labels appear:

TEMPERED GLASS

ABSORBER PLATE

AIR CHANNEL

HOT AIR
140° F

INSULATION

MANIFOLD AREA

COLD AIR   70° F

(B)

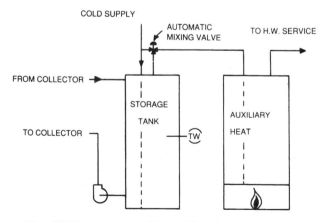

Fig. 20-11 *Schematic of the auxiliary heating equipment.*

Within Fig. 20-11, the following labels appear:

COLD SUPPLY

AUTOMATIC MIXING VALVE

TO H.W. SERVICE

FROM COLLECTOR

STORAGE TANK

TW

AUXILIARY HEAT

TO COLLECTOR

Cold water is blended with hot water in the proper proportion to avoid excessive supply temperature. The mixing valve is sometimes referred to as a tempering valve. Figure 20-12 shows the details of a typical connection for a tempering valve.

You can avoid excessive pressure in the collector loop carrying the antifreeze or heat transfer solution by installing a pressure-relief valve in the loop. Set the valve to discharge at anything above 50 psi. The temperature of the liquid may hit 200 degrees F, so make sure the relief valve is connected to an open drain. The fluid is unsafe and contaminated, so keep that in mind when disposing of it.

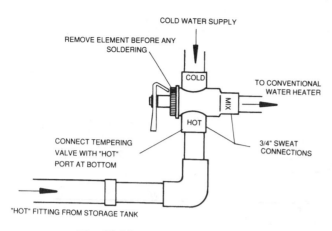

Fig. 20-12 *A typical tempering valve.*

A temperature and pressure-relief valve is usually installed on the storage tank to protect it. Whenever water in the tank exceeds 210 degrees F, the valve opens and purges the hot water in the tank. Cold water automatically enters the storage tank and provides a heating load for the collector loop. This cools down the system. Figure 20-13 gives examples of both safety devices installed in the system.

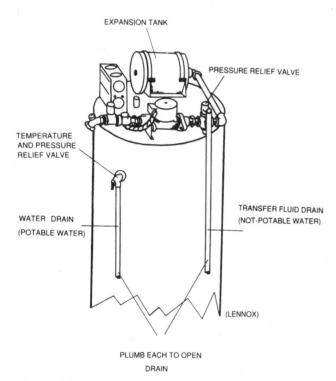

Fig. 20-13 *Note the locations of the safety valves on the heater tank and the expansion tank.*

The collector loop expansion tank (Fig. 20-13) is required to absorb the expansion and contraction of the circulating fluid as it is heated and cooled. Any loop not vented to the atmosphere must be fitted with an expansion tank.

The heat exchanger acts as an interface between the toxic collector fluid and the *potable* (drinkable) water. The heat exchanger must be double-walled to prevent contamination of the drinking water if there is a leak in the heat exchanger. The shell and tube type (Fig. 20-14) does not often meet the local code or the health department requirements.

Air Transfer

Air-heating collectors can be used to heat domestic water. See Fig. 20-15. The operation of this type of system is similar to that of the indirect liquid circulation system. The basic difference is that a blower or fan is used to circulate the air through the collector and heat exchanger rather than a pump to circulate a liquid.

The air transfer method has advantages:

• It does not have any damage due to a liquid leakage in the collector loop.

• It does not have to be concerned with boiling fluid or freezing in the winter.

• It does not run the risk of losing the expensive fluid in the system.

It does have some disadvantages over the liquid type of system.

• It requires larger piping between the collector and heat exchanger.

• It requires more energy to operate the circulating fan than it does for the water pump.

• It needs a slightly larger collector.

Cycle Operation

The indirect and direct water heating systems need some type of control. A differential temperature controller is used to measure the temperature difference between the collector and the storage. This controls the pump operation.

The pump starts when there is more than a 10-degree F difference between the storage and collector temperatures. It stops when the differential drops to less than 3 degrees F.

You can use two-speed or multispeed pumps in a system of this type to change the amount of water being circulated. As solar radiation increases, the pump is speeded up. This type of unit also improves the efficiency of the system.

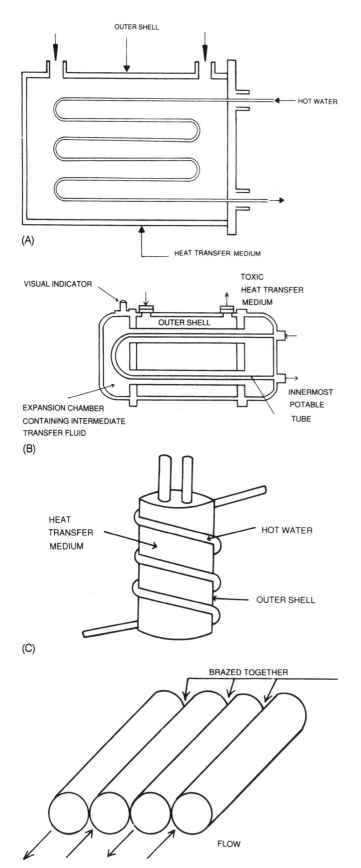

Fig. 20-14 *Heat exchanger designs.*

Designing the Domestic Water Heating System

Any heating job requires that you know the number of Btus (British Thermal Units) needed to heat a space. The type of heating system we are designing is no exception.

Table 20-1 shows minimum property standards for solar systems as designated by the United States Department of Housing and Urban Development. Note that the minimum daily hot water requirements for various residence and apartment occupancy are listed. For example, a two-bedroom home with three occupants should be provided with equipment that can provide 55 gallons per day of hot water. Many designers simply assume 20 gallons per day per person, which results in slightly higher requirements than those listed in Table 20-1.

Another important consideration in sizing the solar domestic hot water system is the required change in the temperature of incoming water. The water supplied by a public water system usually varies from 40 to 75 degrees F, depending on location and season of the year. A telephone call to your local water utility will provide the water supply temperature in your area. Generally, the desired supply hot water temperature is from 140 to 160 degrees F. Knowing these two temperatures and the volume of water required enables you to calculate the Btu requirement for domestic hot water. Figure 20-16 shows how to calculate the Btu requirements for heating domestic hot water.

To find the required collector area needed to provide some portion of the Btu load you can use a number of methods. Figure 20-17 shows the location of an add-on collector. One rule of thumb is: the amount of solar energy available at midaltitude in the continental United States is approximately equal to 2000 Btus per square foot per day. Assuming a collector efficiency of 40 percent, 800 Btus per square foot per day can be collected with a properly installed collector. Using the example in Fig. 20-17, the collector should contain approximately 137 square feet. That is found by 109,956 Btus per day divided by 800, which equals 137. The higher summer radiation levels and warmer temperatures would cause an excess capacity most of the time. A more practical approach is to provide nearly 100% solar hot water in July, which might then average out to 70% contribution for the year. Thus a collector area of 0.7 by 137, or about 96 square feet, might be a more realistic installation. See Fig. 20-18.

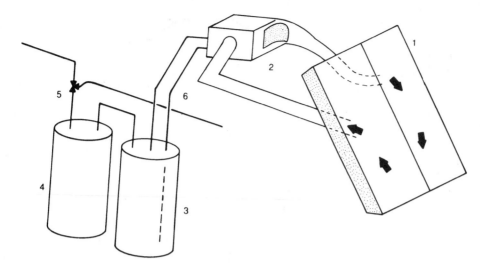

Fig. 20-15 *Schematic of air transfer medium solar water heating system.*

tions. See Fig. 20-19. These eliminate the need to size the storage tank, expansion tank, and pump. If you wish to select individual components, it will be necessary to make the same type of calculations for whole-house heating to determine the sizes of such components. Tank storage would typically be based on one day's supply of energy that is based on the daily Btu load.

Figure 20-20 illustrates a typical piping and wiring arrangement for a solar water heating system.

Table 20-1 *Daily Hot Water Usage (140°F) for Solar System Design*

Category	One- and Two-Family Units and Apartments up to 20 Units				
Number of people	2	3	4	5	6
Number of bedrooms	1	2	3	4	5
Hot water per unit (gallons per day)	40	55	70	85	100

The rule of thumb sizing procedures used here assumes that the collector is installed facing due south and inclined at an angle equal to the local latitude plus 10 degrees. Modification of these optimum collector installation procedures will reduce the effectiveness of the collector. In case the ideal installation cannot be achieved, it will be necessary to increase the size of the collector to compensate for the loss in effectiveness.

Other Components

The components for solar domestic hot water heating systems are available in kits prepackaged with instruc-

IS THIS FOR ME?

The basic question for everyone is: Is this for me? What are the economics of the system? Most manufacturers of packaged solar water heating systems provide some type of economic analysis to assist the installation contractors in selling their customers. For example, pay-back time for fuel savings to equal the total investment may be as little as six to nine years at the present time.

Example Problem:

Given that a family of six people live in a home where the incoming water temperature is 40°F and the requirement is for 150°F hot water, calculate the BTU requirement.

1. Compute: Water requirements = Number of people × 20 gallons per day
   = 6 × 20
   = 120 gallons of hot water used per day

   Because a gallon of water weighs 8.33 pounds, the BTU requirement per day for hot water can be found by:

2. Compute: Heat Required = gallons per day × 8.33 × temperature rise
   = (120 × 8.33) (150 − 40)
   = 109,956 BTU per day

Fig. 20-16 *Circulating Btu requirements.*

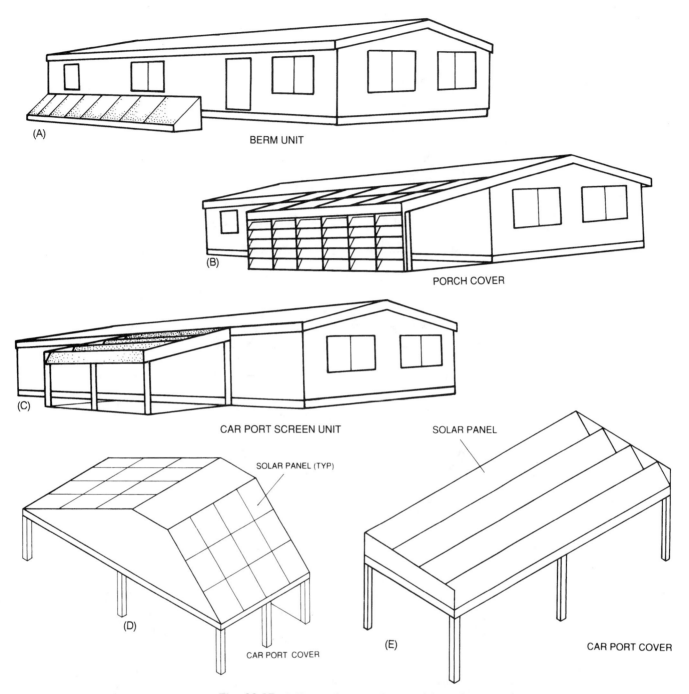

Fig. 20-17 *Collector placement for remodeling jobs.*

A computer service called SOLCOST provides a complete analysis based on the information supplied to the computer service. It includes a collector size optimization calculation that will provide the customer the optimum savings over the life of the equipment. For full details contact:

INTERNATIONAL BUSINESS SERVICES, INC.
Solar Group, 1010 Vermont Avenue
Washington, DC 20005

BUILDING MODIFICATIONS

In order to make housing more efficient, it is necessary to make some modifications in present-day carpentry practices. For instance, the following must be done to make room for better and more efficient insulation:

1. Truss rafters are modified to permit stacking of two 6-inch-thick batts of insulation over the wall plate. The truss is hipped by adding vertical members at

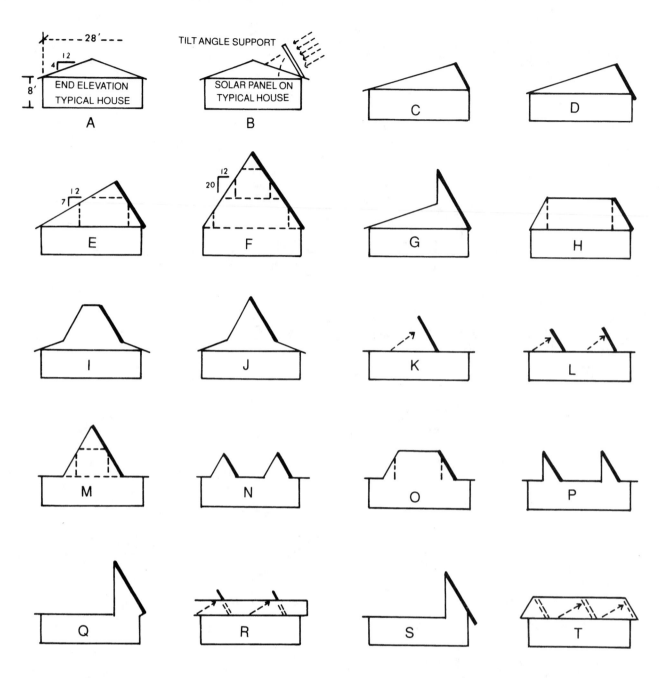

Elevations B, C, D, K, and L are probably the most economical methods of accommodating a steep tilt angle for solar collectors.

Elevations E, H, M, and O are probably the most economical.

Elevations I, J, M, and possibly O, depending on proportions, must be treated carefully.

Elevations G, N, P, Q, and S are possibilities but are more costly than the first group and especially more costly if clearstory space and/or fenestration are to be provided.

Elevations R and T are of some interest in that they provide a method of hiding the collectors completely, but their comparative cost needs special analysis.

Elevation F is obviously very expensive and only applicable under certain conditions. It does provide a three-story opportunity.

Fig. 20-18 *Collector roof configurations.* (National Association of Home Builders)

each end and directly over the 2-×-6 studs of the outer wall.

2. All outside walls use thicker (2 × 6) studs on 2-foot centers to accommodate the thicker insulation batts.

3. Ductwork is framed into living space to reduce heat loss as warmed air passes through the ducts to the rooms.

4. Wiring is rerouted along the soleplate and through notches in the 2 × 6 studs. This leaves the insulation cavity in the wall free of obstructions.

5. Partitions join outside walls without creating a gap in the insulation. Drywall passes between the soleplates and abutting studs of the interior walls. The cavity is fully insulated.

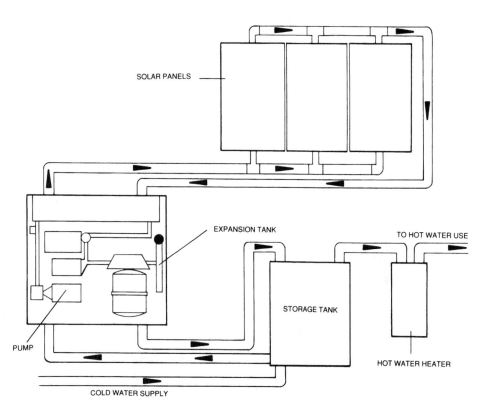

Fig. 20-19 *Prepackaged solar-assisted domestic hot water unit.*

SOLAR PANELS

EXPANSION TANK

TO HOT WATER USE

STORAGE TANK

HOT WATER HEATER

PUMP

COLD WATER SUPPLY

6. Window area is reduced to 8 percent of the living area.

7. Box headers over the door and window openings receive insulation. In present practice, the space is filled with 2 × 10s or whatever is needed. Window header space can be filled with insulation. This reduces the heat loss through the wood that would normally be located here.

BUILDING UNDERGROUND

A number of methods are being researched for possible use in solar heating applications. One of the most inexpensive ways of obtaining insulation is building underground. However, one problem with underground living is psychological. People do not like to live where they can't see the sun or outside. The idea of building underground is not new. The Chinese have done it for centuries. But the problem comes in selling the public the idea. It will take a number of years of research and development before a move is made in this direction.

Advantages

There are a number of advantages of going underground. By using a subterranean design, the builder can take advantage of the earth's insulative properties.

The ground is slow to react to climatic temperature changes. It is a perfect year-round insulator. There is a relatively constant soil temperature at 30 feet below the surface. This could be ideal for moderate climates, since the temperature would be a constant 68 degrees F.

By building underground it would be possible to use the constant temperature to reduce heating and cooling costs. A substantial energy reduction or savings could be realized in the initial construction also.

Figure 20-21 shows a roof-suspended earth home. This is one of the designs being researched at Texas A&M University. It uses the earth as a building material, and uses the wind, water, vegetation, and the sun to modify the climate.

A suitable method of construction uses beams to support the walls. That means structural beams could be stretched across the hole and walls could be suspended from the beams. See Fig. 20-22. The inside and outside walls would also be dropped from the beams. That would allow something other than wood to be used for walls, since they would be non-weight-bearing partitions. The roof would be built several feet above the surface of the ground to allow natural lighting through the skylights and provide a view of outside. This would get rid of some of the feeling of living like a mole.

Another design being researched involves tunneling into the side of a hill. See Fig. 20-23. This still uses the insulative qualities of the earth and allows a southern exposure wall. This way windows and a conventional-type front door could be used. Exposing only the

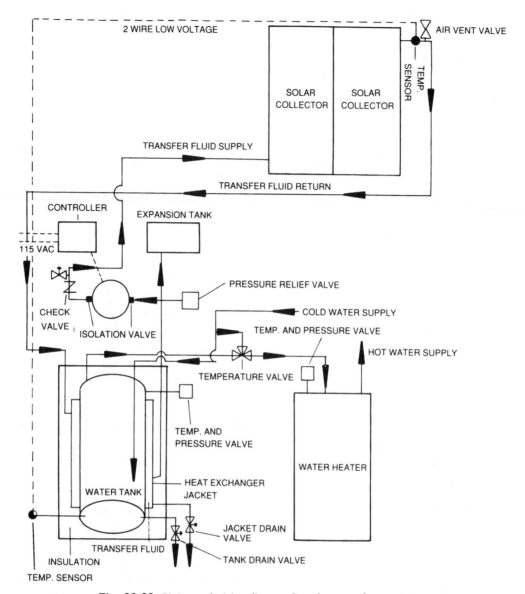

Fig. 20-20 *Piping and wiring diagram for solar water heater system.*

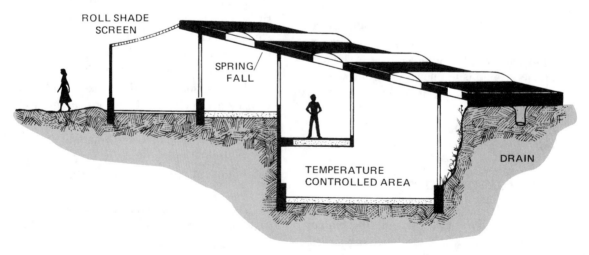

Fig. 20-21 *Roof-suspended earth home.*

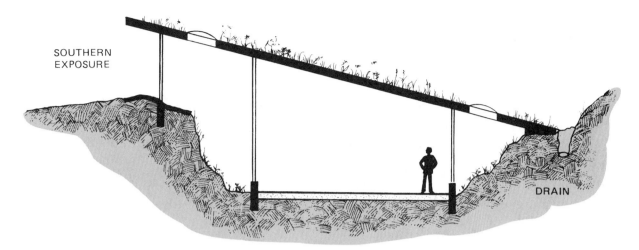

Fig. 20-22 *Another variation of a roof-suspended earth home.*

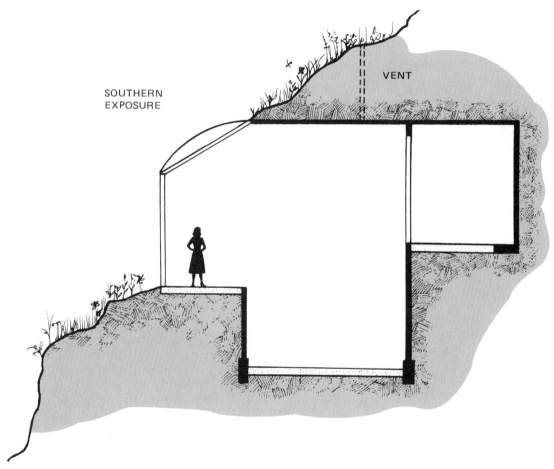

Fig. 20-23 *Hillside earth home.*

southern wall also helps to reduce heating costs. The steep hills in some areas could be used for this type of housing where it would be impossible to build conventional-type housing.

Many more designs will be forthcoming as the need to conserve energy becomes more apparent. The carpenter will have to work with materials other than wood. The new methods and new materials will demand a carpenter willing to experiment and apply known skills to a rapidly changing field.

CHAPTER 20
STUDY QUESTIONS

1. Where are solar heated houses built?

2. Explain how the Tech House air-conditioning system works.

3. Explain how the fireplace works in Tech House.

4. How much can exterior shutters cut heat loss through windows?

5. How does the entry vestibule contribute energy saving in Tech House?

6. What temperature does the air in the attic of a Tech House reach?

7. How much energy do our homes consume each year?

8. What are the advantages of building a house underground?

9. How does a solar collector work?

10. What are the disadvantages of building a house underground?

21

Alternative Framing Methods

THIS CHAPTER ILLUSTRATES NEW METHODS that are presently being used to frame houses. These newer methods are becoming increasingly popular as the price of lumber continues to increase—while its quality decreases. This chapter covers: the most popular alternative framing methods; the advantages and disadvantages to each of these methods; the tools that are associated with each of these methods; the basic sequence for framing a house with these methods; and installation of utilities inside these framing methods.

WOOD FRAMES PREDOMINATE

Approximately 90% of homes in the United States are built with wood frame walls. Because of this heavy use, almost all of the old-growth forests have been harvested. New-growth lumber is typically the only type available to builders. This presents a present-day problem because new-growth lumber is fertilized and watered to make it grow quickly. This produces a larger cell structure, the end result of which is wood that is not as dense or strong and is more susceptible to warping, cracking, and other defects. In addition, the price of lumber has tripled over the past decade and will continue to rise as quality, old-growth lumber becomes less available.

Because of this trend, many home builders use alternative methods for framing houses. These methods are variations of methods used in commercial buildings and therefore meet most local residential building code requirements. In most cases, these framing methods exceed code specifications and make a home more resistant to hurricanes, earthquakes, termites, and other common nuisances that afflict traditionally framed homes.

STEEL FRAMING

Steel-framed homes are not a new phenomenon. Their first debut was in the 1933 Chicago World's Fair. However, it wasn't until the 1960s that the leading steel manufacturers entered the residential housing market. Steel homes have not gained wide popularity in the residential market because of their initial cost. Steel framing had always been more expensive than wood framing. With the rapid increase in the cost of lumber, steel studs are now, in most cases, less expensive than a wooden 2 × 4. In most areas of the country, a steel-framed home costs as much as, or only 15% more than, a wooden-framed home. If the price of lumber continues to rise at the same rate as it has in the past, then steel-framed homes or other methods of framing will become more economical and commonplace.

Advantages and Disadvantages of Steel-Framed Homes

Today, most commercial and industrial buildings are constructed with steel as part of the primary structure. Steel is used because it is the strongest and safest cost-effective building material available. When steel framing is used instead of wood, its advantages include:

- No shrinking, warping, or twisting which allows walls, floors, and ceilings to be straight and square, thereby reducing finish work.
- A higher strength-to-weight ratio so homes can withstand hurricanes, earthquakes, high winds, heavy snow loads, torrential rains, termites, and can span greater distances with less support.
- It is assembled with screws, which eliminates nail-pops and squeaks.
- Steel studs, joists, and trusses are quicker to install.
- Steel-framed homes weigh as much as 30% less, so foundations can be lighter.
- Steel-framed homes provide a better path to ground when struck by lightning, reducing the likelihood of explosions or secondary fires.
- Most steel framing is made to exact size in a factory, so less waste is produced.
- Steel is 100% recyclable.

The main disadvantage to residential steel framing is its initial cost. Other disadvantages include:

- Local building code officials might be unfamiliar with its use.
- Different tools are required.
- Carpenters are not familiar with the building practices associated with steel framing.

Types of Steel Framing

The two main types of steel framing are *red-iron* and *galvanized*. Red-iron framing is based on the standard commercial application in which the structure is built out of a thick red-iron frame at 8-foot spans to distribute the load (Fig. 21-1). Galvanized sheet metal studs, joists, and other framing members are installed at 24 inches O.C. for a screwing base for sheathing.

Galvanized steel framing is based on the traditional wooden "stick-built" method. In this method, the entire home is built using galvanized sheet metal framing members as illustrated in Fig. 21-2. This method is preferred by some builders because its similarity to traditional wooden framing methods makes it easier to

Fig. 21-1 *Red-iron framework at 8-foot intervals with galvanized-steel framing in between to support sheathing.* (Tri-Steel Structures)

retrain carpenters, and cranes are typically not needed to lift heavy red-iron framing members. However, exterior walls require additional horizontal bracing spaced 16–24 inches O.C. and roof framing requires more rafters and purlins.

Tools Used In Steel Framing

Because most metal-framed homes are engineered and sold in kits, most of the framing members are cut to length. However, tin snips or aviation snips are frequently used to cut angles and notches in base track for plumbing manifolds. A chop saw or circular saw with a metal cutting blade is used to make most straight cuts.

Screw guns are used instead of hammers. Powder-actuated nail guns are used to attach base track to concrete foundations. Wrenches and ratchets are used to secure nuts and bolts in red-iron connections. A crane is needed for one day to erect the red-iron framing.

Tape measurers are used to lay out all framing connections and speed squares are used to transfer measurements around corners. Levels and chalk lines are frequently used to keep everything plumb and square. Vice grips are used to clamp framing members together while they are being attached.

Sequence

Most residential steel suppliers are eager for builders to convert to steel framing and will provide extensive training as well as video tapes on framing with steel. In addition, they sell framing kits of precut steel members for standard model floor plans. Detailed plans are provided with each kit that include specifications for stud spacing and other framing members dependent upon the size of the structure. Therefore, only the general sequence for framing with steel is covered in this chapter.

1. Anchor bolts for red-iron framing can only deviate ⅛ inch to ensure proper alignment and accuracy. Therefore, they are installed after the concrete has

Fig. 21-2 *Galvanized-steel framing. Note the horizontal wall bracing and additional rafters that are required for homes made without a red-iron frame.* (Tri-Steel Structures)

hardened. This is done by drilling holes for each anchor bolt with a hammer-drill. Anchor bolts range in length and diameter depending on load specifications. Two anchor bolts are used to fasten each red-iron girder. Anchor bolts are packaged in glass tubes, which break when they are inserted into the hole. When the tube breaks, a two part adhesive mixes together, which bonds the anchor bolt to the concrete foundation. Usually, 24 hours is required for the adhesive to cure before any red-iron framing members are attached.

2. Red-iron framing members are assembled with nuts and bolts on a flat level surface. Each section usually consists of two vertical supports with rafters and bracing. They are erected using a crane.

3. Once the red-iron framing members are attached to the anchor bolts they are straightened by attaching them to come-alongs which are attached to stakes around the perimeter of the home. Come-alongs are tightened and loosened until the framing members are plumb and square. Galvanized sheet metal purlins are then attached every two feet between the eight-foot span of the red-iron rafters to keep them square.

4. Galvanized framing for walls is usually secured to concrete foundations by running two beads of sealant (sill seal) between the concrete and the galvanized base track. A powder-actuated nail gun is then used to drive fasteners every 12 inches O.C. in the base track.

5. Studs are secured by one screw on each side of the base track. Typically, #8 × ½-inch self-drilling low-profile wafer, washer, or pan head screws are used to secure all galvanized framing members. MIG welders also can be used to attach framing members; however, screws are preferred. Studs are attached to red-iron or any other thicker framing members by #12 × ¾-inch self-drilling low-profile head screws. Vice grips should always be used to clamp the framing members together while they are being attached.

6. Furring or hat channel are spaced 16–24 inches O.C. for ceiling framing or for horizontal reinforcement of exterior walls in the galvanized framing method. Steel framing members are identified in Fig. 21-3.

7. Plywood and oriented-strand board (OSB) are commonly used as exterior sheathing for steel-framed homes. Exterior sheathing is staggered every four feet on the roof and on the walls of the galvanized-framing method. Plywood and OSB can be laid horizontally on top of each other without staggering for

Part Identification Guide

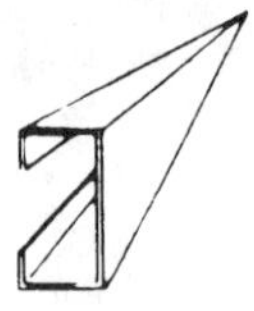	**STUD** - C Channel Type, formed with knurled outer flanges to prevent screw ride, hollow partition framing.
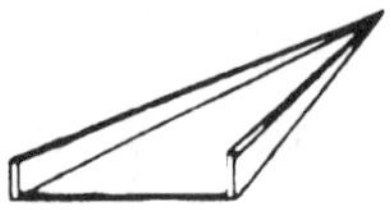	**TRACK** - Channel configuration formed with unhemmed legs sized to receive the corresponding size stud.
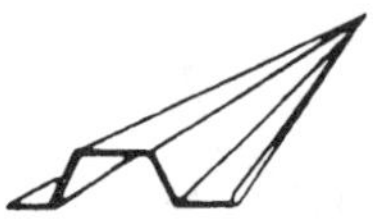	**FURRING - HAT CHANNEL** - Section designed for screw alignment of finished surface material in wall and ceiling furring.
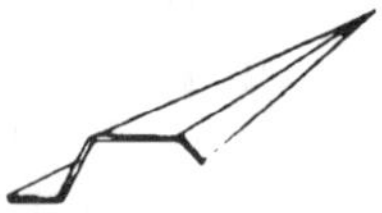	**FURRING - RESILIENT CHANNEL** - is one of the most effective methods of improving sound transmission loss through wood framed partitions and ceilings.
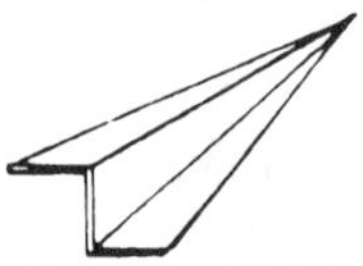	**FURRING** - Z CHANNEL - Designed for vertical placement in wall furring for rigid insulation.
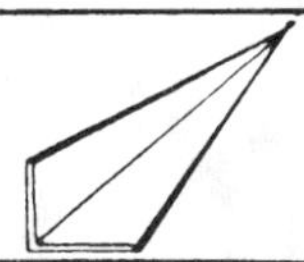	**ANGLE RUNNER** - Angle sections used to secure and brace finish surface material.
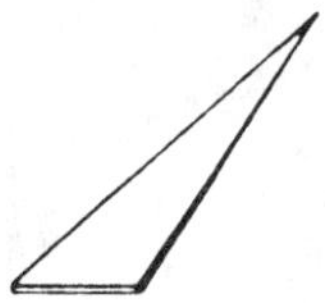	**FLAT STRAPPING** - Strap bracing is for use on internal and external framing to prevent racking and hold proper alignment.
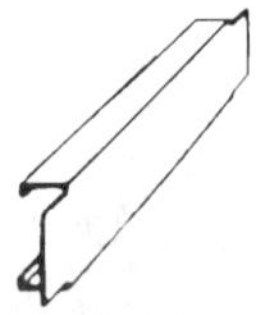	**BRIDGING** - Metal stud wall bridging is used as flush mount backing for mounting cabinets, handrails, grab bars, and mose casework.

Fig. 21-3 *Steel frame part identification guide.* (Southeastern Metals Manufacturing Co.)

walls in the red-iron method. The basic sequence for framing and sheathing a red-iron home is shown in Fig. 21-4. Sheathing should be attached by screws every 24 inches O.C. with five screws securing a 48-inch width. This equates to a minimum of 25 screws per 4' × 8' sheet.

LIFETIME
HOME CONSTRUCTION
Quick, Easy Assembly
Saves Time & Money

❶ The Foundation (slab, pier & beam, pilings or basement) is prepared and exterior wall sections are bolted together on the ground.

Fig. 21-4 *Basic sequence for framing and sheathing using the red-iron method.* (Tri-Steel Structures)

❷ Frame sections are raised, joined together and anchored to the foundation.

❸ Roof trusses are assembled on the ground, lifted into place, attached to the side walls, and braced with furring.

❹ Sheathing is attached to the roof and walls. Doors and windows are installed and dormers are created. Interior framing and insulation are installed.

❺ Exterior finish of your choice is applied and the shell is now complete. The interior can be completed at your own pace.

8. Six screws are used at web stiffening areas such as window sills supported by 6-inch cripple studs. Most other connections require only two screws (one on each side).

9. Utilities, such as electrical wiring and plumbing, can run through the prepunched holes in the studs and joists. However, plastic grommets must first be inserted in the holes so the sharp steel edges of the framing members will not damage the insulation on the wiring. The remaining electrical work usually involves the same techniques used in commercial steel-framing applications.

GALVANIZED FRAMING

When galvanized steel studs are used to frame the entire home, heavier gage studs (12–18 gage)* are used for the exterior load-bearing walls and lighter** gage studs (20–25 gage) are used for the interior curtain walls. Typically, most houses are engineered by a company in advance; depending on the weight of the roof, tile, or composition shingles, or if there is a second story, the gage of the studs will vary from home to home.

*The word *gage* is presently used instead of *gauge*. It is still acceptable to use either form of the word, but it is more "up to date" to use the shorter form.
**Keep in mind that the lighter the gage of the steel, the higher the gage number. That means 12 gage is thicker than 15 gage.

There are three main ways to assemble a steel stud wall. One way is to have all the walls fabricated into panels in a factory that is off-site. See Fig. 21-5. This method, typically called *panelization*, is the fastest-growing segment of residential new construction. The advantages to panelization are:

1. Less material is needed because human errors on the job site are prevented.

2. There is no lost time on the job site because of inclement weather.

Fig. 21-5 *Panelized walls and trusses save material and waste on the job site.* (Steel Framing, Inc.)

3. Fewer skilled laborers are needed on the job site. This means a reduction in the builder's liability and in worker's compensation insurance.

4. Panelized walls produce consistent quality. This is because the factory-controlled conditions include preset jigs and layout tables.

5. Panelized walls can be ordered in advance and shipped to the job site for easy installation. This significantly reduces the project construction time. See Fig. 21-6.

Fig. 21-6 *Factory-made panels being unloaded on the construction site.* (Steel Framing, Inc.)

The second way to assemble a steel wall is called *in-place* framing. This method requires the carpenter to first secure the steel-base track to the foundation. It can be done by bolting it down with anchor bolts or by using powder-actuated nailers. When connecting the base track to a slab using powder-actuated nailers, two low-velocity fasteners are spaced a minimum of every 24 inches. Typically, most exterior walls are connected to the slab with anchor bolts or J-bolts. The base track is usually reinforced with an additional piece of track or stud where the anchor bolt extends through the base track, as shown in Fig. 21-7.

Steel studs are attached to the base track every 16 or 24 inches O.C. with a No. 8 self-tapping screw on either side. Spacing depends on the load placed on each wall and the gage of the stud being used. The corner stud is braced to the ground to prevent movement from the wind or from the stresses during assembly. A spacer stud is placed between each stud thereafter to keep the studs in place. Steel channel (which comes in 16-foot lengths) can be placed between the *cut-outs* of each stud to brace the wall instead of the usual spacers, as shown in Fig. 21-8. Additional steps in fabricating the wall will be discussed in the following.

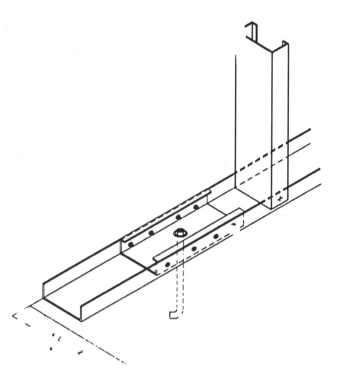

Fig. 21-7 *J-bolt foundation connection.* (Steel Framing, Inc.)

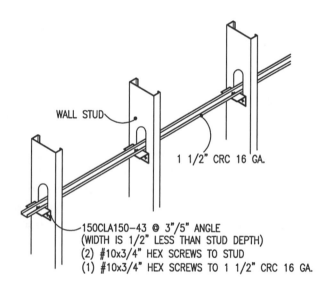

WALL STUD

1 1/2" CRC 16 GA.

150CLA150-43 @ 3"/5" ANGLE
(WIDTH IS 1/2" LESS THAN STUD DEPTH)
(2) #10x3/4" HEX SCREWS TO STUD
(1) #10x3/4" HEX SCREWS TO 1 1/2" CRC 16 GA.

NOTES:
FOR STUD HEIGHTS 10'-0" AND LESS, INSTALL CRC AT MID HEIGHT.
FOR STUD HEIGHTS GREATER THAN 10'-0" INSTALL AT THIRD POINTS.

Fig. 21-8 *Wall bridging detail.* (Nuconsteel Commercial Corp.)

Tilt-up framing is another way to assemble a steel-stud wall. In this method, the wall is assembled flat (horizontally). Then, it is lifted into a vertical position. This method makes it easier for carpenters to assemble a flat straight wall—that is, if the surface they are assembling the wall on is flat. Torpedo levels, which are magnetic, can easily be placed on the steel studs to insure that each is level and square before being installed into the base track with No. 8 self-taping screws.

Steel-strap blocking, wood blocking, or track blocking can be used between studs for reinforcing. This is particularly helpful where overhead cabinets or other heavy fixtures may be attached. Steel straps are also fastened diagonally in the corners of the framing. This keeps the walls square and protects them from shear forces. However, steel strapping is not necessary if OSB or plywood is fastened to the corners of the steel framing.

Details for corners, window and door openings, wall-to-wall panels, intersecting wall panels, and wall-to-roof trusses are shown in Figs. 21-9 through 21-15.

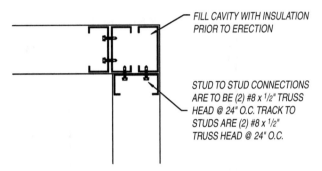

Fig. 21-9 *Corner framing.* (Nuconsteel Commercial Corp.)

INSULATED CONCRETE FORMS

One of the fastest growing framing methods is the insulated concrete form (ICF). In this method, walls are made of concrete reinforced by rebar surrounded by Styrofoam® (expanded polystyrene beads). Most ICFs are hollow Styrofoam® blocks that are stacked together like building blocks with either tongue-and-groove joints or finger joints as shown in Fig. 21-16. Each manufacturer makes the block a different size. Some manufacturers have ties molded in their blocks while others have separate ties that are inserted into the block at the job site. Once the blocks are assembled, reinforced, and braced, concrete is poured into the blocks using a concrete pumper truck. An example of what the concrete and rebar would look like inside the foam is illustrated in Fig. 21-17.

Advantages and Disadvantages to Insulated Concrete Forms

Insulated concrete form homes are framed in concrete, which allows them to have the following advantages over conventional wood-frame homes:

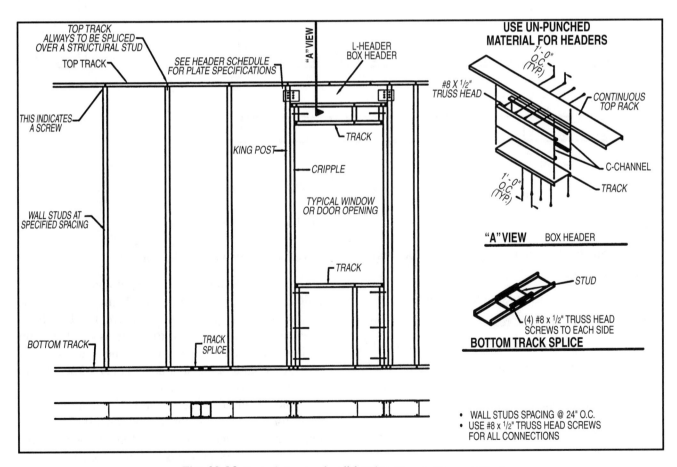

Fig. 21-10 *Typical structural wall framing.* (Nuconsteel Commercial Corp.)

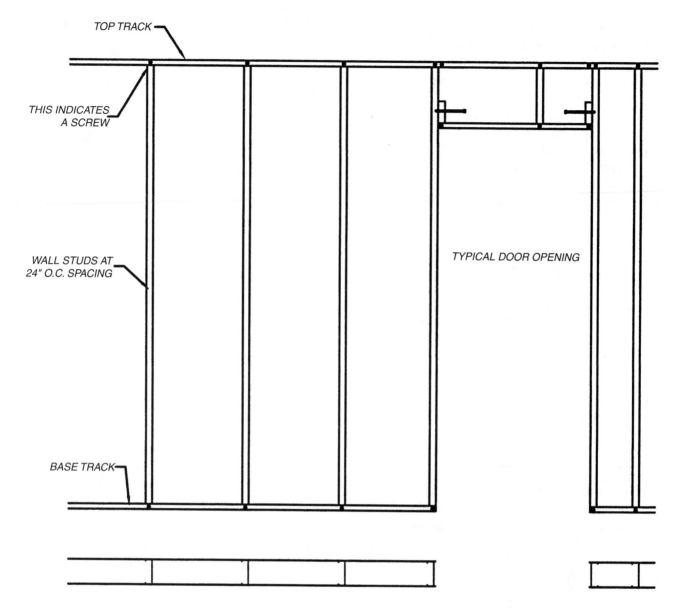

TOP TRACK

THIS INDICATES
A SCREW

WALL STUDS AT
24" O.C. SPACING

TYPICAL DOOR OPENING

BASE TRACK

- WALL STUDS SPACING @ 24" O.C.
- USE #6 x 7/16" SHARP POINT SCREWS FOR ALL CONNECTIONS

Fig. 21-11 *Typical non-structural wall framing.* (Nuconsteel Commercial Corp.)

- It can withstand hurricane force winds (200 mph).
- It is bullet resistant.
- Termites cannot harm its structural integrity.
- It exceeds building code requirements.
- It will lower heating and cooling costs by 50%–80%.
- Load bearing capacity is higher (27,000plf vs. 4,000plf).
- Outside noise reduction is higher.
- Homeowner's insurance rates should decrease 10%–25%.
- Fire rating of walls is measured in hours vs. minutes.

- Reduces pollen inside the home.

An ICF home is much stronger and more energy efficient than a wooden-framed home. However, its major disadvantage is cost. The average ICF home costs about twice as much to frame as its chief competitor. Because the framing of a home is just one of its costs, this figure really amounts to only a 5%–15% increase in the total cost of a home.

Another disadvantage to an ICF is that there is no room for error. If a window or door opening is off, it is much more difficult to cut reinforced concrete than wood. Carpenters must also be retrained to use this framing method correctly.

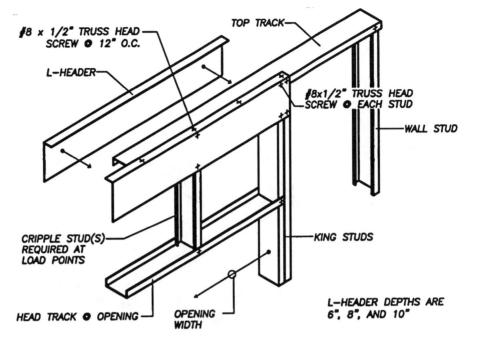

Fig. 21-12 *Double L-header framing.* (Nuconsteel Commercial Corp.)

#8 x 1/2" TRUSS HEAD SCREW @ 12" O.C.

TOP TRACK

L—HEADER

#8x1/2" TRUSS HEAD SCREW @ EACH STUD

WALL STUD

CRIPPLE STUD(S) REQUIRED AT LOAD POINTS

KING STUDS

HEAD TRACK @ OPENING

OPENING WIDTH

L—HEADER DEPTHS ARE 6", 8", AND 10"

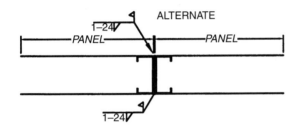

ALTERNATE

1—24

PANEL

PANEL

1—24

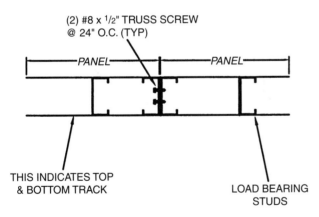

(2) #8 x 1/2" TRUSS SCREW @ 24" O.C. (TYP)

PANEL

PANEL

THIS INDICATES TOP & BOTTOM TRACK

LOAD BEARING STUDS

Fig. 21-13 *Wall panel to wall panel.* (Nuconsteel Commercial Corp.)

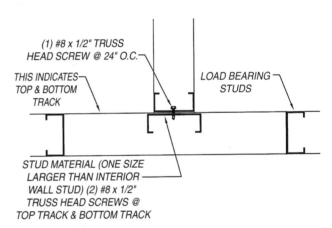

(1) #8 x 1/2" TRUSS HEAD SCREW @ 24" O.C.

THIS INDICATES TOP & BOTTOM TRACK

LOAD BEARING STUDS

STUD MATERIAL (ONE SIZE LARGER THAN INTERIOR WALL STUD) (2) #8 x 1/2" TRUSS HEAD SCREWS @ TOP TRACK & BOTTOM TRACK

Fig. 21-14 *Wall intersection framing.* (Nuconsteel Commercial Corp.)

Tools Used in Insulated Concrete Form Framing

Because concrete can be molded into any shape, it is essential to mold the frame of a home straight and square so walls are plumb and level. In order to achieve this, a carpenter should have a framing square, 2- and 4-foot levels, chalk line, and 30- and 100-foot tape measures. The foam forms can easily be cut with a handsaw or sharp utility knife. In most cases, the foam forms are glued together, so a caulking gun is required.

All the foam forms need to be braced until the concrete hardens. If the construction crew is not using metal braces, then wooden braces need to be constructed. Therefore, a framing hammer, nails, circular saw, and crowbar are needed.

Rebar is run horizontally and vertically throughout the walls of an ICF home. Tools used to work with rebar are rebar cutting and bending tools, metal cutoff saw, rebar twist tie tool (pigtail), 8-inch dikes, 9-inch lineman's pliers, hack saw, tin snips, and wire-cutting pliers. A hot knife (Fig. 21-18) or router can be used to cut notches in the foam to place electrical wiring.

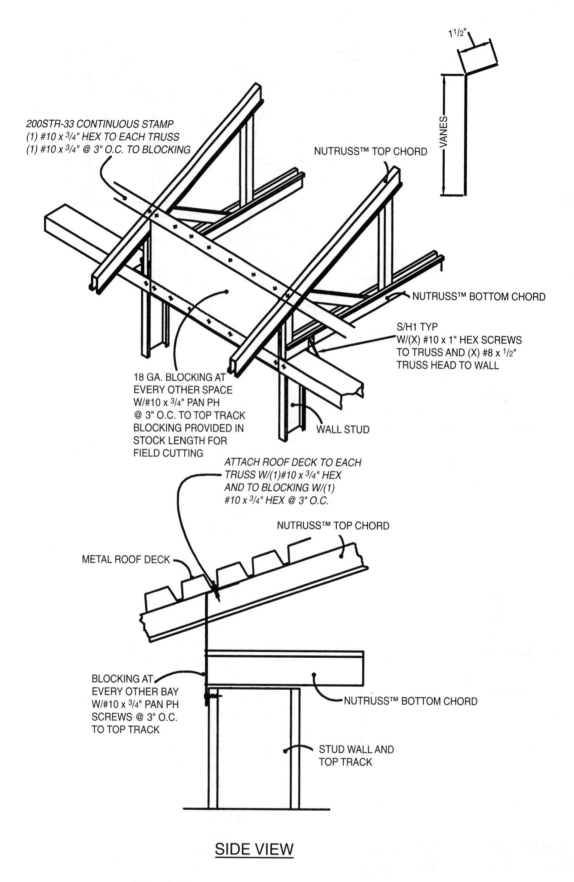

200STR-33 CONTINUOUS STAMP
(1) #10 x 3/4" HEX TO EACH TRUSS
(1) #10 x 3/4" @ 3" O.C. TO BLOCKING

NUTRUSS™ TOP CHORD

VANES

1 1/2"

NUTRUSS™ BOTTOM CHORD

S/H1 TYP
W/(X) #10 x 1" HEX SCREWS
TO TRUSS AND (X) #8 x 1/2"
TRUSS HEAD TO WALL

18 GA. BLOCKING AT
EVERY OTHER SPACE
W/#10 x 3/4" PAN PH
@ 3" O.C. TO TOP TRACK
BLOCKING PROVIDED IN
STOCK LENGTH FOR
FIELD CUTTING

WALL STUD

ATTACH ROOF DECK TO EACH
TRUSS W/(1)#10 x 3/4" HEX
AND TO BLOCKING W/(1)
#10 x 3/4" HEX @ 3" O.C.

NUTRUSS™ TOP CHORD

METAL ROOF DECK

BLOCKING AT
EVERY OTHER BAY
W/#10 x 3/4" PAN PH
SCREWS @ 3" O.C.
TO TOP TRACK

NUTRUSS™ BOTTOM CHORD

STUD WALL AND
TOP TRACK

SIDE VIEW

Fig. 21-15 *Truss to stud wall with blocking.* (Nuconsteel Commercial Corp.)

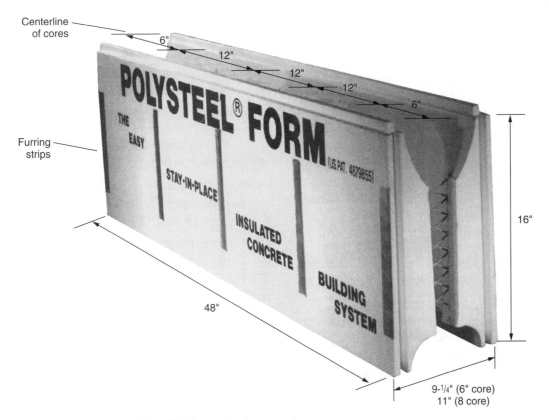

Centerline
of cores

6"

12"

12"

12"

6"

16"

Furring
strips

48"

9-1/4" (6" core)
11" (8 core)

Fig. 21-16 *Insulated concrete form.* (American Polysteel Forms)

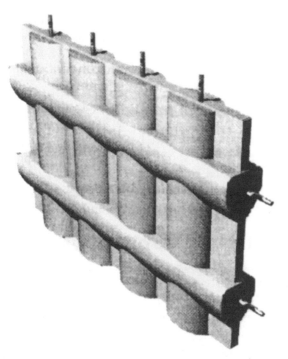

Fig. 21-17 *Concrete and rebar inside an insulated concrete form.*
(American Polysteel Forms)

QC hot knife
with conduit
blade

Fig. 21-18 *A hot knife used to cut foam for electrical wiring and
plumbing.* (Avalon Concepts)

Sequence

Insulated concrete form framing is a radically new approach for carpenters that are familiar with the traditional wood "stick-built" method. The only wood used in this wall framing method is pressure-treated lumber, which is used for sealing rough door and window openings. Because this method is different and each manufacturer has specific guidelines that must be followed to ensure proper installation, training is offered to all individuals constructing ICFs. Many ICFs also send representatives to the job site to supervise those who are building their first home using this method. Therefore, only general guidelines are covered in the following section.

1. Before starting, rebar should extend vertically 2–6 feet from the foundation every one to two linear feet depending on building code requirements.

2. Most manufacturers suggest placing 2 × 4s or some type of bracing around the perimeter of the foundation as a guide for setting the foam forms.

3. Once the first course of foam form blocks is laid, continue to do so placing rebar horizontally every one to two feet (every or every other course of block) as required.

4. Vertical bracing should be tied to the perimeter bracing at 6-foot intervals. Corners should be braced on each intersecting edge with additional diagonal bracing spaced at 4-foot intervals. See Fig. 21-19 for a typical corner bracing diagram. As forms are stacked, vertical bracing can be screwed into the metal or plastic ties built or inserted into the foam form blocks.

5. Openings for windows and doors should be blocked with pressure-treated lumber.

6. After the foam form blocks are stacked to the correct height, place bracing on the top and secure it to the side bracing. Because foam is very light, it will float. Proper bracing is essential to prevent the foam walls from floating and causing blow-outs when the concrete is poured inside them. Additional bracing techniques are illustrated in Fig. 21-20.

7. Walls are usually poured with a boom pump truck in 4-foot increments. Anchor bolts are set in the top of the wall to secure the top plate as a nailing base for roof construction. See Fig. 21-21.

8. Sheathing can be screwed into the metal or plastic ties located every foot in the foam form block, as shown in Fig. 21-22.

9. Electrical wiring and boxes can be installed by gouging out a groove with a router or hot knife. See Fig. 21-23 for electrical wiring tips.

Types of Foam

The most commonly used types of foam utilized in ICFs are the *expanded polystyrene* (EPS), and *extruded polystyrene* (XPS). EPS consists of tightly fused beads of foam. Vending-machine coffee cups, for example, are made of EPS. XPS is produced in a different process and is more continuous, without beads or the sort of "grain" of EPS. The trays in prepackaged meat at the grocery store are made of XPS. The two types can differ in cost, strength, R-value, and water resistance. EPS varies somewhat. It comes in various densities; the most common are 1.5 pounds per cubic foot (pcf) and 2 pounds pcf. The denser foam is a little more expensive, but is a little stronger and has a slightly higher R-value.

Some stock EPS is now available with insect-repellent additives. Although few cases of insect penetration into the foam have been reported to date, some ICF manufacturers offer versions of their product made of treated material. Whether you buy your own foam or you are choosing a preassembled system, you might want to check with the manufacturer about this.

Three Types of ICF Systems

The main difference in ICF systems is that they vary in their unit sizes and connection methods. They can be divided into three types: *plank, panel,* and *block systems* as shown in Fig. 21-24. The panel system is the largest. It is usually 4 by 8 feet in size. That means the wall area can be erected in one step but may require more cutting. These panels have flat edges and are connected one to the other with fasteners such as glue, wire, or plastic channel.

Plank systems are usually 8 feet long with narrow (8- or 12-inch) planks of foam. These pieces of foam are held at a constant distance of separation by steel or

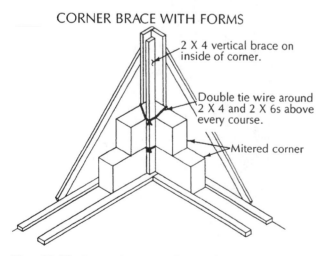

CORNER BRACE WITH FORMS

2 X 4 vertical brace on inside of corner.

Double tie wire around 2 X 4 and 2 X 6s above every course.

Mitered corner

Fig. 21-19 *Corner bracing technique for insulated concrete forms.* (American Polysteel Forms)

2 X 4 TOP RAIL SUPPORT

2 X 4s

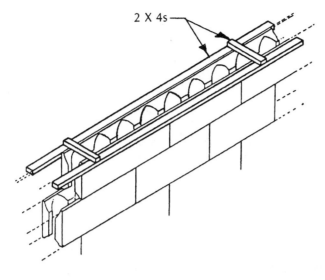

Fig. 21-20 *Additional bracing techniques.* (American Polysteel Forms)

HORIZONTAL TOP BRACE

◆ A horizontal top brace secured to the steel furring strips and supported by a diagonal brace staked to the ground will also provide the alignment and security necessary to keep the Forms in place during the pouring of concrete.

POLYSTEEL FORM® TOP BRACE

Polysteel Forms®

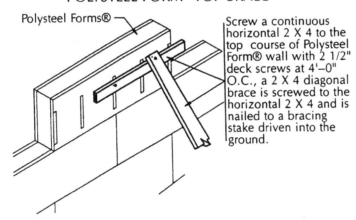

Screw a continuous horizontal 2 X 4 to the top course of Polysteel Form® wall with 2 1/2" deck screws at 4'–0" O.C., a 2 X 4 diagonal brace is screwed to the horizontal 2 X 4 and is nailed to a bracing stake driven into the ground.

Fig. 21-21 *Anchor bolt placement for roof framing.* (AFM)

plastic ties. The plank system has notched, cut, or drilled edges. The edges are where the ties fit. In addition, to spacing the planks, the ties connect each course of planks to the one above and below.

Fig. 21-22 *Attaching sheathing directly to plastic ties in the concrete forms.* (AFM)

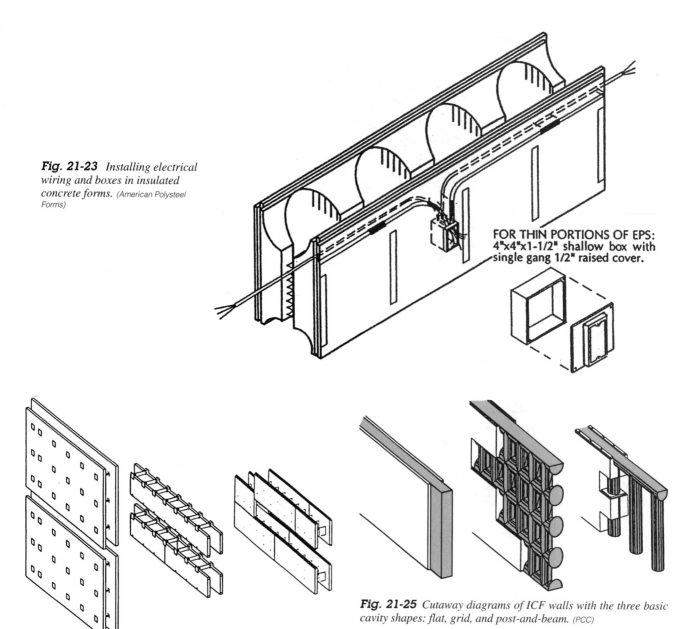

Fig. 21-23 *Installing electrical wiring and boxes in insulated concrete forms.* (American Polysteel Forms)

FOR THIN PORTIONS OF EPS:
4"x4"x1-1/2" shallow box with single gang 1/2" raised cover.

Fig. 21-24 *Diagrams of ICF formwork made with the three basic units: panel on the left, plank in the center, and block on the right.* (PCC)

Fig. 21-25 *Cutaway diagrams of ICF walls with the three basic cavity shapes: flat, grid, and post-and-beam.* (PCC)

Block systems include units ranging from the standard concrete block (8 × 16 inches) size to a much larger 16-inch-high by 4-foot-long unit. Along their edges are teeth or tongues and grooves for interlocking; they stack without separate fasteners on the same principle as children's Lego blocks.

Another difference is the shape of the cavities. Each system has one of three distinct cavity shapes. The shapes are flat, grid, or post-and-beam. These produce different shapes of concrete beneath the foam, as shown in Fig. 21-25.

Note how the flat cavities produce a concrete wall of constant thickness, just like a conventionally poured wall that was made with plywood or metal forms. Cav-

ities are usually "wavy," both horizontally and vertically. If the forms are removed, it can be seen just how the walls resemble a breakfast waffle. The post-and-beam cavities are cavities with concrete only every few feet, horizontally and vertically. In the most extreme post-and-beam systems, there is a 6-inch-diameter concrete "post" formed every 4 feet and a 6-inch concrete "beam" at the top of each story.

Keep in mind that no matter what the shape of the cavity, all systems have "ties." These are the crosspieces that connect the front and back layers of foam that make up the form. If the ties are plastic or metal, the concrete is not affected significantly. However, in some grid systems, they are foam and are much larger. But the forming breaks in the concrete about 2 inches in diameter every foot or so. See Fig. 21-25 for the differences.

Table 21-1 shows eight different systems used in ICF with their dimensions, fastening surface, and various notes with additional details. Note that no matter what the cavity shape is, all systems also have "ties." These ties are the crosspieces that connect the front and back layers of foam. When the ties are metal or plastic, they do not affect the shape of the concrete much. But in some of the grid systems, they are foam and are much larger, forming breaks in the concrete about 2 inches in diameter every foot or so. Figure 21-26 shows the differences.

Table 21-1 *Available ICF Systems*[1] (Portland Cement Assoc.)

	Dimensions[2] (width x height x length)	Fastening surface	Notes
Panel systems			
Flat panel systems			
R-FORMS	8" x 4' x 8'	Ends of plastic ties	Assembled in the field; different lengths of ties available to form different panel widths.
Styroform	10" x 2' x 8'	Ends of plastic ties	Shipped flat and folded out in the field; can be purchased in larger/smaller heights and lengths.
Grid panel systems			
ENER-GRID	10" x 1'3" x 10'	None	Other dimensions also available; units made of foam/cement mixture.
RASTRA	10" x 1'3" x 10'	None	Other dimensions also available; units made of foam/cement mixture.
Post-and-beam panel systems			
Amhome	9 3/8" x 4' x 8'	Wooden strips	Assembled by the contractor from foam sheet. Includes provisions to mount wooden furring strips into the foam as a fastening surface.
Plank systems			
Flat plank systems			
Diamond Snap-Form	1' x 1' x 8'	Ends of plastic ties	
Lite-Form	1' x 8" x 8'	Ends of plastic ties	
Polycrete	11" x 1' x 8'	Plastic strips	
QUAD-LOCK	8" x 1' x 4'	Ends of plastic ties	
Block systems			
Flat block systems			
AAB	11.5" x 16⅝" x 4'	Ends of plastic ties	
Fold-Form	1' x 1' x 4'	Ends of plastic ties	Shipped flat and folded out in the field.
GREENBLOCK	10" x 10" x 3'4"	Ends of plastic ties	
SmartBlock Variable Width Form	10" x 10" x 3'4"	Ends of plastic ties	Ties inserted by the contractor; different length ties available to form different block widths.
Grid block systems with fastening surfaces			
I.C.E. Block	9 1/4" x 1'4" x 4'	Ends of steel ties	
Polysteel	9 1/4" x 1'4" x 4'	Ends of steel ties	
REWARD	9 1/4" x 1'4" x 4'	Ends of plastic ties	
Therm-O-Wall	9 1/4" x 1'4" x 4'	Ends of plastic ties	
Grid block systems without fastening surfaces			
Reddi-Form	9 5/8" x 1' x 4'	Optional	Plastic fastening surface strips available
SmartBlock Standard Form	10" x 10" x 3'4"	None	
Post-and-beam block systems			
ENERGYLOCK	8" x 8" x 2'8"	None	
Featherlite	8" x 8" x 1'4"	None	
KEEVA	8" x 1' x 4'	None	

[1] All systems are listed by brand name.
[2] "Width" is the distance between the inside and outside surfaces of foam of the unit. The thickness of the concrete inside will be less, and the thickness of the completed wall with finishes added will be greater.

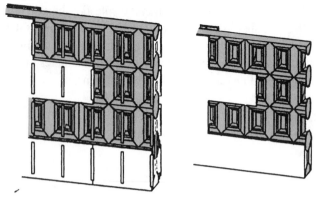

Fig. 21-26 *Cutaway diagrams of ICF grid walls with steel/plastic ties and foam ties.* (PCC)

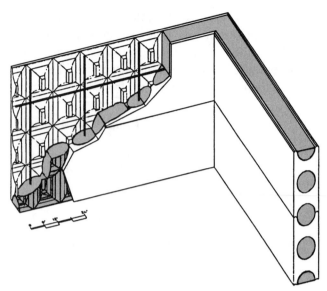

Fig. 21-28 *Cutaway diagram of a grid panel wall.* (PCC)

Another difference is that many of the systems also have a fastening surface which is some material other than foam, embedded into the units that crews can sink screws or nails into the same way as fastening to a stud. Often this surface is simply the ends of the ties; however, other systems have no embedded fastening surface. These units are all foam, including the ties. This generally makes them simpler and less expensive but requires crews to take extra steps to connect interior wallboard, trim, exterior siding, and so on to the walls.

Cutaway views of the wall panels are shown in Figs. 21-27 through 21-34. In Fig. 21-35 the R-Forms panel is being assembled on site. In Fig. 21-36, there is an on-site pile of Ener-Grid panels. Figure 21-37 shows the top view of an Amhome panel that has furring strips embedded. Note how light the panels are. They are easily handled by one person. The worker is carrying a fold-out Fold-Form block before and after it has been spread for adding to the construction.

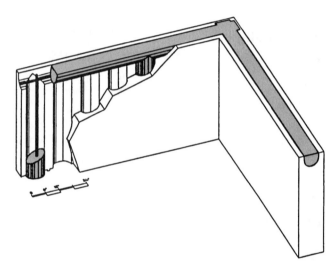

Fig. 21-29 *Cutaway diagram of a post-and-beam panel wall.* (PCC)

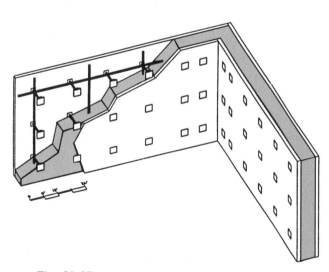

Fig. 21-27 *Cutaway diagram of a flat panel wall.* (PCC)

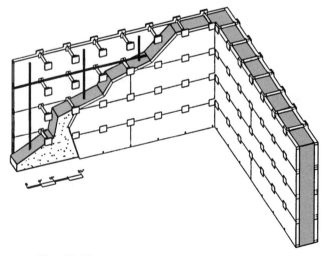

Fig. 21-30 *Cutaway diagram of a flat plank wall.* (PCC)

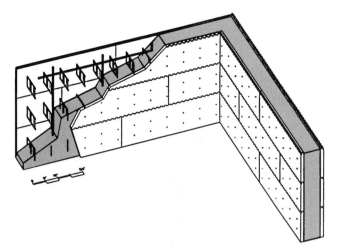

Fig. 21-31 *Cutaway diagram of a flat block wall. (PCC)*

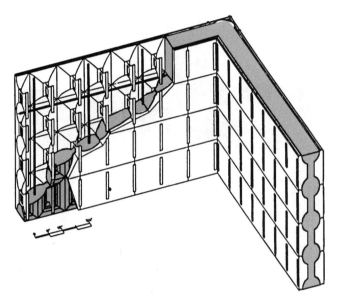

Fig. 21-32 *Cutaway diagram of a grid block wall with fastening surfaces. (PCC)*

Fig. 21-33 *Cutaway diagram of a grid block wall without fastening surfaces. (PCC)*

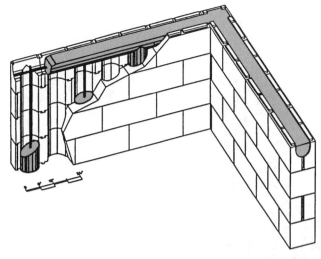

Fig. 21-34 *Cutaway diagram of a post-and-beam block wall. (PCC)*

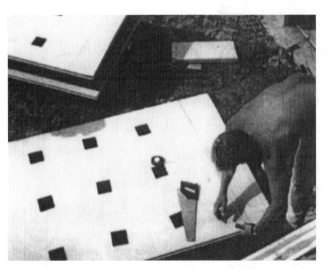

Fig. 21-35 *Site-assembly of R-Forms panels. (PCC)*

Two varieties of SmartBlock are shown in Fig. 21-39. Three of them, A, C, and D, are assembled with plastic ties. A grid block without fastening surfaces is shown in Fig. 21-39B.

Figure 21-40A, B, C, and D shows setting the insulating concrete forms, how the completed forms look, pumping in concrete, and the completed house.

Foam Working Tools

Some of the tools needed for working with foam are not familiar to the usual framer of houses. These tools may include the thermal cutter. This cutter is a new tool that cuts a near-perfect line through foam and plastic units in one pass. Figure 21-41 shows this device. It is made up of a taut resistance-controlled wire mounted on a bench-frame. It is heated with electricity and drawn though the unit while the wire is red hot. It melts a narrow path through foam and plastic

Fig. 21-36 *An ENER-GRID panel.* (PCC)

A

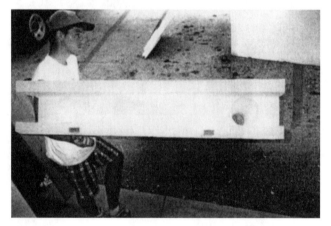

Fig. 21-37 *Top view of an Amhome panel with embedded furring strips.* (PCC)

ties. It is a worthwhile tool to have if you are building a high-volume ICF walled house. It will not cut rough metal ties or the foam-and-cement material of the grid panel systems. However, companies selling thermal cutters are usually located in every community. The grid panel systems can be cut with any of the bladed tools used by a carpenter, but sometimes a chain saw is a handy tool. It goes quickly through the heavier material of these systems and cuts through in one pass, whether cutting on the ground or in place.

Tools needed for cutting and working with foam are shown in Table 21-2.

B

Fig. 21-38 *Folding out a Fold-Form block before use.* (PCC)

Gluing and Tying Units

ICF units are frequently glued at the joints to hold them down, hold them together, and prevent concrete leakage. Common wood glue and most construction

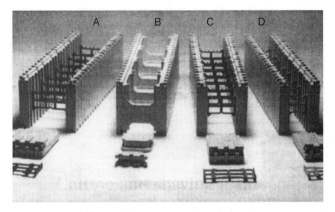

Fig. 21-39 *The two varieties of SmartBlock: a flat block assembled with plastic ties (A,C,D) and a grid block without fastening surfaces in B.* (PCC)

Fig. 21-41 *A thermal cutter. Note the white line that is part of the cutting device. It is attached to the transformer on the left of the framework.* (PCC)

adhesives do the job well. Popular brands are Liquid Nail, PL 200, and PL 400. Some of these can dissolve the foam but, if applied in a thin layer, the amount of foam lost is usually insignificant. Look for an adhesive that is "compatible with polystyrene." Figure 21-42 shows an industrial foam gun.

Rebar is often precut to length and pre-bent but, even if it is, the workers generally have to process a few bars in

the field. Most ICF systems also have cradles that hold the bars in place for the pour, but a few bars need to be wired to one another or to ties to keep them in position.

A

B

C

D

Fig. 21-40 *A. Setting insulating concrete forms. B. Completed formwork. C. Pumping in the concrete. D. The completed home.* (PCC)

Table 21-2 *Useful Tools and Materials* (Portland Cement Assoc.)

Operation or Class of Material	
Tool or Material	**Comments**
Cutting and shaving foam	
Drywall or keyhole saw	For small cuts, holes, and curved cuts.
PVC or mitre saw	For small, straight cuts and shaving edges.
Coarse sandpaper or rasp	For shaving edges.
Bow saw or garden pruner	For faster straight cuts.
Circular saw	For fast, precise, straight cuts. For cutting units with steel ties, reverse the blade or use a metal-cutting blade.
Reciprocating saw	For fast cuts, especially in place.
Thermal cutter	For fast, very precise cuts on a bench. Not suitable for steel ties or grid panel units.
Chain saw	For fast cuts of grid panel units.
Lifting units	
Forklift, manual lift, or boom or crane truck	For carrying large grid panel units and setting them in place. For upper stories, a truck is necessary.
Gluing and tying units	
Wood glue, construction adhesive, or adhesive foam	
Small-gage wire	For connecting units of flat panel systems.
Bending, cutting, and wiring rebar	
Cutter-bender	
Small-gage wire or precut tie wire or wire spool	
Filling and sealing formwork	
Adhesive foam	
Placing concrete	
Chute	For below-grade pours.
Line pump	Use a 2-inch hose.
Boom pump	Use two "S" couplings and reduce the hose down to a 2-inch diameter.
Evening concrete	
Mason's trowel	
Dampproofing walls below grade	
Nonsolvent-based dampproofer or nonheat-sealed membrane product	
Surface cutting foam	
Utility knife, router, or hot knife	Heavier utility knives work better. Use a router with a half-inch drive for deep cutting.
Fastening to the wall	
Galvanized nails, ringed nails, and drywall screws	For attaching items to fastening surfaces. Use screws only for steel fastening surfaces.
Adhesives	For light and medium connections to foam.
Insulation nails and screws	For holding lumber inside formwork.
J-bolt or steel strap	For heavy structural connections.
Duplex nails	For medium connections to lumber.
Small-gage wire	For connecting to steel mesh for stucco.
Concrete nails or screw anchors	For medium connections to lumber after the pour.
Flattening foam	
Coarse sandpaper or rasp	For removing small high spots.
Thermal cutter	For removing large bulges.
Foam	
Expanded polystyrene or extruded polystyrene	Consider foam with insect-repellent additives
Concrete	
Midrange plasticizer or superplasticizer	For increasing the flow of concrete without decreasing its strength. Can also be accomplished by changing proportions of the other ingredients.
Stucco	
Portland cement stucco or polymer-based stucco	

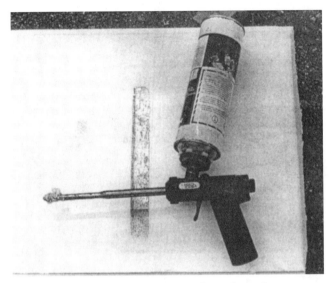

Fig. 21-42 *The industrial foam gun can be used as a glue gun.* (PCC)

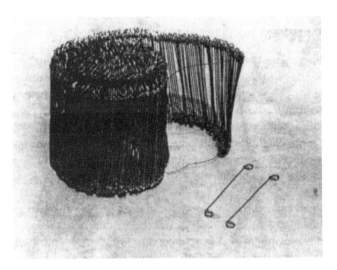

Fig. 21-44 *Roll of precut tie wires.* (PCC)

It is possible to bend rebar with whatever tools are handy and cut it with a hacksaw; however, if you process large quantities, you might prefer buying or renting a cutter-bender. A large manual tool is pictured in Fig. 21-43. It makes the job faster and easier. Cutters-benders are available at steel-supply, concrete-supply, and masonry-supply houses. Almost any steel wire can hold rebar in place, but most efficient are rolls of precut wires shown in Fig. 21-44. These wires make the job faster and easier. Wire coils and belt-mounted coil holders are both sold by suppliers of concrete products, masonry, and steel.

Pouring Concrete

Concrete is best poured at a more controlled rate into ICFs than it is into conventional forms. An ordinary

chute can be used for foundation walls (basement or stem); this is the least expensive option because it comes free with the concrete truck. Precise control is more difficult with a chute. You must pour more slowly than with conventional forms, and you must move the chute and truck frequently to avoid overloading any one section of the formwork.

The smaller line pump pushes concrete through a hose that lies on the ground. See Fig. 21-45. The crew holds the end of the hose over the formwork to drop concrete inside. If possible, use a 2-inch hose. One or two workers can handle it, and it can generally be run at full speed without danger. If only a 3-inch hose is available, you can use it; but pump slowly until you learn how much pressure the forms can take.

Boom pumps are mounted on a truck that also holds a pneumatically operated arm (the boom). The hose from the pump causes the concrete to move along the length of the boom and then hang loose from the end. See Fig. 21-46. By moving the boom, the truck's operator can

Fig. 21-43 *This is a rebar cutter-bender.* (PCC)

Fig. 21-45 *Line pump for concrete.* (PCC)

Fig. 21-46 Boom pump for pouring concrete. (PCC)

dangle the hose wherever the crew calls for it. One worker holds the free end to position it over the form-work cavities. The standard hose diameter is 4 inches.

You will need to have the hose diameter reduced to 2 or 3 inches with tapered steel tubes called "reducers." The narrower diameter slows down and smoothes out the flow of the concrete. Figure 21-47 shows how the reducers are arranged on a boom with hoses. Also, ask for two 90-degree elbow fittings on the end of the hose assembly, as seen in Fig. 21-47. These form an "S" in the line that further breaks the fall of the concrete. The concrete can be leveled off on top of the form by using a mason's trowel.

CONCRETE BLOCK

The second most popular residential framing method is concrete block. Comprised of less than 5% of the residential framing market, concrete block is another traditional method that has been around for a long time. It is mainly used when home owners are interested in a framing method that is resistant to inclement weather and termites. This type of construction is much more costly and time-consuming. Because the techniques associated with masonry and concrete work are numerous and detailed, only the major steps are reviewed.

1. Snap a chalk line on the footing or foundation where the concrete block should be placed.

2. Spread mortar an inch or two thick on the footing or foundation and an inch or so wider and longer than the block.

3. Place a concrete block on each corner of the wall. Make sure both blocks are level and straight.

4. Stretch a string line tightly between the front, top edge of both blocks. This can be done by wrapping a string around a brick and placing it on top of the

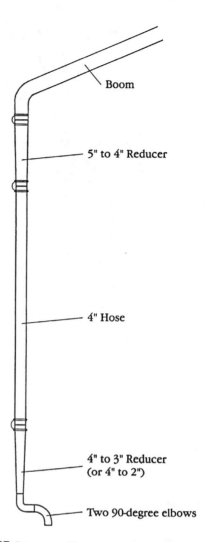

Fig. 21-47 Diagram of boom pump fittings that are suitable for pouring ICF walls. (PCC)

concrete block.

5. Place concrete blocks evenly about seven blocks (if using (8" × 16" block) from the corner. Do this for both corners.

6. Continue to build up the corners by backing up each level of block (course) by half a block. When finished, one block should be on the top of each corner. Each corner is called a *lead*.

7. Place concrete blocks between each lead using a string to keep the top and front edge of the block straight and level. Mortar should be placed on one end of each block. A ⅜-inch-thick layer of mortar is common, because the actual size of most 8" × 16" concrete block is 7⅝" × 15⅝".

8. Allowances should be made for rough openings of windows and doors. Lintels or metal ledges are used to support block over door and window openings.

9. The top course of concrete block is filled with mortar, and anchor bolts are inserted at certain intervals according to building code requirements.

10. Mortar joints are usually smoothed (tooled) with a round bar, square bar, or s-shaped tool.

CHAPTER 21 STUDY QUESTIONS

1. Why are many builders beginning to use alternative methods for framing a house?

2. What are the main advantages and disadvantages to steel framing?

3. List the two types of steel framing and describe their differences.

4. How are anchor bolts attached to a concrete foundation using the red-iron framing method?

5. How is electrical wiring run through a steel-frame home?

6. What does ICF mean and how is it constructed?

7. List at least two advantages and disadvantages to ICFs.

8. How is sheathing installed in an ICF?

9. How is electrical wiring installed in an ICF?

10. Define the following terms: lead, course, lintel, and tooled.

11. What are three ways to build a steel-framed wall?

12. List two advantages of a panelized steel-framed wall.

13. What type of fastener is used to assemble steel and stud walls?

14. What gage steel studs are used for interior and exterior walls?

15. What gage steel is used for studs of weight-bearing walls?

Glossary

access   Access refers to the freedom to move to and around a building, or the ease with which a person can obtain admission to a building site.

acoustical   This refers to the ability of tiles on a ceiling to absorb or deaden sound.

aggregate   Aggregate refers to sand, gravel, or both in reference to concrete mix.

air-drying   Method of removing excess moisture from lumber using natural circulation of air.

alligatoring   This occurs when paint cracks and resembles the skin of an alligator.

anchor bolt   An anchor bolt is a steel pin that has a threaded end with a nut and an end with a 90° angle in it. The angled end is pushed into the wet concrete and becomes part of the foundation for anchoring the flooring or sill plates.

annular ring   There are two colored rings that indicate the growth of a tree. The colors indicate the growth of springwood and summerwood.

apron   A piece of window trim that is located beneath the window sill. Also used to designate the front of a building such as the concrete apron in front of a garage.

arbor   An axle on which a cutting tool is mounted. It is a common term used in reference to the mounting of a circular saw blade.

asphalt shingle   This is a composition-type shingle used on roof. It is made of a saturated felt paper with ground-up pieces of stone embedded and held in place by asphaltum.

asphalt shingles   These are shingles made of asphalt or tar-impregnated paper with a mineral material embedded; they are very fire resistant.

auxiliary locks   Auxiliary locks are placed on exterior doors to prevent burglaries.

awl   An awl is a tool used to mark wood with a scratch mark. It can be used to produce pilot holes for screws.

awning picture window   This type of window has a bottom panel that swings outward; a crank operates the moving window. As the window swings outward, it has a tendency to create an awning effect.

backsplash   A backsplash is the vertical part of a countertop that runs along the wall to prevent splashes from marring the wall.

backsaw   This saw is easily recognized since it has a very heavy steel top edge. It has a fine-tooth configuration.

balloon frame   This type of framing is used on two-story buildings. Wall studs rest on the sill. The joists and studs are nailed together, and the end joists are nailed to the wall studs.

balustrade   A complete handrail assembly. This includes the rails, the balusters, subrails, and fillets.

baluster   The baluster is that part of the staircase which supports the handrail or bannister.

bannister   The bannister is that part of the staircase which fits on top of the balusters.

baseboard   Molding covering the joint between a finished wall and the floor.

base shoe   A molding added at the bottom of a baseboard. It is used to cover the edge of finish flooring or carpeting.

batten   A batten is the narrow piece of wood used to cover a joint.

batter boards   These are boards used to frame in the corners of a proposed building while the layout and excavating work takes place.

batts   Batts are thick pieces of Fiberglas that can be inserted into a wall between the studs to provide insulation.

bay window   Bay windows stick out from the main part of the house. They add to the architectural qualities of a house and are used mostly for decoration.

beam   A horizontal framing member. It may be made of steel or wood. Usually the term is used to refer to a wooden beam that is at least 5 inches thick and at least 2 inches wider than it is thick.

bearing partition   An interior divider or wall that supports the weight of the structure above it.

bearing wall A bearing wall has weight-bearing properties associated with holding up a building's roof or second floor.

benchmark A point from which other measurements are made.

bevel A bevel is a tool that can be adjusted to any angle. It helps make cuts at the number of degrees that is desired and is a good device for transferring angles from one place to another.

bifold A bifold is a folding door used to cover a closet. It has two panels that hinge in the middle and fold to allow entrance.

blistering Blistering refers to the condition that paint presents when air or moisture is trapped underneath and makes bubbles that break into flaky particles and ragged edges.

blocking Corners and wall intersections are made the same as outside walls. The size and amount of blocking can be reduced. The purpose of blocking is to provide nail surfaces at the corners. These are needed at inside and outside nail surfaces. They are a base for nailing wall covering.

blockout A form for pouring concrete is blocked out by a frame or other insertion to allow for an opening once the concrete has cured.

blocks This refers to a type of flooring made of wood. Wide pieces of boards are fastened to the floor, usually in squares and by adhesives.

board and batten A finished wall surface consisting of vertical board with gaps between them. The battens or small strips of wood cover the gaps.

board foot A unit of lumber measure equaling 144 cubic inches. The base unit (B.F.) is 1 inch thick and 12 inches square, or $1 \times 12 \times 12 = 144$ cubic inches.

bonding This is another word for gluing together wood or plastics and wood.

bottom or heel cut This refers to the cutout of the rafter end which rests against the plate. The bottom cut is also called the foot or seat cut.

bow A term used to indicate an upward warp along the length of a piece of lumber that is laid.

bow window A window unit that projects from an exterior wall. It has a number of windowpanes that form a curve.

brace A brace is an inclined piece of lumber applied to a wall or to roof rafters to add strength.

brace scale A brace scale is a table that is found along the center of the back of the tongue and gives the exact lengths of common braces.

bridging Bridging is used to keep joists from twisting or bending.

buck A buck is the same as a blockout.

builder's level This is a tripod-mounted device that uses optical sighting to make sure that a straight line is sighted and that the reference point is level.

building codes Building codes are rules and regulations which are formulated in a code by a local housing authority or governing body.

building paper Also called tar paper, roofing paper, and a number of other terms. It has a black coating of asphalt for use in weatherproofing.

building permits Most incorporated cities or towns have a series of permits that must be obtained for building. This allows for inspections of the work and for placing the house on the tax roles.

built-ins This is a term used to describe the cabinets and other small appliances that are built into the kitchen, bathroom, or family room by a carpenter. They may be custom cabinets or may be made on the site.

bundle This term refers to the packaging of shingles. A bundle of shingles is usually a handy method for shipment.

butt A term that can be used a couple of ways: A *butt hinge* is one where the two parts meet edge-to-edge, allowing movement of the two parts when held together by a pin; to *butt* means to meet edge to edge, such as in a joining of wooden edges.

cantilever Overhangs are called *cantilevers*; they are used for special effects on porches, decks, or balconies.

carpenter's square This steel tool can be used to check for right angles, to lay out rafters and studs, and to perform any number of measuring jobs.

carpet strips These are wooden or metal strips with nails or pins sticking out. They are nailed around the perimeter of a room, and the carpet is pulled tight and fastened to the exposed nails.

carpet tape This is a tape used to seam carpet where it fits together.

carriage A notched stair frame is called a carriage.

casement This is a type of window hinged to swing outward.

casing A door casing refers to the trim that goes on around the edge of a door opening and also to the trim around a window.

cathedral ceiling A cathedral ceiling is not flat and parallel with the flooring; it is open and follows the shape of the roof. The open ceiling usually precludes an attic.

caulk Caulk is any type of material used to seal walls, windows, and doors to keep out the weather. Caulk is usually made of putty or some type of plastic material, and it is flexible and applied with a caulking gun.

cellulose fiber Insulation material made from cellulose fiber. Cellulose is present in wood and paper, for example.

cement Cement is a fine-powdered limestone that is heated and mixed with other minerals to serve as a binder in concrete mixes.

ceramic tile Ceramic tile is made of clay and fired to a high temperature; it usually has a glaze on its surface. Small pieces are used to make floors and wall coverings in bathrooms and kitchens.

certificate of occupancy This certificate is issued when local inspectors have found a house worthy of human habitation. It allows a contractor to sell a house. It is granted when the building code has been complied with and certain inspections have been made.

chair A chair is a support bracket for steel reinforcing rods that holds the rods in place until the concrete has been poured around them.

chalk line A chalk line is used to guide a roofer. It is snapped, causing the string to make a chalk mark on the roof so that the roofers can follow it with their shingles.

chipboard Chipboard is used as an underlayment. It is constructed of wood chips held together with different types of resins.

chisel A wood chisel is used to cut away wood for making joints. It is sharpened on one end, and the other is hit with the palm of the hand or with a hammer to cut away wood for door hinge installation or to fit a joint tightly.

claw hammer This is the common hammer used by carpenters to drive nails. The claws are used to extract nails that bend or fail to go where they are wanted.

cleat Any strip of material attached to the surface of another material to strengthen, support, or secure a third material.

clerestory A short exterior wall between two sections of roof that slope in different directions. The term is also used to describe a window that is placed in this type of wall.

closed-cut For a closed-cut valley, the first course of shingles is laid along the eaves of one roof area up to and over the valley. It is extended along the adjoining roof section. The distance is at least 12 inches. The same procedure is followed for the next courses of shingles.

cold chisel This chisel is made with an edge that can cut metal. It has a one-piece configuration, with a head to be hit by a hammer and a cutting edge to be placed against the metal to be cut.

common rafter A common rafter is a member that extends diagonally from the plate to the ridge.

concrete Concrete is a mixture of sand, gravel, and cement in water.

condensation The process by which moisture in the air becomes water or ice on a surface (such as a window) whose temperature is colder than the air's temperature.

contact cement Contact cement is the type of glue used in applying countertop finishes. Both sides of the materials are coated with the cement, and the cement is allowed to dry. The two surfaces are then placed in contact, and the glue holds immediately.

contractor A contractor is a person who contracts with a firm, a bank, or another person to do a job for a certain fee and under certain conditions.

contractor's key This is a key designed to allow the contractor access to a house while it is under construction. The lock is changed to fit a pregrooved key when the house is turned over to the owner.

convection Transfer of heat through the movement of a liquid or gas.

coped joint This type of joint is made with a coping saw. It is especially useful for corners that are not square.

coping saw This saw is designed to cut small thicknesses of wood at any curve or angle desired. The blade is placed in a frame, with the teeth pointing toward the handle.

corner beads These are metal strips that prevent damage to drywall corners.

cornice The cornice is the area under the roof overhang. It is usually enclosed or boxed in.

course This refers to alternate layers of shingles in roofing.

cradle brace A cradle brace is designed to hold sheetrock or drywall while it is being nailed to the ceiling joists. The cradle brace is shaped like a T.

crawl space A crawl space is the area under a floor that is not fully excavated. It is only excavated sufficiently to allow one to crawl under it to get at the electrical or plumbing devices.

cricket This is another term for the *saddle*.

cripple jack A cripple jack is a jack rafter with a cut that fits in between a hip and a valley rafter.

cripple rafter A cripple rafter is not as long as the regular rafter used to span a given area.

cripple stud This is a short stud that fills out the position where the stud would have been located if a window, door, or some other opening had not been there.

crisscross wire support This refers to chicken wire that is used to hold insulation in place under the flooring of a house.

crosscut saw This is a handsaw used to cut wood across the grain. It has a wooden handle and a flexible steel blade.

cup To warp across the grain.

curtain wall Inside walls are often called curtain walls. They do not carry loads from roof or floors above them.

dado A rectangular groove cut into a board across the grain.

damper A damper is an opening and closing device that will close off the fireplace area from the outside by shutting off the flue. It can also be used to control draft.

deadbolt lock A deadbolt lock will respond only to the owner who knows how to operate it. It is designed to keep burglars out.

deck A deck is the part of a roof that covers the rafters.

decorative beams Decorative beams are cut to length from wood or plastic and mounted to the tastes of the owner. They do not support the ceiling.

diagonals Diagonals are lines used to cut across from adjacent corners to check for squareness in the layout of a basement or footings.

dividers Dividers have two points and resemble a compass. They are used to mark off specific measurements or transfer them from a square or a measuring device to the wood to be cut.

dormer Dormers are protrusions that stick out from a roof. They may be added to allow light into an upstairs room.

double-hung windows Double-hung windows have two sections, one of which slides past the other. They slide up and down in a prearranged slot.

double plate This usually refers to the practice of using two pieces of dimensional lumber for support over the top section or wall section.

double trimmer Double joists used on the sides of openings are called double trimmers. Double trimmers are placed without regard to regular joist spacings for openings in the floor for stairs or chimneys.

downspouts These are pipes connected to the gutter to conduct rainwater to the ground or sewer.

drain tile A drain tile is usually made of plastic. It generally is 4 inches in diameter, with a number of small holes to allow water to drain into it. It is laid along the foundation footing to drain the seepage into a sump or storm sewer.

drawer guide The drawers in cabinets have guides to make sure that the drawer glides into its closed position easily without wobbling.

drop siding Drop siding has a special groove, or edge cut into it. The edge lets each board fit into the next board. This makes the boards fit together and resist moisture and weather.

drywall This is another name for panels made of gypsum.

ductwork Ductwork is a system of pipes used to pass heated air along to all parts of a house. The same ductwork can be used to distribute cold air for summer air conditioning.

dutch hip This is a modification of the hip roof.

eaves Eaves are the overhang of a roof projecting over the walls.

eaves trough A gutter.

elevation Elevation refers to the location of a point in reference to the point established with the builder's level or transit. Elevation indicates how high a point is. It may also refer to the front elevation or front view of a building or the rear elevation or what the building looks like from the rear. Side elevations refer to a side view.

energy Energy refers to the oil, gas, or electricity used to heat or cool a house.

essex board measure This is a table on the back of the body; it gives the contents of any size lumber. The table is located on the steel square used by carpenters.

excavation Excavate means to remove. In this case, excavation refers to the removal of dirt to make room for footings, a foundation, or the basement of a building.

expansion joint This is usually a piece of soft material that is about 1 inch thick and 4 inches wide and is placed between sections of concrete to allow for expansion when the flat surface is heated by the sun.

exposure Exposure refers to the part of a shingle or roof left to the weather.

extension form An extension form is built inside the concrete outer form. It forms a stepped appearance so that water will not drain into a building but drain outward from the slab or foundation slab.

faced insulation This insulation usually has a coating to create a moisture barrier.

faced stapling Faced stapling refers to the strip along the outer edges of the insulation that is stapled to the outside or 2-inch sides of the 2-x-4 studs.

facing Facing strips give cabinets a finished look. They cover the edges where the units meet and where the cabinets meet the ceiling or woodwork.

factory edge This is the straight edge of linoleum made at the factory. It provides a reference line for the installer.

factory-produced housing This refers to housing that is made totally in a factory. Complete units are usually trucked to a place where a basement or slab is ready.

false bottom This is a system of 1-x-6-inch false-bottom or box beams that provide the beauty of beams without the expense. The false beams are made of wood or plastic materials and glued or nailed in place. They do not support any weight.

fascia Fascia refers to a flat board covering the ends of rafters on the cornice or eaves. The eave troughs are usually mounted to the fascia board.

FHA The Federal Housing Administration.

Fiberglas Fiberglas is insulation material made from spun resin or glass. It conducts little heat and creates a large dead air space between layers of fibers. It helps conserve energy.

firebrick This is a special type of brick that is not damaged by fire. It is used to line the firebox in a fireplace.

fire stops Fire stops are short pieces nailed between joists and studs.

flaking This refers to paint that falls off a wall or ceiling in flakes.

flashing Flashing is metal used to cover chimneys or other things projecting through the roofing. It keeps the weather out.

floating The edges of drywall sheets are staggered, or floated. This gives more bracing to the wall since the whole wall does not meet at any one joint.

floating Floating refers to concrete work; it lets the smaller pieces of concrete mix float to the top. Floating is usually done with a tool moved over the concrete.

floorboards This refers to floor decking. Floorboards may be composed of boards, or may be a sheet of plywood used as a subfloor.

flue The flue is the passage through a chimney.

flush This term means to be even with.

folding rule This is a device that folds into a 3-×–6-inch rectangle. It has the foot broken into 12 inches. Each inch is broken into 16 points. Snap joints hinge the rule every six inches. It will spread to as much as 6 feet, or 72 inches.

footings Footings are the lowest part of a building. They are designed to support the weight of the building and distribute it to the earth or rock formation on which it rests.

form A form is a structure made of metal or wood used as a mold for concrete.

Formica Formica is a laminated plastic covering made for countertops.

foundation The foundation is the base on which a house or building rests. It may consist of the footings and walls.

framing Roof framing is composed of rafters, ridge board, collar beams, and cripple studs.

framing square This tool allows a carpenter to make square cuts in dimensional lumber. It can be used to lay out rafters and roof framing.

French doors This usually refers to two or more groups of doors arranged to open outward onto a patio or veranda. The doors are usually composed of many small glass panes.

frostline This is the depth to which the ground freezes in the winter.

furring strips These are strips of wood attached to concrete or stone. They form a nail base for wood or paneling.

gable This is the simplest kind of roof. Two large surfaces come together at a common edge, forming an inverted V.

galvanized iron This material is usually found on roofs as flashing. It is sheet metal coated with zinc.

gambrel roof This is a barn-shaped roof.

girder A girder is a support for the joists at one end. It is usually placed halfway between the outside walls and runs the length of the building.

grade The grade is the variation of levels of ground or the established ground-line limit on a building.

grid system This is a system of metal strips that support a drop ceiling.

grout Grout is a white plaster-like material placed into the cracks between ceramic tiles.

gusset A gusset is a triangular or rectangular piece of wood or metal that is usually fastened to the joint of a truss to strengthen it. It is used primarily in making roof trusses.

gutter This is a metal or wooden trough set below the eaves to catch and conduct water from rain and melting snow to a downspout.

gypsum Gypsum is a chalk used to make wallboard. It is made into a paste, inserted between two layers of paper, and allowed to dry. This produces a plastered wall with certain fire-resisting characteristics.

handsaw A handsaw is any saw used to cut wood and operated by manual labor rather than electricity.

hang a door This term refers to the fact that a door has to be mounted on hinges and aligned with the door frame.

hangers These are metal supports that hold joists or purlins in place.

hardware In this case hardware refers to the metal parts of a door. Such things as hinges, locksets, and screws are hardware.

hardwood The wood that comes from a tree that sheds its leaves. This doesn't necessarily mean the wood itself is hard. A poplar has soft wood, but it is classified as a hardwood tree. An oak has hard wood and is also classified as a hardwood tree.

hardboard A type of fiberboard pressed into thin sheets. Usually made of wood chips or waste material from trees after the lumbering process has been completed.

header A header is a board that fits across the ends of joists.

head lap This refers to the distance between the top and the bottom shingle and the bottom edge of the shingle covering it.

hearth A hearth is the part of a fireplace that is in front of the wood rack.

hex strips This refers to strips of shingles that are six-sided.

hip rafters A hip rafter is a member that extends diagonally from the corner of the plate to the ridge.

hip roof A hip roof has four sides, all sloping toward the center of the building.

hollow-core doors Most interior doors are hollow and have paper or plastic supports for the large surface area between the top and bottom edge and the two faces.

honeycomb Air bubbles in concrete cause a honeycomb effect and weaken the concrete.

insert stapling This refers to stapling insulation inside the 2-×-4 stud. The facing of the insulation has a strip left over. It can be stapled inside the studs or over the studs.

insulation Insulation is any material that offers resistance to the conduction of heat through its composite materials.

Plastic foam and Fiberglas are the two most commonly used types of insulation in homes today.

insulation batts These are thick precut lengths of insulation designed to fit between studs.

interlocking Interlocking refers to a type of shingle that overlaps and interlocks with its edges. It is used in high winds.

in the white This term is used to designate cabinets that are assembled but unfinished.

jack rafter Any rafter that does not extend from the plate to the ridge is called a jack rafter.

jamb A jamb is the part that surrounds a door window frame. It is usually made of two vertical pieces and a horizontal piece over the top.

joist Large dimensional pieces of lumber used to support the flooring platform of a house or building are called joists.

joist hangers These are metal brackets that hold up the joist. They are nailed to the girder, and the joist fits into the bracket.

joist header If the joist does not cover the full width of the sill, space is left for the joist header. The header is nailed to the ends of the joists and rests on the sill plate. It is perpendicular to the joists.

kerf The cut made by a saw blade.

key A key is a depression made in a footing so that the foundation or wall can be poured into the footing, preventing the wall or foundation from moving during changes in temperature or settling of the building.

kicker A kicker is a piece of material installed at the top or side of a drawer to prevent it from falling out of a cabinet when it is opened.

kiln dried Special ovens are made to dry wood before it is used in construction.

king-post truss This is the type of roof truss used to support low pitch roofs.

ladder jack Ladder jacks hang from a platform on a ladder. They are most suitable for repair jobs and for light work where only one carpenter is on the job.

landing A landing is the part of a stairway that is a shaped platform.

lap This refers to lap siding. Lap siding fits on the wall at an angle. A small part of the siding is overlapped on the preceding piece of siding.

laths Laths are small strips of wood or metal designed to hold plastic on the wall until it hardens for a smooth finish.

ledger A ledger is a strip of lumber nailed along an edge or bottom of a location. It helps support or keep from slipping the girders on which the joists rest.

left-hand door A left-hand door has its hinges mounted on the left when viewed from the outside.

level A level is a tool using bubbles in a glass tube to indicate the level of a wall, stud, or floor. Keeping windows, doors, and frames square and level makes a difference in their fit and operation.

level-transit This is an optical device that is a combination of a level and a means for checking vertical and horizontal angles.

load Load refers to the weight of a building.

load conditions These are the conditions under which a roof must perform. The roof has to support so much wind load and snow. Load conditions vary according to locale.

lockset The lockset refers to the doorknob and associated locking parts inserted in a door.

mansard This type of roof is popular in France and is used in the United States also. The second story of the house is covered with the same shingles used on the roof.

manufactured housing This term is used in reference to houses that are totally or partially made within a factory and then trucked to a building site.

mastic Mastic is an adhesive used to hold tiles in place. The term also refers to adhesives used to glue many types of materials in the building process.

military specifications These are specifications that the military writes for the products it buys from the manufacturers. In this case, the term refers to the specifications for a glue used in making trusses and plywood.

miter box The miter box has a hacksaw mounted in it. It is adjustable for cutting at angles such as 45 and 90°. Some units can be adjusted by a level to any angle.

modular homes These houses are made in modules or small units which are nailed or bolted together once they arrive at the foundation or slab on which they will rest.

moisture barrier A moisture barrier is some type of material used to keep moisture from entering the living space of a building. Moisture barrier, vapor seal, and membrane mean the same thing. It is laid so that it covers the whole subsurface area over sand or gravel.

moisture control Excess moisture in a well insulated house may pose problems. A house must be allowed to breathe and change the air occasionally, which in turn helps remove excess moisture. Proper ventilation is needed to control moisture in an insulated house. Elimination of moisture is another method but it requires the reduction of cooking vapors and shower vapors, for example.

moldings Moldings are trim mounted around windows, floors, and doors as well as closets. Moldings are usually made of wood with designs or particular shapes.

monolithic slab Mono means "one." This refers to a one-piece slab for a building floor and foundation all in one piece.

nail creep This is a term used in conjunction with drywall, the nails pop because of wood shrinkage. The nailheads usually show through the panel.

nailers These are powered hammers that have the ability to drive nails. They may be operated by compressed air or by electricity.

nailhead stains These occur whenever the iron in the nailhead rusts and shows through the paint.

nail set Finish nails are driven below the surface of the wood by a nail set. The nail set is placed on the head of a nail, and the large end of the nail set is struck with a hammer.

newels The end posts of a stairway are called newels.

octagon scale This "eight square" scale is found on the center of the face of the tongue of a steel square. It is used when timber is cut with eight sides.

open This refers to the type of roofing that allows a joint between a dormer and the main roof. It is an open valley type of roofing. The valley where two roofs intersect is left open and covered with flashing and roofing sealer.

orbital sander This power sander will vibrate, but in an orbit. Thus causes the sandpaper to do its job better than it would if used in only one direction. An orbital sander can be used to finish off windows, doors, counters, cabinets, and floors.

panel This refers to a small section of a door that takes on definite shape, or to the panel in a window made of glass.

panel door This is a type of door used for the inside of a house. It has panels inserted in the frame to give it strength and design.

parquet Parquet is a type of flooring made from small strips arranged in patterns. It must be laminated to a base.

partition A partition is a divider wall or section that separates a building into rooms.

peeling This is a term used in regard to paint that will not stay on a building. The paint peels and falls off or leaves ragged edges.

penny (d) This is the unit of measure of the nails used by carpenters.

perimeter The perimeter is the outside edges of a plot of land or building. It represents the sum of all the individual sides.

perimeter insulation Perimeter insulation is placed around the outside edges of a slab.

pile A pile is a steel or wooden pole driven into the ground sufficiently to support the weight of a wall and building.

pillar A pillar is a pole or reinforced wall section used to support the floor and consequently the building. It is usually located in the basement, with a footing of its own to spread its load over a wider area than the pole would normally occupy.

pitch The pitch of a roof is the slant or slope from the ridge to the plate.

pivot This refers to a point where the bifold door is anchored and allowed to move so that the larger portion of the folded sections can move.

plane Planes are designed to remove small shavings of wood along a surface. One hand holds the knob in front, and the other hand holds the handle in the back of the plane. A plane is used to shave off door edges to make them fit properly.

planks This refers to a type of flooring usually made of tongue-and-groove lumber and nailed to the subflooring or directly to the floor joists.

plaster This refers to plaster of Paris mixed with water and applied to a lath to cover a wall and allow for a finished appearance that will take a painted finish.

plaster grounds A carpenter applies small strips of wood around windows, doors, and walls to hold plaster. These grounds may also be made of metal.

plastic laminates These are materials usually employed to make countertops. Formica is an example of a plastic laminate.

plate The plate is a roof member which has the rafters fastened to it at their lower ends.

platform frame This refers to the flooring surface placed over the joists; it serves as support for further floor finishing.

plenum A plenum is a large chamber.

plumb bob This is a very useful tool for checking plumb, or the upright level of a board, stud, or framing member. It is also used to locate points that should be directly under a given location. It hangs free on a string, and its point indicates a specific location for a wall, a light fixture, or the plumb of a wall.

plumb cut This refers to the cut of the rafter end which rests against the ridge board or against the opposite rafter.

post and beam Posts are used to support beams, which support the roof decking. Regular rafters are not used. This technique is used in barns and houses to achieve a cathedral-ceiling effect.

prehung This refers to doors or windows that are already mounted in a frame and are ready for installation as a complete unit.

primer This refers to the first coat of paint or glue when more than one coat will be applied.

pulls The handle or the part of the door on a cabinet or the handle on a drawer that allows it to be pulled or opened is called a pull.

purlin Secondary beams used in post-and-beam construction are called purlins.

rabbet A groove cut in or near the edge of a piece of lumber to fit the edge of another piece.

radial-arm saw This type of power saw has a motor and blade that moves out over the table which is fixed. The wood is placed on the table and the blade is pulled through the wood.

rafter scales This refers to a steel square with the rafter measurements stamped on it. The scales are on the face of the body.

rail The vertical facing strip on a cabinet is the stile. The horizontal facing strip is the rail.

rake On a gabled roof, a rake is the inclined edge of the surface to be covered.

random spacing This refers to spacing that has no regular pattern.

rebar A rebar is a reinforcement steel rod in a concrete footing.

reinforcement mesh Reinforcement mesh is made of 10-gage wires spaced about 4 to 6 inches apart. It is used to reinforce basements or slabs in houses. The mesh is placed so that it becomes a part of the concrete slab or floor.

remodeling This refers to changing the looks and function of a house.

residential building A residential building is designed for people to live in.

resilient flooring This type of flooring is made of plastics rather than wood products. It includes such things as linoleum and asphalt tile.

ridge board This is a horizontal member that connects the upper ends of the rafters on one side to the rafters on the opposite side.

right-hand door A right-hand door has the opening or hinges mounted on the right when viewed from the outside.

right-hand draw This means that the curtain rod can be operated to open and close the drapes from the right-hand side as one faces it.

rise In roofing, rise is the vertical distance between the top of the double plate and the center of the ridge board. In stairs, it is the vertical distance from the top of a stair tread to the top of the next tread.

riser The vertical part at the edge of a stair is called a riser.

roof brackets These brackets can be clamped onto a ladder used for roofing.

roof cement A number of preparations are used to make sure that a roof does not leak. Roof cement also can hold down shingle tabs and rolls of felt paper when it is used as a roof covering.

roofing This term is used to designate anything that is applied to a roof to cover it.

rough line A rough line is drawn on the ground to indicate the approximate location of footing.

rough opening This is a large opening made in a wall frame or roof frame to allow the insertion of a door or window or whatever is to be mounted in the open space. The space is shimmed to fit the object being installed.

router A router will cut out a groove or cut an edge. It is usually powered and has a number of different shaped tips that will carve its shape into a piece of wood. It can be used to take the edges off countertops.

run The run of a roof is the shortest horizontal distance measured from a plumb line through the center of the ridge to the outer edge of the plate.

R values This refers to the unit that measures the effectiveness of insulation. It indicates the relative value of the insulation for the job. The higher the number, the better the insulation qualities of the materials.

saber saw The saber saw has a blade that can be used to cut circles in wood. It can cut around any circle or curve. The blade is inserted in a hole drilled previously and the saw will follow a curved or straight line to remove the block of wood needed to allow a particular job to be completed.

saddle A saddle is the inverted V-shaped piece of roof inserted between the vertical side of a chimney and the roof.

saturated felt Other names for this material are tar paper and builder's felt. It is roll roofing paper and can be used as a moisture barrier and waterproofing material on roofs and under siding.

scabs Scabs are boards used to join the ends of a girder.

scaffold A scaffold is a platform erected by carpenters to stand on while they work on a higher level. Scaffolds are supported by tubing or 2 × 4s. Another name for scaffolding is *staging*.

screed A screed may be a board or pipe supported by metal pins. The screed is leveled with the tops of the concrete forms. It is removed after the section of concrete is leveled.

scribing Scribing means marking.

sealant A sealant is any type of material that will seal a crack. This usually refers to caulking when carpenters use the term.

sealer coat The sealer coat ensures that a stain is covered and the wood is sealed against moisture.

shakes This is a term used for shingles made of handsplit wood, in most cases western cedar.

sheathing This is a term used for the outside layer of wood applied to studs to close up a house or wall. It is also used to cover the rafters and make a base for the roofing. It is usually made of plywood today. In some cases, sheathing is still used to indicate the 1 × 6-inch wooden boards used for siding undercoating.

shed In terms of roofs, this is the flat sloping roof used on some storage sheds. It is the simplest type of roof.

sheetrock This is another name for panels made of gypsum.

shim To shim means to add some type of material that will cause a window or door to be level. Usually wood shingles are wedge-shaped and serve this purpose.

shingles This refers to material used to cover the outside of a roof and take the ravages of weather. Shingles may be made of metal, wood, or composition materials.

shingle stringers These are nailing boards that can have cedar shingles attached to them. They are spaced to support the length of the shingle that will be exposed to weather.

shiplap An L-shaped edge, cut into boards and some sheet materials to form an overlapping joint with adjacent pieces of the same material.

side lap The side lap is the distance between adjacent shingles that overlap.

siding This is a term used to indicate that the studs have been covered with sheathing and the last covering is being placed on it. Siding may be made of many different materials—wood, metal, or plastic.

sill This is a piece of wood that is anchored to the foundation.

sinker nail This is a special nail for laying subflooring. The head is sloped toward the shank but is flat on top.

size Size is a special coating used for walls before the wallpaper is applied. It seals the wall and allows the wallpaper paste to attach itself to the wall and paper without adding undue moisture to the wall.

skew-back saw This saw is designed to cut wood. It is hand-operated and has a serrated steel blade that is smooth on the non-cutting edge of the saw. It is 22 to 26 inches long and can have 5½ to 10 teeth per inch.

skilled worker A skilled worker is a person who can do a job well each time it is done, or the person who has the ability to do the job a little better each time. Skilled means the person has been at it for some time, usually 4 to 5 years at the least.

sliding door This is usually a large door made of glass, with one section sliding past the other to create a passageway. A sliding door may be made of wood or glass and can disappear or slide into a wall. Closets sometimes have doors that slide past one another to create an opening.

sliding window This type of window has the capability to slide in order to open.

slope Slope refers to how fast the roof rises from the horizontal.

soffit A covering for the underside of the overhang of a roof.

soil stack A soil stack is the ventilation pipe that comes out of a roof to allow the plumbing to operate properly inside the house. It is usually made of a soil pipe (cast iron). In most modern housing, the soil stack is made of plastic.

soleplate A soleplate is a 2 × 4 or 2 × 6 used to support studs in a horizontal position. It is placed against the flooring and nailed into position onto the subflooring.

span The span of a roof is the distance over the wall plates.

spreader Special braces used across the top of concrete forms are called spreaders.

square This term refers to a shingle-covering area. A square consists of 100 square feet of area covered by shingles.

square butt strip This refers to shingles for roofing purposes that were made square in shape but produced in strips for ease in application.

staging This is the planking for ladder jacks. It holds the roofer or shingles.

stain Stain is a paint-like material that imparts a color to wood. It is usually finished by a clear coating of shellac, varnish or satinlac, or brush lacquer.

stapler This device is used to place wire staples into a roof's tar paper to hold it in place while the shingles are applied.

steel square The steel square consists of two parts—the blade and tongue or the body.

stepped footing This is footing that may be located on a number of levels.

stile A stile is an upright framing member in a panel door.

stool The flat shelf that rims the bottom of a window frame on the inside of a wall.

stop This applies to a door. It is the strip on the door frame that stops the door from swinging past the center of the frame.

storm door A storm door is designed to fit over the outside doors of a house. It may be made of wood, metal, or plastic, and it adds to the insulation qualities of a house. A storm door may be all glass, all screen, or a combination of both. It may be used in summer, winter, or both.

storm window Older windows have storms fitted on the outside. The storms consist of another window that fits over the existing window. The purpose is to trap air that will become an insulating layer to prevent heat transfer during the winter. Newer windows have thermopanes, or two panes mounted in the same frame.

stress skin panels These are large prebuilt panels used as walls, floors, and roof decks. They are built in a factory and hauled to the building site.

strike-off After tamping, concrete is leveled with a long board called a strike-off.

strike plate This is mounted on the door frame. The lock plunger goes into the hole in the strike plate and rests against the metal part of the plate to hold the door secure against the door stop.

striker This refers to the strike plate. The striker is the movable part of the lock that retracts into the door once it hits the striker plate.

stringer A carriage is also called a stringer.

strip flooring Wooden strip flooring is nothing more than the wooden strips that are applied perpendicular to the joists.

strongbacks Strongbacks are braces used across ceiling joints. They help align, space, and strengthen joists for drywall installation.

stucco Stucco is a type of finish used on the outside of a building. It is a masonry finish that can be put on over any type of wall. It is applied over a wire mesh nailed to the wall.

stud This refers to the vertical boards (usually 2×4 or 2×6) that make up the walls of a building.

stump A stump is that part of a tree which is left after the top has been cut and removed. The stump remains in the ground.

subfloor The subfloor is a platform that supports the rest of the structure. It is also referred to as the underlayment.

sump pump This refers to a pump mounted in a sump or well created to catch water from around the foundation of a house. The pump takes water from the well and lifts it to the grade level or to a storm sewer nearby.

surveying Surveying means taking in the total scene. In this case, it refers to checking out the plot plan and the relationship of the proposed building with others located within eyesight.

suspended beams False beams may be used to lower a ceiling like a grid system; these beams are suspended. They use screw eyes attached to the existing ceiling joists.

sway brace A sway brace is a piece of 2×4 or similar material used to temporarily brace a wall from the wind until it is secured.

swinging door A swinging door is mounted so that it will swing into or out of either of two rooms.

table saw A table saw is electrically powered, with a motor-mounted saw blade supported by a table that allows the wood to be pushed over the table into the cutting blade.

tail The tail is the portion of a rafter that extends beyond the outside edge of the plate.

tail joist This is a short beam or joist supported in a wall on one end and by a header on the other.

tamp To tamp means to pack tightly. The term usually refers to making sand tightly packed or making concrete mixed properly in a form to get rid of air pockets that may form with a quick pouring.

taping and bedding This refers to drywall finishing. Taping is the application of a strip of specially prepared tape to drywall joints; bedding means embedding the tape in the joint to increase its structural strength.

team A team is a group of people working together.

terrazzo This refers to two layers of flooring made from concrete and marble chips. The surface is ground to a very smooth finish.

texture paint This is a very thick paint that will leave a texture or pattern. It can be shaped to cover cracked ceilings or walls or beautify an otherwise dull room.

thermal ceilings These are ceilings that are insulated with batts of insulation to prevent loss of heat or cooling. They are usually drop ceilings.

tie A tie is a soft metal wire that is twisted around a rebar or reinforcement rod and chair to hold the rod in place till concrete is poured.

tin snips This refers to a pair of scissors-type cutters used to cut flashing and some types of shingles.

tongue-and-groove Roof decking may have a groove cut in one side and tongue led in the other edge of the piece of wood so that the two adjacent pieces will fit together tightly.

track This refers to the metal support system that allows the bifold and other hung doors to move from closed to open.

transit-mix truck In some parts of the country, this is called a Redi-Mix truck. It mixes the concrete on its way from the source of materials to the building site where it is needed.

traverse rod This is another name for a curtain rod.

tread The part of a stair on which people step is the tread.

trestle jack Trestle jacks are used for low platforms both inside and outside. A ledger, made of 2×4 lumber, is used to connect two trestle jacks. Platform boards are then placed across the two ledgers.

trimmer A trimmer is a piece of lumber, usually a 2×4, that is shorter than the stud or rafter but is used to fill in where the longer piece would have been normally spaced except for the window or door opening or some other opening in the roof or floor or wall.

trowel A trowel is a tool used to work with concrete or mortar.

truss This is a type of support for a building roof that is prefabricated and delivered to the site. The W and King trusses are the most popular.

try square A try square can be used to mark small pieces for cutting. If one edge is straight and the handle part of the square is placed against this straightedge, the blade can be used to mark the wood perpendicular to the edge.

underlayment This is also referred to as the subfloor. It is used to support the rest of the building. The term may also refer to the sheathing used to cover rafters and serve as a base for roofing.

unfaced insulation   This type of insulation does not have a facing or plastic membrane over one side of it. It has to be placed on top of existing insulation. If used in a wall, it has to be covered by a plastic film to ensure a vapor barrier.

union   A union is a group of people with the same interests and with proper representation for achieving their objectives.

utilities   Utilities are the things needed to make a house a home. They include electricity, water, gas, and phone service. Sewage is a utility that is usually determined to be part of the water installation.

utility knife   This type of knife is used to cut the underlayment or the shingles to make sure they fit the area assigned to them. It is also used to cut the saturated felt paper over a deck.

valley   This refers to the area of a roof where two sections come together and form a depression.

valley rafters   A valley rafter is a rafter which extends diagonally from the plate to the ridge at the line of intersection of two roof surfaces.

vapor barrier   This is the same as a moisture barrier.

veneer   A veneer is a thin layer or sheet of wood.

vent   A vent is usually a hole in the eaves or soffit to allow the circulation of air over an insulated ceiling. It is usually covered with a piece of metal or screen.

ventilation   Ventilation refers to the exchange of air, or the movement of air through a building. This may be done naturally through doors and windows or mechanically by motor-driven fans.

vernier   This is a fine adjustment on a transit that allows for greater accuracy in the device when it is used for layout or leveling jobs at a construction site.

vinyl   Vinyl is a plastic material. The term usually refers to polyvinyl chloride. It is used in weather stripping and in making floor tile.

vinyl-asbestos tile   This is a floor covering made from vinyl with an asbestos filling.

water hammer   The pounding sound produced when the water is turned off quickly. It can be reduced by placing short pieces of pipe, capped off at one end, above the most likely causes of quick turnoffs, usually the dishwasher and clothes washing machine.

water tables   This refers to the amount of water that is present in any area. The moisture may be from rain or snow.

weatherstripping   This refers to adding insulating material around windows and doors to prevent the heat loss associated with cracks.

winder   This refers to the fan-shaped steps that allow the stairway to change direction without a landing.

window apron   The window apron is the flat part of the interior trim of a window. It is located next to the wall and directly beneath the window stool.

window stool   A window stool is the flat narrow shelf which forms the top member of the interior trim at the bottom of a window.

wrecking bar   This tool has a number of names. It is used to pry boards loose or to extract nails. It is a specially treated steel bar that provides leverage.

woven   This refers to a type of roofing. Woven valley-type shingling allows the two intersecting pieces of shingle to be woven into a pattern as they progress up the roof. The valley is not exposed to the weather but is covered by shingles.

zoning laws   Zoning laws determine what type of structure can be placed in a given area. Most communities now have a master plan which recognizes residential, commercial, and industrial zones for building.

Index

ABOUT THE AUTHORS

Rex Miller is Professor Emeritus of Industrial Technology at State University College at Buffalo and has taught technical curriculum at the college level for more than 40 years. He is the coauthor of the best-selling third edition of this book, and the author of more than 75 texts for vocational and industrial arts programs. He lives in Round Rock, Texas.

Mark R. Miller is Chairman and Associate Professor of Industrial Technology at Texas A&M University in Kingsville, Texas. He teaches construction courses for future middle managers in the trade. He is coauthor of several technical books, including the best-selling prior edition of this one. He lives in Kingsville, Texas.

Glenn E. Baker is Professor Emeritus of Industrial Education at Texas A&M University and the author of more than 15 books and 100 articles on technical subjects. He lives in College Station, Texas.